17th International Conference on Micro and Nanotechnology for Power Generation and Energy Conversion Applications (PowerMEMS 2017)

Journal of Physics: Conference Series
Volume 1052

Kanazawa, Japan
14 – 17 November 2017

ISBN: 978-1-5108-6823-6
ISSN: 1742-6588

Printed from e-media with permission by:

Curran Associates, Inc.
57 Morehouse Lane
Red Hook, NY 12571

proceedings.com

Some format issues inherent in the e-media version may also appear in this print version.

Printed by Curran Associates, Inc. (2018)

Institute of Physics
Dirac House, Temple Back
Bristol BS1 6BE UK

Phone: 44 1 17 929 7481
Fax: 44 1 17 920 0979

techtracking@iop.org

Additional copies of this publication are available from:

Curran Associates, Inc.
57 Morehouse Lane
Red Hook, NY 12571 USA
Phone: 845-758-0400
Fax: 845-758-2633
Email: curran@proceedings.com
Web: www.proceedings.com

Table of contents

Volume 1052

17th International Conference on Micro and Nanotechnology for Power Generation and Energy Conversion Applications (PowerMEMS 2017)

14–17 November 2017, Kanazawa, Japan

Accepted papers received: 29 June 2018
Published online: 26 July 2018

Preface

Papers

Co-publications in SMS and JMM are available here and here

The 17th International Conference on Micro and Nanotechnology for Power Generation and Energy Conversion Application took place in Kanazawa, Japan, from November 14 to 17 2017. About 220 participants from 17 countries attended this year.

The objective of the PowerMEMS conference is to catalyse innovation in micro- and nano- scale technologies for the energy domain. The scope of the meeting ranges from basic principles, to materials and fabrication, to devices and systems, to applications. The many applications of Power MEMS concern the harvesting, storage, conversion and conditioning of energy, to integrated systems that manage these processes, to actuation, pumping, and propulsion.

Our Conference aims to stimulate the exchange of insights and information, as well as the development of new ideas, in the Power MEMS field. Our goal is to allow the attendees to interact and network within our multidisciplinary community that includes professionals from many branches of science and engineering, as well as energy, policy, and entrepreneurial specialists interested in the commercialization of Power MEMS technologies.

This year's technical program is highlighted by four plenary talks from prominent experts on thermoelectric materials, piezoelectrics, microscale combustion and electrostatic energy harvesters. The contributed program received 163 abstract submissions this year. After careful review by the 44-members of the Technical Program Committee, a total of 145 papers were presented. The 60 contributed oral presentations were arranged in two parallel sessions. The 85 posters were arranged in a "two-in-one" poster session in which the poster session time is divided in two. In addition to oral and poster sessions, we had a special session of "Power MEMS in Action" that is specifically dedicated to table-top demonstrations of micro power sources and small scale energy harvesting systems. With the growing interest in the Internet of Things, there is no better time to bring power MEMS out of the lab and into IoT applications.

The present issue of "*Journal of Physics: Conference Series*" (JPCS) contains the 154 submitted papers that were peer reviewed and accepted in the Conference Proceedings. In addition, selected papers were published in the special issues of "*Journal of Micromechanics and Microengineering*" (JMM) and "*Smart Materials and Structures*" (SMS). The 9 papers were accepted for the publication in the special issues of JMM and SMS, while those papers were not included in this JPCS issue.

Finally, we would like to thank all the people whose careful work contributed to the preparation of this JPCS issue, in particular the session chairs as well as the peer reviewers.

Conference chair: Yuji Suzuki

TPC chair: Isaku Kanno

IOP Conf. Series: Journal of Physics: Conf. Series **1052** (2018) 011002 doi:10.1088/1742-6596/1052/1/011002

Peer review statement

All papers published in this volume of *Journal of Physics: Conference Series* have been peer reviewed through processes administered by the proceedings Editors. Reviews were conducted by expert referees to the professional and scientific standards expected of a proceedings journal published by IOP Publishing.

Novel Materials and Processes to Develop Viable Thermoelectrics

T. Mori[1,2]
[1]MANA, National Institute for Materials Science (NIMS), Tsukuba 305-0044 Japan
[2]Graduate School of Pure and Applied Sciences, University of Tsukuba, Tsukuba 305-8671, Japan
MORI.Takao@nims.go.jp

Abstract. Thermoelectric power generation which represents the direct solid-state conversion of waste heat to electricity is a promising field challenging to reach wide-scale application. There are intrinsic paradoxes in the requirements of physical properties leading to high thermoelectric performance. Namely, high electrical conductivity but low thermal conductivity κ, and large Seebeck coefficient but high electrical conductivity are required. These have made straightforward enhancement of the thermoelectric performance difficult. I will review several novel materials and processes we are developing to overcome the traditional limitations, and which were presented at this Power MEMS 2017 conference.

1. Introduction

Thermoelectric power generation which represents the direct solid-state conversion of waste heat to electricity is a promising field challenging to reach wide-scale application [1,2]. There are intrinsic paradoxes in the requirements of physical properties leading to high thermoelectric performance. Namely, high electrical conductivity σ but low thermal conductivity κ, and large Seebeck coefficient α but high electrical conductivity σ are required. These requirements have made quite difficult the straightforward enhancement of thermoelectric performance, represented by the figure of merit ZT:

$$ZT = \alpha^2\sigma(\kappa_L+\kappa_E)^{-1}T, \qquad (1)$$

where κ_L and κ_E are the lattice thermal conductivity and electrical carrier contribution to the thermal conductivity, respectively. There have been several recent good overall reviews dealing with recent advancements in thermoelectrics [3-5]. I will review several novel materials and processes based on principles whereby we try to develop thermoelectric materials suitable for wide-scale power generation application.

2. Novel materials

2.1. Utilizing magnetism and magnetic materials

We have recently proposed magnetic semiconductors as a way to find high performance thermoelectrics [6-10]. The details of this principle has been noted in a recent review [3]. It was observed in $CuFeS_2$ chalcopyrites that high power factors (~1.1 mW/m/K^2) could be obtained at 300 K at room temperature, despite these materials being comprised of relatively light elements. The carrier effective mass m was estimated to be 4 ~6 times the free electron mass m_0 and magnetic interaction speculated to be the origin of the good thermoelectric properties. Carrier-magnon interaction was also indicated to enhance the Seebeck coefficient α [7]. The thiospinel Sb doped $CuCr_2S_4$ was discovered to be another heavy mass (m=5.9m_0) magnetic semiconductor with ZT~0.43 [9].

One example of the usefulness of this principle was illustrated when a high ZT material $CuGaTe_2$ was doped with magnetic ions of Mn, and effective mass was increased (Fig. 1), resulting in an enhancement of power factor and ZT (40% increase by doping to ZT~0.83 at 870 K) [10]. This indicates that this can be a relatively easy enhancement method possibly applied to other systems also. One issue to be taken into consideration is that while a large effective mass is beneficial for α, it is detrimental to the mobility (i.e. σ) as shown in Fig. 1, so optimization is necessary [3].

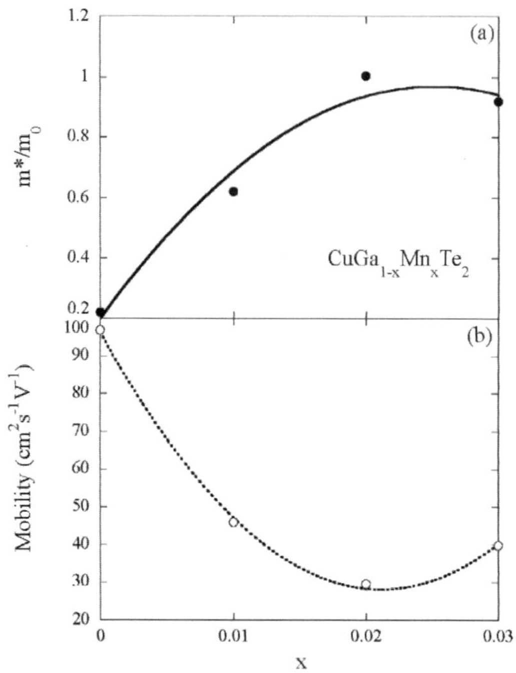

2.2. High temperature borides for topping cycles

There are several attractive thermoelectric power generation applications which require high temperature materials [11]. Foremost is the topping cycle in power plants which can deliver sizable increase in power output with moderate ZT (e.g. ZT=0.7 can lead to 6% power output enhancement) [11,12]. A further advantage is that cascade modules with segments of different temperature range materials are not necessary, but only the high temperature material module.

Borides are attractive because they are high temperature materials with typical melting points much higher than expected working temperatures, but also because they are materials which possess intrinsic "built-in" mechanisms for low thermal conductivity [13,14]. Another advantage is that through incorporation of rare earth or other metal atoms in the voids of the boron network structure, and mixed anions third elements like C, N, Si in the boron network, the physical properties and network structures themselves have been found to be able to be controllable to some degree [15].

Finding matching p, n materials is a usually difficult requirement for thermoelectric application. Borides are predominantly p-type because of the electron deficient nature of the boron cluster crystal structure framework. However, by control of atoms in the voids of the boron framework, excellent p, n control ($|\alpha|$>200 μV/K) was realized even in elemental boron, β-B [16]. This method has advantages in that no foreign elements are involved in the p, n control, and thermal matching is the best that can be envisioned with same constituent elements and same basic crystal structure.

A recent interesting discovery was the large enhancement of thermoelectric properties in REB_{66} for the Sm phase [17]. Compared to other rare earth phases like Y and Er, we observed up to a 30 times enhancement in the power factor. Magnetic measurements indicated mixed valency. The driving force of the TE enhancement was a large increase in σ which is reasonable considering REB_{66} is p-type and divalency would add hole charge carriers. However, in such a case, Seebeck would typically be expected to have a large reduction, but the actual reduction is very small. Because of this huge enhancement in power factor and heavy rare earth mass, ZT has a ~40 times enhancement for SmB_{66} compared to YB_{66} [17]. The explicit mechanism of the huge TE enhancement is not fully solved yet, but is likely related to the indicated mixed valency.

3. Processes

In regards to enhancement of ZT, various kinds of nanostructuring have been shown to be effective to enhance ZT [1]. Theoretically, the lattice thermal conductivity, κ_L, can be approximated by [1]:

$$\kappa_L^{-1} = \kappa_0^{-1} + 2\,R_K\,d^{-1}, \qquad (2)$$

where $\varkappa_0$ is the bulk thermal conductivity, d is the grain diameter, R_K is the Kapitza resistance. Namely, a small grain size can scatter phonons and reduce κ_L. By utilizing the typically different mean free paths (MFP) of electrical carriers and phonons, if one makes nanostructures in the target thermoelectric material on the scale of the phonon mean free path, phonons can be selectively scattered, and thereby enhance ZT.

We have developed a bottom-up nanostructuring method leading to a 100% enhancement (i.e. ZT~1.6) in "empty" rare earth-free skutterudites [18]. This is achieved by utilizing phase diagrams to create surprisingly effective porosity in the materials. Typically, porous materials were considered to be poor thermoelectric materials. However, precipitation and evaporation of a low temperature secondary phase enabled the creation of nano-micropores in the samples, with porosity less than 10% leading to large reduction of κ and only small reduction of σ. A hollow sphere synthesis method to create porous ZnO has also recently shown enhancement of ZT [19].

Processes with such controlled utilization of nano-microporosity should be applicable for the thermoelectric enhancement of other materials also. This method can be expected to be particularly effective for materials with large power factors and relatively high thermal conductivities.

Acknowledgments

The author acknowledges support from Japan Science and Technology Agency (JST) CREST Grant No. JPMJCR15Q6, Japan. Support from Japan Society for the Promotion of Science (JSPS) KAKENHI Grant No. JP16H06441 is also acknowledged. Dr. N. Tsujii and F. Ahmed are thanked for their help in drawing Fig. 1.

References

[1] Koumoto K, Mori T, Thermoelectric Nanomaterials; Materials Design and Applications, Springer Series in Materials Science, vol. 182. Springer, Heidelberg (2013).
[2] Mori T and Nolas G, 2016 *Scr. Mater.* **111** 1.
[3] Mori T, Small 2017 in press doi: 10.1002/smll.201702013.
[4] Zhang X and Zhao Z D 2015 *J. Materiomics* **1** 92.
[5] Liu W, Hu J, Zhang SM, Deng MJ, Han CG, Liu Y 2017 *Mater. Today Phys.* **1** 50.
[6] Tsujii N and Mori T 2013 *Appl. Phys. Express* **6**, 043001.
[7] Ang R, Khan AU, Tsujii N, Takai K, Nakamura R, Mori T 2015 *Angew. Chem. Int. Ed.* **54**, 12909-12913.
[8] Takaki H, Kobayashi K, Shimono M, Kobayashi N, Hirose K, Tsujii N, and Mori T 2017 Appl. Phys. Lett. **110**, 072107.
[9] Khan AU, Orabi RA, Pakdel A, Vaney JB, Fontaine B, Gautier R, Halet JF, Mitani S, Mori T 2017 *Chem. Mater.*, **29**, 2988.
[10] Ahmed F, Tsujii N, Mori T, 2017 *J. Mater. Chem. A* **5** 7545.
[11] Mori T 2016 *JOM* **68**, 2673.
[12] Yazawa K and Shakouri A 2016 *Scripta Mat.* **111** 58.
[13] Cahill D G, Fischer H E, Watson S K, Pohl R O and Slack G A 1989 *Phys. Rev. B* 40 3254.
[14] Mori T, Martin J and Nolas G 2007 *J. Appl. Phys.* 102 073510.
[15] Mori T 2008 *Handbook on the Physics and Chemistry of Rare Earths* vol 38, ed K A Gschneidner Jr, J C Bunzli and V Pecharsky V (Amsterdam: Elsevier) p 105-173.
[16] Sologub O, Salamakha L, Stoger B, Michiue Y, and Mori T 2017 *Acta Mater.* **122** 378.
[17] Sussardi A, Tanaka T, Khan AU, Schlapbach L, and Mori T 2015 *J. Materiomics* **1** 196.
[18] Khan AU, Kobayashi K, Tang D, Yamauchi Y, Hasegawa K, Mitome M, Xue Y, Jiang B, Tsuchiya K, Golberg D, Bando Y, and Mori T 2017 Nano Energy **31**, 152.
[19] Virtudazo RVR, Guo Q, Wu R, Takei T, and Mori T 2017 *RSC Advances* **7**, 31960.

Piezoelectric Energy Harvesting Systems

Kenji Uchino

Int'l Ctr. for Actuators & Transducers, The Penn State Univ., University Park, PA 16802, USA

kenjiuchino@psu.edu

Abstract

Energy harvesting from wasted or unused power has been the topic of discussion for a long time. This paper focuses on harvesting energy from mechanical vibrations using piezoelectric transducers. We will consider comprehensively three major phases/steps associated with piezoelectric energy harvesting: (i) mechanical-mechanical energy transfer, including mechanical stability of the piezoelectric transducer under large stresses, and mechanical impedance matching, (ii) mechanical-electrical energy transduction, relating with the electromechanical coupling factor in the composite transducer structure, and (iii) electrical-electrical energy transfer, including electrical impedance matching, such as a DC/DC converter to accumulate the energy into a rechargeable battery. The problem found in the current research teams is on a narrow research area of each above phase. In order to provide comprehensive strategies on how to improve the efficiency of the harvesting system, I provide a general guideline for piezoelectric energy harvesting systems.

We dealt with detailed energy flow analysis in piezoelectric energy harvesting systems with stiff "Cymbals" (~100 mW) and flexible piezoelectric transducers (~1 mW) under cyclic mechanical load, in order to provide comprehensive strategies on how to improve the efficiency of the harvesting system. Energy transfer rates are practically evaluated for all three steps above. For your information, the former "Cymbal" is to be applied to the automobile engine vibration, while the latter flexible transducer is to the human-wearable energy-harvesting system.

We should also point out here that there is another research school of piezo-energy harvesting; that is, small energy harvesting (mW or lower) for signal transfer applications, where the efficiency is not a primary objective. This school usually treats a burst/pulse load to generate instantaneous electric energy for transmitting signals for a short period (100 ms ~ 10 s) without accumulating the electricity in a rechargeable battery. Successful piezoelectric products in the commercial market belong mostly to this category at present, including "LightningSwitch" [remote switch for room lights, with using a piezoelectric unimorph component], and the 25 mm caliber "Programmable Ammunition" [electricity generation with a multilayer piezo-actuator under shot impact], both of which were originally designed in our group (spin-off company, Micromechatronics Inc., State College, PA).

IOP Conf. Series: Journal of Physics: Conf. Series **1052** (2018) 012002 doi:10.1088/1742-6596/1052/1/012002

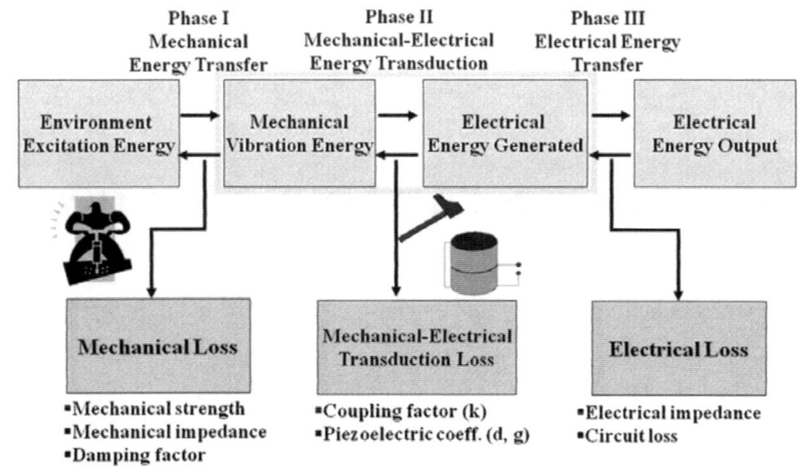

Fig. 1 Three major phases associated with piezoelectric energy harvesting): (i) mechanical-mechanical energy transfer, (ii) mechanical-electrical energy transduction, and (iii) electrical-electrical energy transfer.

Autonomous Power System Using Small Scale Vortex Combustor

Abstract. Micro and meso scale combustors have attracted considerable attention as one of energy conversion media of liquid (or liquefied) hydrocarbon fuels which possess almost 100 times larger energy density than the conventional batteries. Then, in this decade, combustion characteristics in micro and meso scale channels have been extensively studied fundamentally and practically. Based on the obtained knowledge, micro combustion power systems aimed practical use have appeared. Currently, the miniature combustion power system using vortex combustor combined with thermos electric device (TED) is the most powerful system which output attained 18.1W and conversion efficiency 3.0% with 600W thermal input by propane air mixture. As a final step, prototype "autonomous" vortex combustion power system to charge the mobile electrical devices has developed. Using 250g butane cartridge, 10W output can be obtained over 5 hours with running all assisting devices such as micro blower for the combustion air or fans for cooling air of TED.

D Shimokuri[1]

[1] Graduate school of mechanical engineering, Hiroshima University, Kagamiyama, Higashi-hiroshima 739-8527, Japan

cri@hiroshima-u.ac.jp

1. Introduction

Micro and meso scale combustors have attracted considerable attention [1-3] as one of energy conversion media of liquid (or liquefied) hydrocarbon fuels because they possess almost 100 times larger energy density than the conventional chemical batteries. However, the development of micro / meso scale combustors are quite challenging task due to the thermal / chemical effects of combustor wall on the flames. The combustion reactions are chemically terminated or thermally quenched by the combustor walls. In fact, flame cannot propagate through the tube if its inner diameter is less than certain value. The minimum tube inner diameter for the flame propagation is called "quenching diameter d_0", which is important from the viewpoint of fire safety. For example, the quenching diameter for propane / air mixture is 3.1mm under standard condition (see, Table 7.1 and eq.7.1 of [4]). Then, in this decade, to overcome the wall effects and to realize the small scale combustion power systems, a variety of small-scale combustors have been developed adopting a various flame stabilization techniques, such as excess enthalpy combustion, catalytic combustion and so on [1-3,5-7]. And recently, based on the progress of small-scale combustors, miniature combustion power systems have appeared. Table 1 shows the characteristics of the recently developed miniature combustion power systems. Takahashi [8] made a frontier work with a catalytic combustor of 1.2-mm outer diameter and thermo-electric device (TED). The system output attained 210mW with 5.0W thermal input, and the efficiency $\eta = 4.21\%$ was obtained. Yadav [9] developed a power system using a "three step combustor" (maximum 6-mm i.d.) of which output was 2.35W with 51.1W thermal input (η = 4.6%). Jiang [10] developed a power system for higher thermal input which output is 1.3W (η = 0.72%). Merotto [11] developed a considerably powerful system with micro catalytic combustor. The output attained 10W with 2.85% conversion efficiency.

Table 1 characteristics of recently developed miniature combustion power systems.

	Thermal input [W]	Output [W]	Conversion efficiency [%]	Power density [GW/m3]
Catalytic combustion power system by Takahashi (2013)	4.7	0.2	4.21	9.95
Catalytic combustion power system by Merotto (2016)	345	9.86	2.85	-
Plat flame power system by Jiang (2009)	~ 180	2.4	1.3	0.186
Film combustion power system by Li (2009)	~ 300	8.5	2.4 (TPV)	-
Vortex combustion system by Shimokuri (2017)	~ 600	18.1	3	-

2. Vortex combustion power systems

Using a vortex flow is also an effective way to stabilize the flame in narrow channels. Our fundamental investigations revealed that the flame is easily stabilized in small diameter tube if the flow is rotating [4,12,13], and furthermore, a flame of large surface area is established in the tube (see, Fig. 1). The unique flame shape, shown in Fig. 1, contributes to the increase of the mixture consumption rate, and hence, the thermal input through the increase of the flame surface area. Thus, the small scale vortex combustors are expected to be significantly powerful power source.

In fact, A vortex power system for higher thermal input were developed by Li [14] with the film combustion technique [4,15], in which the evaporation and mixing of liquid fuels were enhanced by the vortex flow. In the system, the thermal input reached 560W with liquid fuels, and output of 8.5W, η = 1.52% was obtained with thermo-photovoltaic device. More recently, we develop a powerful miniature power system [16,17] using a vortex combustor and TED (see, Fig.2). In the system, the vortex combustor was embedded in the significantly small heat medium (40 × 40 × 13-mm). The system quickly started with in 3 or 4minutes as shown in Fig. 3 [17]. The maximum output of the system reached 18.1W with 600W thermal input [17], and the system efficiency reached 3.0% (see. Fig. 4). As far as author's knowledge, this is the first miniature combustion power system which output exceeds 15W.

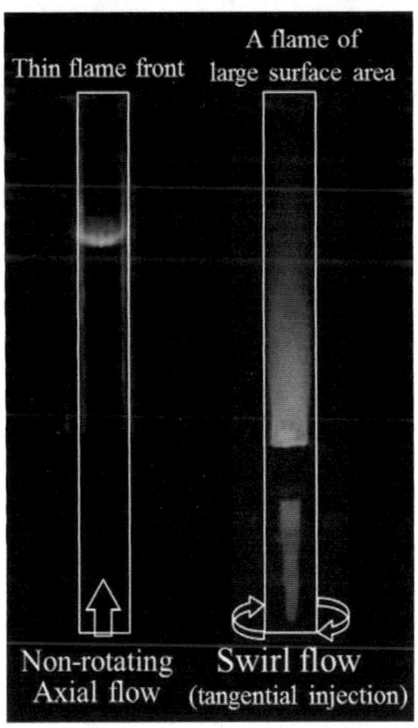

Fig. 1 Flame appearance in non-rotating flow field (left) and vortex flow (right). Tube inner diameter is 3.6mm.

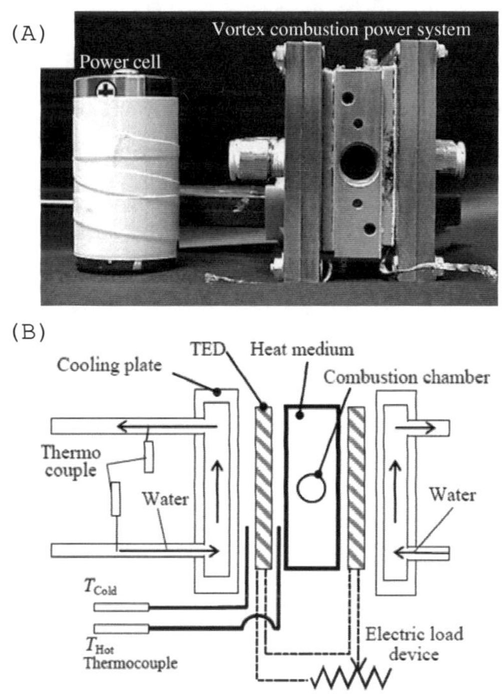

Fig.2 Appearance (A) and schematic (B) of the miniature vortex combustion power system [17].

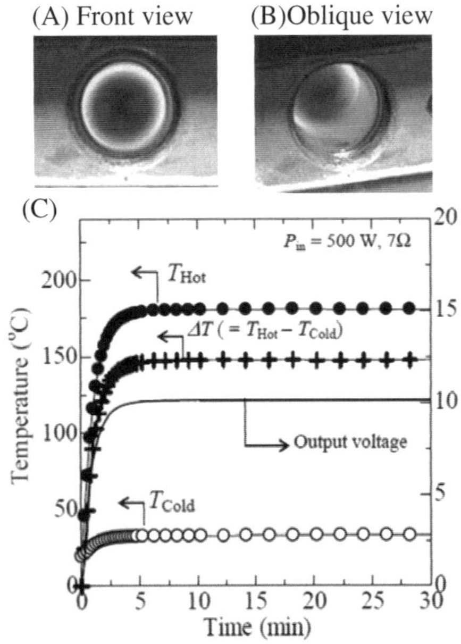

Fig. 3 Flame appearance obtained with opened ignition port (A) (B), and the variations of the wall temperatures and output voltage for $\Phi = 1.0$ of P_{in} = 500W (C) [17].

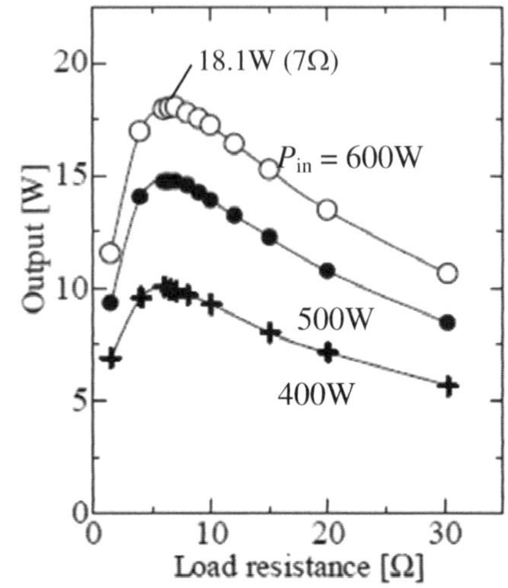

Fig. 4 Output power for various thermal input conditions [17].

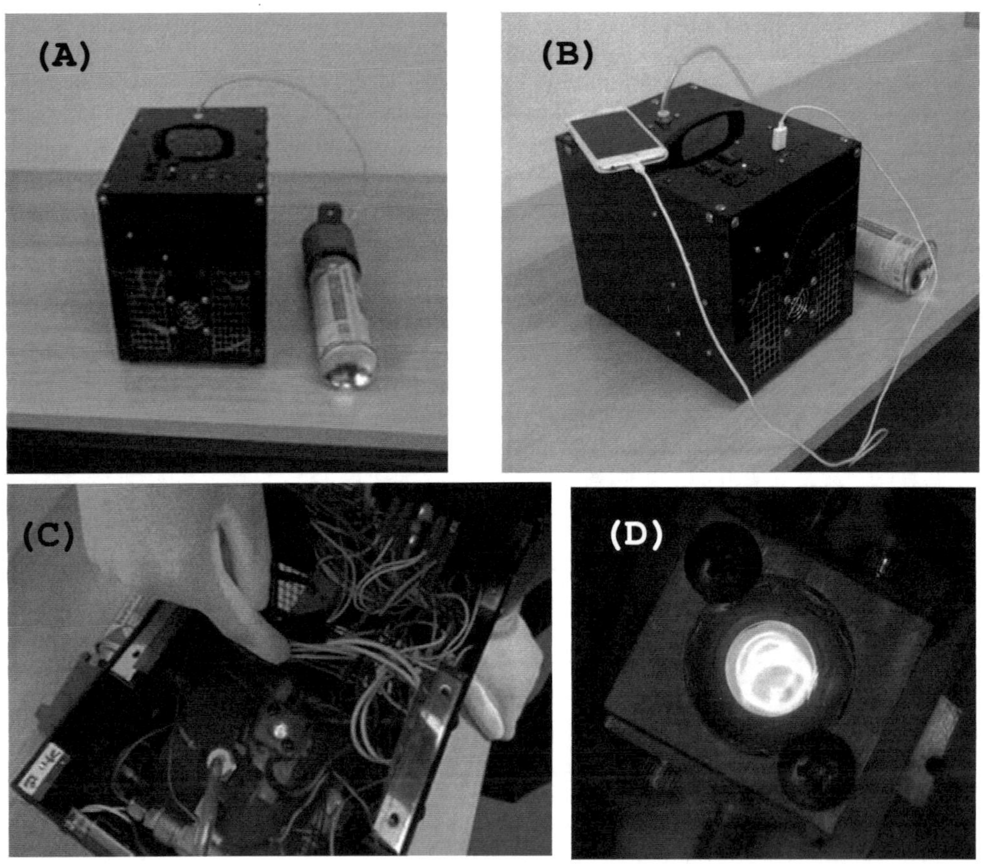

Fig.5 Appearance of autonomous vortex combustion power system ((A), (B)), the system inside (C) and flame appearance during the system run (D). The mobile phone is being charged by the system (B).

IOP Conf. Series: Journal of Physics: Conf. Series **1052** (2018) 012003 doi:10.1088/1742-6596/1052/1/012003

3. Development of Autonomous System

Several miniature combustion power systems have been developed, however, the autonomous, portable system has hardly reported. On Takahashi's autonomous system [8], the combustion air is supplied from a miniature blower, which was exerted by the part of the generated power. Fuel is supplied from bomb, however, the system output attained 223mW and conversion efficiency 1.73%.
Recently, a prototype vortex combustion power system of completely autonomous and much powerful was developed. Figure 5 (A) and (B) shows the appearance of the system, (C) shows the inside and (D) shows the flame appearance during the system run. The system is 200 × 300 × 300 mm and 5 kg. It is relatively large and heavy as it is not fully optimized, but this may be the first portable power system using small scale combustor. On the system, fuel is supplied from butane cartridge while the combustion air was supplied from micro blower which is exerted by the part of generated power. More than 10W can be used from the system over 5 hours by 250g butane cartridge. CO concentration is less than 100 ppm and the noise is less than 40dB.
Although the system output exceeded 10W for the autonomous system, the output power and conversion efficiency must be improved further, which can lead to the system scale down and weight reduction. Other energy conversion devices such as stirring engine are now being applied for the system.

4. References

[1] Fernandez-Pello C A, Proc. Combust. Inst. 2002 29 883-899.
[2] Dunn-Rankin D, Leal E M, Walther D C, 2005 Prog. Energ. Combust. Sci. 31 422-465.
[3] Maruta K, Ju Y, 2011 Prog. Energ. Combust. Sci. 37 669.
[4] Ishizuka S, Dunn-rankin D, 2013 Tubular Combustion (Sustainable Energy) Momentum Press.
[5] Kim N I, Aizami S, Yokomori T, Kato S, Fujimori T, Maruta K, 2007 Proc. Combust. Inst. 31 3243-3250.
[6] Hossain A, Nakamura Y, 2015 Proc. Combust. Inst. 35 3413-3420.
[7] Takahashi S, Wakai K, 2009 45th AIAA/ ASME/ SAE/ ASEE Joint Propulsion Conference & Exhibit, AIAA 2009-5246.
[8] Takahashi S, 2013 J. Combust. Inst. Jpn., 55 57-65
[9] Yadav S, Yamasani S, Kumar S, 2015 Energ. Conv. Manag. 99, 1
[10] Jiang L Q, Zhao D Q, Wang X H, 2011 Ener. Conv. Manag. 52 596-602.
[11] Merotto L, Fanciulli C, Donde R, De Luliis S, 2016 Appl. Energy 162 346.
[12] Shimokuri D, Honda Y, Ishizuka S, 2011 Proc. Combust. Inst. 33 3251-3258.
[13] Shimokuri D, Karatsu Y, Ishizuka S, 2011 Proc. Combust. Inst. 33 3403.
[14] Li Y-H, Lien Y-S, Chao Y-C, Dunn-Rankin D, 2009 Prog. Photovolt: Res. Appl. 17, 327
[15] Sirignano W A, Pham T K, Dunn-Rankin D, 2002 Proc. Combust. Inst. 29 925-931.
[16] Shimokuri D, Hara T, Matsumoto R, 2015 J. Micromech. Microeng. 25, 104004.
[17] Shimokuri D, Taomoto Y, Matsumoto R 2017 Proc. Combust. Inst. 36 4253.

Acknowledgments

This work was supported by Grant-in-Aid for Young Scientists (A) 15H05509. The author gratefully acknowledges the support from Dainichi Co.,Ltd. and KELK Ltd.

Electrostatic energy harvesters and fundamental limits to power

Einar Halvorsen

Department of Microsystems, University College of Southeast Norway, Campus Vestfold, Raveien 215, 3184 Borre, Norway

E-mail: `Einar.Halvorsen@usn.no`

Abstract. Analysis of fundamental performance limits to vibration energy harvesting reveal how damping and electromechanical coupling affect performance both for narrow and wide-band excitations. This talk summarizes the performance limits and presents examples of how electrostatic energy harvesters should be made to perform close to the fundamental limits. It is demonstrated how the overwhelmingly dominant contribution to loss, gas damping, can be understood and limited in an electrostatic harvester. It is seen that for wide-band noise excitations, minimizing loss and maximizing coupling largely suffice to approach the fundamental limits closely. For narrow band harvesting, a successful reduction of damping in a design can make optimization complicated because proof mass displacement increases and displacement limitations become important. Performance optimization then also involves adjusting the electrical load and the mechanical stiffness when the acceleration amplitude changes. Approaches to optimize performance are presented and discussed.

Ever since microscale vibration energy harvesting was proposed as a solution to power wireless microsystems [1, 2], there has been much focus on how much power can be obtained and how to do it. This is of course always a pertinent question for a power system, but it is particularly pressing in this case since the devices are driven by inertia so that minimizing size reduces proof mass and therefore the excitation force.

Much insight into the limits of energy harvester performance has been obtained by optimizing linear models with one mechanical degree of freedom [2, 3]. The mathematical structure of such a model is the same regardless of whether the harvester is electrostatic, piezoelectric or electromagnetic even though the physics represented is significantly different between electromagnetic harvesters and the two other types of devices. The model is an electromechanical two-port transducer driven by the forced motion of a damped proof mass and loaded electrically by a resistor. A major insight is that for linear devices driven by harmonic vibrations, a product of electromechanical coupling factor k^2 and mechanical quality factor Q is decisive for the performance. Below a certain value (≈ 2) of k^2Q, this parameter acts like a figure of merit and above this value, the output power of the linear, unconstrained device levels out at its maximum value. It should be mentioned that different authors differ in whether they prefer to use k^2 or the expedient coupling factor $k_e^2 = k^2/(1-k^2)$ as a measure of coupling, and open-circuit Q, short-circuit Q or short-circuit damping ratio as measure of loss.

For large enough k^2Q there is a choice of optimal load resistance that makes it possible to represent the conversion of energy by a linear mechanical damper, hence the damping force is proportional to velocity and we talk of a velocity damped generator (VDRG) [4]. This results in a simpler, more transparent model with analytical expressions for all important quantities. Hence, the model can be conveniently used to estimate power when it is applicable. When it is not, it is still a useful standard to compare to as when defining energy harvester effectiveness [5]. Two other such canonical architectures are also useful, i.e. the Coulomb-damped resonant generator and Coulomb force parametric generator. These have all been analyzed thoroughly, also when there are displacement limits [4].

With random-noise input it is possible to show that the output power of a linear device is governed by the input power and k^2Q [6, 7]. Quite generally, the input power depends only on the proof mass and spectral density of acceleration for this excitation regardless of whether the device is linear or not. For specific nonlinear architectures, i.e. a device with linear transducer and nonlinear springs, it is possible to make rather explicit and sometimes tight bounds on the output power [8]. However, since the input power is more or less given and must either be delivered out or be lost, it is clear that low loss and high coupling is what is needed for this type of excitation. Experimentally, one can exploit the versatility of MEMS electrostatic energy harvesters to demonstrate that by increasing the coupling by increasing the bias of an externally biased electrostatic device, the output power levels out towards the the input power limit [9].

Loss in vibration energy harvesting devices tend to be dominated by gas-damping, see e.g. [10]. It is therefore important to understand and counteract this mechanism. To this end we have revisited a previous electret-based harvester [11] for which damping as a function of cavity pressure has been measured [12, 13]. At atmospheric pressure we measure $Q = 5.1$. A simple estimate based on incompressible flow and simple formulas for hydraulic resistance R to the flow around the mass gives a damping constant $b = A^2R$ corresponding to $Q = 6.9$. A is the cross-sectional area of the mass. The Q has the same of order of magnitude as the experiment, but is constant with pressure while the experiments give $Q > 200$ at 2.5 mbar. Taking into account compressibility of the gas in the cavities at the ends of the proof mass, gives a squeezed-film-like form of damping with the same damping constant and a cut-off $\omega_c = 2P_0/V_0R$ where P_0 is the nominal cavity pressure and V_0 is the nominal cavity volume. This confirms the qualitative pressure dependency of the quality factor and suggests that squeezing of the gas and flow around the mass is the correct mechanism. Finite element calculations in the hydrodynamic limit improves greatly on the accuracy and confirms that the damping can be explained by gas compressibility and constrained flow. Hence damping can be reduced and performance improved by ensuring low cavity pressure and wide channels for gas flow. As a small gap may be desired for the electrostatic transducer, wider channels could be made outside the transducer area. Packaging would beneficially be hermetic and at low pressure.

With successful reduction of damping, the effect of a finite available displacement range becomes important for harmonic excitation of a resonant harvester. A possible method to improve output power under displacement constrained operation, is to use transducing end-stops [14]. However, it can be difficult to benefit from this approach at intermediate accelerations. For the VDRG and the linear two-port harvester, this limitation can be handled by increasing the damping to contain the motion within the limit [4, 15, 16]. For the linear two-port device, one can optimize both stiffness and load resistance[17]. For sufficiently large values of a figure of merit $M \approx k^2Q$, the result follows the VDRG exactly up to a maximum power determined by M. Furthermore, the VDRG performance coincide first exactly then approximately to fundamental upper bounds on power [18] as acceleration is increased.

Motivated by the above considerations, we have investigated load optimization for an electrostatic harvester [19]. Surprisingly, we find that even under displacement-limited operation, the VDRG behaviour is followed up to a maximum power level which is then sustained. This happens even tough we do not perform stiffness optimization, but rather let the proof mass collide with rigid end-stops. This is contrary to the theory on the linear two-port harvester where stiffness optimization is essential to reach this performance. Analysis shows that the reason is that the first harmonic of the force due to end-stop impacts has a value that coincides with a spring force of optimal stiffness. Hence, it is possible to obtain performance close to the theoretical maximum over a range of accelerations that includes displacement limited operation just by adapting the electrical load.

References

[1] Williams C B and Yates R B 1996 *Sensors and Actuators A: Physical* **52** 8–11
[2] Amirtharajah R and Chandrakasan A P 1998 *IEEE Journal of Solid-State Circuits* **33** 687–695
[3] Renno J M, Daqaq M F and Inman D J 2009 *Journal of Sound and Vibration* **320** 386 – 405
[4] Mitcheson P D, Green T C, Yeatman E M and Holmes A S 2004 *Journal of Microelectromechanical Systems* **13** 429–440
[5] Mitcheson P D, Yeatman E M, Rao G K, Holmes A S and Green T C 2008 *Proceedings of the IEEE* **96** 1457–1486
[6] Halvorsen E 2008 *Journal of Microelectromechanical Systems* **17** 1061–1071
[7] Scruggs J 2009 *Journal of Sound and Vibration* **320** 707 – 725 ISSN 0022-460X
[8] Halvorsen E 2013 *Phys. Rev. E* **87**(4) 042129
[9] Nguyen S, Halvorsen E and Jensen G 2013 *Journal of Microelectromechanical Systems* **22** 892–900
[10] Elfrink R, Renaud M, Kamel T M, de Nooijer C, Jambunathan M, Goedbloed M, Hohlfeld D, Matova S, Pop V, Caballero L and van Schaijk R 2010 *Journal of Micromechanics and Microengineering* **20** 104001
[11] Halvorsen E, Westby E R, Husa S, Vogl A, Østbø N P, Leonov V, Sterken T and Kvisterøy T 2009 An electrostatic energy harvester with electret bias *The 15th International Conference on Solid-State Sendors, Actuators & Microsystems, Transducers 2009* (Denver, Colorado) pp 1381–1384
[12] Husa S, Blystad L C J, Dong T and Halvorsen E 2008 Measurement of gas damping in an energy harvester with nonlinear transduction *Proceedings of PowerMEMS 2008+ microEMS2008* pp 233 – 236
[13] Husa S and Halvorsen E 2009 A study of gas friction in a nonlinear energy harvester *PowerMEMS 2009 Technical Digest* (Washington DC, USA) pp 316–319
[14] Le C P, Halvorsen E, Søråsen O and Yeatman E M 2012 *Journal of Intelligent Material Systems and Structures* **23** 1409–1421
[15] Marboutin C, Suzuki Y and Kasagi N 2007 Optimal design of vibration-driven micro electret generator for energy harvesting *7th Int. Workshop Micro and Nanotechnology for Power Generation and Energy Conversion Applications (PowerMEMS 2007)* (Freiburg) pp 141–144,
[16] Renaud M, Elfrink R, Jambunathan M, de Nooijer C, Wang Z, Rovers M, Vullers R and van Schaijk R 2012 *Journal of Micromechanics and Microengineering* **22** 105030
[17] Halvorsen E 2016 *ArXiv e-prints* (*Preprint* `1603.01909`)
[18] Halvorsen E, Le C P, Mitcheson P D and Yeatman E M 2013 *Journal of Physics: Conference Series* **476** 012026 URL `http://stacks.iop.org/1742-6596/476/i=1/a=012026`
[19] Truong B D, Le C P and Halvorsen E 2016 *Journal of Micromechanics and Microengineering* **26** 124006

Development of a batteryless VHF-Beacon and tracker for mammals

E. Bäumker, F. Schüle and P. Woias

University of Freiburg, Department of Microsystems Engineering (IMTEK), Laboratory for the Design of the Microsystems, Georges-Koehler-Allee 102, 79110 Freiburg

E-mail: eiko.baeumker@imtek.de

E-mail: woias@imtek.de

Abstract. In this paper, we present the design, set-up and testing of a fully energy-autonomous wildlife tracking collar that is supplied by thermoelectric energy harvesting from the animal's body heat. Attached to the neck of a mammal, the device determines the animal's activity, body-temperature and its own energy state. The data is then transmitted in regular short beacons which are also used for triangulation of the animal position. The characterization in laboratory and a field test show that the system design with an ultra-low power switch and a low-voltage step-up-converter with over 42 % efficiency is capable to measure and transmit continuously in intervals of 25 s.

1. Introduction

Wildlife telemetry systems can help here and are one of the most common practices. A subclass of these systems uses relatively simple very-high-frequency (VHF) transmitters that are attached to the animal. A typical tracking collar (e.g. [1]) sends pulses for radio bearing and also encodes temperature or activity into the transmitted signal. Wildlife monitoring becomes more and more urgent in a human-affected world where changed environmental conditions could lead to imbalances in the population of different species [2]. To prevent harm on humans and environment, understanding and monitoring the habits and needs especially of endangered species is one of the main challenges to integrate wildlife. To fulfill this task, typically information on the position, temperature and activity of some tracked animals is required to derive i.e. the migration path or mortality. A drawback of these devices is the limited lifetime of the included batteries. A promising technique is energy harvesting (EH) to get rid of these. Tracking systems that rely fully on energy-harvesting and are carried by the animal still do not exist yet. This paper presents a fully energy-autonomous VHF tracking system that also gathers vital data from the animal carrying it. Section 2 describes the used components and optimizations that were done. In section 3, the performance of the system is determined in the laboratory and in a field test.

2. Design of the system

With the developed system the most common variables such as position, activity and temperature of the animal can be retrieved. The necessary device is directly attached to the animal as a collar. The gathered data is stored internally and sent wireless to a mobile base station as well. The position of the animal is determined via radio bearing used for triangulation. The whole system is powered by thermoelectric energy harvesting only, using the animal's body as a heat resource. The following sections describe in detail each part of the system.

2.1. Energy-harvesting

Our targeted animals are mammals that produce heat, leading to a temperature-gradient between skin and the environment. When a collar is attached to the animal, a certain heat accumulation occurs at the inner side of the collar. Out of this temperature gradient, electric energy can be gathered via a thermoelectric generator (TEG) mounted through the collar. An early prototype has been built and is described in [3], using a small TEG (TEG 083-230-07, Thermalforce) with a footprint of only 18 x 21 mm^2. A field test of this TEG on a sheep is also described in [3], where a mean temperature difference of $\Delta T_{AVG} = 1.5$ K was found. The system presented here uses a similar design, but includes all electronics and now three TEGs in one housing with outer dimensions as shown in Figure 1 and inner parts in 4. The total Seebeck coefficient of the used three TEGs, in a serial electrical connection, is $\alpha = 63\,\mathrm{mV/K}$. The total internal resistance of the three TEGs is $10.8\,\Omega$. At a realistic temperature difference of $\Delta T = 1.5$ K applied to all three TEGs, this results in an open circuit voltage of $V_{OC} = 95$ mV. Up-conversion is needed to bring the voltage to usable values. This is done with a special low-voltage step-up converter described in [4] and improved acc. to [5]. This converter is self-resonant using a transformer-based Meissner oscillator for start-up and a forward converter for higher efficiency convertion. It has, on purpose, a non-regulated output voltage that directly depends on the magnitude of the input voltage and the output loading conditions. The turns ratio of the transformer was chosen such that a usable output voltage of $V_{OUT} = 2.8$ V is reached at an applied $\Delta T = 1$ K at the TEGs at the system's maximum power point, with a supposed efficiency of the step-up of 50% with a starting point at $\Delta T = 0.6$ K.

After up-conversion, the energy is used to charge a 10 mF supercapacitor (BZ05CA103ZSB, Best Cap). A comparator turns on/off the system when reaching certain voltage levels and acts therefore as deadlock protection and power management unit. A very low-power design was chosen for this according to [6]. The voltages for turning on/off the system are set to $V_{ON} = 2.5$ V and $V_{off} = 2.2$ V.

2.2. Microcontroller and sensors

An ultra-low power microcontroller (MSP430FR5969, TI) controls the attached transmitter, two temperature sensors (LMT70A), an accelerometer (ADXL362) and can read out the voltage V_{Cap} of the supercapacitor. The temperature sensors are embedded on the hot and cold side of the TEGs, for determining ΔT. All data is internally stored in a FRAM, which has a capacity for 6400 values. This data can also later be accessed via a built-in programmer interface. Additionally, the data is sent via the wireless transmitter. The duty-cycle for measuring and sending the data is set to a fixed rate. During the inactive time t_{sleep} energy is harvested for the next run. If this time is too short, the comparator will switch off the whole circuit until enough

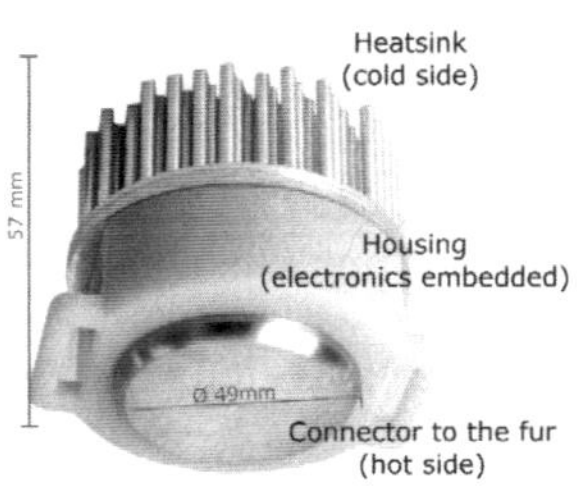

Figure 1. Left: Housing of the tracker with components and dimensions.

Figure 2. Middle: The device attached to the neck of a sheep with a collar.

energy is gathered. The necessary sleep-time dependents on the power income P_{input}, and energy consumption for the measure task $E_{measure}$ and the wireless transmission. Equation 1 predicts the necessary sleep time based on the transmission time $t_{transmit}$, the systems supply current I_{TX} during wireless transmission and the supply current I_{sleep} during the system's power-down time, all at the supply voltage U_{DD}.

$$t_{sleep} = \frac{E_{measure} - P_{input} \cdot (t_{measure} + t_{transmit}) + I_{TX} \cdot t_{transmit} \cdot U_{DD}}{P_{input} - I_{sleep} \cdot U_{DD}} \quad (1)$$

2.3. Wireless Transmission

For maximum flexibility the encoding of the packages was completely done by the microcontroller modulating the transmitter with a simple On-off keying (OOK) scheme. For our system we have tested two Transmitters: The first one transmits at 150 Mhz and is a one-transistor, crystal-controlled RF transmitter according to [7], however modified as described in [3]. The second is a standard 433 MHz AM transmitter. One full data package consists of 8 bytes. It includes a preamble of 4 bit, a device address of 8 bit, two temperature values and the voltage (i.e. charging state) of the supercapacitor, with 12 bit each, and finally the data of a two-axis accelerometer with 8 bit each. This package is manchester-encoded and then send via OOK.

At the receiver side, a HackRF 1 with an omnidirectional antenna (SRH-789, DIAMOND) was attached to a computer. The filtering and decoding was completely done in software (GNU Radio), which has the advantage of great flexibility.

3. System evaluation

3.1. Laboratory tests

To test the efficiency of the step-up converter, its input terminal V_{in} was attached to the three TEGs, whereas the output terminal V_{OUT} was directly connected to the 10 mF supercapacitor. The supercapacitor was pre-charged to V_{off}= 2.2 V at which voltage the system would have been turned off by the comparator. A defined temperature gradient ΔT was then applied to the TEGs. The time interval for charging the supercapacitor to the system's turn-on voltage V_{on} was measured. The efficiency is determined by the mean output-power and the theoretical power of the TEGs at the given temperature for the maximum power point.

Figure 3 shows that the maximum efficiency of 42% is reached at ΔT =1.1 K with an usable output power of 44 μW. This is a quite remarkable result for such low input voltages, especially as the measurement setup does also take discrepancies like the leakage current of the supercapacitor and the changing working range within V_{on} and V_{off} into account. This performance therefore should not differ much in the field tests.

Table 1 shows the characterized current consumption of the main system. Together with the predicted input power P_{input} after the step-up converter, Equation 1 can be used to predict the necessary sleep-time for the system. To verify Equation 1, the duty-cycle of the tracker was set according to the predicted sleep-time at the applied ΔT, thus for the calculated P_{input}. As expected, the tracker was able to keep that duty cycle, and also did not gain any additional energy. Therefore Equation 1 is proven to determine the minimum duty cycle in a sufficiently correct way.

Table 1. Duration an energy consumption of the different system tasks

Task	Energy /μJ	Duration /ms	Power /μW
INIT	0.01	0.03	445
SLEEP			1.25
MEASURE	10.05	10.2	988
TX_{433MHz}	1102	128	8610
TX_{150MHz}	990	512	1935

3.2. Field tests

For a real field test, the device was attached to the neck of a sheep as shown in figure 2. This sheep was kept grazing freely during the week of the experiment in July 2017. For the test, solely the 433 MHz RF transmitter was used. The receiver station was located at a shelter for the sheep. Beside that, a mobile receiver was used for radio bearing and short-term data collection. Unfortunately, the receiver-sensitivity at the base-station was too low to receive data continuously for the whole grazing area. As the sheep preferred to come back to the shelter rarely, data aggregation was difficult. However random inspections with the mobile receiver showed that the system was working and sending data packages approximately every 25 s a package. That interval would match to a $\Delta T \sim 1.2$ K.

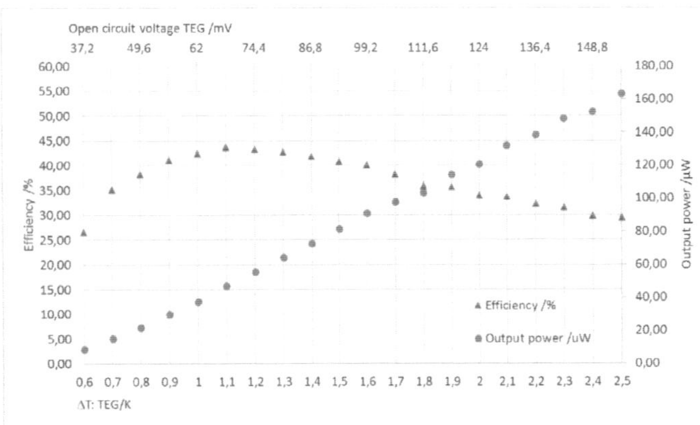

Figure 3. Output power and efficiency of the system's step-up converter as a function of the temperature difference ΔT present at the TEGs

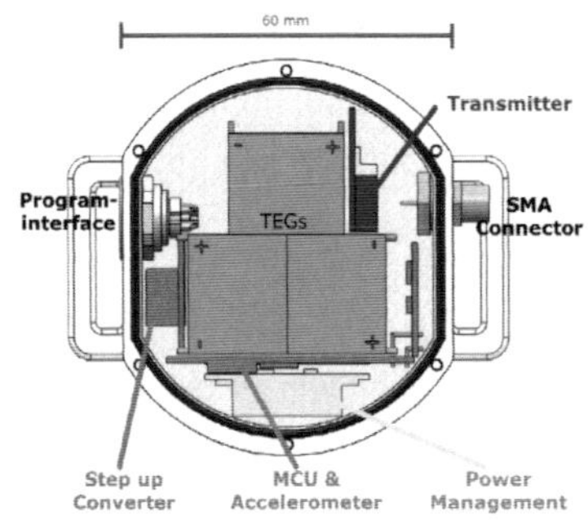

Figure 4. Top view of the tracker – around the three TEGs, the electronics are embedded into the housing

4. Summary and conclusion

We have realized a batteryless VHF tracker that can be used for radio bearing and for monitoring temperature and activity of an animal. This device solely runs on the energy gathered by thermoelectric energy harvesting from the animal's body heat. An improved and fitted design of a low-voltage step-up converter reaches an efficiency of over 42 % in the circuit with an output power around 40 μW, which is enough to supply the systems's microcontroller and all peripherals. Field tests on a freely grazing sheep show that the expected performance is fulfilled with a beacon-interval of about 25 s. To get lower intervals down to 5 s, which is a common value for battery-supplied systems, energy could be saved by sending reduced data package sizes. As the gathered data is normally not necessary in this granularity, short packages could save 90 % transmission time and energy.

References

[1] ATS 2017 Transmitter tracking products online URL `https://atstrack.com/tracking-products/transmitters.aspx`
[2] Reimoser F 2015 *21. Österreichische Jägertagung 2015* 1
[3] Bäumker E, Schüle F and Woias P 2017 Development and performance analysis of an energy-autonomous vhf-beacon for animal-tracking will be published at MikroSystemTechnik Kongress 2017
[4] Woias P, Islam M, Heller S and Roth R 2013 A low-voltage boost converter using a forward converter with integrated meissner oscillator *Journal of Physics: Conference Series* vol 476 (IOP Publishing) p 012081
[5] Stegmeir S 2017 *Selbstversorgte hocheffiziente Spannungswandler für sehr geringe Eingangsspannungen* Master's thesis Albert-Ludwigs-Universität Freiburg - IMTEK
[6] Biancuzzi G, Wischke M and Woias P 2012 Elektrische schaltung dE Patent App. DE201,110,014,512
[7] Kenward R E 2000 *A manual for wildlife radio tagging* (Academic Press)

IOP Conf. Series: Journal of Physics: Conf. Series **1052** (2018) 012006 doi:10.1088/1742-6596/1052/1/012006

A DYNAMIC MODEL OF ARM-EQUIPPED ROTATIONAL ENERGY HARVESTER DURING HUMAN LOCOMOTION

Yuki Tanaka, Tomoya Miyoshi, and Yuji Suzuki

Department of Mechanical Engineering, The University of Tokyo,
7-3-1 Hongo, Bunkyo-ku, Tokyo 113-8656, Japan

Email: ytanaka@mesl.t.u-tokyo.ac.jp

Abstract. A dynamic model of arm-equipped rotational energy harvester (EH) during human walking is proposed. The model describes realistic arm swing based on a two link with shoulder and elbow joints rotation, and includes the electro-mechanical coupling with a rotational electret EH. It is found that output power around 77 μW can be obtained with a rotational electret EH at the walking speed of 1.22 m/s, with minimum outpour power of 40 μW. Optimum design of rotational electret EH for human walking can be made with this type of model.

1. Introduction

Energy harvesting from human walking is suitable for powering battery-less wearable devices. In such applications, rotational energy harvesters (EHs) have advantages over vibration EHs due to the fact that low-frequency 3-D vibration with 3-axis rotation is dominant for human motion. In previous studies [1, 2], sinusoidal external vibration or actual vibration of a specific person is employed to characterize rotational EHs, so that it is not straightforward process to estimate the output power for human walking in general. In the present study, we develop a dynamic model of arm-equipped rotational EH for examining the effect of various parameters of the EH on the output power.

2. Arm Swing Model

Figure 1 shows the present model of rotational EH fixed on an arm modeled with a two link. Segment data of arm is referred to Ref. [3]. In this study, EH is assumed to be fixed at a wrist, near a hand.

Hejrati et al. [4] showed that the shoulder joint angle is approximated with a sinusoidal oscillation. In the present study, based on their statistical analysis, the applied sinusoidal angle in shoulder is given by

$$\theta_{\rm sh} = A_{\rm sh} \sin(2\pi f_{\rm sh} t), \tag{1}$$

where $A_{\rm sh}$ is the amplitude of shoulder joint, $f_{\rm sh}$ is the frequency. Using the parameters given in Ref. [4], $A_{\rm sh}$ and $f_{\rm sh}$ are expressed by human parameters:

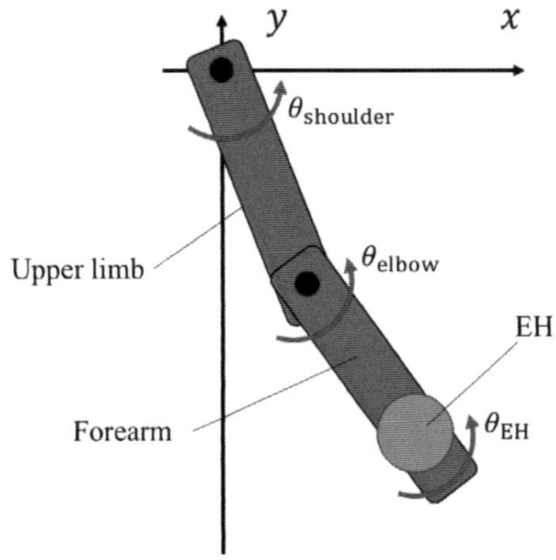

Figure 1. Arm-swing model with rotational energy harvester.

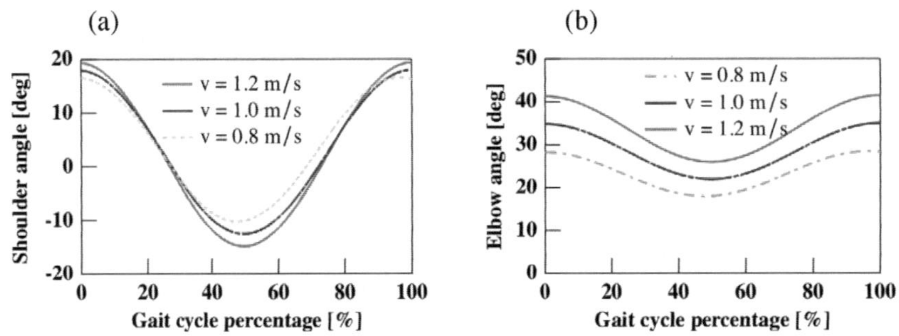

Figure 2. (a) Shoulder angle versus the gait cycle percentage, (b) Elbow angle versus gait cycle percentage at different walking speeds.

$$A_{sh} = -359h + 74.07v + 107h^2 + 0.289mv - 0.267m - 50.1hv + 0.0995s + 353, \quad (2)$$

$$f_{sh} = 0.361v - 0.0141s - 0.561h + 0.00340vs + 1.46, \quad (3)$$

where h is height, m is mass, g is gender ($g = 0$ for females, and $g = 1$ for males), v is waking speed, and s is surface slope on which human walks.

In the present study, the height, weight, and walking speed are assumed to be 1.64 m, 66 kg, and 1.2 m/s, respectively. Also, the gender is assumed to be female. Figure 2 shows temporal change of the modeled shoulder joint angle, which is in good accordance with the actual experimental data [4]. On the other hand, the elbow joint angle exhibits somewhat complex behavior, and it is difficult to approximate with a sinusoid oscillation. However, for the zeroth-order approximation, we employ the following equation for the elbow joint angle as the shoulder angle,

$$\theta_{el} = A_{el} \sin(2\pi f_{sh} t) + \theta_{0el} \quad (4)$$

where A_{el} is the amplitude in degree, and θ_{0el} is the offset angle. Using the parameters in Ref. [4], A_{el} and θ_{0el} are given by

$$A_{el} = 6.39v - 1.59g + 0.117s - 0.182vs + 0.101, \quad (5)$$

$$\theta_{0el} = 26.4v - 6.69g + 0.458s - 0.722vs + 2.155, \quad (6)$$

Figure 2b shows the elbow joint angle versus gait cycle percentage. The temporal profile of the modeled elbow joint angle exhibits similar amplitudes with the data [4] and qualitative accordance. These results indicate that the proposed model describes the arm swing motion during human locomotion with reasonable accuracy. Note that the accuracy of the model can be improved only with more plausible model for the elbow joint angle.

3. Rotational EH Model

In the present study, rotational electret EH [5] shown in Fig. 3a is assumed as an example of rotational EHs. The rotational mass is 15 g, and the eccentricity is 3 mm. Other physical parameters are similar with those in Ref. [5]. Figure 3b shows the one-dimensional electrostatic model [6]. Applying the Gauss's law, the Kirchhoff's law, and the conservations of charges, a differential equation for the output voltage is given by

$$\left\{C_p + \frac{\varepsilon_0\varepsilon_E}{d+\varepsilon_E g}(A_1 + A_2)\right\}\frac{dV}{dt} = -\left\{\frac{\varepsilon_0\varepsilon_E}{d+\varepsilon_E g}\left(\frac{dA_1}{dt}+\frac{dA_2}{dt}\right)+\frac{1}{R}\right\} + \frac{\sigma d}{d+\varepsilon_E g}\frac{dA_1}{dt}, \quad (7)$$

where C_p is the parasitic capacitance, ε_0 and ε_E are the permittivity of vacuum and the relative permittivity of the electret, d and g are the thickness of the electret and the gap distance between the electret surface and the counter electrode. A_1 and A_2 are the overlapping area in between electret and the counter electrode, and in between the guard electrode and the counter electrode. σ is the surface charge density of the electret.

Acceleration in a two-dimensional plane by the arm swing as well as the rotational acceleration is applied to the EH. Estimated mechanical damping of the bearing as well as the electrical damping are assumed as in the following equations:

$$T_{\text{mech}} = D_{\text{mech}}\dot{\theta}_{\text{EH}}, \quad (8)$$

$$T_{\text{elec}} = \frac{V}{R^2}\frac{1}{\dot{\theta}_{\text{EH}}} \quad (9)$$

These damping torques represent the mechanical energy dissipated in the oscillating system and the electrical energy converted from the kinetic energy of rotor.

Using the present arm swing model and the EH model, we calculate the rotational angle and the output voltage. The output power is computed as a time average for 10 s, where transient response still exists.

4. Results

Figure 4a shows the simulation result of rotational angle of EH for different walking speeds. For the walking speeds in this range, the rotor rotates back and forth, and EH sometimes rotates in one direction continuously at higher walking speeds. The output power is shown in Fig. 4b. The peak to

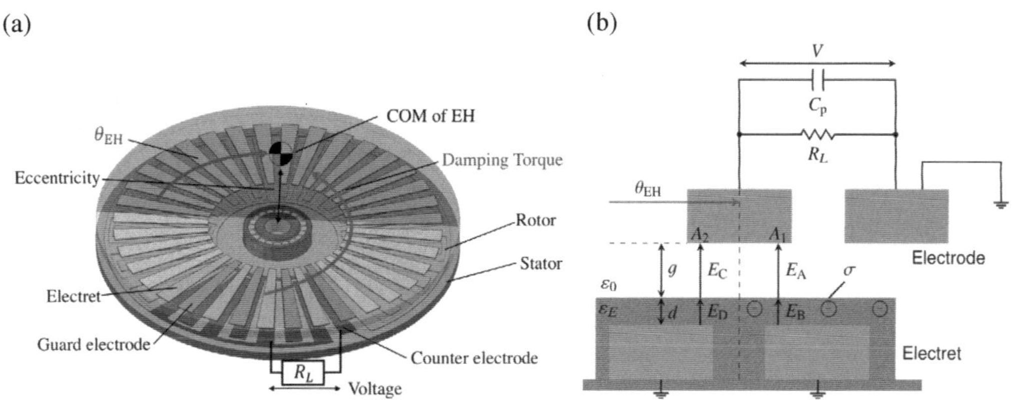

Figure 3. (a) Schematic of rotational electret EH [5], (b) One-dimensional electrostatic model of rotational electret EH.

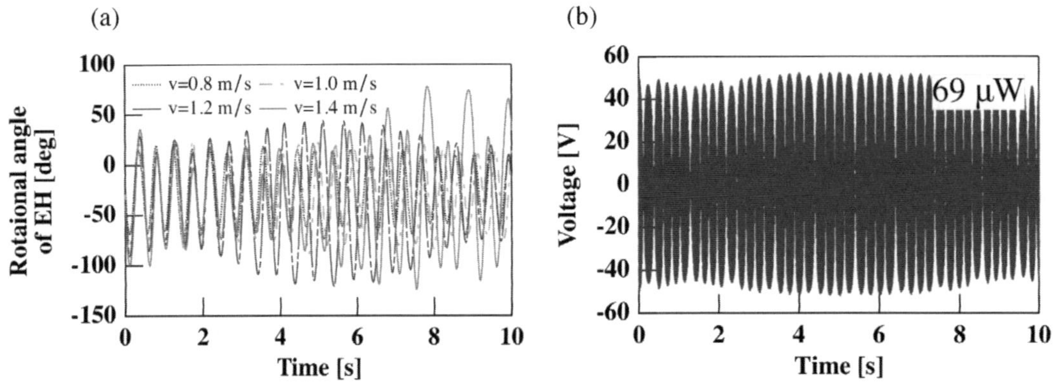

Figure 4. (a) Rotational angle versus time at different walking speed. (b) Simulated output voltage versus time at the walking speed of 1.2 m/s.

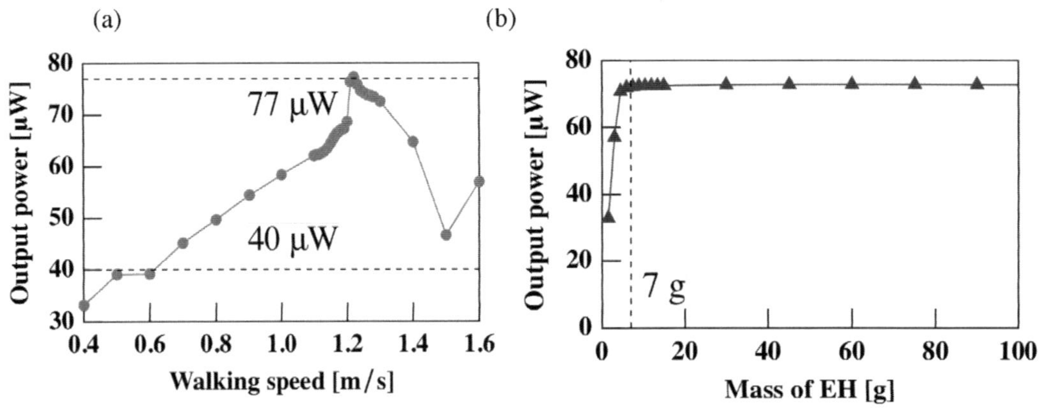

Figure 5. (a) Simulated output power versus walking speed. (b) Simulated output power versus mass of the rotor at the walking speed of 1.3 m/s.

peak output voltage is approximately 100V at the walking speed of 1.2 m/s, and output power of 69 μW is obtained.

Figure 5a shows the output power for different walking speeds. In this range of normal walking speeds, it is shown that the maximum output power 77 μW is obtained at 1.22 m/s, and minimum output power 40 μW at 0.6 m/s. The EH at 1.22 m/s keeps rotating leading to large output power. At a higher speed of 1.5 m/s, on the other hand, the lower output power drops due to smaller rotation angle in the time period of the present simulation.

Figure 5b shows the output power for different rotor mass with constant eccentricity of 3 mm. It is found that the output power is increased with the mass, but levels off at 7 g, showing the existence of minimum rotor mass for an efficient rotational EH.

5. Conclusion

A dynamic model of arm-equipped rotational EH for human walking is proposed based on the knowledge of biomechanics. Under the assumption of this study, up to 77 μW is obtained in the normal walking speed range. In addition, it is shown that the present model can provide information for effective design of high-performance rotational EH for human walking.

This work was supported by JST CREST Grant Number JPMJCR15Q3, Japan.

References

[1] Romero E, Neuman M and Warrington R O 2011 *Proc. MEMS 2011* 1325.
[2] Xue T, Ma X, Rahn C and Roundy S 2014 *J. Phys. Conf. Ser.* **557** 012090.
[3] Leva P 1996 *J. Biomechanics* **29** 1223.
[4] Hejrati B, Chesebrough S, ForeBan K B, Abbott J J and Merryweather A S 2016 *Hum. Movement. Sci.* **49** 104.
[5] Adachi M et al., PowerMEMS 2017 2017 to be presented.
[6] Suzuki Y (2011) *IEEJ Trans. Electr. Electr. Eng.* **6** 101.

Batteryless neural interface using triboelectric nanogenerators (TENGs) to enable a self-sustainable platform for neuromodulation

Sanghoon Lee[1,2,3], Hao Wang[1,3], Nitish V. Thakor[1,2], Shih-Cheng Yen[1,2], and Chengkuo Lee[1,2,3]

[1]Department of Electrical and Computer Engineering, National University of Singapore, Singapore
[2]Singapore Institute for Neurotechnology (SINAPSE), Singapore
[3]NUS Suzhou Research Institute (NUSRI), China

E-mail: elelc@nus.edu.sg

Abstract. The concept of converting human energy into useful power by various mechanisms has been explored as an alternative way to support the operation of implantable bioelectronics. Triboelectric nanogenerators (TENGs) have recently been proposed as a promising technology to harvest energy as an alternative power source for these implantable devices. In this paper, we investigate stacked TENGs as a potential power source for peripheral nerve stimulation. To achieve sufficient electrical energy for the stimulation, we optimize various parameters of the stacked TENGs. We combine the optimized TENGs and neural electrodes to demonstrate batteryless neural interfaces. We conduct direct and selective stimulation of sciatic nerves in rats using this batteryless neural interface to modulate different muscle activation. This prove-concept result indicates that this technology could enable a self-sustainable platform for neuromodulation in the future.

1. Introduction

The emerging field of neuromodulation using implantable bioelectronics involves modulating bodily conditions or control neuroprostheses [1][2]. One of the important challenges for long-term use of these devices is reliable power source with adequate output power. Instead of approaching wired, wireless, or implantable battery, the concept of converting human energy into useful power has been exploring as an alternative approach to support the operation of implantable devices [3]. Triboelectric nanogenerators (TENGs) have recently been proposed as a promising technology to harvest energy for these devices [3]. However, the optimization and design of TENGs for neuromodulation, which may open an attractive research direction of TENGs for direct stimulation of a peripheral nerve, has not been studied yet.

In this paper, we optimized and demonstrated stacked TENGs as a potential power source for neural stimulation of a peripheral nerve. Various parameters, such as electrical connections, the number of stacked layers, and the confinement length, were studied to generate sufficient electrical charges for the nerve stimulation. Furthermore, we demonstrated modulation of muscle functions in rats using batteryless neural interfaces combined the TENGs and neural electrodes. This result provides insight into a self-sustainable platform for neuromodulation (Figure 1). The electrical energy generated by the TENGs is delivered to a peripheral nerve via a flexible neural interface, which allows selective stimulation. Then, it activates target muscle resulting in post-biomechanical energy. This energy may finally be used for the next round of TENGs leading to a self-sustainable platform.

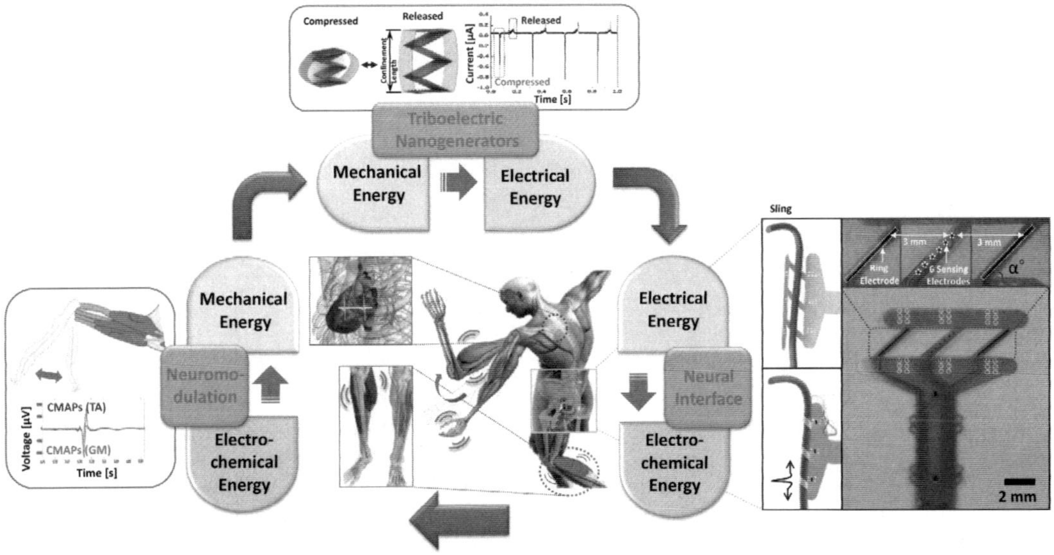

Figure 1. Schematic diagram of a self-sustainable platform for neuromodulation combining neural interfaces and triboelectric nanogenerators (TENGs).

2. Design and Fabrication

2.1 Stacked triboelectric nanogenerators

To support folding and separating layers of the stacked TENGs repeatedly, the zigzag-shaped PET sheet was used. Figure 2 shows the working mechanism of the device. The micropyramid structure of PDMS not only improved the contact area between two layers, but also acted as the first triboelectric layer for the charge generation process. During the layers were folded, the first bottom copper film acted as the bottom electrode (Figure 2a) beneath the patterned PDMS film (Figure 2b). The second copper film served the dual purpose of being the second triboelectric layer and also the top electrode.

For the fabrication of the TENGs, a silicon substrate was etched by KOH for the inverted micropyramid patterns. These micropyramid patterns were transferred onto PDMS films using molding process. After assembling the copper films, the patterned PDMS films were attached on top of alternative facing upwards. For the confinement structure of the TENG devices, different lengths of PET films were used to encompass stacked TENG devices, where the confinement height was designed to have variations ranging from 1 to 6 cm in order to restrict the maximum separation between the top and the bottom layer.

2.2 Flexible sling electrode

The neural interfaces were fabricated using microfabrication technology. It consists of two layers of flexible polyimide with gold sandwiched in between in a sling-shape geometry. It has six active electrodes in the middle bridge and two ring electrodes in the outer two bridges, which enables tripolar or bipolar stimulation configurations [4].

3. Characterization of triboelectric nanogenerators (TENGs)

The characteristics of voltage and power were measured depending on different values of load resistances connected with parallel and series configurations, respectively. The peak power of the layers connected in the parallel was 51.8 μW at the load resistance of 15 MΩ (Figure 3a). The decrement of the maximum power in series configuration was significant when the load resistance value is reduced. It demonstrates that the parallel configuration is suitable for neural stimulation since the impedance between a neural electrode and a sciatic nerve is typically degree of kΩ.

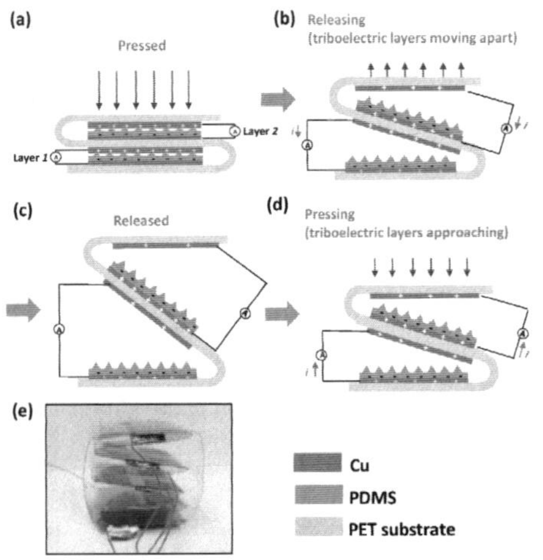

Figure 2. Schematic diagram of working mechanism (a) The device is compressed in contact with triboelectric layers. (b) The layers moving apart from each other as the force is released. (c) Electrostatic equilibrium is reached resulting into no current. (d) The layers approaching each other as external periodic force is applied on multiple stacked devices (e) A picture of the 5 layer device.

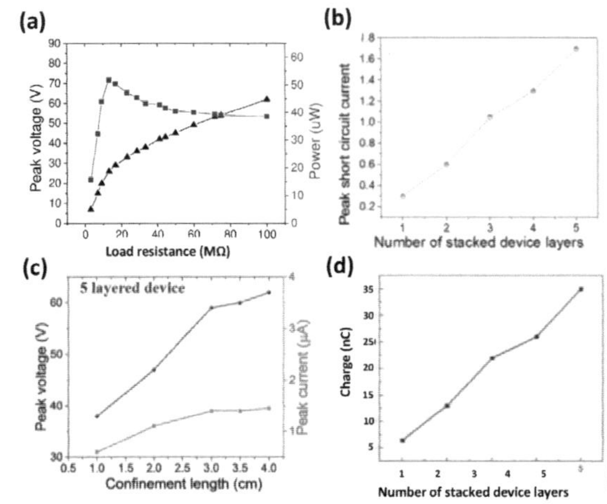

Figure 3. Peak voltage and power characteristics of 5 layers connected (a) in parallel configuration and (b) peak short circuit current for different device layers at 4 Hz. (c) Variation of peak voltage and short circuit current with different confinement lengths of the 5 layered device. (d) Charge generated by stacked TENGs as a result of the number of layers when to fully tap it by a hand at 4 Hz.

In addition, we increased the number of stacked layers in parallel configuration to demonstrate corresponding change in the output voltage and short circuit current at a fixed frequency of 4 Hz and a fixed confinement of 4 cm. The short circuit current increases linearly from 0.3 to 1.7 μA as the number of the layers increase from 1 to 5 (Figure 3b). The output voltage and current of 5 stacked layer depending on the confinement length are shown in Figure 3c. The peak output voltage and short circuit current were saturated with the confinement lengths of 4 cm in the 5 layers. Figure 3d shows calculated charges generated depending on the number of stacked layers. For the stimulation of nerve, charge, which is the product of current amplitude and pulse width, is important. The total charge was achieved in 35 nC with the 5 layers, which is a reasonable threshold charge for stimulating a nerve based on previous study [5]. Finally, the 5 layered device with 4 cm confinement length was used for *in vivo* experiments.

4. In vivo test of batteryless neural interface

The direct stimulation was conducted to activate different patterns of the muscle activations using the batteryless neural interface combined this optimized TENGs and the flexible sling interface (Figure 4a). The two electrodes of the sling interface were selected for the transversal stimulation, which enables to activate a certain part of the sciatic nerve (Figure 4b). The TENG was tapped by a hand to activate the muscles and compound muscle action potentials (CMAPs) of gastrocnemius medialis (GM) and tibialis anterior (TA) muscles were recorded, respectively (Figure 4c). We observed the twitch of the muscle while the stimulation. To fully demonstrate modulation of the TA muscle, we connected the TENG to a pair of Pt/Ir wires implanted on the common peroneal (CP) nerve (Figure 4d). The TENG was tapped by a hand with different frequencies while the CMAPs of GM and TA muscles were recorded (Figure 4e-h). The increased impact force associated with higher tapping frequency leads to higher output voltage. Thereby the muscles were activated more in this case. It demonstrates that the TENG is able to generate sufficient charges to modulate TA muscle via neural interfaces.

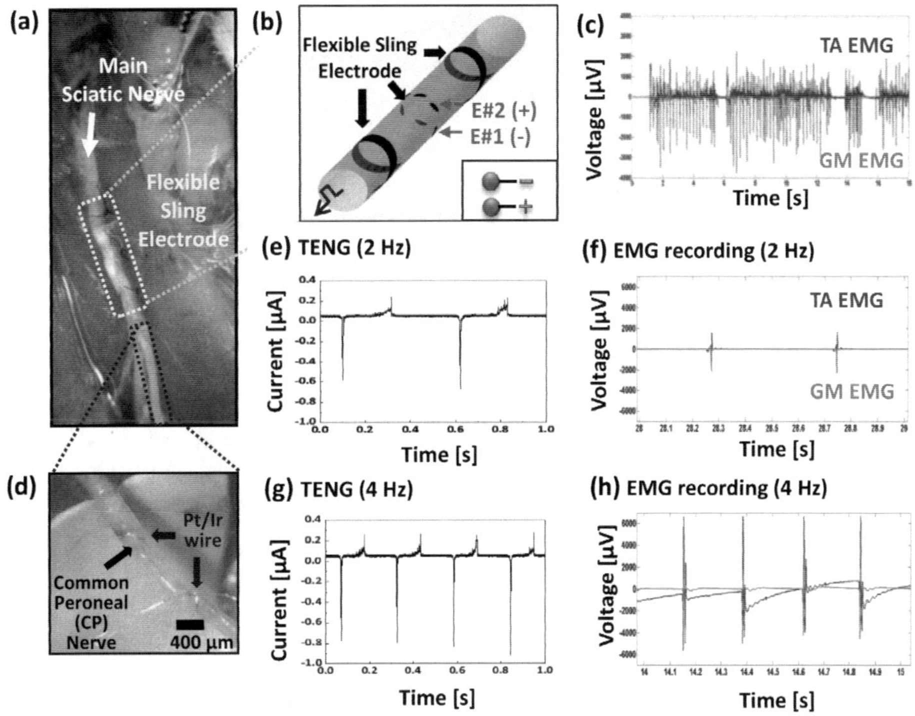

Figure 4. (a) A picture of implanted sling electrode on a sciatic nerve in a rat. (b) Schematic diagram of configuration of implanted sling electrode on a peripheral nerve. (c) Recorded TA and GM EMG signals. (d) A picture of implanted Pt/Ir wires on common peroneal nerve. (e) Current peaks generated by the TENG and (f) the CMAPs recordings at 2 Hz. (g) Current peaks generated by the TENG and (g) the CMAPs recordings at 4 Hz.

5. Conclusion

We demonstrated batteryless neural interface by combining stacked TENGs and neural electrodes. To demonstrate this technology, we conducted direct stimulation of sciatic nerves in rats to selectively activate the TA muscle. Even though the self-sustainable platform could not be achieved yet in this stage, this result shows positive possibility of the self-sustainable technology with this batteryless neural interface in near future.

Acknowledgements

This work is supported by the NRF2011NRF-CRP001-057 Program 'Self-powered body sensor for disease management and prevention-orientated healthcare' (R-263-000-A27-281) from the National Research Foundation (NRF), at the National University of Singapore, Singapore.

References

[1] Famm K, *et al.*, Drug discovery: a jump-start for electroceuticals 2013 *Nature,* 496, 7444, 159.
[2] Lee S, et al., Toward bioelectronic medicine-Neuromodulation of small peripheral nerves using flexible neural clip 2017 *Adv Sci*, 1700149.
[3] Chun J, *et al.*, Boosted output performance of triboelectric nanogenerator via electric double layer effect, 2016 *Nat Commun,*7, 12985.
[4] Lee S, et al., Flexible sling electrode for bidirectional neural signal recording and selective stimulation, 2016 *IEEE MEMS,* 375.
[5] Schiefer M A *et al.*, Selective activation of the human tibial and common peroneal nerves with a flat interface nerve electrode, 2013 *J Neural Eng,* 10, 5, 056006

Organic-Inorganic Thermoelectric Material for a Printed Generator

K Kato[1], K Kuriyama[2], T Yabuki[2], and K Miyazaki[2,3]
[1]Research Center, Lintec Corporation
7-7-3 Tsuji, Minami-ku, Saitama, Saitama 336-0026, Japan
[2]Dpartment of Mechanical and Control Engineering, Kyushu Institute of Technology, 1-1 Sensui-cho, Tobata-ku, Kitakyushu, Fukuoka 804-8550, Japan.

[3]e-mail: miyazaki@mech.kyutech.ac.jp

Abstract. We made a thermoelectric mini-generator by a screen printing method. The sub-milliwatt electric power was generated on a heated plate at 60°C. The composite of Bi_2Te_3 particles and polyimide with ionic-liquid was made as an ink for printing. The printed composite film was annealed at 400 °C for 1 hour in Argon with hydrogen to improve the thermoelectric properties. The diameter of the Bi_2Te_3 particles was about 1-2 μm and polyimide with ionic-liquid was filled between the particles. The Seebeck coefficient of the film was measured by applying a temperature difference along the in-plane direction while measuring the Seebeck voltage. The electrical conductivity was measured by four-point probe method. The cross-plane thermal conductivity was measured by differential 3ω method. The measured Seebeck coefficient of p-type thermoelectric film was 245μV/K, the electrical conductivity was 160 S/cm, and the thermal conductivity was 0.33 W/(m·K). The Seebeck coefficient, the electrical conductivity and thermal conductivity for n-type was -165μV/K, 140 S/cm, and 0.23 W/(m·K), respectively. The non-dimentional figure of merit of the film is 0.87 for p-type, and 0.5 for n-type. Those thermoelectric properties were much improved than those of Bi_2Te_3 with PEDOT:PSS composite. The high thermoelectric properties are achieved by extremely low thermal conductivity of the composites. The thermal conductivities of both p-type and n-type films were lower than that of polyimide film.

1. Introduction
Thermoelectric generator is one of the important technologies for energy harvesting without any moving parts and fuels [1], [2]. In addition, electricity can be generated directly even at low-temperature by thermoelectricity. The thermoelectric micro-generators have been intensively developed by machining process [3], vacuum deposition [4],[5] , electroplating [6], printing [7]-[13] and so on. Thermoelectric performance is evaluated non-dimensional figure of merit ZT, which is defined as $ZT=(S^2\sigma/\kappa)\times T$, where S is the Seebeck coefficient, σ is the electrical conductivity, κ is the thermal conductivity, and T is the absolute temperature. The output power of the thermoelectric generator is increased as increasing ZT, and high-ZT materials have been reported by using nano-structures, such as nano-composite[14], nano-wire[15], nano-crystalline[16], nano-porous[17], [18], superlattices[19], [20] etc. Recently, it has been pointed out that the production cost is also important for the practical use of thermoelectric generator as well as ZT[21][22]. New fabrication process with low cost must be developed. Here, we developed the organic-inorganic thermoelectric materials to make thermoelectric films with high ZT through the low cost processes. The developed materials can be applied to screen printing for making an in-plane

thermoelectric generator. The performance of the screen printed generator was evaluated by output power.

2. Experiment

We prepared for the several printed p-type Bismuth Telluride (Bi_2Te_3) to enhance the thermoelectric properties[7],[11], especially the improvement of the electrical conductivity. The electrical conductivity was measured by four-probe method, and the in-plane Seebeck coefficient of the films was measured at room temperature. One end of the film was heated by a temperature controlled hot plate and the other end was cooled by a heat sink. The temperature difference between the both ends was measured by two K-type thermocouples, and the generated voltage difference was measured simultaneously. The Seebeck coefficient was determined as the ratio of the potential difference along the film to the temperature difference. The cross-plane thermal conductivity was measured by differential 3ω method [23]. The effective thermal conductivity was calculated by the measured thermal resistance of the film in cross-plane. The film thickness was measured by using a stylus profilers (Dektak) to determine both electrical conductivity and thermal conductivity. The film thickness was double-checked from the cross-sectional view of SEM images.

2.1. P-type Bismuth Telluride nano-particles

The nano-particles of p-type Bi_2Te_3 were mixed with the toluene, and the nano-particle solution was dropped onto the Al_2O_3 substrate[7]. The film was sintered to remove the toluene, and to enhance the growth of nano-particles in hydrogen for 60 min at temperature ranging from 300 to 500°C. The samples were cooled down naturally to room temperature after sintering.

2.2. P-type Bismuth Telluride – PEDOT:PSS

Thermoelectric films was prepared by using a mixture of p-type Bi_2Te_3 particles with conductive polymer (PEDOT:PSS), poly acrylic acid and glycerol[11]. The electrical conductivity of the films of Bi_2Te_3 particles was much improved by adding a conductive polymer. It is well known that the conductivity of a random porous material is much lower than that of a bulk material as the percolation effect. The pores between Bi_2Te_3 particles were filled by a conductive material to avoid the huge electrical conductivity reduction. The glycerol was added to the mixture for the effect of the high-boiling solvent for PEDOT:PSS. The growth of the PEDOT-rich particles was enhanced by adding high-boiling solvent to the mixture, and the electrical conductivity of PEDOT:PSS was increased.

2.3. P-type Bismuth Telluride – Polyimide

The composite films of Bi_2Te_3 and polyimide with ionic-liquid were prepared by printing method. Polyimide film was made by heating polyamic acid. Polyamic acid and ionic liquid is solved to NMP (N-methylpyrrolidone) with 2-Methoxyethanol. The Bi_2Te_3 particles were added to the solution of polyamic acid and ionic-liquid. The mixture was spin-coated on the Al_2O_3 substrate, and dried at 120°C for 10 min. The dried films were heated at 300 to 450°C for 60 min in Ar with hydrogen to enhance the thermoelectric properties. The fabricated thermoelectric films were evaluated by the measurements of Seebeck coefficient, electrical conductivity, and thermal conductivity.

3. Results and discussions

The thermoelectric properties of The Bi_2Te_3 nanoparticles films are shown in figure 1. The electrical conductivities were less than 130 S/cm even at high sintering temperature, although the electrical conductivity was improved by increasing sintering temperature. The crystal growth occurs at high temperature as shown in figure 2. The electrical conduction paths were increased by the crystal growth of Bi_2Te_3. The Seebeck coefficient was gradually decreased by increasing annealing temperature due to the Tellurium sublimation. Both high Seebeck coefficient and high electrical conductivity cannot be achieved at any sintering temperature. The highest power factor $S^2\sigma$ of the films was 1.3μW/(cm·K^2) annealed at 350°C, and the highest non-dimensional figure of merit was 0.16 at 300K. Next, the

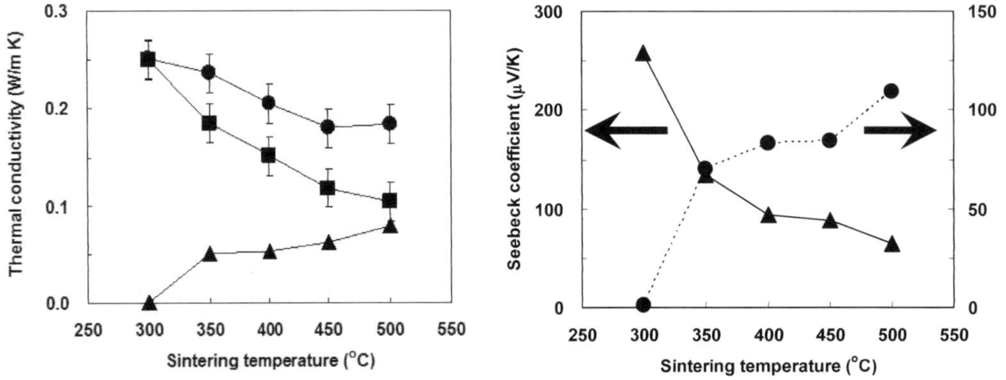

Figure 1. Thermoelectric properties of Bi_2Te_3 nanoparticles films with sintering temperature.

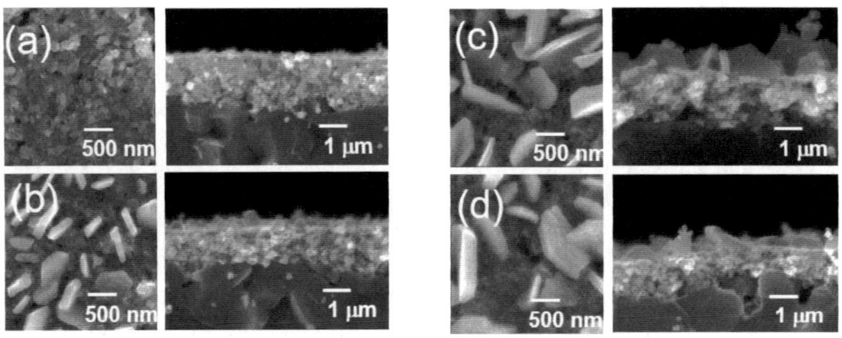

Figure 2. SEM images of Bi_2Te_3 nanoparticles films sintered at (a) 300°C, (b) 350°C, (c) 400°C, (d) 450°C. Left: Top view, Right: Cross-sectional view.

conductive materials were filled between Bi_2Te_3 particles to avoid the huge reduction of electrical conductivity due to percolation effect. PEDOT:PSS was used as the conductive polymer. The measured thermoelectric properties are shown in figure 3. The effective properties of composite calculated the simple models were shown as the solid lines in the figures[9]. The series coupling model is as follows;

$$\sigma = \frac{\sigma_{Bi_2Te_3}\sigma_{Cond.}}{x\sigma_{Bi_2Te_3} + (1-x)\sigma_{Cond.}},\quad S = \frac{S_{Bi_2Te_3}x/\kappa_{Bi_2Te_3} + S_{Cond.}(1-x)/\kappa_{Cond.}}{x/\kappa_{Bi_2Te_3} + (1-x)/\kappa_{cond.}} \tag{1}$$

x is the volume fraction of Bi_2Te_3 in the composite. The parallel coupling model is as follows;

$$\sigma = x\sigma_{Bi_2Te_3} + (1-x)\sigma_{Cond.},\quad S = \frac{S_{Bi_2Te_3}\sigma_{Bi_2Te_3}x + S_{Cond.}\sigma_{Cond.}(1-x)}{\sigma_{Bi_2Te_3}x + \sigma_{Cond.}(1-x)} \tag{2}$$

The higher curve was calculated by a parallel model and the lower curve was calculated by a series model. The measured properties were used for PEDOT:PSS at x=0 in figure 3. The measurements results were expected between two models, but all thermoelectric properties were measured to be lower than the minimum value. The only effective Seebeck coefficient was roughly explained by series connection model. The measured electrical conductivity was well improved to be 400 to 600 S/cm, although the measured electrical conductivity was lower than the expected values. The decrease of thermal conductivity was also huge, therefore the non-dimensional figure of merit is slightly improved

Table 1. Thermoelectric properties of printed films

Materials	Particle Diameter, μm	σ, S/cm	S, μV/K	κ, W/(m·K)	ZT at 300K
P-type Bi_2Te_3 nano particles[7]	0.05	72	135	0.25	0.16
P-type Bi_2Te_3 and PEDOT:PSS [11]	0.66	420	42	0.24	0.09
P-type Bi_2Te_3 and PEDOT:PSS [11]	208	380	79	0.36	0.20
P-type Bi_2Te_3 and Polyimide	1-2	160	245	0.33	0.87
N-type Bi_2Te_3 and Polyimide	1-2	140	-165	0.23	0.5
Single Crystal Bulk p- Bi_2Te_3 [24]	-	780	225	1.37	0.87
Thin film p-Bi_2Te_3[25]	-	640	236	1.05	1.0

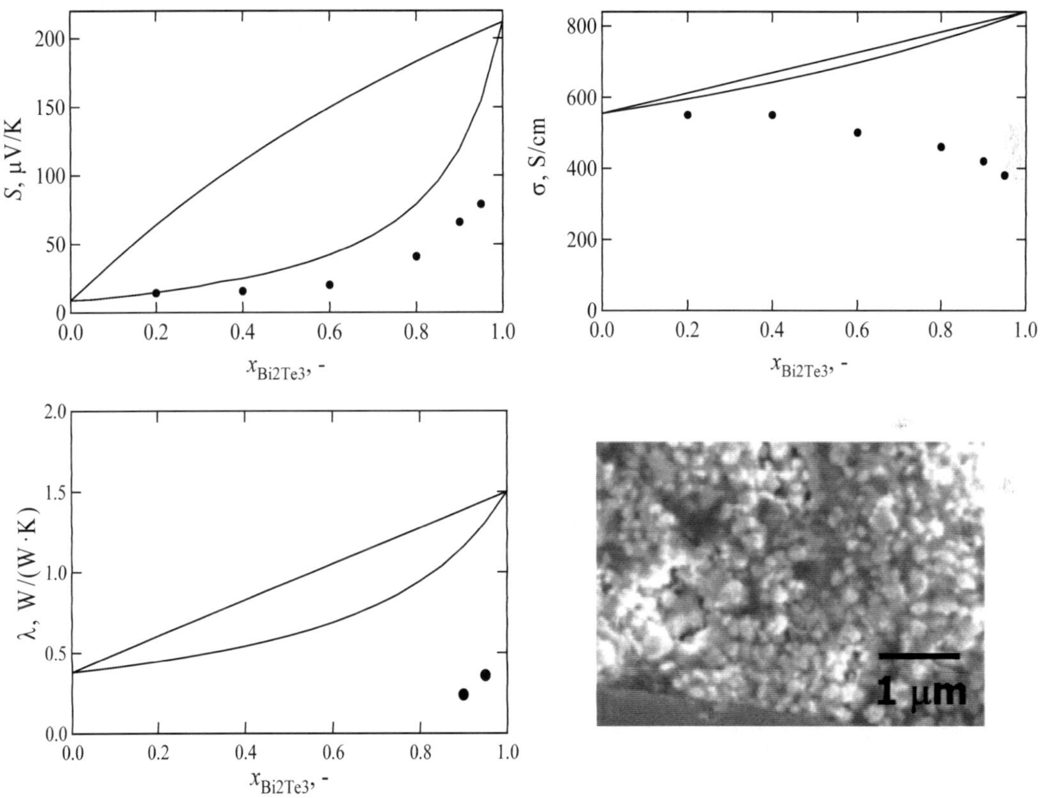

Figure 3. Thermoelectric properties of the composite films of Bi_2Te_3 nanoparticles and PEDOT:PSS in different Bi_2Te_3 volume fractions with simply expected values. Cross-sectional SEM image of a film are shown in lower right.

to be 0.2. Finally, we used polyimide with ionic-liquid for filling materials into the porous Bi_2Te_3 film. The sintering temperature can be increased because both polyimide and ionic-liquid are heat resisting materials. The SEM images of the sintered Bi_2Te_3 films were shown in figure 4. The polyimide with ionic-liquid was filled in pores between Bi_2Te_3 particles. The Seebeck coefficient of the composite film were increased as increasing sintering temperature up to 450°C. The maximum power factor was

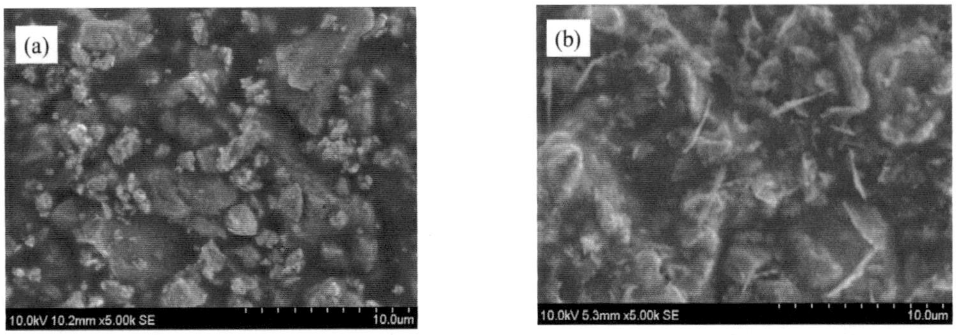

Figure 4. SEM images of composite films of p-Bi_2Te_3 particles and polyimide (a) without ionic-liquid, (b) with ionic-liquid sintered at 400°C.

measured to be 9.6μW/(cm·K^2) in the p-type film, and 3.5μW/(cm·K^2) in n-type films sintered at 415°C. The thermal conductivity of the film was 0.33 W/(m·K). The thermal conductivity of the polyimide film was measured to be 0.39 W/(m·K) by 3ω method. The effective thermal conductivity was lower than thermal conductivities of both polyimide and Bi_2Te_3. The measured thermoelectric properties of the composite films are summarized in table 1. The measured extremely low thermal conductivity resulted in high ZT of the films. The thermal transport can be suppressed due to strong phonon scattering at the interfaces between inorganic and organic materials. The evaluated number of interfaces with 10^{-7} (m^2·K)/W interfacial thermal resistance well explained the measured low effective thermal conductivity of the composite films in the present study.

4. Thermoelectric module

The thermoelectric mini-modules was made by screen printing method as shown in figure 5. The output power of the module was measured during heating up to 60°C of the bottom of the sheet module temperature as well as keeping 17°C of the top surface temperature of the sheet module. The measured open circuit voltage and output power were plotted in Fig. 6. The maximum output power was measured to be 200μW generated by a 5cm×5cm module. The sub-milli-watt output power was generated at 60°C by a printed low cost thermoelectric module.

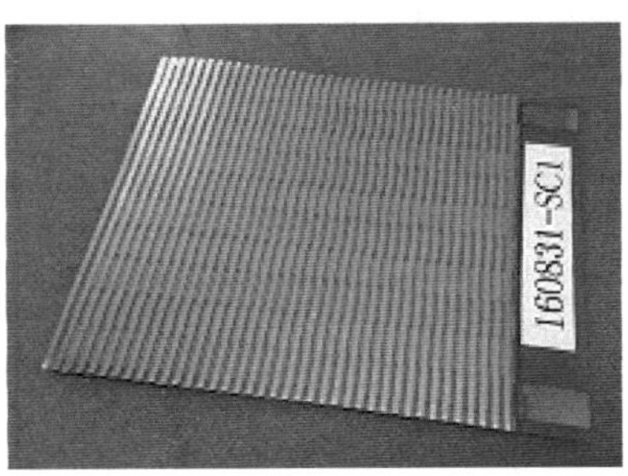

Figure 5. Printed thermoelectric module

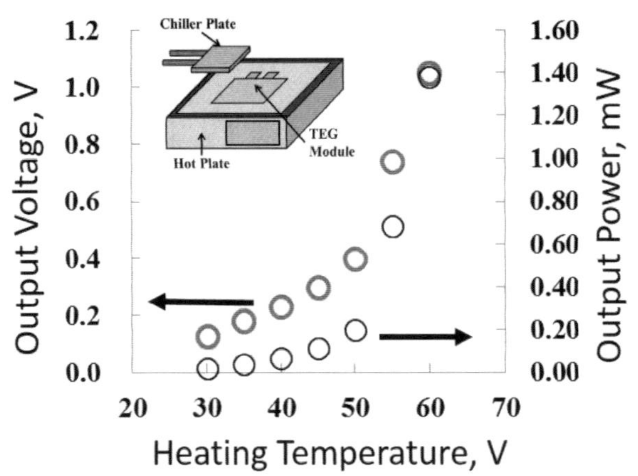

Figure 6. Performance of printed thermoelectric module. Inset is the schematic of the measurement system.

5. Conclusions

The composite of Bi_2Te_3 and conductive polymer were printed on the flexible substrate by a printing method to reduce the production cost as well as keeping the high thermoelectric properties. The maximum ZT was 0.87 for p-type and 0.5 for n-type by using polyimide with ionic-liquid as the conductive polymer. Thermoelectric performance was improved by high temperature sintering using heat-resisting materials. The high ZT was measured due to low thermal conductivity in the present study. The mechanisms of the thermal conductivity reduction should be investigated for the future work. Finally, the thermoelectric mini-generator was made by screen printing, and the performance was measured at near room temperature. The generated power was 200μW by a 5cm×5cm module at 60°C. We demonstrated the high performance of thermoelectric module prepared by low cost processes.

6. References

[1] Priya S and Inman D J 2009 *Energy Harvesting Technologies* Springer pp. v-vi
[2] Harb A. 2011 *Renewable Energy* **36** 2641-2654
[3] Kishi M, Nemoto H, Hamao T, Yamamoto M, Sudou S, Mandai M, and Yamamoto S 1999 *Proc.18th Int. Conf. on Thermoelectrics (Yokohama)*, 301–304
[4] Takashiri M, Shirakawa T, Miyazaki K and Tsukaoto H 2007 *Sens. Actuators A* **138** 329-324
[5] Kurosaki J, Yamamoto A, Tanaka S, Cannon J, Miyazaki K and Tsukamoto H 2009 *J. Electron. Mater.* **38** 1326-1330
[6] Snyder G J, Lim J R, Huang C, and Fleurial J P 2003 *Nature Mater.* **2** 528-531
[7] Kato K, Hagino H, and Miyazaki K 2013 *J. Electron.Mater.* **42** 1313-1318
[8] Madan D, Chen A, Wright P K, and Evans J. W. 2011 *J. Appl. Phys.* **109** 034904
[9] Zhang B, Sun J, Katz H E, Fang F, and Opila R L 2010 *ACS Appl. Mater. Interfaces* **2** 3170-3178
[10] Navone C, Soulier M, Plissonnier M, and Seiler A L 2010 *J. Electron. Mater.* **39** 1755-1759
[11] Takashiri M, Tanaka S, Takiishi M, Kihara M, Miyazaki K, Tsukamoto H 2007 *J. Alloys Compd.* **462**, 351-355
[12] Miyazaki K, Iida T, and Tsukamoto H 2003 *Proc.23rd Int. Conf. on Thermoelectrics (Montpellier)*, 641–643
[13] Ohta T, Kajikawa T, and Kumashiro Y 1990 *Electr. Eng. Jpn.* **110** 213-219
[14] Heremans J P, Jovovic V, Toberer E S, Saramat A, Kurosaki K, Charoenphakdee A, Yamanaka S, and Snyder G J 2008 *Science* **321** 554-557
[15] Hochbaum A I, Chen R, Delgado R D, Liang W, Garnett E C, Najarian M, Majumsar A, and Yang P 2008 *Nature* **451** 163-167
[16] Poudel B, Hao Q, Ma Y, Lan Y, Minnich A, Yu B, Yan X, Wang D, Muto A, Vashaee D, Chen X, Liu J, Dresselhaus M S, Chen G, and Ren Z 2008 *Science* **320** 634-638
[17] Kashiwagi M, Hirata S, Harada K, Zheng Y, Miyazaki K, Yahiro M, and Adachi C 2011 *Appl. Phys. Lett.* **98** 023114
[18] Kato K, Hatasako Y, Uchino M, Nakata Y, Suzuki Y, Hayakawa T, Adachi C, and Miyazaki K 2014 *Adv. Mater. Interfaces* **1** 1300015
[19] Venkatasubramanian R, Siivola E, Colpitts T, and O'Quinn B 2001 *Nature* **413** 597-602
[20] Harman T C, Taylor P J, Walsh M P, and Laforge B E 2002 *Science* **297** 2229-2232
[21] Shannon K Y, LeBlanc S, Goodson K E, and Dames C 2013 *Energy Environ. Sci.* **6** 2561-2571
[22] Yazawa K, and Shakouri A 2011 *Environ. Sci.Tech.* **45** 7548-7553
[23] Takashiri M, Miyazaki K, Tanaka S, Kurosaki J, Nagai D, and Tsukamoto H 2008 *J. Appl. Phys.* **104** 084302
[24] Scherrer H, and Sherrer S 1995 *CRC Handbook of Thermoelectics*, ed Rowe D M (London: CRC Press) p 227
[25] Takashiri M, Tanaka S, and Miyazaki K 2010 *Thin Solid Films* **519** 619-624

Acknowledgments
The authors wish to thank Prof. Takashiri at Tokai University for experimental support. This work was financially supported by JST-ASTEP (AS2621271L).

IOP Conf. Series: Journal of Physics: Conf. Series **1052** (2018) 012009 doi:10.1088/1742-6596/1052/1/012009

Development and optimization of high power density micro-thermoelectric generators

Wenhua Zhang[1], Juekuan Yang[2] and Dongyan Xu[1]

[1] Department of Mechanical and Automation Engineering, The Chinese University of Hong Kong, Shatin, New Territories, Hong Kong Special Administrative Region, China

[2] School of Mechanical Engineering and Jiangsu Key Laboratory for Design and Manufacture of Micro-Nano Biomedical Instruments, Southeast University, Nanjing 211189, China

dyxu@mae.cuhk.edu.hk

Abstract. This paper presents the design, fabrication, characterization, and optimization of micro-thermoelectric generators. The micro-thermoelectric generators are fabricated by a bottom-up approach combining pulsed electroplating and microfabrication techniques. This paper shows a complete set of experimental results of four prototypes and also reports on theoretical analysis that provides design guidelines for optimizing the power density of the prototypes.

1. Introduction

Thermoelectric generators (TEGs) are promising for harvesting waste heat from the environment to power wireless sensor networks in smart buildings. Despite of low energy conversion efficiency, TEGs have many advantages including high reliability, long lifetime, and environmental friendliness. Especially, compared to conventional heat engines, TEGs are compact, scalable, and can be easily driven by small temperature differences. Recently, several high-performance non-flexible and flexible TEGs are reported in the literature [1-3].

Electroplating is commonly used to fabricate TEGs on non-flexible substrates [2, 3]. In comparison with other techniques, electroplating has a number of advantages including efficient usage of raw materials, compatibility with microfabrication processes [4], and low contact resistance. In this work, we developed four non-flexible micro-TEG prototypes by combining pulsed electroplating and microfabrication techniques. Our TEGs demonstrate very low internal resistance and high power density, which can be attributed to the low parasitic contact resistance and high packing density of thermoelectric pillars [3].

2. Design and Fabrication of Micro-thermoelectric Generators

In this work, we designed and fabricated four cross-plane micro-TEG prototypes including TEG27, TEG355, TEG127, and TEG495. The micro-TEGs are fabricated by a bottom-up approach. Figure 1 shows the schematic of the fabrication process. The fabrication was conducted on silicon chips with 300 nm SiO_2 on both sides. First, bottom interconnectors and thin connecting lines are fabricated by sputtering, photolithography, and wet etching (Fig. 1a). Next, a 10 μm thick SU-8 layer is fabricated by photolithography with two open holes patterned on each bottom interconnector (Fig. 1b). One set of holes are covered by photoresist and n-type Bi_2Te_3 is deposited in open holes by the pulsed electroplating (Fig. 1c). Then, another layer of photoresist is used to cover the deposited n-type materials and p-type Sb_2Te_3 is deposited in the rest of open holes (Fig. 1d). Next, top interconnectors are formed by electroplating in a temporary photoresist mold (Fig. 1e-i). After that, the chip is soaked in acetone to remove the photoresist and then partial connecting lines are chemically etched away to break external electrical connections. Then, the micro-TEG is bonded with the top cover.

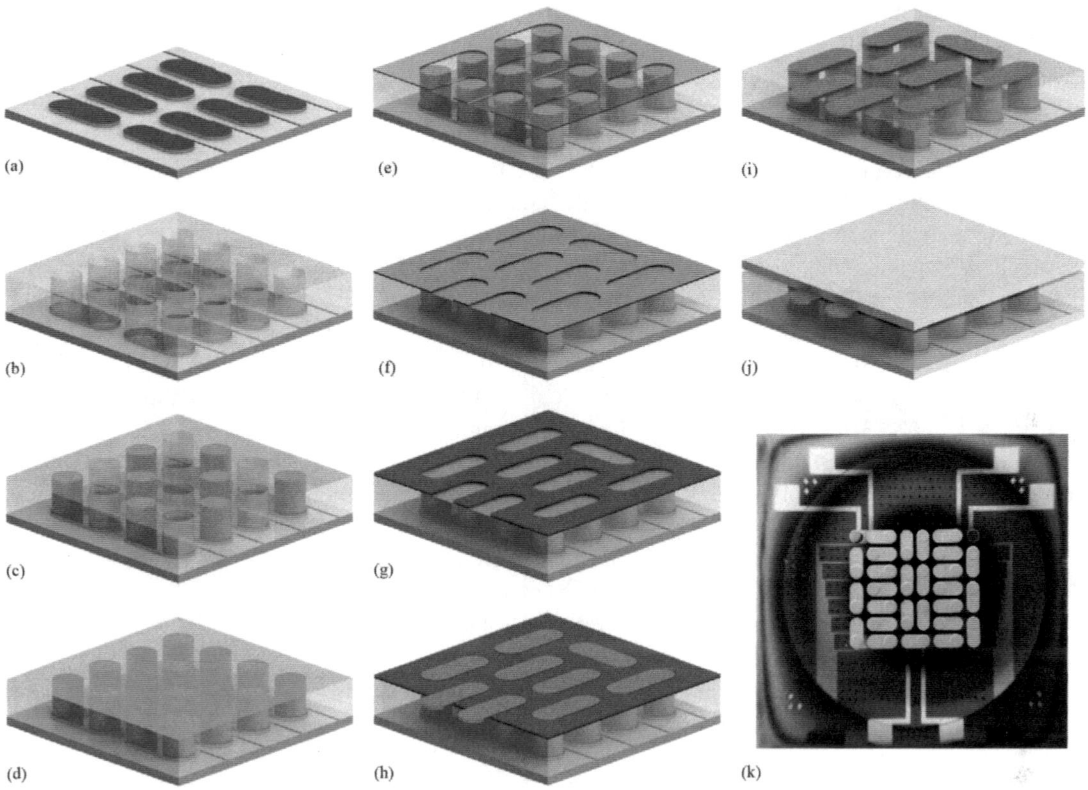

Figure 1. (a-j) The schematic of the fabrication process for the micro-TEGs. (k) The photograph of the TEG27 prototype.

The photograph of TEG27 is shown in Fig. 1k and those of other prototypes are given in Fig. 2. The dimensions of thermoelectric pillars, total number of pairs, and the room temperature internal resistance of all the prototypes are listed in Table 1.

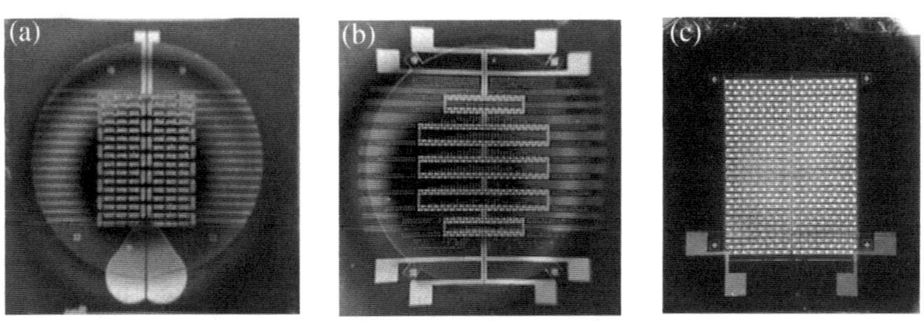

Figure 2. The photographs of the micro-TEG prototypes: (a) TEG127; (b) TEG355; and (c) TEG 495.

Table 1. Dimensions of thermoelectric pillars, total number of pairs, and internal resistance of four micro-TEG prototypes

	TEG 27	**TEG 355**	**TEG127**	**TEG495**
Diameter × Thickness	500 μm × 10 μm	100 μm × 10 μm	200 μm × 10 μm	200 μm × 10 μm
Number of Pairs	27	355	127	495
Internal Resistance	45.2 Ω	180.2 Ω	13 Ω	114.5 Ω

3. Device Performance

The open circuit voltage and the maximum power output of each prototype are characterized by using a home-made setup at different temperature differences and the testing results are summarized in Fig. 3. TEG27 and TEG355 demonstrate very low power output, which is mainly due to the large electrical resistance of thin bottom interconnectors (100 nm Au) and thus large internal resistance as shown in Table 1. In TEG127, bottom interconnectors are thickened by depositing 1.5 μm Au layer via DC electroplating. Meanwhile, thermoelectric properties of both n-type and p-type materials are further optimized. As a result, TEG127 produces a maximum power of 3 mW at $\Delta T = 52.5$ K, corresponding to a power density of 9.2 mW/cm^2, which is the highest value reported for the electroplated TEGs [3]. TEG495 demonstrates a maximum power of 4.4 mW at $\Delta T = 50$ K, which is larger than the counterpart of TEG127 but not as high as we expected. As can be seen in Fig. 3a, the open circuit voltage of TEG495 is 1.36 V at $\Delta T = 50$ K, ~3.4 times the value of TEG127. However, the maximum power of TEG495 at $\Delta T = 50$ K is only ~47% higher than that of TEG127 because of the abnormal increase of the internal resistance as shown in Table 1. The large internal resistance of TEG495 might be caused by the nonuniform current density during the pulsed electroplating, which might lead to larger electrical resistivity of thermoelectric materials.

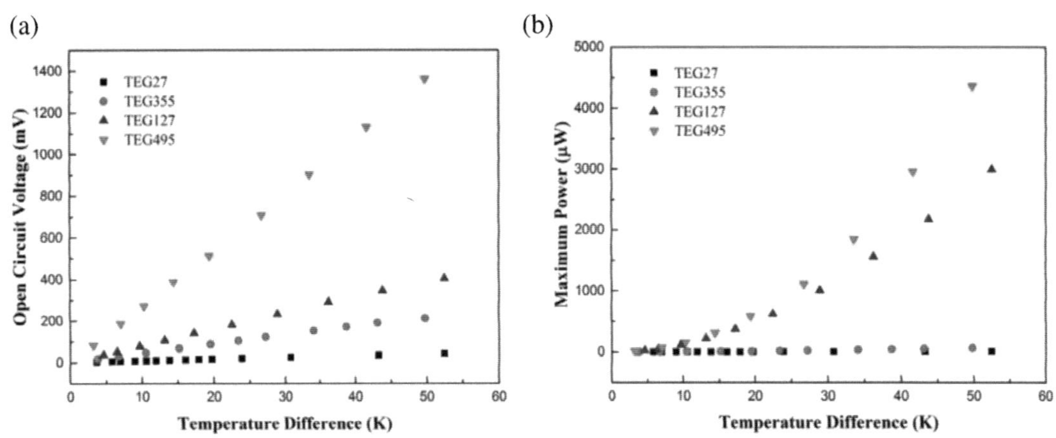

Figure 3. The open circuit voltage (a) and maximum power (b) of four micro-TEG prototypes at different temperature differences.

4. Design Optimization

TEG127 demonstrates the largest power density, which can be attributed to the low parasitic contact resistance and high packing density of the thermoelectric pillars [3]. To explore the possibility of further enhancing the power density of TEG127, we conducted theoretical analysis for this prototype using the model in Fig. 4. The simulation results are given in Fig. 5. Our theoretical analysis shows that the maximum power density of TEG127 might be further enhanced by optimizing the diameter (290 μm) and the thickness (11 μm) of thermoelectric pillars and decreasing the side length of the area containing the thermoelectric pillar.

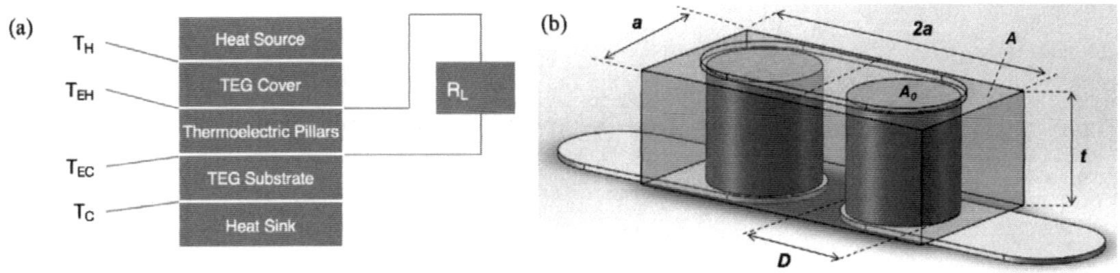

Figure 4. Schematic diagram of a micro-TEG (a) and one pair of n-type and p-type thermoelectric pillars embedded in the filling materials (b).

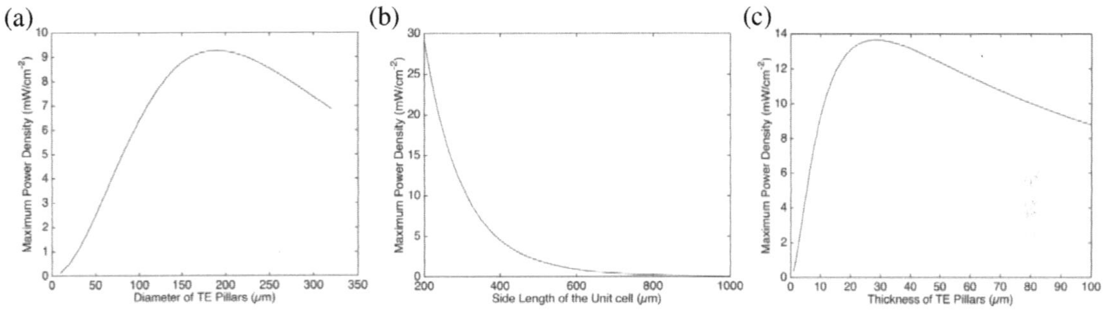

Figure 5. Variation of the calculated maximum power density with diameter of thermoelectric pillars (a), side length of the area containing one pillar (b), and thickness of thermoelectric pillars (c).

5. Conclusions

In summary, we developed four micro-TEG prototypes by combining pulsed electroplating with microfabrication techniques. TEG127 demonstrates the largest power density, ~9.2 mW/cm^2 at a temperature difference of 52.5 K. Theoretical analysis reveals that it is possible to further enhance the power density of this prototype by optimizing the design parameters.

Acknowledgments

The authors acknowledge the financial support from the Research Grants Council under Theme-based Research Scheme (T23-407/13-N) and the Innovation and Technology Commission (ITS/020/16FP) of the Hong Kong Special Administrative Region, China.

References

[1] Kim S J, We J H and Cho B J 2014 *Energy Environ. Sci.* **7** 1959
Kim S J, Lee H E, Choi H, Kim Y, We J H, Shin J S, Lee K J and Cho B J 2016 *ACS Nano* **10** 10851
Kim S J, Choi H, Kim Y, We J H, Shin J S, Lee K E, Oh M-W, Lee K J and Cho B J 2017 *Nano Energy* **31** 258
Park S H, Jo S, Kwon B, Kim F, Ban H W, Lee J E, Gu D H, Lee S H, Hwang Y, Kim J-S, Hyun D-B, Lee S, Choi K J, Jo W and Son J S 2016 *Nat. Commun.* **7** 13403
[2] Roth R, Rostek R, Cobry K, Köhler C, Groh M and Woias P 2014 *J. Microelectromech. Syst.* **23** 961
[3] Zhang W, Yang J and Xu D 2016 *J. Microelectromech. Syst.* **25** 744
[4] Zhou A, Fu Q, Zhang W, Yang B, Li J, Ziolkowski P, Mueller E and Xu D 2015 *Electrochim. Acta* **178** 217

IOP Conf. Series: Journal of Physics: Conf. Series **1052** (2018) 012010 doi:10.1088/1742-6596/1052/1/012010

Importance of grain size for nanostructured poly-Si thermoelectric material

R. Yanagisawa[1], N. Tsujii[2], O. Paul[3], T. Mori[2], and M. Nomura[1,4,5]

[1] Institute of Industrial Science, University of Tokyo, Tokyo, 153-8505 Japan
[2] National Institute of Material Science, Ibaraki, 305-0044 Japan
[3] IMTEK University of Freiburg, Freiburg, 79110 Germany
[4] NanoQuine, University of Tokyo, Tokyo, 153-8505 Japan
[5] PRESTO, Japan Science and Technology Agency, Saitama, 332-0012 Japan

r-yanagi@iis.u-tokyo.ac.jp

Abstract. Silicon thermoelectric device is one of the promising energy harvesters due to its compatibility with current semiconductor technology and environment, and various types of nanostructured silicon are studied to improve thermoelectric efficiency. In this work, we investigate thermoelectric performance of poly-Si membranes with PnC nanopatterning. The results show that PnC patterning reduces the thermal conductivity of membranes by 60 %, and grain size control enables more efficient reduction. These effects of nanostructure and grain can be combined, and important for high thermoelectric material efficiency.

1. Introduction

Thermoelectric devices promise a practical solution to energy harvesting technology required for remote devices developed by recent industries such as the Internet of Things (IoT). In recent years, much attention has been given to the development of Si-based thermoelectric materials with the hope of replacing the current standard material which is expensive and toxic. While the thermoelectric performance of bulk silicon is extremely low, nanostructured Si have reported large thermal conductivity reduction making a good thermoelectric material[1–3]. Furthermore, it is found that phononic crystal (PnC) nanopatterning impedes heat conduction and enhance the thermoelectric performance efficiently[4].

In this work, we experimentally characterized thermoelectric figure of merit ZT ($= S^2\sigma T/\kappa$) of poly-Si membrane nanostructured with PnC heading towards power generating device application. Figure 1 shows the schematic of device which can be fabricated in an SOI wafer by conventional CMOS-MEMS process. The device is composed of several p- and n-dope regions and suspended parts, which create in-plane temperature difference from the environmental temperature difference across the chip. Metal wires are arranged like as pi-module, so that each n- and p-dope regions are connected in series electrically and in parallel thermally. Similar device structure was already studied before[5], but the difference in our structure is that thermoelectric membrane is PnC nanostructured to enhance the temperature difference and output power.

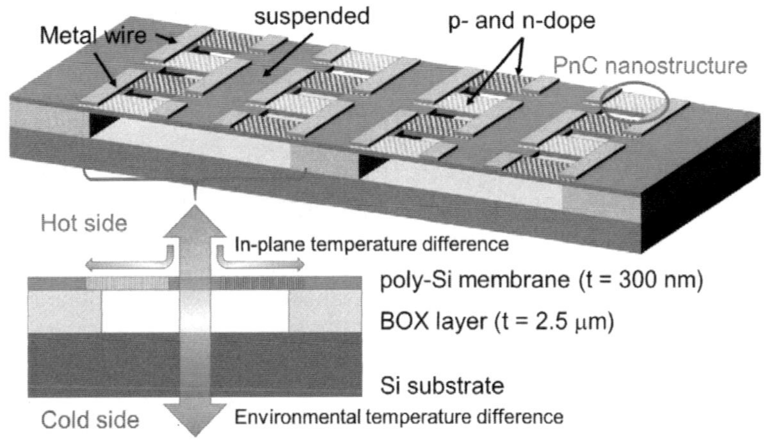

Figure 1. Schematic of membrane based thermoelectric power generator. Suspended parts induce in-plane temperature difference in thermoelectric elements, and PnCs enhance this difference.

2. Methods

We fabricated PnC nanostructures in a 300-nm-thick poly-Si membrane, and measured thermal and electrical conductivities. First, we deposited SiO_2 and poly-Si on Si substrate with LPCVD method. Deposition thickness and temperature is 2.5 μm at 425 °C for SiO_2 and 300 nm at 625 °C for poly-Si. Next, we implanted the dopant ions, Phosphorous and Boron for n- and p-dope respectively, in top poly-Si layer, and annealed the wafers to activate the carriers. Annealing time and temperatures are 30 minutes at 900 °C for n-dope and 950 °C for p-dope, and measured Hall carrier concentrations are 9.2×10^{19} cm^{-3} and 7.2×10^{19} cm^{-3} for n- and p-dope, respectively. Next, we deposited 5 μm × 5 μm Aluminium pads on the poly-Si surface by using the electron beam lithography and electron beam evaporator. Consequently, we made two dimensional PnCs, array of vertical holes, in the poly-Si layer by using electron beam lithography and plasma dry etching with SF_6 and O_2 as gas etchants. At the same time, two slots were also etched which define the width and length of PnC area. Finally, we removed SiO_2 layer under the poly-Si with vapor Hydrofluoric acid for thermal measurement samples, and deposited Gold wires for electrical measurement samples. Figures 2 show the scanning electron microscope (SEM) images of fabricated samples for both thermal and electrical measurement, with different hole radius.

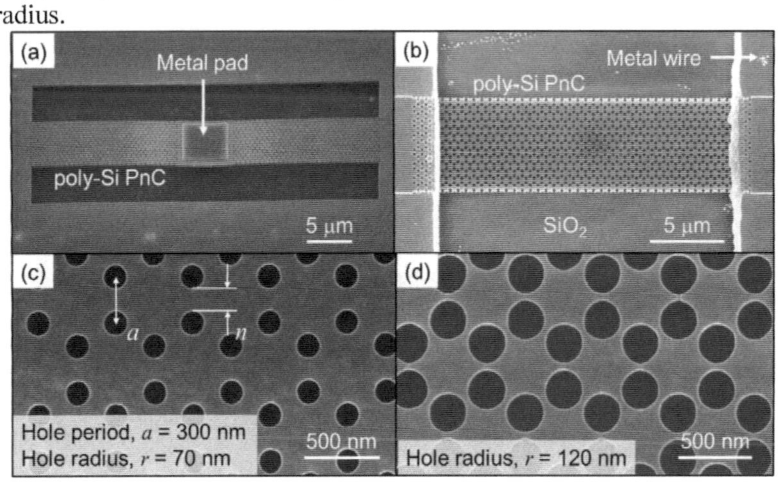

Figure 2. SEM images of fabricated samples (a) for μTDTR measurement and (b) for four probe measurement and detail of PnC with (c) small holes and (d) large holes. The neck size of PnC, distance between adjacent holes, is defined by hole period (a) and radius (r) as $n = a - 2r$.

We measured thermal conductivity of PnCs by means of micro time-domain thermoreflectance (μTDTR). In μTDTR method, we use two different lasers, pulse and probe, and both are focused on the metal pad between PnCs. While pulse laser heats the metal pad, probe signal gives us the reflectivity change of the metal pad caused by temperature rises and drops. From this signal, we obtain the thermal decay time constant of PnCs, and can estimate their thermal conductivities by comparing experiment with finite element model simulation. Detail of measurement can be found in our previous publications[6,7]. We also fabricated the samples for electrical measurement in same wafer, and measured electrical resistance of PnC by four-probe method, and comparing the experimental results with finite element simulation gives us electrical conductivity of PnC. To evaluate the thermoelectric figure of merit, we also measured Seebeck coefficient of a doped poly-Si membrane; measured values are -107 μV/K for n-dope and 147 μV/K for p-dope at 295 K.

3. Measurement results and discussion

Figure 3(a) shows thermal conductivity of nanostructured membrane depending on the PnC hole radius for both n-dope and p-dope samples. Thermal conductivity of membrane without PnC (52.6 W/(m·K) for n-dope and 21.3 W/(m·K) for p-dope) is already reduced from bulk mono crystalline Si value (130 W/(m·K)) because phonons are much scattered by thin film surface and poly crystalline grain boundary. When the PnC is created, thermal conductivity strongly drops from the value of membrane. As hole radius becomes larger, thermal conductivity decreases and reaches more than 60 % reduction from membrane for both n- and p-dope. This big reduction is come from strong phonon back scattering on hole surface, and as neck size (space between adjacent holes) become narrow, the effect become larger because the mean free path (MFP) of phonon is strongly limited by the neck part, where phonons are strongly scattered by the sidewall of the holes.

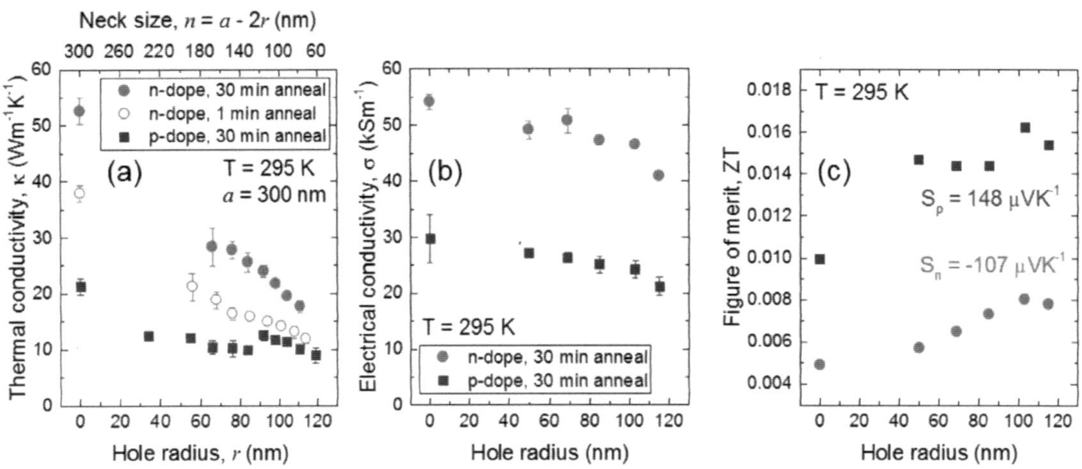

Figure 3. Measured thermoelectric properties of poly-Si membrane with PnC, (a) thermal conductivity, (b) electrical conductivity, and (c) figure of merit, as a function of PnC hole radius. In each figure, red circles and blue squares represent n-dope and p-dope respectively.

On the other hand, as we show in Figure 3(b), electrical conductivity of PnC does not decrease much, if hole radius is not too large. Since electron MFP in silicon is much shorter than phonon MFP, the impact of PnC pattern on electrical conductivity is less. As a result, we improve the figure of merit, ZT for both n- and p-dope. The values we obtain, however, are still too small to apply the device, further improvement is required especially for n-dope, and it can be achieved by reducing thermal conductivity.

It is known that thermal conductivity of poly crystalline material depends on the grain size[8], so we checked grain size with transmission electron microscope (TEM) to clarify the difference between thermal conductivities of n- and p-dope samples. From Figure 4, we found the large difference in grain size, and average grain sizes are estimated as 200 nm and 80 nm for n- and p-dope respectively. Following this, we supposed grain in n-dope Si grew too much during the anneal process causing high thermal conductivity, so we tried short time annealing for another n-dope wafer to make grain size smaller. As shown in Figure 3(a), short annealed n-dope samples have lower thermal conductivity than long annealed ones, it seems we successfully controlled grain growth by changing anneal condition.

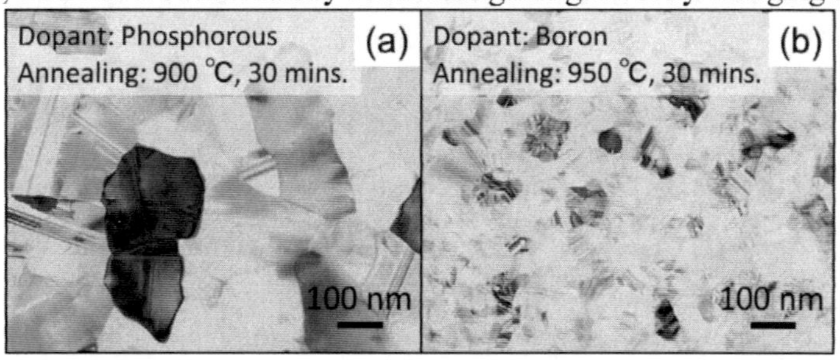

Figure 4. Surface TEM image of (a) n-dope and (b) p-dope poly-Si shows grain size of doped poly-Si strongly depends on the dopant.

4. Conclusion

We investigated thermal and electrical property of nanostructured poly-Si membranes for both n- and p-dope. We achieved to improve thermoelectric figure of merit by using proper PnC nanostructure in which phonons are scattered while electrons keep conductive. Furthermore, we showed controlling grain size is important for poly-Si thermoelectric device to reduce thermal conductivity. Since PnC nanostructure and grain boundary impact on thermal conductivity reduction by different mechanism, it is important to combine them for efficient thermoelectric material.

5. Acknowledgements

This work was supported by the Project for Developing Innovation Systems of the MEXT, Japan, Kakenhi (15H05869 and 15K13270), and PRESTO JST.

References

[1] Boukai A I, Bunimovich Y, Tahir-Kheli J, Yu J-K, Goddard III W A and Heath J R 2008 Silicon nanowires as efficient thermoelectric materials *Nature* **451** 168–71
[2] Hochbaum A I, Chen R, Delgado R D, Liang W, Garnett E C, Najarian M, Majumdar A and Yang P 2008 Enhanced thermoelectric performance of rough silicon nanowires *Nature* **451** 163–7
[3] Lim J, Wang H-T, Tang J, Andrews S C, So H, Lee J, Lee D H, Russell T P and Yang P 2016 Simultaneous Thermoelectric Property Measurement and Incoherent Phonon Transport in Holey Silicon *ACS Nano* **10** 124–32
[4] Nomura M, Kage Y, Müller D, Moser D and Paul O 2015 Electrical and thermal properties of polycrystalline Si thin films with phononic crystal nanopatterning for thermoelectric applications *Appl. Phys. Lett.* **106** 223106
[5] Xie J, Lee C and Feng H 2010 Design, Fabrication, and Characterization of CMOS MEMS-Based Thermoelectric Power Generators *J. Microelectromechanical Syst.* **19** 317–24
[6] Anufriev R, Ramiere A, Maire J and Nomura M 2017 Heat guiding and focusing using ballistic phonon transport in phononic nanostructures *Nat. Commun.* **8** 15505
[7] Maire J, Anufriev R and Nomura M 2017 Ballistic thermal transport in silicon nanowires *Sci. Rep.* **7** 41794
[8] Wang Z, Alaniz J E, Jang W, Garay J E and Dames C 2011 Thermal Conductivity of Nanocrystalline Silicon: Importance of Grain Size and Frequency-Dependent Mean Free Paths *Nano Lett.* **11** 2206–13

Transparent Thin Film for Energy Harvesting

M. Uenuma, J. C. Felizco, D. Senaha and Y. Uraoka

Nara Institute of Science and Technology,
8916-5Takayamacho, Ikoma, Nara 630-0192, Japan

uenuma@ms.naist.jp

Abstract. Amorphous InGaZnO (a-InGaZnO) has already been proved to have good applications in thin-film transistor devices for display application. The a-InGaZnO is n-type semiconductor material and has enormous potential such as transparency, a low temperature fabrication process. Thus, the application of a-InGaZnO to flexible electronics and sensors has been studied. A low thermal conductivity and a low temperature process are merit for a thin film or a flexible thermoelectric module application. In this study, thermoelectric properties of a-IGZO thin film are evaluated. Transparent thermoelectric generator was demonstrated using a-InGaZnO and ITO electrodes.

1. Introduction

Application of thermoelectric conversion of unused waste heat issuing from sources such as the human body and daily lives has recently been studied. It is necessary to investigate new applications such as transparent and flexible thermoelectric devices to incorporate such innovations practically into our daily lives. Therefore, transparent and flexible thermoelectric generators are becoming more attractive for smart windows and wearable systems. To obtain a high thermoelectric property (figure of merit: ZT), a low thermal conductivity is indispensable parameter.

Amorphous InGaZnO (a-InGaZnO) is n-type transparent oxide semiconductor material and has enormous potential such as transparency, a low temperature fabrication process. Thus, the application of a-InGaZnO to flexible electronics for thin film transistor has been studied. The thermal conductivity of the InGaZnO is around 1.0 W/mK due to the amorphous structure [1,2]. Therefore, low thermal conductivity and a low temperature process are merit for a transparent and a flexible thermoelectric module application. Oxide thermoelectric materials can be also used for high temperature applications.

Our group researched thermoelectric properties in the a-InGaZnO thin film [3,4]. We also demonstrated transparent thermoelectric generator (TEG).

2. Experiment

At first, thermoelectric properties of a-InGaZnO film were measured using the sample as shouwn in Fig. 1. The a-InGaZnO films with thickness of 200 nm were deposited on quartz glass substrate (20 mm × 5 mm, 0.5 mm thickness) using sintered targets (In:Ga:Zn _2:2:1, at.%) by RF magnetron sputtering. We employed an input RF power of 100 W in a mixed Ar and O2 atmosphere at fixed total pressure of 0.6 Pa. After the sputtering process, the a-IGZO samples were annealed. After annealing, AU/Mo electrodes were deposited by electron beam evaporation and formed for ohmic contact. Carrier concentrations were modified by changing the sputtering condition. The Seebeck coefficient and electrical conductivity were obtained in the temperature range from 100 K to 400 K using the thermal transport option for the physical property measurement system (PPMS, Quantum Design, Inc.). The thermal conductivity was measured using LFA.

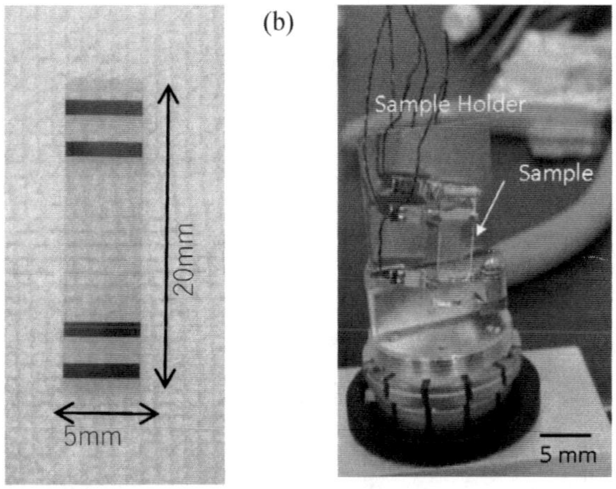

Figure 1. (a) a-InGaZnO/glass sample to evaluate thermoelectric properties. (b) Thermal transport sample puck (PPMS, Quantum Design, Inc.) with a-InGaZnO sample.

3. Measurement and results of a-InGaZnO thin film

Hall mobility was increased with increasing carrier concentration (Fig. 2 (a)). The 13 cm^2/Vs was obtained at the carrier concentration of 9.8×10^{19} cm^{-3} (oxygen flow ratio was 0% in the sputtering process). Figure 2 (b) shows the values of thermal conductivity. The thermal conductivity was around 1.2 – 1.3 W/mK at 300 K. The thermal conductivity was slightly increased by the carrier concentration due to a heat conduction carrier of electrons. The reported values of poly crystal ZnO or AlZnO film thermal conductivity were around 4 – 6 W/mK [5,6]. Therefore, the amorphous oxide has an advantage for thermoelectric application. Electrical conductivity and Seebeck coefficient were shown in Fig. 3. These results the trade-off relation between electrical conductivity and Seebeck coefficient. ZT value was 0.02 ~ 0.03 at 300 K and 0.14 ~ 0.2 at ~800 K (thermal conductivity was assumed 1.3 W/mK). Even amorphous material, the ZT value of a-InGaZnO was almost comparable with InGeO and AlZnO bulk materials owing to the lower thermal conductivity (Fig. 4).

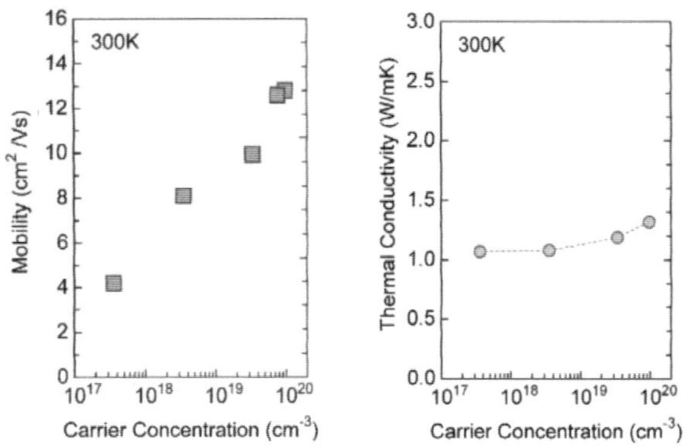

Figure 2. Hall motility and thermal conductivity of amorphous InGaZnO sample with different carrier densities.

IOP Conf. Series: Journal of Physics: Conf. Series **1052** (2018) 012011 doi:10.1088/1742-6596/1052/1/012011

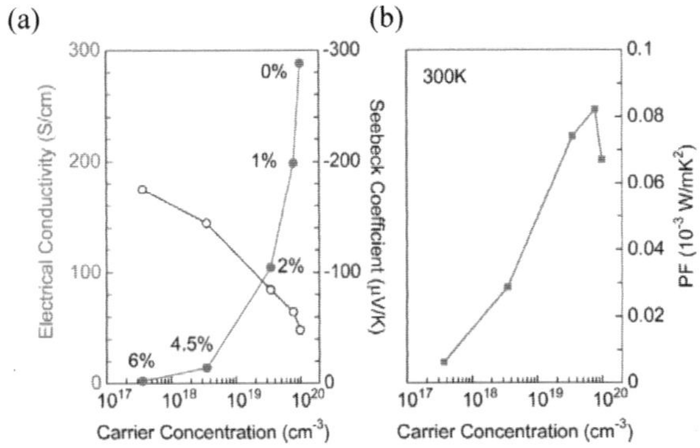

Figure 3. (a)Variations of electrical conductivity, Seebeck coefficient and (b) Power factort, as functions of carrier concentration for the different sputtering oxygen ratios of the a-InGaZnO thin films at 300 K (b) PF as functions of carrier concentration at 300 K.

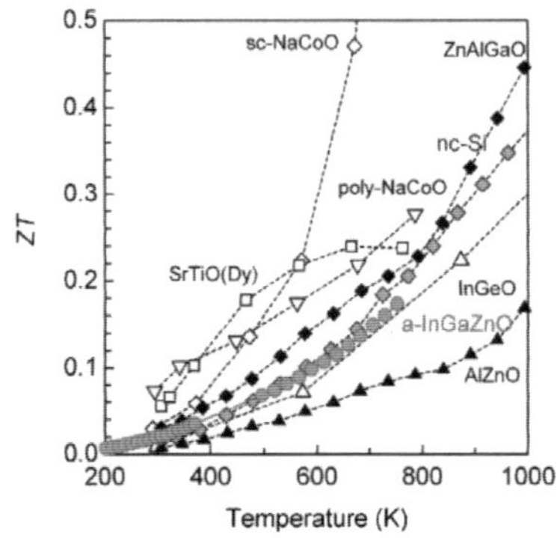

Figure 4. Temperature dependence of ZT values of a-InGaZnO film. Reported values of other bulk oxides were also plotted. [7-10]

4. Demonstration of transparent TEG

Transparent thermoelectric device was demonstrated using a-InGaZnO and ITO electrode uni-leg structure (Fig.5). The sample TEG was fabricated using ITO (155 nm)/Glass substrate with photolithography and etching process. The ITO has a negative Seebeck value (< -10 μV/K). But it was relatively smaller then a-InGaZnO value. Thus we used the ITO as electrode. The 70 pairs of a-InGaZnO and ITO stripes was designed. As shown in Fig.6, the demonstrated TEG was fully transparent. Output power of the demonstrated device was still small. However, the output power will be improved by optimizing the atomic composition of InGaZnO and the device structure.

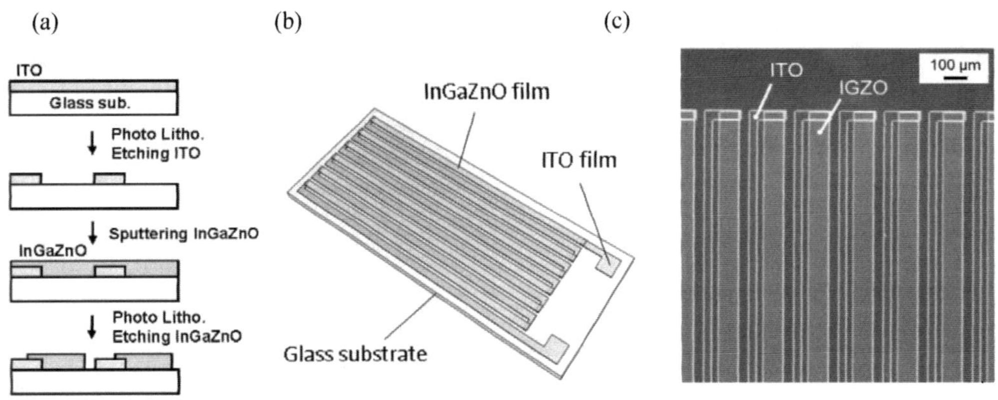

Figure 5. Transparent thermoelectric device. (a) Fabrication process. (b) Schematic image of the sample. (c) Optical microscope image of InGaZnO and ITO uni-leg structures.

(a) 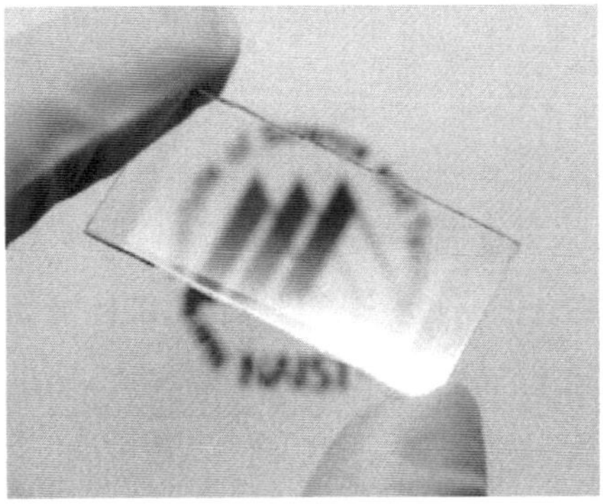

(b) 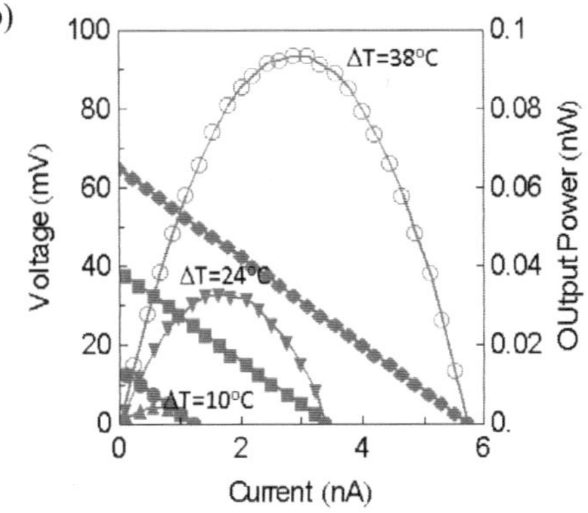

Figure 6. Transparent thermoelectric device of InGaZnO and ITO uni-leg structures. (a) Device image and (b) output characteristics.

5. Conclusion

The thermoelectric properties of amorphous InGaZnO thin film were shown. The low thermal conductivity and relatively high Hall mobility was advantages for thermoelectrical applications. Power factor value was still small. But ZT values of amorphous InGaZnO were almost consistent with other poly-crystal oxide materials even amorphous structure. The transparent TEG was investigated using ITO and a-InGaZnO uni-leg structures. For increase the output power, composition of InGaZnO material and TEG structures needs to be further improvement.

References

[1] T. Yoshikawa et al. 2013 *APEX* **6**,021101
[2] S. W. Cho et al. 2013 *J. Nanomat.* 909786
[3] Y. Fujimoto, M. Uenuma, Y. Ishikawa, and Y. Uraoka, 2015 *AIP advances*, **5**,97209,1-6
[4] Y. Fujimoto, M. Uenuma, Y. Ishikawa, and Y. Uraoka, *34th ICT 2015 and 13th ECT 2015*
[5] S. saini, et al., 2015 *J. Electron. Mater.*, Vol. **44**, No. 6
[6] S. Saini, et al. 2014 *Jpn. J. Appl. Phys.* **53**, 060306
[7] H. Muta, et al., 2003 *J. Alloys Compd.*, **350**, 292–295
[8] M. Ohtaki, et al, 2009 *J. Electron. Mater.*, **38**, 1234–1238
[9] K. Fujita, et al.,2001 *Jpn. J. Appl. Phys.*, **40**, 4644–4647
[10] D. Berardan, et al, 2008 *Solid State Comm.*, **146**, 97

IOP Conf. Series: Journal of Physics: Conf. Series **1052** (2018) 012012 doi:10.1088/1742-6596/1052/1/012012

Pulsed discharge of printed secondary Zn-MnO_2 batteries for IoT and wearable devices

B J Kim[1], J W Evans[2] and P K Wright[1]

[1]Department of Mechanical Engineering, University of California, Berkeley, California, 94720, USA

[2]Department of Materials Science and Engineering, University of California, Berkeley, California, 94720, USA

E-mail: bernard.kim@berkeley.edu

Abstract. This paper reports the manufacturing and testing via pulsed discharging of printed secondary Zn-MnO_2 batteries with ionic liquid gel polymer electrolyte (GPE) for IoT and wearable devices. Printed cells under constant current discharge conditions demonstrated average DC internal resistances of 175-225 Ω/cm^2. Printed cells under pulsed discharge current demonstrated decreasing internal resistances with increasing cycle number, reaching an average plateau of 125 Ω/cm^2 after 50 cycles and remaining beyond 650 cycles. This work improves upon manufacturing methods reported at PowerMEMS 2015 and investigates system-level compatibility with IoT and wearable devices.

1. Introduction

As devices for the Internet of Things continue to develop, there is a need for safe and inexpensive energy storage to power them [1]. These devices generally operate in two regimes: a long period of low-current draw (*e.g.* sleep mode), or a short, high-current pulse (*e.g.* radio transmission) [2].

While battery performance under constant current discharge is a well-understood metric, performance under pulsed discharging, especially for smaller footprint batteries (on the order of 1 cm^2), must be better characterized. Of particular importance is the cell internal resistance, which is proportional to the ohmic (IR) drop observed upon pulse discharging. The magnitude of this drop must be minimized in order to maintain an adequate voltage to power device electronics and avoid brownouts as well as to maximize useable cell capacity (Fig. 1).

This work improves upon a previously presented layer-by-layer manufacturing process in order to decrease cell internal resistance and characterizes fully printed cell performance under a pulsed discharging regime [3].

2. Experimental Methods

Cells utilize a Zn-MnO_2 chemistry with an ionic liquid-based gel polymer electrolyte. The gel polymer electrolyte functions as both a mechanical separator layer as well as an ion permeable layer to facilitate ion transport between the cathode and anode. The ionic liquid [BMIM][OTf] with the dissolved salt $Zn(OTf)_2$ was used as the electrochemically functional electrolyte. This system has been proven to be rechargeable, as has been previously reported [3, 4].

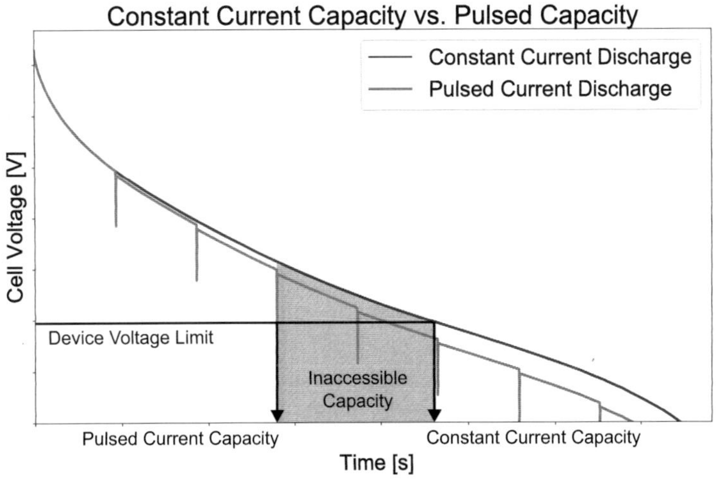

Figure 1. Constant current capacity vs. pulsed current capacity. Due to ohmic drops exceeding the device voltage limit, effective cell capacity and runtime are diminished.

Table 1. Compositions of electrode and electrolyte inks after drying. All materials were measured in wt%.

Cathode	Anode	Electrolyte
91% MnO_2	91% Zn	75% 0.5M ZnOtf in [BMIM][Otf]
2.57% Acetylene Black	2.57% Acetylene Black	–
6.43% PVDF-HFP	6.43% PVDF-HFP	25% PVDF-HFP

Separate inks were made for the cathode, anode, and gel polymer electrolyte, whose compositions are presented in Table 1. The electrode ink compositions were chosen to minimize printed film resistivity. Cells were manufactured using a layer-by-layer additive process. All casting steps took place in ambient at room temperature, and drying steps took place in air or under vacuum at 80°C. After the cells were fully dried, they were removed from the substrate and sealed in stainless steel 2032 coin cells (MTI Corp) for testing. The ink preparation and manufacturing processes were followed as previously reported [3].

Cells were tested on a Bio-Logic BCS 805/810 battery tester. Constant current discharge tests were performed from 1.0-1.8V with 125μA charge/discharge current. For pulsed discharge tests, a 1mA discharge pulse was applied to the cell 5 times for 4ms each, resting 1s between pulses. After discharging, the cell was charged to 1.8V at 125μA and then allowed to rest for 5 minutes before repeating (Fig. 2). To measure cell solution resistance, EIS measurements were also taken.

3. Results & Discussion

3.1. Effect of Drying Environmet on Internal Resistance

EIS results indicate drying the GPE under vacuum produces cells with lower solution resistance compared to cells dried in air (Fig. 3). This is further supported by DCIR measurements during cycling, as cells dried under vacuum exhibit lower DC internal resistance at rest (Fig. 4). Scanning electron micrographs comparing GPE microstructures indicate variations in porosity as

IOP Conf. Series: Journal of Physics: Conf. Series **1052** (2018) 012012 doi:10.1088/1742-6596/1052/1/012012

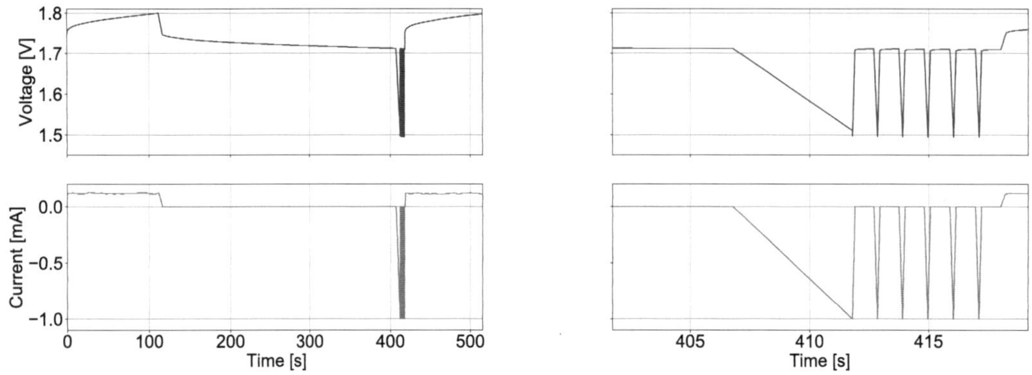

Figure 2. *left*: Full pulse discharge profile; *right*: Inset of pulse discharge region.

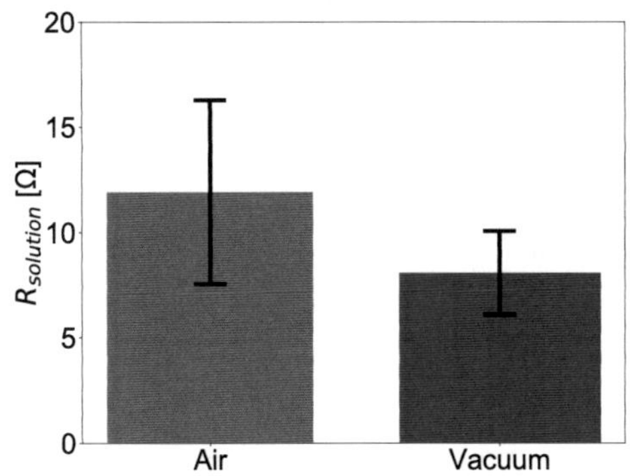

Figure 3. Average EIS measurements of solution resistance by drying atmosphere. Both samples were dried at 80°C.

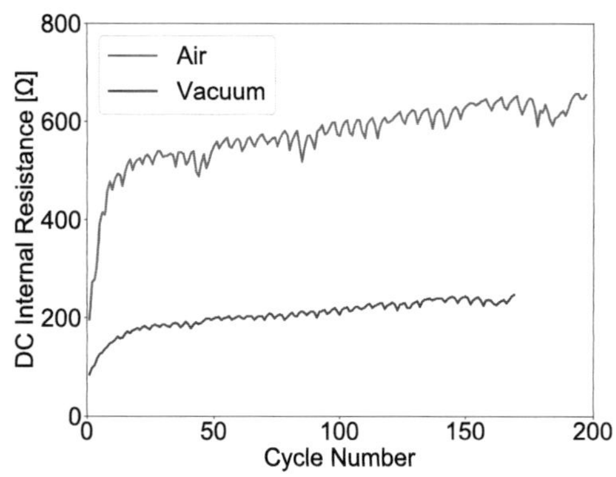

Figure 4. DCIR at rest for cells dried in air and under vacuum cycled under constant current discharge

well as polymer micelle radius (Fig. 5), which could be responsible for this observed difference in internal resistance. The difference in microstructure may affect electrolyte uptake, which could affect ion transport through the separator layer. Further investigation is necessary to characterize the impact of polymer microstructure on cell internal resistance.

3.2. Pulsed Discharging Performance

Figure 6 presents the change in cell internal resistance as a function of cycle life for cells cycled under constant current discharge and pulsed discharge. Cells cycled under pulsed discharge displayed constant internal resistance with increasing cycle number while cells cycled under constant current displayed steadily increasing interal resistance until failure. Furthermore, under pulsed discharging cells, cells cycled repeatable with no loss in performance for over 650 cycles, after which the test was stopped (Fig. 7). This increase in cycle life and plateau in DCIR may be due to the lower magnitude of charge being passed per cycle as well as the shorter time scale for each cycle, inhibiting the number of charge transfer reactions that can occur. Future testing will seek to combine constant current and pulsed discharging to better simulate IoT device operation.

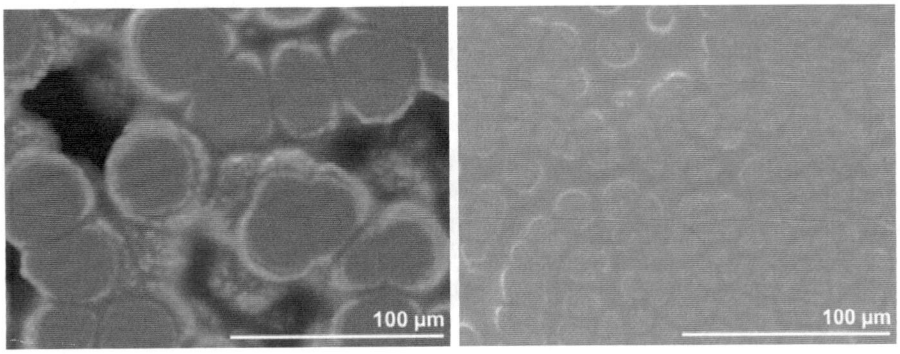

Figure 5. Differences in GPE microstructure as a result of drying atmosphere. Micelle radius and porosity are visually different. Both samples were dried at 80°C. *left*: air ; *right*: vacuum

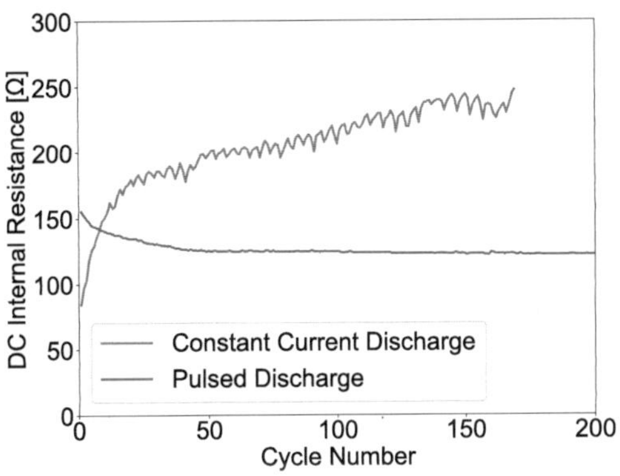

Figure 6. DCIR for cells dried under vacuum under constant current discharge and pulsed discharge

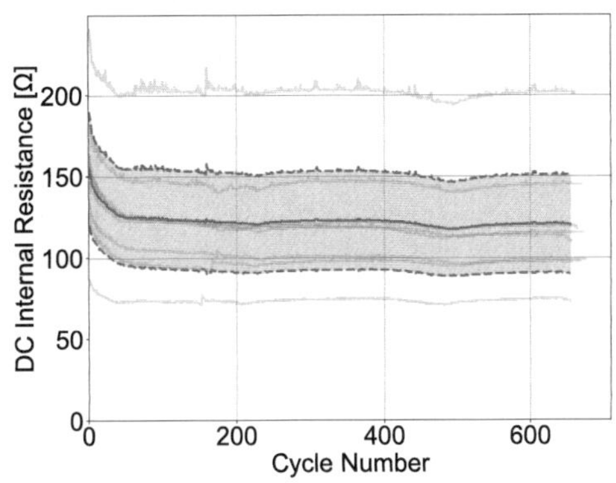

Figure 7. Mean DCIR for all cells tested under pulsed discharging with 95% confidence interval

4. Conclusion

Fully printed cells were manufactured in different atmospheres and tested via constant current and pulsed discharge. The resulting GPE microstructures as a result of drying atmosphere and their effects on cell internal resistance were investigated. GPEs dried under vacuum demonstrated lower solution resistance and cell internal resistance. Cells tested under pulsed discharging exhibited lower and more stable DCIR than cells tested under constant current discharge. Under pulsed discharging, cells were able to provide a 1mA, 4ms pulse current for over 650 tested cycles. Future testing will incorporate both discharging regimes into a single test.

References

[1] Swan M 2012 *J. Sens. Actuator Netw.* **1** 217-253
[2] MacKenzie J D and Ho C 2015 *Proc. of the IEEE* **103** 535-553
[3] Kim B, Winslow R, Lin I, Gururangan K, Evans J and Wright P 2015 *J. Phys.: Conf. Series* **660**
[4] Kumar GG and Sampath S 2003 *Solid State Ionics* **160** 289-300

IOP Conf. Series: Journal of Physics: Conf. Series **1052** (2018) 012013 doi:10.1088/1742-6596/1052/1/012013

HIGH RECYCLABILITY AND POWER PERFORMANCE OF A THIN MICRO LITHIUM-ION BATTERY ANODE

X L Kuang[1,2,3], K D Li[1,2,3], Y F Liu[1,2,3], X J Feng[4] and X H Wang[1,2,3]

[1] Tsinghua National Laboratory for Information Science and Technology, PR China
[2] Institute of Microelectronics, Tsinghua University, Beijing, PR China
[3] Department of Microelctronics and Nanoelectronics, Tsinghua University, Beijing, PR China
[4] GCL System Integration Technology CO., Ltd., Suzhou, PR China

E-mail: wxh-ime@tsinghua.edu.cn

Abstract. We report a high recyclability and fast charging/discharging speed of a thin silicon/carbon anode, which can implement direct prototyping of three dimensional (3D) micro-battery on chip. The silicon/carbon material uses SU-8 as carbon source, and forms Si/void/C structure to provide enough space for volume expansion of the SiNPs during lithium-ion intercalation. By using roll forming, the anode has been demonstrated to enjoy the superior recyclability (563.42 mAh/g in the third cycle, while 472.96 mAh/g after 500 cycles) and the high charging/discharging speed (1.35 hour to finish a whole charge/discharge cycle) with only 60μm thickness.

1. Introduction

Electrochemical energy storage has been demonstrated as one of the most critical technology for a variety of applications, such as portable electronic devices, electric vehicles and renewable energy storage. Silicon has been thought as the most attractive high-capacity anode materials because of its high theoretical capacity (4200 mAh/g), which is more than 10 times to conventional graphite anodes (372 mAh/g) [1]. In order to improve the low conductivity of silicon and enhance the recyclability, which is quite low caused by silicon's large volume expansion (over 300%) during cycling, many silicon and carbon composite materials have been tested [2]. In previous work, X. Li *et al.* has developed a silicon/void/carbon anode by using SU-8 photoresist as carbon source, and so that it is possible to implement direct prototyping of 3D micro-battery on chip as shown in Figure 1 [3].

However, the anode is always broken when carbonizing the graphical SU-8/nanocomposite particles with high depth-to-width ratio structure, just like what we have fabricated in Figure 2 shown. Figure 2a demonstrates that the graphical interdigital Si/void/C anode after lithography. The structure is easily broken as shown in Figure 2c, while carbonizing the SU-8 photoresist because the high stress releasing.

Here, we have fabricated a film anode by using Si/void/C composite and it maybe a new way to form lithium-ion battery on chip. And this study also tests out the best rolling thickness of the 200 μm anode film by balancing capacity retaining and charging/discharging speed.

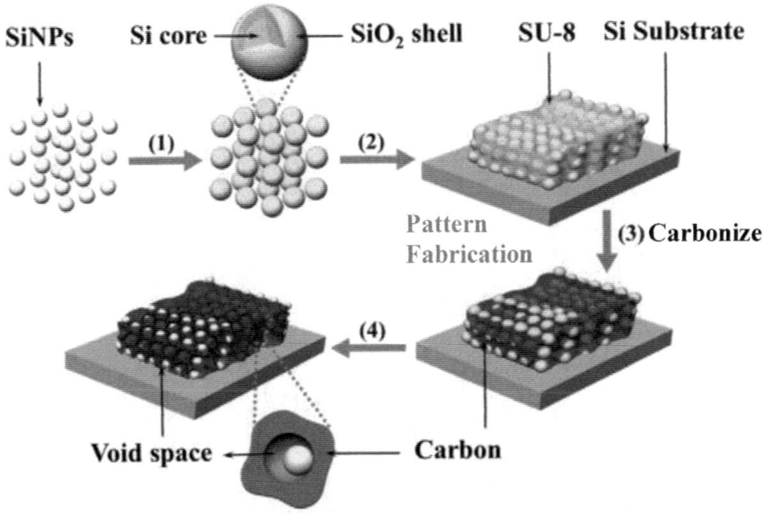

Figure 1. Schematic design and fabrication of the graphical Si/void/C anode based on SU-8. (1) TEOS oxidation process to generate SiO_2 sacrificial layer; (2) Mixing Si/SiO_2 composite particles with SU-8 photoresist and spin-coating on Si substrate; (3) Pattern and pyrolysis process to carbonize SU-8; (4) Etching away SiO_2 layer, forming void space between carbon scaffold and SiNPs. [3]

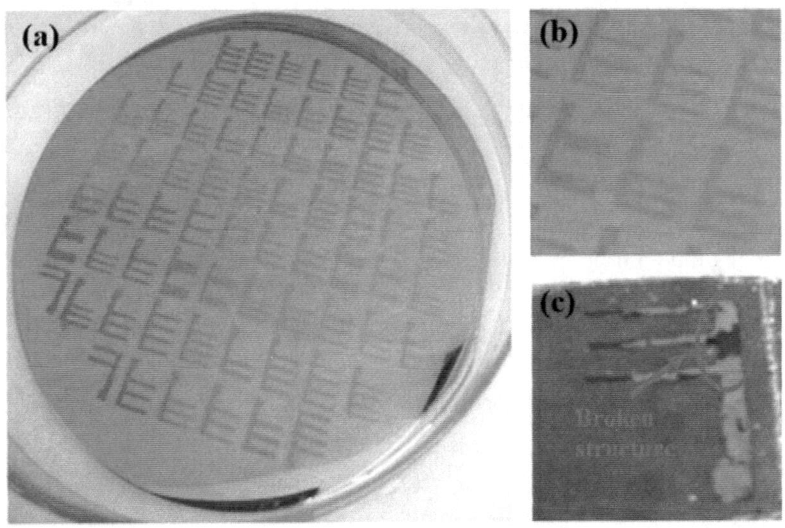

Figure 2. (a) and (b): The images of graphical interdigital Si/void/C anode after lithography with different resolution. (c): The broken interdigital structure of Si/void/C anode after carbonization.

2. Fabrication

The Si/void/C composite material is prepared by using the method in the previous work as shown in Figure 1 [3]. In the first step, SiNPs (about 30~50 nm diameter) coated by SiO_2 layer are fabricated with hydrolysis and condensation of tetraethoxysilane (TEOS) by using typical Stober method [4]. The thickness of SiO_2 shell is 10~15 nm by choosing the weight ratio of silicon and TEOS as 1:10. In the second step, the Si/SiO_2-templated SU-8 galatin are formed by dispersing the Si/SiO_2 particles into the SU-8 photoresist.

Milling and rolling the composite materials, and then filling them into the interdigital electrode maybe another way to form the battery on chip. Also, this method can be used in the industrial production without changing the fabrication process. The silicon/void/C composite is mixed with Super-P and CMC (9:0.5:0.5) by ball milling, and coated on the Cu foil with 200μm thickness. Then, the roll forming is used to improve the volume specific capacity by rolling the anode electrode into thinner thickness.

3. Results and Discussion

Figure 3 shows the SEM images of the Si/void/C anode before and after milling. In Figure 3a, the Si/void/C structure can be seen clearly, and the carbon network is intact. While in Figure 3b, although some structure on the surface are broken during milling, numerous silicon still remains in the carbon shells or holes (broken shells). Therefore, this structure with carbon shells or holes can still contain the volume expansion during the lithium-ion intercalation and deintercalation.

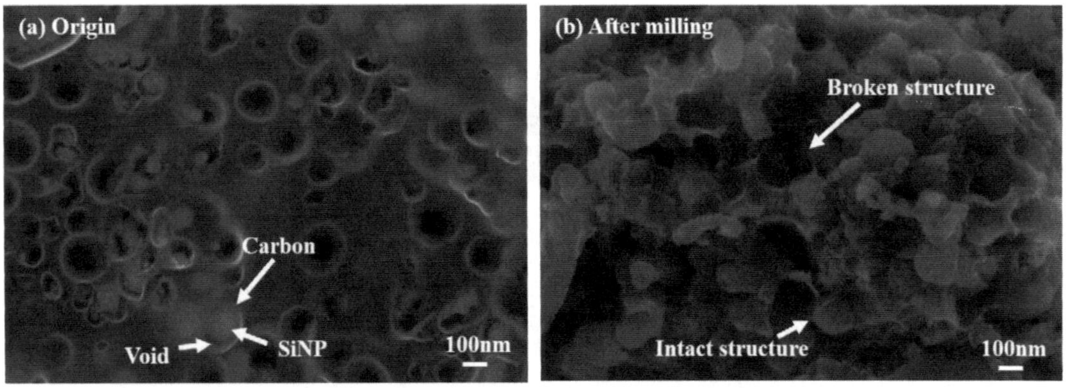

Figure 3. The SEM of Si/void/C anode with different forming process: (a) The direct forming interdigital Si/void/C anode; (b) The Si/void/C anode after ball milling.

The anode material is compacted to different degree by roll forming as shown in Figure 4. The thickness of anode is approximately determined by the thickness of rolling model and Cu foil. For example, the rolling model used in Figure 4a is 40 μm and all the Cu foil is about 20 μm (the Cu foil also is condensed in this step). It can be easily found that the structure is more loose with a thicker anode than the thinner one. In another word, the transmission of lithium-ion will be hampered while increasing the mass per unit volume of the anode material.

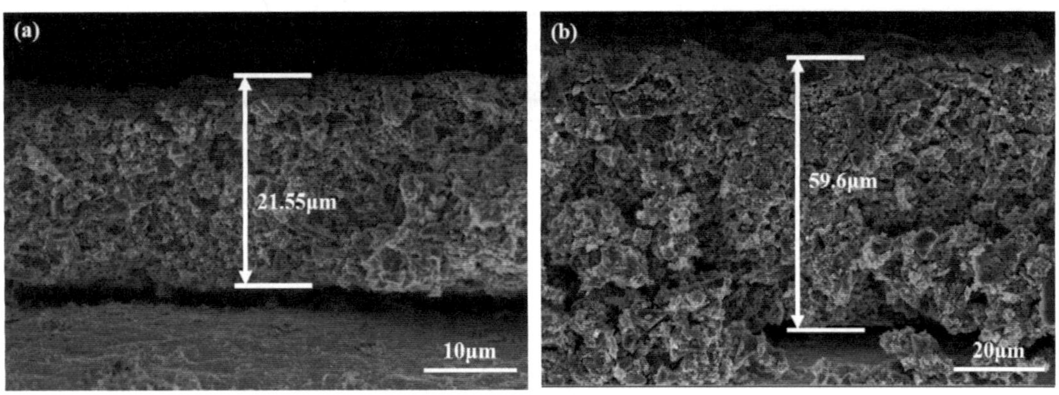

Figure 4. The SEM of the Si/void/C anode with different thickness after rolling forming: (a) The thickness of rolling model is 40μm; (b) The thickness of rolling model is 80μm.
*the thickness of anode = the thickness of rolling model - the thickness of Cu foil

The galvanostatic charge-discharge tests of the anode are performed using a battery testing system (BTS-3000, Neware) at room temperature, with voltage cutoffs set at 0.01V and 1.2V vs. Li/Li+. The charge/discharge rate is calculated with respect to the theoretical capacity of silicon (4200 mAh/g). Therefore, a current density of 1C corresponds to 4200 mA/g.

Figure 5 gives the recyclability test of four samples with different thickness (20μm, 40μm, 60μm, and 80μm). It can be found that the weight specific capacity is decreased while the thickness decreasing, because the void space structure is destroyed more seriously with thinner anode. In all four anodes, the anode with 60μm shows the highest recyclability (563.52 mAh/g in the third cycle, and 472.96 mAh/g after 500 cycles), cause the opportune compression degree. The thicker anode has a lower capacity retain caused by the reaction between more electrolysis and silicon, and the thinner anode shows capacity fading rapidly as there is few void space for silicon expansion.

The charge/discharge profiles of various Si/void/C anode with different thickness are tested to verify the power performance in Figure 6. From these curves, the thinner anode shows the faster charging/discharging speed than the thicker one. for the non-compacted ionic transport channel.

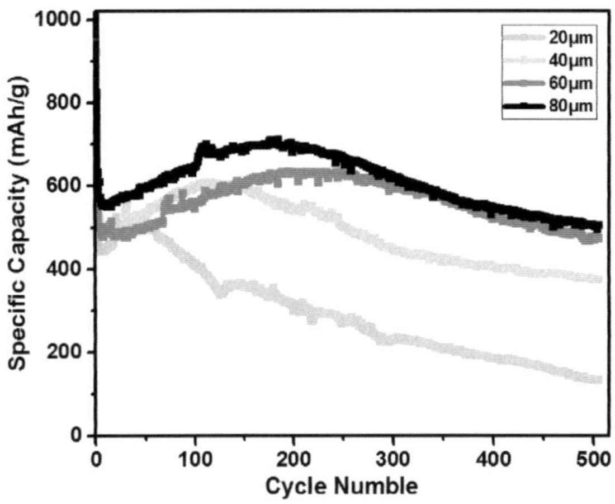

Figure 5. The specific capacities of Si/void/C anode with different thickness after roll forming under 0.2C galvanostatic cycling tests.

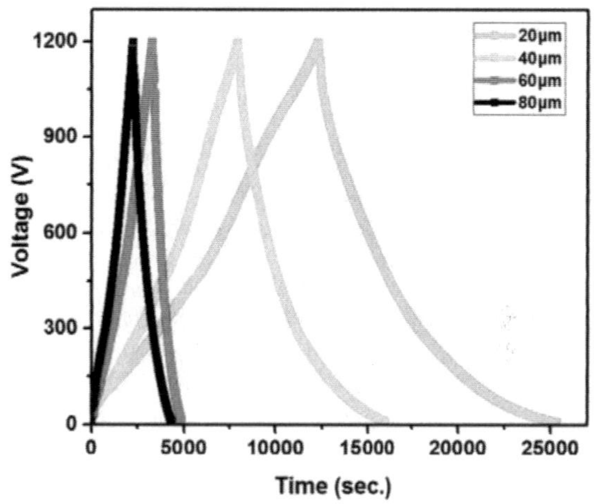

Figure 6. Charge/discharge profiles of various Si/void/C anode with different thickness.

4. Conclusion

We have fabricated a high recyclability and fast charging/discharging speed of a thin anode by using Si/void/C composite, and it can implement direct prototyping of three dimensional (3D) micro-battery on chip in the future. By balancing the capacity retaining and charging/discharging speed, this anode demonstrates a superior recyclability (563.42 mAh/g in the third cycle, while 472.96 mAh/g after 500 cycles) and a high charging/discharging speed (1.35 hour to finish a whole charge/discharge cycle) with only 60μm thickness.

References

[1] A. S. Arico, et al., *Nature Materials*, vol. 4, pp. 366-377, 2005.
[2] M. L. Terranova, et al., *J. Power Sources*, vol. 246, pp. 167-177, 2013.
[3] X. Li, et al., *MEMS 2015*, Estoril, Portugal, 18-22 January, 2015.
[4] W. Stober, et al., *J. Colloid and Interface Sci.*, vol. 26, no. 1, pp. 62-69, Jan, 1968.

Acknowledgement

This work is supported by the National Natural Science Foundation of China (No. 61531166006), Suzhou-Tsinghua Leading Innovation Specific Project (No.2016SZ0302) and 863 program (No. 2015AA043502) of China. The authors also thank Prof Qiu'group in the Department of Chemistry, Tsinghua University, for providing necessary experimental equipments and helpful instructions.

IOP Conf. Series: Journal of Physics: Conf. Series **1052** (2018) 012014 doi:10.1088/1742-6596/1052/1/012014

Photovoltaic module active self-cleaning surface using anisotropic ratchet conveyors fabricated with parylene-C stencil

Di Sun and Karl F. Böhringer

Department of Electrical Engineering, University of Washington, Seattle, WA 98195, USA

E-mail address: karl@ee.washington.edu

Abstract. This paper describes an active self-cleaning surface using an anisotropic ratchet conveyor (ARC) to move a water droplet under orthogonal vibrations. Two different ARC systems were designed and fabricated with self-assembled monolayers and hydrophobic Cytop thin films. A novel way to create micro-sized patterns on Cytop was developed without degrading the original Cytop hydrophobicity. Droplet transport speed is 20.5 mm/s at frequency 52 Hz and acceleration 3.3 g ($g = 9.8\ m/s^2$). Optical transmission measurements show Cytop can improve optical transmittance by 2.5~3.5% over the entire visible wavelength range. Solar cell module power output is measured to verify the output power efficiency gain. The self-cleaning function is demonstrated with ARC patterns.

1. Introduction

Photovoltaic (PV) modules are often installed in remote arid regions with high levels of sand and dust particles. Dust accumulation forms a non-uniform layer on solar panels that block the sunlight and degrade photovoltaic module efficiency over time. The composition of dust particles majorly consists of quartz and silicate minerals, with a distribution size in the 10~100 μm range [1]. In order to reduce maintenance costs and to minimize dust accumulation on the solar panel cover glass, methods have been proposed using superhydrophobic surface modification with micro textures [2], electrodynamic screens [3], photocatalytically active metal oxide to create superhydrophilic surfaces [4], among others.

The ideal properties of self-cleaning coatings and their system design include low adhesion of dust, optical transparency without degradation of light transmission, and the capability to remove dust particles from the top of the solar cell. Water droplets can be directed along the surface using chemical gradients [5], thermal gradients [6], electrowetting [7], and micro textures [8-9] while they are carrying away dust particles along their path and keeping the surface clean. In this paper, we design, fabricate and test an active self-cleaning surface system using anisotropic ratchet conveyors (ARC) by creating periodic micro-sized semi-circular hydrophilic rungs on a hydrophobic background. Two different designs and their corresponding fabrication processes are proposed here: a perfluoro-octyltrichlorosilane (FOTS, contact angle 108°) – trimethylsilanol (TMS, contact angle 53°) system and a Cytop (contact angle 110°)-TMS system. Mico patterning on hydrophobic surfaces (like Teflon, Cytop) is difficult using standard photolithography due to poor adhesion between the photoresist and substrate. Methods have been proposed [10] but the original surface properties will be damaged after treatment with decrease water droplet contact angle. By adopting parylene-C as a stencil mask, we create hydrophilic patterns on top of the Cytop surface without degrading the original surface properties.

2. System design

The design of the ARC surface is shown in Fig. 1. To introduce an anisotropic wetting force, the periodic, semicircular, chemically heterogeneous ratchet rungs are patterned on the silicon/glass surface. Each hydrophilic semicircular rung has 10 μm rung width and 50 μm centre to centre gap width between the adjacent rungs. The radius of curvature is 1000 μm and the curve angle is 120°. The remainder of the silicon wafer surface is treated with a hydrophobic coating. The whole substrate is mounted on a vibration stage.

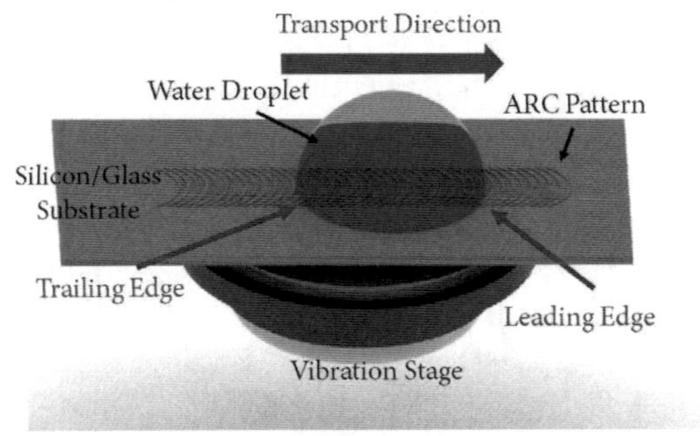

Figure 1. Sessile water droplet sitting on ARC surface. The red periodic rung structures were treated with TMS (water contact angle 53°). The remaining substrate was treated with FOTS or Cytop (water contact angle 108° and 110°, respectively).

As a sessile droplet is pipetted on the ARC surface, the water droplet wets the ratchet patterns along the rungs, creating an air-water-solid three-phase contact line (TPL) along its outer edge. We denote the portion of the water droplet edge that aligns with the rung curvature, which has a mostly continuous TPL, as the leading edge of the droplet, while the other portion, which has only intermittent TPLs across different rungs, is called the trailing edge of the droplet. During each vibration cycle, the leading edge provides higher pinning force than the trailing edge as the water droplet expands and recesses. This asymmetry in pinning forces causes water droplets to move toward the direction of the rung curvature [11].

2.1. Fabrication process

The patterning of FOTS-TMS system could be found from the reference [11]. As for the Cytop patterning process, after Si/glass wafer cleaning, diluted Cytop (Cytop CTL-809M: CTL-Solv.180= 3:1) was spin coated on a silicon wafer and baked under 180°C for 1 hour. Then 2.5 μm parylene was evaporated on the Cytop using a commercial parylene coater (PDS 2010, Specialty Coating Systems, Inc.) under vacuum. 6 μm photoresist (AZ9620) was coated and patterned. The parylene stencil mask and Cytop were etched through with O_2 plasma using reactive ion etching (Vision RIE). Then the parylene stencil mask was peeled off with tweezers. The surface was treated with spin-on HDMS (MicroPrime MP-P20) and baked at 110°C for 2 min.

2.2. Experimental setup

The test wafer was mounted on an aluminum platform, which is attached to the vibration exciter (Brüel and Kjær type 4809) with double adhesive tape. A sinusoidal wave signal was generated by the function generator and amplified by the power amplifier to drive the vibration exciter (Brüel and Kjær type 2718). The vibration amplitude was monitored with a vibrometer (Polytec OFV) via an oscilloscope (Agilent Infiniium). The water droplet movement was captured by high speed camera (FASTCAM Mini UX100) at 2000 fps. 2" by 2" solar cells were sandwiched between the cover glass and the support glass. After the surface treatment and patterning with FOTS-TMS and Cytop-TMS coatings, light transmission data was collected with a Cary 5000 UV-Vis-NIR spectrophotometer (Agilent Technologies, Inc.). Solar I-V curves were measured using the self-custom test bench. We used flood light as the light source and the distance between the solar cell modules and the light source was kept the same for each measurement.

3. Measurement results

3.1. Water droplet movement characterization

A water droplet was pipetted and the drop silhouette was monitored via high-speed camera. The progression of the leading and trailing edge position with time of both FOTS-TMS and Cytop-TMS system were characterized. The droplet transportation on the Cytop-TMS surface, plotted in Fig. 2(a), shows that the leading and trailing edge expanded at 0.57±0.09 mm and 0.36±0.14 mm, respectively, from the rest position during the expansion phase, while the leading and trailing edge recessed 0.28±0.05 mm and 0.74±0.17 mm during the contraction phase. This anisotropy of position change caused the droplet to move forward after each vibration cycle. Fig. 2(b) shows the water droplet contact angle change with time. Pinning anisotropy of the leading and trailing edge can be clearly observed during the recession phase from the contact angle change with time. Compared with the FOTS-TMS system, Cytop-TMS requires lower vibration amplitude and energy to drive the water droplet movement.

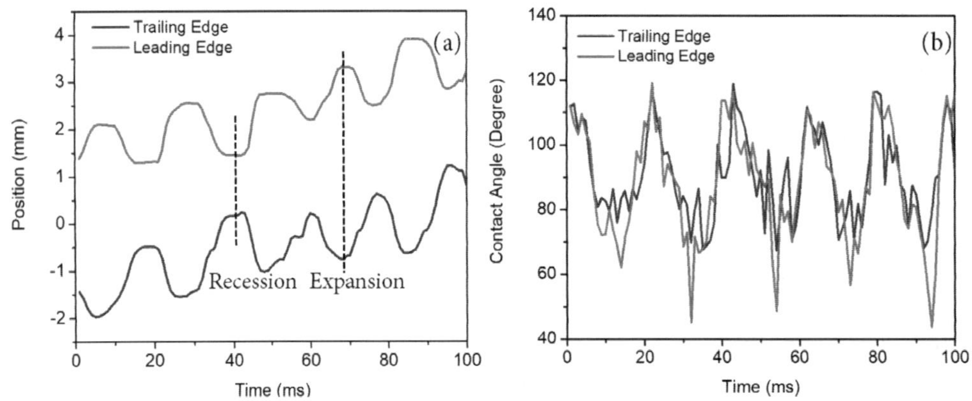

Figure 2. (a) Position change of the leading edge and trailing edge with time. The droplet center at rest is set as the origin. (b) Contact angle change of the leading edge and trailing edge. The water droplet (10 μL) moves forward from its initial position.

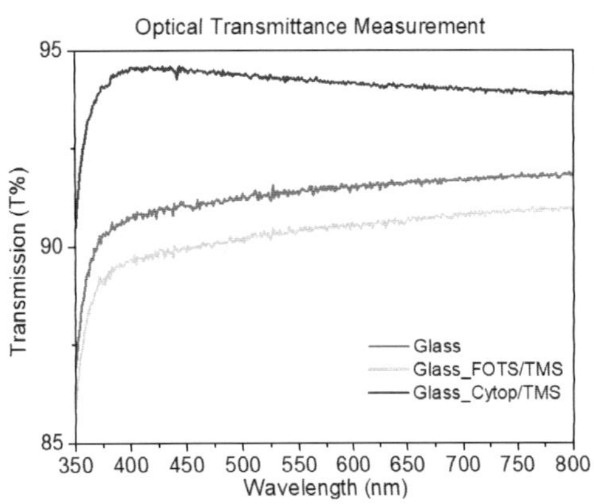

Figure 3. Optical transmittance measurement results with a wavelength range of 350 nm~800 nm. 4" soda lime glass wafers were used as the substrate baseline.

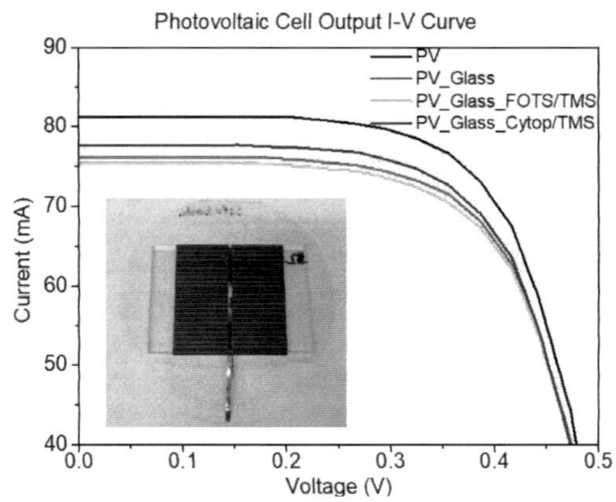

Figure 4. I-V curve measurement results of bare PV cells, PV cells with glass, FOTS-TMS, and Cytop-TMS system. [Insert] Solar cell module with cover glass treated with Cytop-TMS for measurements.

3.2. Light transmission and solar I-V output

Optical transmission measurements within the visible wavelength range are shown in Fig. 3. As for the FOTS-TMS system, the light transmission was degraded due to the added coating of monolayers, but within a range of less than 1%. The glass after FOTS-TMS treatment was transparent and optically flat. Meanwhile, the Cytop-TMS system improved the transmission with an enhancement of 2.5% ~ 3.5% over the visible wavelength range even with an added coating on top of the glass. The reason was that the refractive index of Cytop is ~1.34, which is between air (n_{air}=1.0) and the glass substrate (n_{glass} =

1.5), providing a refractive index match. Like Rayleigh's film, a portion of the incoming light reflects both at the interface of air/Cytopand Cytop/glass but has less reflection as compared to the single reflection at the air/glass interface with a larger refractive index mismatch. Fig. 4 shows the I-V curve measurements for assembled photovoltaic (PV) modules. The ARC structure was patterned in a zig-zag way covering a 2" by 2" area in the upper right of the solar cell. The Cytop-TMS coating generated larger optical output power compared with bare glass and FOTS-TMS surface treatment, in accordance with the light transmission measurements.

3.3. Self-cleaning surface demo

A self-cleaning surface should clean all randomly placed particles on the surface. Fig. 5 presents cleaning of sweetener particles by a droplet as a proof-of-concept. Through additional testing, sand particles, salt grains and airborne dust have all been cleaned by the ARC surface. 10 μL water droplets can carry up to 2 mg of sand particles, representing 20% of the water droplet's original weight.

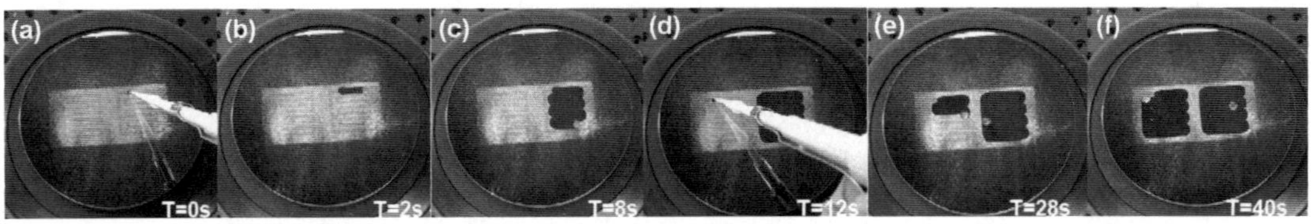

Figure 5. Surface cleaning performance for sweetener (dextrose, maltodextrin, and sucralose) contamination on ARC surface consisting of two ARC tracks. (a)-(f) are subsequent video frames of the experiment. All sweetener on the ARC areas is cleaned by the droplet.

4. Conclusions

In this paper, we demonstrate a proof of concept self-cleaning surface system with ARC tracks using both FOTS-TMS and Cytop-TMS systems. Cytop was successfully patterned using parylene as the stencil mask. Both optical transmission and solar cell power output I-V curves were measured. The Cytop-TMS system can combine both antireflection and self-cleaning functions, which provides an opportunity for PV module applications. For the future plan, we will perform outdoor field tests and explore large-scale, low-cost manufacturing processes creating ARC surfaces on solar panel cover glass.

Acknowledgements

Fabrication was performed at the Washington Nanofabrication Facility (WNF), a member site of the NSF National Nanotechnology Coordinated Infrastructure (NNCI), with support from NSF award ECCS-1308025. We would like to thank the funding and fellowship support from CoMotion and from the Clean Energy Institute at the University of Washington.

References

[1] L.L. Kazmerski, A.S.A.C. Diniz, C.B. Maia, M.M. Viana, S.C. Costa, P.P. Brito, C.D. Campos, L.V.M. Neto, S. De Morais Hanriot, and L.R. De Oliveira Cruz, IEEE J. Photovoltaics **6**, 719 (2016).
[2] R. Blossey, Nat. Mater. **2**, 301 (2003).
[3] R. Sharma, C. Wyatt, J. Zhang, M.K. Mazumder, and N. Mardesich, **1**, 434 (2007).
[4] R. Wang, K. Hashimoto, A. Fujishima, M. Chikuni, E. Kojima, A. Kitamura, M. Shimohigoshi, and T. Watanabe, Nature **388**, 431 (1997).
[5] M.K. Chaudhury and G.M. Whitesides, Science (80). **256**, 1539 (1992).
[6] D.E. Kataoka and S.M. Troian, Nature **402**, 794 (1999).
[7] M.G. Pollack, A.D. Shenderov, and R.B. Fair, Lab Chip **2**, 96 (2002).
[8] T.A. Duncombe, E.Y. Erdem, A. Shastry, R. Baskaran, and K.F. Böhringer, Adv. Mater. **24**, 1545 (2012).
[9] A. Shastry, M.J. Case, and K.F. Böhringer, Langmuir **22**, 6161 (2006).
[10] C. C.CHO, R.M. Wallace, and and L.A. FILES-SESLER, J. Electron. Mater. **23**, (1994).
[11] T.A. Duncombe, J.F. Parsons, and K.F. Böhringer, Langmuir **28**, 13765 (2012).

A T-shaped, plate-type thermoelectric power generator for realizing the higher power density at a small temperature difference

H. Tohmyoh, T. Daimon, N. Ohgi

Department of Finemechanics, Tohoku University, Aoba 6-6-01, Aramaki, Aoba-ku, Sendai 980-8579, Japan

E-mail: tohmyoh@ism.mech.tohoku.ac.jp

Abstract. We report the performance of a T-shaped, plate-type thermoelectric power generator (TPG). The plate-type TPG has been proposed by the authors. A new design of the plate-type TPG realized the higher power density together with the higher power at a small temperature difference. For example, the maximum power of the T-shaped, plate-type TPG was 11.1 W m^{-3} at the temperature difference of 40 K, and this value was over 2 times greater than that of the previous TPG.

1. Introduction

Among the power generation technologies utilizing the renewable energy resources, such as wind, solar, vibration, etc., thermoelectric power generation has been attracted considerable attentions recently because our daily lives are full of waste heat. Seebeck effect is a phenomenon, which convert heat into electricity, can be explained simply by considering an electrical circuit comprising bi-metal wires. If there is a temperature gradient ΔT between two bi-metal junctions, a thermoelectromotive force, ΔV, appears in the circuit and this is given by $\Delta V = S\,\Delta T$, where S (= $\alpha_1 - \alpha_2$) is the Seebeck coefficient of the circuit, and α_1 and α_2 are the absolute thermoelectric powers of each wire metal.

Although the thermoelectric power generators (TPGs) composed of Bi, Te or its related alloys realize greater ΔV [1, 2], such rare metals are the limited resources. Moreover, TPGs composed of Bi, Te or its related alloys have relatively higher internal resistance, which increases the power consumed in the circuit. To realize the efficient power generation from a small ΔT without using such rare metals, a plate-type TPG, which is composed of an Fe plate and an Al layer, has been proposed by the authors [3].

In this paper, a T-shaped, plate-type TPG, which was also composed of Fe and Al, was designed for realizing the higher power density at a small temperature differences. The volume of the T-shaped TPG was 1 / 4 compared with the previous one, and the power densities of the T-shaped TPG measured at the temperature difference of 20 to 60 K were over 2 time greater than that of the previous one.

2. Plate-type thermoelectric power generator

Figure 1(a) shows the photograph and the cross section of the plate-type TPG [3]. The TPG is composed of the Fe plate of 0.1 mm thick. and the Al layer, and the interface between Fe and Al is electrically isolated with the SiO_2 layer expect the part of the bi-metal interface. Because among the common metals, the absolute thermoelectric power of Fe (+16.2 × 10^{-6} V K^{-1}) is greater than that of Al (−1.66 × 10^{-6} V K^{-1}), we selected this materials combination [4]. The size of the TPG was 20 × 30 mm^2. In case that the

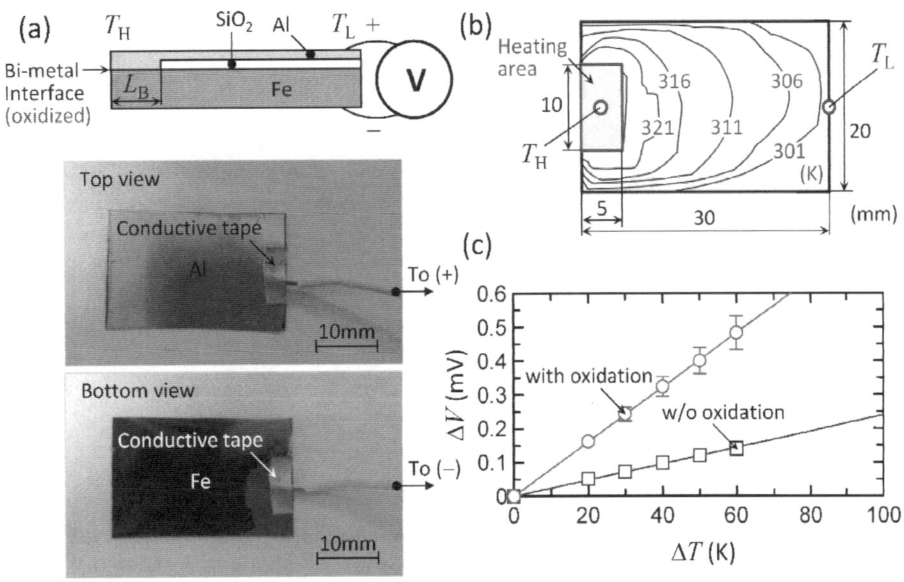

Figure 1. Plate-type thermoelectric power generator (TPG). (a) Schematic of the cross section and overviews. (b) An example of the temperature distribution on the TPG as obtained by an infrared camera. ΔT (= $T_H - T_L$) in this case was 40 K. (c) The relationships between ΔV and ΔT for two types of TPGs with and without oxidation. Reprinted with permission from ref. (3). Copyright (2016) Elsevier.

temperature of the bi-metal side, T_H, becomes larger than that of another side, T_L, ΔV appears in the TPG due to the temperature gradient ΔT (= $T_H - T_L$). An example of the temperature distribution on the TPG, which was measured by an infrared camera, is shown in Fig. 1(b). The heating area was 10 × 5 mm^2, and the value of ΔT in this case was 40 K.

The relationships between ΔV and ΔT for two types of TPGs are shown in Fig 1(c). The surface of the Fe plate of one TPG was oxidized before depositing the Al layer. The value of ΔV of the TPG with oxidized bi-metal interface was greater than that of the TPG without oxidization, and oxidizing the bi-metal interface was found to be effective to enhance ΔV.

3. T-shaped, plate-type thermoelectric power generator

The design of the T-shaped, plate-type TPG proposed in this study is shown in Fig. 2(a) together with that of the previous TPG. Although the size of the bi-metal interface (20 × 5 mm^2) was unchanged, the length of the TPG was decreased to 1 / 2. The cross-sectional structure of the T-shaped TPG was same as that of the previous TPG, see Fig. 1(a). In the results, the volume of the T-shaped TPG was decreased to 1 / 4 compared with the previous one.

Fabrication process of the T-shaped TPG is as follows. The thickness of the used Fe plate was 0.1 mm. Because oxidizing the bi-metal interface was found to be effective to enhance the thermoelectric power generation efficiency, the surface of the Fe plate was initially oxidized at 873 K for 3 h. After that, a part of the oxidized Fe plate was covered with a resist mask, and a tetraethyl orthosilicate (TEOS) layer was formed on the Fe plate for electrical insulation. After removing the resist mask, an Al layer was deposited, and the bi-metal interface was created between the oxidized Fe and Al. The thickness of the Al layer was about 5 μm.

Conductive wires were attached on both surfaces of the Fe plate and the Al layer using conductive tape, and the wires were connected to a voltmeter. A part of the surface of the T-shaped TPG was heated with a ceramic heater, and ΔV was measured for this open circuit. The heating area was 10 × 5 mm^2, and this was same as that of the previous experiment shown in Fig. 1(c).

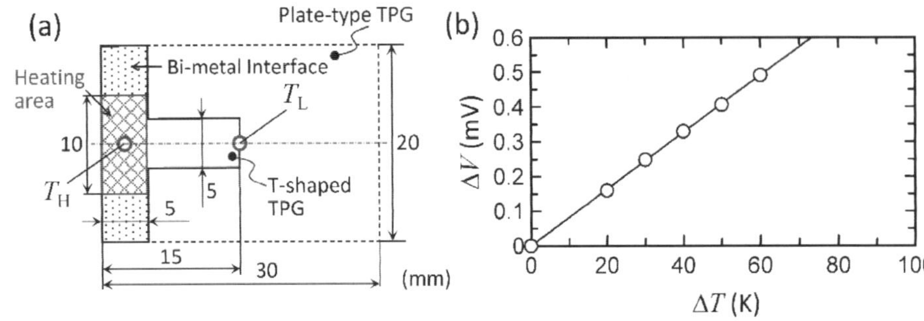

Figure 2. T-shaped, plate-type thermoelectric power generator (a) Schematic. (b) The relationship between ΔV and ΔT.

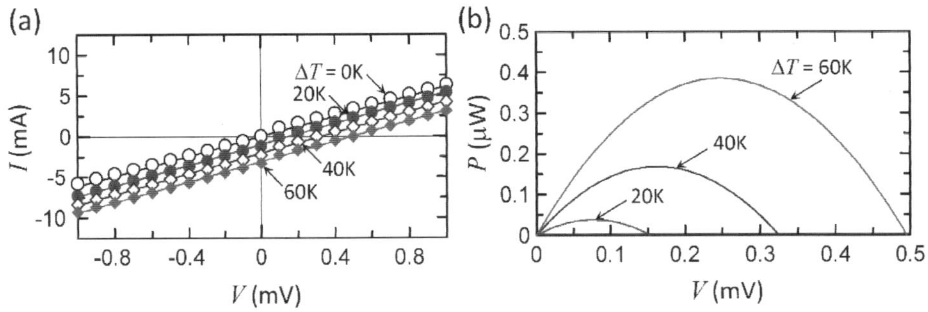

Figure 3. Experiment to confirm the power generation performed for the closed circuit. (a) The $I - V$ characteristics measured at $\Delta T = 0$ to 60 K. (b) P against V at $\Delta T = 20$ to 60 K. Here the values of P were determined from the $I - V$ characteristics in (a).

The relationship between ΔV and ΔT is shown in Fig. 2(b). The value of ΔV was measured for 10 seconds for each condition after the temperature of the hater reached the set value, and the average values of ΔV are plotted in Fig 2(b). ΔV increased linearly against ΔT, and Seebeck effect was clearly observed for the T-shaped TPG fabricated. From the slope of the $\Delta V - \Delta T$ relation, the value of the Seebeck coefficient, S, was found to be 8.2×10^{-6} V K^{-1}, and this was very close to that of the previous one (8.1×10^{-6} V K^{-1}). The results suggest that ΔT induced in the TPG was maintained although the length of TPG was reduced to 1 / 2.

4. Power generation characteristics

To evaluate the power generation characteristics, the current-voltage ($I - V$) characteristics for $\Delta T = 0$ to 60 K were measured. Conductive wires attached on the surfaces of the T-shaped TPG were connected to a source measure unit, and I flowing in the closed circuit was measured under applying V. ΔT in the TPG was induced by a ceramic heater similarly to the experiment for open circuit.

The current (I) – voltage (V) characteristics measured in the closed circuit at $\Delta T = 0$ to 60 K are shown in Fig. 3(a). The slopes of the $I - V$ characteristics for all values of ΔT are almost constant, and this indicates that the internal resistance of the T-shaped TPG was unchanged in the examined temperature range. The internal resistance of the T-shaped TPG was determined to be 0.158 Ω. Although the internal resistance was 28 % larger than that of the previous one (0.123 Ω), this value was relatively lower compared with the reported TPGs [1, 2]. Figure 3(b) shows the power curves determined from the $I - V$ characteristics, and the power P was determined by VI. For all ΔT, P takes its maximum values, P_{max}. The values of P_{max} are summarized in Table 1 together with those of the previous TPG. The values

Table 1. Performance of the T-shaped, plate-type thermoelectric power generator.

TPG	volume (m^3)	S ($\mu V\ K^{-1}$)	R (Ω)	P_{max} (μW) ΔT (K)		
				20	40	60
Previous [a)]	6.0×10^{-8}	8.1	0.123	0.071	0.307	0.669
T-shaped	1.5×10^{-8}	8.2	0.158	0.037	0.167	0.384

[a)] Ref. [3]

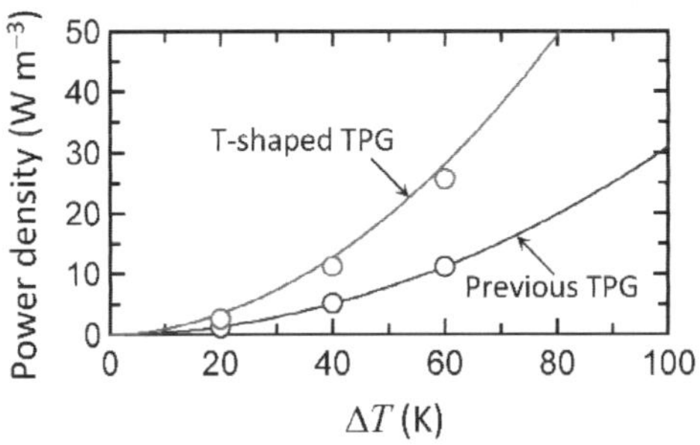

Figure 4. Power density (= P_{max} / volume) against ΔT.

of P_{max} of the T-shaped TPG were decreased to 1 / 2 compared with those of the previous one. The main cause of this decreases might be the increase in the internal resistance of the TPG.

Figure 4 shows the values of power density, which is determined by P_{max} / volume, against ΔT. Although the values of P_{max} became about 1 / 2 of the previous TPG for all the values of ΔT, the values of power density became over 2 times greater than that of the previous one. For example, at $\Delta T = 40$ K, the power density per unit temperature difference is determined to be 0.27 W m^{-3} K^{-1}, and this value is much greater than those of the previously reported TPGs without composed of Bi, Te or its related alloys [5, 6].

5. Conclusions

In this paper, a T-shaped, plate-type thermoelectric power generator (TPG), which is composed of the Fe plate and the Al layer, was reported for power generation form a small temperature difference. Although the volume of the T-shaped TPG became 1 / 4 compared with the previous TPG, the thermoelectromotive forces generated due to the temperature gradient in both generators were same. Moreover, the power density of the T-shaped TPG, which was determined from the current-voltage characteristics measured at the temperature difference of 0 to 60 K, became over 2 times greater than that of the previous TPG.

Acknowledgement

The authors would like to acknowledge Professor M. Saka for valuable discussions throughout this work. This work was supported by JSPS KAKENHI Grant Number 15H03886.

References

[1] S. A. Whalen, C. A. Apblett and T. L. Aselage, J. Power Sources, 180 (2008) pp. 657-663.
[2] F. Ren, P. Menchhofer, J. Kiggans and H. Wang, J. Electron. Mater., 45 (2016) pp. 1412-1418.
[3] H. Tohmyoh and T. Daimon, Microelectron. Eng., 158 (2016) pp. 38-41.
[4] A. Watanabe and H. Tohmyoh, Renew. Energy 83 (2015) pp. 41-46.
[5] A. Yadav, K. P. Pipe and M. J. Shtein, J. Power Sources, 175 (2008) pp. 909-913.
[6] K. J. Norris, M. P. Garrett, J. Zhang, E. Coleman, G. S. Tompa and N. P. Kobayashi, Energy Convers. Manage., 96 (2015) pp. 100-104.

Amorphous Thin Film for Thermoelectric Application

K. Umeda, M. Uenuma, D. Senaha, J. C. Felizco, Y. Uraoka, and, H. Adachi

Nara Institute of Science and Technology,
8916-5Takayamacho, Ikoma, Nara 630-0192, Japan

uenuma@ms.naist.jp

Abstract. Amorphous-InGaZnO is n-type semiconductor material and has enormous potential such as transparency, flexible application owing to a low temperature fabrication process. In this study, Effects of annealing on the thermoelectric properties of a-IGZO thin film are evaluated for a low temperature process. We also demonstrated flexible TEG using a-InGaZnO and PEN substrate.

1. Introduction

Transparent Amorphous Oxide Semiconductor (TAOS) have been well-studied as channel layers for thin film transistors (TFTs), owing to their good transparency, flexibility and room temperature process. Among them, amorphous-InGaZnO (a-IGZO) thin films have been studied for applications other than TFT, which has been reported for a wide range of research results [1]. Our research group had previously reported on the thermoelectric performance of annealed a-IGZO [2, 3].

In this study, the thermoelectric properties of as-deposited and annealing a-IGZO were investigated. Additionally, a flexible a-IGZO thermoelectric generator (TEG) using polyethylene naphthalate (PEN) was fabricated. PEN is commonly used as a flexible substrate and offers good stability up to around 150 °C. Therefore a low temperature processes are needed.

2. Effect of annealing temperature

At first, annealing effects were evaluated. The a-IGZO thin film (200 nm) was deposited on quartz substrates using RF magnetron sputtering using sintered targets (In:Ga:Zn_2:2:1, at %) with an input power of 100 W. The thin film deposition was performed at a pressure of 0.6 Pa under O_2 and Ar atmosphere. In this process, oxygen partial pressure was set at 0%, 1% and 4.5%. Then, Au/Mo electrodes were deposited by electron beam evaporation. Finally, samples were annealed. The Seebeck coefficient and electrical conductivity were obtained at the room temperature.

As a results, annealing process is needed to increase the thermoelectric property (power factor: PF) in the samples with fabricated 1% and 4.5% oxygen partial pressure. On the other hand, the post annealing process is not necessary in the samples with fabricated 0% partial pressure.

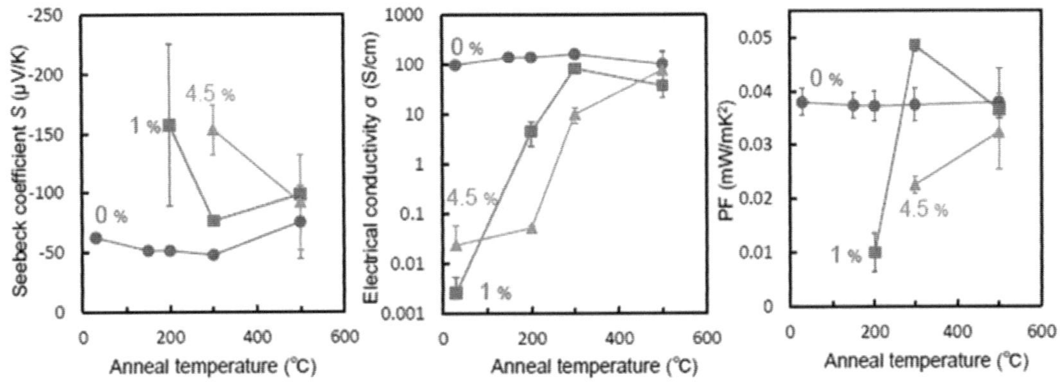

Figure 1. Thermoelectric properties and annealing effect of InGaZnO films on glass substrate formed with different oxygen partial pressure (0%, 1% and 4.5%).

3. Demonstration of flexible InGaZnO-TEG

Then flexible device was demonstrated as follows, the a-IGZO thin film (200 nm) was deposited on PEN substrate (2 cm × 4 cm) with same process (with 0% partial pressure). Au/Mo electrodes (20nm/100nm) were deposited by electron beam evaporation. No heat treatments were performed since a polymer substrate is used (Fig.2).

Output voltage 6.7 mV and output power 0.12 nW was obtained at temperature difference of 53 K (Fig. 3). The thermoelectric performances are stable against bending. From the shape of the fabricated device, it can be seen that the thermoelectric module can generate 0.1nW with only two IGZO parts.

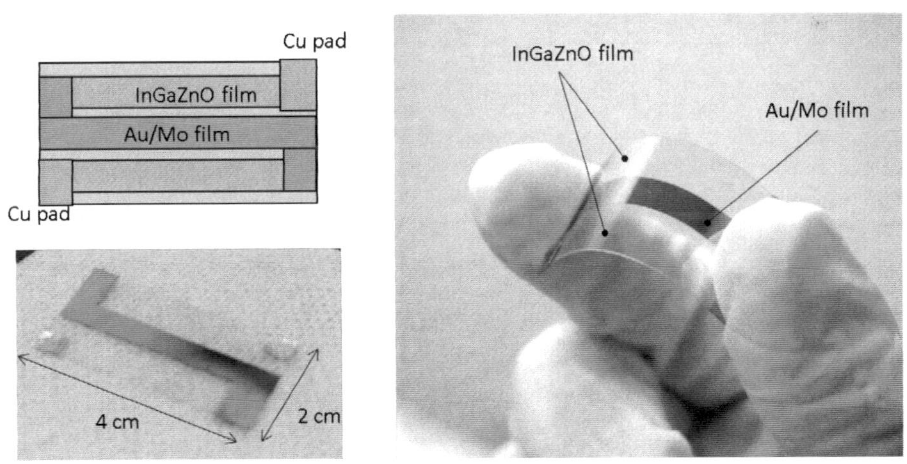

Figure 2. Demonstrated flexibleπ-type device with two-IGZO area and one-electrode area on PEN substrate.

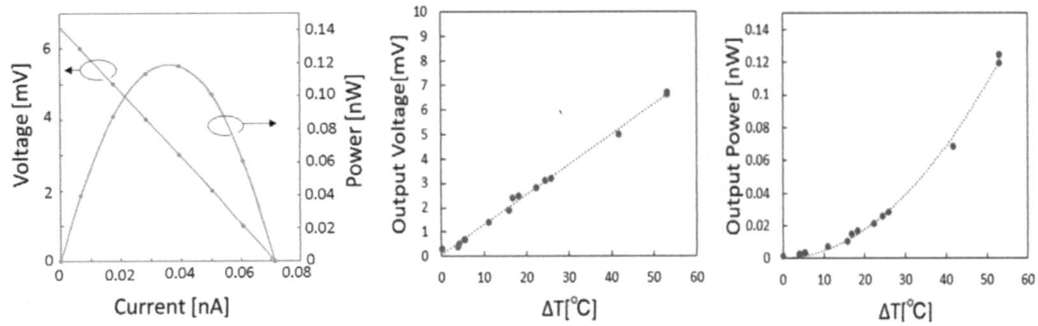

Figure 3. I-V characteristics and output characteristics at ΔT=50 oC. And output voltage at room temperature. Output power at room temperature.

4. Demonstration of flexible InGaZnO-TEG with heat guide

Flexible thermoelectric generator with heat guide was demonstrated using a-InGaZnO and Mo electrode uni-leg structure with heat guide (Fig.4). The sample TEG was fabricated using PEN substrate with metal masks and photolithography process. The heat guides were formed using KMPR photoresist (thickness: 100μm). The 625 pairs of a-InGaZnO and Mo areas was designed. As shown in Fig.5, the demonstrated TEG was fully flexible. Output power of the demonstrated device was still small. However, the output power will be improved by optimizing the atomic composition of InGaZnO and the device structure.

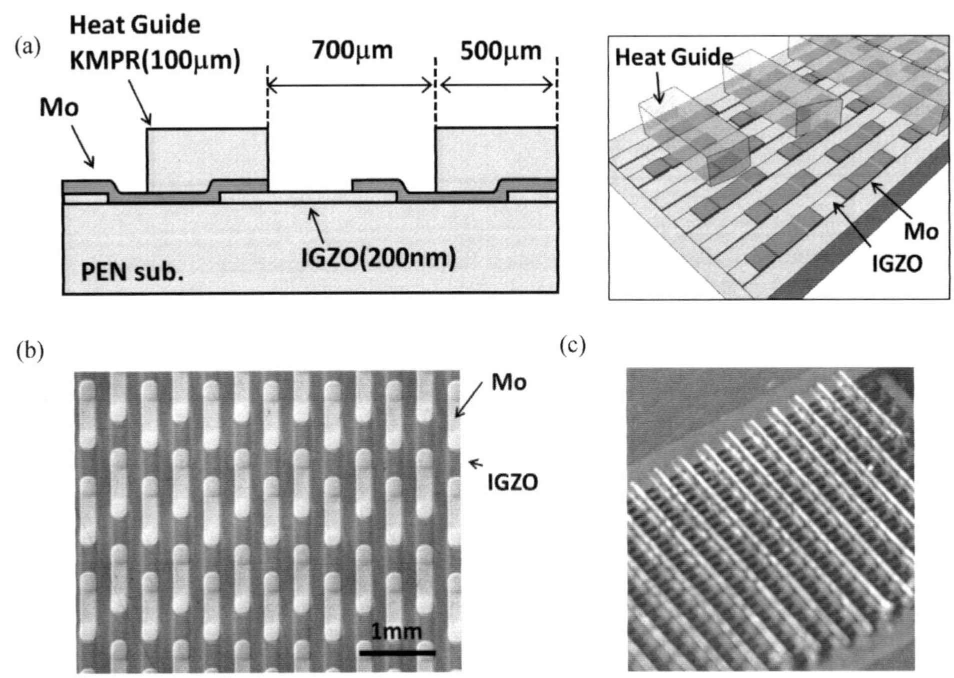

Figure 4. Demonstrated thermoelectric device using hear guide. (a) Schematic image of TEG device. (b) Optical microscope image of TEG device before heat guide formation. (c) Optical microscope image of TEG device after heat guide formation.

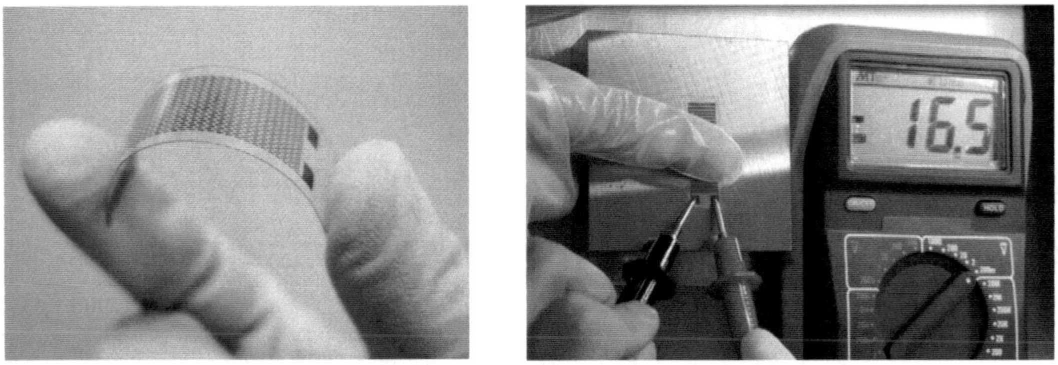

Figure 5. Fabricated TEG device on PEN substrate and demonstrated result (16.5 mV was obtained by finger the TEG).

5. Conclusion

The effects of annealing on thermoelectric properties of amorphous InGaZnO thin film were shown. And we also demonstrated flexible TEG using Mo and a-InGaZnO uni-leg structures on PEN substrate. If the Mo electrode were changed to the transparent conductive materials such as ITO, transparent and flexible TEG is possible with this structures. Additionally, semiconductor fabrication processes of IGZO are already well established. Therefore, it is easy to integrate the hundreds or thousands of pi-structures and it will generates μW scale output power.

References

[1] K. Nomura et al. *Nature*, 432, 2004, 488
[2] Y. Fujimoto, M. Uenuma, Y. Ishikawa, and Y.Uraoka, *AIP advances*, 2015, **5**,97209,1-6
[3] Y. Fujimoto, M. Uenuma, Y. Ishikawa, and Y.Uraoka, 34th ICT 2015 and 13th ECT 2015

IOP Conf. Series: Journal of Physics: Conf. Series **1052** (2018) 012017 doi:10.1088/1742-6596/1052/1/012017

Thermoelectric characteristics of nanocrystalline ZnO grown on fabrics for wearable power generator

Hiroya Ikeda, Faizan Khan, Pandiyarasan Veluswamy, Shota Sakamoto, Mani Navaneethan, Masaru Shimomura, Kenji Murakami, Yasuhiro Hayakawa

Shizuoka University, Johoku 3-5-1, Naka-ku, Hamamatsu 432-8011, Japan

E-mail: `ikeda.hiroya@shizuoka.ac.jp`

Abstract. The present paper reports the fabrication and characterization of low-cost and large-area flexible thermoelectric materials with nanocrystalline ZnO applicable for wearable power generator. The ZnO nanostructures were grown on cotton fabric (CF) and carbon fabric (CAF) by solvothermal synthesis. It was demonstrated that the polarity of Seebeck coefficient in ZnO-covered CF can be controlled by adding Sb and Ag. That is, for the Sb-ZnO/CF sample, the Seebeck coefficient was negative, indicating a n-type semiconductor. On the other hand, its value changed to positive for the Ag-ZnO/CF sample, indicating p-type one. For the ZnO/CAF sample, the Seebeck coefficient showed the same value as a CAF substrate. In this sample, ZnO hardly contributes to the thermoelectromotive force comparing with the CAF, which is due to a lack of the ZnO amount.

1. Introduction

The wearable thermoelectirc power generator, one of the energy harvesting devices, requires flexible thermoelectric materials. To achieve high power-generator efficiency using thermoelectric technology, it is necessary to increase the thermoelectric figure-of-merit (Z), denoted as $Z = S^2\sigma/\kappa$, where S is the Seebeck coefficient, σ is the electrical conductivity and κ is the thermal conductivity of the thermoelectric material [1]. Therefore, it is necessary to achieve an increase in S and σ, and a decrease in κ, simultaneously. One method of overcoming this issue is the introduction of nanostructured semiconductors due to the confinement effect of carriers and phonons [2-4].

We focus our attention on nanostructured ZnO as a thermoelectric material [5-7] while the ZnO nanostructures formed by solvothermal synthesis have been investigated for applying to gas sensor [8], dye-sensitized solar cell [9] and so forth. This is because ZnO is easily obtainable, inexpensive, stable for temperature and humidity, and nontoxic for human skin. Hence, ZnO-related flexible materials are available for clothing. In the present study, with the aim of applying the ZnO nanostructures to a flexible thermoelectric material, we constructed the fabrication method of ZnO nanostructre on cotton fabric (CF) and carbon fabric (CAF), and evaluated the Seebeck coefficient of the obtained ZnO-nanostructure/fabric samples.

2. Fabrication of nanocrystalline ZnO on fabrics

Figure 1 shows a schematic diagram of fabrication procedure of ZnO nanostructures on CF, substrate for flexible material, used in this study [6]. The two-step method was adopted for the fabrication, consisting of seed creation process and nanostructure growth process. Mixture of $Zn(NO_3)_2{\cdot}6H_2O$ and $(CH_2)_6N_4$ was used as precursor solution. A typical scanning electron microscope (SEM) image of ZnO/CF obtained from this fabrication condition is shown in Fig. 2. It is found that white ZnO nanosheet structures can be observed on the cotton fibers. We also confirmed that the UV-transmittance characteristics of the ZnO/CF sample does not change even after 10-times laundry treatment with water. This fact indicates that the nanocrystalline ZnO is chemically bonded to the cotton fiber.

With the aim of controlling the polarity of ZnO, we tried to add Sb and Ag to the ZnO/CF materials. The detailed process is described in Ref. 4. Moreover, instead of the insulator CF substrate, conductive CAF was used as a substrate for flexible thermoelectric materials. In this case, a simple solvothermal process was performed and the detail procedure will be described elsewhere.

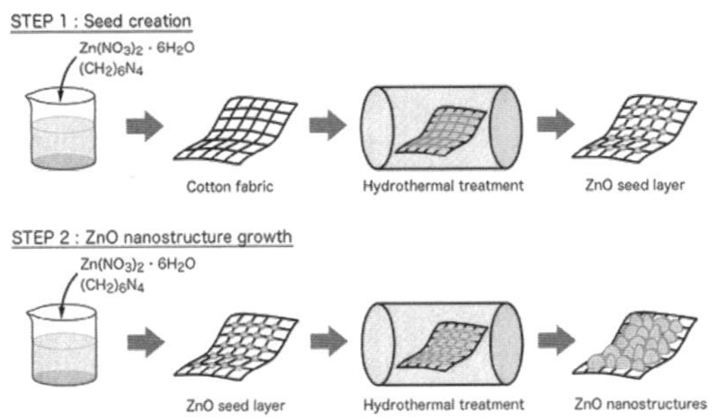

Figure 1. Schematic of fabrication process of ZnO nanostructures on a CF. The two-step method consists of seed creation process and nanostructure growth process.

Figure 2. SEM image of ZnO nanostructures grown on a CF.

Our handmade measurement equipment used in this study is illustrated in Fig. 3. Two Peltier devices connected to heat sinks are placed side by side with a gap of 10 mm between them and can be heated individually. The flexible sample for measuring is placed across the gap, in contact with both of the Peltier devices. Two T-type thermocouples (TCs) buried within Cu electrodes are directly attached to the sample surface. The time evolution of the thermoelectromotive force (TEMF) was measured by a digital multimeter (Keithley 2700) equipped with a switching module (Keythley 7700), simultaneously with the temperatures at the high- and low-temperature regions. The Seebeck coefficient S was evaluated from the TEMF ($\Delta V \equiv V_L - V_H$) and the temperature difference ($\Delta T \equiv T_H - T_L$) by $S = \Delta V/\Delta T$ near room temperature.

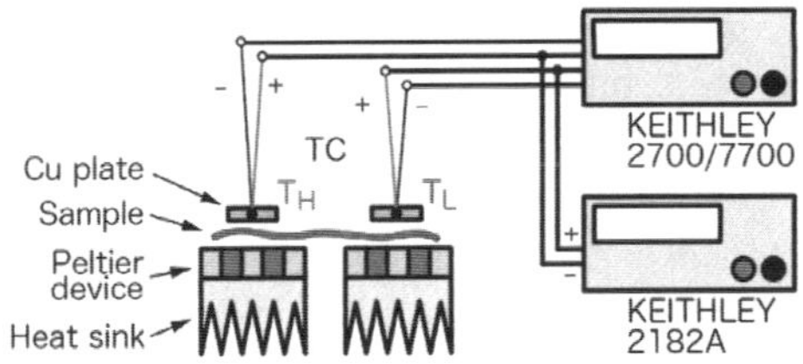

Figure 3. Schematic diagram of the apparatus for Seebeck coefficient measurement.

3. Results and discussion

3.1. Seebeck coefficient of ZnO/CF

Figure 4 shows the Seebeck coefficient evaluated for ZnO/CF, Sb-ZnO/CF, and Ag-ZnO/CF. From the SEM observation, the Sb- and Ag-ZnO/CF samples were confirmed to have composite structure rather than impurity doping [7]. However, it is clearly found that both two Sb-ZnO/CF samples synthesized for different conditions show negative Seebeck coefficient, or n-type semiconductor. As for the difference between these samples, it is worth noting that the Sb-ZnO network on the cotton fiber surface is formed as a freestanding composite, which should be attributed to the Sb incorporation into the Zn sites and O vacancies, resulting in n-type behavior [10-12]. On the contrary, both two Ag-ZnO/CF samples show positive Seebeck coefficient, or p-type semiconductor. In the case of Ag-ZnO composites, the growth process of ZnO is disrupted when the Ag composites are introduced, and the growth process is again favored when enough Ag have recombined with the Zn material which may cause the positive effect on the charge separation efficiency [13,14]. Consequently, the polarity in Seebeck coefficient of ZnO can be controlled by adding Sb and Ag.

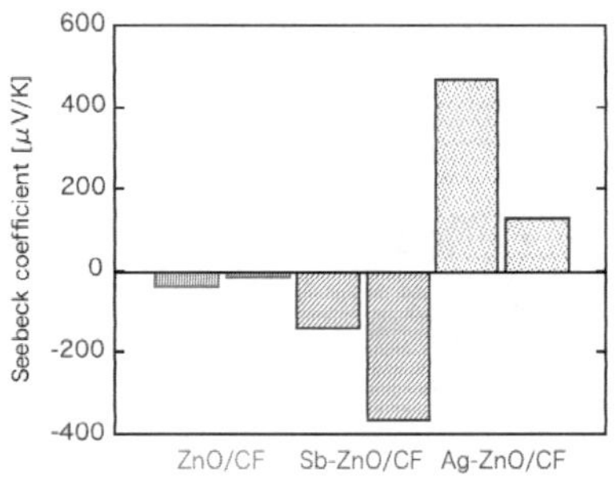

Figure 4. Seebeck coefficients evaluated for ZnO/CF, Sb-ZnO/CF, and Ag-ZnO/CF.

3.2. Nanocrystalline ZnO grown on CAF

Figure 5 shows the measured TEMF as a function of temperature difference for ZnO/CAF and CAF itself. Both graphs make a linear relation, indicating that the Seebeck coefficient is nearly constant in the measured temperature range. The Seebeck coefficient of CAF evaluated from the gradient was 5.0 μV/K. The value is positive, so the CAF used is a p-type material, which was confirmed also by Hall measurement. On the other hand, the Seebeck coefficient of ZnO/CAF sample was evaluated also to be 5.0 μV/K. This fact suggests that the TEMF in the ZnO/CAF sample is produced at CAF since the obtained value is not influenced by the formation of ZnO nanostructures on CAF. Insignificant influence of ZnO nanostructures is considered to be due to a lack of its amount.

4. Conclusions

We investigated the thermoelectric characteristics of ZnO nanostructures formed on CF and CAF by a solvothermal method for the application of wearable power generator device. The ZnO nanosheet structure formed on the cotton fiber. The Seebeck coefficient of nanocrystalline-ZnO/CF was negatively increased by forming the Sb-ZnO composites. In addition, it changed to positive values by forming the Ag-ZnO composites. Therefore, it is found that the ZnO/CF material can be available for n- and p-type thermoelectric materials by adding Sb and Ag. The Seebeck coefficient of ZnO/CAF is about 5 μ V/K and equal to that of CAF itself. The ZnO nanostructures on CAF hardly contributes to TEMF under a given temperature gradient.

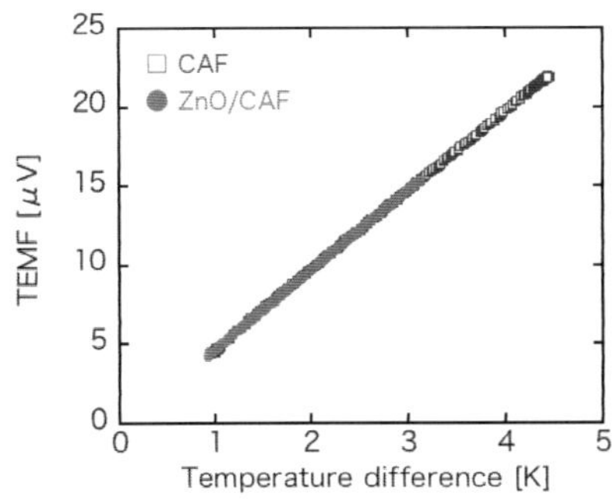

Figure 5. Relation between TEMF and temperature difference for CAF and ZnO/CAF samples. The Seebeck coefficients evaluated from the gradient were 5.0μV/K both for CAF and ZnO/CAF samples.

Acknowledgments
This work was financially supported in part by a Grand-in-Aid for Challenging Exploratory Research (No.16K13717) from the Japan Society for the Promotion of Science and the Cooperative Research Project Program of Research Institute of Electronics, Shizuoka University.

[1] Ioffe A F 1960 *Physics of Semiconductors* (London: Infosearch Lid.)
[2] Hicks L D and Dresselhaus M S 1993 *Phys. Rev.* B **47** 16631
[3] Li D, Wu Y, Kim P, Shi L, Yang P and Majumdar A 2003 *Appl. Phys. Lett.* **83** 2934
[4] Neophytou N and Kosina H 2011 *Phys. Rev.* B **83** 245305
[5] Veluswamy P, Sathiyamoorthy S, Chowdary K H, Muthusamy O, Krishnamoorthy K, Takeuchi T and Ikeda H 2017 *J. Alloy. Compd.* **695** 888
[6] Pandiyarasan V, Suhasini S, Archana J, Navaneethan M, Majumdar A, Hayakawa Y and Ikeda H 2017 *Appl. Surf. Sci.* **418** 352
[7] Veluswamy P, Suhasini S, Khan F, Ghosh A, Abhijit M, Hayakawa Y and Ikeda H 2017 *Carbohydr. Polym.* **157** 1801
[8] Rai P, Kwak W-K and Yu Y-T 2013 *ACS Appl. Mater. Interfaces* **5** 3026
[9] Navaneethan M, Archana J, Arivanandhan M and Hayakawa Y 2012 *Phys. Status Solidi RRL* **6** 120
[10] Zhu B L, Zhu S J, Zhao X Z, Su F H, Li G H and Wu X G 2011 *Physica Status Solidi* A **208** 843
[11] Song X, Bottger P H M, Karlsen O B, Finstand T G and Tafto J 2012 *Physica Scripta* **1-6** 014001
[12] Li Y, Huang L, Zhang P, Ren X and Deng L 2015 *Nanoscale Res. Lett.* **10** 1
[13] Zinberberg L, Mitlin S, Shankar H and Asscher M 2015 *Phys. Chem.* C **119** 28979
[14] Hongi L and Bredas J-L 2016 *Adv. Mater.* **28** 3928

High figure of merit $(MgHf)_xAl_{1-x}N$ thin films for miniaturizing vibrational energy harvesters

H. H. Nguyen[1], L. Van Minh[1], H. Oguchi[2] and H. Kuwano[1]

[1]New Industry Creation Hatchery Center, Tohoku University, Japan
[2]Advanced Institute for Materials Research, Tohoku University, Japan

hung@nanosys.mech.tohoku.ac.jp

Abstract. In this paper, we report our development of piezoelectric MgHf co-doped AlN films with high power figure of merit (FoM) for vibrational energy harvesters (VEHs). The $(MgHf)_xAl_{1-x}N$ films were deposited on Pt/Ti/SOI substrates by reactive ion-beam sputtering AlN and MgHf targets simultaneously. These studies revealed that we achieved a high FoM of 22.3 GPa at x = 0.12, increasing three-fold that of the pure AlN. Our micro-machined energy harvester exploiting the developed $(MgHf)_{0.12}Al_{0.88}N$ films provided the highest normalized power density (NPD) of 12.04 mW.cm^{-3}.g^{-2} among the state-of-the-art piezoelectric VEHs. The results opened the way for the development of high-performance VEHs using in wireless sensors network.

1. Introduction

Piezoelectric VEHs have become one of the novel approaches to self-sustainable power sources for wireless sensor networks (WSN) [1, 2]. High performance and miniaturization of VEHs continue the quest for a better piezoelectric material with higher FoM ($e_{31}^2/\varepsilon\varepsilon_r$).

To date, lead zirconate titanate (PZT) is the most widely used piezoelectric material for sensors and actuators [2, 3]. However, PZT is not the best material for power generation in VEHs due to the high relative permittivity (ε_r of 1118 [4]), that decreases the FoM. Moreover, PZT is recently restricted to be used in society since it contains huge amount of a toxic material, lead (Pb) [3]. For such reasons, lead-free piezoelectric films such as AlN, ZnO, KNN, etc. are introduced to replace the toxic PZT materials [3-8]. Among them, AlN is one of the most promised materials for the high electromechanical-coupling factor (k^2 of ~ 6.5%), low dielectric loss (δ of 0.3%) and low dielectric constant (ε_r of 10) [4]. In addition, the FoM for AlN is comparable to that of PZT (~ 10 GPa for AlN and ~ 13 GPa for PZT [4]).

With the aim of FoM enhancement, AlN piezoelectricity might be improved by tailoring dopants into the Al sites [3, 5-8]. In the previous studies, we proofed that co-doping Mg and Hf into Al sites enhanced approximately 3.5-time longitudinal piezoelectric coefficient (d_{33}) [5, 6]. Since most of piezoelectric VEHs exploit d_{31}-mode structure, measuring the transverse piezoelectric coefficient (d_{31} or e_{31}), dielectric constant (ε_r), and Young's modulus (Y) of MgHf co-doped AlN thin films become crucial.

In this study, we aimed to synthesize high-FoM $(MgHf)_xAl_{1-x}N$ films by reactive ion-beam sputtering with MgHf and AlN targets. We investigated extensively the impact of MgHf dopants concentration on Young's modulus *(Y)*, the transverse piezoelectric coefficient (e_{31}), and the FoM. Also, we developed the first VEH prototype utilizing the successful films.

2. Experimental

The $(MgHf)_xAl_{1-x}N$ films were developed by the reactive ion-beam sputter with MgHf and AlN targets in Ar/N_2 ambient. The 700 nm $(MgHf)_xAl_{1-x}N$ films with the x from 0 to 0.12 were grown on Pt (100 nm)/Ti (6 nm)/SOI substrates. Base pressure of the sputtering chamber was kept at less than 1×10^{-7} Torr. The AlN target was pre-sputtered before each deposition for 20 min in Ar and then for 10 min in the mixture of Ar and N_2. The applied RF powers for AlN and MgHf targets are 140 W and 100 W, respectively. The Ar/N_2 flow rate was kept at 8:32 sccm during deposition process. The detailed deposition process was presented in [6].

$(MgHf)_xAl_{1-x}N$ thin film crystallinity was investigated and reported in our previous report [5]. The c- and a-axis lattice constants were determined from the X-ray diffractometer (Brucker-D8 advance) using θ-2θ scans on the (0002) and $(10\bar{1}2)$ planes in symmetric and asymmetric configurations, respectively [8]. The Y of the developed films was determined from the resonant-frequency shift between $(MgHf)_xAl_{1-x}N$/Si and Si cantilevers [9]. The e_{31} of $(MgHf)_xAl_{1-x}N$ was determined by measuring displacement for a unimorph cantilever under applied voltage. In this measurement, a negative polarized voltage ($|V_{pp}|$= 1 V – 20 V) in sinusoidal waveform at frequency of 1000 Hz, being far away from the resonant frequencies, was applied on the cantilevers. The tip displacement was measured by using a Laser Doppler Vibrometer (Ono-Sokki LV-1710). To determine the FoM, the ε_r was measured using an impedance analyzer (HP-4194A) at 10 kHz.

Figure 1 showed a scanning electron microscopic (SEM) image of the array of $(MgHf)_xAl_{1-x}N$/Si cantilevers with the same width of 200 μm, thickness of 50 μm and lengths of 500 μm, 1000 μm and 1500 μm. The cantilevers were fabricated with various lengths to avoid the geometrical errors.

3. Results and discussions

3.1. MgHf co-doped AlN thin film characterization

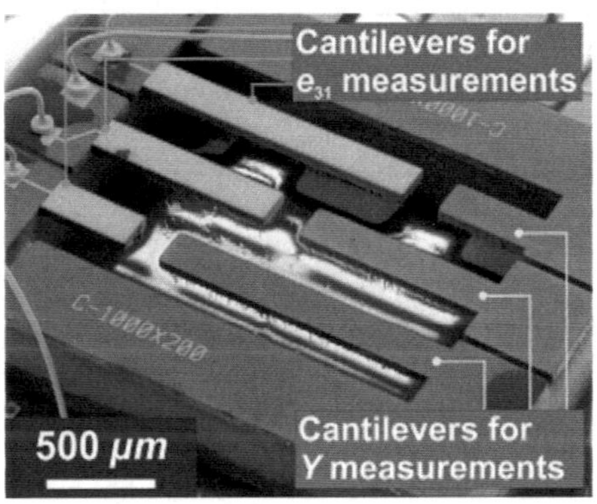

Figure 1. $(MgHf)_xAl_{1-x}N$/Si micro cantilevers for evaluating the Y and the e_{31}

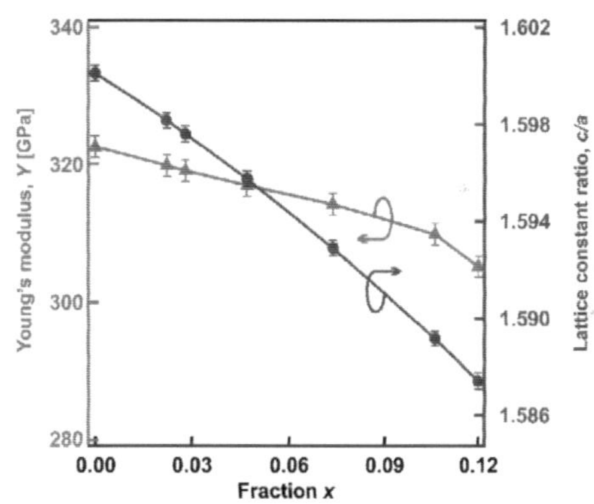

Figure 2. The dependence of the Y and c/a ratio on MgHf dopants concentration

Figure 2 shows the Y and the lattice constant ratio (c/a) depend on the MgHf concentration. The Y of pure AlN decreased from 323 GPa to 305 GPa at x = 0.12. This elastic softening is caused by the decrease in c/a ratio when co-doping Mg-Hf into Al sites [10]. The e_{31} and the ε_r in relation with the MgHf concentration were shown in figure 3. The thin films achieved e_{31} of 1.4 C/m^2 at x = 0.12 that increases of 84 % compared with that of the pure AlN. The ε_r alsoincreased from 10.3 (for x = 0) to 11.6 for (x = 0.12).

The increase in the transverse piezoelectric coefficient led to the high FoM of 22.3 GPa at x = 0.12. Figure 4 exhibits the dependence of the FoM on dopants concentration. The FoM increases three-time that of the pure AlN (from 7.2 GPa for x = 0 to 22.3 GPa for x = 0.12).

A comparison of FoM for $(MgHf)_xAl_{1-x}N$ in this study with PZT and other piezoelectric materials was shown in the figure 5. It demonstrates that pure AlN has experimental FoM comparable to ZnO and PZT. However, the $(MgHf)_xAl_{1-x}N$ provided the highest FoM at x = 0.12 among the investigated piezoelectric materials [2,4,7,11]. At the same dopant concentration, the FoM for $(MgHf)_{0.12}Al_{0.88}N$ was approximately 1.9-fold that of the expensive materials of Sc-doped AlN [7]. In addition, The FoM increases three-time that of the pure AlN, being higher than that of the well-developed PZT [4].

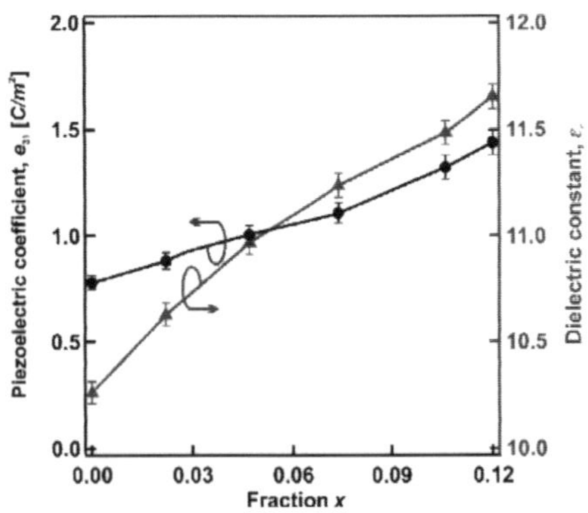

Figure 3. The dependences of e_{31} and ε_r on MgHf dopants concentration

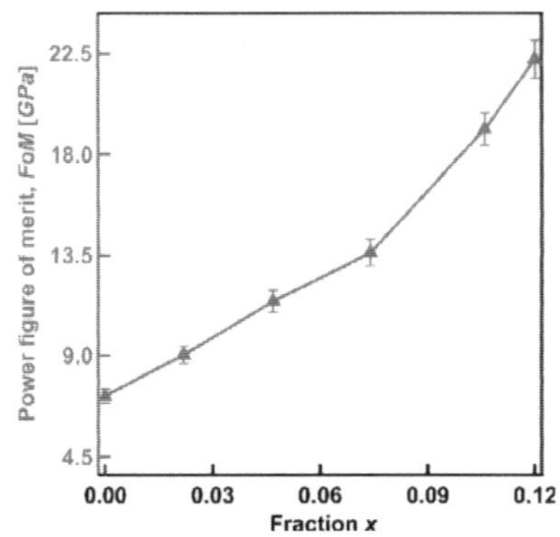

Figure 4. The dependence of FoM on MgHf dopants concentration

3.2. $(MgHf)_{0.12}Al_{0.88}N$-based micro-energy harvesters

A micro VEH consisting of the $(MgHf)_{0.12}Al_{0.88}N$/Si cantilever of 1000 μm × 200 μm × 50 μm and the Si proof-mass of 0.6 mg was fabricated by bulk micromachining. A SEM image of the final device was shown in figure 6. The top and bottom electrodes were connected to separate terminals on a PCB plate by using a ball-wedge wire bonder (TPT-HB16) with Au wire of 25 μm in diameter.

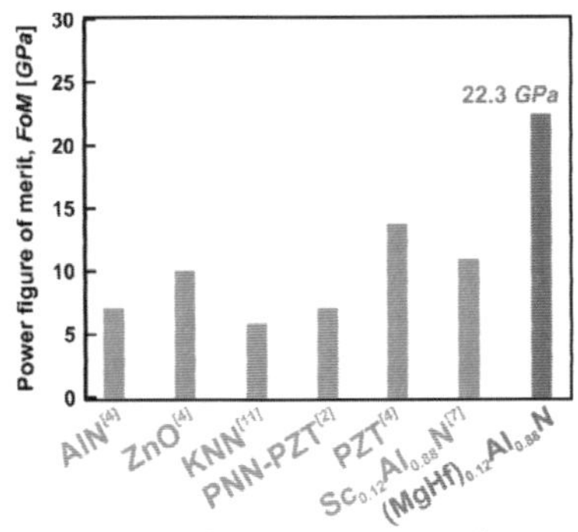

Figure 5. The FoM comparison of piezoelectric materials

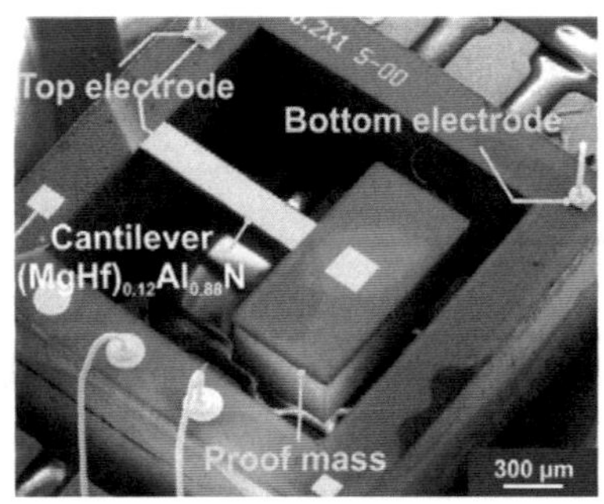

Figure 6. False-colored SEM image of a typical $(MgHf)_{0.12}Al_{0.88}N$-based VEH

To evaluate the device performance, the VEH was fixed on a standard shaker (G-Master APD-200FCG) and the output power spectra were measured under various accelerations as shown in figure 7. The optimal resistance for the highest output power was obtained at 800 kΩ. The device provided a maximum output power of 2.6 μW at the acceleration of 0.9 g (1 g = 10 m/s^2) and the resonance frequency of 2497 Hz. The NPD for this device was 12.04 mW.cm^{-3}.g^{-2}. This NPD increases 5.4-fold that of the best pure AlN-based VEH [12]. In addition, the achievement is the highest value among the published VEHs (figure 8).

4. Conclusion

In summary, we successfully deposited $(MgHf)_xAl_{1-x}N$ thin films by reactive ion-beam sputtering with MgHf and AlN targets. The transverse piezoelectric properties and the Young's modulus depending on the MgHf dopants concentration were investigated. The result showed an 84% increase in e_{31} of $(MgHf)_{0.12}AlN_{0.88}N$ film leads to a threefold increase in the FoM. The VEHs utilizing $(MgHf)_{0.12}Al_{0.88}N$

with the high FoM of 22.3 GPa generated the power of 2.6 μW at the input acceleration of 0.9 g and the resonance frequency of 2497 Hz. Our device provided the highest NPD of 12.04 mW.cm^{-3}.g^{-2}. These results paved the way for development of high-performance $(MgHf)_xAl_{1-x}N$-based VEHs applying in WSN.

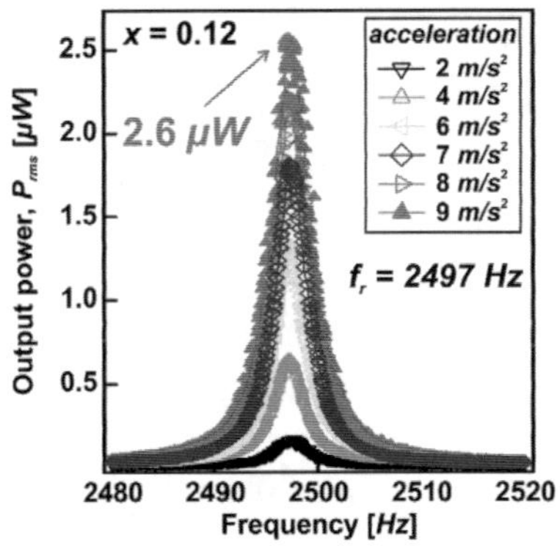

Figure 7. The output power spectra of $(MgHf)_{0.12}Al_{0.88}N$-based VEH

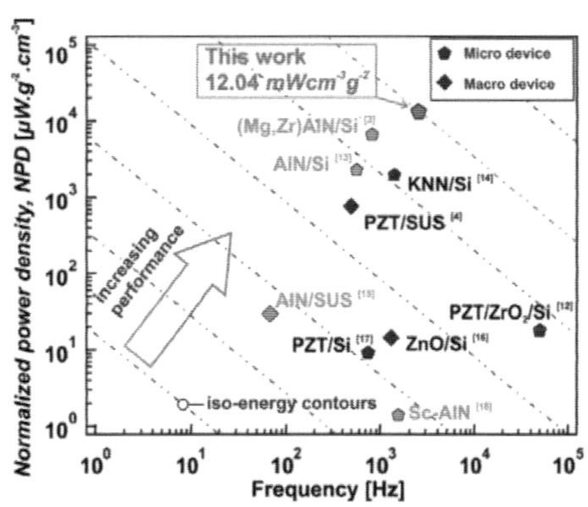

Figure 8. The NPD of $(MgHf)_{0.12}Al_{0.88}N$-based VEH and the state-of-the-art piezoelectric VEHs.

Acknowledgments
The authors would like to thank MNC (Micro/Nano-Machining and Education Center), and Hands-on-access fabrication facility at Nishizawa Research Center, Tohoku University for their valuable supports on experimenting.

References

[1] Kuwano H 1996, *Proc. Int. Symp. Micro Mach. Hum. Sci.*, 21-28.
[2] Baek S H *et al.* 2011 *Science* **334** 958-961.
[3] Minh Le V, Hara M, Yokoyama T, Nishihara T, Ueda M and Kuwano H 2015 *IEEE Trans. Sonics Ultrason.* **62** 2005-2008.
[4] Trolier-McKinstry S and Muralt P 2004 *J. Electroceram.* **12** 7-17.
[5] Nguyen H H, Oguchi H, Minh L V and Kuwano H 2017 *ACS Comb. Sci.* **19** 365-369.
[6] Nguyen H H, Oguchi H and Kuwano H 2016 *J. Phys.: Conf. Ser.* **773** 012075.
[7] Akiyama M, Kano K and Teshigahara A 2013 *Appl. Phys. Lett.* **102** (2) 021915.
[8] Yokoyama T, Iwazaki Y, Onda Y, Nishihara T, Sasajima Y and Ueda M 2015 *IEEE Trans. Sonics Ultrason.* **62** 1007-1015.
[9] Kanno I, Kotera H and Wasa K 2003 *Sens. Actuators, A* **107** 68-74.
[10] Tasnádi F, Alling B, Höglund C, Wingqvist G, Birch J, Hultman L and Abrikosov I A 2010 *Phys. Rev. Lett.* **104** 137601.
[11] Wakasa Y, Kanno I, Yokokawa R, Kotera H, Shibata K and Mishima T 2011 *Sens. Actuators, A* **171** 223-227.
[12] Jeon Y B, Sood R, Jeong J H and Kim S G 2005 *Sens. Actuators, A* **122** (1) 16-22.
[13] Elfrink R, Kamel T M, Goedbloed M, Matova S, Hohlfeld D, Andel Y V and Schaijk R V 2009 *J. Micromech. Microeng.* **19** (9) 094005.
[14] Minh L V, Hara M and Kuwano H 2013 *J. Micromech. Microeng.* **23** (3) 035029.
[15] Cao Z, Zhang J and Kuwano H 2012 *Sens. Actuators, A* **179** 178-184.
[16] Wang P. and Du H 2015 *Rev. Sci. Instrum.* **86** (7) 075002.
[17] Fang H B, Liu J Q, Xu Z Y, Dong L, Wang L, Chen D, Cai B C and Liu Y 2006 *Microelectron. J.* **37** (11) 1280-1284.
[18] Mayrhofer P M, Rehlendt C, Fischeneder M, Kucera M, Wistrela E, Bittner A and Schmid U 2017 *J. Microelectromech. Syst.* **26** (1) 102-112.

MULTILAYER PIEZOELECTRIC MEMS ENERGY HARVESTER BASED ON LONGITUDINAL EFFECT

R. Nakanishi[1], K. Kanda[1], T. Fujita[1], I. Kanno[2], and K. Maenaka[1]

[1]Graduate School of Engineering, University of Hyogo, 2167 Shosha, Himeji, Hyogo, Japan
[2]Department of Mechanical Engineering, Kobe University, 1-1 Rokkodai-cho, Nada-ku, Kobe, Hyogo, Japan

kanda@eng.u-hyogo.ac.jp

Abstract. Piezoelectric MEMS energy harvesters based on multilayer (ML) Pb(Zr,Ti)O_3 (PZT) thin films with interdigitated electrodes (IDE) as their internal and top electrodes are proposed and designed. Thick PZT thin films with good piezoelectric characteristics can be deposited by using multilayer deposition technique. Furthermore, IDE is adopted to the harvester design in order to take advantage of the longitudinal piezoelectric effect, which typically has twice piezoelectric constant as large as transverse effect. The energy harvesters with two electrode configurations, parallel plate electrodes (PPE) and IDE, are designed and estimate these output power. According to the result of finite element analysis, the output power per gravitational acceleration is about 124 μW for the PPE configuration. On the other hand, the output power per gravitational acceleration for the IDE configuration is about three times as large as that for PPE. Moreover, the influence of IDE patterns and the number of PZT layers were investigated. The larger number of layers results in the larger output power.

1. Introduction

Along with the development of the Internet of Things (IoT) society, vast amount of sensors have been used. The sensors should be self-powered in order to avoid troublesome power wiring and exchange cost of butteries. Therefore, energy harvesting devices are important as the alternatives of the sensor butteries. In particular, vibration-type energy harvesting devices have been attracted attention because vibration presents everywhere. For the self-powered sensors, the harvesters and sensors should be integrated as a MEMS device. In addition, piezoelectric harvesters have been the most popular manner because piezoelectric MEMS have features such as simple structures and moderate output impedance.

The typical structure of piezoelectric MEMS energy harvesters are simple beam with a proof mass. The beam is the lamination of piezoelectric thin films and Si supporting layer. The external vibration deforms the beam and provides piezoelectric films with in-plane stress. For the material of the piezoelectric MEMS harvesters, PZT is a promising candidate because of its large electromechanical coupling factor. One of the most important problems is the thickness of PZT for the MEMS energy harvesters. Although thick PZT is preferable to obtain large output power and structural strength, it is difficult to achieve both thick and good piezoelectric characteristics. The typical thickness of PZT for MEMS is 3 μm or less [1]. Multilayer PZT (ML-PZT) thin films with internal electrode layers can achieve 10 μm-thick PZT with good piezoelectric characteristics [2]. In this study, ML-PZT is used for the design of the energy harvesters.

On the other hand, conventional electrode configurations for the piezoelectric MEMS energy harvesters are shown in Figure 1. The most common electrode configuration is parallel plate electrodes (PPE). For PPE, the electric power is generated based on transverse piezoelectric effect. For IDE, the poling direction becomes longitudinal in-plane direction. The longitudinal piezoelectric effect (directions of the stress and poling are same) has about twice piezoelectric constant as large as transverse one. Therefore the output power from the piezoelectric energy harvesters based on the longitudinal effect can be quadruple because that is proportional to the square of piezoelectric constant. However, IDE has several difficulties; (1) the piezoelectric material beneath the electrodes cannot contribute the electric charge generation, (2) the best ratio of IDE gap to width is 2 for optimal poling treatment [3], and (3) the optimum ratio of electrode width to thickness of piezoelectric layer is 0.8 because the pole is not along to longitudinal direction [4]. To lighten these problems, ML-PZT with internal IDE is proposed in this study.

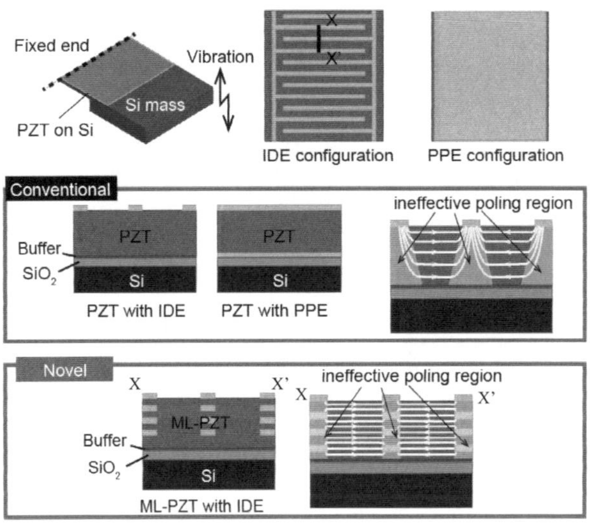

Figure 1. Illustration of ML-PZT with internal IDE.

2. Fundamental design for piezoelectric MEMS harvesters

As a fundamental design, an energy harvester with PPE and single-layer PZT was designed. The harvester has a supporting beam and a proof mass as depicted in Fig. 1. Assuming alternatives to button batteries, device size is configured to 10 × 10 mm^2. Thickness of the mass is configured 525 μm based on the thickness of a 4-inch wafer. Thickness of the PZT is configured to 10 μm based on our conventional study [2]. The length ratio of the beam and mass and substrate thickness is determined to have large output power and sufficient structural strength. The average longitudinal in-plane stress σ in the piezoelectric layer when gravitational acceleration g is applied is expressed as

$$\sigma = \frac{3Y_s Y_p t_s (t_s + t_p) \rho_m g l_m (t_m + t_s)(l_b + l_m)}{Y_s^2 t_s^4 + Y_p^2 t_p^4 + 4Y_s Y_p t_s^3 t_p + 4Y_s Y_p t_s t_p^3 + 6Y_s Y_p t_s^2 t_p^2}, \quad (1)$$

where, Y, t, ρ, and l are Young's modulus, thickness, density, and length, respectively. The subscript "s", "p", "b" and "m" denote Si, PZT, beam, and mass, respectively. The optimum load R is given as

$$R = \frac{t_p}{\omega \varepsilon_p l_e w}, \tag{2}$$

where, ω, w, and ε_p are resonant angular frequency, width, and dielectric constant of piezoelectric layer, respectively. By using R, the output power is given as

$$P_{\mathrm{R\_fr}} = \frac{Q_{\mathrm{m}}^2}{2R}\left(\frac{l_e w d_{31}\sigma}{C}\right)^2, \tag{3}$$

where, d_{31}, C, and Q_{m} are piezoelectric constant, capacitance, and quality factor, respectively. The dimensions were optimized to have maximal output power below the yield strength of the materials under application of 10G (=98 m/s^2) acceleration (Fig. 2), assuming Q_{m} = 100. The dimensions are shown in Table 1. The output power per gravitational acceleration is about 124 μW for the PPE configuration.

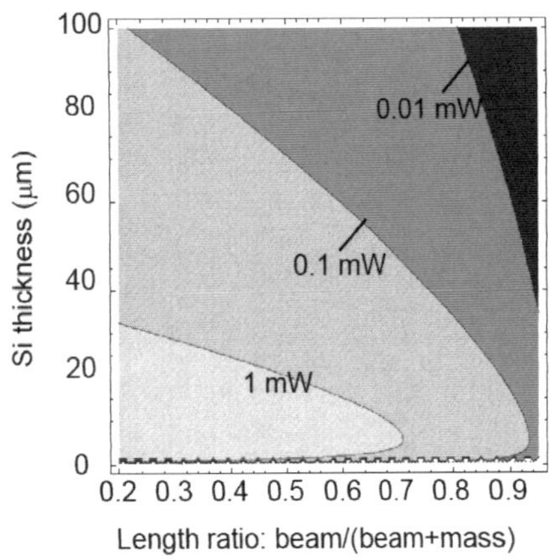

Figure 2. Estimation of output power for designed piezoelectric energy harvester with PPE configuration.(single layer PZT with thickness of 10 μm)

Table 1. Dimensions of designed energy harvester.

Dimensions	Value
Length of beam (mm)	3.33
Thickness of Si on beam (μm)	80
Thickness of PZT on beam (μm)	10
Width of beam (mm)	10
Thickness of Si mass (μm)	525

3. Analysis for IDE configuration

The energy harvester with ML-PZT and internal IDE was designed. For the design, the basic dimensions are same as mentioned in chapter 2. The ML-PZT is composed of four layers of PZT with internal IDE. The thickness of the each PZT layer is 2.5 μm. For actual case with taking the fabrication process into account, the electrodes and Si sandwiching insulation SiO_2 and buffer layers makes stray capacitance because Si substrate is modelled as conductive material. Therefore the influence of these layers (SiO_2 with thickness of 1 μm and TiO_2 buffer layer with the thickness of 75nm) are also modelled. The line width of the IDE is decided to be 5 μm because of convenience in fabrication process. The rest parameter of the dimensions is IDE gap. The poling direction of PZT is assumed as in-plane direction except for the PZT regions underneath the electrode (the underneath regions have out-of-plane poling direction). The output power for the harvester with IDE gap as the parameter was calculated from the results of piezoelectric structure analysis and electromagnetic analysis. The calculation results are shown in Fig. 3. The output power increase with an increase of the IDE gap, because the capacitance between the electrodes pair decrease with the increase of the IDE gap in despite of same generative electric charge. The output power reaches about three times as large as that with PPE. Figure 4 shows the influence of the layer number. The larger number of layers generates larger output power. The out-of-plane poling region underneath the electrodes would influence.

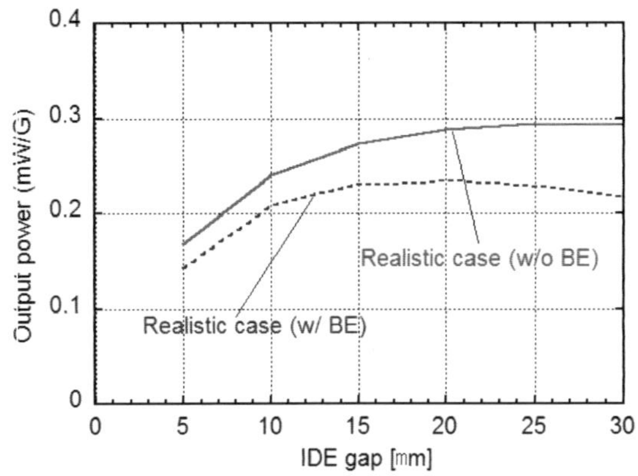

Figure 3. Calculation results for ML-PZT harvester with internal IDE.

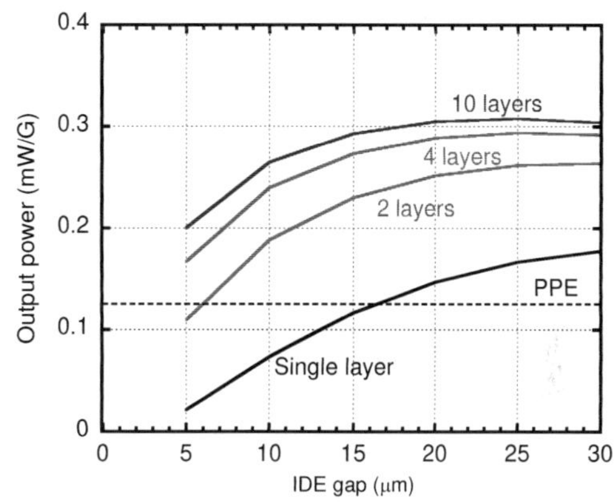

Figure 4. Effect of multilayer

4. Conclusions

This paper proposed and designed piezoelectric MEMS energy harvesters based on multilayer PZT thin films with internal IDE. The output power of the harvester device is estimated about 0.3 mW per gravitational acceleration. This value is about three times larger than that of the conventional harvester with PPE. In addition, it was found that the larger number of layers generates larger output power.

Acknowledgements

This research is supported by JST CREST, JPMJCR16Q4 and JSPS Grant-in-Aid for Scientific Research (C) 17K06269, JAPAN.

References

[1] S. Trolier-Mckinstry and P. Muralt 2004 J. Electroceram Vol.12 pp 7-17
[2] R. Sano et al. 2015 Jpn. J. Appl. Phys. 54 10ND03
[3] N. Chidambaram et al. 2012 IEEE Trans. Ultrason. Ferroelectr. Freq. Ctrl. Vol. 59 pp 1624-1631
[4] R. R. Knight et al. 2011 J. Electroceram. Vol. 26 pp 14- 22

Direct piezoelectric properties of $BiFeO_3$ epitaxial films grown by combinatorial sputtering

T. Yoshimura, K. Kariya, N. Okamoto, M. Aramaki and N. Fujimura

Department of physics and electronics, Osaka Prefecture University, Sakai, Osaka, 599-8531, Japan

tyoshi@pe.osakafu-u.ac.jp

Abstract. We have proposed that $BiFeO_3$ films are suitable for piezoelectric vibration energy harvesting applications, because of the low dielectric permittivity and excellent ferroelectricity. In this paper, the improvement of the piezoelectric properties of the domain-engineered $BiFeO_3$ films was discussed. The (100) epitaxial films were prepared at growth temperatures between 500 and 650°C on Nb-doped (100) $SrRuO_3$ single crystals. The combinatorial sputtering method was employed in order to prevent Bi defects during high-temperature deposition. The domain structure and $e_{31,f}$ piezoelectric coefficients of the films were characterized by a piezoelectric force microscopy and direct piezoelectric response, respectively. The highest $e_{31,f}$ coefficient of -4.3 C/m^2 and figures of merit of 14 GPa were obtained in the film deposited at 600°C, which has the highest domain wall density. This suggests that the introduction of the domain walls is effective to enhance the piezoelectric properties of $BiFeO_3$ films.

1. Introduction

Recently, vibration energy harvesters (VEHs) are receiving a considerable amount of interest [1,2]. While the various transduction mechanisms for VEHs exists such as electromagnetic and electrostatic, the piezoelectric effect has the feature of the direct conversion of the applied strain energy into a useful electric energy [2]. The typical structure of the piezoelectric VEHs is a cantilever with a mass at the free end. The electromechanical coupling factor of the piezoelectric films in this structure is given by

$$k_{31}^2 = \frac{(1-\nu)^2}{Y}\frac{e_{31,f}^2}{\varepsilon_0\varepsilon_r}, \quad (1)$$

where $e_{31,f}$ is the transverse effective piezoelectric coefficient, ε_0 is the permittivity of vacuum, ε_r is the relative permittivity, Y is Young's modulus, and ν is the passion ratio [3]. We have proposed that $BiFeO_3$ films are suitable for piezoelectric VEH (pVEH) applications, [4, 5] because it has a high spontaneous polarization (~100 μC/cm^2) and a low dielectric permittivity (~100) [6, 7]. Piezoelectric MEMS VEH using sol-gel derived $BiFeO_3$ films showed a maximum output power of 10 μW/mm^2/G^2 [3, 8]. This result is comparable to those of the best-performing VEHs using other piezoelectric films [9, 10].

While it is indicated that $BiFeO_3$ films are promising for pVEH applications, the further improvement of the piezoelectric properties of $BiFeO_3$ films needs to facilitate device design. The general approach to enhance the piezoelectric properties is the exploration of the morphotropic phase boundary (MPB). For example, on the (100) epitaxial films of the solid solution system of $BiFeO_3$, $BaTiO_3$, and $SrTiO_3$

reported $d_{33(AFM)}$ coefficient of 83 pm/V was reported [11]. The effect of rare-earth substitution for $BiFeO_3$ epitaxial films was also investigated, and $d_{33(AFM)}$ coefficient of 110 pm/V was reported on Sm^{3+} substituted films [12]. Our approach is the engineering of the domain configuration. In ferroelectric single crystals, the excellent piezoelectric properties are obtained in a direction not parallel to the spontaneous polarization [13]. It was indicated that this mechanism is effective for piezoelectric films also [14]. We have reported that the $e_{31,f}$ coefficient of (100) $BiFeO_3$ epitaxial film (-3.5 C/m^2) is 2.7 times higher than that of the (111) film (-1.3 C/m^2), which indicates that the engineered-domain configuration is effective for $BiFeO_3$ films [4].

For the further improvement of piezoelectric properties, we employed a combinatorial sputtering method [15] to obtain stoichiometric films regardless the growth conditions because it is known that Bi is easily volatile during the deposition. The relationship between the growth condition, crystal structure, domain structure and piezoelectric properties are discussed.

2. Experiment

The schematic illustration of the combinatorial sputtering method is shown in Fig. 1. To fabricate the films with a gradient of Bi/Fe ratio in-plane direction, $Bi_{1.1}FeO_3$ ceramic and pressed powder of Bi_2O_3 were used as the targets, which have a diameter of 2-inches. RF power of 30W and 5W was applied, respectively, at the same time. Nb-doped (100) $SrTiO_3$ single crystals were used as the substrates. The growth temperature was varied from 500 to 650 °C. The films with a thickness of about 300 nm were obtained for the deposition of 5h. The crystal structure of the films was investigated by x-ray diffraction (XRD) of 2θ-ω scan and reciprocal space mapping. The surface morphology and ferroelectric domain structure were observed by scanning probe microscopy (SPM). The effective transverse piezoelectric coefficient, $e_{31,f}$, was characterized by direct piezoelectric response based on substrate bending method [16].

3. Results and Discussion

XRD analysis indicated that the (100) $BiFeO_3$ epitaxial films with cube-on-cube relationship were obtained at the growth temperature from 500 to 650 °C. Significant change in the phase development

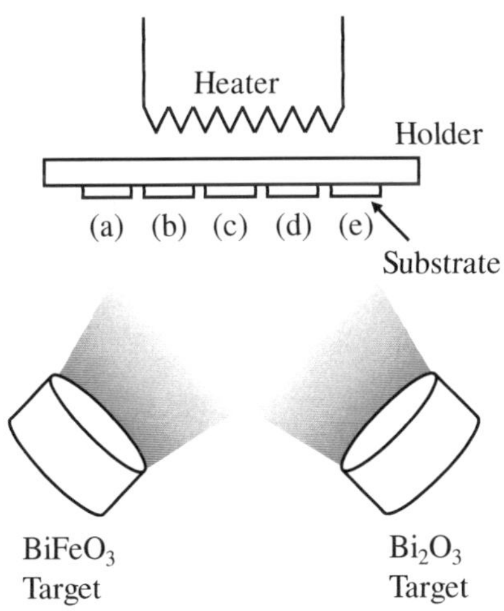

Figure 1. Schematic of the combinatorial sputtering method.

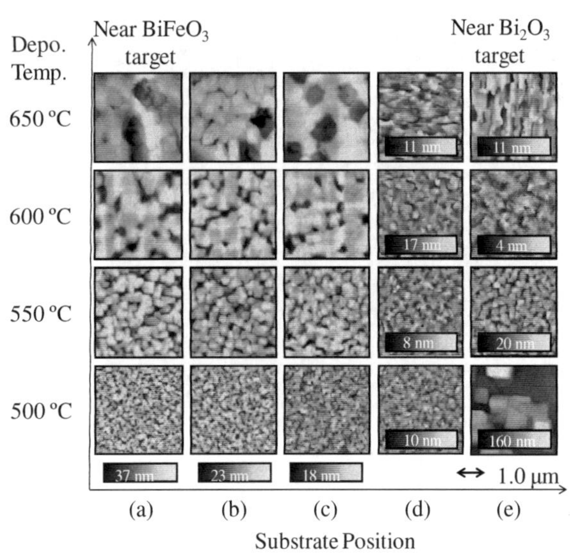

Figure 2. Surface morphology of the $BiFeO_3$ epitaxial films deposited at various temperature.

against the composition was not observed. Figure 2 shows the surface morphology of these films. At the deposition temperature of 500 °C, the smoothest surface is observed at the position (d). Above 600 °C, the smoothest surface is observed at the position (e). These results can be explained by the change of the film composition because it is well known that for $BiFeO_3$ films the surface roughness drastically increases by the off-stoichiometry. Since the volatilization of Bi during the deposition increases with increasing the growth temperature, the position where the film has the stoichiometric composition shifts to the side of the Bi_2O_3 target.

The domain structure of the $BiFeO_3$ films was observed by piezoelectric force microscopy using SPM. The (100) $BiFeO_3$ films have a rhombohedral structure with a spontaneous polarization along the <111> direction. Since the conductive SPM probe was placed parallel to the <110> direction of $BiFeO_3$ in this measurement, the lateral piezoelectric response has four possible directions with respect to the in-plane direction. Figure 3(a) and 3(b) shows the mapping images of the vertical and lateral converse piezoelectric response of the $BiFeO_3$ films with the stoichiometric composition. At the deposition temperature of 500 °C, the domain has a round shape with a size of around 100 nm, which is almost same as the grain size. This indicates that the grain growth is insufficient and the most of the domain walls exist at the grain boundary. At 550 and 600 °C, the domain size observed by the vertical piezoelectric response increases to 100-500 nm and the domains have a polygon shape. The domain size observed by the lateral piezoelectric response is bit smaller than that observed by the vertical piezoelectric response. The (100) $BiFeO_3$ films with rhombohedral structure have the three types of domain walls as shown in figure 3(c). While 180° and 109° domain walls exist between the upward and downward domains, only 71° domain walls exist between the domains with the same vertical component. The domain observation for the films deposited at 550 and 600 °C indicates that the part of the domain walls is 71° walls. At 650 °C, the domain has an elongated shape with a width of 200-300 nm and a length of 500~1000 nm. The domain sizes observed by vertical and lateral piezoelectric response are almost same, which indicates that the domain walls consists of 180° and 109° domain walls.

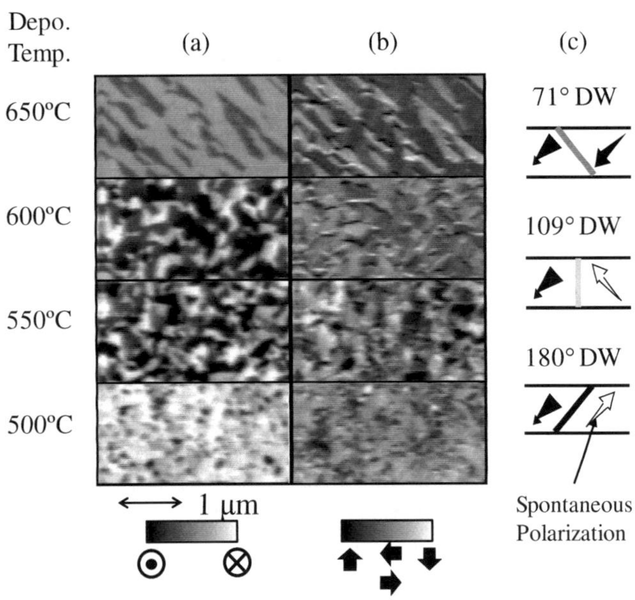

Figure 3. Mapping images of (a) vertical and (b) lateral converse piezoelectric responses of the $BiFeO_3$ films and (c) types of the domain wall (DW).

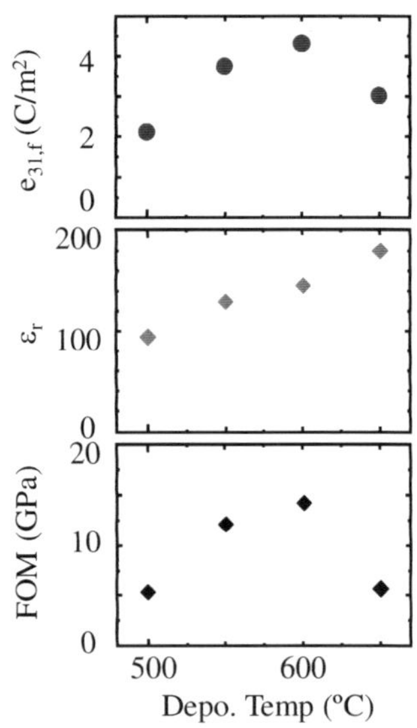

Figure 4. Growth temperature dependence of $e_{31,f}$ coefficient, dielectric constant and FOM of the $BiFeO_3$ films.

The $e_{31,f}$ coefficient and dielectric constant of these films are shown in figure 4. The $e_{31,f}$ coefficient increases with increasing the deposition temperature up to 600 °C, then decreases at 650 °C. The highest $e_{31,f}$ coefficient at 600 °C is -4.3 C/m^2, which is comparable to the best result for $BiFeO_3$ films [6]. We have revealed that the intrinsic contribution in the $e_{31,f}$ coefficient for the (100) $BiFeO_3$ films is about -2 C/m^2, and that the other component in the $e_{31,f}$ coefficient is caused by 71° domain walls. The results presented in this paper are consistent with this. The dielectric constant of the films gradually increases from 100 to 200 with increasing the growth temperature. The figure of merit (FOM) for VEH application, which is given by $e_{31,f}^2/\varepsilon_0\varepsilon_r$, is also shown in figure 4. As can be seen, the highest FOM of 14 GPa was obtained at the growth temperature of 600 °C, at which the highest $e_{31,f}$ coefficient and domain wall density are obtained. Given the thickness of the $BiFeO_3$ films (~300 nm), these results are quite promising, because the obtained FOM (14GPa) is comparable to the typical PZT film with a thickness more than 1 μm. Thus, it can be concluded that the control of the domain structure is the key to obtain high $e_{31,f}$ coefficient and FOM.

Conclusion

Direct piezoelectric properties of (100) $BiFeO_3$ epitaxial films grown by combinatorial sputtering was investigated. The films with smooth surface were obtained at the growth temperature from 500 to 650°C. It was found that the density of 71° domain walls increases between 550 and 600°C. The high $e_{31,f}$ coefficient (-4.3 C/m^2) and FOM (14 GPa) for VEH application are obtained in the films deposited at 600°C. The results obtained in this study suggests that increase of the density of 71° domain walls is effective to improve the piezoelectric properties of (100) $BiFeO_3$ epitaxial films.

Acknowledgments

This work was supported by CREST, JST (JPMJCR16Q4), and the Industrial Technology Research Grant Program in 2011 from New Energy and Industrial Technology Development Organization (NEDO) of Japan.

References

[1] Roundy S, Wright P and Rabaey J 2003 *Comput. Commun.* **26** 1131
[2] Kim S, Priya S and Kanno I 2012 *MRS Bull.* **37** 1039
[3] Yoshimura T, Murakami S, Wakazono K, Kariya K and Fujimura N 2013 *Appl. Phys. Exp.* **6** 051501
[4] Ujimoto K, Yoshimura T, Ashida A and Fujimura N 2012 *Appl. Phys. Lett.* **100** 102901
[5] Kawahara Y, Ujimoto K, Yoshimura T and Fujimura N 2012 *Jpn. J. Appl. Phys.* 51 09LB04
[6] Wang J *et al.* 2003 *Science* **299** 1719
[7] Yun K, Ricinschi D, Kanashima T, Noda M and Okuyama M 2004 *Jpn. J. Appl. Phys.* **43** L647
[8] Murakami S, Yoshimura T, Satoh K, Wakazono K, Kariya K and Fujimura N 2013 *J. Phys.:Conference Series* **476** 012007
[9] Shen D, Park J, Noh J, Choe S, Kim S, Wikle H and Kim D 2009 *Sens. Actuators A*, **154** 103
[10] Defosseux M, Allain M, Defay E and Basrour S 2012 *Sens. Actuators A* **188** 489
[11] Sasaki T, Hirabayashi Y, Kobayashi H and Sakashita Y 2011 *Jpn. J. Appl. Phys.* **50** 09NA08
[12] Cheng C, Kan D, Anbusathaiah V, Takeuchi I and Nagarajun V 2010 *Appl. Phys. Lett.* **97** 212905
[13] Park S and Shrout T 1997 *J. Appl. Phys.* **82** 1804
[14] Yoshimura T, Trolier-McKinstry S 2002 *J. Appl. Phys.*, **92** 3979
[15] Kurokawa F, Tsujiura Y, Hida H, Kanno I 2014 *Jpn. J. Appl. Phys.* **53** 09PA06
[16] Yoshimura T, Sakiyama H, Oshio T, Ashida A, Fujimura N 2010 *Jpn. J. Appl. Phys.* **49** 021501

Bimorph vibration energy harvester with flexible 3D mesh structure

T Tsukamoto[1], Y Umino[1], S Shiomi[1], K Yamada[1], T Suzuki[1,2],

[1]Gunma University, 1-5-1 Tenjin cho, Kiryu-shi, Gunma, Japan
[2]JST PRESTO, Japan

E-mail: suzuki.taka@gunma-u.ac.jp

Abstract. In this paper, we propose a piezoelectric bimorph vibration energy harvester (VEH) with flexible 3D mesh elastic layer for the purpose of improving output power and lowering resonance frequency. Due to the high void ratio of the 3D mesh structure, it is possible to lower the bending stiffness of the cantilever. So, the deflection of the harvester and the strain in the piezoelectric layer increase. As a result of a vibration test, the resonance frequency is 26% lower and the output power is 1.5 times higher than the conventional solid flat plate VEH. Compared with the flat plate VEH, the proposed mesh VEH has 1.3 times larger tip deflection and the total output power is 20.4 μW that is possible to use as a power supply of low-power consumption sensor nodes for WSNs.

1. Introduction

Recently, self-powered sensor nodes for wireless sensor networks (WSNs) has been developed using VEH [1]. Vibration energy harvesting is classified as electromagnetic, electrostatic, triboelectric, and piezoelectric. Among the four types of harvesting methods, piezoelectric VEH has attracted much attention because of the higher energy density, the simplicity of configuration, and the compatibility with the MEMS technology [2,3]. In the piezoelectric VEH, a large variety of cantilever beams with an elastic layer and one or two piezoelectric layers have been presented [4-6].

A bimorph configuration has two piezoelectric layers, so it has a potential to achieve large electric power. However, the smaller the size of the cantilever-type VEH becomes, the more difficult it is to lower resonance frequency and to achieve large electric power that is enough for sensor node battery. Here, in this study, we propose a bimorph VEH with a flexible 3D mesh structure in order to control bending stiffness: to lower resonance frequency and to achieve large electric power simultaneously.

2. Design of bimorph VEH with 3D mesh structure

2.1. Configuration of VEH

The proposed vibration energy harvester is designed as bimorph configuration. Figure 1 shows a schematic of the proposed bimorph cantilever composed of an elastic layer having a flexible 3D mesh structure and piezoelectric layers on both sides. The dimensions of the proposed VEH are shown in table 1. In the bimorph configuration, the neutral axis is located at the center of the elastic layer. Since the strain of the piezoelectric layer is proportional to the distance from the neutral axis, increasing the distance causes large strain under identical cantilever deflection. On the other hand, when the thickness of elastic layer increases, the deflection and the strain are reduced because the bending stiffness is increased. Moreover, the resonance frequency increases. Regarding these demerits, we aim to design and fabricate a flexible VEH to achieve large strain and low resonance frequency simultaneously. Therefore, in this study, we propose a method to reduce bending stiffness while increasing the film thickness of the elastic layer by applying a 3D mesh structure with voids as the elastic layer and evaluate the validity of the structure.

2.2. FEM structural analysis of VEH

Figure 2 represents the effect on the bending stiffness of the 3D mesh structure. The horizontal axis of the graph is the line spacing of the mesh structure in figure 1, and the vertical axis represents the ratio of the bending stiffness of mesh VEH to the flat plate VEH. The normalized bending stiffness is defined by dividing bending stiffness of a mesh harvester by that of a flat plate VEH. The FEM analytical model is made by using COMSOL Multiphysics 5.2a, and the result is calculated from the ratio of tip displacement when the static load is applied to the tip of a cantilever. The bending stiffness decreases with the increase of line spacing of mesh structure. From this result, it is found that the bending stiffness can be controlled and lowered without changing the device size, configuration, and basic fabrication process by adopting the mesh structure to the elastic layer of the bimorph VEH.

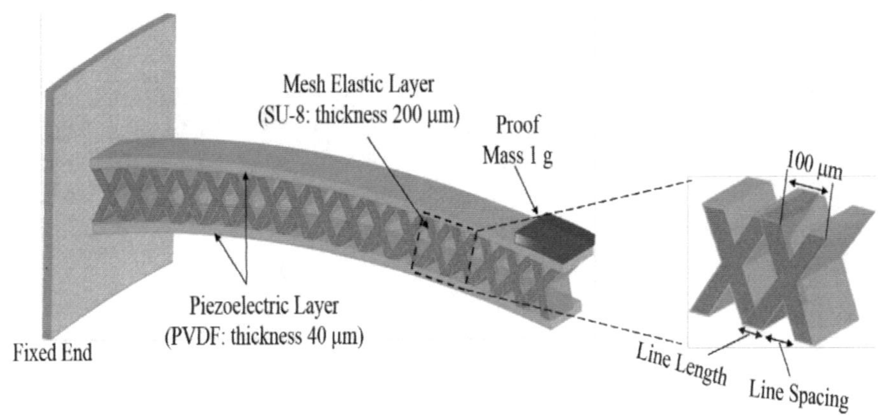

Figure 1. Schematic of a proposed VEH with 3D mesh structure (10mm×21mm×280μm). The device consists of two PVDF piezoelectric layers bonded on both sides of a 3D mesh elastic layer and a proof mass at the tip. A sum of line spacing and line length is 100 μm.

Table 1. Dimensions of the proposed VEH. In the table, the subscripts p and e mean a piezoelectric material and an elastic layer, respectively.

Width	10 mm
length	21 mm
Thickness: t_p	40 μm
Thickness: t_e	200 μm
Young's modulus: Y_p	0.95 GPa
Young's modulus: Y_e	3.25 GPa
Piezoelectric constant d_{31}	22 pC/N
Piezoelectric constant d_{33}	35 pC/N
Proof mass	1.0 g

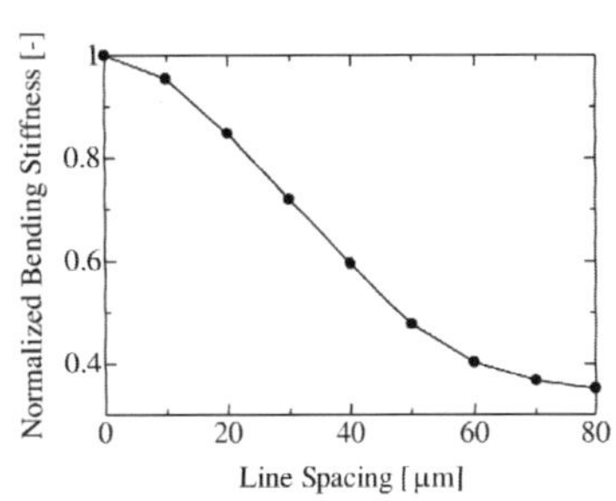

Figure 2. Normalized bending stiffness as a function of line spacing of mesh structure analyzed by FEM. The normalized bending stiffness is calculated by dividing bending stiffness of a harvester with the mesh elastic layer by that of a harvester with a solid flat plate elastic layer.

3. Experiments

3.1. Fabrication of VEH with 3D mesh structure

The proposed VEH is fabricated by inclined photolithography [7] and bonding of piezoelectric PVDF sheet (KF piezo film: Kureha Corporation) and elastic SU-8 layer (Nihon Kayaku: SU-8 3005). Inclined photolithography is one of the fabrication technique to fabricate 3D microstructure by exposing UV light from the oblique direction for photoresist on a substrate. Here, the mesh structure is fabricated by inclined exposure twice from different oblique directions. The fabrication process of the 3D mesh structure is shown in figure 3. SEM images of the fabricated mesh structure using the inclined photolithography are shown in figure 4. In this paper, we fabricated the VEH prototype with a 3D mesh structure, which is designed as the line length and the line spacing of the mesh structure are both 50 μm.

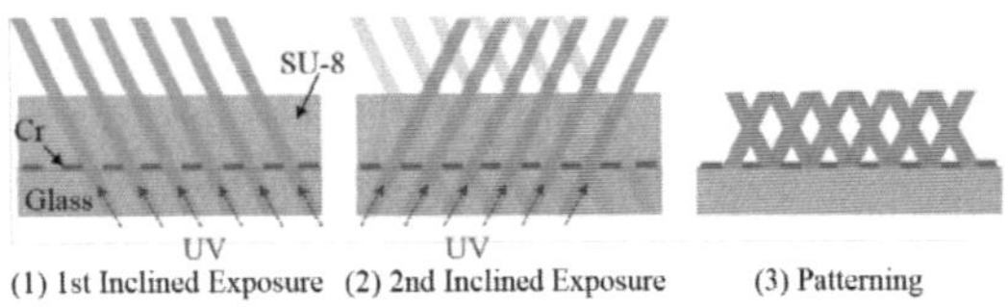

Figure 3. Fabrication principle of a photolithography process for mesh structure. Inclined exposure is used 2 times to fabricate 3D mesh structure.

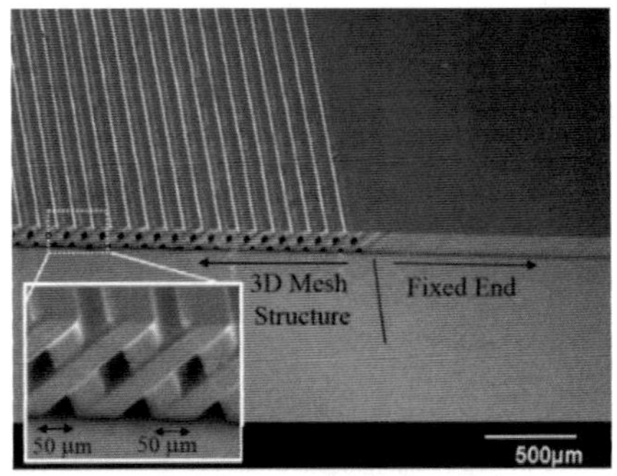

Figure 4. SEM images of the fabricated 3D mesh elastic layer. Both line spacing and line length of mesh structure are 50 μm.

3.2. Vibration Test

We evaluated the power generation performance by vibration test, and compared the proposed mesh type VEH with the device having a flat plate structure of the same size and thickness in the elastic layer. Experimental setup for vibration test is shown in figure 5. A load resistance of 10 MΩ is connected to the VEH, and the amount of output power is calculated from the measured output voltage and load resistance under oscillation. The acceleration monitored by an acceleration sensor becomes constant as 4.9 m/s^2.

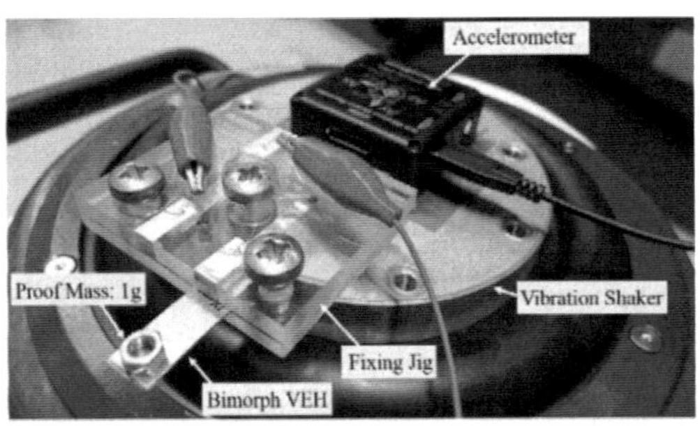

Figure 5. Photograph of vibration test setup with a VEH, a vibration shaker, and an accelerometer.

4. Results and Discussion

The frequency response is shown in figure 6. In the mesh type VEH, the output power is 1.5 times higher than the flat plate one and resonance frequency is reduced by 26% at the same time. Since the bending stiffness decreases due to the mesh structure, the deflection of the bimorph cantilever increases and the amount of strain of the piezoelectric film also increases. Thus, the output power increases because of the large strain in piezoelectric layer. In addition, figure 7 shows the output voltage waveform during the sinusoidal oscillation. In both the mesh and the flat types, we observed that sinusoidal output is synchronized with the oscillation frequency. As shown table 2, the output voltage of mesh type VEH is 1.3 times higher than that of the flat type one. Therefore, from the results of vibration test, mesh structure shows the validity for the elastic layer for bimorph VEH.

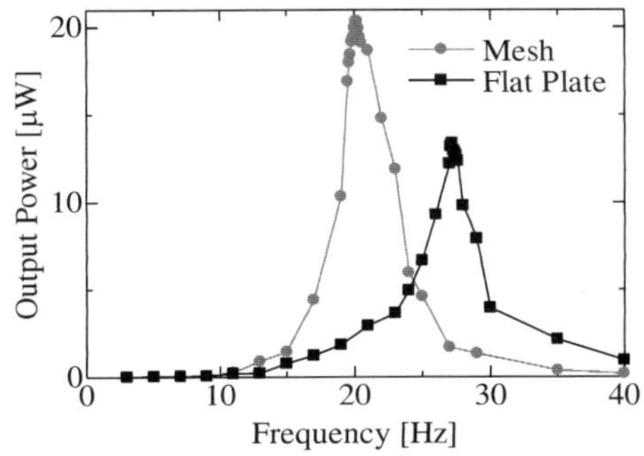

Figure 6. Frequency response of output power. In the proposed mesh VEH, the resonance frequency is lowered by 26 % and the output power is increased by 1.5 times compared with the flat plate VEH.

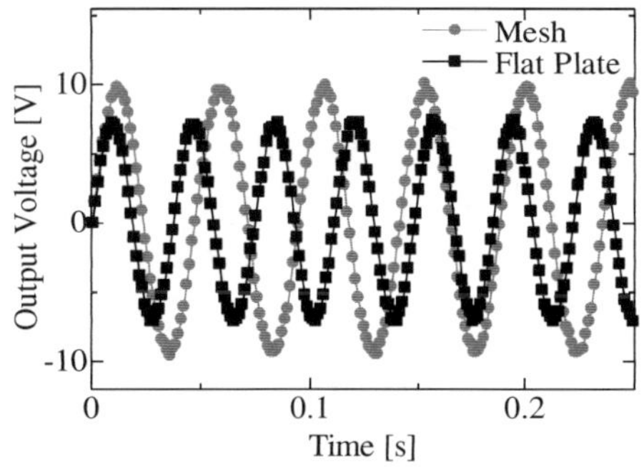

Figure 7. Time response of output voltage of the proposed mesh and the flat plate VEHs under each resonant condition. Voltage amplitude generated by the mesh VEH is 1.3 times higher than one of the flat plate VEH.

Table 2. Experimental results of vibration test at each resonant condition.

	Resonant frequency [Hz]	Total electric power [μW]	Tip displacement [mm]
Mesh	20.1	20.4	6.0
Flat	27.4	13.3	4.6

5. Conclusion

In this study, we confirmed the validity of the bimorph VEH with 3D mesh structure as an elastic layer. First, we showed that the bending stiffness of the mesh harvester can be controlled by changing mesh line spacing. Second, As a result of the vibration test, the resonance frequency of mesh harvester is 26% lowered and the tip deflection is 1.3 times larger compared with the flat plate VEH. Finally, the total output power is 20.4 μW that is possible to use as a power supply of low-power consumption sensor nodes for WSNs.

Acknowledgement

This research was partially supported by JSPS Science Research Grant JP17H03196, JST PRESTO Grant Number JPMJPR15R3, and MEXT Nanotechnology Platform Project (The University of Tokyo Microfabrication Platform).

References

[1] F Karim, S Zeadally 2016 *Renew. Sustain. Energy Rev* 55 1041-54
[2] Janphuang. P and Mishra B et al 2012 *Proc. PowerMEMS 2012* 145-148
[3] G Ferin and C Bantignies et al 2015 *J. Phys. Conf. Ser.* **660** 12085
[4] I Kanno 2015, *J. Phys. Conf. Ser* **660** 12011
[5] H Liu, C J Tay et al 2011 *J. Micoelectromech. Syst* **20** 1131-1142
[6] H S Kim, J H Kim, and J Kim 2011 *Int. J. Precis. Eng. Manuf* **12** 1129-41
[7] T Suzuki, H Kotera, I Kanno, and D Hiramaru 2014 *US Patent* 8871433

Self-powered SSHI for Electret Energy Harvester

Yiran Liu[1] and Yuji Suzuki[1]

[1]Department of Mechanical Engineering, The University of Tokyo,
7-3-1 Hongo, Bunkyo-ku, Tokyo 113-8656, Japan

Email: yliu@mesl.t.u-tokyo.ac.jp

Abstract. A nonlinear power management circuit for electret-based energy harvesting is designed based on the synchronized switch harvesting on inductor (SSHI) technique. With the present circuit, the output power of rotational electret energy harvester is increased by 75% if compared with the conventional full-bridge rectifier.

1. Introduction

Among various vibration energy harvesters, electret energy harvester (EH) is free of external bias-voltage, and has merit of large output power in the low frequency range [1]. It is well known that electret EH suffers from parasitic capacitance, which, even in pF range, leads to substantial power loss due to the shunting effect [2]. Therefore, eliminating the negative impact of parasitic capacitance is crucial to enhance the output power. On the other hand, synchronized switch harvesting on inductor (SSHI), which deals with clamped capacitance in piezoelectric EH, is often proposed to increase harvested power by several hundred percent [3]. However, if compared with piezoelectric EHs, electret EHs exhibit much higher output voltage and much lower current, making electret EH sensitive to the leakage current and parasitic capacitance introduced by the SSHI circuit itself. The objectives of the present study are to design self-powered SSHI for electret EH and to examine its performance using a prototype electret EH.

2. Self-powered SSHI Interface Circuit for EEH

2.1. Circuit model of rotational EEH

Figure 1a shows a schematic of our rotational electret EH [4]. Figure 1b shows its circuit model based on the 1-D model [1]. The model includes an equivalent current source I_s, an internal resistance R_i, and an internal variable capacitance C_i. Their expressions are given by the following equations:

$$I_s = \begin{cases} \dfrac{2\varepsilon_2\varepsilon_1\varepsilon_0 nAf_v V_s}{d\varepsilon_1 + g\varepsilon_2} & 0 \le t < \dfrac{1}{2nf_v} \\ -\dfrac{2\varepsilon_2\varepsilon_1\varepsilon_0 nAf_v V_s}{d\varepsilon_1 + g\varepsilon_2} & \dfrac{1}{2nf_v} \le t < \dfrac{1}{nf_v} \end{cases}$$

$$R_i = \left[2Anf_v \left(\frac{\varepsilon_2\varepsilon_1\varepsilon_0}{d\varepsilon_1 + g\varepsilon_2} - \frac{\varepsilon_1}{g+d} \right) \right]^{-1}$$

$$C_i(t) = \begin{cases} 2A\left(\dfrac{\varepsilon_2\varepsilon_1\varepsilon_0}{d\varepsilon_1 + g\varepsilon_2} - \dfrac{\varepsilon_1}{g+d}\right) nf_v t + \dfrac{\varepsilon_1 A}{g+d} & 0 \le t < \dfrac{1}{2nf_v} \\ 2A\left(\dfrac{\varepsilon_2\varepsilon_1\varepsilon_0}{d\varepsilon_1 + g\varepsilon_2} - \dfrac{\varepsilon_1}{g+d}\right)\left(\dfrac{1}{nf_v} - t\right) + \dfrac{\varepsilon_1 A}{g+d} & \dfrac{1}{2nf_v} \le t < \dfrac{1}{nf_v} \end{cases}$$

Table 1. Definition of parameters in electret EH.

f_v	Rotational frequency	1 Hz
n	Number of electrode pairs	109
A	Maximum overlapping area	5 cm^2
V_s	Surface potential of electret	840 V
ε_2	Relative permittivity of electret	2.1
ε_1	Relative permittivity of air	1
d	Thickness of electret	15 μm
g	Air gap	85 μm

Table 2. Comparison between EEH and PEH

	Electret EH [5]	Piezoelectric EH [6]
I_s (μA)	8.79	87
R_i (MΩ)	1200	2.08
C_i (pF)	44~48 (avg. 46)	33740
C_p (pF)	111	Neglected

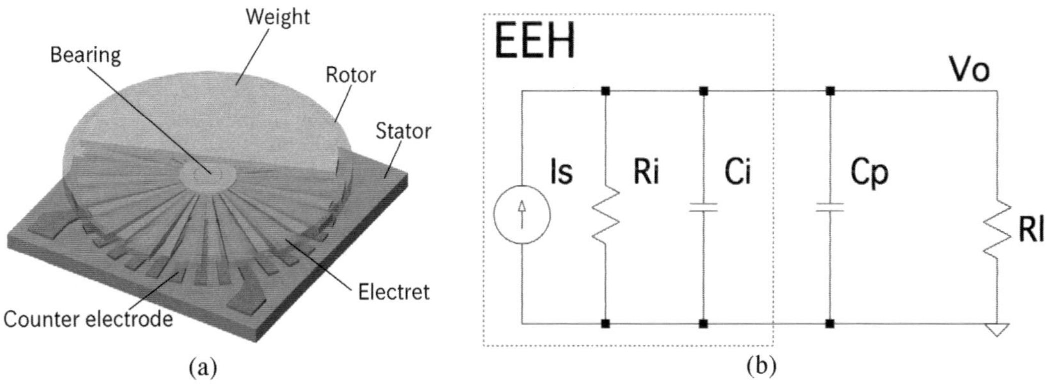

Figure 1. Rotational electret energy harvester [5]. (a) Schematic, (b) Circuit model.

Table 2 gives a comparison between our electret EH [5] and the piezoelectric EH [6]. It is clear that the electret EH shows much lower I_s and parallel capacitance. In this paper, we use the average value of C_i for simplicity since the influence of its little variation is negligible.

2.2. SSHI interface circuit

Figure 2a shows the external powered SSHI circuit design modified from [7] to verify the effectiveness of SSHI on electret EH. V_{sense}, an attenuated version of the rectifier input voltage, is extracted by R_{sense} and connected to the peak detector. The voltage peak is detected by comparing V_{sense} and its delayed signal made by the RC filter with R_{ed} and C_{ed}. The comparator therefore outputs positive (resp. negative) voltage V_{drive}, when V_{sense} is at the maximum (resp. minimum) values. At this instant, the NMOS M1 (resp. PMOS M2) transistor turns on initiating the inversion process. The latter is stopped when the current tends to reverse, as the diode D1 (resp. D2) stops conducting, thereby inverting V_o synchronously with I_s [7]. Due to high voltage and low power concerns, high voltage small signal MOSFETs BSP89 and BSP92P from Infineon are chosen for M1 and M2, respectively. Note that the comparator is externally powered by source meters in this case.

Figure 3b shows the voltage waveforms simulated with LTspice. The load is 7 MΩ. In a standard circuit, the EH is connected to load through a bridge rectifier and a reservoir capacitor. In the present SSHI, switching actions guarantee that output voltage V_{sshi} and I_s have the same sign; therefore, their product is always positive, indicating more power extracted from the mechanical part to the electrical part. Meanwhile, with the same load, V_{rec} in SSHI exceeds that in the standard circuit, indicating more power harvested on load. Neglecting R_i, we can derive the output power as

$$P_o = I_s V_{rec} - 2(1-\gamma)(C_i + C_p) f_s V_{rec}^2$$

where $\gamma = V_{inv}/V_{rec}$. Note that γ =-1 cooresponds to the standard circuit.

IOP Conf. Series: Journal of Physics: Conf. Series **1052** (2018) 012022 doi:10.1088/1742-6596/1052/1/012022

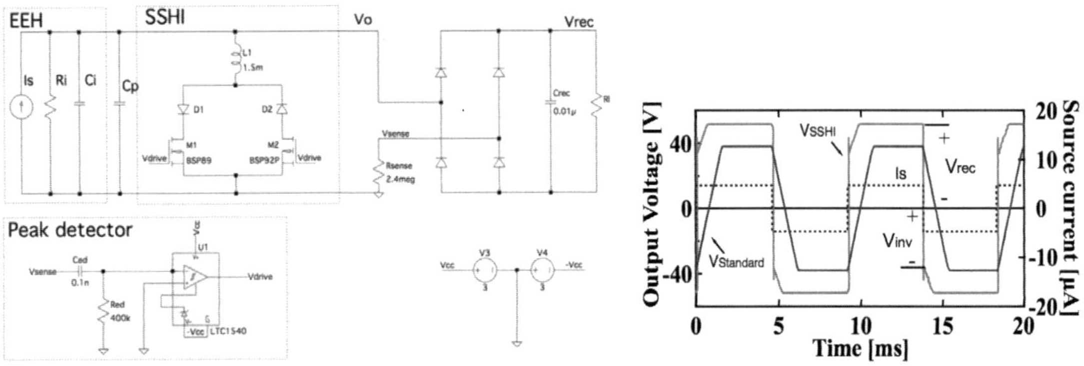

(a) (b)

Figure 2. External powered SSHI. (a) Circuit design, (b) Simulated waveforms.

2.3. *Self-powering scheme*

Figure 3a shows a self-powered SSHI scheme to make the system truly closed in terms of energy. Due to high voltage, the comparator which requires ±3V cannot be directly powered by the electret EH by adding a pair of diode-capacitor branches as in [7]. Here we propose a scheme using two DC/DC converters. The idea is to use a portion of the low-voltage output of the DC/DC converter to reversely power the comparator. On the other hand, creating a dual-polarity power supply by commercial inverter can be power consuming, hence we employ two sets of identical NMOSs and control blocks to take charge of the switching actions respectively. In this case, however, positive control signal and negative control signal may overlap (Fig. 3b) due to inconformity of electric components. Here an enabling signal is introduced to avoid overlapping. Taking the M1-M2 branch for instance, the enabling signal (enp) generated synchronously with the output voltage of electret EH prevents the branch from conducting when output voltage of electret EH is negative, even if M1 is turned on by *ctrlp*. Assuming that the DC/DC converter exhibits the same conversion efficiency in both self-powered SSHI and standard cases, the power output is calculated as

$$P_o = (V_{ccp}^2 + V_{ccn}^2) / R_{load}$$

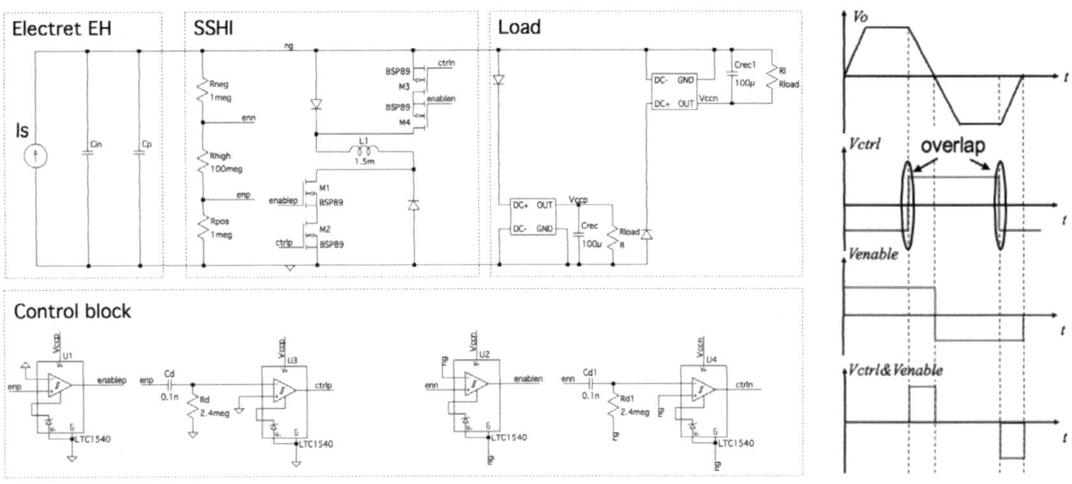

(a) (b)

Figure 3. Self-powered SSHI. (a)Circuit design, (b)Voltage waveforms in control block

3. Experimental results

Figure 4a shows the experimental setup of self-powered SSHI. A pair of DC/DC converters with over 50% conversion efficiency for electret EH [8] is employed in the present study. The rotational EEH is mounted on a test bench, in which the rotational speed is kept constant at 1 rps. The output voltage is measured by oscilloscope (DSO6014A, Agilent) through a 100:1 probe. The output voltages of the DC/DC converters V_{ccp} and V_{ccn} are measured through a 10:1 probe. Figure 4 shows the measured output voltage of the electret EH and V_{ccp} on load (500kΩ) in the standard and the SSHI circuit. Thanks to the SSHI, V_{ccp} in SSHI exceeds that in the standard circuit, indicating more power harvested on load.

Figure 5 shows the output power versus load. With the present self-powered SSHI circuit, 90μW has been obtained on a 500 kΩ load, which is 1.75 times as high as the harvested power with the standard bridge circuit. The performance of SSHI can be further enhanced in two ways: one is to reduce power consumption of the control block by, for instance, increasing resistance of the voltage divider, adjusting R_{ed} and C_{ed}; the other is to increase the inversion ratio, which can be realized by reducing the resistance and current amplitude in the SSHI switching branches.

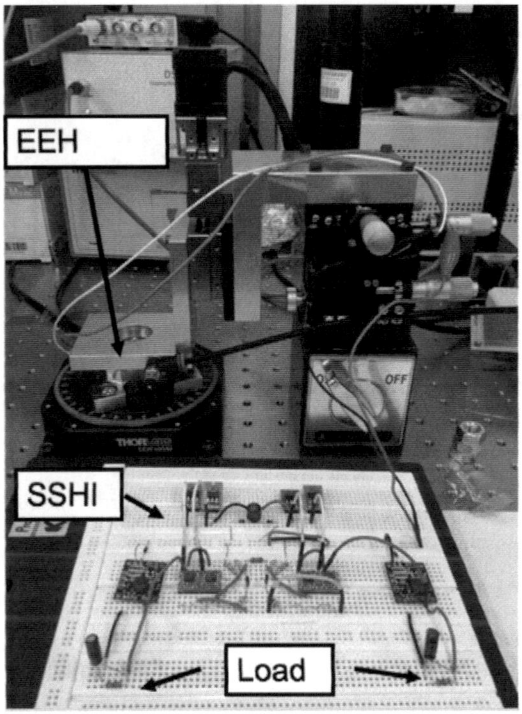

(a)

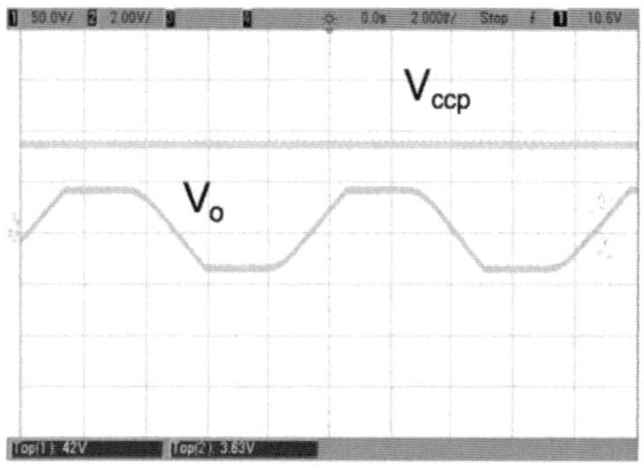

(b)

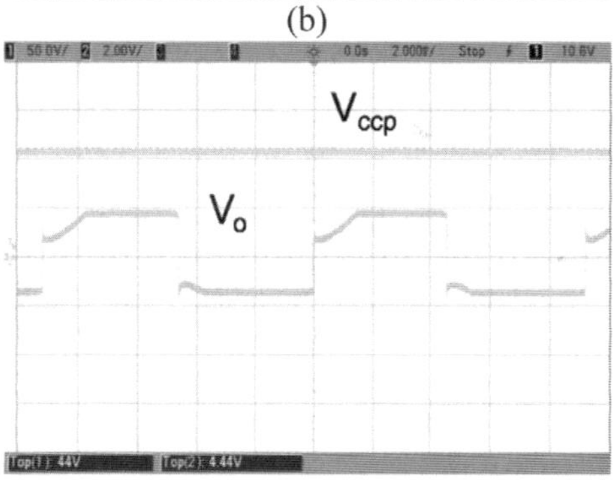

(c)

Figure 4. (a)Experimental setup, (b)Measured voltages in the standard circuit, (c)Measured voltages in the present SSHI circuit.

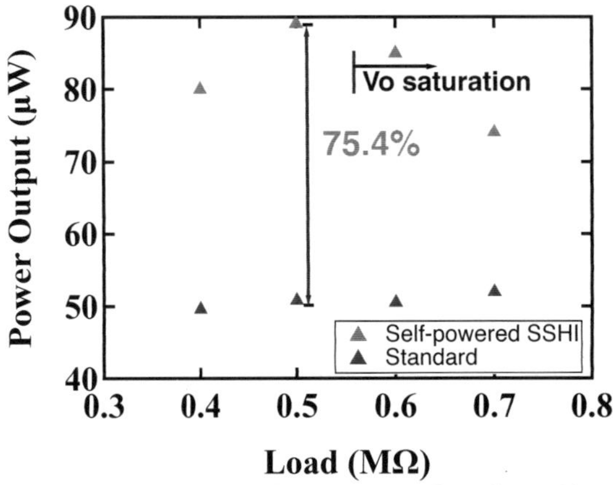

Figure 5. Harvested power as a function of load.

4. Conclusion

Self-powered SSHI for electret EH is proposed for the first time. Based on the current design, over 70% more power can be harvested on load with self-powered SSHI. Our future work will focus on enhancing the performance of self-powered SSHI and on integrated design of circuit.

This work was supported by JST CREST Grant Number JPMJCR15Q3, Japan.

References

[1] Y. Suzuki, "Recent progress in MEMS electret generator for energy harvesting." *IEEJ Trans. Electr. Electr. Eng.* 6 (2011) 101-111.

[2] U. Bartsch, et al. "Influence of parasitic capacitances on the power output of electret-based energy harvesting generators." *Proc. PowerMEMS 2009,* (2009) 332-335.

[3] D. Guyomar, et al. "Toward energy harvesting using active materials and conversion improvement by nonlinear processing." *IEEE Trans. Ultrason. Ferroelectr. Freq. Control,* 52 (2005) 584-595.

[4] J. Nakano, et al. "MEMS rotational electret energy harvester for human motion." *J. Phys.: Conf. Ser.*, 660, (2015) 012052.

[5] M. Adachi, et al. "Development of rotational electret energy harvester using print circuit board" *PowerMEMS 2017*, (2017), to be presented.

[6] J. Liang, and W.-H. Liao. "Improved design and analysis of self-powered synchronized switch interface circuit for piezoelectric energy harvesting systems." *IEEE Trans. Ind. Electr.*, 59 (2012) 1950-1960.

[7] M. Lallart, "Nonlinear technique and self-powered circuit for efficient piezoelectric energy harvesting under unloaded cases." *Energy Convers. Manag.*, 133 (2017) 444-457.

[8] http://www.owl-sol.com/product.html

Extending the range of wireless power transmission for bio-implants and wearables

N Garraud, D Alabi, S Chyczewski, J. D. Varela, D P Arnold, A Garraud
Interdisciplinary Microsystems Group, Department of Electrical and Computer Engineering, University of Florida, Gainesville, FL, 32611, USA

E-mail: agarraud@ufl.edu

Abstract. This paper reports the latest achievements on an electrodynamic wireless power transmission (EWPT) system employing low-frequency (~17 Hz), low-amplitude (< 1 mT) magnetic fields. Compared to inductively or resonantly coupled coils, the EWPT approach facilitates power transmission to multiple receivers in a cluttered environment with little restriction on their orientation. Using a 5.5 cm^3 receiver, a maximum of 22 mW power transfer (4 mW/cm^3 power density), and a maximum transmission distance of 60 cm are independently demonstrated. These capabilities are achieved using one of two transmitter technologies: a coil-based transmitter, and a rotating-magnet transmitter. Through-body transmission and multi-receiver transmission are also demonstrated, opening the way for applications targeting bio-implants and wearables.

1. Introduction

While wireless power transmission (WPT) has been heavily researched in the last few years, there remain numerous challenges for efficiently transferring useful power levels to small power receivers over extended distances. Four interdependent and competing parameters are important for assessing various WPT technologies: the output power, the end-to-end power efficiency, the transmission distance and the receiver size.

In terms of near-field (non-radiative) WPT technologies, inductive coupling is well suited for efficient (>80 %) power transmission at close ranges, where the gap between the two coils is much less than the coil diameter [1]. However, the transmitted power and the system efficiency decrease dramatically while increasing the distance between the transmitter and the receiver (theoretically, the power decreases with the distance to the sixth power, i.e. $P \propto 1/d^6$). Resonant inductive coupling overcame the low mutual coupling problem by using high-Q resonators, demonstrating high power transmission over meters [2]. However, the receiver antenna needs to be finely tuned with the transmitter antenna, and the system efficiency is highly sensitive to their orientation and axial mismatch.

Other WPT constraints to consider include human exposure to electromagnetic fields, receiver orientation independence, transmitter and receiver volume and weight, transmission to multiple receivers, and the ability to transmit power through a "cluttered" environment where conductive and/or ferromagnetic objects may be in between the transmitter and receiver.

The work presented in this paper focuses on mid-range transmission (i.e. 10 to 60 cm) with electrodynamic WPT (EWPT) using a permanent magnet embedded in the receiver that acts as a resonant torsional oscillator. Relying on a high mechanical Q-factor, the EWPT approach is shown to transmit power using safe, low-frequency, low-amplitude magnetic fields at mid-range distances (i.e. several diameters of the transmitter antenna).

2. System Design

2.1. EWPT Working Principle

In the EWPT working principle, a time-varying magnetic field emitted by the transmitter imparts a magnetic force or torque on the receiver magnet, inducing its motion (Figure 1a) [3,4]. In the system demonstrated here, a torsion oscillator is used in the receiver. The mass-spring system exhibits a maximum rotation amplitude at the resonance frequency, which is chosen as the system operating frequency. The magnet rotation within the receiver windings generates electrical power by induction and power is delivered to an external load. Resonant-type EWPT relies on a high mechanical Q-factor of the oscillator to maximize responsivity to low-amplitude magnetic fields.

EWPT generally operates at much lower frequencies (10 – 1000 Hz) than inductively or resonantly coupled systems (100 kHz – 10 MHz) [1]. The lower frequencies facilitate higher safety margins for magnetic field amplitudes [5,6] and better penetration through electrically conductive media (e.g. metal, human body, seawater). This enables wireless power delivery to multi-receivers in a crowded, cluttered environment, such as a home/office environment, an automobile, or within the human body for deep-biomedical-implants.

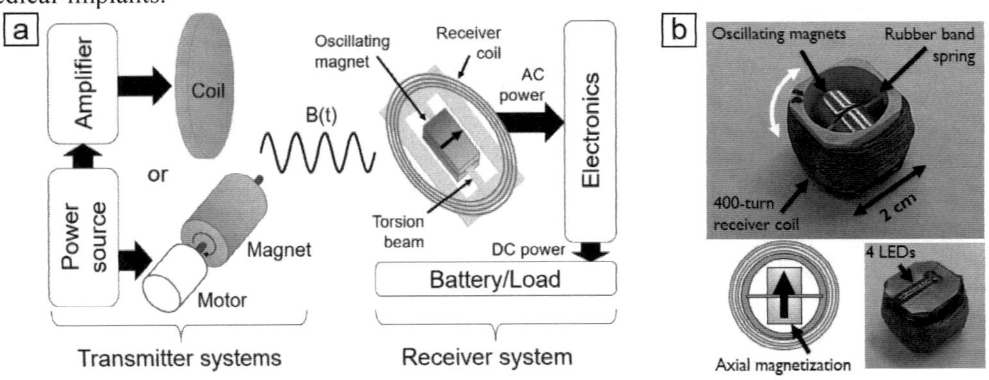

Figure 1: *(a) General schematic of EWPT (electrodynamic wireless power transmission). (b) Prototype of resonant receiver built for demonstration.*

2.2. Receiver Design (Torsion Oscillator)

A macroscale torsion oscillator receiver is built to demonstrate and optimize the EWPT system at mid-range (Figure 1b). It is worth noting that the primary focus of this paper is on the transmitter and not the receiver. The receiver is composed of two cylindrical Nd-Fe-B magnets (8.5 mm diameter, 6 mm long) supported on a rubber band spring, which facilitates torsional resonance around 17 Hz with a Q-factor of 36. The receiver coil is made of 400 turns of AWG-36 wire wrapped around a rigid plastic housing. The receiver is 5.5 cm^3 and 15 g.

2.3. Transmitter Design and Optimization

The alternating magnetic field (17 Hz in this case) can be generated either by a coil transmitter supplied with an AC current [4,7], or by a rotating permanent magnet actuated by a DC motor [8], as illustrated in Figure 2a. While a coil generates a time-varying uniaxial field (meaning at any given point around the transmitter, the field only increases and decreases in time, but does not change direction), the rotating magnet creates a more complex time-varying rotating magnetic field, resulting in increased flexibility in the receiver orientation. The rotating magnet also provides a large magnetic moment for "free," and is especially efficient at long distances. However, it is limited in rotation rate, relies on moving parts, and the static field cannot be turned off, which may be undesirable for certain applications.

In order to compare the transmission technologies and their range of transmission, three coils (denoted coil #1, coil #2 and coil #3) and a rotating magnet are built (Figure 2a). A transmitter performance can be analyzed through a figure of merit $fm = m^2 f^2 / P_{in}$ combining important characteristics of the magnetic source: m the magnetic moment of the transmitter (units of A·m²), f its operating frequency, and P_{in} its input power. This figure of merit quantifies how effective the magnetic source is to create a magnetic moment for a given input power and is independent of the operating current (for current-carrying coils). For annular, pancake-type coils, it can be shown that fm is maximum when the coil inner diameter is about 0.36 times its outer diameter (the inner loops burn proportionally more power than the magnetic moment they generate). The maximum outer diameter is generally limited by the application, but also by the maximum reactive power deliverable by the power amplifier at the operating frequency, since the coil inductance increases with the diameter. Compared to un-optimized coil #1 with a calculated value of fm of 5.2×10^6 W/T², coil #2 and coil #3 are optimized with fm reaching 13.4×10^6 W/T² and 84×10^6 W/T², respectively. The more compact rotating-magnet transmitter exhibits a fm of 72×10^6 W/T².

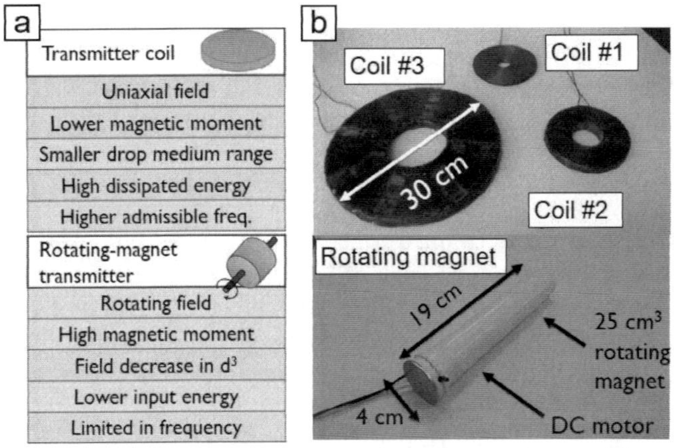

c

	Dimensions	Number of turns	AWG	Magnetic moment m (A.m²)	fm (10⁶ W/T²)
Coil #1	15-cm Ø 6 mm thick	200	16	6.2	5.2
Coil #2	15-cm Ø 15 mm thick	169	12	15	13.4
Coil #3	30-cm Ø 17 mm thick	350	12	36	84
Rotating magnet	4-cm Ø 9-cm long	N/A	N/A	38	72

***Figure 2**: (a) Comparison between the coil transmitter and the rotating-magnet transmitter technologies. (b) Built transmitter coil #1, #2 and #3 and built rotating-magnet transmitter. (c)Main characteristics of each fabricated transmitter.*

3. Experimental Setup and Results

Measurements are conducted to compare the different transmitters. First, each transmitter is characterized by measuring the magnetic field surrounding it for a given input power. Then, the wireless power delivery to the resonant receivers is measured as a function of the distance from the source.

3.1. Magnetic Field Mapping

An AC input signal generated by a waveform generator (Agilent 33120A) is sent to a linear power amplifier (Crown XLS2500), supplying current to the transmitter coil. The coil current is measured using a current probe connected to an amplifier (Tektronix TCP305A/ TCPA300 respectively), while the output voltage is measured using two 10x oscilloscope probes, all of which are connected to a 4-channel oscilloscope (Agilent DSO-X-2004A) to estimate the coil input power. In the following experiments, the input power to the coils is fixed at 10 W. The rotating-magnet transmitter is supplied by a 19 V wall transformer. The 5 W motor input voltage, regulating the rotation frequency, is modulated by a potentiometer and a TO-220AB MOSFET.

The magnitude of the magnetic field is measured using a 3-axis probe (F.W. Bell, model ZOA99-3208) connected to a 3-axis gaussmeter (F.W. Bell, series 9950). From the measurements, Figure 3a shows various iso-envelopes of magnetic field from 0.05 mT to 5 mT for the four transmitters (the results from each transmitter plotted in just one quadrant). The transmitter coil input powers are kept constant at 10 W for fair comparison (input current of 3.3 A_{rms} for coil #1, 5.8 A_{rms} for coil #2, 2.8 A_{rms} for coil

#3). For low magnetic fields, a sharp improvement in range is noticed moving from coil #1 to #2 and again from #2 to #3. The performance of coil #3 is quite similar to the rotating-magnet transmitter: both 0.05 mT envelopes have a radius around 40 cm. Compared to coil #1, the volume of the 0.05 mT envelope increases by 1.6 fold for coil #2, by 4.2 fold for coil #3, and by 4 fold for the rotating-magnet transmitter. It is shown later than 0.05 mT is plenty large enough to deliver power via EWPT.

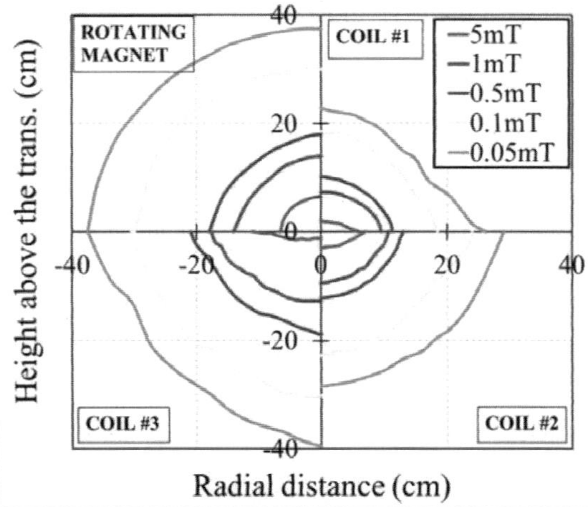

Figure 3: *Iso magnetic field envelopes for the four different transmitters (10 W input for transmitter coils).*

3.2. Wireless Power Performance

For wireless power measurements, the receiver is connected to a variable resistive load box. The ac power delivered to the load is determined by measuring the time-varying load voltage and load current (via current probe). Figure 4a shows the evolution of the output power with the load resistance for different magnetic field amplitudes. The optimal load resistance corresponds to the maximum load power. Simulations ran with a lumped-element model under Simulink (dashed lines) fit the experimental data (solid lines) for low magnetic fields up to 0.2 mT. For higher magnetic fields, the mismatch corresponds to very large magnet rotation reaching the maximum range (±90°), and correspondingly less well-controlled torsional motion.

The matched-load output power versus the transmitter-receiver distance is reported on Figure 4b, the receiver being along the central axis of the transmitter coils (or perpendicular to the rotating magnet). For mid-range distances beyond 10 cm, coil #3 and the rotating-magnet transmitter outperform the two other coils. 1.5 mW power transmission is achieved up to 60 cm, as well as 140 μW up to 100 cm, which is still useful in powering some electronic devices. Note that the transmitter coils have an advantage in close range (<10 cm) compared to the rotating-magnet transmitter. In fact, at close range, the strong force attraction between the rotating magnet transmitter and the receiver magnet inhibits rotational motion of the receiver.

3.3. Wireless Transmission in Cluttered Environments

Four 2-mW LEDs are connected to multiple receivers for demonstration. Figure 5a demonstrates the power transmission through a body (upper torso), lighting LEDs at 30 cm from the rotating-magnet transmitter. Moreover, Figure 5b demonstrates simultaneous power transmission to three arbitrarily-placed/oriented receivers in environment cluttered with metallic objects.

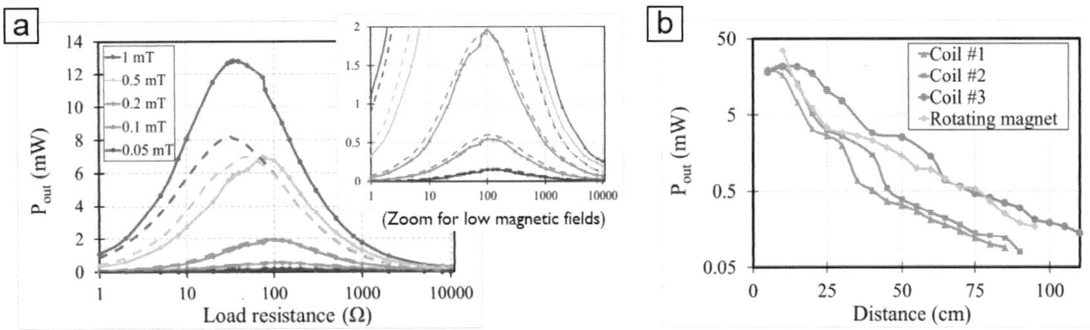

Figure 4: *(a) Measurements of receiver power at different magnetic field amplitudes vs. the load resistance (experimental data: solid lines, simulation: dashed lines). (b) Matched-load ac output power vs. distance for the different types of transmitters.*

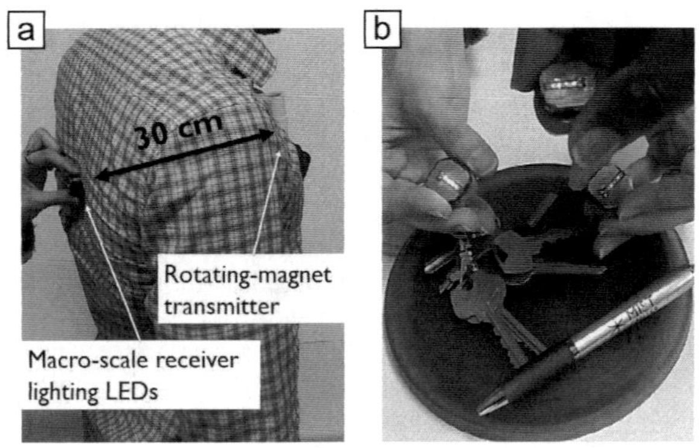

Figure 5: *Demonstration of WPT (a) through body for bio-implants and wearables, and (b) to multiple receivers in a cluttered environment.*

Conclusion

In this paper, two different WPT transmitter technologies are compared: a coil supplied by an AC current or a rotating magnet operated by a motor. The optimized coil #3 and the rotating-magnet transmitter show similar magnetic field and power transmission performances. These two transmitters can transmit non-negligible power (i.e. 1.5 mW) up to 60 cm with optimized coils and power density up to 4 mW/cm^3, a 28-fold improvement compared to previous work [3,7]. Future work will focus on miniaturizing the receiver while keeping a high-power density.

Acknowledgments

This work was supported in part by the NSF I/UCRC on Multi-functional Integrated System Technology (MIST) Center (NSF Grant IIP-1439644).

References

[1] Wei X, Wang Z and Dai H 2014 A Critical Review of Wireless Power Transfer via Strongly Coupled Magnetic Resonances *Energies* **7** 4316–41

[2] Barman S Das, Reza A W, Kumar N, Karim M E and Munir A B 2015 Wireless powering by magnetic resonant coupling: Recent trends in wireless power transfer system and its applications *Renew. Sustain. Energy Rev.* **51** 1525–52

[3] McEachern K M and Arnold D P 2013 Electrodynamic wireless power transmission to a torsional receiver *J. Phys. Conf. Ser.* **476** 12004

[4] Garraud A, Munzer D J, Althar M, Garraud N and Arnold D P 2015 Watt-level wireless power transmission to multiple compact receivers *J. Phys. Conf. Ser.* **660**

[5] C95.6 2002 *IEEE Standard for Safety Levels With Respect to Human Exposure to Electromagnetic Fields, 0-3 kHz*

[6] C95.1 2010 *IEEE Standard for Safety Levels With Respect to Human Exposure to Radio Frequency Electromagnetic Fields, 3 kHz to 300 GHz*

[7] Challa V R, Mur-Miranda J O and Arnold D P 2012 Wireless power transmission to an electromechanical receiver using low-frequency magnetic fields *Smart Mater. Struct.* **21** 115017

[8] Li W 2009 *High efficiency wireless power transmission at low frequency using permanent magnet coupling* (Master's Thesis, University of British Columbia)

IOP Conf. Series: Journal of Physics: Conf. Series **1052** (2018) 012024 doi:10.1088/1742-6596/1052/1/012024

A highly sensitive and ultra-low-power wake-up receiver for energy-autonomous embedded systems

P Woias, S Heller, U Pelz

Albert-Ludwigs-University of Freiburg, Department of Microsystems Engineering (IMTEK), Laboratory for Design of Microsystems, Georges-Koehler-Allee 102, 79110 Freiburg, Germany

E-mail: woias@imtek.de

Abstract. This article presents the design, optimization and characterization of an ultra-low-power wake-up receiver for embedded wireless sensor nodes. The RF part of the circuit is based on an optimized diode detector for an amplitude-modulated (AM) 315 MHz RF carrier, followed by an ultra-low power low-frequency (LF) amplifier for the demodulated signal, and a commercial ultra-low-power wake-up receiver IC. Special emphasis has been put into optimizing the RF part with respect to its input sensitivity and selectivity. With two different versions of this circuit part, RF reception sensitivity could be raised from -52 dBm to -63 dBm. Power consumption is 7.4 μW at a minimum supply voltage of 2 V.

1. Introduction

Today, the development of embedded wireless sensor systems is driven by a number of up-coming - or already present - target applications, like medical assistive devices and implants, Internet-of-Things (IoT) technologies and distributed sensor systems in smart buildings, structural health monitoring, industrial sensing, automotive or aircraft applications. In many of these application areas a large number of sensor nodes are deeply embedded into their respective application sites. This calls, inevitably, for selective wireless communication between nodes and respective base stations, as any wired connectivity would severely reduce flexibility in application, deployment, modification or service. Also, communication is frequently bidirectional and non-synchronized: A base station must, for instance, be able to address an individual wireless sensor node (WSN) at deliberate times to reprogram the WSN's tasks or to retrieve information on demand.

In this situation a serious problem arises from the fact that every WSN has to listen, more or less continuously, to potentially incoming messages. In a conventional system design this would require, at least, regular and frequent up-times of the WSN and its RF transceiver. The system will wake up for a short time, activate its wireless communication interface and listens to potentially in-coming messages. If no message arrives, this effort is fruitless and wastes energy. The unnecessary power drain from the system battery or energy harvesting unit is considerable and will reduce the WSNs operational lifetime and reliability, especially if it is not easily accessible for service.

In this situation, energy-autonomous as well as battery-operated WSNs do greatly benefit from the integration of a wake-up receiver. This unit listens continuously for incoming wireless communication while consuming an absolute minimum of electrical power and wakes up the system from its power-saving sleep mode only if a valid communication request is detected. Its task is merely to receive basic information at a low data rate, e.g. the address of a specific WSN to be accessed. Therefore, energy-saving receiver concepts are conceivable.

2. Wake-up receivers: Requirements and state-of-the-art

The premier requirement for a wake-up receiver is ultra-low-power operation. Moreover, its RF frequency range should allow for long-range communication in order to reach a WSN over a larger distance. Receiver sensitivity must be high from the same reason. With a widespread commercial application in mind, the use of a license-free frequency range, like one of the available ISM bands, is preferable. As these frequency bands may be highly covered, selectivity must be good enough to suppress interferences from neighbour RF channels. Finally, the wake-up receiver should be available in a highly compact form, e.g. as integrated circuit.

Taking the measures of that time, low-power receivers have been available as integrated circuits in the 1970s already: For instance, an AM medium wave receiver of the company Ferranti was operational from a supply voltage of 1.1 V only, with 200 μA supply current [1]. High RF sensitivity (typ. 50 μV) was achieved through the integration of a multi-stage amplifier for the demodulated audio signal. Automatic gain control was realized for receiving signals from both weak and strong transmitters.

These design elements are also present in all modern commercial wake-up receiver ICs [2,3,4]. Simi-lar to [1] they operate at low frequencies (LF, typ. 125 kHz) for the sake of low power consumption (typ. a few μW). This does, however, require large electrical antennas or magnetic pickup coils. High-frequency wake-up receivers usually consume much higher power if realized as ICs. Gamm et al. [5] have shown a solution for this design trade-off: By modulating a high-frequency carrier with a pulsed LF signal, they combine the advantages of high frequency communication (smaller antennas, larger distances) with the low power draw of integrated LF wake-up receivers. The same concept is found in [6], others use the WSN's microcontroller for data decoding [7,8].

In all cases RF transmission is based on amplitude modulation (AM), and all receivers use a diode detector as RF frontend. The main reason for this choice, despite its limited channel resolution and noise immunity, is again power consumption: Frequency or phase modulation are more reliable and tolerant against interferences, however would require a much more power-hungry circuitry, whereas a detector radio is using the energy present in the RF signal itself for demodulation.

3. Design concept

Combining the benefits of the concepts outlined above, the wake-up receiver realized within this study uses a diode detector as RF stage with a LF amplifier for improving overall reception sensitivity [1] and a commercial LF wake-up receiver IC for further signal amplification and data decoding [5]. The receiver operates within the 315 MHz ISM band and requires a pulsed, amplitude-modulated signal. Figure 1 shows the circuit diagram with two different implementations of the RF detector circuit.

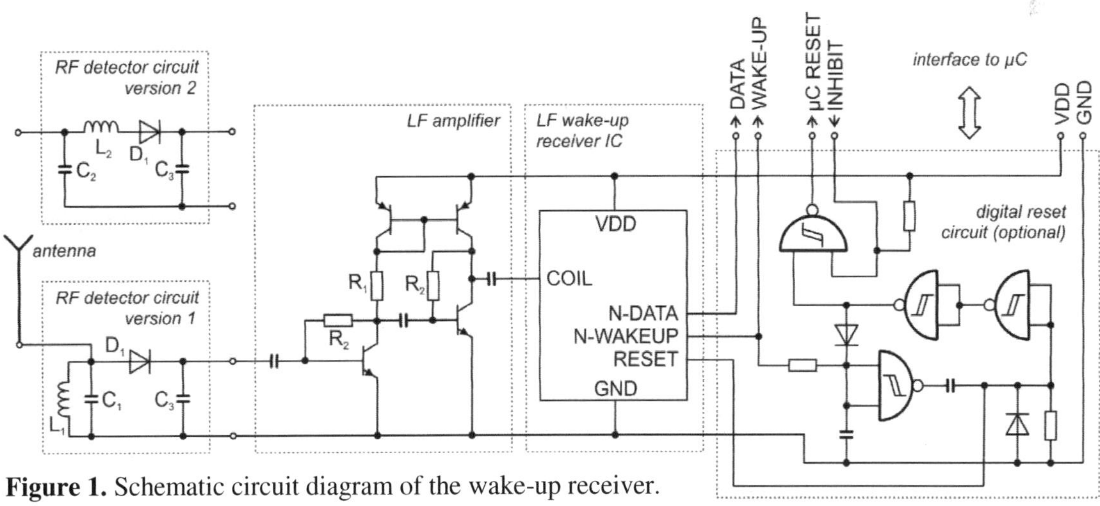

Figure 1. Schematic circuit diagram of the wake-up receiver.

The detector output signal is fed into a two-stage LF transistor amplifier and, from there, into the input of the LF wake-up receiver IC [10]. The LF amplifier is made from two high-gain bipolar transistors (BC 860 C) with a current mirror at their collector terminals. Gain and bias set-point of both transistor

stages are adjusted via respective collector-emitter feedback resistors R_2. Capacitive AC coupling is realized between all stages. The amplifier and all other active circuit parts operate at VDD = 2V.
The commercial wake-up receiver IC in use is designed for ultra-low-power and low-frequency operation within upper and lower corner frequencies of 180 kHz and 30 kHz, respectively [4]. To match this frequency range, a pulsed modulation frequency of 32.768 Hz was used for digital data transmission: At the transmitter every data bit representing a logical "1" will turn on the 32 kHz modulation, whereas a logical "0" will lead to the transmission of an unmodulated carrier signal. After demodulation and amplification, a pulse train of 32 kHz bursts will appear at the input COIL of the wake-up receiver IC to be further amplified, decoded and presented as serial data stream at the output N-DATA. The maximum power consumption for this IC is approximately 2.8 μW at 2 V.
A global or an address-dependent wake-up can be performed as required: Global wake-up occurs after a valid signal has been present at the input COIL for a fixed latency period. Then, a digital wake-up detection signal appears at the output WAKE-UP for e.g. activating a connected microcontroller. Now, additional data, e.g. an address, can be transmitted and will appear as serial data stream at the output DATA to be decoded by the activated microcontroller. The WAKE-UP signal is latched and has to be cleared via the ICs input RESET. This could be done easily by the microcontroller. As an alternative and as safety feature, a logic circuit has been implemented instead with four NAND Schmitt trigger gates to enable - after a pre-set time - an automatic reset of the wake-up receiver and the system's microcontroller (output μC RESET). Microcontroller reset can be prevented via the input INHIBIT. This allows the automatic reset of a non-responding microcontroller by simply sending a wake-up signal. As this logic circuit is optional, its power consumption is not taken into account in table 1.

4. Implementation, optimization and characterization

The RF detector circuit and the LF amplifier have been optimized with regard to their performance and power consumption. Fig. 2 shows the frequency-dependent gain of the LF amplifier as a function of its supply current, with an obvious trade-off between both. Finally, voltage gain was set to appr. 10 at 32 kHz, with a power consumption of 4 μW at a supply voltage of 2 V. Therefore, a demodulated signal with 100 μV_{RMS} amplitude is sufficient for wake-up, taking the input sensitivity of the wake-up receiver IC (1 mV_{RMS}) into account.
The RF detector was optimized step-wise towards higher sensitivity. Characterization was done by applying an amplitude-modulated RF signal (100% modulation depth) directly at the antenna input and by recording the minimal RF input power required for detecting a valid wake-up. As a starting point, the non-optimized version 1 uses a LC tank (L_1, C_1) for frequency tuning, a RF Schottky diode (Agilent HSMS 285) for demodulation and the capacitor C_3 for filtering. Although not optimized, the RF sensitivity of version 1 does already reach -52 dBm.

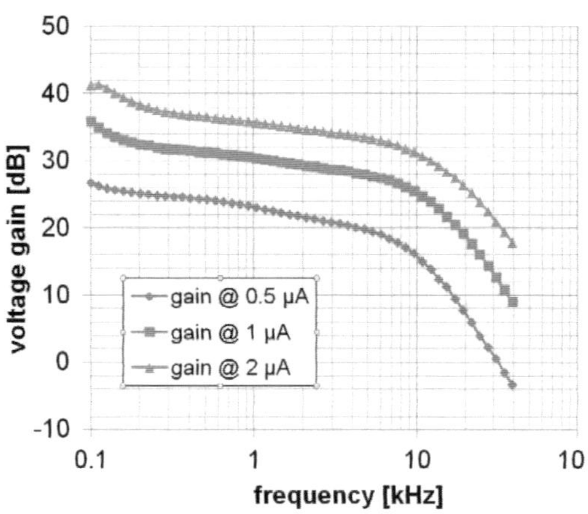

Figure 2. Voltage gain of the LF amplifier as a function frequency and of supply current, with a supply voltage of 2 V. Resistor R_1 is used to adjust the supply current (see fig. 1).

In contrary to that, version 2 was optimized with respect to maximal power transfer from the antenna to the detector, using an antenna matching network composed from L_2 and C_2. As a result, the input impedance of version 2 is close to the desired 50 Ohm, which represents the typical impedance of an ideal whip antenna. The input reflection coefficient s11 of version 2 is given in figure 3. At resonance, version 2 of the detector circuit has an input sensitivity of -63 dBm.
Practical implementation was made on standard double-sided printed circuit boards (PCB, FR4 with 0.8 mm thickness). For comparison of RF detectors, exchangeable test boards were built with a common shield layer on one side and the SMD components placed on the other side. SMA connectors were used for the antenna input and an edge connector to connect these boards to the rest of the circuit.

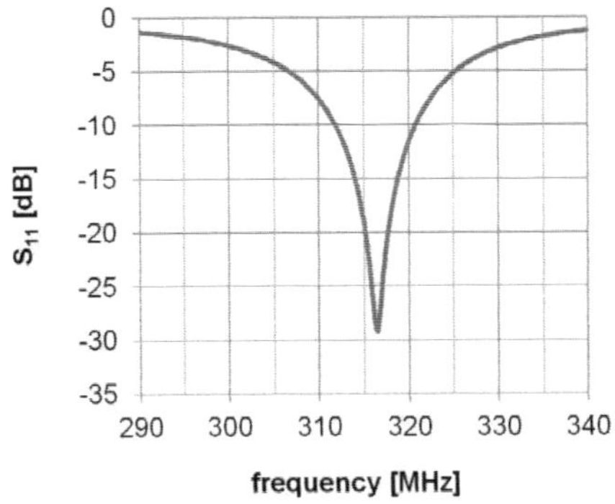

Figure 3. S_{11} plot of the RF detector version 2

Figure 4. Fully integrated version 1 of the wake-up receiver (PCB size: 15 mm x 24 mm)

In addition, an integrated version was built with the wake-up receiver on one PCB side, as visible in Figure 4. All digital logics are placed onto the backside for the sake of small board dimensions.

5. Discussion and Conclusions

As shown in table 1, the input sensitivity of the non-optimized version 1 is already in the typical range known from a number of other publications. The sensitivity of version 2 is significantly higher. A power consumption of 7.4 µW is typical for the majority of wake-up receivers presented so far. This value could, however, be reduced by operating the LF amplifier at lower power: Taking the same input impedance at the RF input, a LF voltage gain of only 5 at 32 kHz, with a total power consumption of approximately 3.5 µW, would still result in a calculated RF sensitivity of approximately -60 dBm. This leaves the user with the freedom to decide whether low power consumption or high RF sensitivity is preferable. Also, a further optimization, e.g. by choosing a lower modulation frequency in conjunction with suitable wake-up detection, leaves additional room for a reduction of power consumption.

	This work		[5]	[6]	[7]*	[8]*
	V. 1	V. 2				
RF frequency [MHz]	315	315	868	868	300	868
RF sensitivity [dBm]	-52	-63	-52	-60	---	-55
supply voltage [V]	2	2	3	2.85 - 3.6	3	2
supply current [µA]	3.7	3.7	2.59	2.37	1.6	---
supply power [µW]	7.4	7.4	7.77	---	4.8*	1.2*

Table 1. Comparison of this work with other research. (*) pure RF receivers and signal conditioners, without data decoding and/or wake-up detection, (---) no values given.

6. Acknowledgement

Sebastian Stöcklin from the IMTEK's Laboratory for Electrical Instrumentation is gratefully acknowledged for professional assistance in all RF measurements.

7. References

[1] ZN 414 - An AM Radio Receiver, datasheet, Ferranti Limited, 1973.
[2] AS 3930, Single Channel Low Frequency Wakeup Receiver, datasheet, Austria Micro Syst., 2014.
[3] AS 3933, 3D Low Frequency Wakeup Receiver, datasheet, Austria Micro Systems, 2011.
[4] ATA 5283 Interface IC for 125 kHz wakeup function, datasheet, Atmel, 2007.
[5] G. U. Gamm, M. Sippel, M. Kostic, L. M. Reindl, *Proc. 6th Int. Conf. on Intelligent Sensors, Sensor Networks and Information Processing*, Dec. 7-10, 2010, Brisbane, Australia, 121-126.
[6] S. Bdiri, F. Derbel, O. Kanoun, *Journal of Sensors and Sensor Systems* **5**, 2016, 433-446.
[7] M. Malinowski, M. Moskwa, M. Feldmeier, M. Laibowitz, J. A. Paradiso, *Proc. 5th Int. Conf. on Embedded Networked Sensor Systems*, Nov. 6-9, 2007, Sidney, Australia, 145-159.
[8] M. Magno, V. Jelicic, B. Srbinovski, V. Bilas, E. Popovici, L. Benini, *IEEE Trans. Ind. Informatics* **12** (2), 2016, 633-644.

Continuous Machine Health Monitoring Enabled Through Self-Powered Embedded Intelligence and Communication

J Cornett, A O'Grady, A Vouaillat, J Michaud, F. Muret, W. Weatherholtz J Bai, M Dunham, P Riehl, B Chen, M Farrington and T Galchev
Analog Devices, Inc., 804 Woburn St, Wilmington, MA, 01887, USA

tzeno.galchev@analog.com

Abstract. This paper presents a fully self-powered machine health monitoring wireless node. A bismuth telluride thin film thermoelectric generator (TEG) is used to convert the temperature difference between the hot machine and the cooler ambient into electrical energy. By providing an autonomous power supply to the wireless condition-based monitoring (CBM) system, the cost of ownership is significantly reduced and this technology can become readily accessible to a much wider variety of applications. The paper discusses the system operation, component choices, and thermal transport issues that must be addressed to make an autonomous system viable. The system power consumption can range from 12 – 200 μW depending on the refresh rate which spans the range of 30 s to 30 min. A temperature difference of 2.4 – 10 °C is sufficient to cover this operating span, considering the performance of typical thin-film thermoelectric harvesters.

1. Introduction

Autonomous condition-based monitoring (CBM) has the potential to eliminate downtime, reduce maintenance costs, and improve worker safety. In addition to vibration sources that exist by design, factory equipment can produce telling signs in their vibration signature that warn of things like bearing defects, imbalance and gear meshing, and provide an opportunity for timely corrective actions. Permanent CBM, often prohibitively expensive, is currently reserved for only the most critical pieces of equipment and only in certain industries like energy generation and petrochemical production, which can afford this technology. Other machines are monitored intermittently (or not at all) using hand tools and/or long time intervals that leave much to be desired with respect to the quantity of data, its repeatability, and the cost of having dedicated technicians. Connected devices and the Internet of Things have the potential to fundamentally reshape how CBM is performed and who has access to this type of technology. However, to make it truly cost competitive, wireless devices should not require maintenance themselves; therefore, an energy source that lasts the lifetime of the sensor node is required. This paper presents a fully autonomous CBM node that makes use of thermoelectric energy harvesting technology to supply its own power requirements. A bismuth telluride-based thin film thermoelectric generator (TEG) in the CBM node is used to convert the temperature difference between a hot machine it is installed on and the ambient into electrical energy. Few self-powered system implementations have been presented that use thermoelectric harvesting (for example [1, 2]) and even fewer have made it into the marketplace. Analog Devices seeks to reduce the technological burden of implementing self-powered systems by: 1) providing a wide portfolio of harvesters, power management ICs, and ultra-low-power sensors, microcontrollers and other core components, 2) easing the design and analysis of self-powered systems with tools that help in understanding the energy ambient, streamlining the planning and design process, and 3) offering reference designs such as this one that cover important details in both the physical implementation as well as the system design and optimization. The paper describes the system architecture and component selection to enable its ultra-low-power operation and thereby compatibility with energy harvesting.

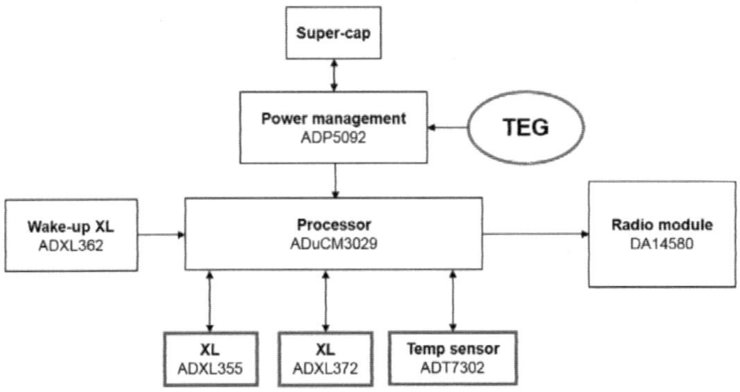

Figure 1. Block diagram of the machine health monitoring system.

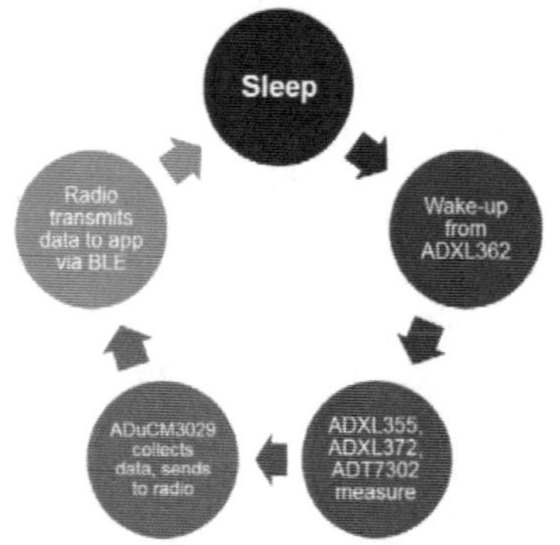

Figure 2. Typical task sequence for continuous self-powered monitoring.

2. Design

A block diagram of the self-powered machine health monitoring system is given in Figure 1. Power is supplied by a thermoelectric generator that can convert waste heat into useable electrical energy. In addition, this system makes use of the Analog Devices ADP5092 power management chip and ADuCM3029 microcontroller. The ADP5092 is designed for efficient conversion of dc power from energy harvesters (typically solar cells and TEGs) to charge storage elements (in this case, a supercapacitor) and power small battery-free devices. The ADP5092 employs a two stage charging process that takes care of system start-up and other cold-start events that may arise when the stored energy has been fully depleted. The ADuCM3029 is designed with 4 programmable active and sleep modes, allowing for excellent control over power consumption. SPI, I2C and UART interfaces allow for communication with a wide range of sensors. Four sensors are included in this system: (1) ADXL362, powered constantly and used only to wake up the microcontroller, (2) ADXL355 and ADXL372, low-power 3-axis accelerometers chosen to cover a dynamic range of 25 μg to 200 g, and (3) ADT7302, a low-power digital temperature sensor. Data from these sensors is collected by the microcontroller and transmitted via Bluetooth.

2.1. System design

A typical task sequence for the continuous machine health monitoring system is shown in Figure 2. When in sleep mode, only the ultra-low-power ADXL362 and the ADP5092 (continuously storing energy generated by the TEG in the supercapacitor) remain powered. When a wake-up signal is generated from the ADXL362 (either due to a set vibration limit on the board or a fixed timer), measurements are collected from the remaining accelerometers (1024 samples of 9 bytes) and the temperature sensor (one sample of 2 bytes). After performing an FFT on the accelerometer data (reduced down to the 50 largest 12-bit coefficients), it is sent to the radio module for transmission. All components, except the ADXL362 and ADP5092, then return to sleep mode. How often data is transmitted ultimately depends on the particular use case—how much the TEG can generate from the available waste heat, whether data is required on a fixed timer or only on perturbation, etc.

2.2. Thermal considerations

Section 3.1 highlighted the "electrical system" constituting the machine health monitoring sensor node, powered entirely by thermoelectric energy harvesting. Since the TEG output depends strongly on the availability of waste heat and how efficiently that waste heat is accessed, careful consideration must also go into design of the "thermal system".

A schematic of the system is shown in Figure 3. In this configuration, the sensor node is attached to the machine (heat source) using a magnet and heat is dissipated through an anodized aluminum fin array to the ambient (heat sink); in this way, a temperature gradient is maintained across the TEG. The key to optimizing TEG power generation is to minimize parasitic thermal losses, directing all heat flow through the TEG. Thermal resistances of interfaces and materials along the path of heat flow, in *series* with the TEG, must be minimized. This includes the magnet, the aluminum block standoff, the thermal grease and adhesive layers, and the heat fin. In addition, thermal resistances along heat flow paths in *parallel* with the TEG must be maximized. These parallel heat paths include the plastic standoffs on either side of the TEG, the plastic housing enclosing the entire sensor node, as well as the air around the TEG. Optimization of the thermal system requires careful choice of materials and geometries—typically, at least half of the available temperature drop (between the machine and ambient) should fall across the TEG.

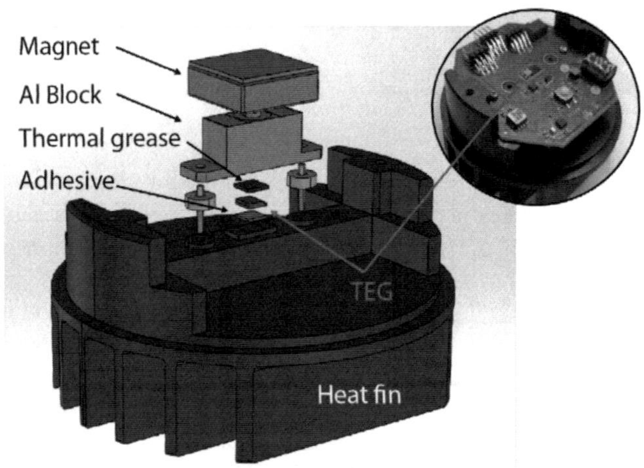

Figure 3. Schematic of the machine health monitoring system. The TEG sits in a hole in the PCB (shown in the inset photograph), directly attached to the heat fin. The plastic housing surrounding the sensor node is not shown.

3. Results

Estimates for the power consumption by component are given in Table 1 for data transmission every 30 seconds and every 30 minutes. For data updates every 30 s, nearly all of the 200 μW is consumed by the radio module. For data updates every 30 min, on the other hand, power consumption is dominated by the wake-up accelerometer (which as mentioned above is always on).

Table 1. Average power consumption in μW by component for two data update rates.

	DA14580	*ADuCM3029*	*ADXL355*	*ADXL372*	*ADXL362*	*ADT7302*	***Total***
30 s	172	16.1	5.56	0.685	4.89	0.240	**199**
30 min	5.82	0.433	0.093	0.011	5.39	0.004	**11.8**

Figure 4. Photograph of the completed demo. A retrofitted speaker serves as the heat and vibration source in lieu of a machine.

Optimized thermoelectric energy harvesters can easily generate power on the order of 10 μW to 200 μW [3-5], depending on the temperature gradient available for a given use case. Using a particularly conservative estimate for TEG power generation, a temperature gradient ΔT of ~10 °C would be required to power the sensor node with a 30 s update rate and a ΔT of only ~2.4 °C required for a 30min update rate. Further optimization of the machine health monitoring system could mean a reduction in the required ΔT and/or an increase in the data update rates possible—making self-power continuous machine health monitoring available for a wider range of use cases.

A photograph of the completed demo is shown in Figure 4. The CBM system is shown resting on a retrofitted Bluetooth speaker which has been converted into a stand-in for a hot, vibrating piece of factory equipment by integrating a heat source and using the speaker to provide the vibration.

4. Conclusion

Autonomous condition-based monitoring has the potential to change the way in which factories operate machinery today, all the while reducing costs and improving safety. The technology necessary to achieve this is already available today, including sensing (ADXL355, ADXL372, ADXL362, ADT7302), processing (ADuCM3029), and finally power management (ADP5092). While one must take care of a number of issues related to thermal transport and making sure the maximum temperature gradient is seen across the TEG, a wide range of operating conditions and duty cycles are enabled with easily achievable temperature differentials between 2.4 – 10 °C.

References

[1] A. Elefsiniotis, et. al., *Sensors and Actuators A: Physical,* **206**, pp. 159-164, 2014.
[2] W. Wensi, et. al., *International Journal of Distributed Sensor Networks,* **9**, p. 232438, 2013.
[3] J. Cornett, et al., *Journal of Electronic Materials,* **46**, pp. 2844-2846, 2017.
[4] Micropelt GmbH 2012, "TGP-751, TGP-651 ThermoGenerator-Package (TGP)" *TGP651 datasheet*
[5] Nextreme Thermal Solutions, Inc "eTEG HV56 Power Generator" *eTEG HV56 datasheet*

IOP Conf. Series: Journal of Physics: Conf. Series **1052** (2018) 012026 doi:10.1088/1742-6596/1052/1/012026

Speed vs Efficiency and Storage Type in Portable Energy Systems

M E Kiziroglou[1,2], M Cowell[3], B T Kumaravel[3], D E Boyle[2], J W Evans[3] P K Wright[3] and E M Yeatman[2]
[1]ATEI Thessaloniki, Greece
[2]Imperial College London, U.K.
[3]University of California at Berkeley, CA, USA

Abstract. Portable power management systems must optimise power interfacing, storage and routing, to meet application specific functionality requirements. Two key aspects are reliability and efficiency. For reliable operation, it is required that powering on/off the system must occur in a planned manner. For efficient operation, it is desired that the system is powered for an optimal amount of time. maximizing its useful operational outcome per unit of energy consumed. This can be achieved by optimizing energy usage based on the anticipated energy income and power demand of duty-cycled power consumers. Both battery and supercapacitor storage can be employed to meet energy and power density demand, on both sides, and to enable fast transition from cold-starting to active power management. A simplified model is used to calculate the reliability of a simple solar-powered microsystem. The modelling of dynamically configurable interfacing and storage may enable a new generation of power management, providing reliable power from irregular and small energy sources.

1. Introduction

The management of energy delivery, storage and usage in wireless devices is critical for emerging IoT technologies, including mobile phones, wearable electronics, small electric vehicles and wireless sensor networks. Key performance aspects of portable power management include the efficiency of energy conversion and transfer, energy storage density, the ability to store and deliver energy at high rates (power density) but also the efficiency, speed and safety of energy transfer between storage media. Depending on the application, these requirements can be fulfilled by using more than one type of energy storage and a diverse range of power flow interfaces. In parallel, the exploitation of energy sources with challenging features such as low signal level and irregular availability is highly desirable, including indoor lighting, stray motion and stray heat flow. To regulate this increasingly multidimensional energy ecosystem, a holistic design approach is required [1]. In addition, the reliability of energy availability must be quantified.

In this paper, three important aspects of portable power management are discussed: the trade-off between fast versus efficient energy transfer, the use of combined battery and supercapacitor storage and the definition of a figure-of-merit for the quantification of portable power quality. A case study on solar energy availability is analysed against a deterministic energy demand scenario and a control – oriented approach for system level modelling and optimization is proposed.

2. Challenges in Portable Power Management

2.1. *Speed vs Efficiency*

In the case of energy transfer/transduction from one stored form to another, such as when charging a smaller portable system from a moderately bigger one or when using a fuel, the available energy is finite and it is reduced by usage. There, conversion and transfer efficiency is the main parameter of interest. In contrast, when exploiting a power source such as sunlight, or when using a practically infinite energy source such as the motion or heat of a large industrial machine to deliver energy to sensor systems, any energy that is not converted is practically lost. In such systems maximization of power delivery, instead of efficiency is desirable. This difference is illustrated in Fig.1 (left), using the equivalent output (R_{GEN}) and input (R_{PM}) resistance of the source and the target system respectively. The indices refer to a generator output resistance and the Power Management input resistance, corresponding, as an example, to a generator – power management interface. Maximum energy transfer efficiency is obtained when $R_{PM} >> R_{GEN}$, whereas maximum power transfer is obtained at $R_{PM} = R_{GEN}$.

While this model serves well as a general description of the two different cases, it must be noted that in generators such as solar cells, the output response to different loads may be non-linear and a more detailed circuit model is typically used. Overall, power management systems need to operate either at optimized efficiency or maximum power, depending on their position in the power supply chain, and also on specified usage priorities. As an example, if a portable device is to be charged by an also portable power bank, the choice between efficient and rapid energy transfer may depend on the available connection time window.

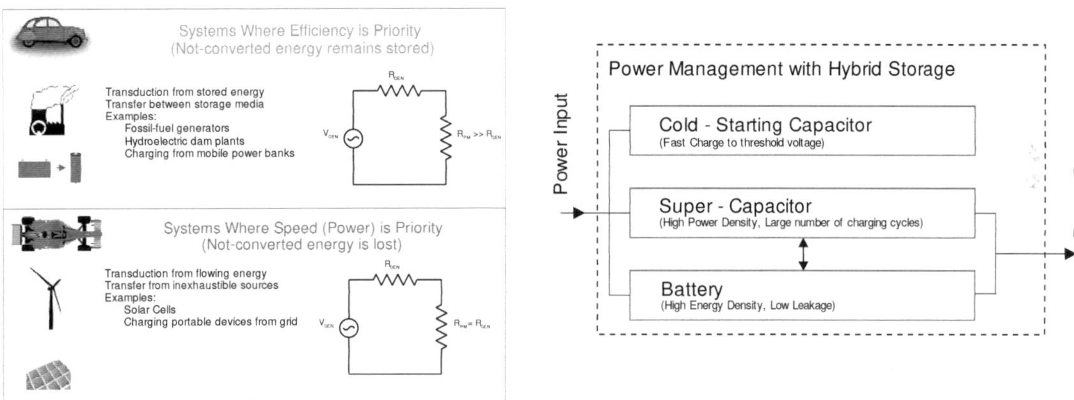

Figure 1. Left: Illustration of the difference between cases where maximization of efficiency or net power transfer is more important. Right: Block diagram of the proposed hybrid energy storage architecture for portable systems.

2.2. The Requirement for Hybrid Energy Storage

The ability of a power management and storage system to optimize efficiency, charging speed and reliability of power availability is largely determined by the storage type that is employed. It is possible to employ both a battery and a super-capacitor, to combine high energy density and low leakage, with high input and output power. Such hybrid systems have been proposed and demonstrated, mainly for electric vehicle applications [2, 3]. In addition, in systems relying on low-voltage power sources such as thermoelectric energy harvesters, a smaller capacitor can be used for fast transition from passive to active power management during cold-starting. A block diagram of a system with a fast-charging, a high power and a high energy storage unit is illustrated in Fig.1 (right).

While the latter technique of using a small capacitor which can charge quickly during cold-starting has been successfully adopted in state-of-the-art power management integrated circuits, the combination of battery and super-capacitor arrays depends on the specific characteristics of power availability and usage applications. The optimum balance between super-capacitor and battery storage can be determined by statistical analysis of anticipated incoming and outgoing energy, based on experimental data. This requirement for storage type customization is a challenge to manufacturing general purpose portable power supplies. The development of printed hybrid energy storage technology may provide a cost-effective solution to this challenge [4].

2.3. Quantification of Portable Power Quality
One of the major challenges of portable power is its reliability. This is due to the- irregularity of incoming energy and power demand, but also to the uncertainty of storage state. Optimisation requires the definition of a quantitative measure of reliability for such systems. A simple metric would be the percentage of desirable power-on time that is successfully covered by the power supply:

$$R = \frac{Successfully\ covered\ powering\ time}{Total\ desirable\ powering\ time}$$

This reliability metric can be directly calculated by simulating the behaviour of a storage system with a given storage capability, power management efficiency and power routing strategy, under given incoming and outgoing power profiles. In addition, stochastic models and distributions can be used, depending on the application scenario. This approach reveals the requirement for dynamic control of power management within the system, taking into account the history and anticipated future of both power availability and demand. An example architecture will be shown in section 4 of this work.

3. Reliability of a solar-powered power supply

As an example, the behaviour of a solar-powered power supply was simulated, using a simplified model and solar irradiance data in order to provide a quantitative assessment of powering reliability. The data, representing 13.8 years of solar irradiance from the National Renewable Renergy Lab, South Park, Co, USA are illustrated in Figure 2. The Simulink model receives as input, time-series irradiance data and simulates the behaviour of the WSN. The energy storage levels and system voltages are monitored throughout the simulation to determine the fraction of the day in which the WSN is functional.

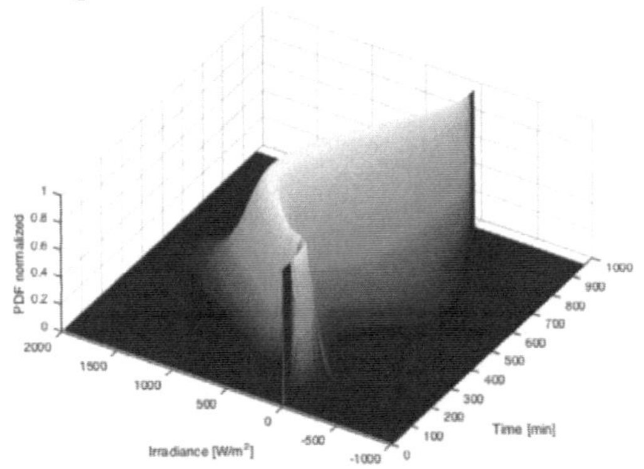

Figure 2. The probability density function of solar input to a Simulink Model. The data represent 13.8 years of solar irradiance from the National Renewable Energy Lab station in South Park, Colorado, USA.

By feeding temporally varying energy availability into the equivalent circuit model of the WSN and applying the voltage triggering algorithm, a model that predicts the behaviour of our energy autonomous WSN over the course of a day is obtained. Because the model applies a voltage triggering algorithm, this model can dynamically adjust its duty cycle to adapt to the input energy availability. Starting the day before sunrise and with a depleted energy reservoir, the WSN is completely off. As the sun rises, the photovoltaic system begins to create a net positive energy flow and the energy reservoirs begin to charge. At the first threshold voltage, the system enters sleep mode and periodically wakes to check the power bus voltage. Once the second threshold voltage is reached, the system has adequate stored energy to undergo a data acquisition and transmission procedure followed by power bus voltage measurement and sleep.

As this model accepts solar availability as its input, it is possible to observe the sensor node responding to periodic cloud cover throughout the day (Figure 3). By considering the ratio of the WSN's actual up time to the length of the day, the fraction of the day in which the WSN operates is obtained. This is a measure of the WSN's reliability, with a perfectly reliably WSN achieving 100% up time. It corresponds to the reliability metric as defined in section 2.3, for a requirement for continuous sensor node functionality. By including larger energy harvesters and more energy reservoirs, it is possible to improve the operation of the simulated WSN toward 100% reliability.

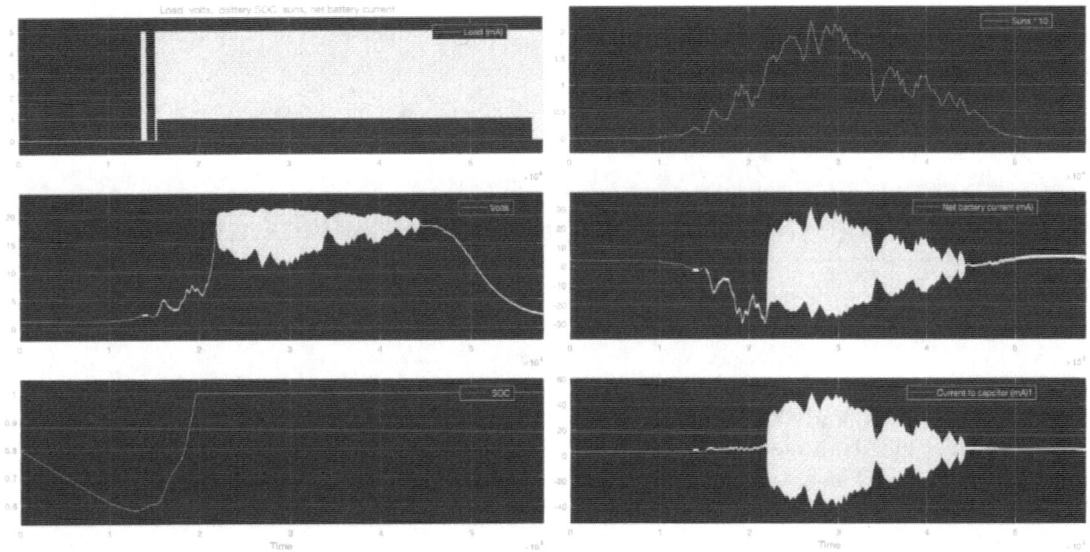

Figure 3. Simulated performance of an example power management system. Storage levels and system voltages are monitored to determine the fraction of the day in which the node is functional.

4. Conclusion

In conclusion, portable power supply systems can improve their performance by selecting the balance between power transfer speed and efficiency either at the design stage or dynamically, according to conditions and power demand specifications. In addition, hybrid energy storage systems, i.e. balancing use of both a supercapacitor and a battery unit, can expand their incoming and outgoing power and energy capabilities. Active power routing and dynamic interfacing may be required to provide flexibility and optimal exploitation of small amounts of energy, especially for energy harvesting systems. A block diagram of such a system is illustrated in Figure 4. The expansion of hybrid storage models to account for stochastic input and duty cycled power demand, and to allow dynamic interfacing, could allow the design of advanced power management systems for a range of contemporary electronic devices.

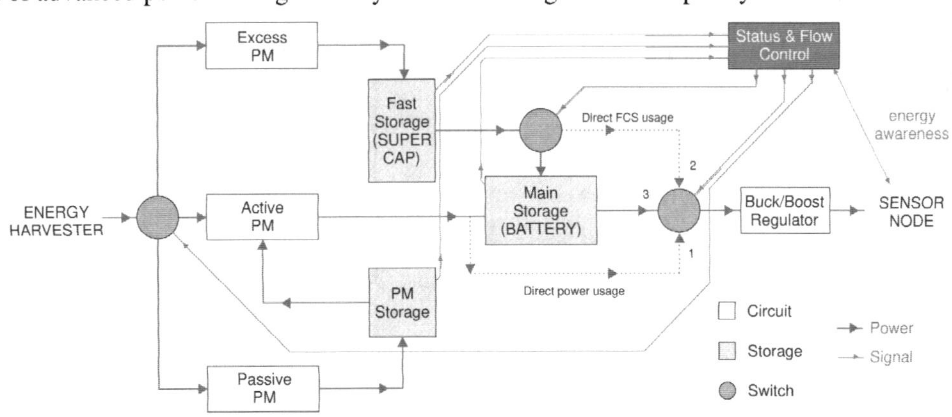

Figure 4. Block diagram of an example hybrid storage architecture with dynamic power routing control for portable power supplies. energy harvesting systems. This system includes a battery main storage, a super-capacitor high-power storage (FCS) and a separate storage unit for passive power management.

References

[1] Boyle et al, IEEE Pervasive Computing, 15 (2016), pp. 28-35, 2016
[2] Cao et al, IEEE Trans. Power Electron., 27 (1), pp. 122-132, 2012
[3] Magno et al, , IEEE Trans. Ind. Electron., 61 (4), 1871 – 1881, 2014
[4] Cowell et al, PowerMEMS 2014, J. Phys.: Conf. Ser., 557, 012061, 2014

Bennet's charge doubler boosting triboelectric kinetic energy harvesters

A Ghaffarinejad[1,2], Y Lu[1], R Hinchet[3], D Galayko[4], J Y Hasani[2], P Basset[1]
[1]Univ. Paris-Est, France / ESYCOM / ESIEE Paris, France
[2]Iran University of Science and Technology, Tehran 1684613114, Iran,
[3]Sungkyunkwan Univ., Korea,
[4]UPMC-Sorbonne Universités / LIP6, CNRS, France

E-mail: ghaffarinejadali@elec.iust.ac.ir

Abstract We present for the first time how a Bennet's doubler conditioning circuit can boost the electrical energy extracted from an electrostatic Kinetic Energy Harvester (e-KEH) based on triboelectricity, commonly called Tribo-Electric Nano-Generator (TENG). After 1000 charging cycles, the harvested power is increased by more than 2 orders of magnitude with the Bennet doubler compared to a half-wave diode bridge. The maximum energy extracted by a 3×3 cm² device made of PFA and aluminum is no less than 4.6 μJ per tap through the Bennet's doubler circuit.

1. Introduction

Electrostatic Kinetic Energy Harvesters (e-KEHs), along with piezoelectric and electromagnetic ones, have been under extensive research in the recent years as a new source of energy [1],[2],[3]. E-KEHs convert power from mechanical vibrations into electrical power by varying the capacitance of an electromechanical transducer. For the energy conversion, the capacitive transducer needs an electrical charge that can be obtained by successive triboelectric contacts. Triboelectric charging of the capacitance's electrodes is based on the principle of charge transfer between two contacting surfaces due to the difference in the electronegativity of the materials they are made of [4]. Various applications are introduced for triboelectric KEH such as powering implantable medical devices[5], powering of sensors [6], energy scavenging from water motion[7] or human daily activities[8].

Compared to electret-based e-KEHs, triboelectric e-KEHs do not carry a pre-imposed charge with them. Therefore, the bias of the capacitive transducer comes from repeated tapping or sliding motions of its electrodes. Since triboelectric charging is not uniform on the surface, it is very difficult to reach the theoretical maximum charge density on the material. Thus, tribo-based e-KEHs usually produce less energy per cycle in comparison with electret-based harvesters. Hence, it is important to have a conditioning circuit that has the ability to increase the charge on the transducer's terminals during its operation like the Bennet's doubler family[9] [9],[10] to boost the harvested power. Several instances of application of the Bennet's doubler in energy harvesting applications have been reported previously that emphasize the importance of this circuit configuration as a conditioning circuit for energy conversion purposes[11].

2. Test and Measurement Setup

Figure 1 displays the setup used to actuate the capacitive harvester. The variable capacitance is composed of two electrodes in aluminium and a dielectric layer made of Polymer-PFA

(Perfluoroalkoxy) covering one of the electrodes . The PFA is provided by GoodFellow [12] and has a thickness of 50μm, a relative permittivity of 2.1 and a melting point of 315°C.

2.1. sample preparation and actuation

To build the variable capacitance, the PFA is cut in squares of 3×3 cm². One polymer piece is attached firmly on the bottom electrode, which is a one-sided aluminum tape with the same dimensions as the polymer. The bottom electrode is glued on the supporting table through a rigid insulator. The top electrode is glued on the force sensor with the same type of rigid insulator.

A magnetic shaker is adjusted to produce a vertical sinusoidal vibration with frequency of f=5Hz that applies a normal maximum force on the polymer surface in contact mode, measured with a force sensor as 0.3±0.05N .The vertical motion serves two purposes: (i) the variation of the capacitance value of the harvester in order produce electric power, (ii) the application of a pressure contact between the PFA and the aluminum surface that is hard enough to induce triboelectric charges.

2.2. Conditioning circuits

Figure 2 shows two conditioning circuits working with a e-KEH: a Bennet's doubler and a half-wave diode bridge. Both are employed to store electrical energy in a reservoir capacitance C_{res}. Each diode on the schematic is actually composed of two diodes with a breakdown voltage of 200V and a reverse saturation current of 1 nA. To overcome the limited span of the high-impedance follower used for recording the output signal, C_{res} is the series combination of C_{res1} =100nF and C_{res2} =10nF, which works as a capacitance divider. An additional voltage source of V_b=25V is also employed to extend the output saturation range of the amplifier. The storage voltage V_{res} is measured through this capacitive divider and the follower loaded with a resistive divider (R_1=R_2=1kΩ). Evolution of V_{res} over time with respect to recorded V_{out} is defined by the following formula:

$$V_{res} = k_1(k_2 V_{out} + V_b) \tag{1}$$

Where $k_1 = 1 + \frac{C_{res1}}{C_{res2}}$ and $k_2 = 1 + \frac{R_2}{R_1}$.

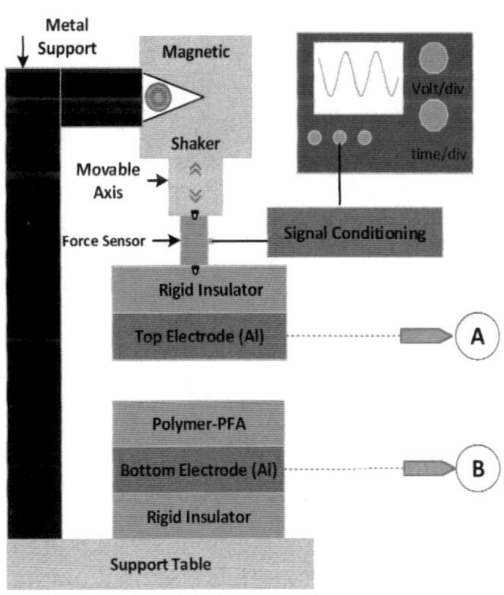

Figure 1. Setup of the experiments. A magnetic shaker, force sensor and variable capacitor are the basic parts of the tests.

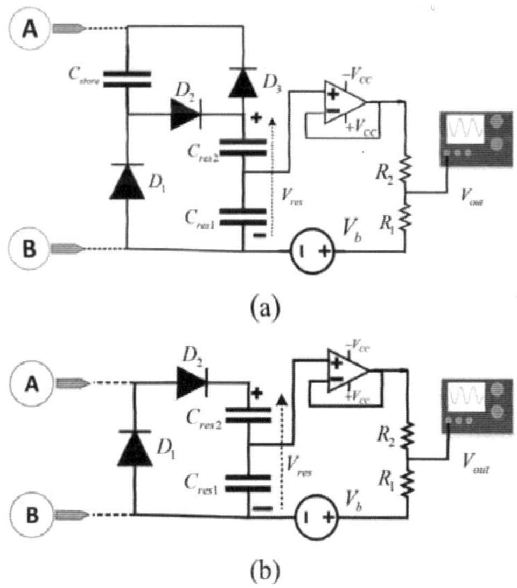

Figure 2. Conditioning circuits used for the triboelectric e-KEH. (a) Bennet's doubler. (b) Half-wave rectifier. C_{res1} =100nF, C_{res2} =10nF, $\pm V_{cc}$=±30, R_1=R_2=1kΩ, V_b=25V, C_{store} =100nF.

By setting proper values for k_1 and k_2, large value of V_{res} can be measured. Since the AOP is powered with +/- 30V, in the following experiments the maximum value of V_{res} is 605 V. Connecting points A and B in figure 1 and figure 2 indicate the proper connection of the output of the e-KEH to the input of conditioning circuits.

3. Experimental results and discussion

To evaluate the capacitance variation of the device, we used the technique introduced in [13][13] during the oscillation of the top electrode. According to figure 3, extreme values of the transducer variation in one cycle under the applied excitation are C_{max}=78pF and C_{min}=26pF , plus 16pF of parasitic capacitance introduced by the setup. Figure 3 also depicts the magnitude of the contact force between top and bottom electrode during one cycle of oscillation. As we expect, the maximum capacitance occurs at the same time with the maximum of the force.

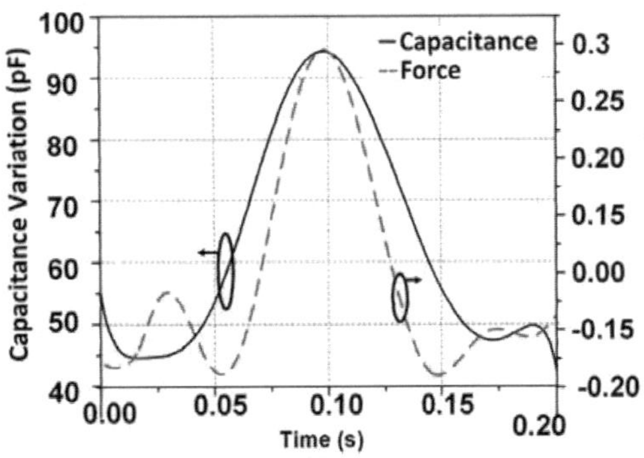

Figure 3. Measurements of the applied force and capacitance variation of the tribo e-KEH (including a 16 pF parasitic capacitance).

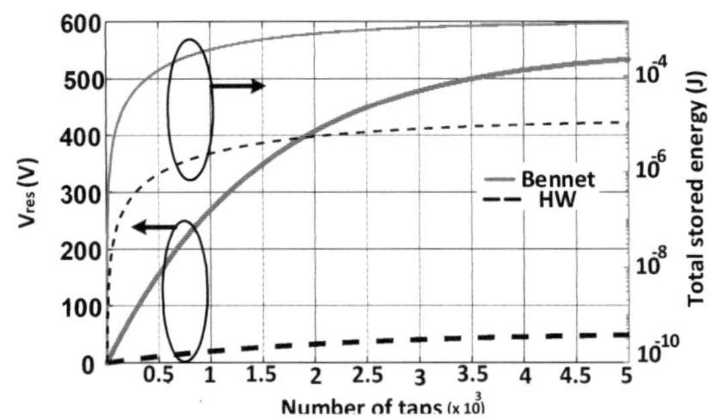

Figure 4 Measurements of the voltage and the total stored energy in C_{res} for the half-wave (HW) rectifier and the Bennet's doubler.

As shown in figure 4, after 5000 taps a saturation occurs with V_{res}, even with the Bennet's doubler conditioning circuit that is supposed to generate an exponential increase[14]. However, the maximum value of V_{res} is 533V with the Bennet circuit, to be compared with 48V for the half-wave circuit. Consequently, the total storage energy is 1.2mJ and 10μJ respectively. On the contrary to the experiments in [11], the origin of this saturation for the experiment with the Bennet's doubler is not due to the spring-softening effect but to the saturation of the operational amplifier at the output.

Figure 5 compares the harvested power for the two conditioning circuits. The maximum power for both circuits occurs around 200 s. Then 23μW are extracted with the Bennet's doubler and only 138 nW with the half-wave bridge.

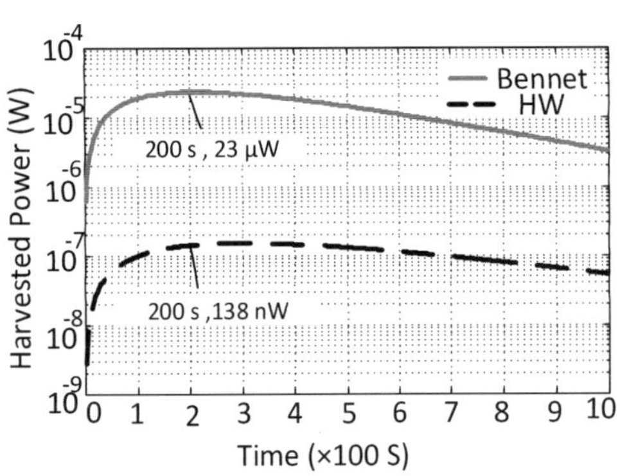

Figure 5. Harvested power doubler and half-wave rectifier.

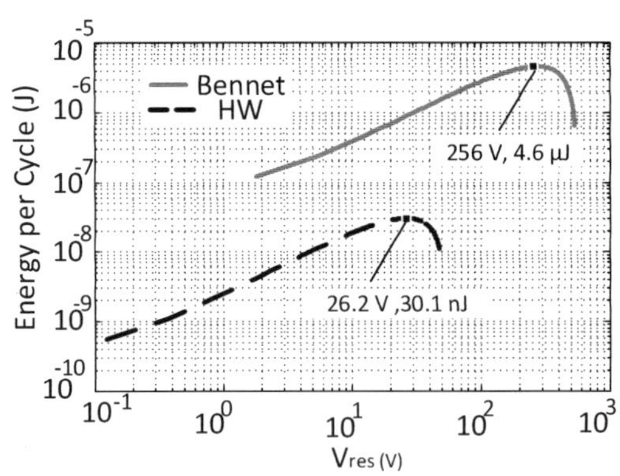

Figure 6. Converted energy per tap as a function of voltage across reservoir capacitance with Bennet doubler and half-wave rectifier.

Figure 6 depicts the converted energy per cycle as a function of stored voltage across C_{res}. From this figure, it appears that the highest energy per tap extracted from the e-KEH using the half-wave conditioning circuit is 30 nJ, compared to the 4.6 µJ for the Bennet's circuit, which shows more than 2 order of magnitude improve in the harvested power. The reason for this much better performance is the self-charging behavior introduced by the Bennet's doubler, which in theory double the charges in the transducer's terminals during each tap. However, it is also noticeable that the peak harvested energy for the Bennet's doubler is at V_{res} =256V while it is 26V for the half-wave circuit.

4. Conclusion:

The much improved performance of the Bennet's doubler circuit with triboelectric e-KEHs is due to the self-increase of the triboelectric charge generated at each contact. By the bennet's circuit, it was possible to reach up to 550 V across the capacitive transducer, much higher than the 48 V provided by half-wave circuit. Using the Bennet's doubler the maximum obtained energy per cycle of 4.6 µJ is calculated, compared to 30 nJ for the half-wave circuit. The limiting factor that prevents the Bennet's circuit to reach values above 4.6µJ/per tap is the high reverse current of the HV diodes used in the circuit, while the maximum voltage of 605V is imposed by the monitoring setup.

5. References

[1] Lu Y, Cottone F, Boisseau S, Marty F, Galayko D and Basset P 2015 A nonlinear MEMS electrostatic kinetic energy harvester for human-powered biomedical devices *Appl. Phys. Lett.* **107** 253902

[2] Lefeuvre E, Badel A, Richard C and Guyomar D 2005 Piezoelectric Energy Harvesting Device Optimization by Synchronous Electric Charge Extraction *J. Intell. Mater. Syst. Struct.* **16** 865–76

[3] Beeby S P, Torah R N, Tudor M J, Glynne-Jones P, O'Donnell T, Saha C R and Roy S 2007 A micro electromagnetic generator for vibration energy harvesting *J. Micromechanics Microengineering* **17** 1257–65

[4] Fan F-R, Tian Z-Q and Lin Wang Z 2012 Flexible triboelectric generator *Nano Energy* **1** 328–34

[5] Zheng Q, Shi B, Li Z and Wang Z L 2017 Recent Progress on Piezoelectric and Triboelectric Energy Harvesters in Biomedical Systems *Adv. Sci.* **4** 1700029

[6] Rasel M S, Halim M A and Park J Y 2015 A PDMS based triboelectric energy harvester as self-powered, active tactile sensor system for human skin (IEEE) pp 1–4

[7] Rodrigues C R S, Alves C A S, Puga J, Pereira A M and Ventura J O 2016 Triboelectric driven turbine to generate electricity from the motion of water *Nano Energy* **30** 379–86

[8] Proto A, Penhaker M, Conforto S and Schmid M 2017 Nanogenerators for Human Body Energy Harvesting *Trends Biotechnol.* **35** 610–24

[9] de Queiroz A C M and Domingues M 2011 The Doubler of Electricity Used as Battery Charger *IEEE Trans. Circuits Syst. II Express Briefs* **58** 797–801

[10] Dorzhiev V, Karami A, Basset P, Marty F, Dragunov V and Galayko D 2015 Electret-Free Micromachined Silicon Electrostatic Vibration Energy Harvester With the Bennet's Doubler as Conditioning Circuit *IEEE Electron Device Lett.* **36** 183–5

[11] Karami A, Galayko D and Basset P 2017 Series-Parallel Charge Pump Conditioning Circuits for Electrostatic Kinetic Energy Harvesting *IEEE Trans. Circuits Syst. Regul. Pap.* **64** 227–40

[12] Anon Goodfellow Inc, Wholesale and distribution company, website: http://www.goodfellow.com

[13] Basset P, Galayko D, Paracha A M, Marty F, Dudka A and Bourouina T 2009 A batch-fabricated and electret-free silicon electrostatic vibration energy harvester *J. Micromechanics Microengineering* **19** 115025

[14] Basset P, Blokhina E and Galayko D 2016 *Electrostatic kinetic energy harvesting* (Hoboken, NJ: Wiley)

Zinc oxide nanowire-parylene nanocomposite based stretchable piezoelectric nanogenerators for self-powered wearable electronics

A S Dahiya[1*], F Morini[1], S Boubenia[1], C Justeau[1], K Nadaud[1], K P Rajeev[1], D Alquier[1], G Poulin-Vittrant[2]
[1] GREMAN UMR 7347, Université de Tours, CNRS, 16 rue Pierre et Marie Curie, 37071 TOURS Cedex2, France.
[2] GREMAN UMR 7347, CNRS, Université de Tours, INSA-CVL, 3 rue de la Chocolaterie, CS 23410, 41034 BLOIS Cedex, France.

*abhishek.dahiya@univ-tours.fr

Abstract. Sustainable flexible micro power supplies for wearable and mobile electronics is a strategic field in today's energy research, which could offer a viable solution to provide the energy required by microsystems. In this interest, we report a facile, cost-effective and industrially scalable process flow for the fabrication of a high performance stretchable piezoelectric nanogenerator (SNG) on polydimethylsiloxane (PDMS) substrate. The present proposed fabrication process is substrate independent. The proof-of-concept is shown by fabricating SNGs on bank cards, free standing PDMS and on silicon substrates. The active layer of the present SNG device is realized by encapsulating ZnO nanowires (NWs), synthesized by hydrothermal method, in parylene C polymer matrix. The SNG devices exhibit excellent performance with peak power density of ~0.7μW/cm^2 under a 13 N vertical compressive force at 5 Hz. The obtained electricity from the SNG is used to drive electronic devices such as liquid crystal displays (LCDs) without employing any storage unit, implying the device significance in the field of consumer electronics.

1. Introduction

Wearable systems and flexible electronics have gained extensive research interest recently, as they enable a range of applications including health monitoring, military and prosthetics [1,2]. The stable real-time operation of electronic devices on the wearable systems pretty much depends on the energy, which is currently supplied through standard batteries which have limitations, such as large size and weight, non-flexibility, and finite reservoir of available charges. To be effective, especially for above mentioned applications where portability is necessary, it is critical for wearable systems to be energy autonomous. Smart nanogenerators (NGs) for harvesting energy sources including solar, thermal and mechanical energy have been intensively investigated for the realization of autonomous electronic systems. Solar energy, for example, is one of the important green energy sources for energy autonomous electronic systems such as electronic skin [1]. However, solar cells application to power wearable systems may have limitations such as non-availability of light (at night and/or indoor). Converting thermal energy using pyro- and/or thermoelectric generator can be an interesting solution to power bulky electronic systems but face serious limitations to drive micro-/ nanosystems, as it is difficult to have large temperature difference across thin devices and thus the output power is very low. Alternatively, kinetic/mechanical energy is a power source that is ubiquitous and can be a potential solution to supply green energy for the development of autonomous electronics and has potential to be fabricated over fully flexible and stretchable platform [3].

Mechanical energy harvesting systems that can be realized on highly flexible, stretchable and foldable substrates, and further the possibility to transfer them on any surface topology, could be used for the development of energy autonomous wearable systems. For example, piezoelectric NGs (PENGs) have been used to scavenge body movements to power micro-/nano devices [3]. Since the first demonstration of the PENGs in the scientific domain in 2006 [4], they have been developed and evolved in a variety of configurations. Flexible PENG device structures, such as vertically integrated NG (VING) [5,6], laterally integrated NG [7], and/or nanocomposite generators (NCG)[8] have been demonstrated using diverse piezoelectric nanomaterials. Each device configuration has its advantages and limitations in terms of the average power generated, ease of integration into the devices and stretchability. In this work, we exploit advantages of both VING and NCG device prototypes and developed a facile, cost-effective and industrially scalable process flow for the fabrication of high performance, mechanically robust, stretchable nanogenerator (SNG) on PDMS substrate.

2. Experimental work

2.1 Zinc oxide nanowire synthesis

The ZnO NWs are synthesized at around 90 °C on ZnO thin-film (deposited by sputtering) coated PDMS substrate by the hydrothermal process, as reported elsewhere [9]. The reaction nutrients *i.e.* equimolar ratio (50 mM) of zinc nitrate hexahydrate (30 ml) and Hexamethylenetetramine (30 ml) and 5 mM polyethylenimine (1 ml) in DI water are added. The substrates are immersed face down in a Teflon lined stainless steel autoclave reactor and placed in a preheated convection oven for 6 hours.

2.2 SNG characterization test-bench

The experimental test bench is a mechanical shaker equipped with an actuator arm. The amplitude and frequency of the shaker are controlled *via* an Agilent 33 250A function generator and a LDS PA100E power amplifier is used to drive the shaker. The actuator arm is attached to the reciprocating shaker platform at one end and a height adjustable screw is inserted through it at the opposite end. To avoid excessive damage to the SNGs during testing, the screw tip is also insulated by soft styrene–butadiene rubber. The SNGs are then firmly fixed onto a solid aluminium block using double sided adhesive tape. The screw tip is positioned over the active region of the SNG, without making contact.

3. Results and Discussions

The SNG device fabrication process is as follows: First, a thin PMMA layer (~ 100 nm) is spin-coated over a rigid substrate used as a sacrificial layer. Then, a ~250 μm thick PDMS layer, as a stretchable device substrate, is spin-coated (250 rpm, 100 rpm/sec and 15 sec). Subsequently, in a next step, the bottom electrode (Ti (100nm) / Pt (20nm)) and ZnO thin-film (50 nm) are sputtered over the cured PDMS. A low temperature hydrothermal process is used for the growth of vertically aligned ZnO NWs on various device substrates such as bank cards, free standing PDMS and Si substrates (Figure 1a-c). Following, the substrate with NWs is transferred into a chemical vapour deposition (CVD) system for the deposition of parylene C film (~600 nm). The decision of choosing parylene C as the encapsulation matrix has been made while considering its effectiveness of infiltration between the ZnO NWs, conformity and thickness controllability, and compatibility of this coating technique with flexible substrates [10]. The active area of the device is defined by depositing top metal electrode of ~ 100/400 nm Ti/Al layers using a shadow mask. Finally, the copper wires are connected to top and bottom electrodes using epoxy silver paste and the whole device is packaged in PDMS, to improve the robustness of the SNG device and to prevent any moisture and/or dust particle to stick onto the top metal. At the end, the complete device is lifted-off from the rigid substrates making it highly foldable, flexible and stretchable. A schematic of the final device structure is shown in Figure 1d. To confirm the parylene C infiltration and conformity, respectively in between and around the NWs, we captured

IOP Conf. Series: Journal of Physics: Conf. Series **1052** (2018) 012028 doi:10.1088/1742-6596/1052/1/012028

SEM pictures from a device fabricated on Si substrates. As shown in Figure 1e, NWs are completely encapsulated in parylene C film with high degree of infiltration between the NWs and conformity over the top surface of the NWs array.

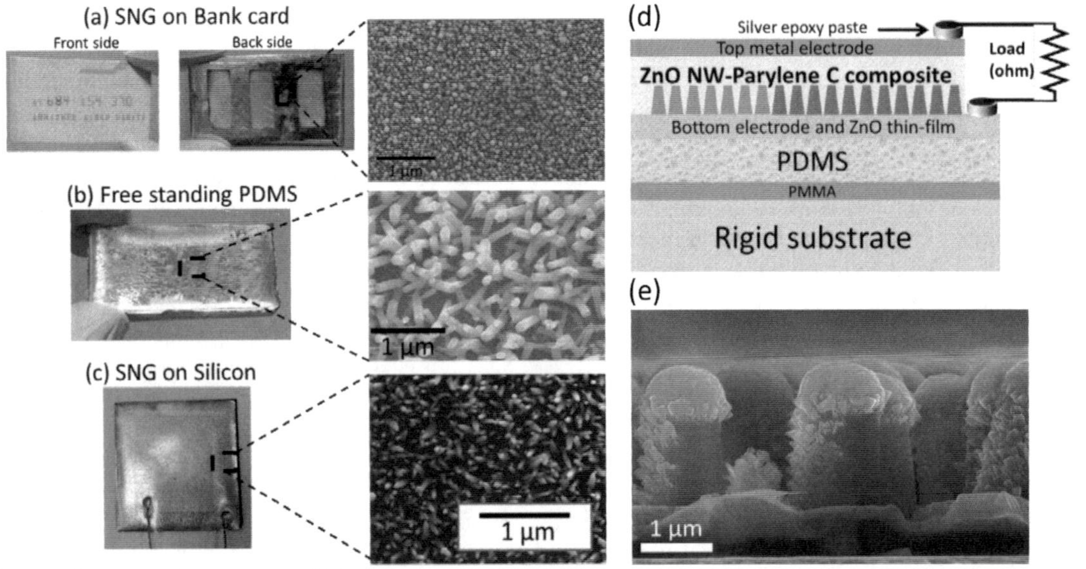

Figure 1. Proof-of-concept for SNGs fabricated on: (a) bank cards, (b) free standing PDMS and (c) silicon substrates. The magnified image in each panel shows the as-grown ZnO NWs on respective substrates. (d) The final device structure and (e) SEM image after parylene C and metal deposition on NWs grown on silicon.

The fabricated SNG devices are characterized using a custom built test bench in order to assess their energy harvesting performance and the results are shown in Figure 2. Typical optical micrograph of the fabricated SNG device with an active area $1.2cm^2$ (free standing PDMS after lifting-off the whole device from rigid substrate) is shown in Figure 2a. A schematic of the characterization system is shown in Figure 2b. The description of the test bench can be found in the experimental section. The energy harvesting performances of SNG devices are measured under different amplitudes of compressive force (6-13 N). It is well know that for PENG devices, the voltage generation is directly proportional to the applied stress (magnitude) and stress rate. The measurement results shown here are performed at fixed frequency of 5Hz. As seen from Figure 2c, the open circuit voltage (V_{OC}) amplitude increases from 3 to 9.1 V with the increase of compressive force amplitude from 6 to 13 N. The force sensitivity of the present device is ~0.7±0.1 V/N at 5Hz. It is worth to mention that the observed output voltage signal may arise because of piezoelectric properties of parylene C polymer [11]. However, its piezoelectric coefficient (d_{33} ~ 0.2 pm/V) is 60 times lower than bulk ZnO (d_{33} ~ 12 pm/V) and further, poling step is needed to have any piezoelectric response from it [11]. It is to note that we have not applied any voltage bias to the SNG devices before or during the measurements. Therefore, we can rule out any contribution from parylene C to the observed piezoelectric response from the SNG device. The short-circuit current (I_{SC}) is also measured under vertical compressive force (with 13 N amplitude and 5 Hz frequency). The obtained output signal is shown in Figure 2d. A peak I_{SC} of ~0.6 μA is obtained from the SNG devices. To assess the impact of resistive loading on peak surface power density ($P_{max} = (V_{RL} \times I_{RL})/A$), we measure the device performances across a wide range of resistive loads (R_L), where V_{RL} and I_{RL} are the instantaneous peak voltage and current at a given resistive load respectively, and A is the device active area. The experimental data resulting from the various R_L values can be found in Figure 2e. From Figure 2e, we can see that the SNG device can produce P_{max} ~0.7 μW/cm² at the matched load of 8 MΩ. In terms of peak volume power density, the performance exhibited by the present SNG device (~4 mW/cm^3) is comparable to the state-of-the-art value for VING devices [12]. Furthermore, to demonstrate the potential utilization of the generated electric output power, the SNG is further evaluated by connecting it to consumer electronic devices such as LCDs. For this, a larger area (~8 cm²) SNG device on bank cards is excited using a mechanical shaker by applying compressive forces with 13N amplitude. As shown in Figure 2f, many digits of a LCD screen can be driven without any external circuits, using the SNG only, implying the device significance in the field of consumer electronics energy supply.

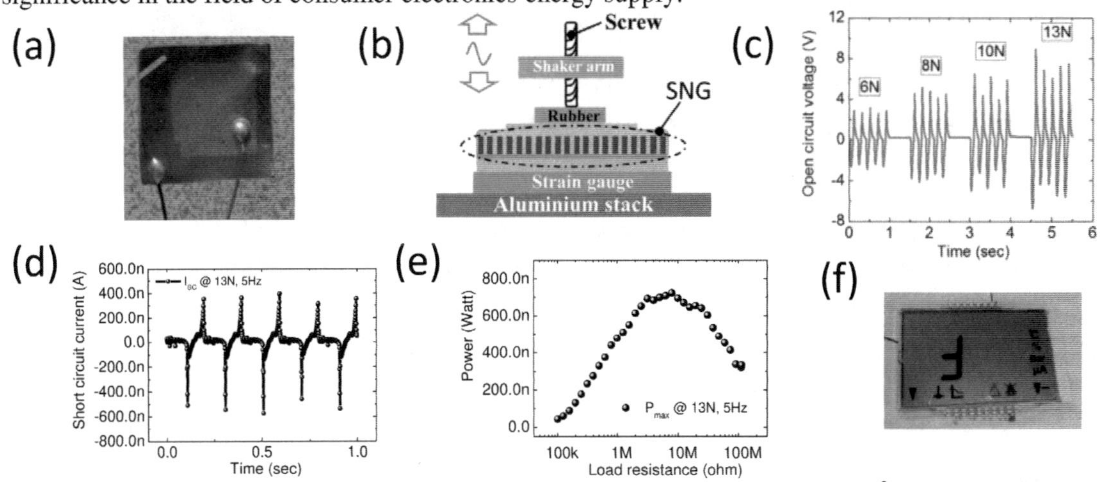

Figure 2. Functional characterization of the fabricated SNG devices (active area ~1.2 cm^2): (a) an optical image of the SNG on free standing PDMS. (b) Schematic of the custom-built test bench. (c) Output open circuit voltage w.r.t. applied compressive force. (d) The short circuit current w.r.t. time at constant amplitude (13N) and frequency (5Hz) of force. (e) Dependence of the output power density on the external load resistance and (f) optical image of the continuous display of LCDs driven by the SNG device.

4. Conclusions

In summary, we have demonstrated an industrially scalable process flow for the fabrication of high performance mechanically robust stretchable nanogenerators (SNG) on PDMS substrate. The SNG devices exhibit excellent performance with a high piezoelectric open circuit voltage of 9V, output short-circuit current density of ~0.5 μA/cm², and peak power density of ~0.7 μW/cm^2, under a 13N dynamic vertical compressive force at 5 Hz. The obtained electricity from the SNG device was used to drive LCDs without employing any storage unit. The results obtained from the fabricated SNG devices clearly demonstrate the interest of our novel fabrication process for the development of autonomous wearable electronics for personal health care and human-machine interfaces.

References

[1] Núñez C G, Navaraj W T, Polat E O and Dahiya R 2017 *Adv. Funct. Mater.* **27** 1606287
[2] Coyle S, Wu Y, Lau K-T, De Rossi D, Wallace G and Diamond D 2007 *MRS Bull.* **32** 434–42
[3] Proto A, Penhaker M, Conforto S and Schmid M 2017 *Trends Biotechnol.* **35** 610–24
[4] Wang Z L and Song J 2006 *Science (80-.).* **312** 242–6
[5] Rivera V F, Auras F, Motto P, Stassi S, Canavese G, Celasco E, Bein T, Onida B and Cauda V 2013 *Chem. - A Eur. J.* **19** 14665–74
[6] Wang X, Song J, Liu J and Wang Z L 2007 *Science* **316** 102–5
[7] Xu S, Qin Y, Xu C, Wei Y, Yang R and Wang Z L 2010 *Nat. Nanotechnol.* **5** 366–73
[8] Park K Il, Lee M, Liu Y, Moon S, Hwang G T, Zhu G, Kim J E, Kim S O, Kim D K, Wang Z L and Lee K J 2012 *Adv. Mater.* **24** 2999–3004
[9] Opoku C, Dahiya A S, Oshman C, Daumont C, Cayrel F, Poulin-Vittrant G, Alquier D and Camara N 2015 *Nanotechnology* **26** 355704
[10] Dahiya A S, Opoku C, Poulin-Vittrant G, Camara N, Daumont C, Barbagiovanni E G, Franz G, Mirabella S and Alquier D 2017 *ACS Appl. Mater. Interfaces* **9** 573–84
[11] Kim J Y, Cheng A and Tai Y 2011 *2011 IEEE 24th Int. Conf. Micro Electro Mech. Syst.* 473–6
[12] Zhu G, Wang A C, Liu Y, Zhou Y and Wang Z L 2012 *Nano Lett.* **12** 3086–90

A self-powered triboelectric sensor for wide-range pressure detection in wearable application

M S Rasel, H O Cho, J W Kim, and J Y Park[1]
Micro/Nano Devices and Packaging Laboratory, Department of Electronic Engineering, Kwangwoon University, 447-1 Wolgye-dong, Nowon-gu, Seoul, 139-701 Korea.

E-mail: jaepark@kw.ac.kr (J Y Park)

Abstract. This paper reports the structural design, fabrication, and experimental validation of a self-powered triboelectric sensor for wide-range pressure detection based on as-fabricated micro-structured conductive elastomer composite. To design a biocompatible Triboelectric Nanogenerator (TENG) for pressure sensing, micro-structured Polydimethylsiloxane (PDMS) and Multiwall Carbon Nano Tube MWCNT/PDMS films were fabricated through sandpaper template based inexpensive fabrication process. MWCNTs has been doped into PDMS to tune the internal resistance of the TENG and to optimize its output power. A range of forces was applied on the device and corresponding performances were analyzed to calibrate the device as wide-range pressure sensor. Our as-fabricated prototype has wide range of mechanical force sensitivity starting from 2N to 80N and it works at higher forces as well, resulting in output peak-to-peak voltages and currents of 1~17V and 10~80nA, respectively.

1. Introduction

In recent years, energy harvesting technology has been investigated to realize the potential energy supply for micro-actuators, micro-sensors, especially wearable and implantable medical devices. A variety of harvesters have been demonstrated to convert ambient energy into the electrical energy employed by different mechanisms, including electromagnetic [1-2], electrostatic [3], piezoelectric [4] mechanisms. But these devices are not strong enough to endure long-time exposure to the applied excitations because most of them are not fabricated with organic materials. Triboelectric energy nanogenerators (TENGs) [5–7], by coupling triboelectrification and electrostatic induction have been popularly explored very recently due to relatively high output performance and easy fabrication process. The output voltages of TENGs can reach very high through optimized structure and proper materials selection from triboelectric series [8]. Recently, the applications of the TENGs have been explored for self-powered sensors and wearable devices [9–12], which is our area of interest in this work. To show the application of TENG as wide-range pressure sensor, we proposed a self-powered device with double-side tribological layers of Polydimethylsiloxane (PDMS) and Multiwall Carbon Nanotube (MWCNT)/PDMS. A range of forces was applied on the device under vertical contact-separation mode and corresponding voltage and current performance were analyzed to calibrate the sensibility of the device as pressure sensor. This paper will show a complete set of experimental results; including the effect of surface modification of the triboelectric layer, influence of force on TENG output performance, output power characteristics, and reliability of the device.

[1] To whom any correspondence should be addressed.

2. Design, fabrication and operation

Figure 1 shows the 2-D schematic view and detailed diagram of the proposed pressure sensor. The pressure sensor includes PDMS film with backside metal electrode and a composite film of CNT/PDMS. Both PDMS and CNT/PDMS were patterned with micro structures to increase the effective contact area, which in turns enhances the electrical output of the device. For the fabrication of micro-structured PDMS and CNT/PDMS layer, we used sandpaper (CAMI grit designation: 100, average grit diameter: 140µm) as template. The sandpaper template was attached on Si wafer by using adhesive tape. After pouring the PDMS (and CNT/PDMS) mixture on sandpaper template, spin coating (SCS 6800 spin coater, Specialty Coating System) of PDMS and CNT/PDMS mixture was performed individually at 200rpm for 30sec to achieve around 250µm thickness for each. After a degassed process for 30 minutes under vacuum, samples were incubated at 85°C for 1 hour. Au/Ti electrode (100nm) is deposited on the plane side of PDMS surface using e-beam evaporation process. Finally, PDMS film with electrode covering and CNT/PDMS film were peeled off from the sandpaper template and were diced into required device size. The area of PDMS film and PDMS/MWCNT film is 20mm × 20mm each while the thickness is 0.25mm each. The maximum internal gap between two triboelectric layers is 3 mm. The total dimension of the final fabricated prototype is 20.8cm^3 (4cm×4cm×1.3cm).

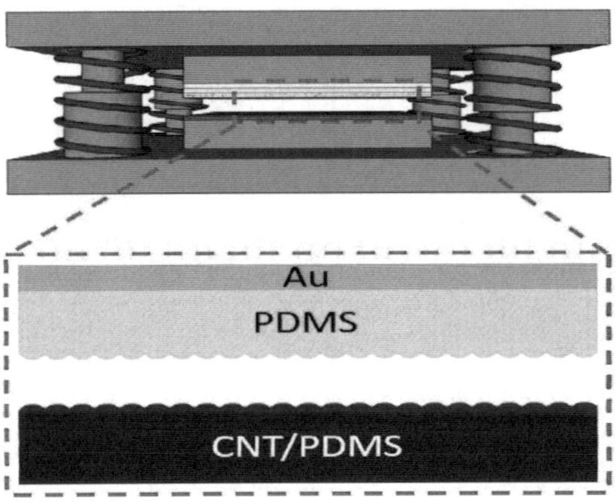

Figure 1. Schematic structure of the proposed self-powered triboelectric pressure sensor.

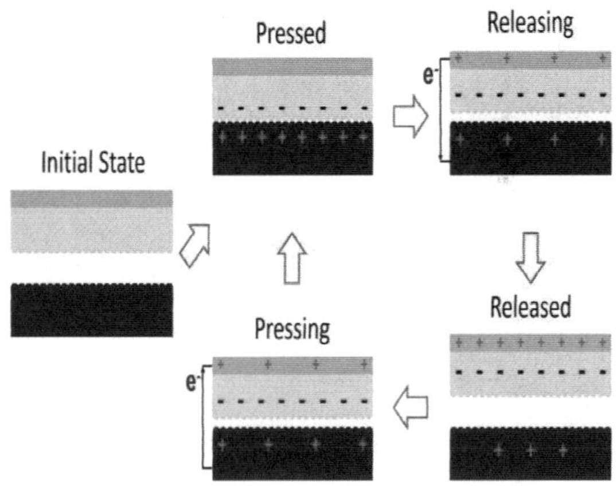

Figure 2. Working principle of reported sensor based on triboelectric charge transfer process.

Figure 2 demonstrates the working principle of the triboelectric pressure sensor. Triboelectric output is caused by the coupling between the triboelectric effect and electrostatic induction under periodic contact and separation of the two materials surfaces. The CNT/PDMS surface becomes positively charged and the PDMS surface becomes negatively charged when they are brought into contact. The PDMS film was pressed downward until it contacts to the bottom CNT/PDMS when a vertical force was applied. While the force is releasing, electric potential difference between the PDMS and CNT/PDMS drives electrons from PDMS electrode to CNT/PDMS because they have different charge affinity. The larger affinity difference between two triboelectric materials contributes to the larger amount of transferred charges. When the applied force was released, electrons started accumulating on the CNT/PDMS layer. A small separation gap between two layers of the prototype was utilized to facilitate the steady equilibrium state.

3. Experimental results and discussions

In order to investigate the performance of proposed triboelectric pressure sensor, we used an electrometer (Keithley 6514) to measure the output voltage and current of the proposed pressure sensor at applied vertical mechanical force from force gauge. Figure 3 shows the voltage for flat and micro-structured PDMS based pressure sensor at 50N mechanical force was 5V and 10V, respectively. The peak-peak

IOP Conf. Series: Journal of Physics: Conf. Series **1052** (2018) 012029 doi:10.1088/1742-6596/1052/1/012029

short-circuit current for flat and micro-structured PDMS based pressure sensor at 50N mechanical force was 8nA and 25nA, respectively, as shown in Figure 4. For applied mechanical forces ranging from 2N to 80N, the resultant peak-to-peak voltages (Figure 5) and currents (Figure 6) were 1~17V and 10~80nA respectively. Results show the proposed pressure sensor requires at least 2N for contact between PDMS and CNT/PDMS layer. During experiments, the output voltages increase with the increase of applied forces. The reason is that large forces cause large deformation in the tribological layers when they come in contact, which increases the effective contact area. In addition, the large force increases the kinetic energy in the tribological layers, which causes more electrons transfer between two tribological layers.

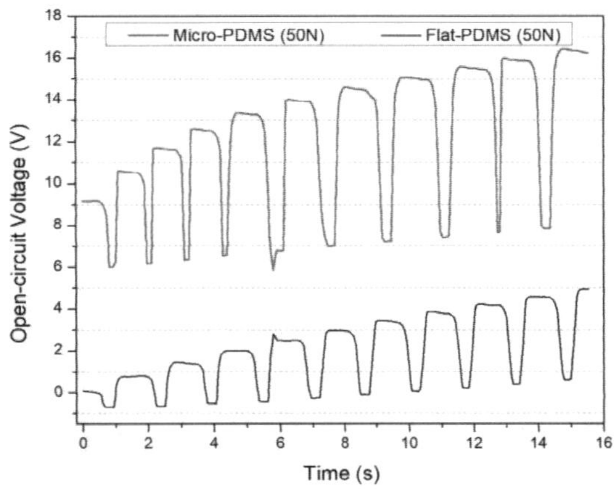

Figure 3. Output voltage performance for flat and micro-structured PDMS layers as triboelectric material.

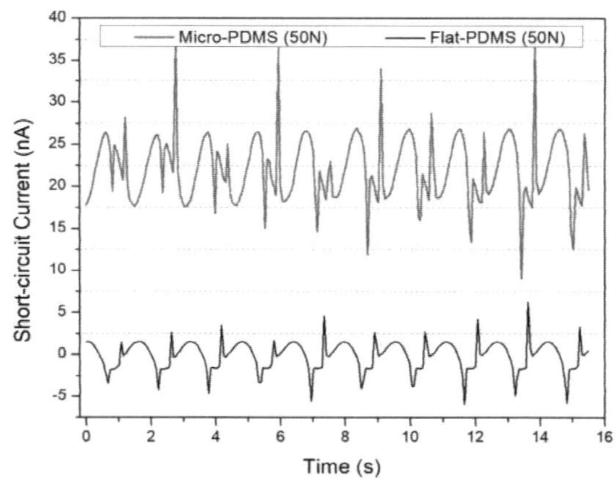

Figure 4. Output current performance for flat and micro-structured PDMS layers as triboelectric material.

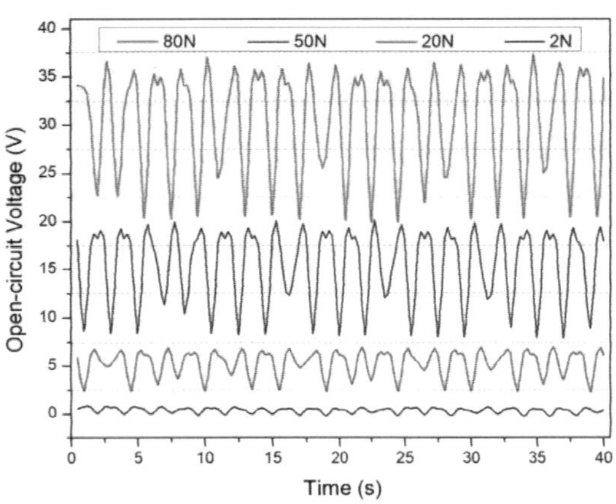

Figure 5. Instantaneous open-circuit voltage waveforms under different mechanical pressing forces on the sensing device.

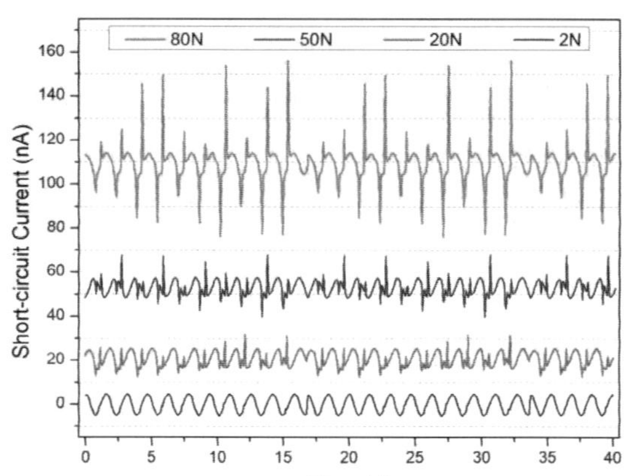

Figure 6. Instantaneous short-circuit current waveforms under different mechanical pressing forces on the sensing device.

We measured voltages for long repetitive working cycles to validate the reliability of the proposed device. Result shows the working prototype has high stability up to thousands of repetitive working cycles. The peak-to-peak voltage is almost constant at 14V for 50N force (Figure 7). To investigate the triboelectric close-circuit output performance, the peak-peak output voltages and peak powers under different loading resistances are. The optimum loading resistance of the triboelectric pressure sensor prototype was found to be 10MΩ (Figure 8). The optimum triboelectric output peak power was around 0.18μW.

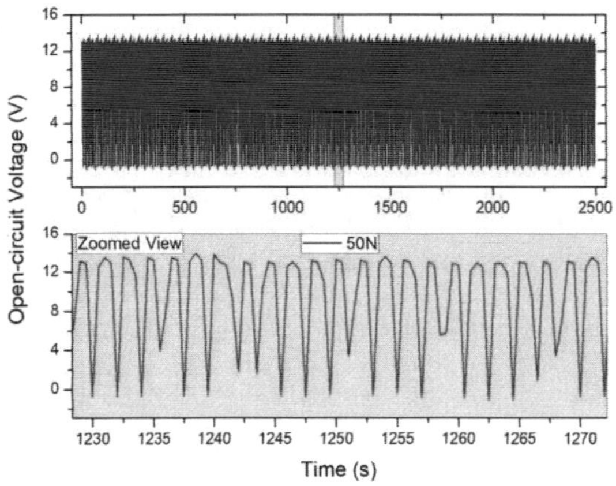

Figure 7. Reliability testing of the sensor under a constant mechanical force of 50N.

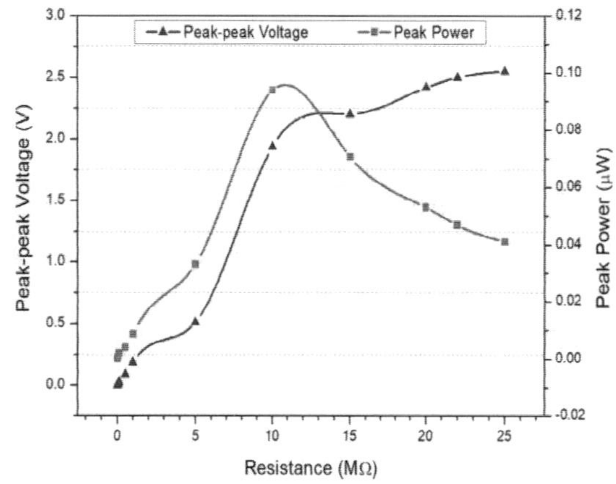

Figure 8. Load voltage and peak power output under a constant mechanical force of 50N.

4. Conclusions

We developed a self-powered triboelectric pressure sensor which can work on vertical contacting mode operation. The output performance was systematically studied in relation to surface patterns and applied mechanical pressing forces. To enhance the effective contact area between triboelectric layers (PDMS as positive and CNT/PDMS as negative) which in turns increases the output performance, sandpaper based inexpensive and feasible fabrication process of micro-structured PDMS and CNT/PDMS films were utilized. The device can be assembled inside shoe (in-sole) and has the potential of wearable force sensing application for different human weights and walking motions. Thus, we believe this work pushes forward a significant step towards practical wearable applications of self-powered triboelectric pressure sensors.

Acknowledgements

This work was supported by the Technology Innovation Program (10065696) funded by the Ministry of Trade, Industry & Energy (MI, Korea). The authors are grateful to MiNDaP (Micro/Nano Devices and Packaging Lab.) group members of Kwangwoon University.

References

[1] Salauddin M and Park J Y, 2017 *Smart Mater. Struct.* **26** 035011
[2] Salauddin M, Halim M A, and Park J Y, 2016 *Smart Mater. Struct.* **25** 095017
[3] Chen G, Li Y F, Xiao H M, and Zhu X, , 2017 *J. Mater. Chem. A,* **5** 4150-4155
[4] Toyabur R M, Salauddin M, and Park J Y, 2017 *Ceram. Int.* **43** S675-681
[5] Gu G Q et al., 2017 *ACS Appl. Mater. Interfaces* **9** 11882-11888
[6] Zhang S L et al., 2017 *Adv. Funct. Mater.* **27** 1606695
[7] Rasel M S and Park J Y, 2017 *Appl. Energy* **206** 150-158
[8] Pi Z et al., and Wu D, 2014 *Nano Energy* **7** 33–41
[9] Rasel M S, Miah M A, and Park J Y, 2015 *Proc. IEEE SENSORS 2015* 1-4
[10] Gu H J et al., 2017 *ACS Nano* **11** 856-864
[11] Rasel M S and Park J Y, 2017 *Proc. TRANSDUCERS 2017* 1804-1807
[12] Wang J et al., 2016 *Nat. Commun.* **7** 12744

IOP Conf. Series: Journal of Physics: Conf. Series **1052** (2018) 012030 doi:10.1088/1742-6596/1052/1/012030

Self-Powered Triboelectric Inertial Sensor Ball for IoT and Wearable Applications

Qiongfeng Shi[1,2,3], Hao Wang[1,2,3], Tianyiyi He[1,2,3] and Chengkuo Lee[1,2,3]*

[1] Department of Electrical and Computer Engineering, National University of Singapore, 4 Engineering Drive 3, Singapore 117576
[2] Center for Intelligent Sensors and MEMS, National University of Singapore, E6 #05-11F, 5 Engineering Drive 1, Singapore 117608
[3] Hybrid-Integrated Flexible (Stretchable) Electronic Systems Program, National University of Singapore, E6 #05-4, 5 Engineering Drive 1, Singapore 117608
*E-mail: elelc@nus.edu.sg

Abstract. Motion monitoring can provide useful information for rehabilitation or diagnostics purpose. Recently, triboelectric nanogenerator (TENG) has developed rapidly as a promising technology for energy harvesting and self-powered sensing. However, research effort of using TENG for multi-axis acceleration sensing is very limited. What is more, rotation sensing has not been achieved by TENG to date. Here we propose a 3-dimensional (3D) symmetric triboelectric nanogenerator based inertial sensor ball (TENG-Ball) for self-powered multi-axis acceleration and rotation sensing. The TENG-Ball has great potential in motion monitoring, internet-of-things (IoT) and wearable applications.

1. Introduction

Human motion information monitoring (i.e. biomechanical parameter monitoring) can provide useful physical and mental status information of our body, which can be further used for rehabilitation or diagnostics purpose [1,2]. Nowadays, the achievement of motion monitoring is normally by traditional inertial sensor module, such as accelerometer and gyroscope. However, the large power consumption of these sensor modules remains as one major concern towards green and sustainable developing future. In the past few years, enormous self-powered sensors based on triboelectric nanogenerator (TENG) have been extensively investigated for battery-less applications in internet-of-things (IoT), wearable electronics, healthcare and harsh environment monitoring. However, research effort of using TENG for multi-axis acceleration sensing is very limited. For instance, hybrid acceleration sensor (by triboelectric and electromagnetic) and spherical TENG based acceleration sensor was reported for only one-axis acceleration sensing [3-5]. To enable multi-axis sensing, a 3-dimensional (3D) acceleration sensor was proposed [6]. Yet the realization of multi-axis sensing is indeed from the three orthogonal integrated tube shaped TENGs, which greatly increases the device size and complexity. Furthermore, rotation sensing by TENG has not been reported up to now. Therefore, in this work, a 3D symmetric triboelectric nanogenerator based inertial sensor ball (TENG-Ball) is proposed for self-powered multi-axis acceleration and rotation sensing. The device also demonstrates the ability of hand motion recognition for virtual reality, game control and smart control interface. With the great sensing capability, the TENG-Ball can be used for a wide range of IoT and wearable applications.

2. Device design

2.1. Device configuration

The schematic diagram of a conventional gyroscope and the proposed TENG-Ball is shown in Figure 1a,b. The structure of the TENG-Ball is shown in Figure 1c,d. It consists of a spherical shell frame (65 mm diameter), four Al electrodes covering with polytetrafluoroethylene (PTFE) thin film and multiple steel balls with diameter of 6.3 mm. The multiple steel balls are adopted as movable mass and positive triboelectric layer. The fabrication of the TENG-Ball started from two hemi-spherical frames by 3D printing. Then Al and PTFE were attached on the inner surface. Last, steel balls were enclosed inside and the device was sealed by polydimethylsiloxane (PDMS). The four electrodes are marked as Ex+, Ey+, Ex- and Ey-. Then Ex+ and Ex- are connected as Ex, while Ey+ and Ey- are connected as Ey. The photograph of the TENG-Ball before and after encapsulation is shown in Figure 1e,f.

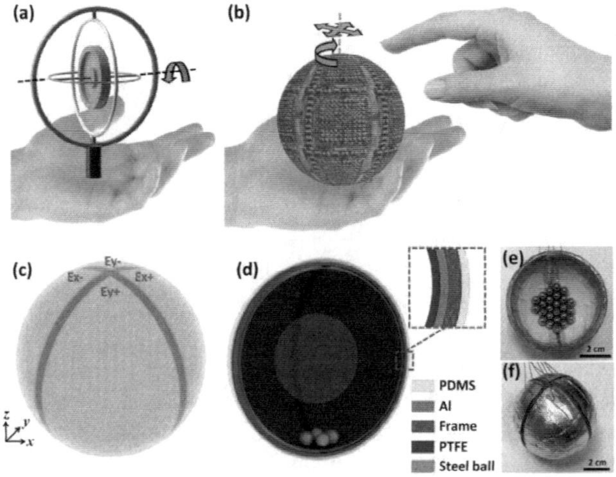

Figure 1. Schematic diagram of (a) a gyroscope and (b) the TENG-Ball. (c) Tilted view and (d) cross-sectional view of the device. Photograph (e) before and (f) after encapsulation.

2.2. Working principle

The TENG-Ball is able to operate under various usage scenarios. The working principle of the device under in-plane vibration is shown in Figure 2a by side view. Only the inner electrodes and the PTFE thin film are shown in the schematic diagram. After contact, steel balls end up with positive charges and PTFE with negative charges due to different electron affiliation. When the steel balls are on the left electrode, electric potential difference arises and current flow appears. Then when the steel balls are on the right electrode, current flow is in the opposite direction. The working principle of the device under spinning or rotation motion is shown in Figure 2b by top view. Electric potential difference is sequentially induced on all the four electrons, resulting in current flow in the respective electrodes.

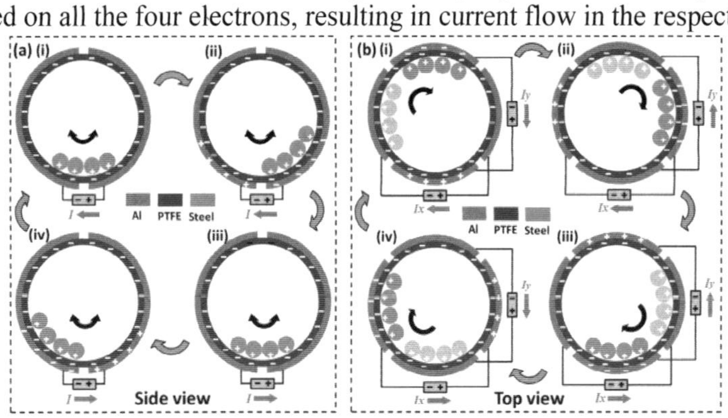

Figure 2: Working principle of the device under (a) vibration and (b) spinning or rotation motion.

3. Results and discussion

3.1. Acceleration sensing

The symmetric structure enables the TENG-Ball to function as a self-powered multi-axis acceleration sensor. Figure 3a-c shows the output from Ex and Ey for x, y and z acceleration sensing with 5, 15 and 25 steel balls encapsulated inside. For x-axis acceleration, voltage from Ex increases significantly at the beginning and gradually saturates. The result of 25 steel balls shows the best performance. The linear sensing range and sensitivity for x axis acceleration sensing is 4.87 g and 6.08 V/g. Meanwhile, voltage from Ey is negligible since no electric potential difference on Ey. Similar results are achieved for y axis with the linear sensing range and sensitivity of 5.06 g and 5.87 V/g. For z axis, voltage is generated from both Ex+ and Ey+ and the sensitivity is 3.62 V/g. The TENG-Ball shows multi-axis sensing capability that exhibits great potential for various complex motion monitoring.

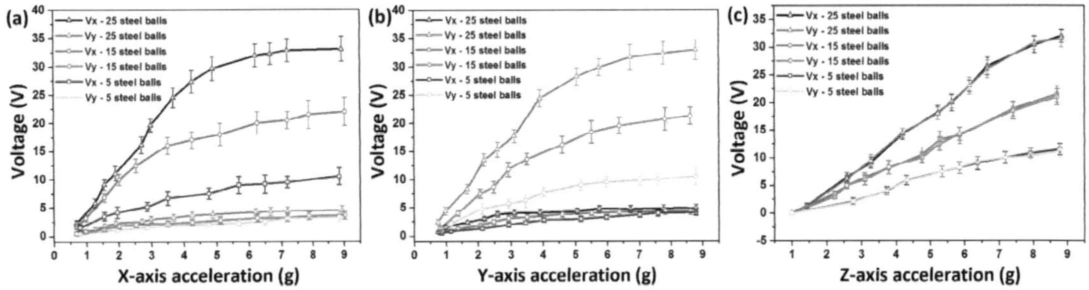

Figure 3: Output voltage versus acceleration in (b) x, (c) y, and (e) z direction.

3.2. Rotation sensing

The TENG-Ball can also operate as a self-powered rotation sensor. As shown in Figure 4a, the TENG-Ball is fixed on a servo motor to generate rotation. Figure 4b shows the results of 5, 15 and 25 steel balls for rotation angle sensing from 10° to 180° at 280 °/s. The output increases with the rotation angle due to higher electric potential difference. Figure 4c shows the results for rotation rate sensing at 180°. The inset depicts the waveform for rotation rate of 280 °/s. The voltage also increases with the rotation rate. The device with 25 steel balls shows the rotation rate sensitivity of 3.5 mV/°/s.

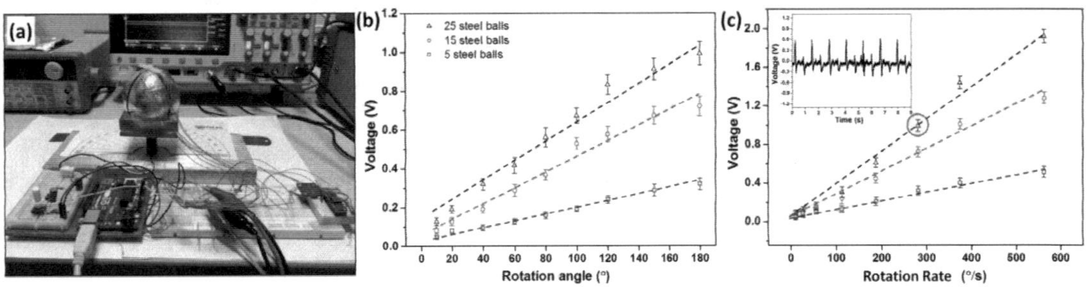

Figure 4: (a) Rotation sensing setup. Output voltage versus (b) rotation angle and (c) rotation rate.

3.3. Hand motion recognition

Due to the capability of acceleration and rotation sensing, the TENG-Ball can be used for hand motion or gesture recognition. Figure 5a shows the photograph of the TENG-Ball held by a human hand for movement and rotation monitoring. Commercial accelerometer (ADXL325, Analog Devices) and gyroscope (ADXRS622, Analog Devices) are fixed on top to detect the actual acceleration and rotation. Output voltage is connected to oscilloscope for waveform recording when human hand is moving left, right, forward and backward or rotating clockwise and anticlockwise, as in Figure 5a. Figure 5b depicts the typical waveform for movement monitoring. For moving left, voltage from Ex first shows a positive peak and then several vibrating peaks while voltage from Ey is almost 0. On the other hand, voltage from Ex first shows a negative peak for moving right. This is mainly because when the TENG-Ball first moves to the left side, the steel balls are swung on Ex+ first, generating a positive peak in Ex at the beginning. For moving right, the steel balls are swung on Ex- first, generating a negative peak in Ex at the beginning. Similar results are observed for moving forward and backward. Figure 5c shows the results for rotation monitoring. For clockwise rotation, voltage from Ex first shows a negative peak and voltage from Ey first shows a positive peak. While for anticlockwise rotation, voltage from Ex first shows a negative peak and voltage from Ey also first shows a negative peak. This is due to the sequence of steel balls on the corresponding electrodes. In practice, the output pattern can serve as control signal for game control or smart control system.

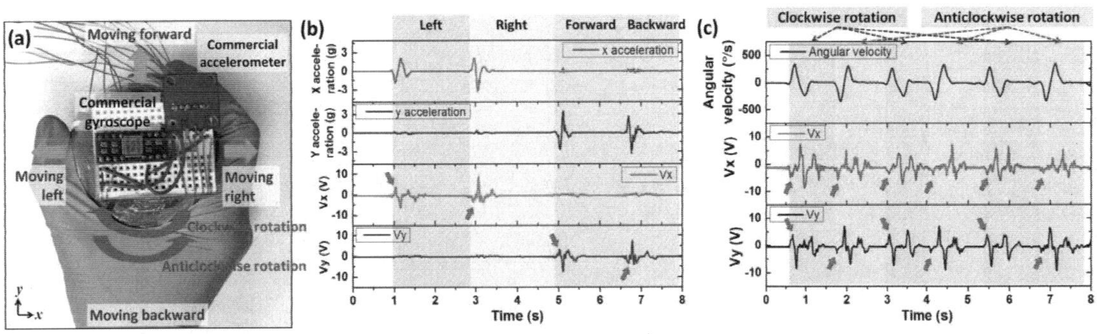

Figure 5: (a) Photograph and (b) measurement results of moving forward, backward, left and right. (c) Photograph and (d) measurement results of rotating clockwise and anticlockwise.

4. Conclusion

In summary, a TENG-Ball is proposed for self-powered multi-axis acceleration and rotation sensing. The 3D symmetric structure is highly adaptable in various usage scenarios and can harvest energy from a wide range of ambient energy sources (vibration, rotation and human motions). The TENG-Ball shows multi-axis acceleration sensing capability with sensitivity of 6.08 V/g, 5.87 V/g and 3.62 V/g in x, y and z axis. Besides, the TENG-Ball can also serve as a self-powered rotation sensor with rotation rate sensitivity of 3.5 mV/°/s. Moreover, the TENG-Ball exhibits great performance in hand motion recognition. Looking forward, the proposed TENG-Ball can be a key component for the healthcare motion monitoring, smart control system, IoT and wearable applications.

Acknowledgements

This work was supported by grants from the National Research Foundation (NRF) Competitive Research Programme (CRP) project "Self-Powered Body Sensor Network for Disease Management and Prevention Oriented Healthcare" (R-263-000-A27-281), NRF CRP project "Peripheral Nerve Prostheses: A Paradigm Shift in Restoring Dexterous Limb Function" (R-719-000-001-281), Ministry of Education (MOE) Faculty Research Committee (FRC) Grant (R-263-000-B56-112) "Thermoelectric Power Generator (TEG) Based Self-Powered ECG Plaster – System Integration (Part 3)" and National Natural Science Foundation of China under Grant No. 61474078 at NUS (Suzhou) Research Institute, Suzhou, China.

References

[1] Sveistrup H 2004 *J. NeuroEng. Rehabil.* **1** 10.
[2] Baker R 2006 *J. NeuroEng. Rehabil.* **3** 4.
[3] Gupta R K, Shi Q, Dhakar L, Wang T, Heng C H and Lee C 2017 *Sci. Rep.* **7** 41396.
[4] Zhang H, Yang Y, Su Y, Chen J, Adams K, Lee S, Hu C and Wang Z L 2014 *Adv. Funct. Mater.* **24** 1401-1407.
[5] Wang X, Niu S, Yin Y, Yi F, You Z and Wang Z L 2015 *Adv. Energy Mater.* **5** 1501467.
[6] Pang Y K, Li X H, Chen M X, Han C B, Zhang C and Wang Z L 2015 *ACS Appl. Mater. Inter.* **7** 19076-19082.

IOP Conf. Series: Journal of Physics: Conf. Series **1052** (2018) 012031 doi:10.1088/1742-6596/1052/1/012031

Autonomous power supply for aeronautical health monitoring sensors

M Bafleur[1], V Boitier[1], D Bramban[2], J-M Dilhac[1], X Dollat[1], J Féau[2] and S Jugé[3]

[1]LAAS-CNRS, Université de Toulouse, CNRS, INSA, UPS, Toulouse, France, [2] Airbus, [3] Altran Technologies

Abstract. New applications of embedded systems are envisioned in the context of aeronautics. However, the increasing burden of on-board cabling requires wireless solutions. Moreover, concerns such as safety or system lifetime often preclude the use of electrochemical energy storage. Ambient energy capture, storage and management are therefore key topics. In this paper we illustrate these concepts by describing an actual implementation in the engine harsh environment of an Airbus A380. A thermogenerator was used to harvest ambient energy and the electronics associated with it allows the local system being energy autonomous. All items were devised and tested in order to meet the criteria of DO-160G aeronautical standard. The set of equipment was mounted on aircraft, and submitted to tens of test flights during roughly a year, in very various flight conditions.

1. Introduction

Numerous new applications of embedded systems, and more specifically of sensor networks are today envisioned in the context of aeronautics. Generally speaking, sensor networks will play an increasingly important role: among others, monitoring the aging of the structure of the aircraft (Structural Health Monitoring - SHM) making it possible to only carry out repairs when necessary, thus saving time, weight (by decreasing mechanical margins) and therefore fuel.

However, the deployment of new wired sensor networks comes up against the complexity of electric networks already present in commercial aircrafts: as an illustration the total cable length is around 500 km for an Airbus A380. This praises in favour of wireless networks, but wireless means that each node of the network has to be energy autonomous. The simplest solution is therefore to rely on an energy reserve provided by an electrochemical battery featuring a high energy density and a weak self-discharge. However usage restrictions such as the required regular replacement of the battery, and risks (fire or explosion) associated with their operation in harsh environments, cannot be ignored. Consequently it is worth to investigate ambient energy harvesting combined with electrostatic energy storage using supercapacitors, in order to provide the required energy autonomy.

2. Requirements

The goal of this project was to arrange SHM sensors (temperature and crack detectors) deployed in the engine area (figure 1), and a local acquisition system (datalogger) transmitting data to a concentrator [1]. A module to harvest ambient energy and the electronics associated with it are, on the other hand, aimed to make the local system energy autonomous. The presence of permanent thermal gradients, found as soon as the engine is started, made sense to choose thermogeneration as the source of electric energy. Power has to be supplied to the datalogger under 5 V. The power consumption of the specific datalogger to be built for this project by another partner [2] was initially estimated around 2.5 mW, with 300 mW power spikes during data transmission. The system is not active when engines are stopped and energy storage has to be scaled only for such power surges. In addition, all system has to adhere to DO-160G aeronautical standard setting minimum performance requirements in terms of environmental conditions such as temperature, pressure, acceleration, vibrations, fire, electromagnetic interferences (EMI), etc. The corresponding certification process is not fully presented in this paper.

3. System architecture

3.1. Thermogenerator

The global structure of energy management system is depicted in figure 2: a limited storage based on two supercapacitors is used to deal with power surges without oversizing the thermogenerator (TEG).

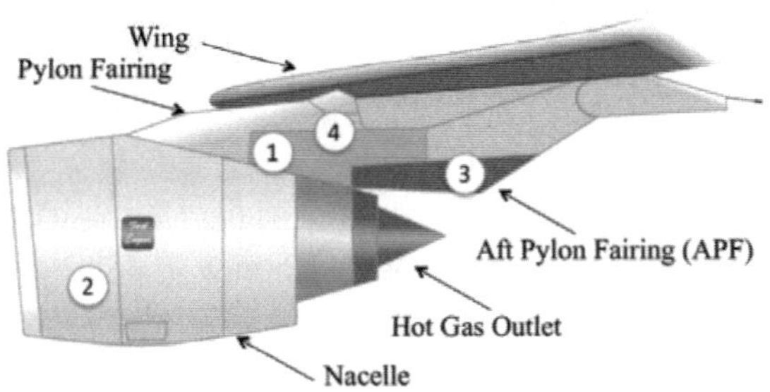

Figure 1. Engine's immediate environment: (1) thermogenerator (2, 3) sensors (4) energy and signal processing electronics.

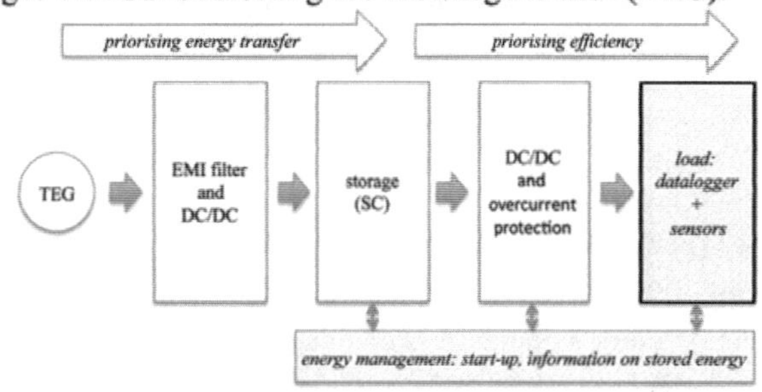

Figure 2. Architecture of energy management system (TEG stands for thermogenerator, and SC for supercapacitors).

As shown in figure 1, the system is surprisingly cabled locally over large distances. The reason is as follows: basically the sensors are placed in close proximity to mechanical parts (nacelle and aft pylon fairing) that they monitor. These parts are by nature submitted to aggressive environments (temperature-wise here), incompatible with electronic devices. The thermogenerator itself obeys another constraint: it is placed along the path of a thermal flux not necessarily present neither at the sensor nor electronic system locations. These local cabling raises the issue of using a wire connection to an already existing power cable installed nearby. However, in this paper, we will consider the above installation as a proof of concept of energy harvesting usage in the very specific context of aviation.
We first designed and manufactured a module incorporating the thermogenerator (figures 3 and 4) to be placed in a cell of the pylon. This cell was identified as likely to provide a sufficient in-flight thermal gradient between cell wall and inside air. The thermogenerator is in contact with the pylon wall through the intermediary of a stainless steel bearing plate. The thermogenerator is sandwiched between this plate and a copper exchanger, and is placed under pressure by the intermediary of two titanium flanges. The electric output of the thermogenerator is linked to a high-temperature electric connector certified for aeronautics. This connector is set in the plate by an intermediary part made from a high temperature polymer support, to lower the temperature of the connector. This polymer is also the material used for the inserts in flange-holders to lower thermal bridges.
The assembly was mechanically designed in such a way that in case of fire (the standard assumes that in this case the wall of the pylon is brought to 1100°C), as soon as the thermogenerator's semi-conducting elements (Bi_2Te_3, melting point 573 °C) and materials made of polymers will have melted or burned, the module remains attached to the wall of the pylon, and that especially all of its metallic pieces remain interdependent. Digital simulations have shown that the copper exchanger will not melt under conditions imposed by the DO-160G standard. For the selected position for the thermogeneration module, the thermal gradient between the wall of the pylon and the air inside was initially evaluated at cruise level between 120 and 180°C, according to weather conditions and the flight altitude. The wall temperature could itself exceed 250°C on the hottest days (Maximum Hot

Day). The selected thermogenerator (*KELK Komatsu KTGM161-18*) was tested in a climatic chamber. In a worst-case scenario, i.e. for a 110°C gradient, the device provided 39 mW at maximum power point, power sufficient enough for our design brief.

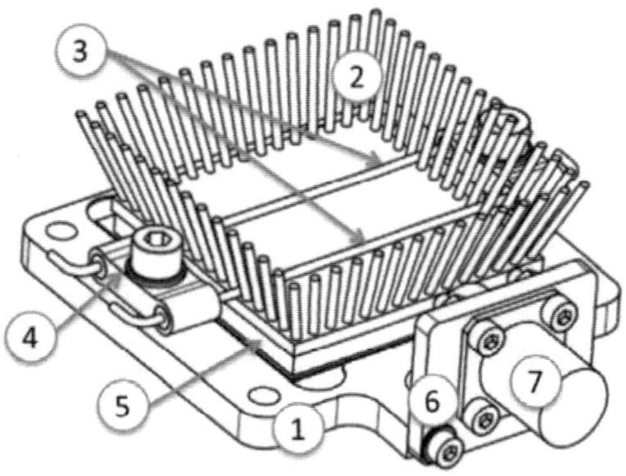

Figure 3. Perspective view of the thermogeneration module. (1) plate (2) exchanger (only one part of its shafts is shown) (3) flanges (4) flange-holders (5) thermogenerator (6) connector-holder (7) electric connector.

Figure 4. Detailed view of the thermogeneration module. The total weight is 497 g. Overall dimensions are 9 x 12 x 3 cm.

3.2. Electronic energy management system

Already shown functionally in its concept in figure 2, the more detailed layout of the electronic energy management system is presented in figure 5. We very briefly describe it below.

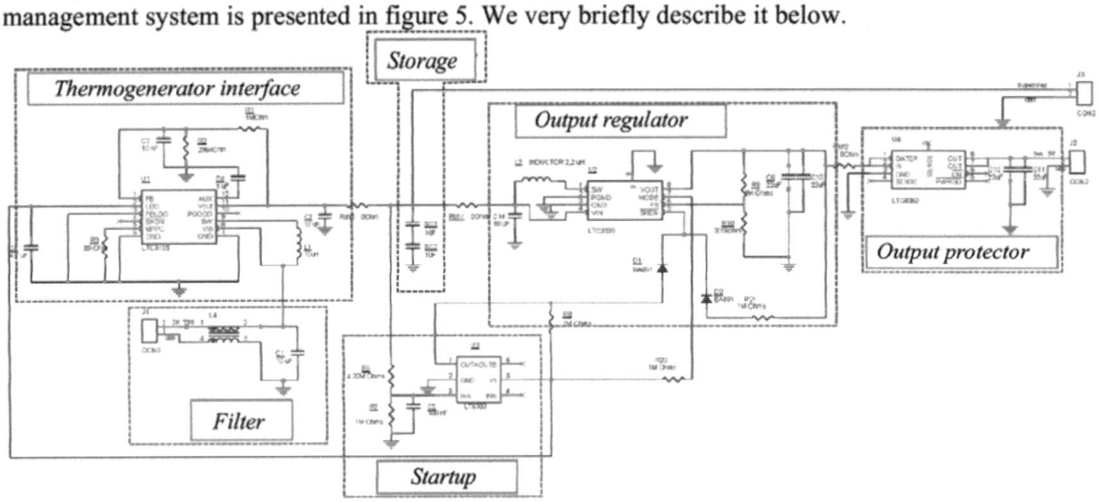

Figure 5. Overall structure of the electronic energy management system.

A common mode filter aimed at protecting the system against conducted electromagnetic interferences was placed at the electronic card input. Then there is a continuous boost regulator operating generally at a maximum power point (using a user predefined voltage setpoint). It also protects the supercapacitors from over voltage and provides a regulated auxiliary voltage for the startup management circuit. Storage is fulfilled by two BCAP0003 ultra-capacitors by Maxwell, connected in series (equivalent to 1.65 F). Regulating the output voltage is fulfilled by a boost regulator (step-up), whose output voltage in our case is set at 5 V. Finally, and although the regulator already has integrated safeguards against over currents, we added, in an interface with the load, a specific protection circuit. Figure 6 gives the electronic card built. To fulfil the DO 160G standard it was coated (Electrolube DCA SCC3) in order to protect against fluids, and to prevent electric arcs (dielectric rigidity of the coating: 90 kV/mm). The card was integrated into a 316L stainless steel housing to comply with the fireproofing standard. Aeronautical high-temperature connectors (Souriau, series 2997 and 3649) are finally used in the connection between the thermogenerator and the load.

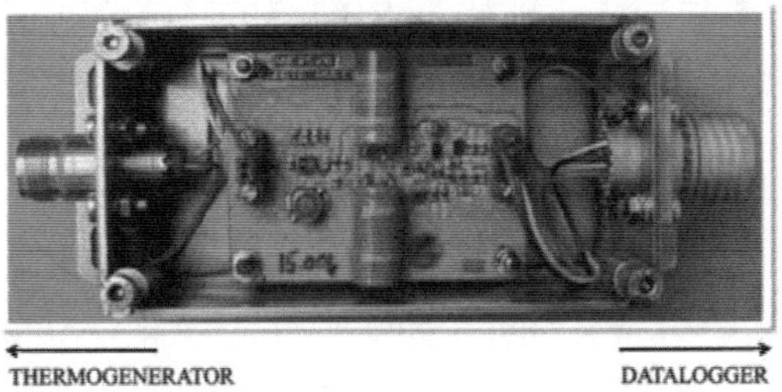

Figure 6. Electronic energy management system in its housing. The total weight is 504 g.

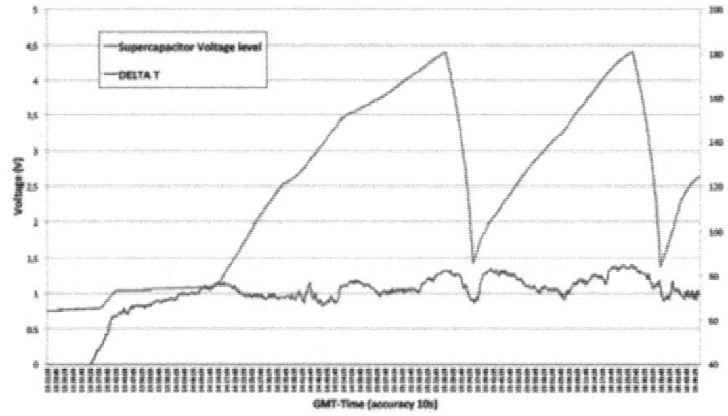

Figure 7. Thermal gradient (bottom) and charges and discharges (top) of supercapacitors during flight, induced by energy storage and periodical powering of the load.

4. Flight testing

The above energy management system has been correctly operating in flight for 1 year. The datalogger was cabled to be either powered by aircraft power network (to guarantee continuous operation) or by the harvester as soon as the supercapacitor voltage was high enough. Unfortunately, the datalogger mean consumption initially supposed to be around 2.5 mW was in practice 10 times higher and the harvester could not power permanently the datalogger as shown in figure 7 (supercapacitor voltage decreases when the harvester supplies the datalogger as the harvested power cannot compensate the consumption). In effect the datalogger was powered by the harvester 15% of the time for flights at low or medium altitude. At high cruising levels (typically10 000m) air pressure was too low to permit sufficient thermal exchanges by convection.

5. Conclusion

The above system (thermogenerator and associated electronics) is robust and has been operating properly for over a year although temperatures and vibrations are significant. Unfortunately, due to imprecise initial specifications and overestimated temperature gradients, it could not power the load permanently in all flight phases. Nevertheless the above work is a good proof of concept. Future work should include a mechanical design to operate with force convection (using relative wind) to foster thermal exchanges at high flight levels.

Acknowledgments
Authors wish to acknowledge assistance from Airbus technical teams and financial support from *Investissements d'Avenir* within the framework of CORALIE project.

References

[1] Dilhac JM and Boitier V 2016, *Energy Autonomy of Batteryless and Wireless Embedded Systems – Aeronautical Applications* (ISTE Press, London, Elsevier, Oxford)

[2] Dramitinos G, Fragkopoulos D, Papadas C, Vagiannis P, Makris K, Brambran D and Feau J 2017, *A Fully Integrated Health Monitoring Smart Sensor, Proc. 11th International Workshop on Structural Health Monitoring (Stanford),* (DEStech Publications, Inc., Lancaster, Pennsylvania, USA).

FABRICATION OF COPPER/COPPER-NICKEL THIN-FILM THERMOELECTRIC GENERATORS WITH ENERGY STORAGE DEVICES

Y Shimizu[1], M Mizoshiri[1], M Mikami[2], J Sakurai[1] and S hata[1]

[1] Department of Graduate School of Engineering, Nagoya University, Furo-cho, Chikusa, Nagoya 464-8693, Japan

[2] National Institute of Advanced Industrial Science and Technology, 2266-98, Anagahora, Shimo-shidami, Moriyama-ku, Nagoya 463-8560, Japan

Abstract. We demonstrated the fabrication of Cu/Cu-Ni thin-film thermoelectric generators for energy harvest. The thin-film generator consisted of 50 pairs of Cu/Cu-Ni thermocouples. The widths of Cu and Cu-Ni thermoelectric elements were 75 μm and 225 μm, respectively. The length of the both elements was 5 mm. Cu and Cu-Ni thin-films were deposited on polyimide substrates with low thermal conductivity using radio-frequency magnetron sputtering method and formed by lift-off techniques. Then, the thin-film generators were coated with polyimide thin films. The generation properties of the fabricated device, which consisted of serially-connected 15 thin-film generators, were evaluated. When the hot side of the device was heated at 203°C using a hot plate as a thermal source, the temperature difference between the hot and cold sides was approximately 70°C. The open-circuit voltage and maximum power were 2.18 V and 21 μW, respectively. A commercially-available light emission diode was successfully operated by connecting the device to a capacitor for energy storage.

1. Introduction

Energy harvesting devices using light, vibration and thermal energy have attracted attention in Internet of Things (IoT) society because these devices can supply electric energy to stand-alone sensors in the networks [1-2]. Thermoelectric microgenerators create power in micro watts by converting temperature gradient into electric energy using the Seebeck effect [3], and can utilize relatively low temperature (less than 300°C) waste heat which has not been widely used so far. We have fabricated Bi-Te thin-film thermoelectric generators that exhibit excellent figure of merit [4]. However, there are disadvantages that the materials consist of scarce metals, and the toxicity and difficulty of microfabrication due to brittleness.

In this study, we demonstrated the fabrication of thin-film thermoelectric generators using Cu-Ni thermoelectric alloys that are well-known as thermoelectric materials and exhibit excellent microfabrication properties. Then, the generation properties of the fabricated device that consisted of serially-connected 15 thin film generators were evaluated. The figures of merit of these materials are generally lower due to their high thermal conductivities, resulting that it is difficult to generate temperature differences in the bulk Cu-Ni alloys. Our proposed device enabled to generate a large temperature difference between the hot and cold sides of the pn junctions by using Cu-Ni thermoelectric elements that are sufficiently thinner than the substrates with low thermal conductivity.

2. Experimental methods

2.1. Design of thin-film thermoelectric generator

A design of thin-film thermoelectric generator is shown in figure 1. The generator consisted of 50 pairs of thermocouples, Cu (p-type) and Cu-Ni (n-type) thin-film thermoelectric elements. The widths of the Cu and Cu-Ni elements were 75 μm and 225 μm, respectively. The thickness and length of the both elements were 200 nm and 5 mm, respectively. The elements were fabricated on polyimide substrates with low thermal conductivity (0.29 W/mK), compared to that of Cu (386 W/mK) and Cu-Ni (23 W/mK). The thickness of polyimide substrates was 50 μm.

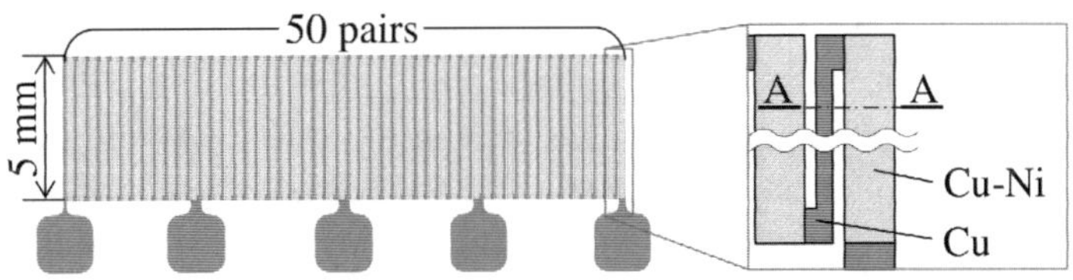

Figure 1. Schematic of thin-film thermoelectric generator.

2.2. Fabrication process of thin-film thermoelectric generator

Figure 2 shows the fabrication process of the thin-film thermoelectric generator. Cu and Cu-Ni thin films were deposited by radio-frequency (RF) magnetron sputtering method and were formed by lift-off techniques. Cu and $Cu_{60}Ni_{40}$ alloy targets were used for the film depositions. Then, polyimide thin films were coated on the thin-film generators using a photosensitive polyimide (Toray, Photoneece, PW-1500). The RF power and Ar process gas pressure for Cu and Cu-Ni thin-film depositions were 30 W and 1.0 Pa, and 40 W and 0.5 Pa, respectively. The base pressure and deposition time was 3.0×10^{-3} Pa or less and 10 min, respectively. The photosensitive polyimide was baked at 140°C, 240°C, and 340°C for 5 min each. Electrical resistivities of the both films were measured by a four-probe method. Cu-Ni thin-film was examined using X-ray diffraction (XRD) analysis (Rigaku, RINT RAPID-S).

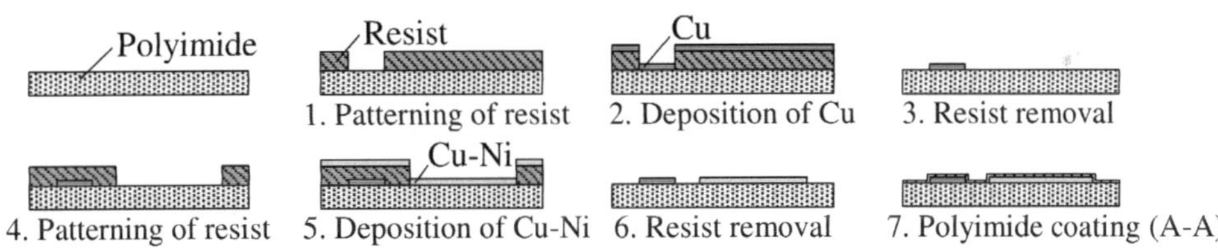

Figure 2. Schematic of fabrication process of the thin-film thermoelectric generator.

2.3. Evaluation method of thermoelectric device

The generation properties of the fabricated device (750 pairs), which consisted of serially-connected 15 thin-film thermoelectric generators, were evaluated. Generation voltage and output power were evaluated by heating the hot side of the device at approximately 200°C using a hot plate as a thermal source. In addition, the device was connected to a capacitor (220 μF) for energy storage, following that a commercially-available light emission diode (LED) was operated by using the energy.

3. Results and discussion

3.1. Fabrication of thin-film thermoelectric generators

The electrical resistivities of Cu and Cu-Ni thin-film were 7.7 μΩ cm and 67 μΩ cm, respectively, which were almost the same as the electrical resistivities of the bulk materials. Figure 3 shows XRD spectrum of Cu-Ni thin film deposited on polyimide substrate. The diffraction peaks corresponding to Cu-Ni were observed although those of the oxidation of the thin-film were not observed. This result indicates that high purity Cu-Ni thin films were deposited by the RF sputtering method.

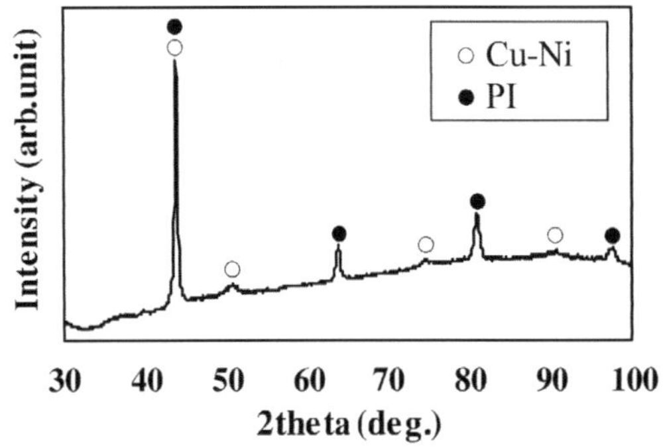

Figure 3. XRD spectrum of Cu-Ni thin-film deposited on polyimide substrate.

The thin-film thermoelectric generator and the device containing the 15 thin-film generators are shown in figures 4 and figure 5, respectively. The thermoelectric thin-film elements were successfully formed on the polyimide substrates without large deformation of the substrate. The internal resistance of the thin-film generator was 3.7 kΩ and the their serially-connected device was 56 kΩ. These values indicate that the contact resistance of each thin-film thermoelectric generator was negligibly small. The size of the thermoelectric device shown in figure 5 was 10×25×3 mm^3.

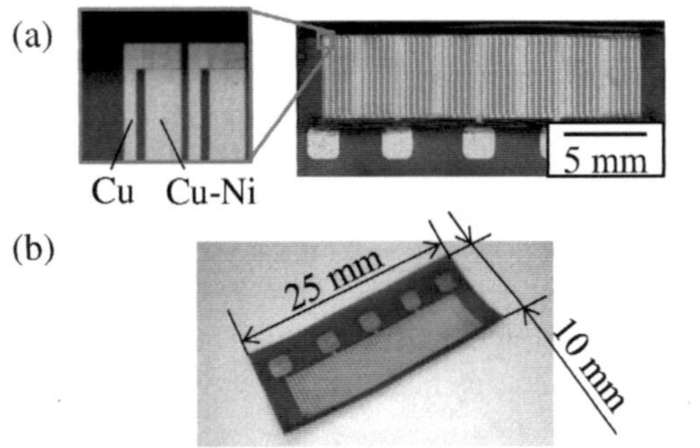

Figure 4. (a)Optical microscope image and (b)photograph of the thermoelectric generator.

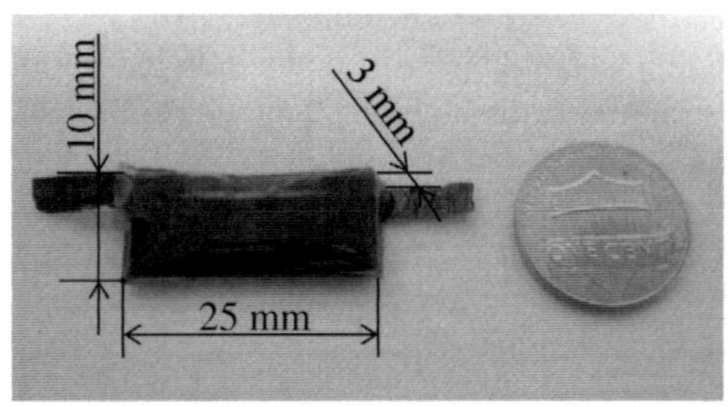

Figure 5. The thermoelectric device that consisted of 15 thin-film generators.

3.2. Evaluation of thermoelectric device

Figure 6 shows generation properties of the device. When the hot side of the device was heated at 203°C, the temperature difference between the hot and cold sides of pn junctions was approximately 70°C. The open-circuit voltage and maximum power were 2.18 V and 21 μW, respectively.

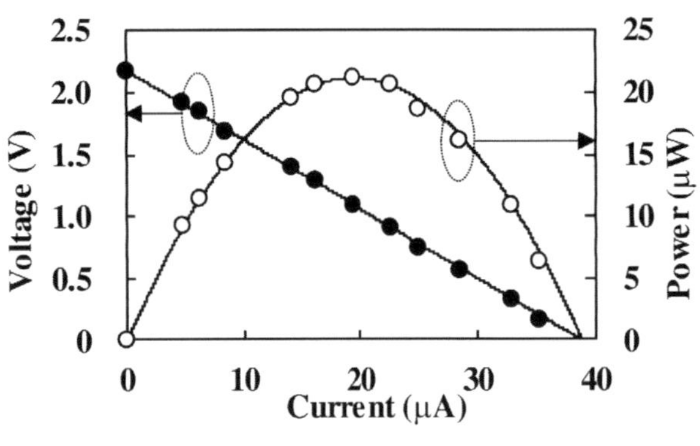

Figure 6. Dependences of generation voltage and output power on current. The device consisted of 15 thin-film generators (750 pairs).

The device was connected to the capacitor and LED. Figures 7 show the circuit diagram and photograph of actual circuit. The charge characteristic of the capacitor which was connected to the device is shown in figure 8(a). The power of 4.6×10^2 μJ was charged into the capacitor for 60 s. Furthermore, the LED was successfully operated using the charging energy as shown in figure 8(b). The thermoelectric generators with storage devices is expected to be used to supply electric energy to stand-alone devices.

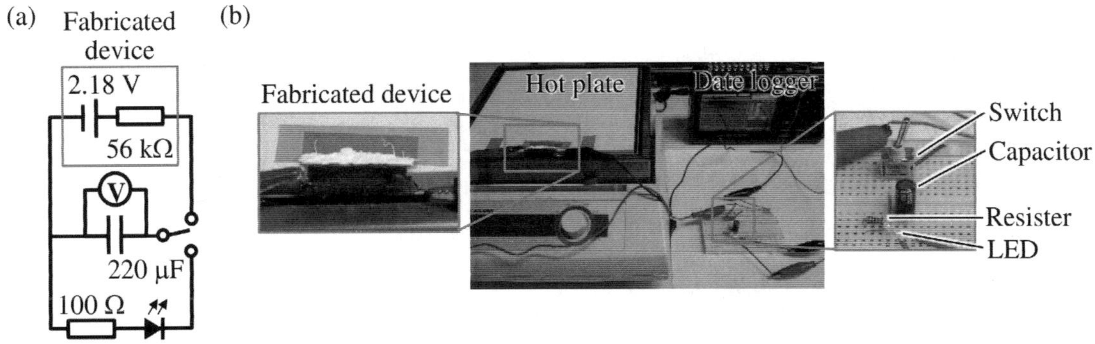

Figure 7. (a)Circuit diagram and (b)photograph of the fabricated device with energy storage circuit.

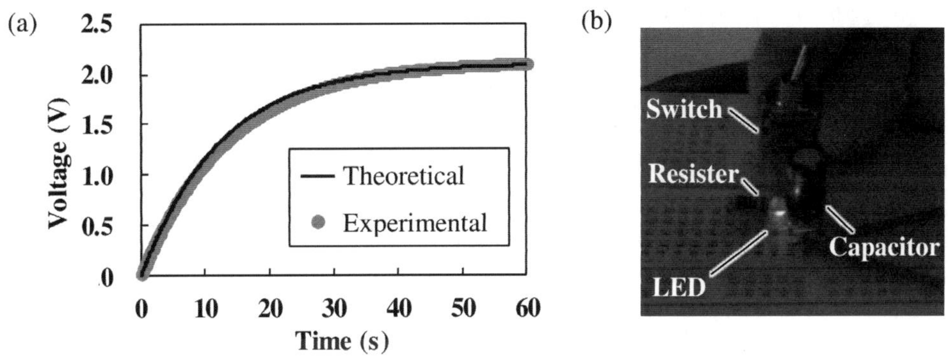

Figure 8. (a)Charge characteristics using a capacitor (220 μF) and (b)LED operation by the serially-connected device (750 pairs).

4. Conclusions

Cu/Cu-Ni thin-film thermoelectric generators were fabricated by sputtering method and lift-off techniques. The generation properties of the fabricated device (750 pairs), that consisted of 15 thin-film generators, were evaluated. When the temperature difference between the hot and cold sides was created as approximately 70°C using the heat source at 203°C, the open-circuit voltage and maximum power were 2.18 V and 21 μW, respectively. The LED were successfully operated by connecting the serially-connected device to the capacitor for energy storage.

5. References

[1] P Kamalinejad, C Mahapatra, Z Shaeng, S Mirabbasi, V.C.M. Leng and Y.L. Guan 2015 *IEEE commun. Mag.* **53 6** 102-108
[2] R J M Vullers, R V Schaijk, I Doms, C V Hoof, R Mertens 2009 *Solid-state Electron.* **53 7** 684-693
[3] A Date, A Date, C Dixon, A Akbarzadeh 2014 *Renew. Sust. Energ. Rev.* **33** 371-381
[4] M Mizoshiri, M Mikami and K Ozaki 2012 *Jan. J. Appl. Phys.* **53 6S** 6FL07

Acknowledgments

This work was partly supported by the Asahi Glass Foundation for Industrial Technology and the Nanotechnology Platform Program (Micro–Nano Fabrication) of the Ministry of Education, Culture, Sports, Science and Technology, Japan (MEXT).

Electrostatic Unsteady Thermal Energy Harvester Using Nematic Liquid Crystal

Hong Xie[1], Kenichi Morimoto[1], and Yuji Suzuki[1]
[1]Dept. of Mechanical Engineering, The University of Tokyo, Tokyo, Japan

Email: hxie@mesl.t.u-tokyo.ac.jp

Abstract: A novel electrostatic unsteady thermal harvester is proposed using nematic liquid crystal as the temperature-sensitive dielectric. With its large anisotropic permittivity and relatively high resistivity, an early prototype is made using fluorinated liquid crystal BCH-5F.F.F. Temperature change of the liquid crystal induces its permittivity change, thereby output voltage is obtained. An effective circuit model is developed and the simulation results are compared with the experimental data for designing harvester with better performance.

Keywords: Electrostatic, Thermal energy harvesting, Nematic liquid crystal, Resistivity

1. Introduction

Energy supply for low-power electronics is a big challenge with increasing demand of portable and wireless devices. Thermal energy harvesting can be a promising solution for self-powered electronics such as wireless sensing node [1], which is one of the essential components of the Internet of Things. Conventional thermal energy harvesting mostly utilizes constant temperature gradient based on the Seebeck effect [2]. More recently, time-dependent fluctuating ambient temperature has been reported to be an alternative thermal energy source for energy harvesting [3, 4].

We previously proposed electret-based unsteady thermal generator utilizing permittivity change of ferroelectric material [5]. By using potassium tantalate niobate (KTN) crystal, output power of 572 nJ has been obtained in 300 s [6]. However, relatively high operation temperature and long recovery time are the main drawbacks of the KTN-based system, which limit the actual application of unsteady energy harvester working in the room temperature range with rapid but small temperature fluctuations.

In this paper, nematic liquid crystal with anisotropic permittivity is, for the first time, studied as temperature sensitive dielectric in the electret-based unsteady thermal energy harvester to improve the temperature sensitivity and fast-response to temperature variation.

2. Working Principle

The basic working principle of electret-based unsteady thermal harvester using liquid crystal is shown in figure 1a, where the electret, the liquid crystal cell and the external load are connected in series. The electret serves as the permanent voltage source. There will be induced charge in both the electret capacitor C_e and the temperature-sensitive capacitor $C(t)$. Figure 1b depicts the phase change of the nematic liquid crustal from the nematic to isotropic phase over the clearing temperature T_{NI}. Temperature variation of surrounding environment around liquid crystal's clearing temperature will lead to the permittivity change from phase change. The permittivity change during the phase transition can be utilized as the source of thermal-to-electrical energy conversion. The induced charge of capacitor will be redistributed as the result of capacitance change and output power can be obtained. Compared with our previous model based on KTN [6], the leakage current through the temperature-sensitive dielectric cannot be neglected due to the limited resistivity of liquid crystal. As shown in figure 1c, the liquid crystal cell can be modeled with the dielectric and the resistor. Based on the Kirchhoff's law, and the charge conservation, we get

$$V_{lc} = I_{leak} R_{In} , V_E = I_{load} R_{load}, \tag{1}$$

$$I_{total} = I_{load} + I_{leak}, \tag{2}$$

$$\Delta Q_{lc} = \int I_{total} dt , \Delta Q_E = \int I_{load} dt, \tag{3}$$

where the V_{lc} and $R_{In}$are the voltage and internal resistance of the liquid crystal cell. V_E and R_{load} are the voltage and resistance of the external load, while Q_{lc} and Q_E are the induced charge of liquid crystal capacitor and electret. I_{total}, I_{load} and I_{leak} are the current of liquid crystal capacitor, external load and internal resistance, respectively. Output voltage can be estimated based on this model.

In order to reduced energy loss by the leakage current, liquid crystal with high resistivity is required. Fluorination of liquid crystal has been reported as a sufficient method to increase the resistivity and to maintain the anisotropic permittivity [7]. Compared with commonly-used nematic liquid crystal such as 5CB, BCH-5F.F.F is a typical fluorinated liquid crystal with low polarizability and high polarity, which result in high resistivity and high anisotropic permittivity. The detailed dielectric property will be discussed in the next section.

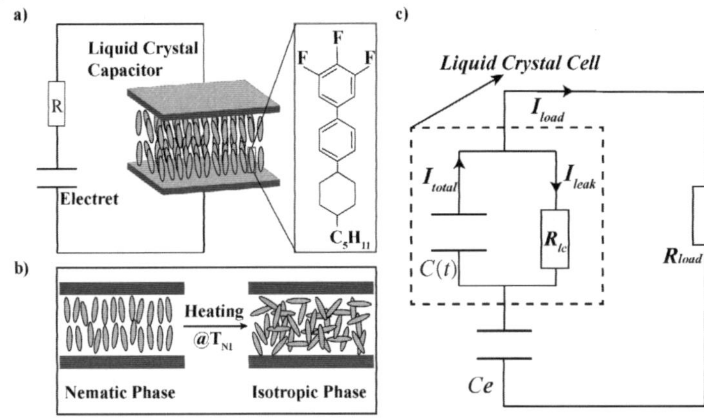

Figure 1. a) Working principle of electret-based thermal energy harvester, b) Circuit model of unsteady harvester considering the leakage current of liquid crystal.

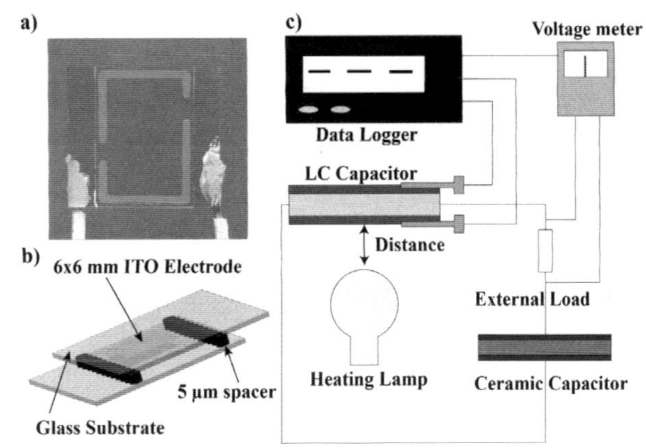

Figure 2. a,b) Photo and structure of liquid crystal cell used in the current study. c) Setup for power generation experiment.

3. Material Characterization and Experimental Results

As shown in figures 2ab, a cell with a 6 mm x 6 mm ITO electrode area and with a 5 μm gap is used for both material characterization and power generation experiments. Detailed experiment setup is shown in figure 3c. A light bulb is used to heat up the liquid crystal cell. In this report, the electret is replaced with a ceramic capacitor with a priming voltage for preliminary experiments. The components are firstly connected in series, and then the liquid crystal cell is heated up. The temperature information and the output voltage of the external load are recorded via a data logger. An impedance analyzer from Solartron is used to precisely measure the impedance of the liquid crystal cell. It is found that the volume resistivity of BCH-5F.F.F at around 26 °C is $7.4\times10^9\ \Omega\cdot m$. Since the resistivity of 5CB is around $3\times10^7\ \Omega\cdot m$, BCH-5F.F.F has 250 times higher resistivity than 5CB.

The transverse and axial permittivities of BCH-5F.F.F at different temperatures from 25 – 60 °C are measured by using a liquid crystal cell with homeotropic alignment. The permittivity measured with low AC voltage (0.5 V_{RMS}) corresponds to the transverse permittivity, while, with high AC voltage (20 V_{RMS} = 4 V_{RMS}/μm), the liquid crystals should be aligned perpendicularly to the substrate, so that the axial permittivity is measured. As shown in figure 3, the axial permittivity is as large as 13 at 25 °C, and decreased with increasing temperature. The clearing temperature where phase transition occurred is around 56 °C.

In the present study, the priming voltage of ceramic capacitor is chosen as 50 V. The external load is 810 MΩ. Figure 4 shows the permittivity change of BCH-5F.F.F when heated up with the lamp during power generation experiment. It is found that the permittivity drops at around 50 s when the liquid crystal phase changes from nematic to isotropic.

Figure 5 shows the output voltage for different capacitances of the ceramic capacitor. As the temperature of liquid crystal increases, voltage increases due to decreasing internal resistance, and clear positive voltage peak around 50 s is observed as the result of steep permittivity change during the phase transition of BCH-5F.F.F. Then, the voltage drop follows because of the charge dissipation. Also, higher voltage output can be found for the higher initial charge with higher capacitance. For the capacitance of 4.7 nF, voltage peak around 2.3 V is obtained during heating. Based on the experimental results, we can verify the electrostatic working mechanism utilizing permittivity change of liquid crystal phase transition.

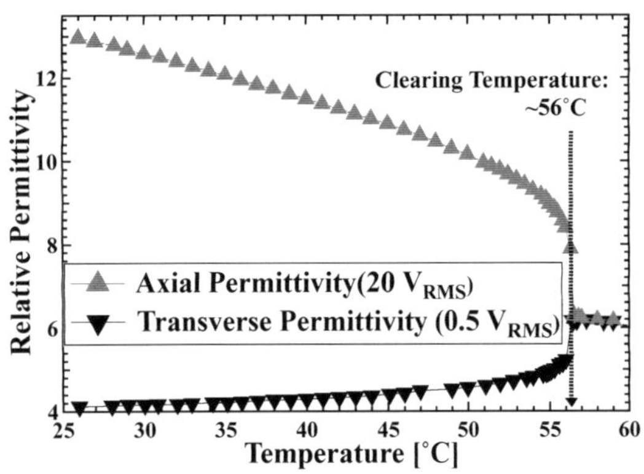

Figure 3. Permittivity change for both transverse and axial direction for 25-60 °C.

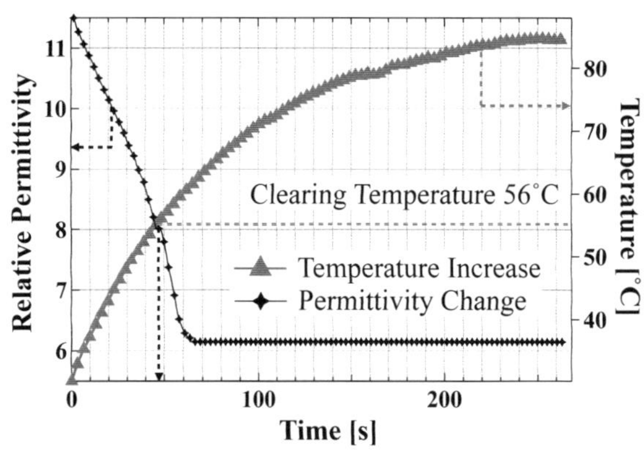

Figure 4. Permittivity and temperature change of liquid crystal cell during heating in experimental setup.

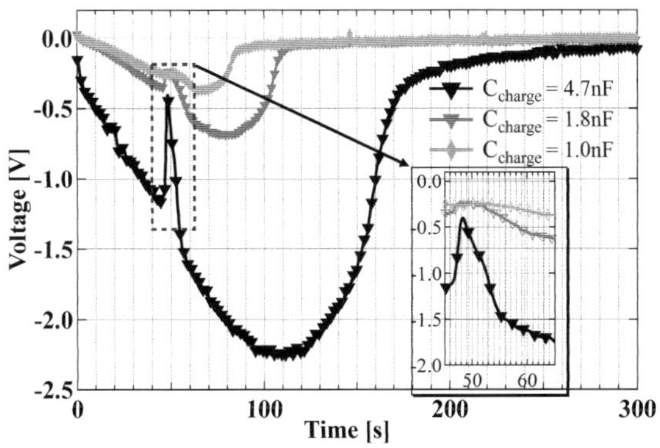

Figure 5. Voltage output during heating with different charged capacitor

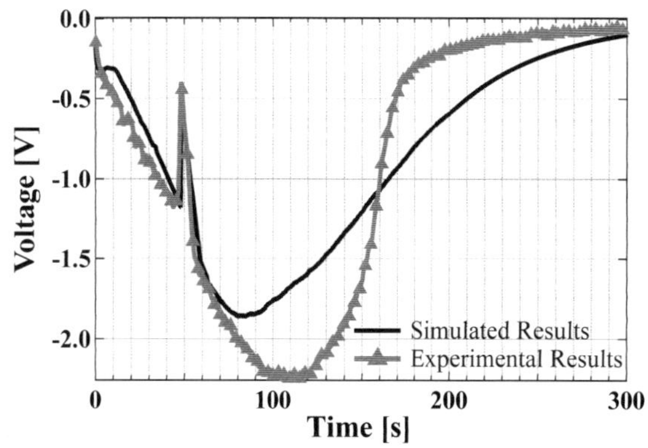

Figure 6. Comparison between simulated and experimental results in the case of 4.7 nF ceramic capacitor

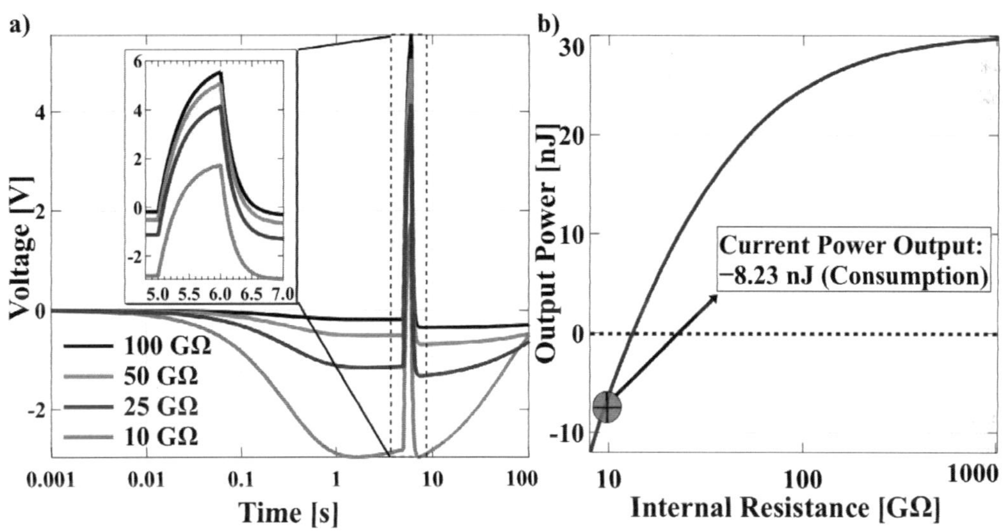

Figure 7. a) Simulated voltage output with increasing internal resistance of liquid crystal cell. b) Simulated output power with increasing internal resistance of liquid crystal cell

Figure 6 shows comparison between the experimental data and the simulation results for C_{charge}=4.7nF. The simulation results especially the positive voltage peak during the phase transition are in good accordance with the experimental data. The shift of leakage peak around 100 s between simulation and experimental is possibly from the measurement error of resistance of liquid crystal cell.

Based on the results above, it is found that, even with BCH-5F.F.F, the leakage current is still a dominant part of the current generated from the temperature variation; the priming charges are leaked out through the liquid crystal layer. Figure 7a shows the output voltage for higher internal resistance of the liquid crystal cell. With increasing internal resistance, the negative output voltage due to the leakage current decreases dramatically, and the positive voltage peak increases. As shown in figure 7b, maximum of 23 nJ can be expected during one heating cycle when the internal resistance is as large as 100 GΩ.

4. Conclusion

A novel electret-based unsteady thermal energy harvester using nematic liquid crystal is proposed for the first time. Fluorinated nematic liquid crystal BCH-5F.F.F is chosen as the temperature-sensitive permittivity material with high resistivity. Output voltage by the temperature fluctuations is examined and compared with an effective circuit mode. A sharp output current peak is observed, while the leakage current is still dominant. Much higher internal resistance is needed to reduce the energy loss. Maximum of 23 nJ can be expected with a liquid crystal with 100 GΩ internal resistance.

This work was supported by JST CREST Grant Number JPMJCR15Q3, Japan.

References:

[1] J. A. Paradiso and T. Starner, "Energy scavenging for mobile and wireless electronics," *IEEE Pervasive Comput.*, vol. 4, pp. 18–27, 2005.
[2] D. M. Rowe, "Thermoelectrics, an environmentally-friendly source of electrical power," *Renew. Energy*, vol. 16, pp. 1251–1256, 1999.
[3] A. Cuadras, M. Gasulla, and V. Ferrari, "Thermal energy harvesting through pyroelectricity," *Sens. Actuators Phys.*, vol. 158, pp. 132–139, 2010.
[4] A. Moser, et al., "Thermoelectric energy harvesting from transient ambient temperature gradients," *J. Electr. Mater.*, vol. 41, pp. 1653-1661, 2012.
[5] J. Yoshida, K. Morimoto, and Y. Suzuki, "Electrostatic Thermal Energy Harvester Using Unsteady Temperature Change," *J. Phys. Conf. Ser.*, vol. 476, 012079, 2013.
[6] H. Xie, K. Morimoto, and Y. Suzuki, "Electret-based Unsteady Thermal Energy Harvester using Potassium Tantalate Niobate Crystal," *J. Phys. Conf. Ser.*, vol. 773, 012023, 2016.
[7] M. Hird, "Fluorinated liquid crystals – properties and applications," *Chem. Soc. Rev.*, vol. 36, pp. 2070–2095, 2007.

Less gives more: on the optimal filling fraction of piezoelectric acoustic power receivers

M Gorostiaga, M C Wapler and U Wallrabe

Department of Micro-actuators, IMTEK-University of Freiburg, Georges-Koehler Alle 102-02, D-79110, Freiburg im Breisgau, Germany

E-mail: wallrabe@imtek.uni-freiburg.de

Abstract. In this paper, we investigate the suitability of 1-3 composite transducers as receivers in acoustic power transfer applications. In contrast to the commonly used PZT transducers, the composites allow us to modify the acoustic impedance, the losses and the electromechanical coupling factor, and hence analyse their influence on the efficiency of acoustic power receivers. Comparing composite receivers with 40%, 60% and 80% filling factors, we find that the lowest filling fraction gives the highest efficiency. This is due to its lower acoustic mismatch with water, which compensates its lower electromechanical coupling and mechanical quality factor.

1. Introduction

Wireless acoustic power transfer has attracted some attention recently as an alternative to the inductive approach to, for example, recharge the depleted batteries of biomedical implants. In the literature, PZT pure plate transducers have been usually used as acoustic power receivers [1], but so far, their suitability as receivers has not been discussed.

Pure hard PZT plates have a very large mechanical quality factor Q_m, and are fully made of piezoelectric material, which might be ideal to harvest as much energy as possible from the incoming acoustic waves. However, their reflection coefficient with soft-tissue is very large ($R \approx -0.92$), and they require an aspect-ratio of 20 (diameter/thickness) to show a clear thickness resonance [2], which is crucial to accurately predict the optimal electric loads. As seen in [3, 4], the optimal loads at the resonances are purely ohmic when the transmission medium is water or human tissue, and therefore, their values can be easily determined by experiments. By contrast, the optimal loads are inductive for the full frequency range when the transmission medium is steel. In this case, finding the loads experimentally is not trivial any more, and consequently, a good prediction is needed.

An alternative to pure PZT transducers are the so-called 1-3 composite transducers that consist of small piezoelectric pillars embedded in a polymer. These composite transducers offer a clear thickness resonance with low aspect-ratios, which makes them ideal for miniturization, and therefore have no spurious modes (see figure 1)[5], which could compromise the receiver performance and should be avoided for a good optimal load prediction [6]. Furthermore, their acoustic impedance can be tuned with their PZT-polymer ratio and their electromechanical coupling factor k_t is higher than that of PZT plates [7]. These properties make them ideal as acoustic power receivers, but the advantages come at the cost of higher dielectric losses δ_{tn} and a lower mechanical quality factor Q_m [5].

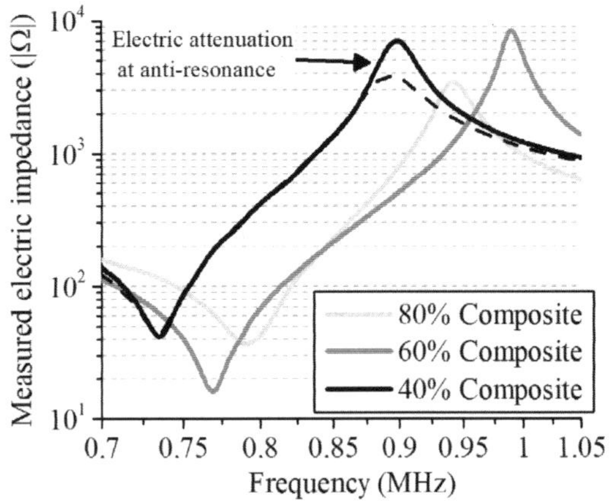

Figure 1. Measured electric impedance of the test-transducers, which showed a clear thickness resonance.

Table 1. Acoustic and piezoelectric properties of the tested composite transducers: k_t=electromechanical coupling; Q_m=mechanical quality factor; δ_{tn}=dielectric losses; R=reflection coefficient with water; Z_p=acoustic impedance.

Transducer	80%	60%	40%
r_p (mm)	7.95	7.95	7.95
Aspect-ratio	7.6	8.1	8.1
ρ (kg/mm^3)	6472	5144	3816
f_s (kHz)	789	761	737
k_t	0.58	0.68	0.64
Z_p (MRayl.)	25.6	19.9	13.5
ϵ_{33}^s	554	434	293
R_{trans}^{water}	-0.89	-0.86	-0.80
δ_{tn}	0.026	0.029	0.030
Q_m	27	51	42

Exploring these properties, we will perform acoustic power transfer experiments with composite transducers that will allow us to view the influence of the acoustic and piezoelectric parameters on the efficiency of acoustic power receivers. As we will see from the measurements in water, less piezoelectric material in the receiver may lead to higher efficiencies and improved acoustic power transfer, contrary to intuition.

2. Experimental set-up

To prove that less piezoelectric material can give a higher efficiency, we tested three 1-3 composite transducers with different PZT-polymer fillings in water as a transmission medium at the front side and air at the back side. The properties of the transducers can be found in Table 1, which we extracted by fitting the measured electric impedance to the emitter model in [8]. As we can see, the acoustic impedance mismatch with water decreases as the piezoelectric content in the transducer becomes smaller. Furthermore, the three transducers show low mechanical quality factors Q_m and high dielectric losses δ_{tn}, but their electromechanical coupling k_t is approximately 25% higher than that of pure PZT plates [9].

In figure 2, we can see a schematic drawing of the experimental set-up. We performed frequency sweeps between 0.70 MHz and 1.05 MHz with wave bursts of 50 cycles sent from a distant emitter to avoid standing waves, and obtained from the measured voltage V the power

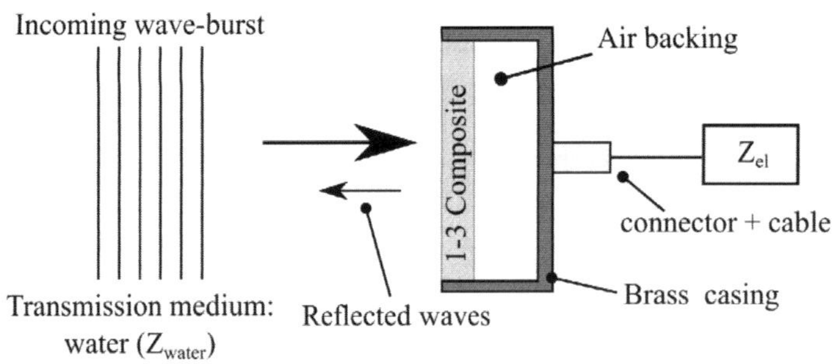

Figure 2. Schematic drawing of the receiver with the attached electric loads Z_{el}, and definition of the incoming power W_{in} and the dissipated power W_{load} to obtain the efficiency $\eta = W_{load}/W_{in}$.

$W_{load} = V^2/Z_{el}$ dissipated at the attached electric loads Z_{el} as explained in [8]. For simplicity, we only considered ohmic loads between $12\,\Omega$ and $5.6\,\mathrm{k}\Omega$, but we saw in [8] that the optimal power maximization loads are complex away from the resonances. Finally, we measured the pressure P of the incoming wave bursts with a needle hydrophone at the same position without the receivers of surface area A to obtain the incoming acoustic power $W_{in} = (AP)^2/Z_{water}$, and hence the efficiency of the receivers $\eta = W_{load}/W_{in}$. Due to the non-planar shape of the incoming waves, we corrected the efficiency to account for the actual force on the receiver as explained in [8].

3. Measurement results

In figure 3, we see on the one hand that the 40% transducer achieves the highest efficiency (0.67) at the resonance despite having less piezoelectric material and higher dielectric losses than the 60% (0.59) and 80% (0.48) composite receivers. Furthermore, it has a lower Q_m and electromechanical coupling k_t than the 60% composite, but its performance is still better.

On the other hand, the 40% receiver should also have a local maximum at the anti-resonance with a higher efficiency than the other receivers, but this is attenuated, probably due to leakage of water vapour into the brass casing. This leakage modified the electric impedance at the anti-resonance (see figure 1), which also lead to a decrease of the efficiency at this frequency.

The better performance of the 40% transducer is due to its lower acoustic reflection with water, which we can also see in figure 4. There we show the numerical prediction of the maximum efficiency at the resonance of the transducers with the optimal load according to the model in [3, 4]. The predictions suggest that the transducer with the closest acoustic impedance to the transmission medium will have the best performance for a set of receivers with equal losses and electromechanical coupling. Furthermore, a lower acoustic impedance mismatch can compensate the efficiency decrease caused by a slightly lower Q_m, as it is the case of our 40% transducer compared to the 60% composite. However, a receiver with a much lower Q_m, such as our 80% composite, will perform worse for the whole front reflection spectrum. Additionally, simulations of the 60% and 40% composites suggest that higher Q_m values will increase the front reflection

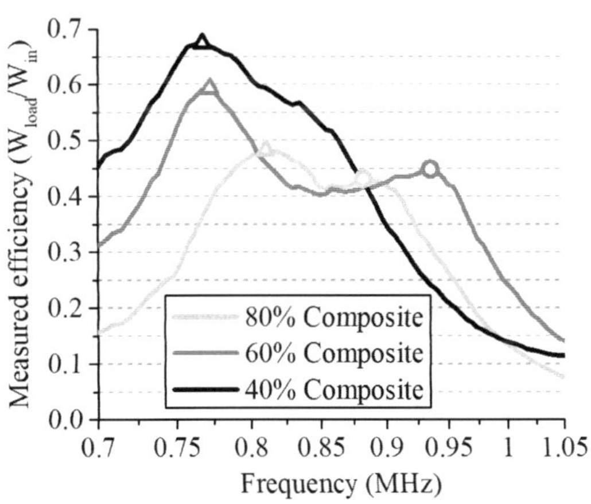

Figure 3. Measured efficiency of the test-transducers with ohmic loads. Triangles mark the efficiency at the resonance and the circles at the anti-resonance. The 40% receiver showed no anti-resonance due to the leakage of water-vapour during testing.

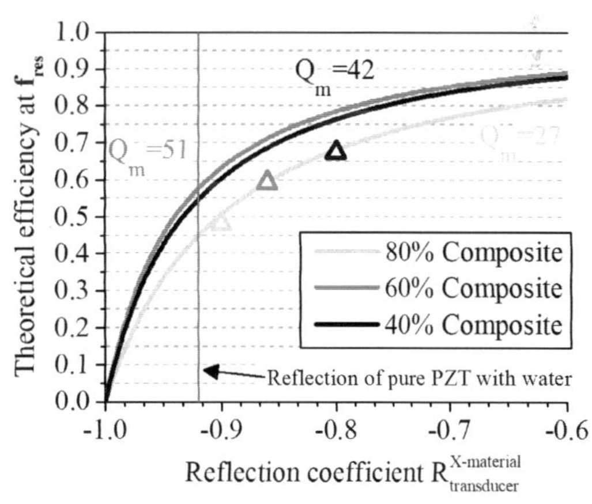

Figure 4. Predicted efficiency (lines) at the resonance of the tested transducers with the optimal loads attached as a function of the front-material and transducer reflection coefficient from the model in [3, 4]. Triangles: measured efficiency in water.

coefficient range, at which the efficiency shows high values ($\eta \geq 0.7$).

In the figure, we have also plotted the reflection coefficient (gray line; R=0.92) of pure PZT with water, which initially suggests that pure PZT transducers would achieve a lower efficiency than the composites with a lower acoustic impedance mismatch. However, we must stress that their mechanical quality factor can be higher than that of our test-composites ranging from $Q_m \approx 80$ for soft to $Q_m \approx 2000$ for hard PZT materials [9]. A higher Q_m will lead to higher efficiencies as we can conclude by comparing the predicted efficiency curve of the 80% composite to the other two transducers.

Finally, we have seen a mismatch between the measured and the predicted efficiency of the 40% and 60% composites, which we believe is due to the influence of the brass casing to which the composites are clamped. Furthermore, the power transfer is also very sensitive to the positioning of the receiver inside the narrow central beam of the acoustic field as we already saw in [8], which can also reduce the measured efficiency if there is a slight misalignment.

4. Conclusions

In this paper, we have investigated the influence of the filling fraction of 1-3 composite acoustic power receivers on the maximum efficiency. In our acoustic power transfer experiments, the 40% composite receiver performed best due to its low reflection coefficient with water, despite having less PZT content and higher dielectric losses δ_{tn} than the 60% and 80% receivers, and a lower Q_m and electromechanical coupling k_t than the 60% composite. Finally, we conclude that 1-3 composite transducers with a low acoustic impedance mismatch with the transmission medium and despite their low Q_m are very capable acoustic power receivers due to their high efficiency ($\eta \geq 70\%$).

Acknowledgments

We gratefully acknowledge the financial support from the *Vector Stiftung* for the project *Energiebertragung mittels Ultraschall* and the BrainLinks-BrainTools Cluster of Excellence funded by the German Research Foundation (DFG, grant no. EXC 1086). Furthermore, we also like to thank *Smart Materials Gmbh* for providing us 1-3 composite transducer samples.

References

[1] Basaeri H, Christensen D B and Roundy S 2016 A review of acoustic power transfer for bio-medical implants *Smart Mater. Struct.* **25** 123001

[2] Kunkel H A, Locke S and Pikeroen B 1990 Finite-element analysis of vibrational modes in piezoelectric ceramic disks *IEEE Trans. Ultrason. Ferroelectr. Freq. Control* **37** 31628

[3] Gorostiaga M, Wapler M C and Wallrabe U 2017 Analytic model for ultrasound energy receivers and their optimal electric loads *Smart Mater. Struct.* **26** 085003

[4] Gorostiaga M, Wapler M C and Wallrabe U 2017 Corrigendum: Analytic Model for Ultrasound Energy Receivers and their Optimal Electric Loads (2017 Smart Mater. Struct. 26 085003) *Smart Mater. Struct.* https://doi.org/10.1088/1361-665X/aa8b24

[5] Safari A, Janas V F and Bandyopadhyay A 1997 Development of fine-scale piezoelectric composites for transducers *AIChE J.* **43** 28492856

[6] Gorostiaga M, Wapler M C and Wallrabe U 2017 On the Optimal Electric Loads for Ultrasound Energy Receivers IEEE ISAF-IWATMD-PFM 2017

[7] Smith W A and Auld B A 1991 Modeling 1-3 composite piezoelectrics: thickness-mode oscillations *Ultrason. Ferroelectr. Freq. Control IEEE Trans. On* **38** 4047

[8] Gorostiaga M, Wapler M C and Wallrabe U 2017 Analytic model for ultrasound energy receivers and their optimal electric loads II: Experimental validation *Smart Mater. Struct.* **26** 105021

[9] PI Ceramic GmbH *https://www.piceramic.com/en/*

Pull-in actuation in hybrid micro-machined contactless suspension

K V Poletkin, R Shalati, J G Korvink, V Badilita

Spin&Photon Applications (SPA-) Lab, Institute of Microstructure Technology, Karlsruhe Institute of Technology, Eggenstein-Leopoldshafen D-76344, Germany

E-mail: kirill.poletkin@kit.edu, jan.korvink@kit.edu, vlad.badilita@kit.edu

Abstract. This paper presents a result of study of the pull-in phenomenon in the hybrid micro-machined contactless suspension (µ-HCS), combining inductive suspension and electrostatic actuation, reported at PowerMEMS 2015 [1]. Assuming the quasi-static behavior of a levitated proof mass, a non-linear analytical model describing the pull-in actuation along the vertical direction is developed. The developed model allows us to predict the static pull-in parameters of the suspension and to show a dependence of these parameters on suspension design. It is shown that the pull-in displacement can be larger by almost a factor of two than one occurring in a spring-mass system with constant stiffness (classic pull-in). The model is verified by using numerical estimations as well as experimental data and agrees well with measurements and calculations.

1. Introduction

Micro-machined contactless suspensions (µ-CS), employing the phenomena of electromagnetic levitation, eliminate mechanical attachments between stationary and moving parts in Micro-Electro-Mechanical Systems (MEMS) and as a result provide the solution of fundamental issue in MEMS related to the domination of friction over inertial forces in the micro-world. As a result, a new generation of micro-sensors and actuators based on levitation have been demonstrated.

Depending on a source of force field, µ-CS can be simply classified as electrostatic, magnetic, and hybrid. For instance, electrostatic suspensions (µ-ECS) were successfully used in micro-inertial sensors [2]. Magnetic suspensions (µ-MCS) can be also further classified as inductive, diamagnetic and superconducting suspensions, which found applications in micro-bearings [3, 4], micro-inertial sensors [5, 6], bistable switches [7] and nano-force sensors [8]. Hybrid suspensions (µ-HCS) combine different force fields, which make the main difference of µ-HCS from both µ-ECS and -MCS.

In particular, capabilities of μ-HCS were demonstrated in applications as micro-motors [9, 10] and micro-accelerators [11]. A wide range of different operation modes such as the linear and angular positioning, bistable linear and angular actuations and the adjustment of stiffness components of μ-HCS were demonstrated and experimentally studied in the prototypes reported in [1, 12, 13]. Thus, μ-HCS establish a promising direction for further improvements in performance and operation capabilities of micro-sensors and -actuators.

In this work, the pull-in phenomenon in μ-HCS shown in Fig. 1 is analytically and numerically studied. In order to model micro-machined inductive CSs, the qualitative technique developed in [14], where the induced eddy current within a levitated micro-object is approximated by a magnetic dipole, is used. Note that this technique has been recently generalized in [15, 16], where the induced eddy current is approximated by a system of dipoles. Then, a reduced analytical

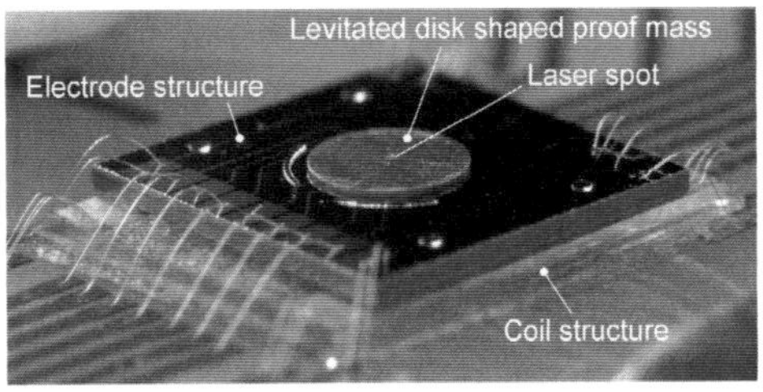

Figure 1. Fully assembled prototype of hybrid suspension under testing (the detailed description of the device can be found in [1, 12]).

model of μ-HCS, which describes the behaviour of levitated proof mass in a vertical direction, is derived. Using the obtained analytical model, the static pull-in behavior of μ-HCS is studied and verified by experimental result published in [1, 12].

2. A reduced model

In order to develop a model of μ-HCS shown in Fig. 1, let us consider its scheme shown in Fig. 2. It consists of stabilization and levitation coils to provide the stable levitation of a dick shaped proof mass (PM) and the embedded electrodes as shown in Fig. 2. Changing the equilibrium position of PM by means of electrostatic force acting on its bottom surface and keeping the same current in the coils the dynamics of μ-HCS is adjusted. Also, applying the pull-in voltage to the electrodes the bi-stable actuation is performed [1, 12].

Assigning an origin to the equilibrium point O and assuming that the resistivity of conducting PM, and its linear and angular velocities are small, then the exact quasi-static nonlinear model, which describes the behavior of PM along the vertical y-axis, is [16, 17]

$$m\frac{d^2y}{dt^2} + mg + \frac{I^2}{L}\frac{dM}{dy}M + \frac{A}{4}\frac{U^2}{(h+y)^2} = F, \tag{1}$$

where m is the mass of levitated PM, g is the gravity acceleration, I is the amplitude of a harmonic current i of coils, L is the self inductance of the PM, M is the mutual inductance between the PM and coils, U is the applied voltage to the electrodes both of each which has the same area of A_e, $A = \varepsilon_0 A_e$, ε_0 is the permeability of free space, F is the force acting on the PM along the y-axis.

In a general case, the mutual inductance, M, is a complex nonanalytic function. This fact becomes the main difficulty for analytical study of suspension model (1). However, accounting for particularities of micro-machined performance of device [15] such that the linear sizes of coils and PM is much larger than the levitation height, h_l, and the distribution of density of induced eddy current is not homogenous. Due to these particularities of the device, the force interaction along vertical direction is reduced to interaction between eddy current, i_{el} and levitation coil current [18]. Considering both the levitation coil and the eddy current circuit as filamentary circles, the mutual inductance between the levitation coil and eddy current can be described by the Maxwell formula.

Moreover, upon holding a certain condition described in [13], the Maxwell formula can be well approximated by the logarithmical function [14]. Hence, accounting for (1), the following

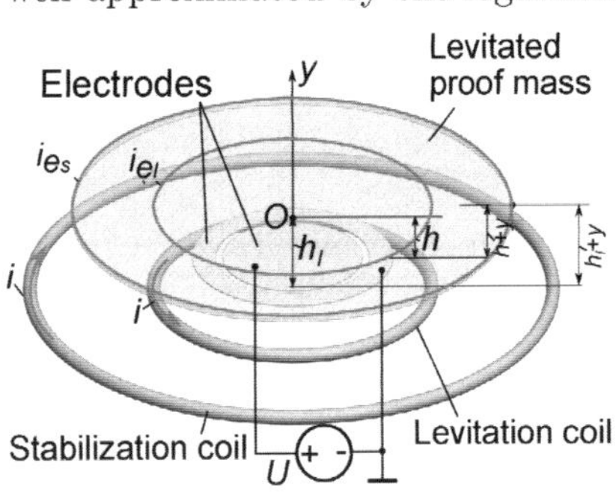

Figure 2. Scheme for modelling hybrid contactless suspension: U is the potential applied to electrodes, h is the space between an electrodes plane and equilibrium point of proof mass, h_l is the levitation height between a plane of coils and equilibrium point of proof mass, i_{el} and i_{es} are the eddy currents corresponding to the maximum current density [16].

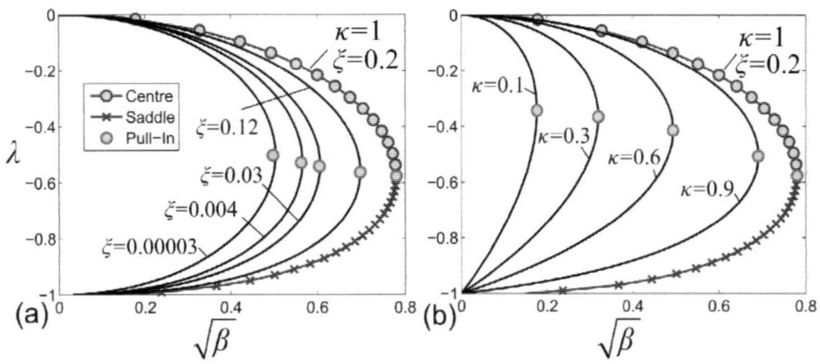

Figure 3. Bifurcation diagrams and their evolution in depending on the suspension design parameters: a) the effect of ξ changing in a range from 3×10^{-5} to 0.2; b) the effect of κ in a range from 0.1 to 1.0 (centre and saddle are corresponding to stable and unstable equilibrium, respectively).

reduced analytical model of suspension in the dimensionless form can be proposed as follows

$$\frac{d^2\lambda}{d\tau^2} + 1 - \eta\frac{\kappa}{1+\kappa\lambda}\left[\ln\frac{4}{\xi(1+\kappa\lambda)} - 2\right] + \frac{\beta}{(1+\lambda)^2} = \tilde{F}, \tag{2}$$

where $\eta = I^2a^2/(mghL)$, $a = r_l\mu_0$, μ_0 is the magnetic permeability of free space, r_l is the radius of levitation coil, $\xi = h_l/(2r_l)$, $\tau = \sqrt{g/h}t$, $\lambda = y/h$, $\beta = AU^2/(4mgh^2)$, $\kappa = h/h_l$ and $\tilde{F} = F/mg$. As it is shown in [14], the accuracy of approximation of modelling the electromagnetic force is dependent on the parameter ξ. If the parameter ξ is less than 0.3, the electromagnetic force is approximated by the logarithmic function with the error less than six percentages. Upon trending the parameter ξ to the zero, the error between the exact equation and approximation as well trends to the zero. Worth noting that in all known prototypes of μ-HCS published in the literature the parameter ξ is less than 0.25. This fact provides the applicability of the reduced model for further analytical study of μ-HCS.

3. Static pull-in

At the equilibrium point ($\lambda = \beta = 0$), η must be equal to $D = \ln 4/\xi - 2$. Hence, the equilibrium state is defined by the following equation

$$f(\lambda,\beta) \equiv -\frac{\kappa\lambda}{1+\kappa\lambda} - \frac{\ln(1+\kappa\lambda)}{D(1+\kappa\lambda)} - \frac{\beta}{(1+\lambda)^2} = 0. \tag{3}$$

Using (3), the bifurcation diagram can be mapped as shown in Fig. 3, which depicts the distribution of centre and saddle as well as pull-in points depending on the design parameters. The pull-in points correspond to the transient state, in which the sign of $f(\lambda,\beta)$ is unchangeable in the vertical direction [19]. For a case of $\kappa = 1$, pull-in has the following parameters: displacement is $\lambda_{pi} = (1 - e - D)/(2D + e)$, the square of voltage is $\beta_{pi} = -(\lambda_{pi} + \ln(1 + \lambda_{pi})/D)(1 + \lambda_{pi})$. For a case, when κ is small ($\kappa \ll 1$), pull-in parameters can be approximated as $\lambda_{pi} \approx -1/(3-2\kappa)$ and $\beta_{pi} \approx \kappa(1+1/D)(2-2\kappa)^2/(3-2\kappa)^3$. Once κ tends to zero, the pull-in displacement becomes the same as in classic static pull-in occurring in the spring-mass system with electrostatic actuation [20] and corresponds to $\lambda_{pi} = 1/3$. However, the square of pull-in voltage is different from the classic static pull-in and becomes $\beta_{pi} = \kappa(1 + 1/D)4/27$.

Fig. 4 provides comparison of results of experimental measurements of static pull-in behavior along vertical direction reported in [1] and modelling. Table 1 shows the parameters of the device

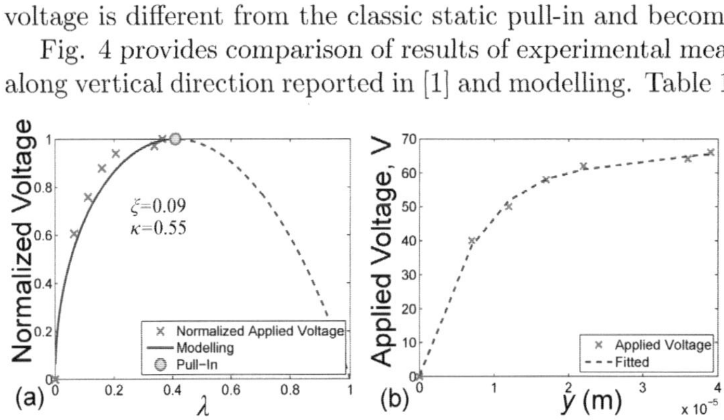

Figure 4. Static pull-in behavior for experiment reported in [1]: a) comparison of modelling and experiment (parameters of device are shown in Table 1; b) measurement of applied voltage vs the proof mass displacement [1].

Table 1. Results of measurements and modelling of the static pull-in displacement.

	Parameters	Device [1]	Device [12]
Measured	Levitation height, h_l	180 μm	200 μm
	Spacing, h	100 μm	120 μm
	Diameter of levitation coil, d_l	2 mm	2 mm
	Pull-In displacement	**39** μm	**45** μm
Modelled	ξ	0.09	0.1
	κ	0.55	0.6
	Pull-In displacement	**41** μm	**49** μm

and sums up results of study of the static pull-in displacements taking from Fig. 4. In addition, Tab. 1 provides results for μ-HCS reported in [12]. As it is seen from analysis of Fig. 4 and Tab. 1 the modelling agrees well with experimental data.

4. Conclusion

In this work the static pull-in behavior of μ-HCSs have been studied. Pull-in parameters were defined analytically based on the developed model. Results of modelling were verified by experimental data. This fact provides the applicability of the reduced model for further analytical study of pull-in phenomenon in μ-HCS. Moreover, we showed that the pull-in displacement in μ-HCS is larger in comparing with classic static pull-in displacement. In particular, for a μ-HCS having design parameters such as $\kappa = 1$ and $\xi = 0.2$ the pull-in displacement can be larger by almost a factor of two in compared to classic one.

Acknowledgments

KVP acknowledges with thanks the support from German Research Foundation (Grant KO 1883/26-1). JGK acknowledges support from the European Research Council (ERC) under contract 290586.

References

[1] Poletkin K V, Lu Z, Wallrabe U and Badilita V 2015 *J. Phys.: Conf. Series* **660** 012005
[2] Murakoshi T, Endo Y, Fukatsu K, Nakamura S and Esashi M 2003 *Japan. J. Appl. Phys* **42** 2468–72
[3] Lu Z, Poletkin K V, den Hartogh B, Wallrabe U and Badilita V 2014 *Sens. Actuator A-Phys.* **220** 134–143
[4] Poletkin K V, Moazenzadeh A, G Mariappan S, Lu Z, Wallrabe U, Korvink J G and Badilita V 2016 *IEEE Magn. Lett.* **7** 1–3 URL `http://dx.doi.org/10.1109/LMAG.2016.2612181`
[5] Zhang W, Chen W, Zhao X, Wu X, Liu W, Huang X and Shao S 2006 *Sens. Actuator A-Phys.* **132** 651–57
[6] Garmire D, Choo H, Kant R, Govindjee S, Sequin C, Muller R and Demmel J 2007 *Transducers 2007 - 2007 Int. Solid-State Sensors, Actuators and Microsystems Conf.* (IEEE) pp 1203–06
[7] Dieppedale C, Desloges B, Rostaing H, Delamare J, Cugat O and Meunier-Carus J 2004 *Proc. IEEE Sensors* vol 1 pp 493–96
[8] Abadie J, Piat E, Oster S and Boukallel M 2012 *Sensor. Actuat. A-Phys.* **173** 227–37
[9] Liu W, Chen W Y, Zhang W P, Huang X G and Zhang Z R 2008 *Electron. Lett.* **44** 681–683
[10] Xu Y, Cui Q, KAN R, Bleuler H and Zhou J 2017 *IEEE-ASME Trans. Mechatron.* **PP** 1–1 ISSN 1083-4435
[11] Sari I and Kraft M 2015 *Sensor. Actuat. A-Phys.* **222** 15–23
[12] Poletkin K V, Lu Z, Wallrabe U and Badilita V 2015 *J. Microelectromech. Syst.* **24** 1248–50
[13] Poletkin K V 2017 *2017 19th Int. Conf. on Solid-State Sensors, Actuators and Microsystems (TRANSDUCERS)* pp 934–37 URL `https://doi.org/10.1109/TRANSDUCERS.2017.7994203`
[14] Poletkin K V, Chernomorsky A, Shearwood C and Wallrabe U 2014 *Int. J. Mech. Sci.* **82** 110–21
[15] Poletkin K V, Lu Z, Wallrabe U, Korvink J and Badilita V 2017 *Int. J. Mech. Sci.* **131-132** 753–66
[16] Poletkin K V, Lu Z, Wallrabe U, G Korvink J and Badilita V 2017 *Solid-State Sensors, Actuators and Microsystems (TRANSDUCERS), 2017 Transducers-2017 19th Int. Conf. on* (IEEE) pp 528–31
[17] Poletkin K V, Chernomorsky A I and Shearwood C 2012 *IEEE Sens. J.* **12** 2407–2413
[18] Lu Z, Poletkin K V, Wallrabe U and Badilita V 2014 *Micromachines* **5** 1469–1484
[19] Andronov A A, Vitt A A and Khaikin S E 2013 *Theory of Oscillators* vol 4 (Elsevier)
[20] Elata D and Bamberger H 2006 *J. Microelectromech. Syst.* **15** 131–140

STUDY OF DYNAMIC RESPONSE OF SILICONE ELASTOMER MICROFABRICATED HYBRID MEMBRANES VERSUS TEMPERATURES AND AGING TIME

A. Diallo, R. Chutani, M. Barthès, S. Bégot, S. Khadraoui, M. De Labachelerie and F. Lanzetta

FEMTO-ST institute, Univ. Bourgogne Franche-Comté, CNRS, France.

E-mail: dassimou.diallo@gmail.com

Abstract. This study concerns the dynamic behaviour of microfabricated Room-Temperature-Vulcanizing (RTV) silicone elastomers based Hybrid Membranes (HMs) when subjected to different operating temperatures and after aging. The micro fabricated HMs structures aimed at being implemented in micromachines, such as for instance a micro-Stirling engine. The hybrid membranes consist of a silicon spiral structure embedded with a thin layer of an elastomeric RTV-silicone. The different process temperatures are consistent with clean room process, such as anodic bonding. In addition to an experimental study, Finite Element Model (FEM) numerical simulations are presented. The simulations include the location of the stress concentration points during a solicitation and membrane dynamic frequency, which are consistent with the experimental test results.

1. Introduction

When miniaturizing, sliding piston are replaced by membranes in micro engines to avoid friction losses and fluid leakage through the cylinder–piston gap, designated as one of the main limitations [1, 2]. However, the stability of their natural frequency is a critical aspect for optimization. A design of a thick fluidic diaphragm based on silicone Hybrid Membranes (HMs) has been proposed in [3]. Knowing that the temperature has an influence on the mechanical properties of silicone membranes [4], one of main issues of these membranes could be their dynamic behavior under severe temperature conditions, such as during wafers assembly by high temperature anodic bonding process or during the operation of a thermal microengine. The applied temperature during anodic bonding is usually in the range [200-500] °C [5]. In the field of microtechnologies, it is particularly used for sealing silicon (Si) and glass wafers, leading to an irreversible bonding. In this paper, we propose a study of the dynamic behavior of those microfabricated silicone-based HMs when subjected to different operating temperatures.

2. Membrane design

The HMs geometry is based on the work of Chutani *et al.* [6]. They demonstrated the microfabrication of thick hybrid fluidic membranes with a low resonance frequency consisting on a planar Si spiral with a central disc (cf. Figure 1), embedded in a Room Temperature Vulcanizing elastomeric silicone (RTV-silicone). The RTV-silicone is a single component that has good adhesion to Si and cross-links when exposed to air moisture to form a tough, flexible, silicone rubber [7]. The planar Si spiral spring with its central disk (embedded in the RTV-silicone) increases the stiffness of the membrane, and allowed a controlled deformed shape to attain a larger swept volume for microengines.

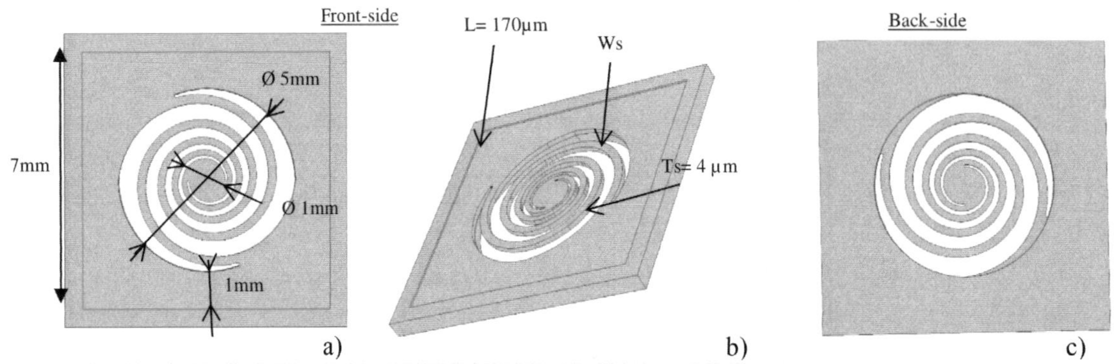

Figure 1. Silicon spiral spring architecture and geometrical parameters; L is the location of the RTV-silicone layer deposited by a squeegee, Ws is the width of the turns of the upper part of the spiral.

The logarithmic spirals geometry (defining the planar springs) is shown in Figure 1. On this figure, "Ts" is the chosen thickness of the spiral, "L" is the thickness of the RTV-silicone layer above the spiral and "Ws" is the width of the spiral. To aim at uniform stress distribution along the structure, their width (Ws) varies proportionally with the spiral angle [5].

3. Microfabrication

The microfabrication process flow for HMs is described in figure 2a. The SEM and optical images of the fabricated components are shown in figure 2b and 2c respectively. Embedded RTV-silicone allows here hermetical membrane to be obtained and a drastic improvement of the Si spring robustness [6]. A Si spring with a 1mm diameter central disc was used for both experimental tests and FEM modelling.

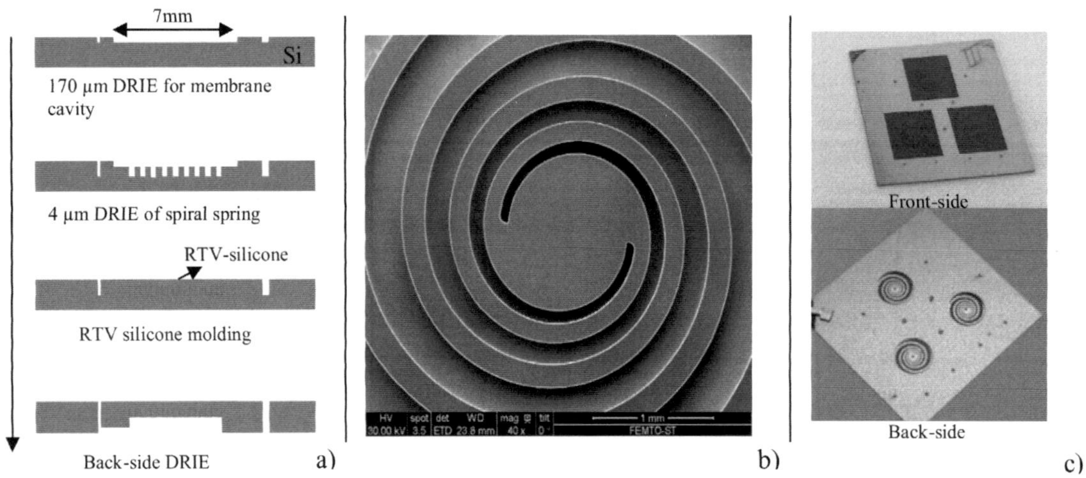

Figure 2. Microfabrication a) HMs microfabrication process flow b) SEM images of the silicon spiral c) Optical images of front and back-side of HMs sample.

4. FEM modelling

A COMSOL Multiphysics mechanical-structural model of the HMs was simulated, including the planar spring alone (cf. figure 3a) and with RTV-silicone (cf. figure 3b) at room temperature. Geometrical parameters and properties of the RTV-silicone used for the FEM model are given in table 1. The 3D FEM model can be used to study the effects of HM thickness, and/or the effect of the type of elastomer, while designing the spiral spring to aim at the desired stiffness value or frequency. The geometric parameters of the structure can be easily adjusted to provide inertial effects by increasing the thickness and / or width of the central disc. This adjustment allows an increase in the mass of the HM.

Table 1. RTV-silicone parameters. D_{cd} is the diameter of the central disc; h_{HM} and h_{sp} are respectively the thickness of the HM and the spiral; E is the Young modulus; ν is the Poisson ratio; ρ_m is the density

	Geometry			Material		
RTV-Silicone layer	h_{HM} = 174 µm	D_{cd} = 1 mm	h_{sp}= 4 µm	E = 1.5 MPa	ν= 0.49	ρ_m = 1100 kg/m^3

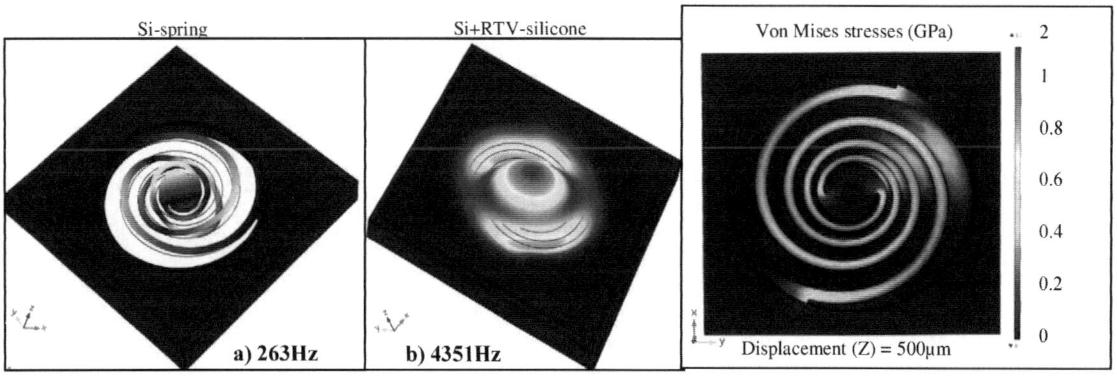

Figure 3. FEM model: a) Resonance frequency (conic deflection, 6th mode) of the Si spiral. b) Resonance frequency of the Si spiral and RTV-silicone assembly. c) Location of the Von Mises stresses concentration points during a 500 µm prescribed displacement on the HM.

According to the FEM, the sixth mode of deflection corresponds to the maximum displacement of the membrane, therefore we considered only this mode. FEM model of the Si spring alone shows a frequency of 263 Hz (figure 3a) whereas the HMs exhibits a resonance of 4351 Hz (figure 3b). Figure 3c shows the zones of strong mechanical constraints and a maximal displacement of the central disc of 500 µm while remaining within the yield strength of the two materials [8]. The zones of strong mechanical constraints were maximal at the salient points of the spring.

5. Characterization, results and discussions

A Polytec laser vibration sensor was used to evaluate HMs frequencies (figure 4) when the structure was submitted to harmonic excitation. These dynamic characterizations were performed for six different membranes, denoted m1 to m6. At room temperature, the HMs presented a 6th harmonic frequency around an average of 4673 ± 120 Hz. The Figure 4a presents the comparison between experimental and numerical results. A discrepancy of 7% is observed between the FEM model frequency when compared to the experimental one: the model is thus consistent with experimental results.

The samples characterized at room temperature were then subjected to given temperatures (100, 200 or 300°C) during given time periods consistent with those of an anodic bonding total process time. Further, membranes were also submitted to aging at 200°C. Results of both experiments are given on the figure 4b. After heating, frequencies of membranes were measured at room temperature. For each membrane, we present the difference between the mean value of the initial frequencies before temperature treatment, and the ones found after being submitted to given operating conditions. Taking into account the reproducibility of the membranes natural frequencies, no significant change was found after heating at 100, 200 or 300°C: therefor the cleanroom process does not damage or modify significantly the membranes properties. Moreover, the resonance frequency of HMs remained almost the same after 113h, 168h, 337h and 503h of aging at 200°C. The HM mechanical properties does not seem to change very much after high temperature exposure. The HMs should thus be suitable for applications at rather high temperature conditions.

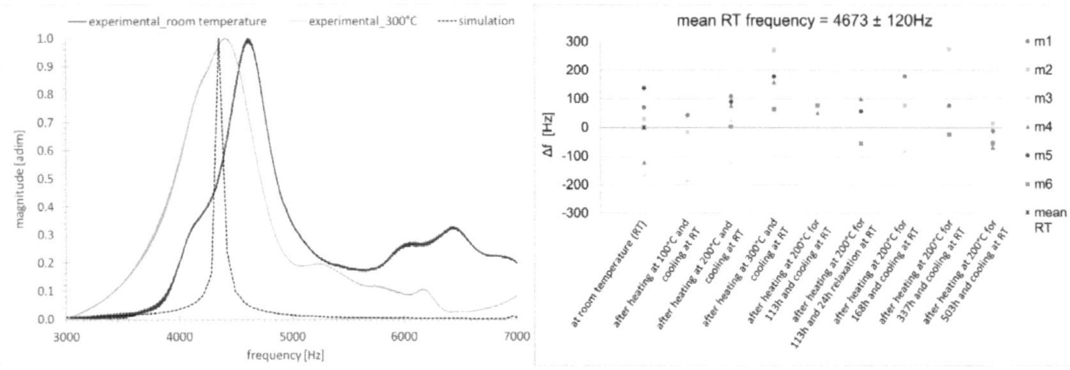

a) b)

Figure 4. a) Experimental (at room temperature and after heating at 300°C) and numerical resonance frequencies, b) Variations of Δf versus operating conditions for each HM (mi, with $1 \leq i \leq 6$). With Δf = [mean RT (room temperature) frequency – measured HM frequency].

6. Conclusion

We have modelled, and experimentally demonstrated the stability of hybrid membranes resonance frequency (6th mode), which are thus suitable for micromachines, especially thermal microengines. The numerical simulation made it possible to highlight the zones of strong mechanical constraints and the expected piston-mode resonance frequency. Numerical results were in agreement with the experimental test results. Experimental results showed that high temperature processes, like anodic bonding, should therefore not change the mechanical properties of the hybrid membranes.

Acknowledgments

This work has been supported by the Labex ACTION project (contract "ANR-11-LABX-0001-01"), the Region Bourgogne Franche-Comté STIP'MEMS project and the French National Research Agency (ANR) under grant no. ANR-12-SEED-0005-01 (MISTIC Project). It has been partly supported by the French RENATECH network and its FEMTO-ST technological facility.

References

[1] S. K. Chou, W. M. Yang, K. J. Chua, J. Li, and K. L. Zhang, "Development of micro power generators – A review," *Appl. Energy*, vol. 88, no. 1, pp. 1–16, Jan. 2011.

[2] H. K. Bardaweel, M. J. Anderson, R. F. Richards, and C. D. Richards, "Optimization of the dynamic and thermal performance of a resonant micro heat engine," *J. Micromechanics Microengineering*, vol. 18, no. 10, p. 104014, 2008.

[3] F. Formosa, A. Badel, and H. Favrelière, "Development of low frequency, insulating thick diaphragms for power MEMS applications," *Sensors Actuators A Phys.*, vol. 189, pp. 370–379, Jan. 2013.

[4] T. Rey, G. Chagnon, J-B. Le Cam and D. Favier, Polymer Testing, 32 (2013), pp. 492-501.

[5] M. Argoud, "Mécanismes de collage et de transfert de films monocristallins dans des structures à couches de polymères," *LETI du CEA Grenoble* (2012), pp. 24-25

[6] R. Chutani, F. Formosa, M. de Labachelerie, A. Badel and F. Lanzetta, "Microfabrication of hybrid fluid membrane for microengines," *J. Micromechanics Microengineering*, vol. 26, 124009, 2016.

[7] Permatex® High-Temp Red RTV-Silicone; http://www.permatex.com

[8] K. E. Petersen, "Silicon as a mechanical material," *Proceeding of the IEEE 1982*, vol. 70, pp. 420-457.

IOP Conf. Series: Journal of Physics: Conf. Series **1052** (2018) 012037 doi:10.1088/1742-6596/1052/1/012037

High flexible piezoelectric PZT thin films deposited on stainless steel mesh

T Nishi, T Ito, T Umegaki and I Kanno

Department of Mechanical Engineering, Kobe University, 1-1 Rokkodai-cho, Nada, Kobe, Japan

E-mail: kanno@mech.kobe-u.ac.jp

Abstract. This paper describes high flexible piezoelectric vibration energy harvesters (PVEHs) composed of PZT thin films on stainless steel (SS) mesh. The PZT thin films were deposited on Pt/Ti-coated SS mesh substrates which were plain-woven with the SS wire diameter of 15 μm and 30 μm, respectively. Relative dielectric constant of the PZT thin films on the SS mesh of φ15 and φ30 was measured to be 475, 263, respectively. In φ15 device, the remnant polarization and the coercive field was 11.9 $\mu C/cm^2$ and 48.7 kV/cm, respectively. For the measurement of the piezoelectric response, we attached conductive rubber sheets to the rectangular top electrode. The maximum output voltage of about 30 mV was measured in each bending motion. We confirmed the capability of the PZT thin films on SS mesh for the PVEH applications.

1. Introduction

In recent years, wearable devices have attracted considerable attention as the next generation portable electronics, however, power supply to those devices are one of the critical issues for the practical applications. Piezoelectric vibration energy harvesters (PVEHs) have been thought to be the key technologies as the power sources to practical wearable products [1,2]. One of the most distinct features of the mechanical human motion is the large displacements [3]. To utilize such mechanical energy effectively, the required specification of PVEHs for wearable devices are not only high-efficient power generation but also reliability, light weight and flexibility. We confirmed the reliability of PVEH using stainless substrates [4], for that reason it is useful to design stainless substrate structures which reduce its rigidity.

In this study, we tried to prepare piezoelectric PZT thin films on flexible metal texture with sufficient flexibility. The PZT thin films were deposited on mesh substrates composed of small SS fibers. We evaluated their electric properties including piezoelectric response by bending motion and examined the capability for the wearable PVEH applications.

2. Experiments

The PZT thin films with a composition of Zr/Ti = 53/47 were deposited on the Pt/Ti-coated SS mesh by rf-magnetron sputtering. The mesh substrates used were plain-woven meshes with the SS wire diameter of 15 μm (φ15) and 30 μm (φ30), respectively. The number of φ15 and φ30 wires in mesh is 640 and 300 per inch, respectively. They were heated up at around 600 °C during the deposition. Prior to the PZT deposition, (Pb,La)TiO_3 (PLT) seed layers were deposited at the interface of PZT and Pt/Ti/SS mesh to improve the crystal growth of the PZT [5]. The thickness of PZT thin films was about 2.2 μm.

After the PZT deposition, post annealing was conducted to complete the crystallization of the PZT thin films to perovskite phase. After the PZT deposition, Pt top electrode was prepared through a shadow mask and evaluated their dielectric, ferroelectric and piezoelectric properties. Fig. 1 shows the schematic illustration and an optical image of PZT thin films deposited on SS mesh substrates.

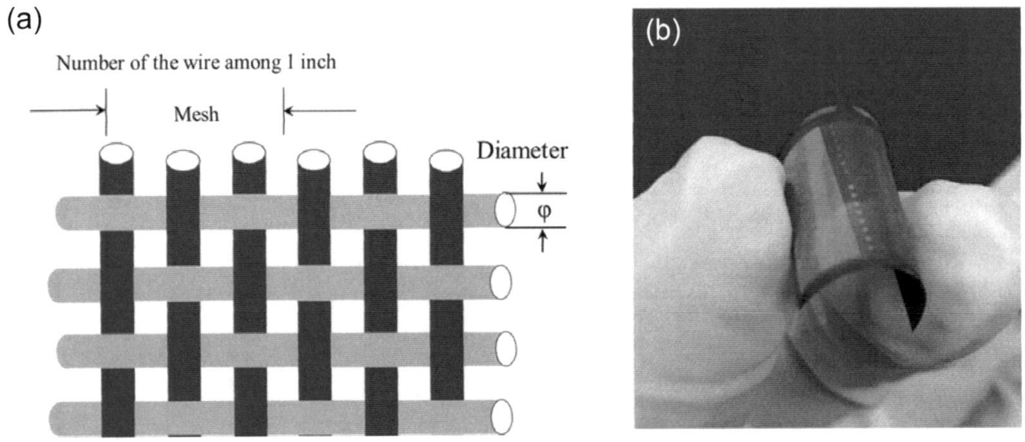

Figure 1. (a) Schematic illustration of mesh substrate, and (b) an optical image of PZT thin films on stainless mesh substrates. Large rectangular Pt electrode and circular electrodes of 0.3 mm and 0.5 mm in diameter were deposited on PZT thin films

3. Results and discussion

3.1 PZT layer on SS mesh

Crystal structure of the PZT thin films on mesh substrates was examined by x-ray diffraction (XRD) measurements. Fig. 2 shows the XRD patterns of the PZT thin films on φ15 and φ30 mesh substrates, respectively. The measurements indicated that the pyrochlore-free polycrystalline PZT layers were successfully grown on the SS mesh substrates. Fig. 3 shows the surface and cross-sectional SEM micrographs of the PZT thin films on SS mesh substrates. We confirmed that the surface of SS wires in meshes were coated by the dense PZT layers of about 2.2 μm in thickness without any cracks or pin-holes.

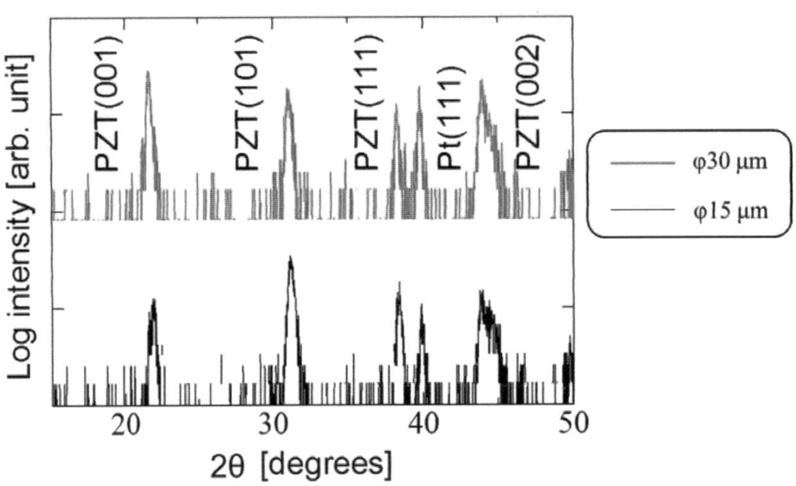

Figure 2. XRD patterns of PZT thin films on φ15 and φ30 mesh substrates.

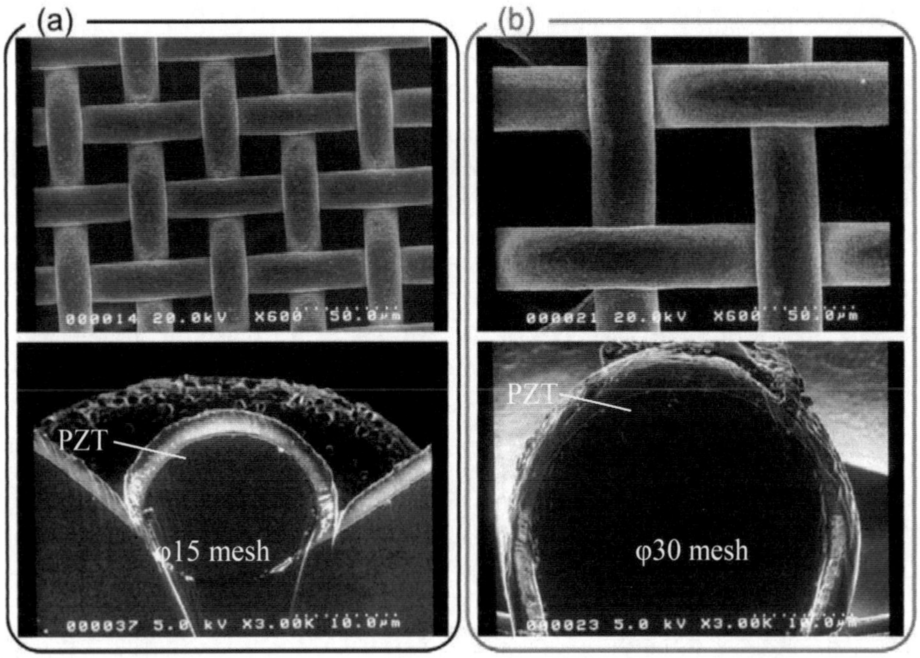

Figure 3. Surface and cross-sectional SEM images PZT layers (a) φ15 mesh and (b) φ30 mesh.

3.2 Dielectric and ferroelectric properties

Relative dielectric constant of the PZT layers on the SS mesh of φ15 and φ30 was calculated from the capacitance of PZT layers coated by Pt dot electrodes of 0.3 mm. In the measurements, we took into account of the wire density of each mesh substrate, while the surface area was assumed to be flat plane neglecting round surface net area of the wires. The apparent relative dielectric constant of the PZT layers on φ15 and φ30 meshes were measured to be 475 and 263, respectively. The difference of the electric properties is attributed to the difference between apparent and net electrode areas. Dielectric loss of the PZT layers on the SS mesh of φ15 and φ30 was measured to be 0.33 and 0.20, respectively. Although the value was slightly higher than that of PZT thin films on the flat SS substrates, we could not observe apparent short circuit between top and bottom electrodes [6, 7].

Fig. 4 shows the *P–E* hysteresis loops for PZT layers on φ15 and φ30 SS mesh. In the φ15 PZT device, the remnant polarization and coercive electric field were 11.9 μC/cm^2 and 48.7 kV/cm, respectively. Although the ferroelectric properties of the φ30 PZT device were almost same as that of the φ15 PZT device, the remnant polarization of φ30 SS mesh was smaller than that of φ15 SS mesh because of the difference net electrode area.

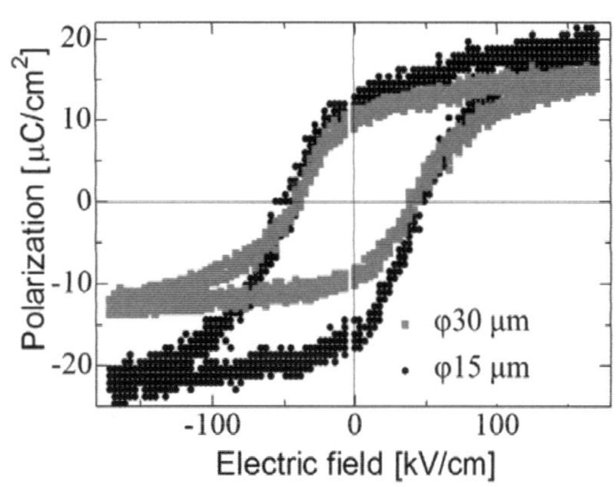

Figure 4. P-E hysteresis loops of stainless steel mesh.

3.3 Power generation

Power generation performance was measured by bending PZT layers on SS meshes. In order to extract the generated power on the rectangular Pt top electrodes, conductive rubber sheet was bonded on it as shown in

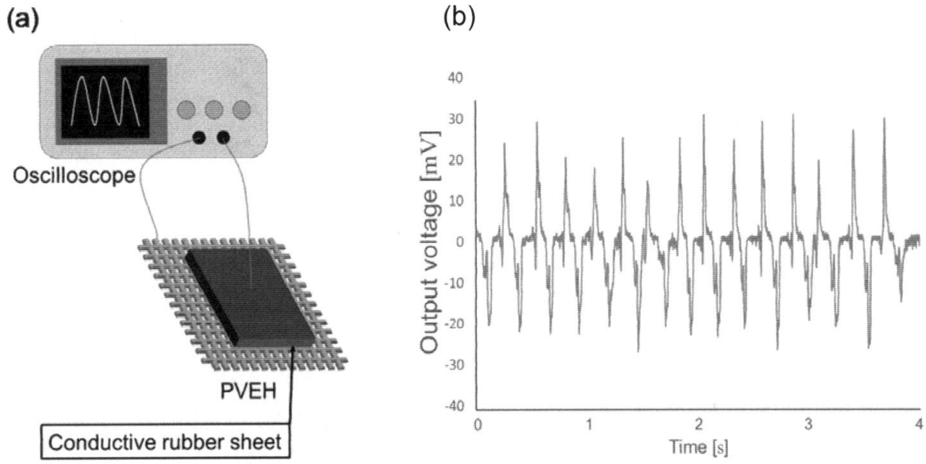

Figure 5. (a) The schematic illustrations of measurement set up and (b) the output voltage of φ30 stainless steel mesh.

Fig. 5 (a). The output voltage of the mesh structure PVEHs in φ30 device are shown in Fig. 5 (b). The output voltage of approximately 30 mV was measured by each bending motion. It was lower than expected because of relatively high resistance (about 20 Ω) of the conductive rubber sheets, however, we could measure stable power generation from the PZT layers on flexible SS mesh substrates.

4. Conclusions

We successfully deposited high flexible piezoelectric PZT thin films on SS mesh of 15 μm and 30 μm in diameter by rf-magnetron sputtering. The PZT layers on SS mesh substrates showed typical ferroelectric properties. By each bending motion, we confirmed stable output voltage of 30 mV from the PZT layers. The results suggests that the PZT thin films on the flexible SS mesh are promising power sources for wearable devices.

References

[1] Bowen, C R, Kim H A, Weaver P M and Dunn S 2014 *Energy Environ. Sci.* **7** 25-44
[2] Kanno I 2015 *J. Phys. Conf. Ser.* **660** 012001
[3] Jung W S, Lee M J, Kang M G, Moon H G, Yoon S J, Baek S H and Kang C Y 2015 *nano energy* **13** 174-181
[4] Tsujiura Y, Suwa E, Kurokawa F, Hida H and Kanno I 2014 *J. Phys. Conf. Ser.* **557** 012096
[5] Kanda K, Kanno I, Kotera H and Wasa K 2009 *J. Microelectromech. Syst.* **18** 610-615
[6] Bose A Sreemany M and Bysakh S 2013 *Appl. Surf. Sci.* **282** 202-210
[7] Suzuki T, Kanno I, Loverich J J, Kotera H and Wasa K 2006 *Sens. Actuators A Phys.* **125** 382-386

Acknowledgments
The authors acknowledge the financial support of Japan Science and Technology Agency (JST) A-STEP (AS272S002a) and CREST (JPMJCR16Q4).

IOP Conf. Series: Journal of Physics: Conf. Series **1052** (2018) 012038 doi:10.1088/1742-6596/1052/1/012038

A High Performance Piezoelectric Micro Energy Harvester Based on Stainless Steel Substrates.

W.H. Tang[1], T.K. Lin[1], C.T. Chen[1], Y.H. Fu[1], S.C. Lin[1], W.J. Wu[1*]

[1]Department of Engineering Science and Ocean Engineering Science, National Taiwan University, Taipei,
*Email: wjwu@ntu.edu.tw

Abstract. This paper presents a high performance piezoelectric micro energy harvester (PMEH) fabricated on stainless-steel substrate with metal MEMS process. The PMEHs fabricated in this study are with simple unimorph or bimorph cantilever structure with one or two layers of high quality lead zirconate titanate (PZT) piezoelectric films deposited by aerosol deposition method (ADM) and a glued tungsten proof mass (4 mm* 6 mm* 1 mm). The thickness of the PZT active layer are 18μm for unimorph PMEH and 10μm thick on both sides for bimorph PMEH. The length and width of the cantilever structure is 9mm and 6mm. With all the experimental process optimized, the results show that unimorph PMEH has a maximum output power of 122μW tested with optimal load under 0.5 g acceleration vibration level in resonant frequency around 120Hz. The corresponding value of bimorph PMEH is 304μW. The normalized power density (NPD) for unimorph PMEH and bimorph PMEH are $18.9 mW\cdot cm^{-3}\cdot g^{-2}$ and $46.81 mW\cdot m^{-3}\cdot g^{-2}$ respectively, which outperformed all previous published PMEHs based on either silicon or stainless steel substrates.

1. Introduction

The output performance of piezoelectric micro energy harvesters (PMEH) with area smaller than 1 cm^2 has great improvement over the past decade. The performance of the piezoelectric materials, the device structure design, the fabrication process, and the substrate materials are all crucial factors to optimized the output perfomance of PMEHs. Early studies of PMEHs are mostly based on silicon substrate with conventional MEMS processes[1-4], the chip area of these PMEHs are usually small and with higher resonant frequencies in around kHz range. Although the normalized power density (NPD) of silicon based PMEHs can be high, the devices are difficult to find real field applications because the mechanical vibration frequencies are usually in low frequencies. Furthermore, silicon and ceramic piezoelectric material are both brittle material which tends to break easily under strong vibration levels. There are growing studies of PMEHS based on metal substrates, especially stainless steel [5-8]. Metal substrate materials like stainless steel are ductile materials show not only much better mechanical strength over silicon but can be pre-stressed to avoid tensile cracks of ceramic piezoelectric materials. The metal MEMS processes of cantilever stucture are also simpler and cheaper because the metal substrates are usually chosen as the final beam thickness and buck micromachining backside etching for silicon is not required. The output power of metal substrate based PMEHs therefore showed better performance in lower frequency but the NPD are in general smaller than silicon based devices with high performance thin piezoelectric layers. In this study, high performance unimorph and bimorph PMEHs with around 10 μm and 20 μm high quality lead zirconate titanate (PZT) deposited by aero deposition method (ADM) on 60 μm stainless steel substrates are fabricated and tested. The device is design to have natural frequency around 120 Hz to harvest vibration from motors. The resonant frequency is much lower than typical silicon based devices but with the optimized ADM deposited PZT film, the PMEHs in this study shows even higher NPD than conventional silicon devices. The material characterization of the deposited PZT films, the device fabrication and characterization are all presented in this paper.

2. Material Characterization

Fig 1 shows x-ray diffraction (XRD) patterns of the PZT powder and the deposited films annealed with different temperature. The crystalline phase with different annealing temperature remains pure perovskite phase as the PZT powder. Although the deposited films annealed with higher temperatures shows stronger perovskite phase, the stainless-steel substrate will be heavily oxidized along with degraded mechanical strength, and the PZT films also shows higher dielectric losses. The annealing temperature for later device fabrication is chosen below 650°C in this study. Fig 2 shows the cross section SEM image of the PZT film deposited by ADM with thickness around 20μm. The PZT film deposited by ADM is quite dense and strong bonded to the stainless steel substrate. In order to analyze the ferroelectric characteristic of the PZT films, ferroelectric hysteresis loop is measured by ferroelectric analyzer (aixACCT TF analyzer 2000). Fig 3 shows the ferroelectric hysteresis loop on polarization-electrical field (P-E) plane of the PZT film deposited on stainless-steel substrate. The measured electrical field varies from 20 to 300V at 10Hz. The P-E loop shows good symmetry and the coercive field Ec = 5.41 V/μm and remnant polarization Pr = 6.08 $\mu C/cm^2$. To sum up, the series material characterizations indicate the PZT films deposited on stainless steel has excellent ferroelectric properties.

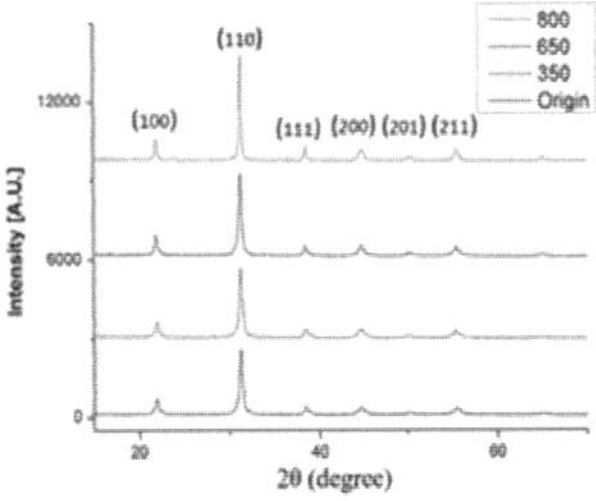

Fig 1. XRD scanning pattern of the PZT powder at different annealing temperature.

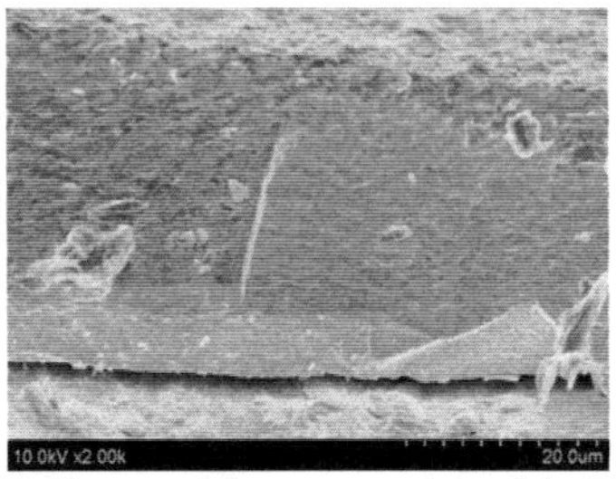

Fig 2. SEM image of the cross section of the PZT film.

3. Device Fabrication

The PMEH device fabrication process is based on metal MEMS process detailed in our previous report [9]. After the cantilever beam is released with wet etching process, a tungsten proof mass is glued to the tip of the cantilever beam with epoxy resin to lower the resonant frequency. The PZT layers are poled under high electric field and high temperature to activate piezoelectricity. Fig 4 shows the photo of the bimorph PMEH device. The dimension of the cantilever beam is 6mm x 8mm, and the total thickness of the bimorph PMEH is about 80μm composed of two 10μm of PZT films and 60μm stainless steel substrate. The fabrication process of the unimorph PMEH is similar to the bimorph PMEH. The difference between is all the deposited layers only need to be done in one side of stainless steel substrate.

4. Experimental Setup

The experimental setup of the bimorph or unimorph PMEH device characterization is described below. The PMEH is mounted on the shaker to simulate the base excitation from a vibration source. The shaker is driven by a sinusoidal wave with a function generator through a power amplifier. An accelerometer is mounted side by side with the energy harvester to measure the acceleration level. The acceleration level and corresponding open-circuit output signal are visualized on the oscilloscope.

5. Measurement Results

5.1. Unimorph piezoelectric micro energy harvester

Fig 5 shows the frequency response of output voltage (V_{p-p}) versus the excitation frequency under open circuit condition at 0.5g driven acceleration level. The device shows non-linearity and has different frequency response curve for frequency forward and reverse sweeping. The resonant frequency of forward sweep and reverse sweep slightly shifts from 111.3Hz to 110.5 Hz. The result shows that the peak voltages of forward sweep and reverse sweep are 21.4V_{p-p} and 25.4V_{p-p}. The output power and voltage is then tested with different resistive load at 0.5

g acceleration level and showed in Fig 6. The optimal load resistance of the unimorph PMEH is around 100kΩ, and corresponding maximum output power is 122μW.

5.2. Bimorph piezoelectric micro energy harvester

The bimorph PMEH is tested with the same setup and condition. Fig 7 shows the frequency response plots with forward and reverse sweeping. The resonant frequencies of forward sweep and reverse sweep are both 126.1 Hz which means then device shows almost no non-linearity. The peak voltage is 34.6V_{p-p}, which is much better to the unimorph device thanks to symmetric structure layout of bimorph devices. Fig 8 shows the output power versus load resistance at 0.5 g acceleration level. Under 0.5 g, the output power shows the peak value of 304μW corresponding to the 150kΩ optimal load.

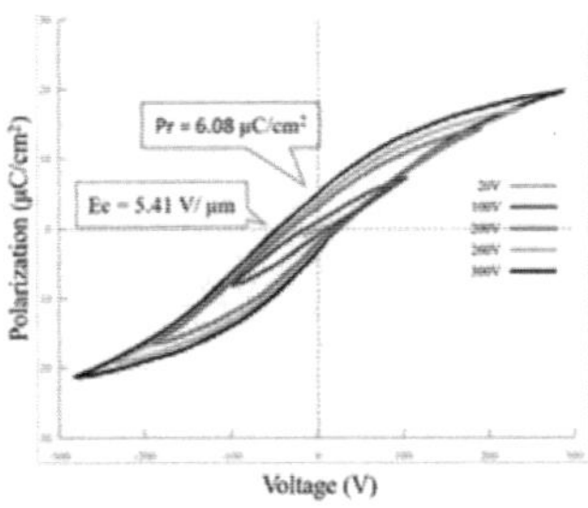

Fig 3. Ferroelectric hysteresis loop of the PZT layer at different measured voltages.

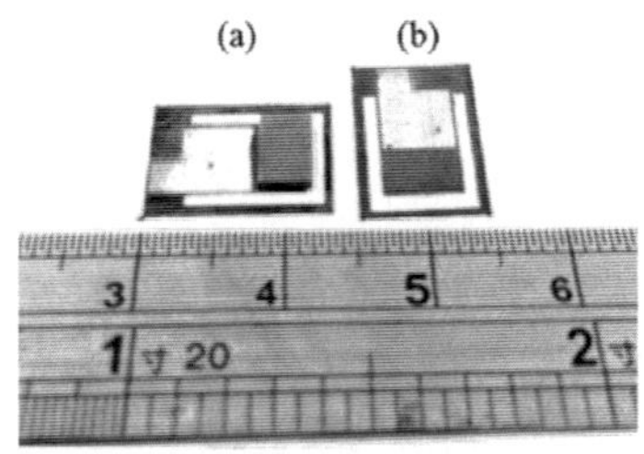

Fig 4. Photo of bimorph PMEH from (a) bottom view and (b) top view.

5.3. Comparison of normalized power density

With complete fabrication process optimized, the unimorph PMEH and bimorph PMEH show excellent output performance. The performances of the PMEHs are benchmarked to state-of-the-art PMEH based on silicon and stainless steel substrates published before with the device area smaller than 1 cm^2. The normalized power density (NPD) are calculated and presented in Fig 9. The NPD for the unimorph PMEH and bimorph PMEH in this study are 18.9mW·cm^{-3}·g^{-2} and 46.81mW·m^{-3}·g^{-2}, which outperformed all previous published PMEHs based on either silicon or stainless steel substrates.

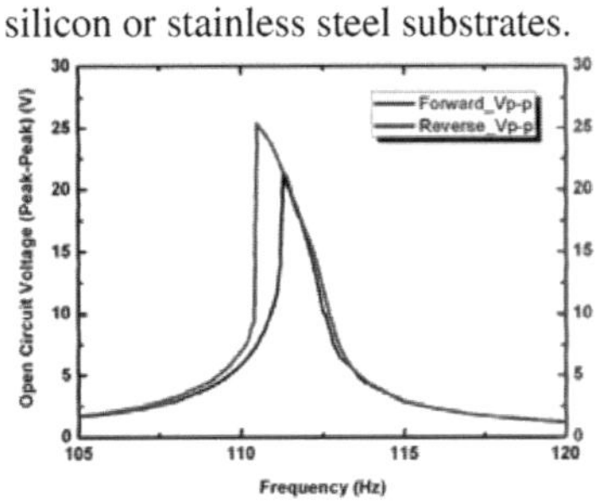

Fig 5. Frequency response of unimorph PMEH with output voltage versus excitation frequency under open circuit at 0.5g acceleration level.

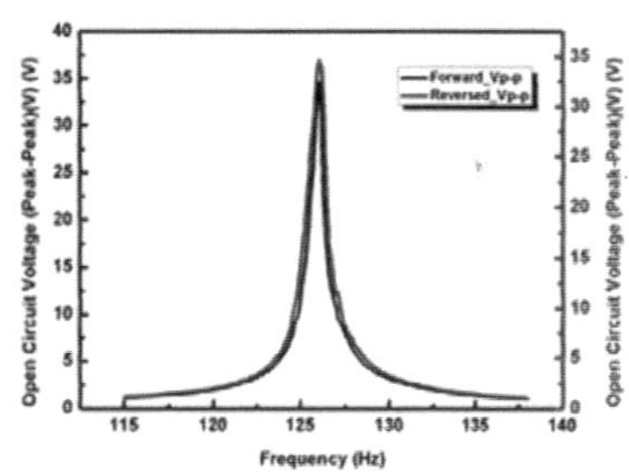

Fig 7. Frequency response of bimorph PMEH with output voltage versus excitation frequency under open circuit at 0.5g acceleration level.

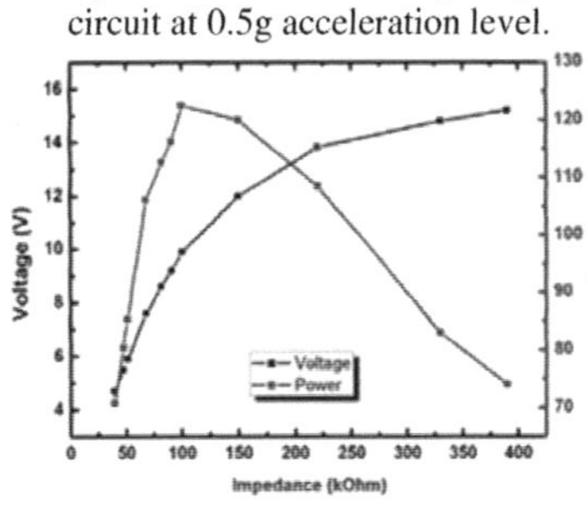

Fig 6. Experimental output performance of bimorph PMEH under different loading impedance at 0.5g acceleration level.

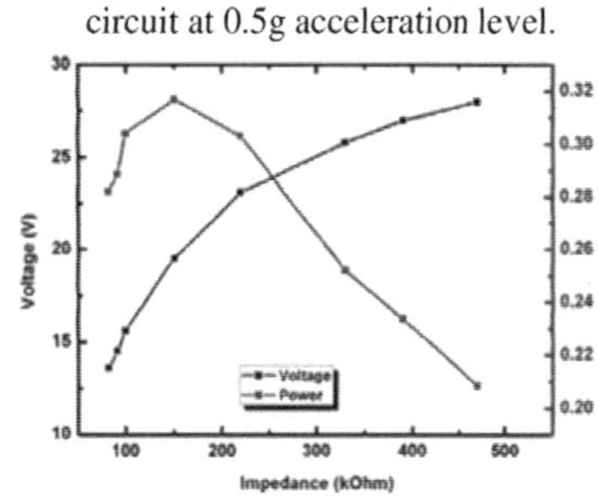

Fig 8. Experimental output performance of bimorph PMEH under different loading impedance at 0.5g acceleration level.

6. Conclusion

In this paper, we presented a unimorph PMEH and bimorph PMEH fabricated with metal MEMS processed on stainless steel substrates. The high-quality PZT layers deposited by ADM shows promising piezoelectric characteristics. The experimental result shows that the output power of unimorph PMEH is 122μW tested with 100kΩ optimal load under 0.5 g acceleration level around 120Hz. As the bimorph PMEH, the corresponding value is up to 304μW. The normalized power density (NPD) for the unimorph PMEH and bimorph PMEH are $18.9 mW \cdot cm^{-3} \cdot g^{-2}$ and $46.81 mW \cdot m^{-3} \cdot g^{-2}$ respectively. The NPD outperformed previous published energy harvester based on either silicon or stainless steel substrates, which makes the high performance stainless steel based PMEH more feasible to the practical applications.

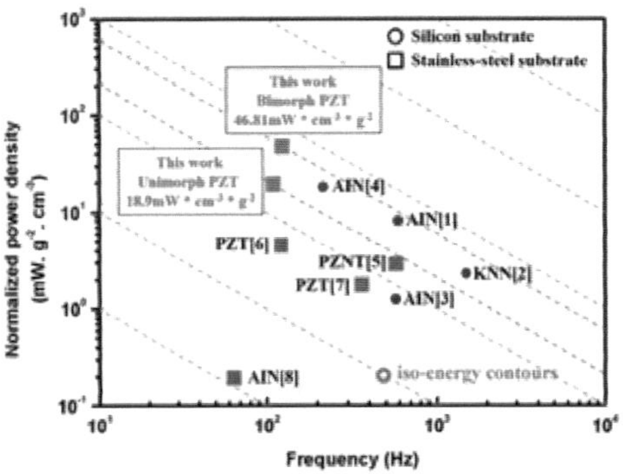

Fig 9. Benchmark of NPD with the state-of-the-art PMEHs based on silicon and stainless steel substrates.

Acknowledgment

The authors gratefully acknowledge the facility support from the Nano-Electro-Mechanical-Systems Research Center at National Taiwan University and the funding support gratefully acknowledge the support of Agence Nationale de la Recherche, France and Ministry of Science and Technology, Taiwan through respective grants ANR-15-CE22-0015-01 and 105-2923-E-002-006 -MY3 (BESTMEMS project).

References

[1] M. Le Van, M. Hara, T. Yokoyama, T. Nishihara, M. Ueda, and H. Kuwano, "Highly piezoelectric MgZr co-doped aluminum nitride-based vibrational energy harvesters [Correspondence]," *IEEE Transactions on Ultrasonics, Ferroelectrics, and Frequency Control,* vol. 62, pp. 2005-2008, 2015.

[2] L. V. Minh, M. Hara, F. Horikiri, K. Shibata, T. Mishima, and H. Kuwano, "Bulk micromachined energy harvesters employing (K, Na)NbO3thin film," *Journal of Micromechanics and Microengineering,* vol. 23, p. 035029, 2013.

[3] R. Elfrink, T. M. Kamel, M. Goedbloed, S. Matova, D. Hohlfeld, Y. van Andel, *et al.*, "Vibration energy harvesting with aluminum nitride-based piezoelectric devices," *Journal of Micromechanics and Microengineering,* vol. 19, p. 094005, 2009.

[4] E. Arroyo, Y. Jia, S. Du, S. T. Chen, and A. Seshia, "High temperature performance of a piezoelectric micro cantilever for vibration energy harvesting," *Journal of Physics: Conference Series,* vol. 773, p. 012001, 2016.

[5] L. Van Minh, T. Sano, T. Fujii, and H. Kuwano, "Vibrational micro-energy harvesters utilizing Nb-doped Pb(Zr,Ti)O3films on stainless steel substrates," *Journal of Physics: Conference Series,* vol. 773, p. 012003, 2016.

[6] K. Morimoto, I. Kanno, K. Wasa, and H. Kotera, "High-efficiency piezoelectric energy harvesters of c-axis-oriented epitaxial PZT films transferred onto stainless steel cantilevers," *Sensors and Actuators A: Physical,* vol. 163, pp. 428-432, 2010.

[7] Y. Tsujiura, E. Suwa, F. Kurokawa, H. Hida, K. Suenaga, K. Shibata, *et al.*, "Lead-Free Piezoelectric MEMS Energy Harvesters of (K,Na)NbO3Thin Films on Stainless Steel Cantilevers," *Japanese Journal of Applied Physics,* vol. 52, p. 09KD13, 2013.

[8] Z. Cao, J. He, Q. Wang, M. Hara, H. Oguchi, and H. Kuwano, "High output power AlN vibration-driven energy harvesters," *Journal of Physics: Conference Series,* vol. 476, p. 012034, 2013.

[9] S.-C. Lin and W.-J. Wu, "Piezoelectric micro energy harvesters based on stainless-steel substrates," *Smart Materials and Structures,* vol. 22, p. 045016, 2013.

Fabrication and characterization of micromachined piezoelectric energy harvesters exploiting flexible Pb(Nb,Zr,Ti)O_3/SUS

T. Takahashi[1], L. Van Minh[1], K. Umeda[2], T. Fujii[2] and H. Kuwano[1,3]

[1]New Industry Creation Hatchery Center, Tohoku University, Japan
[2]Fujifilm Corp., 577 Ushijima, Kaisei-machi, Kanagawa 258-8577, Japan
[3]Sendai Smart Machines Co., Ltd., Japan

minhlv@nanosys.mech.tohoku.ac.jp

Abstract. This work presents the fabrication and characterization of micromachined piezoelectric energy harvesters (EHs) utilizing flexible Pb(Nb,Zr,Ti)O_3(PZNT)/SUS foils. Micromachining stainless steel (SUS) foils using photoresist mask was crucial for MEMS energy harvesters and was addressed by electrochemical etching (EE). We achieved the high etching rate of 1.3 μm/min, vertical-to-lateral etching ratio of 2.6:1, and small roughness of 250 nm by using ($4H_3PO_4$:$1H_2SO_4$) electrolyte. The technique enabled us to scale down PNZT/SUS-based EHs with a total dimension of 2.54 mm × 1.75 mm × 0.5 mm. The EH provided the power of 5.12 μW at 10 $m.s^{-2}$. This results in the best normalized power density (NPD) of 3.13 $mW.cm^{-3}.g^{-2}$ from the flexible ferroelectric-on-metal substrates.

1. Introduction

Compelling research on vibrational EHs has mainly focused on piezoelectric materials grown on silicon substrates. The silicon-based piezoelectric EHs have achieved substantial progress on the output power density [1,2]. However, from practical perspectives of exploiting ubiquitous ambient vibrations, they have to face the silicon brittleness.

Exploiting piezoelectric materials on the metallic substrates that excel in mechanical properties such as large fracture toughness and high elastic strain limit provides a solution to the issue. Unfortunately, low piezoelectricity on metallic substrates and difficulties in micromachining contribute to low performance of metal-based energy harvesters [3-5]. Cao *et al.* had used the stainless-steel substrate for fabricating piezoelectric macro-energy harvesters with the NPD of 40-400 $\mu W.g^{-2}.cm^{-3}$ [4,5]. More than 10-fold decrease is compared with those of Si-based harvesters [1,2].

To address the problem, we introduced the piezoelectric micro-energy harvesters that exploited the highly piezoelectric Nb-doped PZT films grown directly on the stainless-steel substrate [6,7]. The first prototype shows that the PNZT energy harvester provided the normalized power density of 2.8 $mW.g^{-2}.cm^{-3}$, being comparable with those of Si-based piezoelectric EHs [7]. However, micromachining PNZT/SUS-based EHs had difficulty in microfabricating precisely SUS [4,5,7,8].

In this work, we addressed the issue by optimizing the EE technique. Then, the established technique was to fabricate micro-EHs from PNZT films grown directly on thin SUS foils. The evaluation of the fabricated device is also presented.

2. SUS electrochemical microfabrication

SUS microfabrication plays a crucial role to realize SUS-based micro-EHs. Conventionally, SUS is etched by wet etching with $FeCl_3$ solution [8]. The side etching and the etched surface roughness are too large for releasing the micro-patterns. For improvement the etching quality, fine-grained stainless steels are used for fabricating microstructures [8]. For high etching rate, precise micropattern and independent raw SUS material, EE provides unique advantages over competing technologies [9]. Here we investigated EE for SUS materials and compared the achievement to the traditional wet etching with $FeCl_3$. A typical stainless steel (Nilaco, Japan, grain size: 20-30 µm) was used for investigating the SUS microfabrication.

EE set-up is shown in Figure 1 (a). The SUS sample was connected to the anode that was parallel to the counter SUS electrode (cathode). The electrochemical system (VersaSTAT 4, Princeton Applied Research) was used to supply voltage power to these electrodes and to measure the electrochemical current. Photoresist (OFPR-800 LB, Tokyo Ohka Kogyo Co., Ltd.) was utilized for the mask. Electrolyte solutions were H_3PO_4 (98.08%, Wako) and H_2SO_4 (98.08%, Wako) that were mixed together at different volume ratios (1:1 and 4:1).

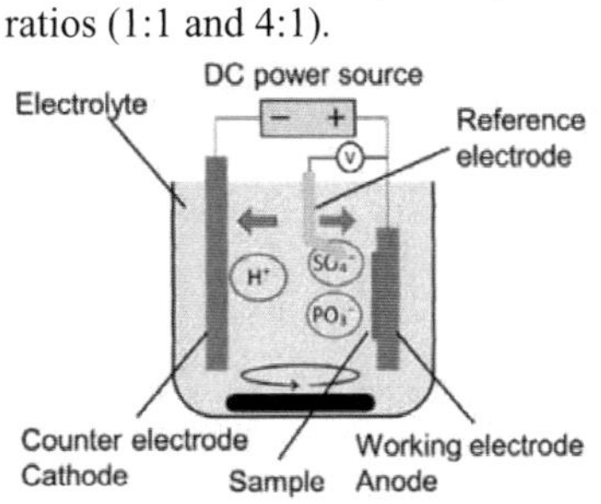

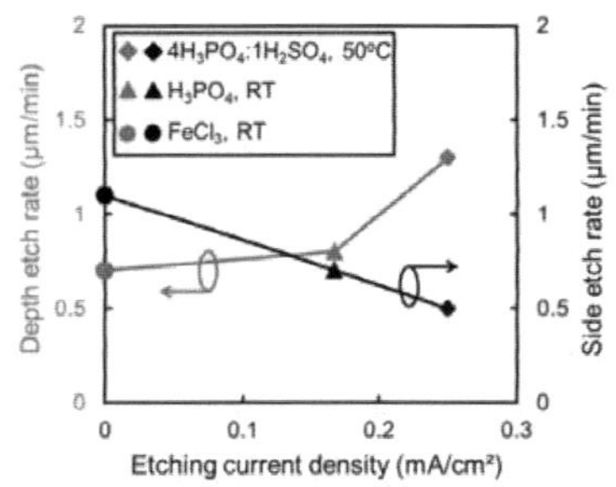

Figure 1. Electrochemical etching for SUS microfabrication: (a) electrochemical etching set-up, (b) depth (red) and side (black) etch rate of stainless steel foils with various electrolytes ($4H_3PO_4$:$1H_2SO_4$) @ 50 °C and H_3PO_4 @ RT and non-electrochemical etching $FeCl_3$ @ RT.

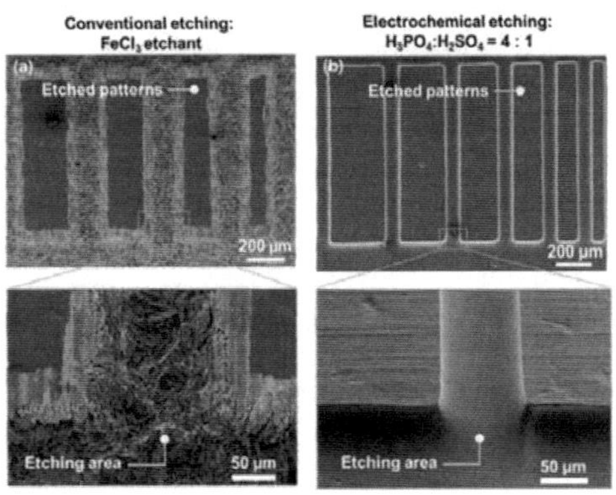

Figure 2. Scanning electron microscopic (SEM) images of SUS430-etched micropatterns with etching time of 30 min: (a) conventional $FeCl_3$ wet etching, (b) electrochemical etching (EE) with ($4H_3PO_4$:$1H_2SO_4$) @ 50 °C and 2.25 mA/cm^2.

For comparison, SUS wet etching by $FeCl_3$ (38%, RT) was also conducted. The etching with $FeCl_3$ caused larger side etching rate than depth etching rate. In contract, the EE provided smaller side etching rate than depth etching rate (figure 1 (b)). Especially, using ($4H_3PO_4$:$1H_2SO_4$) electrolyte with the optimal condition at 50 °C and the current density of 2.25 mA/cm^2 led to high etching rate of 1.3 µm/min and high depth-to-side etching rate ratio of 2.6. The 5-fold increase of the depth-to-side etching rate was compared to the traditional $FeCl_3$ wet etching. Figure 2 shows the SEM images of SUS patterns that were etched by the $FeCl_3$ and the EE of ($4H_3PO_4$:$1H_2SO_4$ @ 50 °C). Our EE with the optimized condition offered the clear patterns and obtained the peak roughness of 250 nm while the former did remove the smallest pattern of 50 µm and had the peak roughness of 10 µm.

3. Micromachined PNZT/SUS-based EHs

The PNZT-based micro-EHs were developed from the established EE and wafer-to-wafer bonding with polymer adhesive. The devices were fabricated from two separate wafers: a PNZT/SUS wafer named a device wafer and a 500 µm Cu thick wafer named a proof mass wafer.

The fabrication process of a device wafer was shown in figure 3 (i)-(iv). First, PNZT films of 3 µm thick were grown directly on SUS430 foils of 100 µm thick at the same sputtering conditions compared with the PNZT grown on SUS substrates of 500 µm in our previous publication (figure 3 (i)) [7]. Second, PNZT film was etched to shape piezoelectric element and open contact window of the device by using the fast-atomic beam (FAB) with SF_6 gas (figure 3 (ii)) [1]. Third, the top and bottom electrodes of Au(200 nm)/Cr(10 nm) were formed on the cantilever by sputtering and wet etching with Au/Cr etchants, respectively (figure 3(iii)). Forth, to achieve the SUS beam thickness of 50 µm, the wafer was etched from the backside by the developed EE (figure 3 (iv)) with OFPR-800 LB photoresist mask.

For the proof-mass wafer, Cu was etched from one-side by $FeCl_3$ (38%). The etching was stopped when the etching depth approached approximately 400 µm (figure 3 (v)).

After finishing the processes of the two separate wafers, they were bonded together by a wafer-to-wafer bonder (EVG 520) with 50 µm SU-8 adhesive (figure 3 (vi)). After that, the SUS beam was shaped by the proposed EE. Finally, Cu-remained layer was released by the aforementioned Cu wet etching to complete the device. Figure 4. Shows SEM image of the fabricated micro-EH. The device consisted a PNZT/SUS cantilever of 740 µm × 1750 µm × 50 µm and a Cu proof mass of 1800 µm × 1750 µm × 500 µm.

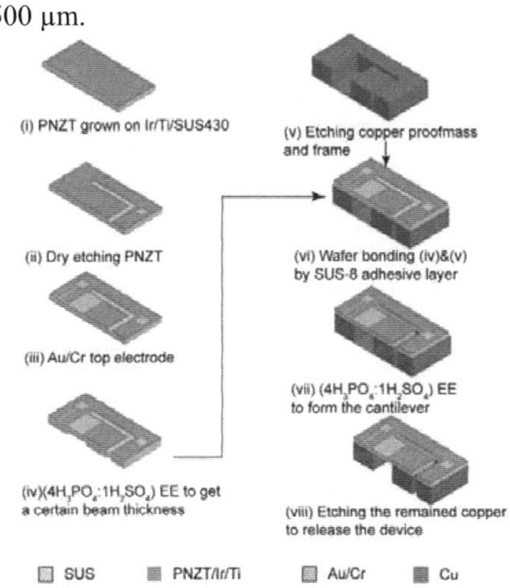

Figure 3. Microfabrication process flow of PNZT/SUS-based micro-energy harvester.

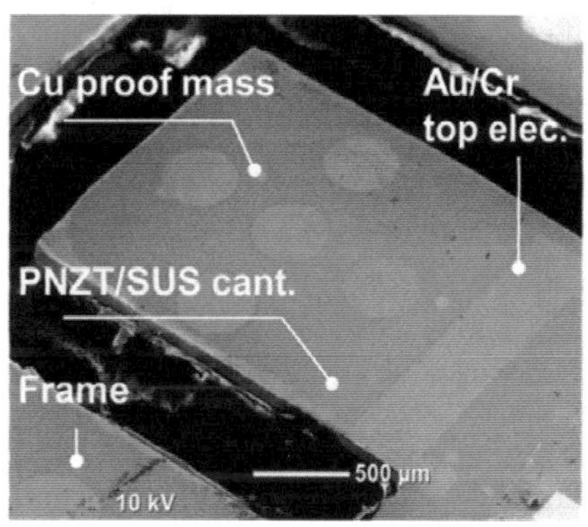

Figure 4. SEM image of PNZT/SUS-based micro-energy harvester, consisting of PNZT/SUS cantilever of 740 µm × 1750 µm × 50 µm and the copper proof mass of 1800 µm × 1750 µm × 500 µm

4. Evaluation of PNZT/SUS micro-energy harvesters

The fabricated PNZT/SUS based energy harvester was evaluated power characteristics using the resistive connection for matching impedance and the shaker for controlling the harmonic acceleration. Figure 5 displays the output power of the PNZT/SUS-based EHs under various accelerations at the resonance of 2040 Hz and the optimal resistance of 80 kΩ. The *Q*-factor decreased from 148 to 94 when the input acceleration increased from 1 $m.s^{-2}$ to 30 $m.s^{-2}$. The device provided the output power of 5.12 µW at 10 m/s^2. The NPD of 3.13 $mW.cm^{-3}.g^{-2}$ increased of 9.1% that of the previous report [7].

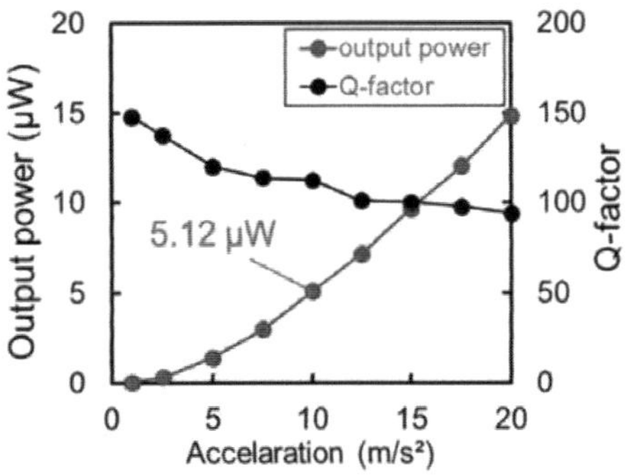

Figure 5. Output power (red) and *Q*-factor (black) of PNZT/SUS-based energy harvester at the resonant frequency of 2040 Hz.

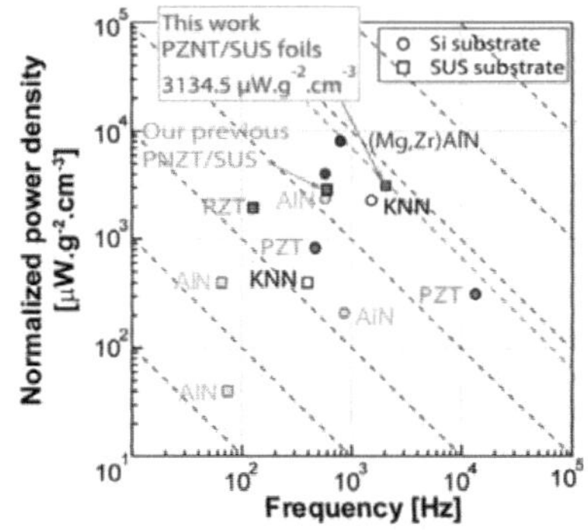

Figure 6. The NPD of this work PNZT/SUS-based and the state-of-the-art EHs in relation with the resonant frequency.

For comparison, the performance of the state-of-the-art piezoelectric micro-energy harvesters has been expressed on the NPD-frequency graph consisting of the iso-energy contours (figure 6). The NPD in this work was 8-fold those AlN/SUS harvester [4,5]. Since all devices operated at different frequencies, in term of generated energy, the PNZT/SUS-based harvester is comparable to the best of AlN/Si and KNN/Si based energy harvesters [1].

5. Conclusion

This work presents the fabrication and characterization of micromachined piezoelectric EHs exploiting flexible $Pb(Nb,Zr,Ti)O_3$(PZNT)/SUS foils. SUS microfabrication was established by EE. The electrolyte of ($4H_3PO_4$:$1H_2SO_4$) provide the high SUS etching rate of 1.3 μm/min and large vertical-to-lateral etching ratio of approximately 3:1. The micromachining process of the EHs was established from the developed EE and wafer-to-wafer bonding with SU-8 adhesive. The output power was 5.12 μW at the acceleration of 10 $m.s^{-2}$. The EH was confirmed with the high NPD of 3.13 $mW.g^{-2}.cm^{-3}$. This achievement is comparable to the best of AlN/Si and KNN/Si based EHs.

Acknowledgments

The authors would like to thank Japan Science and Technology Agency (JST). This work was supported by A-STEP project of JST (AS2815014R).

References

[1] Minh Le V, Hara M, Yokoyama T, Nishihara T, Ueda M and Kuwano H 2015 *IEEE Trans. Ultrason., Ferroelect., Freq. Control* **62** (11) 2005-2008
[2] Van Minh L, Hara M and Kuwano H 2014 *IEEE 27th International Conference on MEMS,* San Francisco, 397-400
[3] Moriwaki N, Minh L V, Ohigashi R, Kitayoshi H and Kuwano H 2016 *J. Phys.: Conf. Ser.* **773** 012049
[4] Cao Z, He J, Wang Q, Hara M, Oguchi H and Kuwano H 2013 *J. Phys.: Conf. Ser.* **476** 012034
[5] Cao Z, Zhang J and Kuwano H 2012 *Sensors and Actuators A* **179** 178-184
[6] Fujii T, Hishinuma Y, Mita T and Naono T 2010 *Sensors and Actuators A* **163** 220-2256
[7] Van Minh L, Sano T, Fujii T and Kuwano H 2016 *J. Phys.: Conf. Ser.* **773** 012003
[8] Asanuma H, Hara M, Oguchi H and Kuwano H 2014 *J. Phys.: Conf. Ser.* **557** 012124
[9] Datta M and Landolt D 2000 *Electrochi. Acta* **45** 2525-2558

IOP Conf. Series: Journal of Physics: Conf. Series **1052** (2018) 012040 doi:10.1088/1742-6596/1052/1/012040

H-TALIF measurement for wall radical quenching modelling in microscale combustion

Yong Fan[1], Yu Saiki[2], Sangeeth Sanal[1] and Yuji Suzuki[1]

[1]The University of Tokyo, Hongo 7-3-1, Bunkyo-ku, Tokyo 113-8656, Japan
[2]Nagoya Institute of Technology, Gokiso-cho, Showa-ku, Nagoya 466-8555, Japan

fan@mesl.t.u-tokyo.ac.jp

Abstract. Two-dimensional hydrogen atom (H) distribution in a CH_4/air premixed flame in a narrow channel was measured using a newly-developed two-photon absorption LIF (TALIF) setup with high efficiency 205 nm laser. The wall chemical effect of SUS321 and quartz surfaces on the H distribution in a methane flame was examined. Based on the measured H atom concentration near the wall, a radical quenching model focused on the H adsorption is proposed and the initial sticking coefficient (S_0) of H on SUS321 and quartz surfaces are estimated as 0.1 and 0.01, respectively. Sensitivity analysis of CO, HRR and species distributions with varying S_0 of H, O, OH, CH_3 were performed. It is found that the H adsorption should be dominant in the wall chemical effect on the methane flame.

1. Introduction

Flame-wall interaction (FWI) plays an important role in the near-wall flame behavior of internal combustion engines and gas turbines. The ignition process and emission can be greatly changed due to the FWI. Dreizler and Bohm [1] reviewed advanced laser diagnostic techniques to assess near-wall thermal, fluidic and chemical properties for the investigation of FWI. In the case of micro combustors for the portable power generation, FWI becomes more significant due to the large surface to volume ratio, resulting in flammability and flame stability problems. Flame quenching due to the thermal FWI in micro channels has been intensively studied to understand wall heat loss effects [2-6]. It was found that the flammability region shrinks as the wall heat loss increases with the downscaling, and the quenching distance, defined as the minimum channel height to hold a flame, decreases when the wall heat loss is reduced by external heating [4]. Also, unwanted flame instability phenomena such as the oscillating flame (also called FREI) and the spinning flame were observed in various micro combustors having different geometries [2-6].

Unlike the intensively-studied wall thermal effect, it was only recently that strong wall chemical effect in microscale combustion was observed [7-8] and examined quantitatively [9-10]. Miesse et al. [7] found that the quenching distance strongly depends on the wall material at high wall temperature (1000 °C). Saiki et al. [9-10] conducted OH planar laser induced fluorescence (OH-PLIF) measurements of a methane/air premixed flame in a quartz micro channel with/without alumina or metal coatings. They found that the near-wall OH concentration is significantly affected by the surface materials at high wall temperatures (≥800 °C). Raimondeau et al. [11] numerically examined the effect of wall radical quenching by arbitrarily setting all the initial sticking coefficients (S_0) for the adsorption of H, O, OH, CH_3 to 0~1.0. Saiki et al. [9-10] obtained S_0 over alumina (S_0=0), quartz (S_0=0.01) and metal (S_0=0.1) surfaces based on their OH-PLIF measurements.

Up to now, S_0 for the adsorption of different radicals is assumed to be the same for different radicals. This is because of limited quantitative information on the near-wall radical distribution other than OH. In the present study, the adsorption of H in the wall radical quenching was examined through the measurement of H atom concentration using newly-developed two-photon absorption LIF (H-TALIF) and numerical simulations assuming different S_0 for H, O, OH, CH_3.

2. Experimental Setup & Numerical Methods

Figure 1a shows the experimental setup for the examination of wall radical quenching using H-TALIF. A lifted CH_4/air premixed flame from the honeycomb burner is held in 3 mm-wide gap between two quartz plates. The bulk mean velocity and equivalence ratio ϕ for the mixture are 0.22 m/s (at 20 °C) and 0.95, respectively. A black quartz layer is fused to the back side of each quartz plate for the adsorption of thermal radiation from the IR heater. The wall temperature, T_w, is monitored by thin sheathed type-R thermocouple (ϕ 0.5 mm) in the plate. The wall surface is alternatively coated with ~100 nm-thick SUS321 film to change the chemical boundary while keeping the thermal boundary identical.

Observing H atom using the single-photon absorption LIF is difficult due to the strong absorption of VUV photons at 102 nm by other species in the air. Instead, the H atom in the flame is excited by two photons of 205.144 nm laser in TALIF, and the fluorescence at 656 nm is captured by the ICCD camera with a bandpass filter. Figure 1b shows newly-developed 615 nm→205 nm third harmonic generation (THG) setup for H-TALIF measurements. Unlike previous setups having poor THG efficiency (~2%) due to separate 615 nm and 307.5 nm optical paths [12-13], all THG optics in the present setup are arranged in a single axis through the implement of a dual-wavelength wave plate and a walk-off compensation BBO crystal. As a result, alignment is much easier, and a THG efficiency as high as 9 % has been obtained, which makes the 2-D H-TALIF measurement possible. A laser sheet of 0.2 mm×3.0 mm is formed. The energy distribution in the laser sheet is calibrated beforehand by taking the TALIF signal of Kr gas with uniform concentration as the reference.

Two-dimensional numerical simulations were carried out in ANSYS Fluent 18.0 with employing GRI-Mech 3.0 as the gas phase reaction mechanism and the wall radical quenching model as the surface reaction mechanism. Beside the wall radical quenching model assuming equal S_0 for all H, O, OH, CH_3 adsorptions, the model assuming the adsorption of H only is also applied. The computation domain is half of the 3-mm-wide channel due to the symmetry and the boundaries are set to the same as in the present experiment. Details are described in our previous work [10]. Sensitivity analysis of the exhaust CO, heat release rate (HRR) and species distributions to the independently varying S_0 for H, O, OH, CH_3 was also performed, where the channel width is set to a smaller value of 0.7 mm to obtain stronger chemical effect.

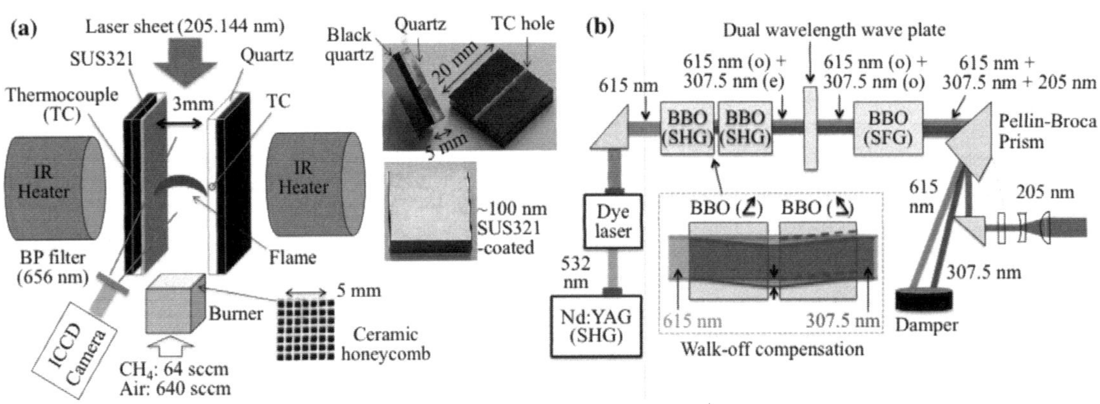

Figure 1. (a) Experimental setup of H-TALIF measurement of CH_4/air premixed flame in a 3 mm-wide channel. (b) Schematic of 205 nm laser generation.

3. Results and Discussion

Figure 2 shows contours of normalized H molar concentration ($\chi_H/\chi_{H,max}$) measured through the H-TALIF being calibrated by the Kr-TALIF. Measured wall normal distributions of $\chi_H/\chi_{H,max}$ at T_w =1000 °C are plotted in Fig. 3 and compared with simulated distributions assuming either all H, O, OH, CH_3 adsorptions or H adsorption only. It is shown that $\chi_H/\chi_{H,max}$ over SUS321 is lower than that over quartz, so that SUS321 surface is more active than quartz in the adsorption of H. Simulated near-wall $\chi_H/\chi_{H,max}$ is very sensitive to S_0 and decreases gradually from S_0 =0 (inert) to 0.1 (active). $\chi_H/\chi_{H,max}$ with assuming all radical adsorptions and only the H adsorption shows no big difference, which implies that H adsorption should be dominant in the wall chemical effect. By matching measured and simulated $\chi_H/\chi_{H,max}$, S_0 for SUS321 and quartz are estimated as 0.1 and 0.01, which agrees well with our previously obtained S_0 values based on the OH-PLIF measurement and the radical quenching model assuming equal S_0 of H, O, OH, CH_3 [9-10].

The importance of H adsorption is obvious through the above comparison, yet it is of concern that how sensitive the combustion field is to the adsorption of H and other radicals. Figure 4 shows the sensitivity results through simulations with varying S_0 of H, O, OH, CH_3 separately. It is found that both CO and heat release rate (HRR) are extremely sensitive to the adsorption of H, but not O, OH, CH_3. 35 % reduction in CO and 15 % improvement in HRR are obtained with decreasing S_0 of H from 0.1 to 0.001. Figure 5 compares CO and HRR in cases of all radical adsorption and single radical adsorption of H, O, OH, CH_3 each. Even if only the H adsorption is assumed, CO and HRR are deteriorated to the same level as all radical adsorptions.

As shown in Figs. 6ab of χ_H contours with/without the H adsorption, χ_H near the wall decreases drastically. It is shown in Fig. 6c of wall normal distributions that not only χ_H but also χ_{OH} and χ_O

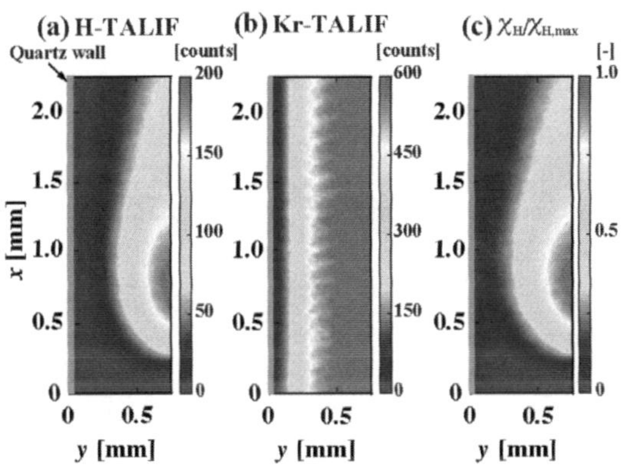

Figure 2. Contours of (a) H-TALIF and (b) Kr-TALIF signals, and (c) computed $\chi_H/\chi_{H,max}$ near quartz surface at T_w = 600 °C.

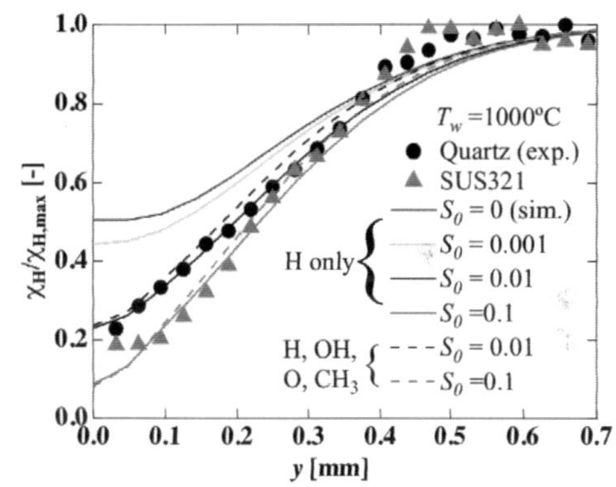

Figure 3. Measured wall normal distributions of $\chi_H/\chi_{H,max}$ near quartz and SUS321 surfaces and simulated results for S_0 = 0, 0.001, 0.01, 0.1.

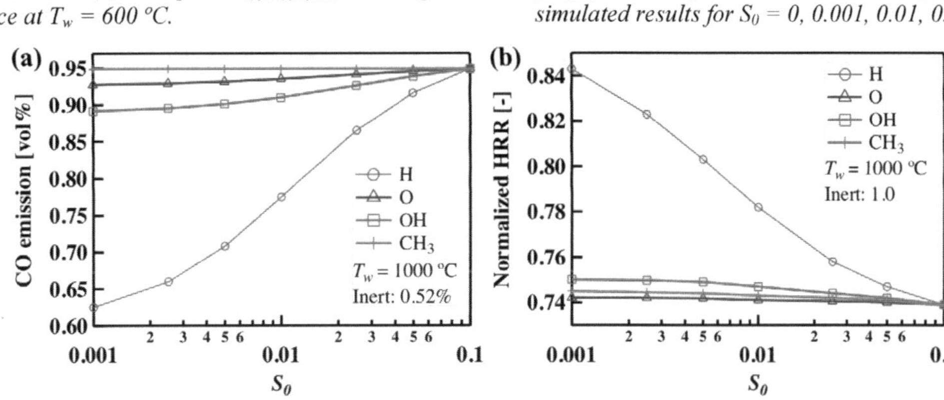

Figure 4. Sensitivity of H, OH, O, CH_3 adsorptions to (a) CO emission and (b) HRR.

decreases with assuming only the H adsorption. This explains why our previously-obtained S_0 based on the OH-PLIF measurement and the equal S_0 model [9-10] could have the same value as here. The adsorption of H has larger impact than other radicals on combustion characteristics due to its large diffusivity and suppression of a chain branching reaction ($H+O_2=O+OH$).

4. Conclusion

Two-dimensional H atom distribution of the CH_4/air premixed flame in a narrow channel is successfully measured with a newly-developed TALIF setup having high efficiency 205 nm laser. Based on the measured H atom concentration near the wall, a wall radical quenching model focused on H adsorption is proposed and the initial sticking coefficient of H on SUS321 and quartz surfaces are estimated as 0.1 and 0.01, respectively. Sensitivity analysis shows that the H adsorption should be dominant in the wall chemical effect.

This work is supported through JSPS KAKENHI Grant Number 15H03927 and 15H05508.

References

[1] A. Dreizler and B. Bohm, Proc. Combust. Inst. 35 (2015) 37-64.
[2] K. Maruta, Proc. Combust. Inst. 33 (2011) 125-150.
[3] Y. Ju and K. Maruta, Prog. Energy Combust. Sci. 37 (2011) 669-715.
[4] Y. Fan et al., Proc. Combust. Inst. 32 (2009) 3083-3090.
[5] Y. Fan et al., Proc. Combust. Inst. 33 (2011) 3267-3273.
[6] B. Xu and Y. Ju, Proc. Combust. Inst. 31 (2007) 3285-3292.
[7] C.M. Miesse et al., AIChE J. 50 (2004), 3206-3214.
[8] K.T. Kim et al., Combust. Flame 146 (2006), 19-28.
[9] Y. Saiki et al., Proc. Combust. Inst. 34 (2013), 3395-3402.
[10] Y. Saiki et al., Combust. Flame 162 (2015), 4036-4045.
[11] S. Raimondeau et al., Proc. Combust. Inst. 29 (2002) 901-907.
[12] W.D. Kulatilaka et al., Appl. Opt. 47 (2008), 4672-4683.
[13] Z. Yin et al., Proc. Combust. Inst. 35 (2015), 3455-3462.

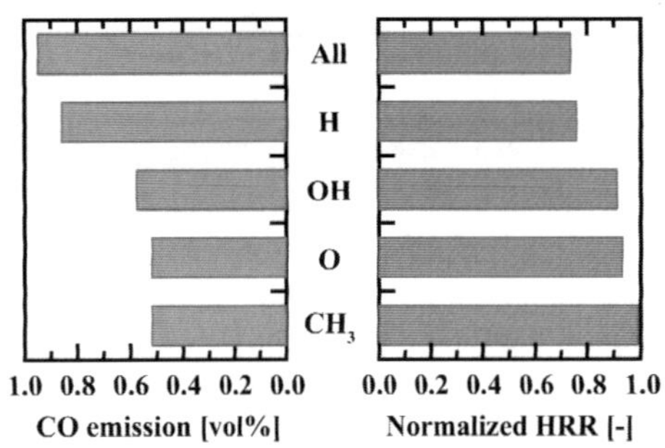

Figure 5. CO and HRR in cases of H, OH, O, CH_3 adsorptions and single radical adsorption.

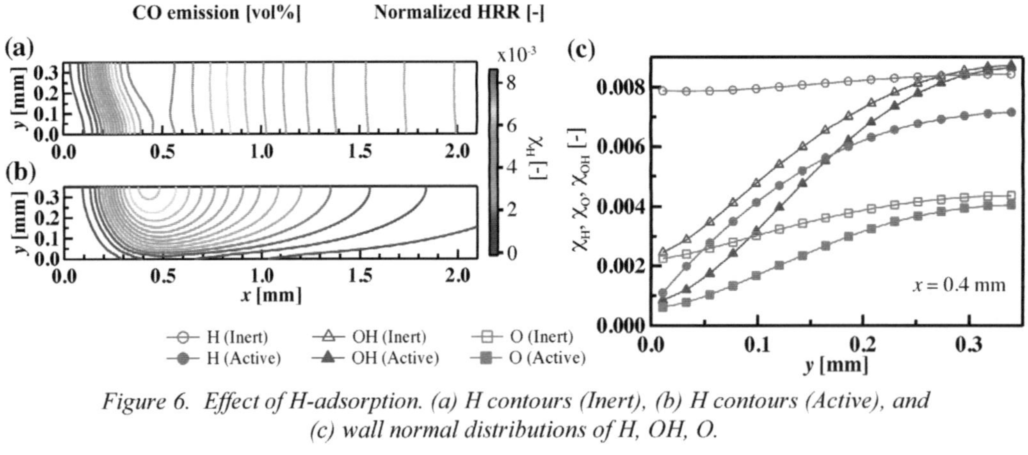

Figure 6. Effect of H-adsorption. (a) H contours (Inert), (b) H contours (Active), and (c) wall normal distributions of H, OH, O.

Towards a portable mesoscale thermophotovoltaic generator

Walker R. Chan[1], Veronika Stelmakh[1], Sunny Karnani[2], Christopher M. Waits[2], Marin Soljacic[3], John D. Joannopoulos[1,3] and Ivan Celanovic[1]

[1] Institute for Soldier Nanotechnologies, Massachusetts Institute of Technology, 77 Massachusetts Avenue, Cambridge, Massachusetts 02139, USA

[2] Sensors and Electron Devices Directorate, US Army Research Laboratory, 2800 Powder Mill Rd., Adelphi, Maryland 20783, USA

[3] Department of Physics, Massachusetts Institute of Technology, 77 Massachusetts Ave., Cambridge, Massachusetts 02139, USA

E-mail: wrchan@mit.edu

Abstract. Thermophotovoltaics (TPV) is the conversion of fuel to electricity with heat and light as intermediaries, and is a promising source of high energy density power at the mesoscale. This work describes our transition from our bench-top experiments to a fully-integrated portable generator. Specifically, we redesigned the microcombustor for propane-air combustion from the previous propane-oxygen design. Next, we validated vacuum package of the microcombustor, necessary to preserve the photonic crystal, in a 50+ day experiment in which there was no degradation of vacuum level. Finally, we vacuum packaged a microcombustor with integrated photonic crystals in a housing with two infrared-transparent windows to transmit the thermal radiation to external PV cells. Although considerable challenges remain, this work demonstrates the feasibility of a mesoscale TPV system.

1. Introduction

Thermophotovoltaics (TPV) is a promising technology that can generate watts of electricity from a heat input similar to that of a lighter, in a volume that is only a fraction of a cubic inch. In our system, a propane-fueled microcombustor heats a photonic crystal emitter to incandescence and the resulting tailored thermal radiation drives low-bandgap photovoltaic (PV) cells to generate electricity. Previously, we achieved unprecedented efficiencies of above 4% in a mesoscale form factor by employing new high-temperature material systems, processing, and fabrication techniques, and showed computationally that 12% efficiency should be achievable with engineering optimization [1]—exceeding the 2–3% was previously thought to be the practical limit [2]. (For comparison, an efficiency of 1.5% is equivalent to lithium batteries.) This work describes our transition from our bench-top experiments to a fully-integrated portable generator. Specifically, we redesigned the microcombustor for propane-air combustion from the previous propane-oxygen design. Next, we validated vacuum packaging of the microcombustor, which is necessary to suppress convective heat loss and prevent degradation of the photonic crystal. Finally, we vacuum packaged the microcombustor with integrated photonic crystals

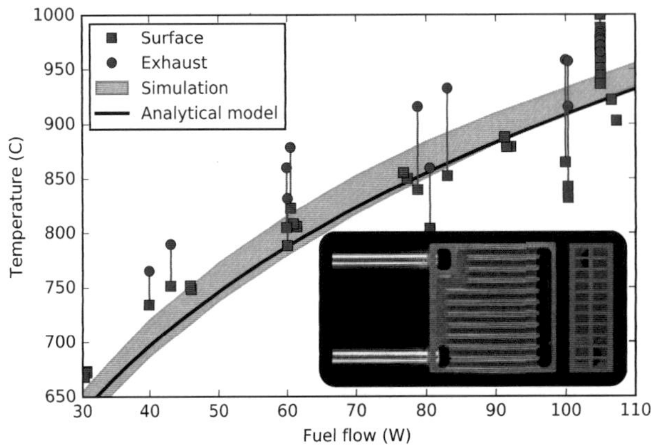

Figure 1. Simulated and measured operating temperature for the propane-air microcombustor as a function of fuel flow (calculated from the lower heating value). Inset: microcombustor structure.

(described in Ref. [3]) in a package with two infrared-transparent windows to transmit the thermal radiation to external PV cells.

2. Propane-Air Microcombustor

The microcombustor catalytically reacts propane with air to bring the photonic crystal emitter to incandescence. It is a 20×20 mm Inconel slab with two internal layers of catalyst-loaded channels, as shown in Fig. 1. Propane and air react in the channel and the resulting heat is conducted through the channel walls to the photonic crystal emitters on the top and bottom surfaces. The device was fabricated by diffusion brazing stacked layers. A two layer design was required in order to attain sufficiently long residence time and sufficiently short diffusion time to react 100 W of propane flow with air. We modified the propane-oxygen microcombustor used in the bench-top experiment for propane-air operation [1], using the design and fabrication methodology detailed in Ref. [4].

Two simulations were used. First, we used a simple analytical heat balance model, which assumed a uniform temperature across the burner surface. Second, we developed a reduced-order modeling tool that enables the evaluation of different reactant compositions and geometry modifications on system thermal efficiency and surface temperature uniformity. These results are shown in Fig. 1, where the shaded region represents the range of surface temperatures predicted by the second model.

The microcombustor was tested in the vacuum chamber described in the next section. A thermocouple was spot welded to the surface to measure temperature and another thermocouple was inserted into the exhaust tube during some experiments to measure exhaust gas temperature. The microcombustor was ignited by bubbling the air through methanol, which reacted at room temperature over platinum, until heated to around 300°C, at which point propane-air operation was possible and the methanol bubbler was bypassed. Exhaust and surface temperatures for a range of propane flows are shown in Fig. 1.

3. Vacuum Packaging

In order to prevent degradation of the photonic crystal emitter and to suppress convective losses, vacuum packaging is required. To test the feasibility of vacuum packaging we assembled

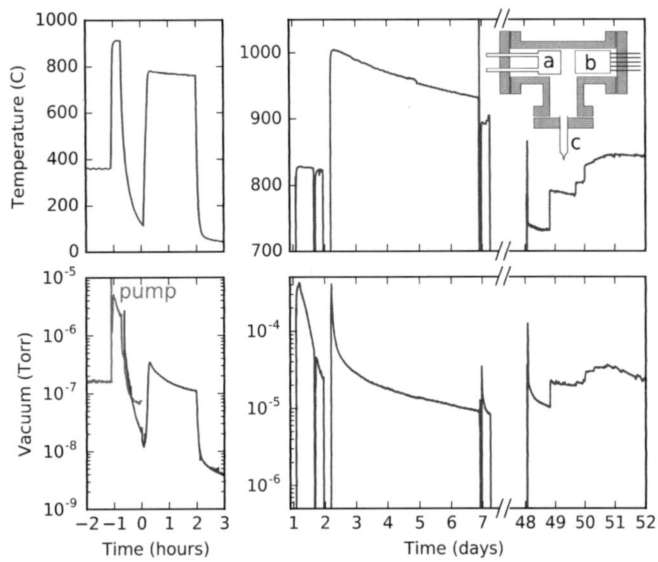

Figure 2. Temperature and vacuum during a 50+ day experiment, with time measured relative to pinch-off ($t = 0$). Inset: the experimental setup was composed of a microcombustor (a), ion gauge (b), and copper pinch-off (c).

a microcombustor and hot filament ion gauge in a Conflat (CF) tee, as shown in the inset of Fig. 2. A zirconium based getter (Saes) was also placed in the chamber. A soft copper tube connected the test chamber to a turbomolecular pump and residual gas analyzer. With the turbomolecular pump running, the chamber was heated to 350°C to degas and to activate the getter. In addition, the burner was ignited and the ion gauge was electrically heated. At the beginning of the bakeout, water vapor dominated; at the end, hydrogen and carbon monoxide dominated. The hydrogen was determined to be permeating through the walls of the microcombustor by injecting pulses of deuterium into the combustion reaction and observing the isotope ratio in the vacuum chamber. At the end of the bakeout, the copper tube was pinched off with a hydraulic crimper, isolating the vacuum chamber from the pump. Temperature and vacuum near pinch-off are shown in Fig. 2.

After pinch-off the microcombustor was run for six days without degradation of the vacuum. In fact, an improvement in vacuum was observed because the getter continued to act as an internal pump. The apparatus was left for nearly 40 days, during which the vacuum level was $\sim 10^{-8}$ Torr. The microcombustor was successfully reignited and run without vacuum degradation, and stepped through several fuel flows.

4. System integration

After validating the microcombustor and vacuum packaging strategy, we designed and fabricated an integrated unit with a photonic crystal emitter and sapphire windows to transmit the infrared radiation to PV cells. The package, shown in Fig. 3, was fabricated from commercial sapphire viewports and custom components by welding and brazing. Although vacuum level could not be directly measured, no discoloration of the microcombustor or photonic crystal (an indication of oxidation) was observed.

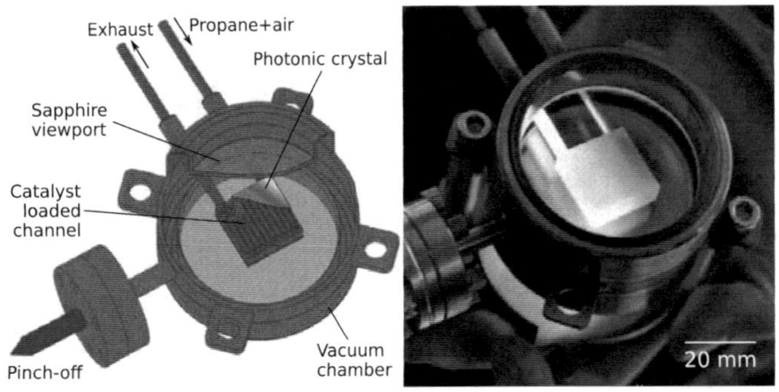

Figure 3. Cutaway rendering and photograph of the vacuum packaged microcombustor with integrated photonic crystals in operation.

5. Conclusion

This work advances our bench-top TPV experiment towards a portable system by design and fabrication of a propane-air microcombustor and of a vacuum packaged microcombustor with integrated photonic crystal and infrared transparent windows to allow for PV cells. Although considerable engineering efforts are required in the area of fuel and air delivery, cell cooling, and packaging, this initial effort demonstrates that mesoscale thermophotovoltaic power generation is practical. This technology would have applications reducing the weight of batteries carried by soldiers, extending the range of drones, and enabling new applications by untethering portable electronics from bulky power sources.

Acknowledgments

The authors would like to thank: J. Daley and T. Savas of MIT, and R. Geil of the University of North Carolina at Chapel Hill for assistance with fabrication of the photonic crystal; and the MIT Central Machine Shop for assistance with fabrication of the microcombustor and vacuum package. Fabrication of the photonic crystal was done in part at NSL at MIT and at CNS at Harvard University, a member of the National Nanotechnology Infrastructure Network (NNIN), supported by the National Science Foundation (NSF) under NSF Award No. ECS-0335765. Photonic crystal fabrication was supported as part of the Solid-State Solar-Thermal Energy Conversion Center (S3TEC), an Energy Frontier Research Center funded by the U.S. Department of Energy (DOE), Office of Science, Basic Energy Sciences (BES), under Award No. DE-SC0001299 / DE-FG02-09ER46577. This work was additionally supported by the Army Research Office through the Institute for Soldier Nanotechnologies under Contract No. W911NF-13-D-0001 and the Micro Autonomous Systems and Technology Collaborative Technology Alliance under Contract No. 892730.

References

[1] Chan W R, Stelmakh V, Ghebrebrhan M, Soljačić M, Joannopoulos J D and Čelanović I 2017 *Energy Environ. Sci.* **10** 1367–1371
[2] Chan W R, Bermel P, Pilawa-Podgurski R C N, Marton C H, Jensen K F, Senkevich J J, Joannopoulos J D, Soljacic M and Celanovic I 2013 *Proceedings of the National Academy of Sciences of the United States of America* **110** 5309–14
[3] Stelmakh V, Rinnerbauer V, Geil R D, Aimone P R, Senkevich J J, Joannopoulos J D, Soljacic M, Celanovic I, Soljačić M and Celanovic I 2013 *Applied Physics Letters* **103** 0–4
[4] Chan W R, Stelmakh V, Allmon W R, Waits C M, Soljacic M, Joannopoulos J D and Celanovic I 2016 *Journal of Physics: Conference Series* vol 773 p 012108 p 012108

Development of Powerful Miniature System with Heat Transfer Controlled Vortex Combustor and Thermo Electric Device

D Shimokuri[1], Y Taomoto[1], H Satou[2], N Yokoo[2]

[1] Graduate school of mechanical engineering, Hiroshima University, Kagamiyama, Higashi-hiroshima 739-8527, Japan
[2] Dainichi Co.,Ltd., 780-6 Kitatanaka, Niigata Minami-ku, Niigata, 950-1237, Japan

cri@hiroshima-u.ac.jp

Abstract. One of the advantage of the miniature vortex combustor is high heating efficiency which comes from rapid heat transfer by the vortex flow, however, this may attribute to the decrease of the wall temperature, and hence, the flame temperature. Then, in this study, a new heat transfer controlled vortex combustor is proposed in which the combustion cylinder is installed to separate the positions of heat extraction and heat release. Results showed that the output of the system with a new vortex combustor reached 20.0W. And the high output power was confirmed to be maintained to Φ = 0.8, and furthermore, CO emission is significantly inhibited less than 500ppm for fuel lean conditions. These results clearly indicate that the flame temperature is maintained enough high and the flame stability is increased in the heat transfer controlled vortex combustor.

1. Introduction

Due to strong demands for small-scale power sources, meso- and micro-combustors have received considerable attention [1-3]. In our previous work [4,5], we have successfully developed miniature vortex combustion power system, which scale is comparable to the power cell. The system consists of combustor and TED (Thermo-Electric Device), and its output and conversion efficiency were confirmed to reach 18.1W and 3.0% respectively [5]. As far as author's knowledge, this is the first miniature combustion power system which output exceeds 15W.

However, further increase in the output power and conversion efficiency is strongly desired for the practical use of the miniature combustion power system. The problem of the vortex combustor developed in previous work (hereafter, called as "conventional combustor" or "proto type") is that the inner / outer combustor wall temperature is significantly low. On the prototype combustor, the combustion chamber was fabricated inside the heat medium to minimize the system scale as possible, then, the heat transfer region (combustor surfaces of TEDs were placed on) was designed too close to the heat release region (combustion chamber), which attributes to the low temperature, "cold" combustor wall. The cold combustor wall must affect to the flame stability and combustion efficiency by the thermal quenching effects which lead to the decrease of the stable combustion range in Φ as well as the large amount CO emission due to the incomplete combustion. Then, in this study, to increase the combustor wall temperature, a new concept miniature combustor was developed in which the heat of combustion is extracted at separated position from heat release. The output power and CO concentration of new combustor were examined for various equivalence ratio conditions.

2. Experimental

Figure 1 shows the schematic of the vortex combustion power system. The system consists of a combustor, thermo-electric devices (TEDs) and cooling plates. The combustor is made of duralumin, and 52 × 52 × 76-mm (205cm^3) which is slightly larger than the power cell (D-cell, 60cm^3). A vortex combustion chamber of 10.0-mm inner diameter is fabricated inside the heat medium.

Table 1 shows the details of TED used in this study. The TEDs are connected in line and its hot sides are place on the heat medium whereas the cold sides are contacted to the cooling plates, which are cooled by water of 1.2 L/min. The system output power (P_{out}), voltage and current are measured by an electric load device (TEXIO, LSA-165) for various load resistance (1~30 Ω). The temperatures on the hot / cold side of TED are monitored by K-type thermocouples (0.25-mm) and data logger (Graphtec, MT-100). The inner wall temperature of the combustion cylinder was also measured by thermography (Nippon AVIONICS, TSV-200EX).

Figure 2 shows the schematic of the combustor (a), appearance of the combustion chamber installed inside the heat medium (b), appearance of the combustor (c) and the swirl generator (d). As shown in Figs 2(a) and (b), the combustor consists of heat medium, heat transfer plate, combustion cylinder, base plate and swirl generator. The combustible mixture is ejected from the swirl generator, and if it is ignited, a flame can be established inside the combustion cylinder. The heat of burned gas is transferred to the combustion cylinder, and then, conducted to the heat medium through the thin "heat transfer plate" and "base plate" (Fig. 2(a),(b)). Finally, the heat is transferred to a couple of TEDs placed on the surface of the heat medium. On this combustor, the combustion chamber is designed not to directly contact to the surface of the heat medium where TEDs are placed on. That is, the region of heat release (combustion cylinder) is separated from the region of heat extraction (the surface of the heat medium). Those regions are connected only by the thin heat transfer plate and base plate. Thus, the combustion cylinder can be kept high temperature, which may lead to the high flame stability and CO inhibition.

Table 1. specifications of the thermo-electric device used in this study.

Type	KELK, KTGM-161-18
Scale	52 × 52 × 4.4 mm, 48g
Limit temperature	Cold side 180°C Hot side 280°C
Output / conversion efficiency [a]	24W / 7.2% @ ΔT = 250K 7W / 4.2% @ ΔT = 150K

a: obtained by supplier with electric heater

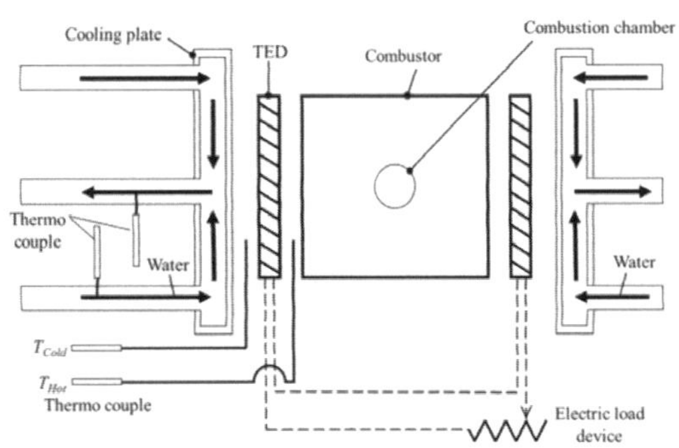

Fig.1 schematic of the heat transfer controlled vortex combustion power system

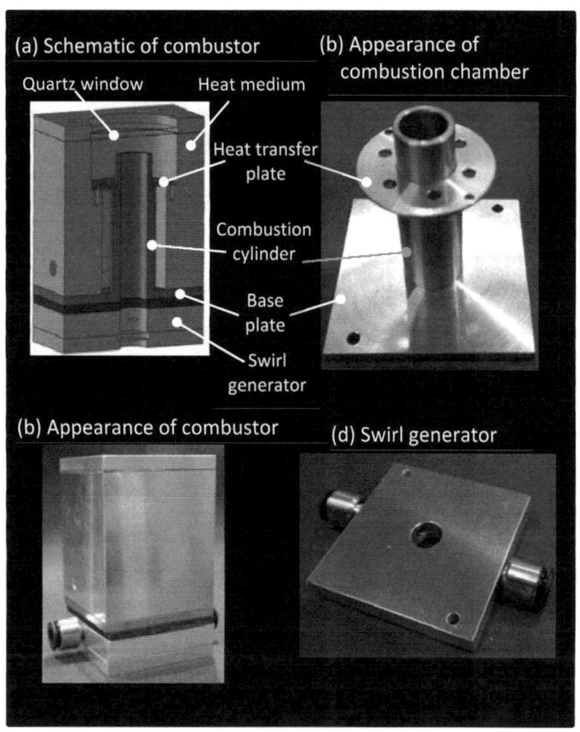

Fig.2 schematic of the combustor

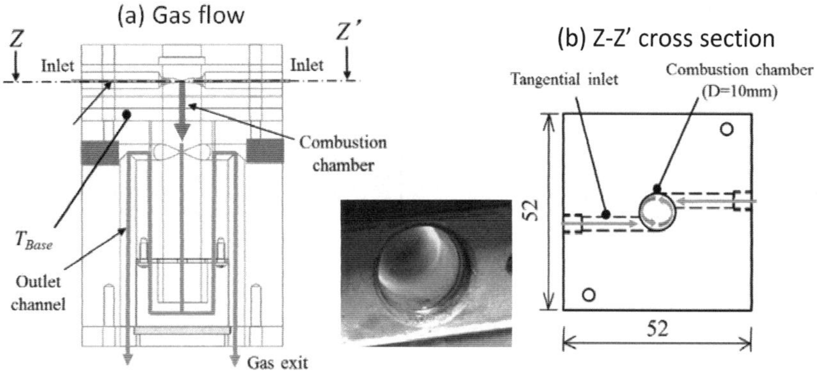

Fig.3 schematic of the heat transfer controlled miniature vortex combustor

Figure 3 shows the schematic of the gas flow inside the combustor. The combustible mixture is injected into the combustion chamber of 10-mm inner diameter (Fig. 3(a) "combustion chamber") through the injection hole of 3.8mm in inner diameter (Fig. 3(a) "Inlet"). Figure 3(b) shows Z-Z' cross section of Fig. 3(a). As shown in Fig. 3(b), the mixture is tangentially injected into the chamber to form the vortex flow. If the mixture is ignited, a flame can be established in the chamber (see, small inserted picture in Fig. 3). The burned gas flows inside the combustion cylinder, and turn the direction at the edge of the chamber and flows the outside the cylinder. As the combustion cylinder is covered by the hot burned gas, the cylinder wall is expected to be kept at high temperature. The burned gas enters into the outlet channel installed inside the heat medium, and flows out from the two exit (Fig. 3 "Gas exit").

3. Results

Figure 4 shows the output power of the conventional (△) and new (●)vortex combustion power systems for various load resistance. The data was obtained for Φ = 1.0 and 600W thermal input. Propane was used as fuel. In Fig. 4, it was confirmed that the output of the new system reached 20W, which is 10% higher than the conventional one (18.1W). As far as author's knowledge, the maximum output power as well as the system efficiency of this system (20W, 3.33%) is prominent compared to the other combustion power systems, such as Li's system [6] (8.2W output, 2.4% conversion efficiency) or Merotto's system [7](9.86W providing 417W input, 2.4% efficiency).

Figure 5 shows the system output for various equivalence ratio. Thermal input and load resistance is kept at 600W and 7Ω, respectively. On the conventional system, the output tends to decrease with decreasing the equivalence ratio, and drastically decreases when Φ became less than 0.85. On the other hand, the output of the new system is maintained high value, larger than 19W even for Φ = 0.81. On the new combustor, the inner wall of the combustion chamber is expected to be enough high to enhance the flame stability, which attributes to the high flame temperature, and hence, high output even under the fuel lean conditions. Then, the inner wall temperature of combustion chamber was examined with thermography. Figure 6 shows the results of the inner wall temperature measurement. The image was obtained through the window (see Fig. 2(a), the part of quartz window). To obtain the infrared emission, the quartz window was removed after the output was reached steady state. Immediately after that, the images were obtained. As can be seen in the Table below the Figs 6(a) and (b), the maximum temperature of inner wall of combustion chamber reaches to 450°C, which seems enough high than the water cooled wall temperature in the conventional combustor.

Finally, CO concentration was examined for the conventional and new combustor during power generation. Results are shown in Fig. 7. On the conventional combustor, CO concentration exceeds1.0% around stoichiometric condition. With decreasing the equivalence ratio, CO concentration once decreases, however, increases again when Φ = 0.81. The increase of CO at Φ =

IOP Conf. Series: Journal of Physics: Conf. Series **1052** (2018) 012042 doi:10.1088/1742-6596/1052/1/012042

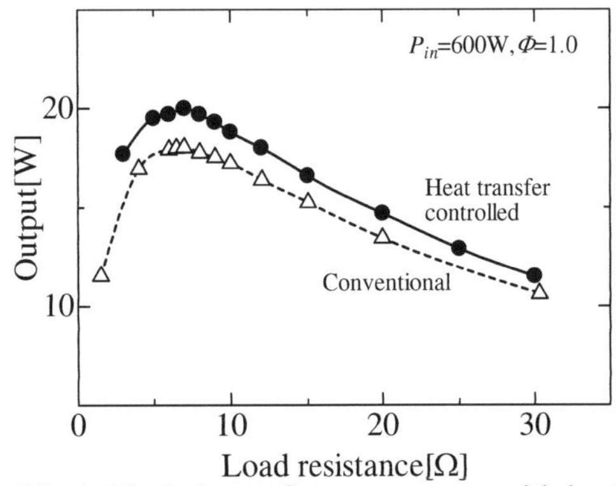

Fig.4 Variations of output power with load resistance

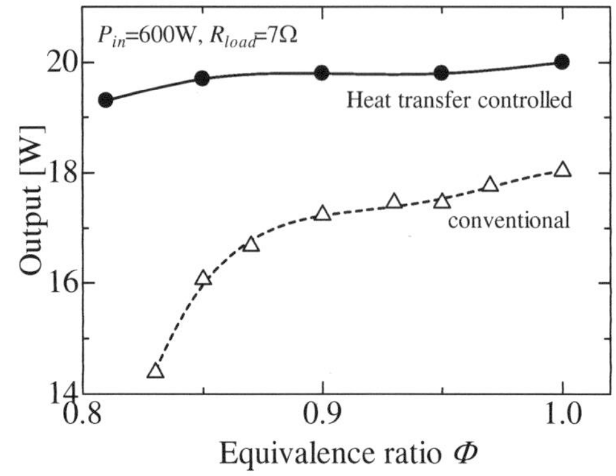

Fig.5 Variations of output power with equivalence ratio

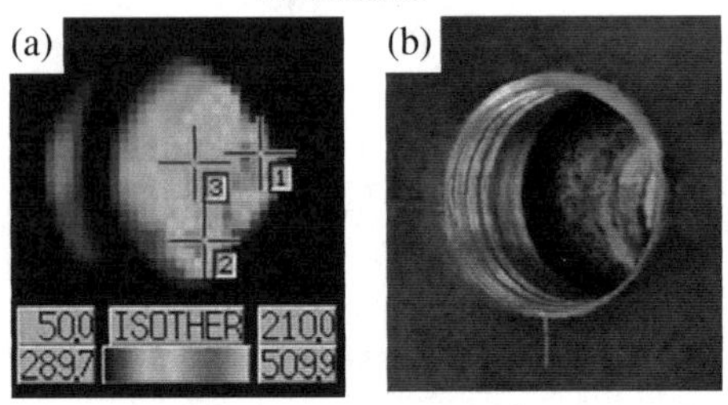

Positon	Temperature (°C)
1	450.0
2	406.1
3	381.9

Fig.6 Appearance of combustion chamber using thermography (a) and direct image (b) and temperature of each position

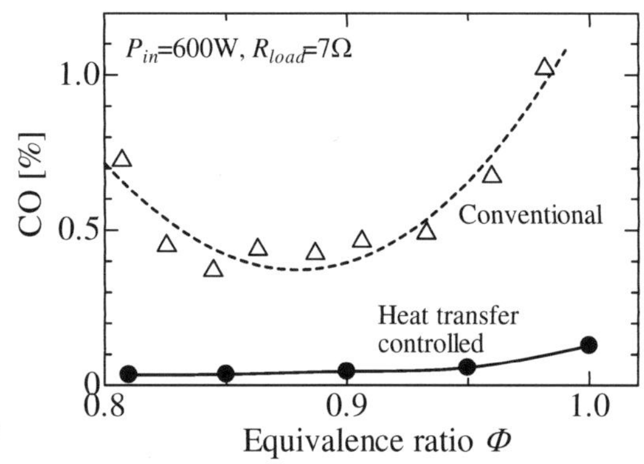

Fig.7 CO concentration for various equivalence ratio

0.81 may come from the incomplete combustion due to the cold wall effects. On the other hand, CO concentration of new combustor is significantly low for all equivalence ratio conditions. For $\Phi = 0.81$, CO concentration is suppressed to 330ppm with 18.3W of high output. These results clearly indicates that the new combustor is much effective compared to the conventional combustor.

4. References

[1] Fernandez-Pello C A 2002 *Proc. Combust. Inst.* **29** 883.

[2] Dunn-Rankin D, Leal E M and Walther D C 2005 *Prog. Energ. Combust. Sci.* **31** 422.

[3] Maruta K, Ju Y, 2011 Prog. Energ. Combust. Sci. 37 669.

[4] Shimokuri D, Hara T and Matsumoto R 2015 *J. Mecromech. Microgen.* **25** 104004.

[5] Shimokuri D, Taomoto Y, Matsumoto R 2017 *Proc. Combust. Inst.* **36** 4253.

[6] Li Y H, Li H Y, Dunn-Rankin D, Chao Y, 2009 *Proc. 7th AsPaCC* ID: 10098.

[7] Merotto L, Fanciulli C, Donde R, De Luliis S, 2016 *Appl. Energy* **162** 346.

Acknowledgments

This work was supported in part by Grant-in-Aid for Young Scientists (A) 15H05509. The authors gratefully acknowledge the support from KELK Ltd.

Improvement of effectiveness and output of electret energy harvester by symmetric comb-drive structures

H Honma[1], H Mitsuya[2], G Hashiguchi[3], H Fujita[1], and H Toshiyoshi[1]

[1]The University of Tokyo, Meguro-ku, Tokyo, Japan
[2]Saginomiya Seisakusho, Shinjuku-ku Tokyo, Japan
[3]Shizuoka University, Hamamatsu, Shizuoka, Japan.

honma-hh@iis.u-tokyo.ac.jp

Abstract. In this paper, we experimentally demonstrate a new method to improve the effectiveness of energy harvester that is based on solid-state electret. A symmetric three-ports electrostatic comb-electrode mechanism is used to reduce the binding electrostatic force that constraints the movable electrodes, thereby allowing a formation of a higher electret potential without electrostatic collapse. Two energy harvesters having different electret potentials are prepared to compare the harvester effectiveness and the electrical damping. The harvester effectiveness is increased by 4 times by increasing the electret potential from the -60 V to -250 V. At the same time, the electrical quality factor is also decreased from 700 to 500. Therefore, this new mechanism provides the energy-harvesting device for a wireless sensor node which can compatibly be made compact and can generate large output.

1. Introduction

Energy harvester is a vital component for Internet-of-Things (IoT) applications [1] that must operate without batteries to be free from frequent replacement. Amongst various harvesters, vibration type is thought to be a most potential for powering wireless sensor nodes that are deployed in various areas such as monitoring system for social infrastructures, where further of miniaturization and high power are needed to drive wireless sensor nodes. Amongst various principles, electret-based energy harvester can effectively convert the ambient vibrational energy owing to the flexibility in MEMS design to adapt to the environmental frequency range. Use of a large stroke of a moving mass and improvement of conversion effectiveness are a straightforward method to increase the output power [2, 3]. In general, the output power of the electret-based energy harvester increases by increasing the electrical damping [4]. However, electrostatic constraint force also increases with increasing the electret density, resulting in small mechanical amplitude of the moving mass in a small acceleration range [5].

To solve this problem, we newly introduce a symmetric electrodes layout that cancels the binding electrostatic force despite a high-density electret. At the same time, the output current under low acceleration condition is increased [6] owing to the large force factor. In this paper, we experimentally demonstrate the improvement of the harvester effectiveness by adopting a three-port comb-electrode mechanism. Performance is studied on two different electret potentials to compare the harvester effectiveness and the electrical damping.

2. Design and fabrication

When the vibrational energy harvester operates at resonance, generating power is given by [7, 8]

$$P_{rms} = \frac{1}{4} m\omega^3 X_{lim} y E_H, \tag{1}$$

where m and ω are the mass and the resonance angular frequency of the movable electrode, respectively. Variables y and X_{lim} are the oscillation displacements of the movable electrode and the chip, respectively, and E_H is the energy conversion efficiency defined by

$$E_H = \frac{1}{1+\frac{Q_e}{Q_m}} \tag{2}$$

which depends on the mechanical Q_m and electrical Q_e-factors [7].

The harvester effectiveness is increased because the electrical Q-factor is also reduced with the increase of the electret potential [4]. At the same time, large displacement of the oscillator mass is required to convert kinetic energy into high electrical power.

Figure 1 shows the schematic structure of the energy harvester based on the symmetric three-ports mechanism to cancel the electrostatic binding force. The electret formed by potassium doping is deployed on the comb fingers [9], while the moving shuttle is electrically grounded. Figure 2(a) shows a developed energy harvester. Scanning electron microscope (SEM) images are shown in Fig. 2(b) and (c). Table 1 shows measured displacements of the moving shuttle. Total 900 pairs of comb-fingers are made in a 100-μm thick SOI, and finished with the electrets charged at -60 V and -250 V with respect to the moving shuttle [6]. The suspensions are designed at a resonant frequency of 125 Hz. Maximum displacements are almost equal, 347 μm and 348 μm for the electrets of -60 V and -250 V, respectively, thereby the design of the mechanical and electrical dumping is independently performed by the symmetric structure.

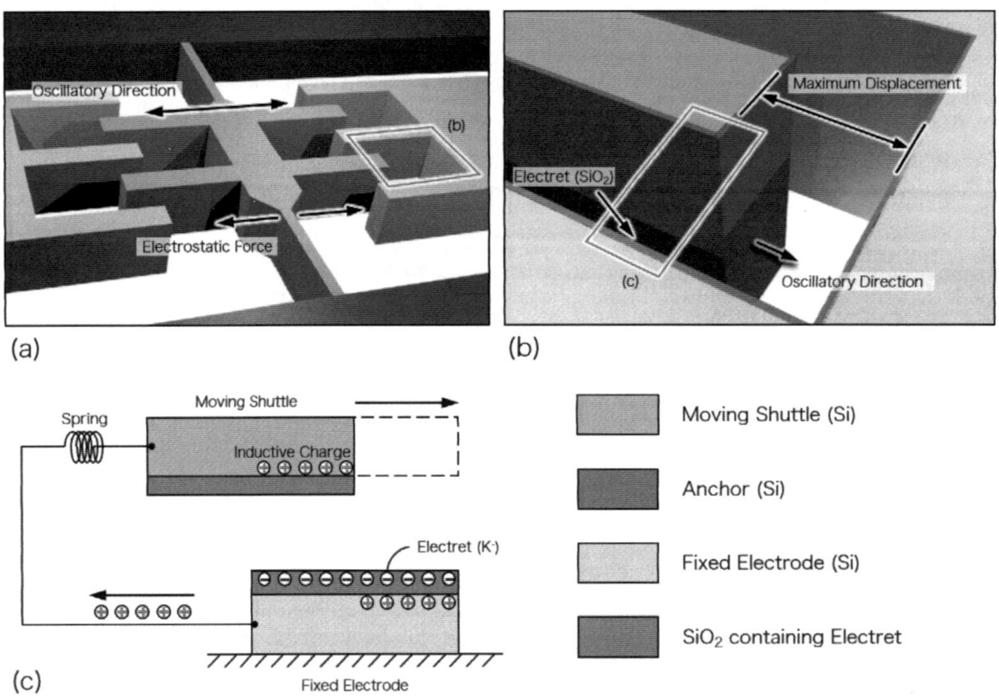

Figure 1. (a) Schematic structure of the electrostatic energy harvester based on the three-ports comb electrode mechanism. (b) Schematic picture of the electret formed on the side wall of comb fingers. (c) The relationship between the mechanical operation and the inductive charge in the electret-based energy harvester.

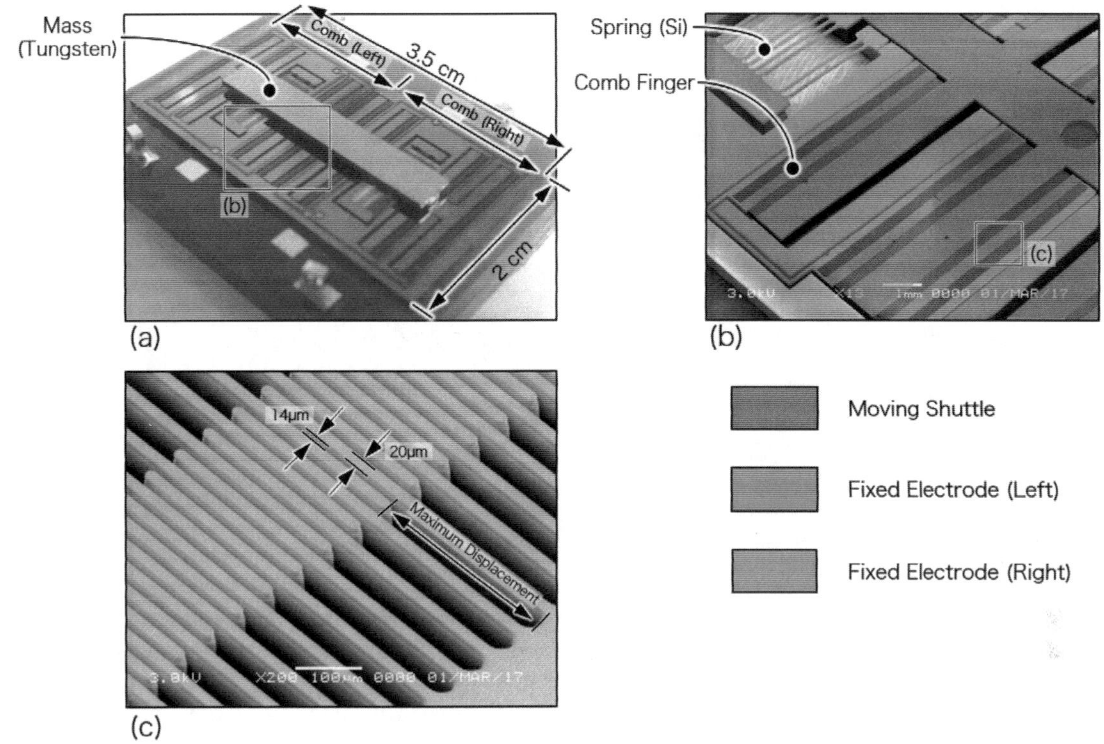

Figure 2. (a) Developed energy harvester composed of the three-ports comb-electrode mechanism. (b) SEM image of the electrode layout, and (c) close-up view of the comb electrodes.

Table 1. Experimentally observed displacement of the energy harvesters prepared with two different electret potentials.

Electret Potential (V)	Maximum Displacement (μm)
-60	346.8
-250	347.5

3. Experimental results

The developed energy harvesters were put on the vibration-testing equipment and mechanically shaken at a constant acceleration and oscillation stroke. At the same time, the oscillation displacement of the moving electrode was measured by the high-speed camera.

Figure 3(a) shows the experimental setup to measure the output power by using the induction current when the device is shaken. Figure 3(b) is the harvester effectiveness as a function of the load resistance. The moving mass was operated at the maximum operating condition. Maximum effectivenesses at the resonance were 23% and 86% for the electret potentials of -60 V and -250 V, respectively. From this measurement, use of the higher electret potential effectively improves the harvester effectiveness in a small acceleration range. Figure 3(c) shows the experimentally obtained displacement of the moving mass as a function of the load resistance. The input displacement of 0.1 μm is used to keep the oscillator mass at a constant amplitude at resonance. Depression of stroke was 10 μm and 36 μm for the electret potentials of -60 V and -250 V, respectively. The electrical Q-factor was also decreased from 700 to 500 by increasing the electret potential value.

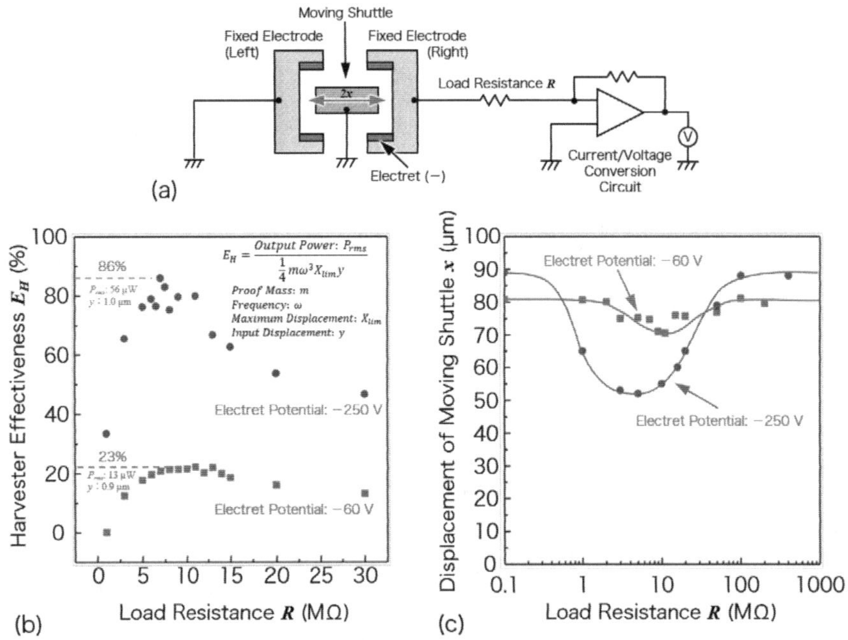

Figure 3. (a) Harvester effectiveness of the electrodes charged at -60 V and -250 V with respect to the moving mass. (b) Harvester Effectiveness as a function of the load resistance at the maximum operating condition in resonance. (c) Measured stroke of the moving shuttle as a function of the load resistance under the input displacement of 0.1 μm in resonance.

4. Conclusions

In conclusion, a symmetric comb-electrode structure was found to have improved the electret potential while maintaining the stable mechanical operation of the energy harvester. A use of higher electret potential was also found to increase the electrical damping, which also increases the power converted into electrical output. Therefore, maximum output power was further increased by using higher acceleration for input.

References

[1] Atzori L, Iera A and Morabito G 2010 *Comput. Networks,* **54** 2787-2805
[2] Mitcheson D P, Yeatman M E, Rao K G, Holmes S A and Green C T 2008 *Proc. THE IEEE,* **96** 1457-86
[3] Sakane Y, Suzuki Y and Kasagi N 2008 *J. Micromech. Microeng.*, **18** 104011
[4] Mitcheson D P, Reilly K E, Toh T, Wright K P and Yeatman M E 2007 *J. Micromech. Microeng.*, **17** S211-S216
[5] Suzuki Y, Miki D, Edamoto M and Honozumi M 2010 *J. Micromech. Microeng.*, **20** 104002
[6] Honma H, Mitsuya H, Hashiguchi G, Fujita H and Toshiyoshi H 2017 *Proc. 19th Int. Conf. Transducers 2017*, June 18-22, 2017, Kaohsiung, Taiwan.
[7] Beeby P S, Tudor J M and White M N 2006 *Meas. Sci. Technol.*, **17** R175-R195
[8] Mitcheson D P, Green C T, Yeatman M E and Holmes S A 2004 J. Microelectromech. Sys., **13** 429-440
[9] Koga H, Mitsuya H, Sugiyama T, Toshiyoshi H and Hashiguchi G 2016 *Proc. 16th Int. Conf. PowerMEMS 2016,* Dec. 6-9, 2016, Paris, France.

Acknowledgement

This work was supported by JST CREST Grant Number JPMJCR15Q4, Japan.

Characterization of fluorinated nematic liquid crystal for high-power electrostatic energy harvester

K. Kittipaisalsilpa[1], T. Kato[2], and Y. Suzuki[1]

[1]Department of Mechanical Engineering, The University of Tokyo, Japan
[2]Department of Chemistry and Biotechnology, The University of Tokyo, Japan

Email: kasidis@mesl.t.u-tokyo.ac.jp

Abstract. The use of fluorinated nematic liquid crystal (NLC) as the medium in between the stator and the rotor of electrostatic vibration energy harvester is proposed for output power enhancement. Fluorine terminal group of 3,4,5-Trifluoro-4'-(trans-4-pentylcyclohexyl)-1,1'-biphenyl (BCH-5F.F.F) offers high resistivity that is two-order-of-magnitude higher than that of 5CB used in our previously study. In a power generation experiment, output power as high as 490 μW is obtained at 10 Hz and 1.00 $\text{mm}_{\text{peak-peak}}$ vibration, while keeping the leakage current as low as a few μA.

1. Introduction

Energy harvesting [1, 2] attracts much attention for powering low-power-consumption electronics. Electrostatic/electret vibration energy harvester (EH) has good compatibility with MEMS technologies, and is advantageous in power generation under low vibration frequencies and in small dimensions [2-4]. However, the output power is very sensitive to the parasitic capacitance (C_p) between interdigitated electrodes, and its reduction is crucial for the electrostatic/electret EH [5-7].

Chen et al. [7] propose a model of C_p for current-collector electrodes, and demonstrate improvement of the output power by using a MEMS-based suspended electrode structure with low effective permittivity of the substrate. In general, C_p consists of 2 parts; the capacitance through the substrate (permittivity: ε_1) and through the dielectric in the gap (permittivity: ε_2) as shown in Figure 1. When a high permittivity fluid is filled inside the gap between the rotor and the stator, the vertical permittivity ($\varepsilon_{||}$) is increased, so that the capacitance change during the vibration is increased. However, with isotropic permittivity material ($\varepsilon_{||} = \varepsilon_{\perp}$), C_p also becomes large, which leads to deterioration of the output power. To address this issue, we previously proposed the use of nematic liquid crystal (LC) with positive anisotropic permittivity ($\Delta\varepsilon = \varepsilon_{||} - \varepsilon_{\perp} > 0$) [8]. Figure 2 shows the concept of liquid-crystal-enhanced electrostatic generator. The rod-shaped molecules filling the gap between the rotor and the stator align its longitudinal axis parallel with electric field. With high $\varepsilon_{||}$ and low $\varepsilon_{\perp}$, the power output is much increased, while the increase of C_p is suppressed. We demonstrated this concept using conventional nematic LC (4-Cyano-4'-pentylbiphenyl, 5CB). However, 5CB has large amount of ionic impurities as high as 10^{14} cm^{-3} [9], such that the resistivity tends to be low and the leakage current is very large. This makes it very difficult to obtain repeatable experimental data.

To reduce the leakage current, nematic LC with high resistivity and reasonable anisotropic permittivity is needed. According to Hard [10] and Kirsch [11], fluorine substituent in LC molecule modifies the LC properties depending on the number of substituents and their location such as, the melting point, the mesophase range, the anisotropic permittivity, and the resistivity. In addition, fluorine has the highest electronegativity among elements in periodic table, making high polarity for C-F bonds. In the present study, we propose fluorinated NLC as a new anisotropic permittivity liquid for power enhancement of electrostatic energy harvesters.

2. Fluorinated Nematic Liquid Crystal

In this study, we employ 3,4,5-Trifluoro-4'-(trans-4-pentylcyclohexyl)-1-1'-biphenyl (BCH-5F.F.F, Ark Pharm) as the representative of fluorinated NLC. Figure 3 shows the molecular structures of 5CB used in our previous study, and BCH-5F.F.F. The major difference between these two NLC is their terminal groups. Dipole moment by the cyano substituent (C≡N) in 5CB exhibits high polarity and high polarizability. On the other hand, three fluoro substituents at the end of aromatic rings in BCH-5F.F.F exhibits low polarizability, while maintaining high polarity. High polarity ensures the anisotropic permittivity. Polarizability describes the ease of electron cloud distortion by external influences. In other words, lower polarizability confers higher resistivity [10-11]. The resistivity of BCH-5F.F.F in our impedance measurement is 7.4×10^9 Ωm, which is two-order-of-magnitude higher than that of 5CB (1.3×10^7 Ωm).

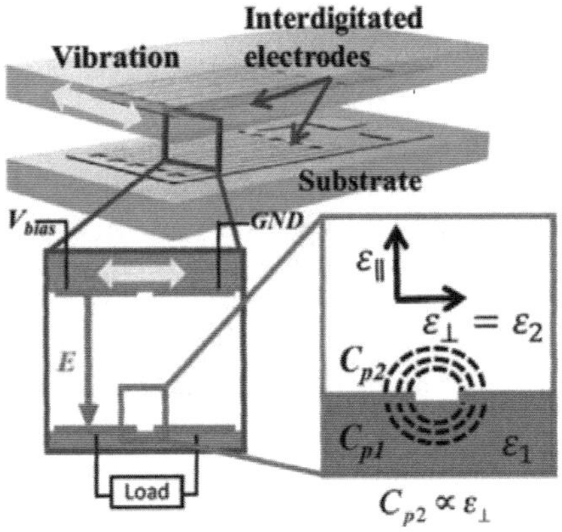

Figure 1. Schematic diagram of electrostatic energy harvester and origin of parasitic capacitance.

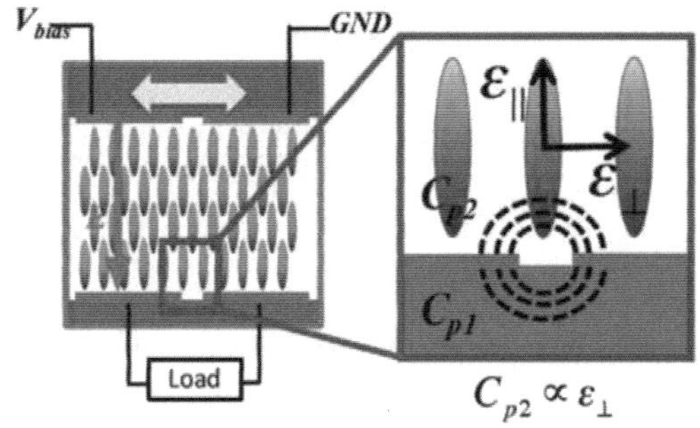

Figure 2. Concept of liquid-crystal-enhanced electrostatic energy harvester.

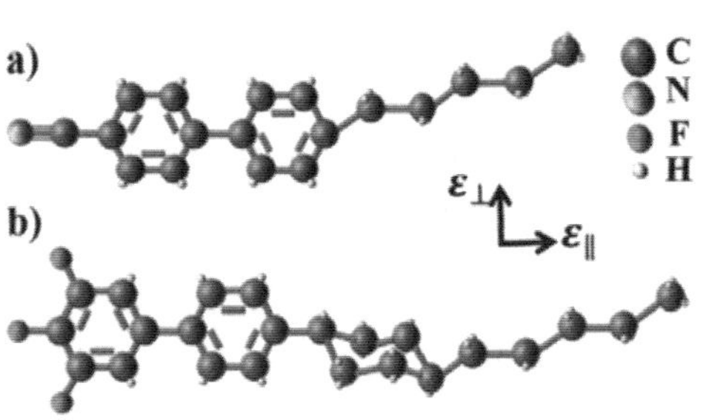

Figure 3. Molecular structures of (a) 5CB, and (b) BCH-5F.F.F.

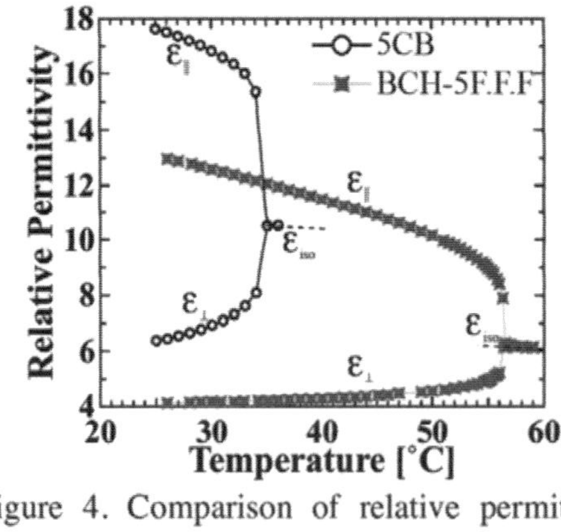

Figure 4. Comparison of relative permittivity between 5CB and BCH-5F.F.F as a function of temperature.

Figure 4 shows the comparison of relative permittivity as a function of temperature. A glass cell with a 5 μm gap is used, and the permittivity is measured at 1 kHz. For both NLCs, the horizontal alignment is used, and the transverse permittivity $\varepsilon_{\perp}$is measured with a low applied AC voltage. When the applied AC field is higher than 2 V/μm, the orientation of NLCs is changed from horizontal to vertical, so that the axial permittivity $\varepsilon_{||}$ can be measured. Although the axial permittivity $\varepsilon_{||}$ of BCH-5F.F.F is lower than that of 5CB, its $\varepsilon_{\perp}$ is also lower. Thus, the ratio of $\varepsilon_{||}/\varepsilon_{\perp}$ is 3 in both NLCs. The nematic range of BCH-5F.F.F is 29 – 56 °C, which is wider than that of 5CB (22 - 34°C), leading to wider operation temperature range of liquid-crystal-enhanced electrostatic energy harvester.

3. Power Generation Experiment Setup

The experimental setup for power generation experiments can be found in Ref. [8]. The interdigitated Cr/Au electrode was patterned on a Tempax substrate using MEMS technologies. The width of a pair of electrode fingers is 0.5 mm with the interdigitated gap of 0.05 mm. The electrode area is 20 mm × 20 mm. Figure 5a shows the finished substrates with the interdigitated Cr/Au electrodes.

Schematic of the experimental setup and its photo are shown in Figure 5b and 5c, respectively. Briefly, two substrates with the interdigitated electrodes are set in parallel on the top and the bottom stages with the gap of 100 μm. The top electrode is connected to a source meter unit (2410, Keithley) for the bias voltage of -400 V, while the bottom one is connected to an external load. The vibration frequency and amplitude are 10 Hz and 1.00 $mm_{peak-peak}$, respectively. BCH-5F.F.F is filled in the dielectric gap with an estimated amount of 40 μL, and its temperature is kept at around 32 °C. The leakage current with BCH-5F.F.F is as low as 6 μA, which is much lower than that for 5CB.

To avoid an unwanted effect of the parasitic capacitance of the read-out circuit, the output current is connected to a programmable current amplifier (CA 5350, NF) through an external load, by which the current is converted to a DC voltage with the gain of 10^6 V/A. Then, the output voltage is measured and the data are stored by a data logger unit (NR-600, Keyence).

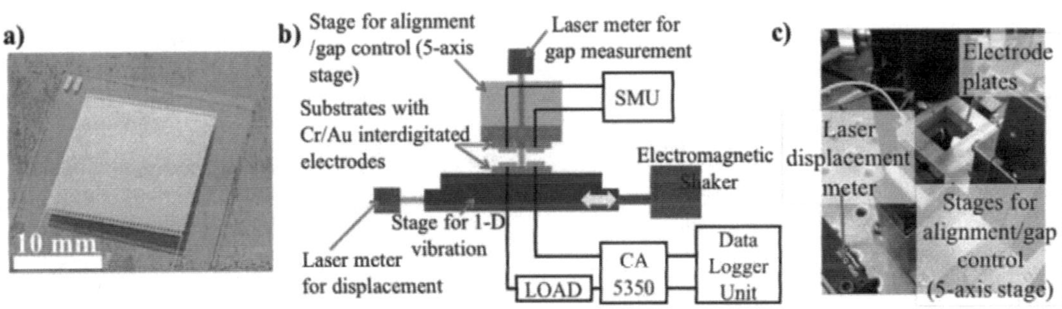

Figure 5. (a) Microfabricated Cr/Au interdigitated electrode on Tempax glass, (b) Schematic diagram of experimental setup for power generation, and (c) photo of the setup.

4. Experimental Results

Figure 6 shows the output voltage waveforms at optimal load resistance of the air and the fluorinated-NLC-filled gap. The number of positive and negative peaks corresponds to number of times that top and bottom electrodes fully overlap with each other in one oscillation period. The peak-to-peak voltage for BCH-5F.F.F reaches 210 V, which is much higher than that of the air gap (50 $V_{peak-peak}$). The output power versus the load resistance is shown in Figure 7. For the air gap, the optimal load is 35 MΩ, and the maximum output power P_{max} at the matched impedance is 4.98 μW. On the other hand, for the BCH-5F.F.F-filled gap, P_{max} is as high as 490 μW at 7 MΩ, which corresponds to 100 times higher output power compared to that of the air gap.

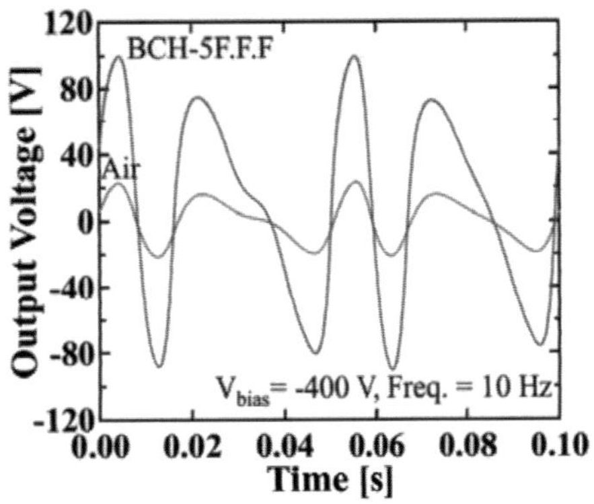

Figure 6. Output voltage waveform at the optimal load resistance.

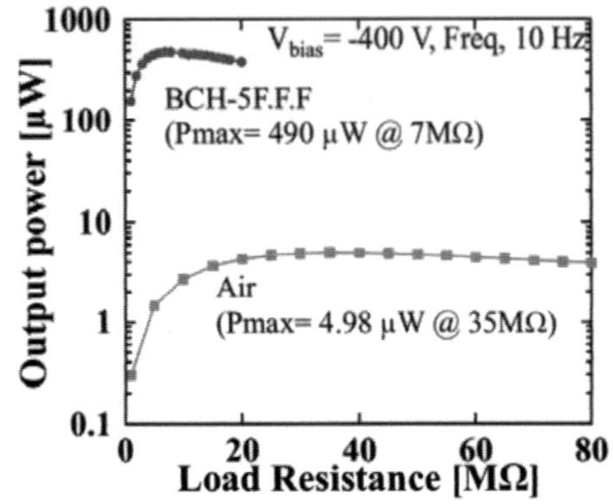

Figure 7. Output power for the air and the BCH-5F.F.F-filled gaps.

5. Conclusion

In this paper, we have proposed the use of fluorinated NLC for liquid-crystal-enhanced electrostatic energy harvester, which has higher resistivity than conventional NLC with cyano terminal. With BCH-5F.F.F, which has wide temperature range of the liquid crystal phase, the resistivity has been increased by 570 times if compared with 5CB. On the other hand, the ratio between the permittivities in the two orthogonal directions is almost the same at around 3. In preliminary vibration power generation experiment, the output power has been increased by 100 times (P_{max}= 490 μW) compared with the device with an air gap.

This work was supported by JST CREST Grant Number JPMJCR15Q3, Japan.

References

[1] S. Roundy, P. K. Wright, and J. Rabaey, "A study of low level vibration as power source for wireless sensor nodes," *Comput. Commun.*, vol. 26, pp. 1131-1144 (2003).

[2] Y. Suzuki, "Recent progress in MEMS electret generator for energy harvesting," *IEEJ Trans. Electr. Electr. Eng.*, vol.6, pp 101-111 (2011).

[3] C.P. Le, and E. Halvorsen, "MEMS electrostatic energy harvesters with end-stop effects," *J. Micromech. Microeng.*, vol. 22, 074013 (2012).

[4] P. Basset et al., "Electrostatic vibration energy harvester with combined effect of electrical nonlinearities and mechanical input," *J. Micromech. Microeng.*, vol. 24, 035001 (2014).

[5] T. Masaki et al., "Power output enhancement of a vibration-driven electret generator for wireless sensor applications," *J. Micromech. Microeng.*, vol. 21, 104004 (2011).

[6] U. Bartsch et al., "Influence of parasitic capacitances on the power output of electret-based energy harvesting generators," *Power MEMS 2009*, Washington DC, pp. 332-335 (2009).

[7] R. Chen, and Y. Suzuki, "Suspended electrode for reducing parasitic capacitance in electret energy harvesters," *J. Micromech. Microeng.*, vol. 23, 125015 (2013).

[8] K. Kittipaisalsilpa, T. Kato, and Y. Suzuki, "Liquid-crystal-enhanced electrostatic vibration generator," *IEEE MEMS 2016*, Shanghai, pp. 37-40 (2016).

[9] H. Mada, and M. Ryuzaki, "Ion influence on nematic liquid crystal cell impedance at low frequency," *Jpn. J. Appl. Phys.*, vol. 34, pp. 1134-1136 (1995).

[10] M. Hard, "Fluorinated liquid crystals – properties and applications," *Chem. Soc. Rev.*, vol. 36, pp. 2070-2095 (2007).

[11] R. Kirsch, "Fluorine in liquid crsytal design for display applications," *J. Fluorine Chem.*, vol. 177, pp. 29-36 (2015).

Batch-fabrication and characterization of miniaturized axisymmetric electropermanent magnets

C Velez, L P Tatum, B S Herstein, C P Becker and D P Arnold

Interdisciplinary Microsystems Group, Department of Electrical and Computer Engineering, University of Florida, FL, 32611, United States

Email: darnold@ufl.edu

Abstract. This work describes the simulation, batch fabrication, functional demonstration, and magnetic field/force characterization of highly miniaturized ~3.8 mm^3 axisymmetric electropermanent magnets (EPMs). Two different bonded high-coercivity, rare-earth permanent magnet powders (~15 µm Sm_2Co_{17} particles and 6 µm NdFeB particles) were evaluated for the fabrication of the fixed magnets in the EPM, reaching latching force on/off ratios of 191:1 and 303:1 respectively. Compared to conventional side-by-side architecture, the axisymmetric design provides: 1) symmetric magnetic field along the magnetization axis, as opposed to the asymmetric fields in the typical transverse configuration, 2) higher and controllable latching force in a smaller volume, and 3) a path for further batch-microfabrication and system integration.

1. Introduction

An EPM is a switchable magnet [1] that only consumes power during transition between the on/off states. The EPM structure comprises a switchable magnet and a fixed magnet, which are sandwiched between two soft magnetic cap plates. The switchable magnet switches its magnetization direction to turn "on" or "off" a magnetic latching field. Application drivers for miniature EPMs include actuators/latches for microrobots and programmable matter [2], switchable magnets for microfluidic devices [3], microgrippers [4] and more. The EPM makes use of a pulse of current in a coil to flip the magnetization of one of the pair [5]. Figure 1 compares the traditional side-by-side EPM (transversal field [2–4]) vs. the new (axisymmetric field) architecture. The axisymmetric architecture aims to generate stray B fields and latching forces along the axis, using two concentric permanent magnets and two soft magnetic cap plates. This work focuses on a design and fabrication of a novel sub-millimeter axisymmetric EPM architecture and a fabrication technique that is amenable to miniaturization, building on the theory and design principles previously proposed by the authors in [6].

2. Simulations of the Axisymmetric EPM

Using 2D simulations in COMSOL Multiphysics, the B fields and latching forces of both architectures are studied as a function of several design variables: outer radius, permanent magnet aspect ratio and radius ratio, and cap plate thickness. For this investigation, the following materials were assumed: grade N52 NdFeB as fixed inner magnet, grade 5 Alnico for the outer switching magnet, and mild/low-carbon steel (AISI 1018) for the cap plates. The cap plate thickness was found to be an important variable for tuning the on/off latching force ratio Figure 1b shows the final proposed dimensions as a function of the magnetic layer thickness (T). These dimensions were selected to maximize the on/off latching force ratio for a given cap plate thickness.

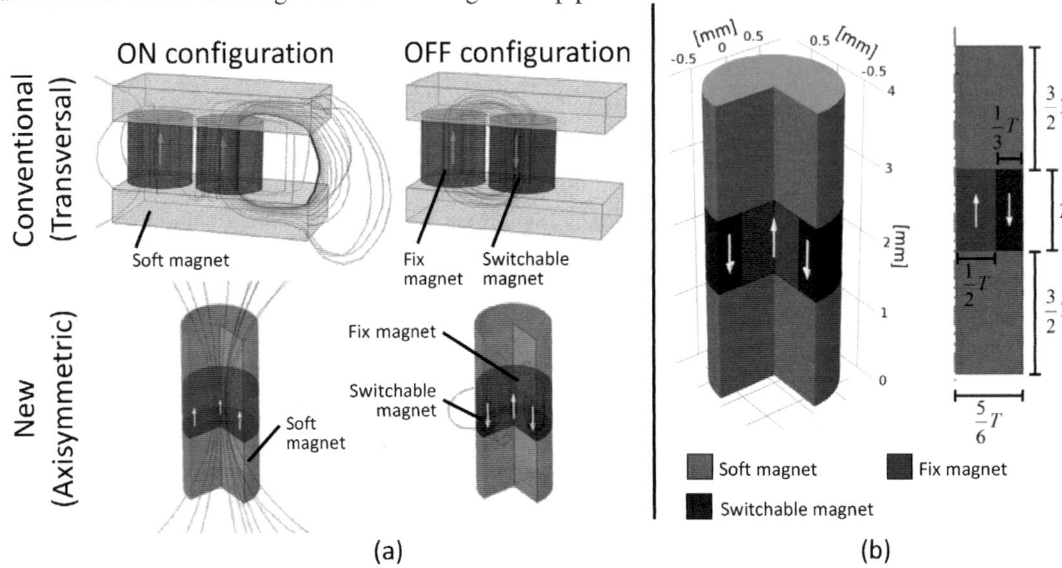

Figure 1. (a) Operational concept of the conventional and new axisymmetric EPM architectures showing the off and on states. (b) Example EPM configuration yielding maximum latching force on/off ratio (a 3T/2 thickness in cap plates will yield 450 force ratio).

Simulations of the stray magnetic field of the optimal EPM evidence ~10x difference in B field near the EPM poles in the on/off states (Figure 2a). Figure 2b shows the magnitude of the on/off B-fields along the central axis with and without the cap plates. Additional modeling is used to predict the latching force to a semi-infinite steel plate. Figure 2c shows how the on/off latching force ratio can be tuned by changing the thickness of the cap plates. Simulations of the new configuration (dimensions used in Figure 1b suggest an on/off force ratio of 450, with the ability for tuning this ratio from 1 up to 784, for sizes between 8 mm^3 and 34 mm^3, respectively. These results suggest the new axisymmetric design may afford hundred-fold higher latching force ratio than conventional EPM architectures in 1/10th the volume.

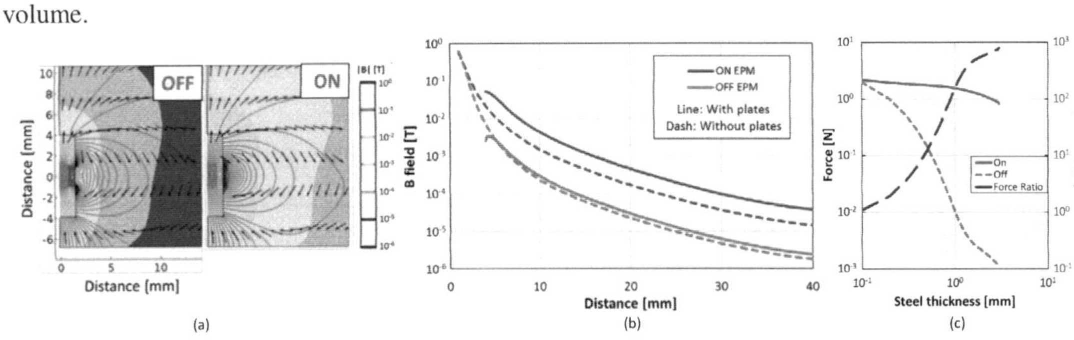

Figure 2. 2D axisymmetric COMSOL simulation. (a) B field norm contour plots. (b) Comparison of on/off fields for EPM with and without cap plates. (c) Latching force generated by different EPM's (on and off states) vs. thickness of the steel plates.

3. Fabrication of the Axisymmetric EPM

In the axisymmetric EPM, a ring-shaped switchable magnet surrounds a cylindrical fixed magnet. The concentric magnets are fabricated by punching ~1 mm diameter holes (using a cutting cannula) in

0.8-mm-thick rubber-bonded iron oxide substrates (which forms the switchable magnetic material), previously demagnetized. Fixed magnets are fabricated on these holes by bonding high-coercivity, rare-earth permanent magnet powders (~15 μm Sm_2Co_{17} particles or 6 μm NdFeB particles) and cyanoacrylate glue as bonding agent (applied on both sides), following technique described in [7]. Finally, a ~2 mm outer ring is punched, resulting in two-magnet assembly called EPM core (Figure 3). For the fully assembled EPMs, two steel cap plates (shunt) 125 μm thick were glued top and bottom of the magnet before punching the EPM core to guaranty self-align and uniform diameter of cap plates.

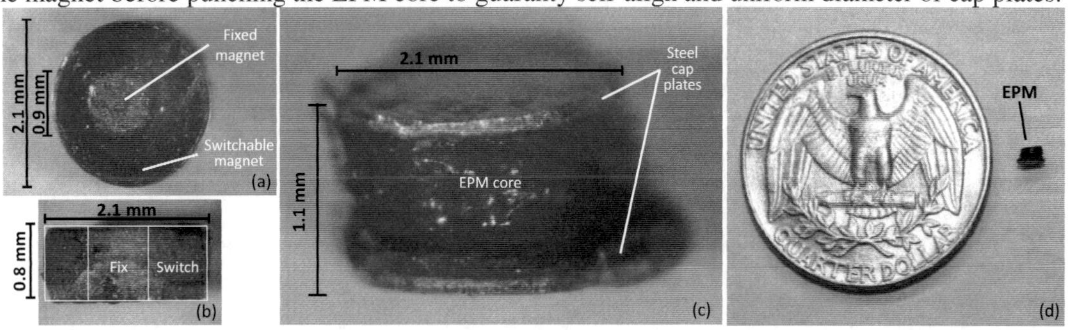

Figure 3. Fabrication of the EPM core, (a)Top view and (b) cross section. (c) Frontal view of the EPM including cap plates. (d) Size comparison.

4. Characterization of the Axisymmetric EPM

EPM cores are magnetically characterized by using vibration sample magnetometer (VSM, ADE Technologies EV9). Figure 4 presents a comparison of the magnetization curves for the EPM core (without caps) and the individual components (switchable and fixed bonded magnets). The SmCo EPMs yielded higher remanence (μ_0Mr = 117 mT) and lower intrinsic coercivity (H_{ci}=210 kA/m) than the NdFeB EPMs (75 mT and 252 kA/m).

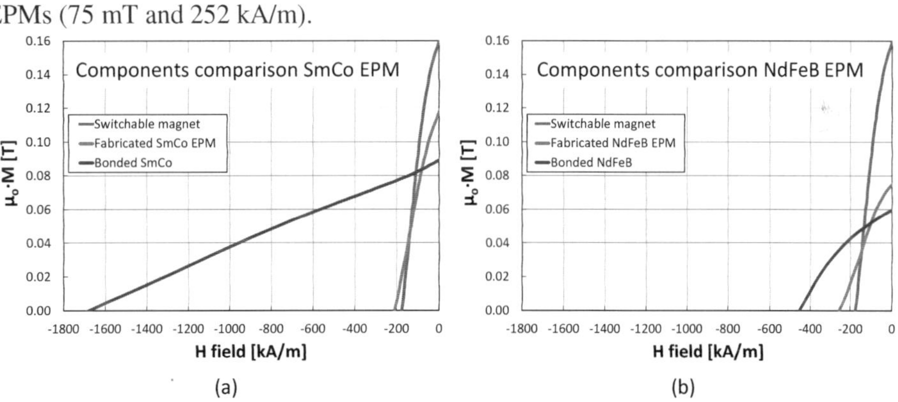

Figure 4. Second quadrant magnetization curve (VSM) of two EPM core configurations and its components. Each graph represents the magnetization of the bonded magnet, the switchable magnet and the fabricated EPM core for: (a) SmCo and (b) NdFeB EPM cores.

Additional magnetic characterization is performed by magneto optical images (MOI) and using a pulse magnetizer to switch the EPM cores from on state (pulsing 7 T in the axial direction) to off state (by pulsing -700 mT for SmCo or -440 mT for NdFeB EPMs). MOI characterization (Figure 5a) of EPMs at on/off states, demonstrates there exists a reversal magnetic field (>300 mT for SmCo or >430 mT for NdFeB) capable to reverse the magnetization of the switchable magnet without affecting the fixed magnet. Cross section measurements of the B field (Figure 5b) suggest that the average on/off ration of the magnetic cores for the SmCo (27.3 mT/-6 mT) and NdFeB (23 mT/8.3 mT) EPM cores are 4.6 and 2.8 respectively.

From the MOI images is it possible to calculate the magnetic flux (in units of nWb) produced by the EPM core when magnetized at different reversal magnetic fields (Figure 5c) and to report the magnetic flux on/off ratios. A full magnetization of (7 T) was always applied before applying different reverse magnetic fields (using VSM) to each EPM core before measurement. Figure 5c demonstrates that it exists a reversal magnetic field that cancel completely the magnetic flux of the EMP. This is a prove of concept of the feasibility of obtaining an "off" state from the EPM. This reversal magnetic field to turn the EPM off is >430 mT for the NdFeB sample and >300 mT for the SmCo sample.

Fully assembled EPMs with the steel cap plates were also used to measure the magnetic flux in the on/off state (by applying the reversal magnetic fields described above). The proof of concept of a EPM that can be turned "on" and "off" was demonstrated and magnetic flux on/off ratio of ~2 were achieved for both samples (NdFeB and SmCo).

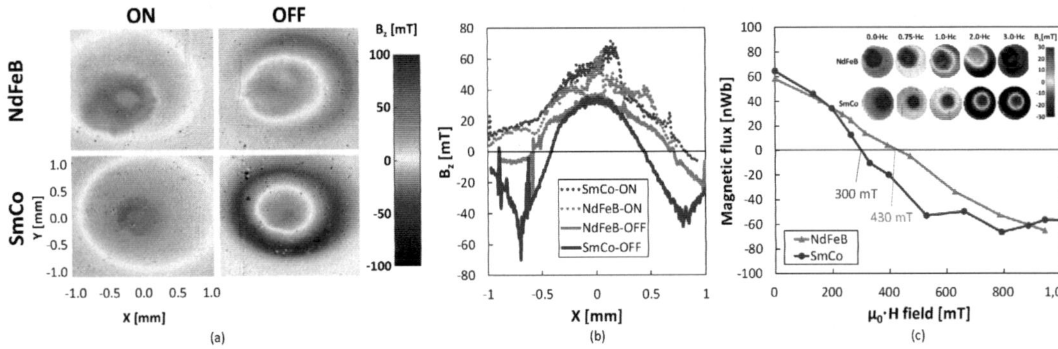

Figure 5. Stray magnetic field (MOI) for on and off states of the EPM. Two configurations with different bonded magnets were measured: NdFeB and SmCo. (a) top view MOI images and (b) cross section measurement over the X axis (c) Calculated net magnetic flux produced by two EPM configurations (NdFeB and SmCo) for different reversal fields. Inset shows selected MOI images from which the magnetic flux was calculated. Magnetization values were selected as a function of each fixed magnet coercivity (H_c).

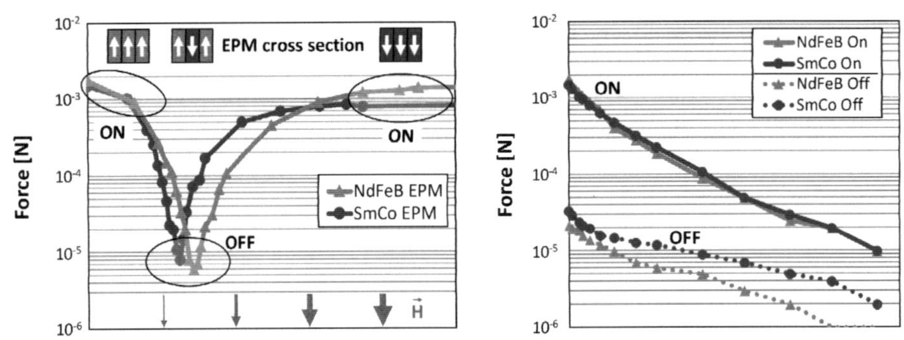

Figure 6. The demagnetization strength that turns each of the electropermanent magnets off was found from this graph by identifying the demagnetization state of each one that generated the lowest surface force. (b) Measured force at different distances from the EPM surface for on/off states.

An assemble of microbalance (Explorer 2, Ohaus) with an automated 3D micro positioner (build on house with Newport DC servo controllers) was implemented as a variation of experiments proposed by [8], to measure the latching force between the EPM and an approximate infinite ferromagnetic plate (mild/low-carbon steel AISI 1018). By cautiously lowering the EPM over the ferromagnetic plate, the latching force will raise the plate away from the balance and the force will be register as weigh in the balance. Figure 6a illustrate the latching force of the EPMs after applying different reversal magnetic fields (using VSM), demonstrating that latching on/off ratios of 191:1 (SmCo) and 303:1 (NdFeB) can be achieved. Figure 6b illustrates how the latching force will vary with the distance from the magnet to the ferromagnetic plate.

5. Conclusions

A novel axisymmetric configuration of EPM was proposed. Using simulations and optimization of sub-millimeter axisymmetric EPM it was predicted that the latching force ratio can be tuned from 1 up to 784, for sizes between 8 mm^3 and 34 mm^3 by varying the thickness of the cap plates. Compared to conventional architectures, the axisymmetric design also provides better performance in a smaller form factor and symmetric magnetic field along the magnetization axis, as opposed to the asymmetric fields in the transverse direction.

A fabrication technique that enables the batch microfabrication of this components was also demonstrated. Two fully assembled EPM SmCo and NdFeB with latching force ratio of 191:1 and 303:1 were demonstrated. Magneto optical images (MOI) characterization showed the it was possible to cancel the magnetic flux of the EPM cores, applying external reversal fields >300 mT for SmCo or >430 mT for NdFeB. Force characterization using a build on house measuring technique proved that the external reversal fields to turn the EPMs off were >245 mT for SmCo and >282 mT for NdFeB. These are fundamental advances in the small-scale fabrication of EPMs and an important demonstration of the working principle for the axisymmetric EPM design.

References

[1] Israelson A F 1965 Turn-off permanent magnet, US Patent 3452310 3

[2] Knaian A 2010 *Electropermanent magnetic connectors and actuators: devices and their application in programmable matter. PhD Dissertation, MIT*

[3] Padovani J I, Jeff S S and Howe R T 2016 Electropermanent magnet actuation for droplet ferromicro fluidics *Technology* **4** 110–9

[4] Velez C, Arnold D P, Gonzalez Z I and Osma J F 2017 Sub-millimeter Electropermanent Magnets for Microgrippers *International Conference on Manipulation, Automation and Robotics at Small Scales (MARSS)* pp 20–2

[5] Jurish F 1996 "Electrical permanent-magnet system for manouvring a magnetic, particularly a ferromagnetic, load" International Patent WO1997003911 24

[6] Velez C, Becker C and Arnold D P 2016 Investigation of axisymmetric sub-millimeter electropermanent magnets (Late News) *PowerMEMS* (Paris - France)

[7] Oniku O D, Bowers B J, Shetye S B, Wang N and Arnold D P 2013 Permanent magnet microstructures using dry-pressed magnetic powders *J. Micromechanics Microengineering* **23** 75027

[8] Laumann D and Heusler S 2017 Determining magnetic susceptibilities of everyday materials using an electronic balance *Am. J. Phys.* **85** 327–32

Acknowledgments

The authors acknowledge Dr. Alexandra Garraud, Nicolas Garraud, and Xiao Wen for valuable technical discussion.

IOP Conf. Series: Journal of Physics: Conf. Series **1052** (2018) 012046 doi:10.1088/1742-6596/1052/1/012046

High-temperature compatible, monolithic, 3D-printed magnetic actuators

Anthony P Taylor[1] **and Luis F Velásquez–García**[2,*]

[1] Edwards Vacuum LLC, Sanborn, NY, 14132, United States of America
[2] Microsystems Technology Laboratories, Massachusetts Institute of Technology, 77 Massachusetts Ave., Cambridge, MA, 01239, United States of America

* Corresponding author, Velasquez@alum.mit.edu

Abstract. We report the design, fabrication, and characterization of the first miniature 3D-printed, monolithic magnetic actuators that are compatible with high temperature (>200 °C) operation. The actuator is a cylindrical frame that holds a coil and a 24 mm diameter, 150 μm-thick membrane connected at its center to a piston with a cavity containing an embedded SmCo magnet. The fused filament fabrication (FFF) method is used to print the non-magnetic portion of the actuator out of pure nylon 12. The displacement of a single-layer membrane actuator is characterized for various coil bias voltages; a maximum displacement equal to 302 μm was obtained with 20V DC applied to the driving coil. Data analysis demonstrates that the magnetic force is proportional to the square of the current drawn by the coil –as expected from theory.

1. Introduction

Various PowerMEMS devices require large displacement and large force actuation to be efficient, e.g., diaphragm vacuum pumps [1],[2]. However, hardware made with standard i.e., cleanroom, microfabrication typically cannot deliver such performance without the added device complexity of hydraulic amplification of large-force, small-displacement actuation, e.g., piezoelectric [3]. Magnetic actuation is an attractive choice to deliver large displacement and large force in a compact form factor [4]; nonetheless, state-of-the-art magnetic actuation for microsystems focuses on monolithic integration of the fabrication of the magnets within the cleanroom-bound process flow of the microsystem [5], highly constraining the choice of materials, geometries, and substrates.

Additive manufacturing, i.e., the layer-by-layer fabrication of objects using a computer-aided design (CAD) file [6], has recently been explored as a processing toolbox for MEMS; researchers have reported 3-D printed microsystems with performance on par or better than counterparts made with standard microfabrication [7], as well as demonstrated microsystem designs not previously implemented due to fabrication complexity or inherent system three-dimensionality [8],[9]. Proof-of-concept stereolithography 3D-printed valves with NdFeB magnets have been reported [10], but the devices required a large external electromagnet for actuation, had magnets inserted after printing, and the NdFeB magnets should demagnetize at temperatures above ~80 °C. In this work, miniature actuators were printed in pure nylon 12 –a high-performance plastic, using the fused filament fabrication (FFF) method –where a thermoplastic filament is extruded from a hot nozzle to create layer by layer a solid object. The actuators have embedded SmCo magnets that are not demagnetized by the heated nozzle (@ 250 °C) while being sealed in place midstream in the printing process.

2. Characterization of printing method and feedstock

A CreatBot model DX Plus 3D printer (Henan Suwei Electronics Tech., Zhengzhou City, China) was used to manufacture test structures and devices in nylon 12 (Orbi-Tech, Leichlingen, Germany) with a 600 μm-diameter stainless steel nozzle extruding feedstock at 250 °C. The printer has a glass bed that was coated with a thin layer of PVA-based adhesive before printing to promote adhesion of the object to the bed. Nylon is hydrophilic and water absorption adversely impacts FFF printing; therefore, to minimize filament moisture content, the spooled filament was baked at 80 °C for 2 hours in a vacuum oven prior to use, and the printer was housed in a 3DPrintClean Model 600 enclosure (Mountainside, NJ) with recirculating HEPA filtration.

To FFF-print an object, a CAD model in *stl* format is created (SolidWorks 2015, Dassault Systemes, Waltham, MA) and exported to a slicer software (Simplify3D), which transforms the 3-D model into a set of horizontal cuts, i.e., *slices*, creating a *gcode* file that contains the traveling path that rasters each slice with associated conditions (e.g., nozzle temperature, bed temperature, feedstock feed rate, nozzle speed). The gcode file is transferred to the 3-D printer using a flash memory card, to then print the object by following the instructions of the gcode file. By trial and error, it was determined that the thinnest single-layer membranes that were repeatable, robust, and appeared leak tight had a nominal thickness equal to 150 μm. In addition, it was also determined by trial and error that the thinnest layers that could reliably be printed with the setup had a nominal thickness equal to 38 μm.

Step pyramids were printed and characterized to assess the capabilities of the printing process and printer; the objects were printed in either 38 μm-thick or 150 μm-thick layers. In these experiments X and Y correspond to the in-plane dimensions (i.e., across the glass bed), while Z corresponds to the out-of-plane dimension; the size of any in-plane feature is a multiple of the layer height. Metrology of the structures conducted with a white light interferometer shows good correspondence between differential measurements in the printed object and the CAD file (Figure 1). The offset in Z is on the order of a few microns, while the offsets in X and Y are around 360 μm; the X and Y offsets can be easily compensated for when generating the gcode file used to run the printing job.

Knowing the elastic properties of the printed material is essential to model the structural performance of the actuator. Uniaxial tension tests were conducted with a PASCO Materials Testing Machine ME-8236 (Pasco, Roseville, CA) on five coupons to characterize the Young's modulus and tensile strength of test structures printed in 150 μm thick layers with one 600 μm-wide outer shell and infill layers printed at alternating angles of -45° and +45° with respect to the major-axis, creating a crosshatch pattern. The structures are 1.3 mm thick, have a narrow 2 mm wide beam in the center, and have tabs at either end with ridges that interface with the flat coupon fixture Pasco ME-8238. The tests were conducted at 8 mm/min constant pulling rate and 20 Hz sampling frequency. From data analysis, the Young's modulus of the crosshatched FFF-printed nylon 12 is estimated at 606 ± 10 MPa (R^2=0.99) and the tensile strength is estimated at 37.9 ± 3.5 MPa. It is anticipated that the crosshatched structure is more flexible than the bulk material; the calculated Young's modulus is about half the lower bound reported in the literature (1270 MPa [11]); nonetheless, the estimated tensile strength is close to the reported minimum value in the same reference.

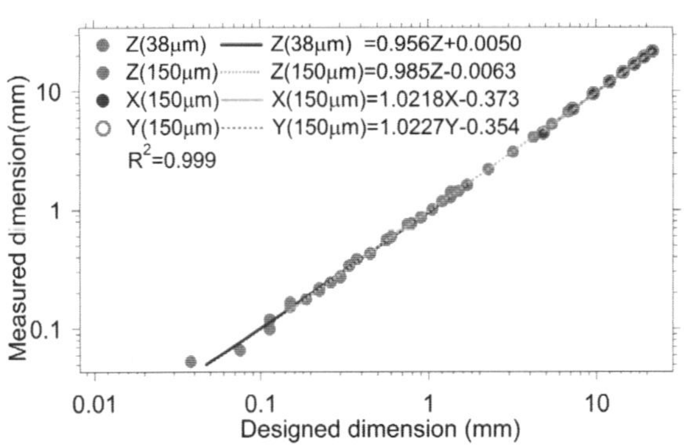

Figure 1. In-plane (X, Y) and out-of-plane (Z) printed features vs. corresponding CAD features for FFF-printed objects in nylon 12 using 38 μm-thick and 150 μm-thick layers. For objects printed with 150 μm layers there is a ~2% mismatch between the dimensions of a printed object and those of the corresponding CAD file (in-plane printed features are larger than those in the CAD model, while out-of-plane printed features are smaller than those in the CAD model).

3. Device design and characterization

The actuator is a frame that holds a coil (from an Omron LY2-DC24 relay, iron core removed) and a 24 mm diameter, 150 μm-thick membrane connected at its center to a 12.4 mm diameter piston with a 9.5 mm diameter cavity containing the embedded magnet (Figure 2); a 150 μm-thick layer secures the magnet in place with a 600 μm wide lip around the edge. Finite element analysis of the actuator (SolidWorks 2015, Dassault Systemes, Waltham, MA) predicts that a force of 0.12 N on the piston results in a 302 μm maximum membrane displacement (i.e., at the center, where the piston holding the magnet is connected) and a maximum stress equal to 1.9 MPa (Figure 3) –which is well below the limit for fatigue infinite life of nylon 12 (~19 MPa) [12].

To print an actuator, the printer bed is heated at 60 °C and then coated with a thin layer of adhesive. After that, the temperature of the bed is raised to 80 °C and the membrane is printed directly onto the adhesive film, followed by the piston and frame. Single 600 μm-wide inner and outer shells are defined, and the infill layers are printed alternately at ±45° with respect to the X-axis of the printer until completion. A SmCo magnet 3/8" in diameter and 1/8" thick from Dura Magnetics Inc. (Sylvania, OH; part # SC3712-500, 26 MGOe maximum energy product) is embedded into the cavity formed within the piston while the printer is paused, right before printing the sealing lip on top of the magnet to hold it in place. Once completed, the actuator is removed from the printer bed and cleaned with isopropanol to remove any remaining adhesive. Four holes printed in the frame are tapped with 6-32 threads so nylon set screws can be attached to grip the coil. The design sets the distance between the closest coil windings and the top surface of the embedded magnet in the piston at 1.85 mm, i.e., there is a 0.85 mm air gap and 1 mm-thick nylon part comprising a spool plate holding the windings in place. The membrane of the devices tested were made of one 150 μm-thick layer (Figure 4, right).

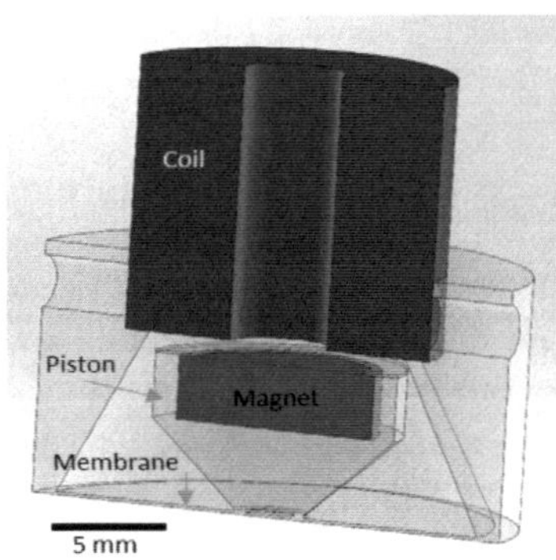

Figure 2. Cross-section of magnetic actuator with FFF-printed nylon 12 body, embedded SmCo magnet, and off-the-shelf driving coil.

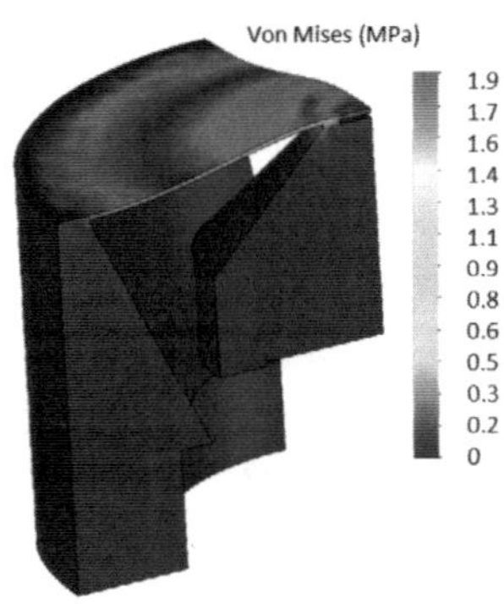

Figure 3. Finite element solution of the actuator's stress field while a 0.12 N force is applied to the piston.

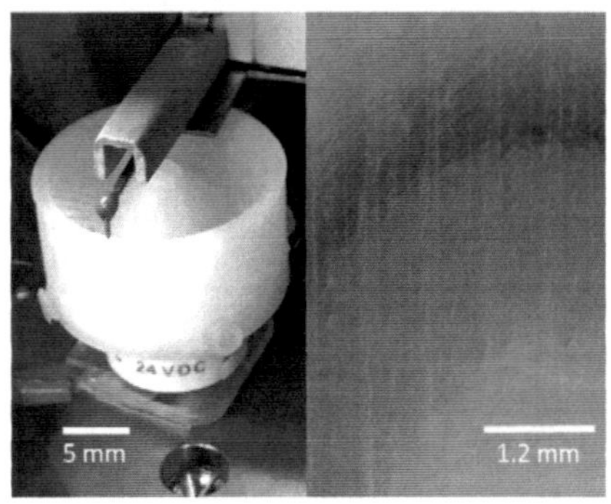

Figure 4. A single-layer, 150 μm-thick membrane actuator while being tested (left) and close-up of the membrane showing the striations due to the rastering of the nozzle (right).

4. Experimental results

The displacement of a single-layer membrane actuator was characterized for various coil bias voltages in a Dektak stylus profilometer with ± 500 μm vertical range (Figure 4, left); the maximum displacement recorded was 302 μm with 20V DC applied to the driving coil (Figure 5). Both negative and positive voltages were applied to the coil; the shape of the displaced membrane was approximately symmetric. The magnetic force versus the square of the current drawn by the coil was estimated from current and displacement measurements in the Dektak experiments and finite element analysis (Figure 6); a linear fit to the data indicates that the magnetic force generated by the coil is proportional to the square of the coil current –as expected from theory.

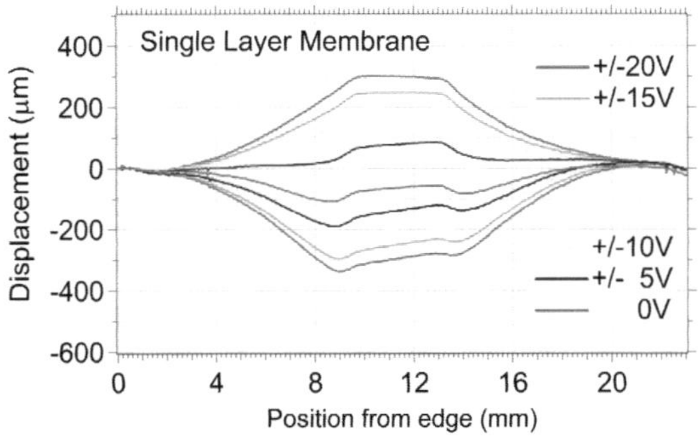

Figure 5. Membrane displacement vs. radial position from the edge for various DC bias coil voltages for an actuator with a 150 μm-thick single-layer membrane. The membrane sags at 0V due to the weight of the magnet.

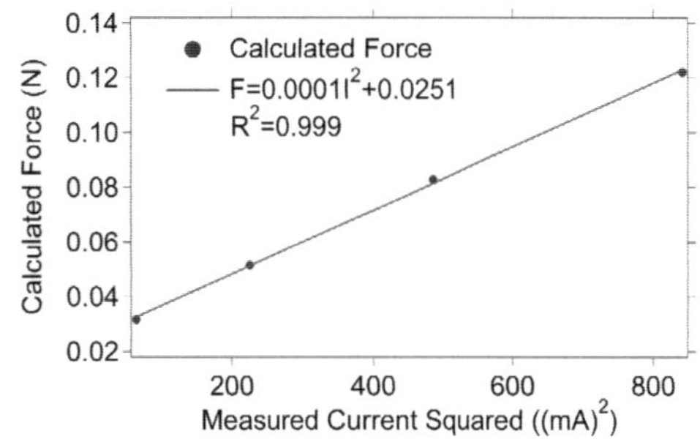

Figure 6. Magnetic force acting on the SmCo magnet (from finite element simulations) versus the square of the current drawn by the driving coil (from experimental measurements).

5. Conclusions

We have demonstrated the first miniature, 3D-printed, monolithic magnetic actuators that are compatible with high temperature (>200 °C) operation. The displacement of a 150 μm-thick, single-layer membrane actuator is characterized for various DC coil bias voltages, resulting in a maximum membrane displacement equal to 302 μm when 20V DC are biased across the driving coil. Current research efforts focus on achieving larger stroke actuation and higher force.

Acknowledgements

The authors would like to thank Ashley Beckwith and Zhumei Sun, Velásquez Research Group, Massachusetts Institute of Technology (MIT), for helping conduct metrology and photography of printed structures, and David Terry and Kurt Broderick, Microsystems Technology Laboratories, MIT, for their help with induction coils and metrology.

References

[1] Zhou H, Li H Q, Sharma V and Schmidt M A 2011 *Proc. IEEE 24th Int. Conf. Micro Electro Mech. Syst.* (Cancun, Mexico) 1095–8
[2] Taylor A P and Velásquez–García L F 2017 *J. Microelectromech. Syst.* 10.1109/JMEMS.2017.2743020
[3] Roberts D C, Li H Q, Steyn J L, Yaglioglu O, Spearing S M, Schmidt M A, and Hagood N W 2003 *J. Microelectromech. Syst.*, **12** (1) 81–92
[4] Sadeghi M M, Kim H, Peterson R and Najafi K 2016 *J. Microelectromech. Syst.* **25** 557–69
[5] Arnold D P and Wang N 2009 *J. Microelectromech. Syst.* **18** (6) 1255–66
[6] Vaezi M, Seitz H and Yang S 2013 *Int. J. Adv. Manuf. Technol.*, **67** (5–8) 1721–54
[7] Velásquez-García L F 2015 *J. Microelectromech. Syst.* **24** (6) 2117–27
[8] Olvera-Trejo D and Velásquez-García L F 2016 *Lab Chip* **16** (21) 4121–32
[9] García-López E, Olvera-Trejo D and Velásquez-García L F 2017 *Nanotechnology* **28** (42) 12pp 425302
[10] Kim S, Lee J and Choi B 2016 *Int. J. Precision Engineering and Manufacturing* **17** 937–42
[11] https://www.matbase.com/material-categories/natural-and-synthetic-polymers/thermoplastics/engineering-polymers/material-properties-of-polyamide-12-nylon-12-pa-12.html#properties
[12] Van Hooreweder B, Moens D, Boonen R, Kruth J-P, and Sas P 2013 *Polymer Testing* **32** 972–81.

Novel concept of a series linear electromagnetic array artificial muscle

R Shalati, K V Poletkin, J G Korvink and V Badilita

Spin&Photon Applications (SPA-) Lab, Institute of Microstructure Technology (IMT), Karlsruhe Institute of Technology (KIT), 76131 Karlsruhe, Germany

r.shalati@stierblick.com, kirill.poletkin@kit.edu, vlad.badilita@kit.edu

Abstract. We propose an electromagnetic artificial muscle (EMAM) actuator whose smallest functional unit/cell consists of an electromagnet created by a solenoidal coil. These cells are coaxially arrayed, being separated by a spring. When actuated by means of an electrical current, the fiber formed by these cells arrayed in parallel collapses, as the gap created by the spring closes in relation to the strength of the current. Fibers are designed to be joined, forming a muscle-like structure. The system is fully customizable in terms of displacement and actuation force, as the stroke can be enlarged by adding more cells to the fiber, while the force can be adjusted with the thickness of the bundle. A mathematical model to describe the operation of a functional unit of the EMAM is given. It predicts the relation between the applied current and the displacement, and is in agreement with the experiment. This novel concept of an EMAM carries the benefits of a linear solenoid array actuator, such as high stroke and uniformity of motion, into the micro domain. It is moreover extended with the capability of precise displacement in micro domain but also fast actuation.

1. Introduction

As modern day robotics, prosthetics and positioning technologies make increasingly higher demands on the properties of actuators such as large stroke, constant force and low power consumption, classical solenoid-based actuators have been discarded so far [1,3].

Concepts of arraying single solenoid actuators to meet these requirements are emerging and pointing to a common architecture that arranges single actuators to a fiber that bypasses the typical disadvantages of classical solenoid-based electromagnetic actuation such as low stroke and uneven force distribution on the stroke length. Challenges are conquered by operating a multitude of similar actuators simultaneously in a chain-like arrangement within their efficiency maximum to overcome the limitations of both possible displacement and uniformity of actuation in terms of maximum force and acceleration.

Obata [1] presents a macroscopic "Muscle Motion Solenoid Actuator" that consists of several identical macroscopic solenoidal actuators arrayed to a fiber to enlarge the total stroke. The number of turns of the solenoidal actuators is adjusted along the coil length to further support a constant force along the actuator stroke. Each cell has a stroke of 40mm and can deliver about 14N at a cylindrical stroke of 20mm, while having an initial length of about 100mm and a diameter of 48mm. In a further study [3], Obata shows how these solenoidal actuators can be designed in order to provide a maximum force when arrayed to a fiber, by incorporating high permeability casings and neodymium

magnets to maximize the density of the flux and minimize the flux loss. Li et al. [2] characterizes the design of a "multi-class series artificial sarcomere array" that transfers the general microstructure of a skeletal muscle into a macroscopic device. The structure is described as a fiber consisting of consecutive planes, each housing several axially arranged identical electromagnetic actuator cells working in parallel. Each cell incorporates a locking mechanism, various coils and permanent magnets. While both concepts mentioned above reveal the advantages of a serial axial coupling of similar actuators, they constitute complex designs. This complexity limits the transfer of the fiber arrangement and its clear benefits into the micro domain. The required fabrication steps, the amount of parts, the wide diversity of materials and manufacturing processes impede a miniaturization.
In a simpler approach targeted on supporting human motion, Takai et al. [4] introduce a concept where all cores (actuators) are mechanically connected, though not attached to the housing that inherits multiple coaxially arranged solenoids (stators). Takai clearly points out the necessity "to investigate by how much the size can be reduced while retaining a reasonable amount of force".

As such concepts have only used macroscopic unit cells, i.e., several cm/cell [4] and focused on the actuation force that can be generated, our design is also targeting a high actuation precision. We propose a design that is comparable to the fiber concept of Obata [1], but aims at miniaturization to make the array concept and its advantages available for microscopic application in actuation and in contrast to mentioned concepts for positioning technology (without the need of an antagonistic fiber).

a) 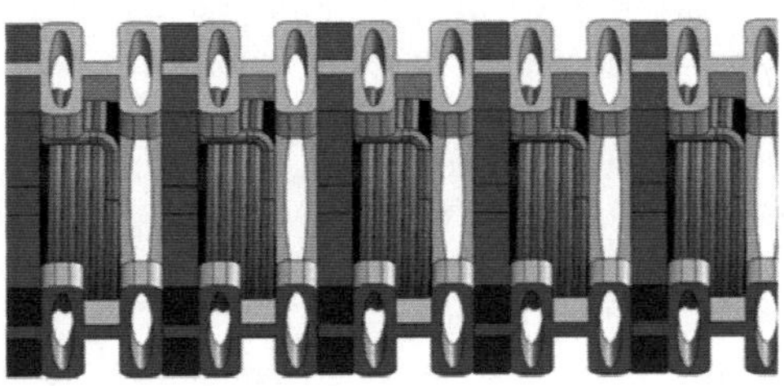

b) 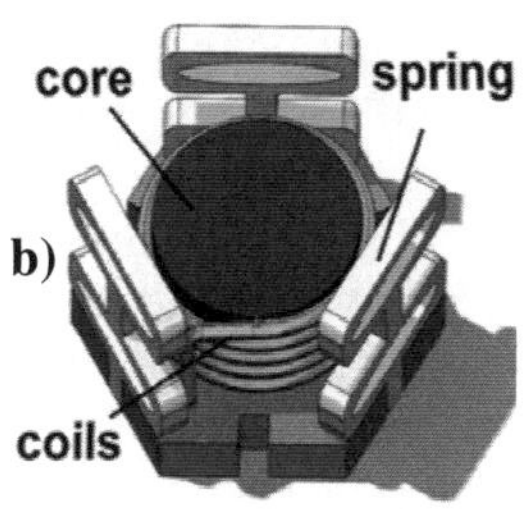

Figure 1. a) Concept of the EMAM muscle fiber composed of a series of individual electromagnetic actuators ;

b) The building block of the EMAM muscle fiber - individual half-cell.

The introduced concept limits the architecture to its vital elements. It can be described as a fully scalable electromagnetic artificial muscle (EMAM), which is supposed to consist of a multitude of fibers, each of them obtained by arraying (Fig. 1a) a large number of individual identical electromagnetic actuators (Fig 1b). As each cell is separated from its consecutive cell by a spring, the actuation motion becomes regardless of outer forces controllable and therefore definite and precise.
This is supported by further contributing an exact mathematical model for the behavior of a functional cell unit to increase both the accuracy of displacement in the micro domain and the speed and force of actuation within macroscopic spaces. Thereby a novel field of solutions for actuation and displacement challenges in micro world will be accessed.

The proposed design and its features are described in section 2. A mathematical model for a controlled actuation is given in section 3, while in section 4 a simplified calculation method is provided. Section five explains the experimental verification and section 5 draws the conclusion.

2. The electromagnetic artificial muscle (EMAM)

The EMAM concept presents a design that can meet the requirements of micro sized fabrication however retaining the benefits of this architecture in general.

The EMAM's basic building block consists of a solenoidal coil wound around a ferromagnetic core, and a spring resetting the elements after attraction and leading to a controlled actuation as it delivers a counterforce for the magnetic actuation.

As this building block/smallest functional unit needs a second equivalent unit to function, it is considered a half-cell. Coaxially arranged and separated by mentioned springs, the cells form a fiber that inherits electrically parallel electromagnets. An application of a current induces the electromagnetic attractive forces that overcome the restoring forces of the springs and collapsing the fiber as a whole while delivering controllable, bidirectional actuation force. Experiments with a 1:20 scale prototype of a functional unit revealed the exact positioning capabilities as a specific stroke is repeatedly related to a certain current with a precision of up to 50 micrometer.

The EMAM draws its thrust force from the high ratio of diameter of solenoids to air gap between the consecutive electromagnets. Furthermore this design particularity raises the fiber density, as the cells have a height to diameter ratio of over 2. Therefore the EMAM fiber surpasses previous attempts [1, 3, 5] in terms of cell density of a fiber, leading to a more uniform actuation needed for precise positioning.

As the targeted cell has a diameter of about 3 mm and a height of 1,5 mm (+0.5mm actuation gap) a fiber with the length of 400mm is capable to inherit 200 cells. This will lead to a theoretically deliverable stroke of 100mm, while retaining a high precision section of about 50mm of the stroke due to the pull-in phenomena discussed in section 3.

Specific demands can be met by elongating the fiber as desired to alter the maximum stroke, while the actuation force is customizable through the amount of fibers bundled up and working in parallel. This is furthermore facilitated by the proposed hexagonal architecture that is designed to enable a high fiber density when joined to a bundle.

3. A mathematical model for describing the operation of the actuation unit of the EMAM

An understanding of the behavior of such an actuator is established by providing an exact mathematical model of the static pull-in point of a functional EMAM cell unit (mathematically described as two coaxially arranged solenoids with radius r, separated by a spring with the length l, while one solenoid is fixed to the ground and the other one sits congruently above.

The pull-in point is well-known within electrostatics, but poorly examined within the field of linear electromagnetic arrays. This point marks the verge of a stable and controllable vs. a powerful and fast displacement where the attraction grows exponentially and the two electromagnets collide with maximum force and velocity.

Utilizing the Maxwell formula for the mutual inductance between two coaxial wire loops, the inductance can be described dimensionless as

$$\hat{L}_m = \frac{1}{k}\left(1 - \frac{k^2}{2}\right)K(k) - k^{-1}E(k) \tag{1}$$

where $\hat{L}_m = L_m \cdot (2\mu_0 r)^{-1}$, $k^2 = (1 + \xi\varepsilon^2)^{-1}$, $\varepsilon = x/l$ and $\xi = l^2(4r^2)^{-1}$while K(k) and E(k) are complete elliptic integrals of the first and second kind, $\mu_0 = 4\pi \cdot 10^{-7}\, H/m$ is the magnetic permeability of the vacuum, x the displacement of the movable coil and l the distance in-between.

Applying the Lagrange formalism, the energy of the system was exactly modeled. Hence, the stability thresholds were derived from this nonlinear, analytical approach and the pull-in behavior has been predicted (Fig. 2). This direct dependence between the current and the displacement is given by

$$I^2 = (1 - \varepsilon)\left(\frac{\delta L_m}{\delta k}\frac{\delta k}{\delta \varepsilon}\right) \tag{2}$$

which can be derived from the equilibrium, when

$$F_{total} = F_{el.mag} - F_{spring} = \frac{\delta L_m(x)}{\delta x} I^2 - c(1 - x) \tag{3}$$

where c is the spring constant and thereupon exclusion of I.

4. Simplified model for easy calculation of the magnetic pull-in point

Taking into account that the radii of the solenoids exceed their separating gap/distance of displacement by far (r>>l) the parameter k becomes practically one, leading to a highly simplified, calculation method, avoiding complex elliptical integrals (Fig.2) with an accuracy that grows with the ratio r to l:

$$L_m = \mu_0 r\left[ln\frac{4}{\xi(1+x/l)} - 2\right] \tag{4}$$

5. Experiment

Thereupon the model was verified by experimental data obtained by a prototype that represents a scale model of a functional EMAM cell unit. It is constituted of 2 axial solenoidal coils with a diameter of

40mm and 220 windings each, separated by 3 band springs. Using a laser distance sensor the displacement was measured against the applied current, leading to the data points in diagram Fig. 2 including the pull-in point, that are in agreement with the modeling. In spite of being tailored for the observation of the pull-in phenomena, the coreless prototype is able to deliver a maximum thrust force of 0.4 N overcoming the collective stiffness of the springs of 1109 N/m while having a maximum stroke of 3.5 mm.

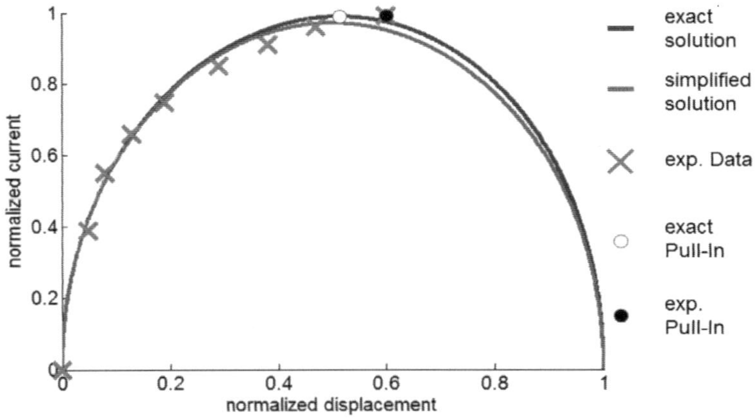

Figure 2. Pull-in diagram.

Figure 3. Example of a 3D wire-bonded solenoidal micro-coil.

6. Conclusion

Utilizing a reliable fabrication of solenoidal micro-coils by wirebonding technology (Fig. 3) previously demonstrated in our group [5,6] and developing a micro fabrication compatible design of both reset/spring and carrier elements, we obtained the potential for further miniaturization of the EMAM, consequently increasing the displacement accuracy. The advantages of this miniaturization combined with the applied understanding of the pull-in behavior open up a novel field of solutions where time-saving, fast macro displacement and accurate, controllable micro positioning is required within the same actuator. Hence an application within the field of pick and place of MEMS-components utilizing this technology for micro grippers as well as various leverage systems becomes highly beneficial.

As the results of our modeling and prototyping are encouraging to go further we will be going to theoretically investigate the pull-in phenomena for n elements. From a constructive point of view we will examine the behavior of a complete fiber as well as the bundling up of multiple fibers with yet to build scale prototypes.

References

[1] Obata S, 2014 New Linear Solenoid Actuator for Humanoid Robot *Mecatronics-2014-Tokyo* DOI: 367-370. 10.1109/MECATRONICS.2014.7018558.

[2] Li J et al. 2014 Design and optimization of multi-class series-parallel linear electromagnetic array artificial muscle *Biomed Mater Eng.* 24(1):549-55. doi: 10.3233/BME-130841.

[3] Obata S, 2012 A Muscle Motion Solenoid Actuator *IEEJ Transactions on Fundamentals and Materials* 132. 63-71. 10.1541/ieejfms.132.63.

[4] Takai A et al. Oct. 2013 Prototyping the flexible solenoid-coil artificial muscle for exoskeletal robots *Proc. IEEE 13th Int. Conf. Control, Autom., Syst. (ICCAS)* pp 1046–1051

[5] Kratt K et al. 2010 A fully MEMS-compatible process for 3D high aspect ratio micro coils obtained with an automatic wire bonder *J. Micromech. Microeng* 20 015021

[6] K. V. Poletkin et al. 2016 Polymer Magnetic Composite Core Boosts Performance of Three-Dimensional Micromachined Inductive Contactless Suspension *IEEE Magnetics Letters* vol. 7 no. pp 1-3 DOI: 10.1109/LMAG.2016.2612181

IOP Conf. Series: Journal of Physics: Conf. Series **1052** (2018) 012048 doi:10.1088/1742-6596/1052/1/012048

Energy-aware 3D micro-machined inductive suspensions with polymer magnetic composite core

K V Poletkin[1], Z. Lu[2], A. Moazenzadeh[3], S. G. Mariappan[3], J G Korvink[1], U. Wallrabe[2], V Badilita[1]

[1]Spin&Photon Applications (SPA-) Lab, Institute of Microstructure Technology, Karlsruhe Institute of Technology, Eggenstein-Leopoldshafen D-76344, Germany
[2]Laboratory for Microactuators, Department of Microsystems Engineering – IMTEK, University of Freiburg, D-79110 Freiburg, Germany
[3]Voxalytic GmbH, D-76228 Karlsruhe, Germany

E-mail: `kirill.poletkin@kit.edu, vlad.badilita@kit.edu`

Abstract. This paper addresses the issue of Joule heating in micromachined inductive suspensions (MIS) and reports a significant decrease of the operating temperature by using a polymer magnetic composite (PMC) core. The PMC material has a high resistivity, thus inhibiting the formation of eddy currents, and a high permeability, thus guiding the magnetic field more efficiently within the MIS structure. We experimentally study the distribution of the PMC material inside the MIS structure and evaluate the effect of the core from the dependence of the levitation height on the excitation current. The experiments carried on in ambient room temperature demonstrate that the temperature inside the micromachined inductive suspension is reduced to 58°C, which is a record-low temperature compared to other MIS structures reported before.

1. Introduction

Micromachined inductive suspensions operate by exciting a levitation coil with an alternative current, therefore creating an alternative magnetic field, which induces a current in a conductive proof mass, the proof mass being levitated on top of the MIS structure. However, the energy transfer between the excitation current of the coil and the current induced in the proof mass is rather inefficient, and this translates in relatively high excitation currents with respect to the achievable levitation height. The first MIS prototype used single turn coils obtained by planar lithography [1] and reported a temperature as high as 600°C. A significant reduction of the operating temperature down to 120°C has been achieved by replacing the planar coils with 3D wirebonded microcoils [2]. Following the trend of approaching room temperature operation of micromachined inductive suspensions, we have recently demonstrated further improvement by introducing a polymer magnetic composite core material [3].

In this work, we investigate the distribution of the PMC core material inside the MIS structure, focusing on the dependence between the levitation height and the excitation current. A further reduction of the operating temperature is demonstrated for operation in ambient room temperature, i.e., 58°C, which is a record-low value to date. At the same, time, this work suggests solutions for further improvement in terms of energy efficiency of micromachined inductive suspensions.

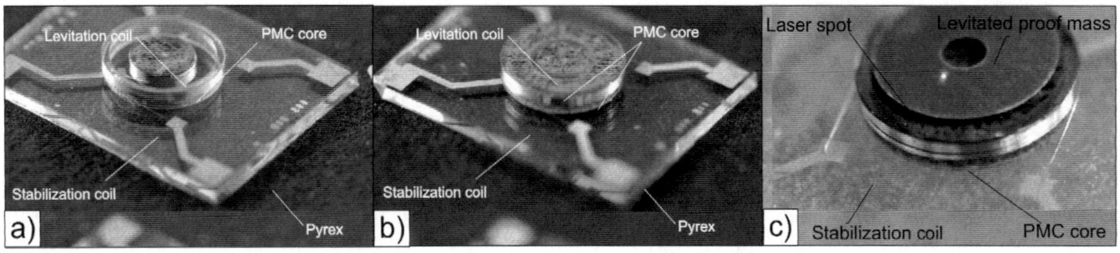

Figure 1. a) Design A: only the inner (levitation) coil is filled with PMC material; b) Design B: both the inner (levitation) and the outer (stabilization) coils are filled with PMC material; c) An Al proof mass is successfully levitated on a design B MIS structure.

2. Fabrication and operating principle

The MIS structures used in this work have been presented in detail in previous reports [2] and consists of two concentrical solenoidal wirebonded microcoils obtained by wirebonding. The role of the inner "levitation" coil (2 mm in diameter and 20 windings) is mainly to levitate the conductive proof mass, whereas the role of the outer "stabilization" coil (3.9 mm in diameter and 12 windings) is to achieve a stable levitation of the proof mass. The height of the coils is 650 μm. In order to investigate the influence of the PMC core material to the MIS performance, two different designs have been investigated experimentally: design A - only the levitation coil is filled with PMC material Figure 1a), and design B - the entire volume of both coils is filled with PMC material Figure 1b). The performance of these structures is compared to the performance of the structure without core. Figure 1c) shows a MIS prototype with an integrated PMC core successfully levitating an Al proof mass (3.2 mm in diameter and 13 µm in thickness). The PMC core [4] consists of a NiFeZn ferrite soft magnetic powder (CMD5005, National Magnetics Group, USA) dispersed in a mixture of two epoxies (AW4510 and HW4804, Huntsman Advanced Materials GmbH, Switzerland). My means of a dispenser (DX-250, OKI, USA), the MIS structure has been filled with the PMC mixture.

The research hypothesis of this work is that the PMC core focuses the magnetic field lines around the body of the two coils, thus minimizing their air-path and increasing the magnetic induction in the region of stable levitation. This translates in lower excitation currents required to achieve the same levitation height, therefore a lower operating temperature.

The field, B, generated by the coil without core can be found by solving the following set [3]

$$\begin{cases} B + \mu_0 H = \mu_0 \dfrac{(n_l - n_s)I}{l'_{air}}; \\ B = \mu_0 H, \end{cases} \tag{1}$$

where n_l and n_s are the number of windings for levitation and stabilization coils, respectively, H is the magnetic field, l'_{air} is an effective path of magnetic field passing through the air, and I is the current through the coil. Solving set (1), we obtain the magnetic field in the center of the levitation coil without core as follows

$$B_a = \mu_0(n_l - n_s)I/(2l'_{air}). \tag{2}$$

For coils with core the relationship between B and $\mu_0 H$ becomes as

$$Hl_{core} + Bl''_{air}/\mu_0 = (n_l - n_s)I, \tag{3}$$

where l''_{air} is the effective path of magnetic field passing through the air for coils with the magnetic core. Accounting for $l_{core} < l'_{air}$ and $l''_{air} \leq l'_{air}$, and the fact that μ_{core} is large [4], the magnetic induction generated by coils with core is

$$B_b \approx \mu_0(n_l - n_s)I/l''_{air}. \quad (4)$$

As a result, even if $l''_{air} \approx l'_{air}$ (due to the fact that the linear size of coils is one order of magnitude larger than their height), still the magnetic field generated by the coils with integrated core is larger than by coils without core.

3. Experimental Results

The experimental setup consists of a current amplifier (LCF A093R) delivering a square wave excitation current to the coils. A function generator (Arbstudio 1104D) was used to control the amplitude and the frequency of the excitation current, while a laser distance sensor (LK-G32) was measuring the levitation height. Finally, an IR (Infrared) camera PI-160 (Optris GmbH, Berlin, Germany) was used to measure the temperature of the MIS prototype.

The dependence between the levitation height and the excitation current in the coils at a frequency of 10 MHz has been measured experimentally and is being shown in Figure 2a) for the following MIS prototypes: design A, design B and, in order to enable direct comparison, a prototype without core. The position of the three curves in Figure 2a) indicate that, for the same excitation current, higher levitation heights are being achieved for a higher degree of filling the coil volume with PMC material. Conversely, for the same levitation height, e.g., 110 µm, the highest actuation current is obtained for the structure without core - 110 mA, and the smallest current value - 70 mA, is demonstrated for design B, i.e., for the prototype having the entire coil volume filled with PMC material.

In order to check the prediction of the equations above, we choose one specific operation point for the design B prototype of the suspension: for an excitation current of 60 mA, the levitation height of the proof mass is 110 µm. In order to achieve the same effect for the suspension without core, i.e., 110 µm, the excitation current must be 110 mA, which is a factor of 1.8 higher. Equations (2) and (4) above also predict that the magnetic induction doubles for a prototype with a magnetic core, therefore we can conclude a relatively good agreement between

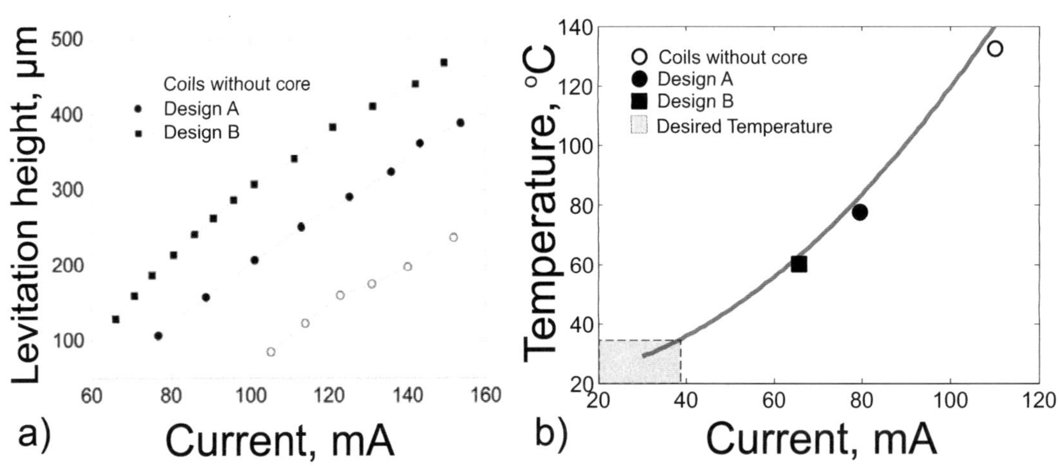

Figure 2. a) Dependence of the levitation height on the excitation current (at 10 MHz) as a function of the PMC distribution inside the MIS; b) Temperature vs. current dependence for a levitation height of 110 µm.

the theoretical prediction and the experimental results. It means that managing the distribution of magnetic field within coils by designing magnetic core the required current for levitation can be reduced together with operating temperature. Figure 2b) sums up this fact by showing the dependence of operating temperature for considered coils design on the excitation current. Figure 2b) predicts that operating temperature can be comparable with ambient one, when the excitation current will be less than 40 mA. This can be reached for instance by adding a PMC backing plane to the MIS design, could decrease the operating temperature to around room temperature values. A significant reduction of the current leads to a significant reduction in the Joule heating within the MIS structure. This is indeed confirmed by the temperature map of the design B prototype shown in Figure 3, which demonstrates that the temperature on top of the structure is lowered to 58°C. Moreover, the temperature distribution is relatively uniform within a range of 10°C, whereas the prototype without core shows a large temperature gradient from the coil windings to the center of the structure [5, page 1482].

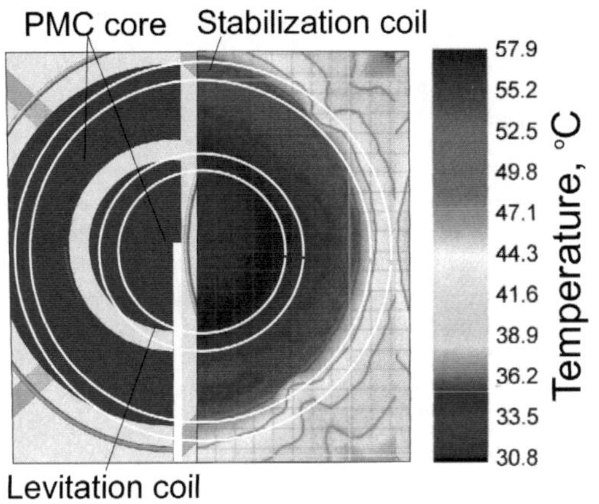

Figure 3. Temperature map for design B MIS prototype operating at 60 mA.

4. Conclusion

This work continues the efforts in our group in the field of MIS based on 3D microcoils obtained by automatic wirebonding, by further integrating a polymer magnetic composite material as a core to improve the MIS performance. Due to its high resistivity, the PMC core hinders the formation of eddy currents. Due to its high magnetic permeability, the PMC core increases the magnetic field in the region of stable levitation. Both properties lead to a more efficient energy transfer from the excitation, i.e., actuation current to the proof mass to be levitated. We expect that further improvements can be achieved by exploiting the flexibility of the dispensing method for the PMC core to cover the bottom and the outermost surface of the coils, i.e., embedding the entire MIS structure in the PMC material. However, the very top of the MIS structure must not be convered by the magnetic composite material in order to allow the lines of the magnetic field to "escape" the structure in order to induce eddy currents inside the conductive proof mass. Approaching room temperature in the operation of 3D micromachined inductive suspensions is a pragmatic step in the direction of integrating these elements into levitated micro-systems.

References

[1] Shearwood C, Ho K, Williams C and Gong H 2000 *Sensor. Actuat. A-Phys.* **83** 85–92
[2] Lu Z, Poletkin K V, den Hartogh B, Wallrabe U and Badilita V 2014 *Sens. Actuator A-Phys.* **220** 134–143
[3] Poletkin K V, Moazenzadeh A, G Mariappan S, Lu Z, Wallrabe U, Korvink J G and Badilita V 2016 *IEEE Magn. Lett.* **7** 1–3
[4] Mariappan S G, Moazenzadeh A and Wallrabe U 2016 *Micromachines* **7** 60
[5] Lu Z, Poletkin K V, Wallrabe U and Badilita V 2014 *Micromachines* **5** 1469–1484

IOP Conf. Series: Journal of Physics: Conf. Series **1052** (2018) 012049 doi:10.1088/1742-6596/1052/1/012049

Feasibility of vibration energy harvesting powered wireless tracking of falcons in flight

Maisie M. Snowdon, James Horne, Buck Gyr and Yu Jia

Department of Mechanical Engineering, University of Chester, Chester CH2 4NU, UK

E-mail: yu.jia.gb@ieee.org

Abstract. The use of wireless tagging of birds has been widely used for monitoring or tracking purposes. This include over 10 thousand wireless tracking devices currently used by the UK falconers alone. However, due to the concern of not burdening the birds with a heavy battery, the existing lightweight telemetry tracking systems can only last for days, if not hours. Falcons can have top flight speeds in excess of a hundred miles an hour, which makes it a near impossible task to track a missing falcon after the battery has been depleted. This paper investigates the feasibility of incorporating a piezoelectric vibration energy harvesting system to act as a secondary power source for the wireless tracking of falcons. The ultimate aim is to both extend the primary battery life and enable periodic burst transmissions of telemetry after the depletion of the primary battery. The presented tracking and harvesting system is lightweight and has been field trialled on a gyrfalcon at the Chester Cathedral Falconry.

1. Introduction

Wireless tagging of avian species has existed for decades, including the monitoring of birds during migration and the tracking of birds of prey by sanctuaries and falconries. Falcons can have top flight speeds in excess of a hundred miles an hour, making it extremely difficult to track outside the visual range, without the assistance of a telemetry system. There are over 25 thousand falconers in the UK alone and an estimated 10 thousand wireless tracking devices are used on birds of prey in the UK to-date [1]. This typically included radio-transmitters that have limited transmission range, but conservationists have increasingly moved towards the more accurate Global Positioning System (GPS) for telemetry [2].

However, due to the critical need to minimise the overall weight of the tag in order to minimise burden on the birds, the battery life of the wireless telemetry system is very limited. Despite the more accurate telemetry capabilities, GPS require more power budget than radio transmitters. Furthermore, this technological challenge has resulted in the limited weight budget of the wireless tracker for birds (typically in the order of few grams to few tens of grams) been overwhelmingly occupied by the battery, with the active wireless device only able to occupy a small fraction of the overall device weight [3].

In the meantime, energy harvesting has been an emerging technology in recent years that has been demonstrated as a means to recover ambient energy to replenish the batteries for microelectronics [4]. The use of solar and vibration energy harvesting methods to power wireless tracking has been previously investigated in the literature [5]. However, the mean concern on minimising the parasitic effects and burden on birds still linger. Solar cells require large surface area and is only suitable under direct solar irradiation. On the other hand, the use of backpack-like mechanisms to recover vibration energy [3] can be a source of parasitic burden for the body and the wings that cause distress for the birds.

This paper reports a new approach to harvest the vibration energy of birds, with a specific focus on birds of prey due to collaboration and access to field trails with the Chester Cathedral Falconry [6]. The proposed system employs a lightweight piezoelectric cantilever-based mechanism attached to the tail feather of a falcon. In an interview with the Chester Cathedral Falconry, the falconers provided an end-user design specification of achieving an add-on system weighing less than 10 grams to their existing wireless tracker.

2. Method

The Chester Cathedral Falconry used off-the-shelf radio transmitters with wire antenna on their falcons. These tracking devices weighed just under 10 grams and were clipped into a tail feather harness using a spring clip mechanism. The lightweight fixture on the tail feather resulted in minimal stress for the birds. The battery life of the radio transmitter lasts 3 weeks and the range of transmission does not extend far beyond the visual range ($\sim$ 100 m). While the more accurate and range-independent GPS tracking devices were available for the falconry too, the battery life does not last more than 1 day.

In order to extend and complement the battery powered trackers, a piezoelectric vibration energy harvesting (VEH) system was integrated. A key objective of the study was to validate the feasibility of applying piezoelectric VEH to the tail feather of falcons. MIDE's PPA-1022 piezoelectric beam was employed, which had a 0.18 mm thick PZT-5H layer. The FR4 encapsulated beam was 53.0 mm in length, 10.3 mm in width, 0.7 mm in thickness and weighed 0.8 grams. The beam was attached to the casing of the radio transmitter in a cantilever-based arrangement. However, since the radio transmitter itself is only clipped onto the tail feather harness, the PZT cantilever beam is not strongly anchored to a fixed plane.

Design of the experimental setup in a lab environment is shown in figure 1, where the harvester is clipped into the tail feather harness and is fixed on a mount that interfaces with a mechanical shaker for vibration testing.

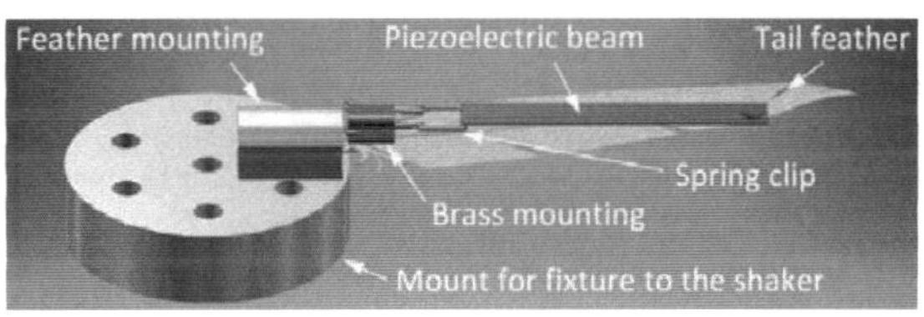

Figure 1. Design of the proposed system for laboratory experimentation on a shaker table. The harvester recovers vibration of the tail feather and radio antenna (not shown here) primarily through cantilever beam bending oscillation.

An image of the transducer is shown in figure 2 and the assembled setup on a mechanical shaker is shown in figure 3. The PZT beam is able to respond to the base point excitation as well as the oscillation of the tail feather and the radio antenna. A full bridge rectifier circuit using die level Schottky diodes were assembled on a small circuit board was used to interface the PZT element and an AVX super-capacitor (measured at 5 μF) as illustrated in figure 4. The overall assembled system that has been used during the field trial is shown in figure 5.

A specific gyrfalcon by the name of Buck was chosen for the field trials. Video image analysis of the falcon's flight estimated a tail feather motion of approximately 4 Hz to 6 Hz with an average acceleration attaining $\sim$ 1 g near the tail feather harness. The frequency of the harvester beam can be tuned to $\sim$ 5 Hz by adding 3 grams of end mass.

Figure 2. PZT beam.

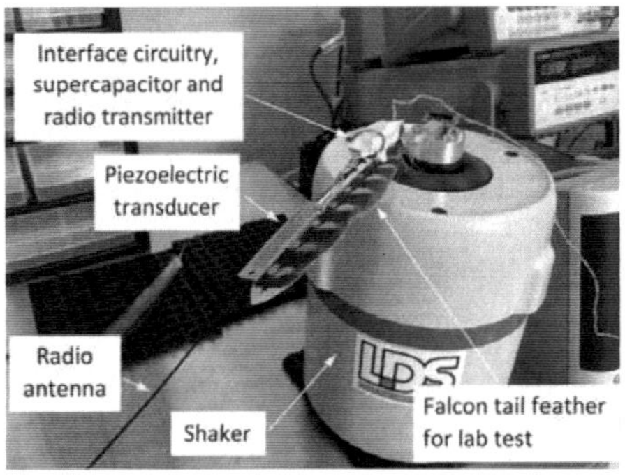

Figure 3. Experimental apparatus setup on a mechanical shaker. The system can be securely clipped onto the tail feather harness. Representative vibration tests were carried out.

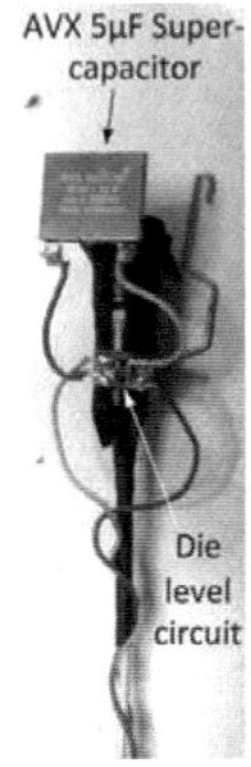

Figure 4. Rectification and storage circuitry.

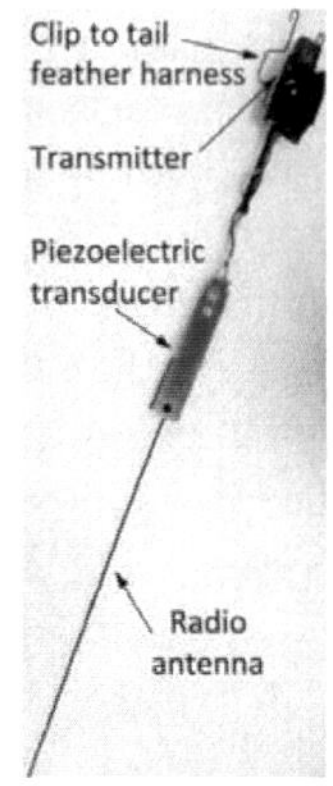

Figure 5. Assembled tracking and harvester system.

3. Experimental results

The frequency responses of the harvester with and without end mass, when subjected to various levels of base-point input acceleration, are shown in figures 6 and 7. An average power in excess of 0.7 mW was observed at 0.5 g acceleration at 5 Hz. However, real tail falcon vibration during flight is unlikely to be purely sinusoidal and continuous, and the frequency is likely to be varying in nature. Therefore, the device is not expected to operate in steady resonance.

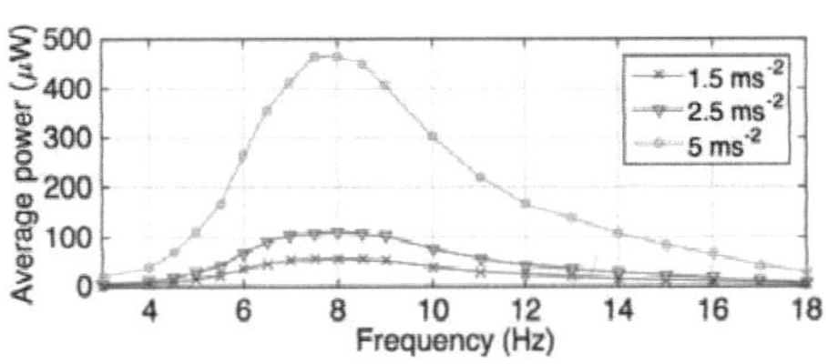

Figure 6. Frequency domain power output of the PZT harvester with no end mass.

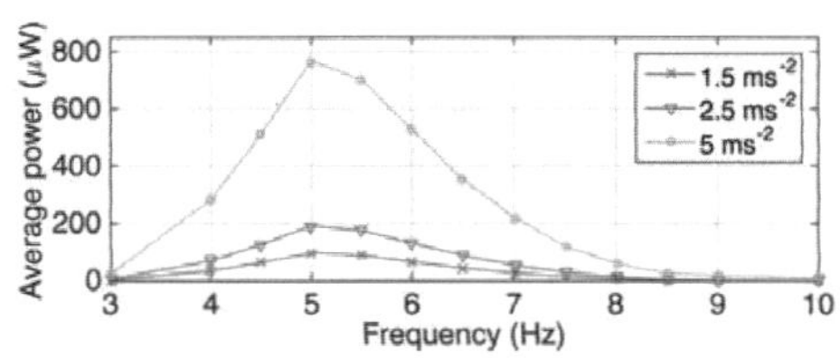

Figure 7. Frequency domain power output of the PZT harvester with 3 grams end mass.

4. Field trial

The assembled prototype device seen in figure 5 was clipped onto the existing falcon tail harness as shown in figure 8. The overall weight of the add on system (including the harvester, end mass, circuit, storage capacitor and fixture) was about 5 grams. Figure 9 show the gyrfalcon in flight with the prototype device.

A number of tests were carried out across various dates. Voltage across the super-capacitor was measured before and after the falcon flight show on the day. Due to logistic limitations, only 1 measurement was possible per show. Table 1 summarises the measurements taken from 3 different field trials. A parasitic leakage of about 0.8 μW from the circuitry was measured. This further disadvantaged the field trial measurement experiment, as the falcon was on land during most part of the show and flight time only lasted approximately 5 minutes.

The maximum average power recovered was ~3 μW. This deviation from the laboratory testing is due to the additional continuous parasitic leakage from the circuitry when the bird is

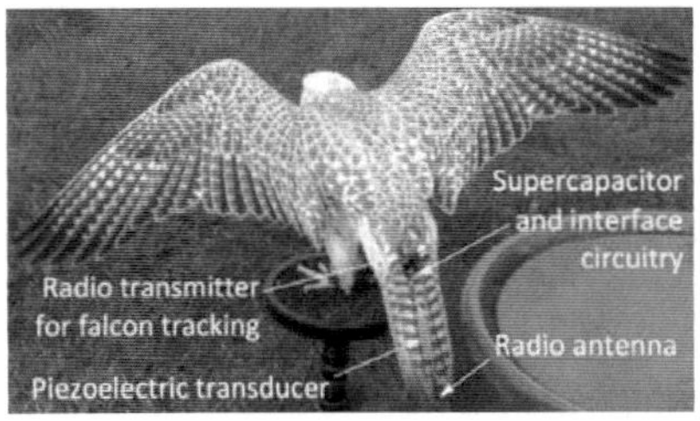

Figure 8. Radio tracking system and piezoelectric vibration harvester on a gyrfalcon.

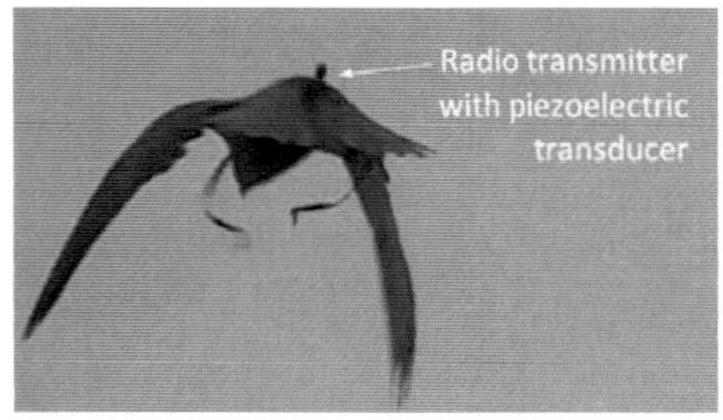

Figure 9. Gyrfalcon in flight with the radio tracking system and piezoelectric harvester.

Table 1. Field trial results of recovered energy into a 5 μF supercapacitor during 3 different tests on a gyrfalcon in flight. All tests lasted 5 minutes. The circuit has an average parasitic leakage of 0.79 μW. The average generated power presented here compensates for this loss.

	Start voltage (V)	End voltage (V)	Energy stored (mJ)	Average generated power (μW)
Test 1	0.005	0.057	0.008	0.82
Test 2	1.223	1.302	0.499	2.45
Test 3	1.230	1.344	0.734	3.23

not in flight. Nonetheless, despite the significant parasitic losses and non-continuous excitation conditions, the field trials revealed positive results in terms of net energy gain from the harvesting of tail feather motion during flight.

Conclusion and future work

A prototype device that incorporates piezoelectric VEH onto a radio tracking device was developed. While a number of challenges still persist, such as parasitic losses and optimal frequency tuning, the field trails validated the feasibility of recovering energy from falcon tail feathers during flight. The result holds promise to extend the battery life of tracking devices for birds, as well as the VEH system being used as a back up power source for wireless transmitters long after the primary battery has been depleted. Future work involves further dynamic analysis of bird tail motion in order to better understand the vibration spectrum for design optimisation.

Acknowledgments

Authors wish to thank the Chester Cathedral Falconry, especially falconers Tommy McNally and Kieran Morrison; the science and engineering technical team at the University of Chester; as well as sincere gratitude to Michael S. Horne for providing additional technical support.

References

[1] House of Commons - Environment Audit Committee: Written evidence submitted by the Hawk Board, Publications Parliament UK, 2017.
[2] W. Bouten, E.W. Baaij, J. Shamoun-Baranes and K.C.J. Camphuysen, A flexible GPS tracking system for studying bird behavior at multiple scales, *Journal of Ornithology*, **154**(2), 2013, pp. 571-580
[3] M.W. Shafer, R. MacCurdy, J.R. Shipley, D. Winkler, C.G. Guglielmo and E. Garcia, The case for energy harvesting on wildlife in flight, Smart Mater. Struct., 24 (2015) 025031 (12pp).
[4] S.P. Beeby, M.J. Tudor and N.M. White, *Meas. Sci. Technol.*, **17**, 2006, R175-R195
[5] R. MacCurdy, T. Reissman, E. Garcia and D. Winkler, A methodology for applying energy harvesting to extend wildlife tag lifetime, Proc. IMECE2008-68082, 2008, (10pp).
[6] Chester Cathedral Falconry and Nature Gardens, [available online] `https://chestercathedral.com/falconry-nature-gardens/`

EXPERIMENTAL VALIDATION OF WIDEBAND PIEZOELECTRIC ENERGY HARVESTING BASED ON FREQUENCY-TUNING SYNCHRONIZED CHARGE EXTRACTION

A. Brenes[1], E. Lefeuvre[1] and C.-S. Yoo[2]

[1]Centre for Nanoscience and Nanotechnology (C2N) – Site d'Orsay, Univ. Paris-Sud, Université Paris-Saclay, Bât. 220, Rue Ampère, 91405 Orsay, France

[2]Electronic Convergence Materials & Device Research Center, KETI, Rep. of Korea

E-mail: alexis.brenes@c2n.universite-paris-saclay.fr

Abstract. This paper reports, for the first time, experimental evidence of the effectiveness of the frequency-tuning synchronized charge extraction technique (FTSECE) for wideband piezoelectric energy harvesting. We also present results proving that, for a single-degree-of-freedom (SDOF) resonator, the theoretical bandwidth of the system is infinite and discuss major practical limitations to that ultimate performance.

1. Introduction

Amongst the topologies based on synchronized switching for piezoelectric energy harvesting, a first technique called "Synchronous Charge Extraction" (SECE) has initially been studied by Lefeuvre et al. [1] However, this technique is known to yield relatively poor performance for resonators exhibiting a strong electromechanical coupling coefficient. To improve the performance of classical SECE in terms of maximum power and/or bandwidth, recent works have proposed simple modifications to the original circuit [2] [3] [4]. However, to optimize the performances further, one should simultaneously tune the instant and duration of switch closing depending on the oscillation frequency. This last statement has led to the FTSECE technique, shortly described in [5]. No experimental evidence proving its feasibility has been published so far. Moreover, no study has yet determined the analytical solution of the optimal control of FTSECE architectures.

2. System under study

In this paper, we focus on a previously-published architecture [5] where the energy stored in the piezoelectric layer is temporarily transferred to a parallel inductor when a controlled switch S in turned on (see Fig. 1).

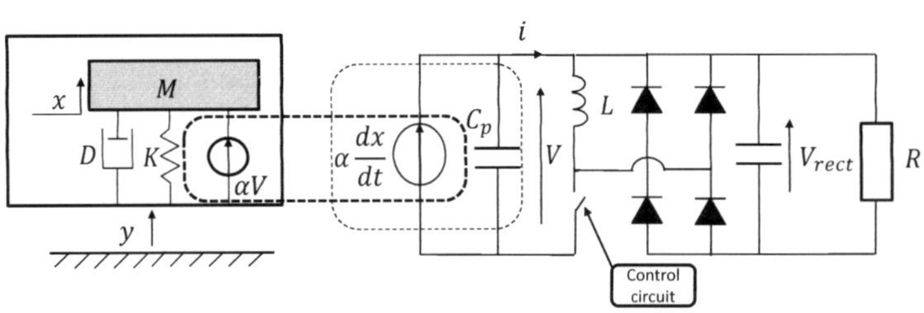

Figure 1. Interface circuit used for FTSECE technique [5].

The inductor is chosen so that $T_{LC} = 2\pi\sqrt{LC}$ is largely inferior to the period of the mechanical oscillations and the charge extraction is supposed to be instantaneous. The generator is a single-degree-of-freedom linear piezoelectric resonator of stiffness K, mass M, coupling factor α, capacitance C_p and quality factor Q, clamped to a base. The displacement of the base is written y and its acceleration γ. The motion x of the resonator is assumed to be sinusoidal (first-harmonic analysis). The coupled system is described by (1), where V is the piezoelectric voltage and i the current flow at the output of the resonator. The dynamics of the system is given by (1).

$$\begin{cases} \frac{d^2x}{dt^2} + \frac{\omega_0}{Q}\frac{dx}{dt} + \omega_0^2 x + \frac{\alpha V}{M} = \frac{d^2y}{dt^2} \\ i = \alpha\frac{dx}{dt} - C_0\frac{dV}{dt} \end{cases} \quad (1)$$

We also define the electromechanical coupling coefficients $k_m^2 = \alpha^2/KC_p$ and $k^2 = \frac{\alpha^2}{KC_p+\alpha^2}$.

3. Optimal operating conditions of FTSECE

In classical SECE, the charges are extracted from the generator when reaching its maximum voltage until the capacitance of the piezoelectric transducer is fully discharged [3]. Classical SECE has two main drawbacks. First, it does not provide the maximum available power P_{lim} (see Table I) except for a very specific value of the coupling coefficient ($k^2Q = \pi/4$). Secondly, in the case of strongly coupled generators ($k^2Q > \pi$), the -3dB bandwidth of a classical SECE energy harvester is quite poor compared to other well-known techniques such as linear load adaptation [5]. Two parameters can be controlled in this technique: the voltage drop and the instant of extraction. With the tuning factor β and the phase shift ϕ defined in Figure 2, the average power extracted from the piezoelectric transducer can be obtained via the method of harmonic balance :

$$P = \frac{16}{\pi}\frac{M\gamma^2}{8\omega_0}\frac{1-\beta}{1+\beta}\frac{k_m^2\Omega\cos^2\phi}{\left[1-\Omega^2+k_m^2\left(1+\frac{2}{\pi}\frac{1-\beta}{1+\beta}\sin 2\phi\right)\right]^2+\left[\frac{\Omega}{Q}+k_m^2\frac{4}{\pi}\frac{1-\beta}{1+\beta}\cos^2\phi\right]^2} \quad (1)$$

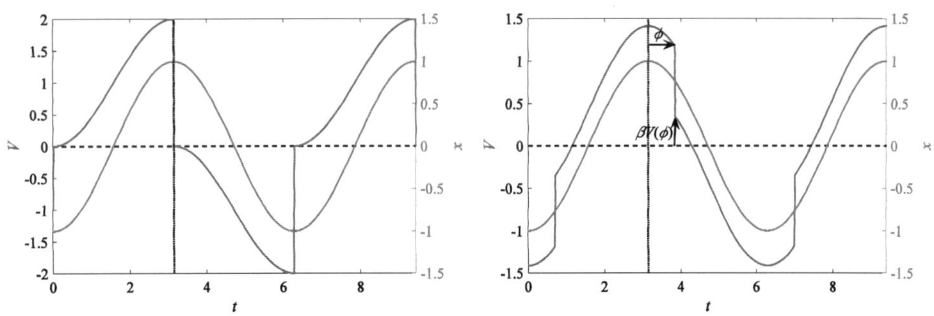

Figure 2. Typical waveforms in the SECE technique (left) and FTSECE technique (right).

To optimize the harvested power at a given frequency Ω, β and ϕ must be tuned properly. From (1), the optimal conditions (2) on β_{opt} and ϕ_{opt} leading to the maximal harvested power can be obtained, for instance with a symbolic calculation software (Wolfram Mathematica). Under these conditions, the harvested power is maximum and constant at any frequency, with a theoretically-infinite bandwidth.

$$\begin{cases} \phi_{opt} = -\text{Arctan}\left[\dfrac{Q(1+k_m^2-\Omega^2)}{\sqrt{\Omega^2+Q^2(1+k_m^2-\Omega^2)^2}}\right] \\ \beta_{opt} = \dfrac{4k_m^2Q\Omega - \pi[\Omega^2+Q^2(1+k_m^2-\Omega^2)^2]}{4k_m^2Q\Omega + \pi[\Omega^2+Q^2(1+k_m^2-\Omega^2)^2]} \cdot \\ P(\Omega) = Q, \ \forall\Omega \end{cases} \tag{2}$$

4. Experimental measurements and comparison with theoretical predictions

In our setup (see Fig. 6), the piezoelectric energy harvester is composed of a 36x36x0.2 mm3 PZT-5H plate bonded on a 60x37x0.5 mm3 stainless steel plate. This composite plate is clamped to a rigid aluminum base, and a magnetic inertial mass is attached at the free-end of the plate. The control of the switch is operated by a waveform generator delivering voltage pulses two times per period. The mechanical quality factor is $Q \simeq 50$ and the coupling coefficient is $k^2 \simeq 0.042$. The natural frequency is 69.5Hz. $L = 150\text{mH}$ and $C_p \simeq 100\text{nF}$.

Experimental measurements are reported in Fig. 6. They highlight that the -3dB bandwidth of the FTSECE harvester is multiplied by around 3.5 times compared to classical impedance matching.

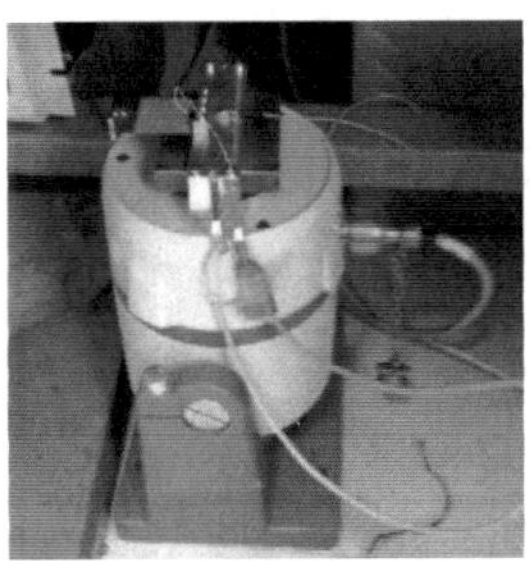

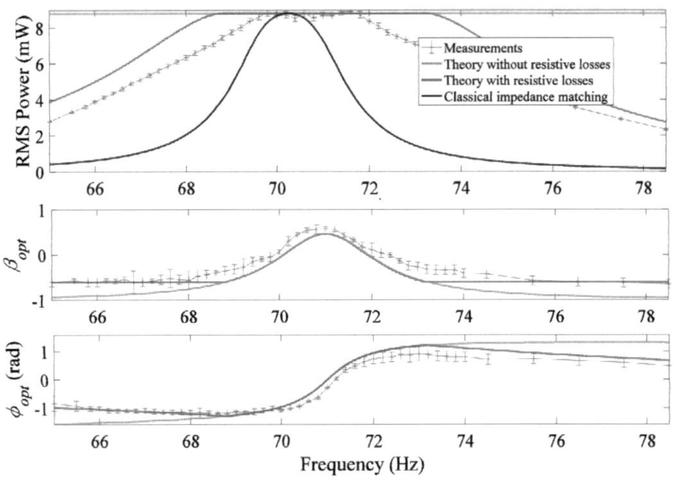

Figure 3. Picture of the harvester in the experimental setup, experimental results of FTSECE and comparison with theoretical results and with classical impedance matching.

The FTSECE system bandwidth is theoretically infinite but limited, in practice, by the resistive losses in the circuit (inductor, piezoelectric materiel, switches, etc.). These losses limit the possibility of voltage inversion through LC-oscillation. Hence, β is bounded by $\beta_{min} \leq 0$ which verifies :

$$\beta_{min} = -\exp\left[-\frac{\pi}{2}R\sqrt{\frac{C_p}{L}}\right] \tag{3}$$

where R is the series resistance of the imperfect inductance (in addition to the usually small resistance of the switch). The evolution of $\beta_{min}(L)$ in the case of a piezoelectric resonator with $C_p = 100\text{nF}$ is reported in Figure 4 for commercially-available inductors with volume lower than 1cm^3 (data from [6]). β_{min} mainly limits the bandwidth of the harvester. In order to optimize the performance of FTSECE, one should use a very small inductance (here lower than 1mH) to ensure $\beta_{min} \leq 0.95$ but this increases the risk of actuating parasitic resonances.

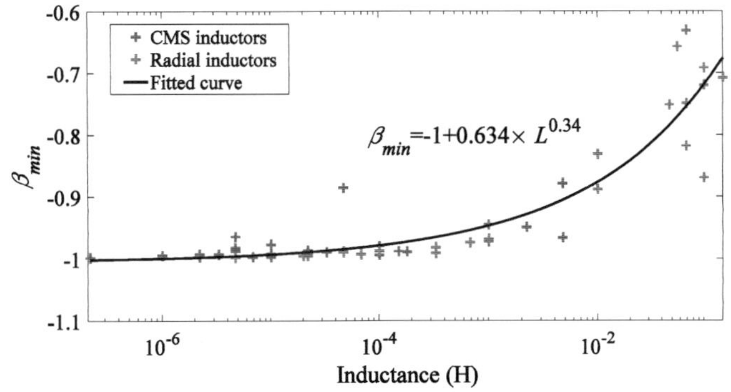

Figure 4. Lower bound of the tuning factor for commercially available inductors (data from [6]). Example for a resonator with $C_p = 100\text{nF}$.

5. Conclusions and future perspectives

The so-called FTSECE technique has been experimentally-validated. For a single-degree-of-freedom (SDOF) resonator, the bandwidth of such a system is theoretically infinite but limited, in practice, by the resistive losses in the circuit (inductor, piezoelectric materiel, switches, etc.). Diminishing the parasitic resistance by lowering the inductor value would partially solve this problem but increases the risk of actuating high-frequency parasitic modes of the resonator, resulting in additional energy losses. Manufacturing a single-mode resonator to allow the reduction of the inductor losses while avoiding parasitic resonances is the subject of ongoing research.

Acknowledgements

This work was supported by the Korea Institute of Energy Technology Evaluation and Planning (KETEP) and the Ministry of Trade, Industry & Energy (MOTIE) of the Republic of Korea (No. 20158510060040).

References

[1] E. Lefeuvre, A. Badel, C. Richard, and D. Guyomar, "Piezoelectric energy harvesting device optimization by synchronous electric charge extraction," *J. of Intelligent Material Systems and Structures*, vol. 16, pp. 865-876, 2005.

[2] E. Lefeuvre, A. Badel, A. Brenes, S. Seok, and C.-S. Yoo, "Power and frequency bandwidth improvement of piezoelectric energy harvesting devices using phase-shifted SECE interface circuit," *J. of Intelligent Material Systems and Structures*, 2017.

[3] A. Richter et al., "Tunable Interface for Piezoelectric Energy Harvesting," in *Systems, Signals & Devices (SSD), 2014 11th International Multi-Conference on*, Barcelona, 2014.

[4] P. Gasnier et al., "An Autonomous Piezoelectric Energy Harvesting IC based on a Synchronous Multi-Shots Technique," *IEEE Journal of Solid-State Circuits*, vol. 49, no. 7, pp. 1561-2014, July 2014.

[5] A. Badel and E. Lefeuvre, "Nonlinear Conditioning Circuits for Piezoelectric Energy Harvesters," in *Nonlinearity in Energy Harvesting Systems*, E. Blokhina et al., Eds. Cham, Switzerland: Springer, 2016, ch. 10, pp. 353-357.

[6] Farnell. (2017, March) farnell.com. [Online]. http://fr.farnell.com/inductances

[7] A. Badel and E. Lefeuvre, "Wideband piezoelectric energy harvester tuned through its electronic interface circuit," in *PowerMEMS*, Hyogo, Japan, 2014.

[8] Y. C. Shu and I. C. Lien, "Analysis of power output for piezoelectric energy harvesting systems," *Smart Materials and Structures*, vol. 15, pp. 1499-1512, 2006.

A hybrid piezoelectric and electromagnetic energy harvester for scavenging low frequency ambient vibrations

R M Toyabur, J W Kim, and J Y Park[1]
Micro/Nano Devices and Packaging Lab., Department of Electronic Engineering, Kwangwoon University, 447-1 Wolgye-dong, Nowon-gu, Seoul, 139-701 Korea.

E-mail: jaepark@kw.ac.kr (J Y Park)

Abstract. This work presents a hybrid type piezoelectric and electromagnetic multi-degree of freedom (MDOF) energy harvester for low-frequency ambient vibrations. The proposed device is designed with unique MDOF mechanical structure that helps to obtain multiple resonant modes operation of the harvester at low frequency range. The mechanical design comprises of a clamped-clamped beam as primary beam and four secondary beams as cantilever array. During external excitation, the secondary cantilever beams vibrate simultaneously along with the magnetic proof mass that causes mechanical strain to the piezoelectric elements as well as relative motion between magnets and coils placed underneath of it. A prototype of the hybrid MDOF energy harvester has been fabricated, simulated, and experimentally verified. The frequency response of the prototype implies that the proposed hybrid MDOF harvester has four resonant modes which concentrate around 12, 16, 17, and 21 Hz, respectively. On the other hand, a maximum of 218μW and 204μW peak powers have been obtained across the respective optimal load resistances from a single piezoelectric and electromagnetic generator of the array at 3rd resonance (17Hz) under 0.1g base acceleration.

1. Introduction

During last few decades, harvesting energy from ambient mechanical vibration has drawn much interest due to its ubiquitous existence in the environment and unlimited lifetime. Many efforts have been made to research and develop the vibration energy harvesters based on various energy conversion mechanisms, for instance, electromagnetic [1], electrostatic [2], piezoelectric [3], and triboelectric [4]. Scavenging energy from mechanical vibration can be utilized in many interesting applications including low-power electronic devices, wireless sensor networks(WSN), biomedical applications and remote hostile environment sensors, etc. [5-7]. Most conventional vibration energy harvesters comprise of single-degree-of-freedom (SDOF) systems and perform well at single resonant frequency which makes them less efficient for lower frequency vibrations [8]. Most of the vibration based piezoelectric energy harvesters have been employed a single resonant frequency of the structure which generates maximum power when the excitation frequency matches the resonant frequency of the device. Moreover, the power (P) generated by a low frequency (ω) resonant generator decreases dramatically because $P \propto \omega^3$ and too low to be utilized [9]. In order to overcome these constraints, we have proposed a MDOF vibration energy harvester where multiple high-mode resonant frequencies are shifted down to practical ambient frequency

[1] To whom any correspondence should be addressed.

ranges. Subsequently, hybridizing two transduction mechanisms from one single mechanical motion increase the power generation capability of the energy harvester. Lead-based polycrystalline ceramic Lead Zirconate Titanate (PZT) is used as piezoelectric material in piezoelectric (PE) mechanism which is most widely used because of their several mechanical and electrical advantages. On the other hand, electromagnetic (EM) mechanism utilizes the permanent magnets and conductive coil to generate electric potential according to Faraday's law of electromagnetic induction. Recently, we have a proposed a MDOF vibration energy harvester where only piezoelectric mechanism is utilized to convert mechanical vibration energy into electrical energy [10]. In this work, we have approached to hybridize our previous work with electromagnetic mechanism without increasing the harvester volume. The proof mass of the piezoelectric generator was replaced with permanent magnets to add the electromagnetic generators in the harvester. This hybrid approach generates significant power than that of the individual energy harvesting unit due to the combinational operation of PE and EM mechanisms under the same mechanical movement.

2. Hybrid energy harvester design

Figure 1 shows the schematic structure of the proposed hybrid vibration energy harvester. It consists of a primary beam which clamped at both sides of the harvester's housing whereas four secondary beams are engaged with the primary beam at the middle. The secondary beams are considered as an individual cantilever beam with proof mass attached at the free end of the beam that works as the spring-mass-damper system. The beams are made of flexible material where the primary beam acts as a linear first order transmission beam connected with secondary cantilever beam and vibrates transversely by external excitation. In the meantime, the permanent magnets are used as proof mass of the secondary cantilever and conductive coils are placed below the cantilevers to make the electromagnetic generator. During external vibration, the whole system works as MDOF system and the secondary beams vibrate in according to their modal resonant frequency as well as the permanent magnets move respectively, as shown in figure 2 (Simulated by COMSOL Multiphysics software). The piezoelectric elements are attached on the secondary beams that convert induced mechanical strain into electrical potential, whereas, the relative motion between the magnetic proof mass and coil induces voltage accordingly.

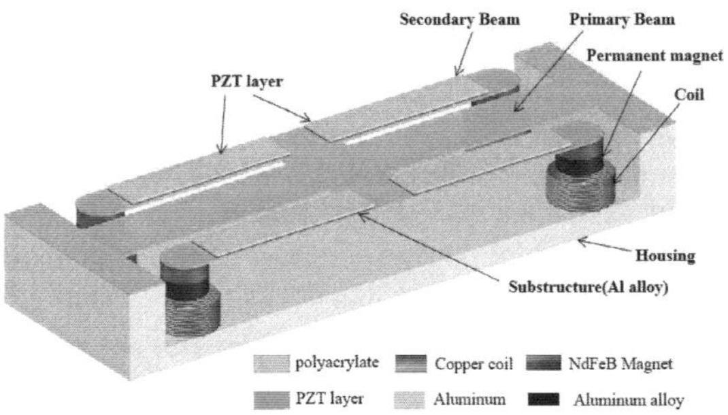

Figure 1. 3D schematic of the proposed hybrid energy harvester.

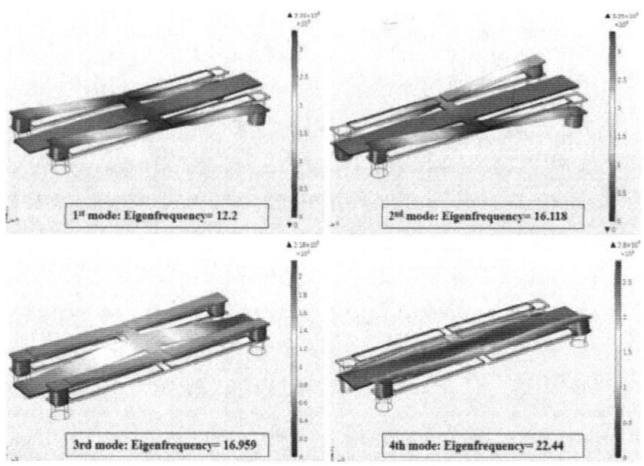

Figure 2. Finite element modelling and simulation of proposed multimodal shaped structure.

3. Prototype fabrication

A macro-scale hybrid prototype was fabricated that consists of two transductions parts as shown in figure 3. A low Young's modulus polyacrylic sheet (thickness: 0.5mm) is used to make primary (dimension: 90 × 10 × 0.5 mm^3) and secondary (dimension: 40 × 5× 0.5 mm^3) beams with laser cutter. PZT ceramic material (5 H Navy Type VI) was used as piezoelectric material which was attached on the secondary beams with substructure (Aluminium alloy) using adhesive. NdFeB (N35) cylindrical magnets (Ø6×5) are employed as proof mass of the secondary cantilevers where 500-turns bobbinless hollow cylindrical coil

(Ø0.2mm copper wire) is placed underneath of the magnets which constitute the EM transduction part. The entire system is placed inside a housing made of aluminium.

4. Experimental results and discussion

Figure 4 illustrates the schematic diagram of the complete experimental setup. The proposed energy harvester was mounted on the electrodynamic shaker (B & K 4809) to characterize the frequency response and output performance of the harvester at different external excitation. The outputs of the harvester were measured and stored in a digital oscilloscope (Tektronix TDS5052B). In order to understand the output characteristics of the proposed energy harvester, we have analyzed the output voltage and power of a single PE and EM part of the harvester.

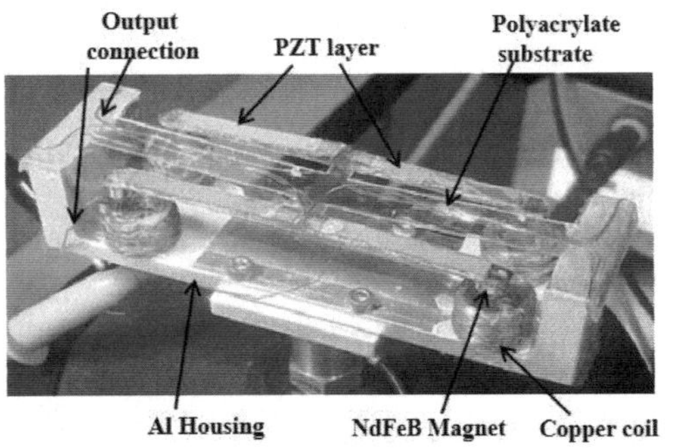

Figure 3. Photograph of the fabricated prototype hybrid energy harvester under test.

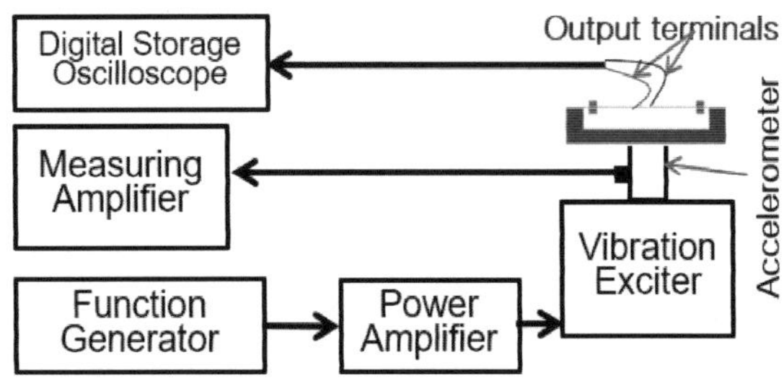

Figure 4. Schematic diagram of the experimental test setup.

Figure 5 shows the frequency response of a single PE and EM generator under constant 0.1g acceleration. Inspection of the voltage curve as a function of frequency indicates that the harvester has four resonant modes below 30Hz range as we predicted from Finite element method (FEM) simulation. The four resonant modes of the multimodal energy harvester are 12, 16, 17, and 21 Hz respectively. However, the voltage levels are varied at different resonant frequencies, because each resonant frequency has a unique modal shape. During 3rd resonant mode (17Hz), the generators produce the highest voltage than other three resonant modes below 30Hz. Additionally, figure 6 shows the instantaneous voltage waveform of the EM and PE generators across the respective optimal load resistance at 17Hz (0.1g acceleration). The generated voltage of the PE generator is higher than EM generator whereas the current generation is higher in EM generator than PE generator corresponding to the various output loads. Therefore, power generation is almost similar for two kinds of mechanisms. Figure 7 shows the maximum peak-peak voltage and power generated by EM and PE generator across various load resistances at 3rd resonant mode (17 Hz) under 0.1g base acceleration. A single EM generator delivers maximum of 204μW peak power to an 8.5Ω optimal load resistance at 3rd resonant mode whereas a PE generator delivers maximum of 218μW peak power to an 80KΩ optimal resistive load at same condition.

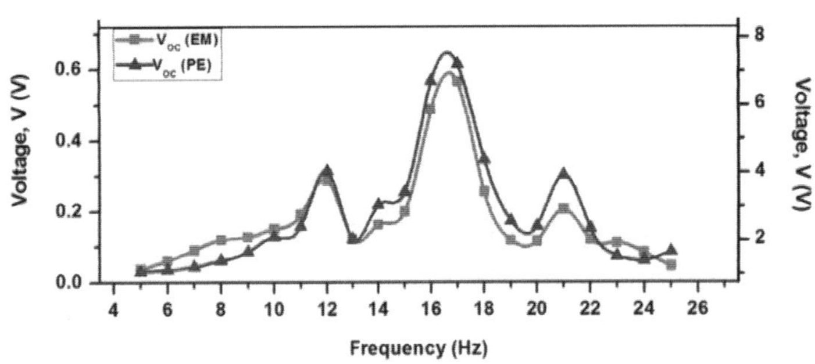

Figure 5. Frequency response of a single secondary beam of the proposed hybrid energy harvester under 0.1g acceleration.

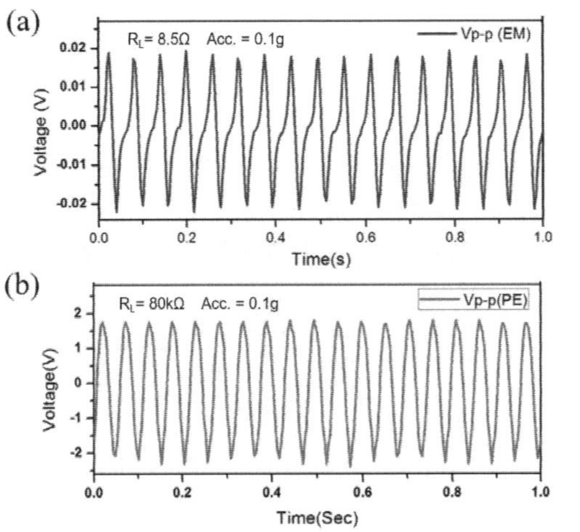

Figure 6. Instantaneous peak-peak voltage waveform across the optimum load resistance of single primary beam at 3rd resonant mode (17 Hz) under 0.1g base acceleration: (a) electromagnetic generator (Load: 8.5Ω) (b) piezoelectric generator (Load: 80 KΩ).

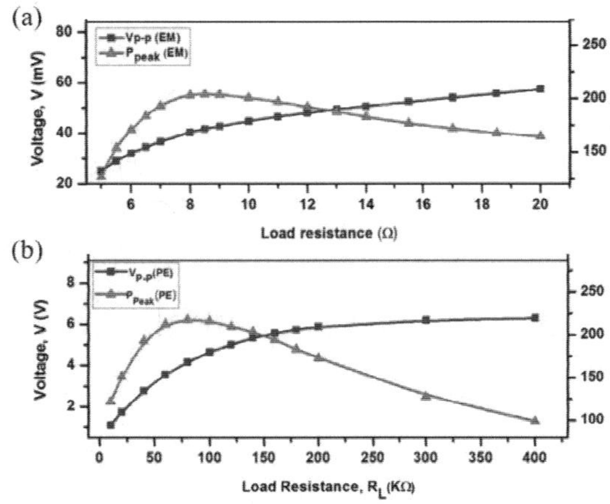

Figure 7. Load voltages (p-p) and powers (peak) versus load resistances generated by (a) magnet-coil assembly (EM) and (b) single primary PZT cantilever beam of fabricated prototype at 3rd resonant mode(17Hz) (PE) under 0.1g base acceleration.

5. Conclusions

In this paper, we have proposed and demonstrated a vibration energy harvester in which piezoelectric transduction and electromagnetic transduction mechanism have been hybridized to scavenge electrical energy from low-frequency mechanical vibrations. This hybrid approach increases the overall power generating capability of the energy harvester. By using multiple degrees of freedom technique, we have four effective resonant frequencies at low frequency range which increases the bandwidth of the energy harvester. A macroscale prototype has been fabricated and tested in the laboratory environment. In this harvester, single piezoelectric transducer generated 218μW peak power whereas the single EM transducer generated 204μW peak power corresponding to their optimum load resistances at 17Hz under 0.1g acceleration. Future work will include more optimized design, miniaturization, and implementation of the proposed hybrid vibration energy harvester.

Acknowledgements

This work was supported by the Technology Innovation Program (10065696) funded by the Ministry of Trade, Industry & Energy (MI, Korea). The authors are grateful to MiNDaP (Micro/Nano Devices and Packaging Lab.) group members of Kwangwoon University.

References

[1] Salauddin M, Halim M A, and Park J Y, 2016 *Smart Mater. Struct.* **25** 095017
[2] Nguyen S D and Halvorsen E, 2011 *J. Microelectromech. Syst.* **20** 1225-1227
[3] Halim M A and Park J Y, 2015 *Ceram. Int.* **41** S702–S707
[4] Rasel M S and Park J Y, 2017 *Appl. Energy* **206** 150-158
[5] Salauddin M and Park J Y, 2017 *Smart Mater. Struct.* **26** 035011
[6] Halim M A and Park J Y, 2014 *J. Appl. Phys.* **115** 094901
[7] Halim M A, H. Cho and Park J Y, 2015 *Energ Convers Manage* **106** 393-404
[8] Gatti G, Brennan M, Tehrani M, Thompson D, 2016 *Mech. Syst. Signal Process.* **66** 785–792.
[9] Ashtari W A, Hunstig M, Hemsel T, Sextro W, 2013 *Sens. Actuators A,* 200 138–146.
[10] Toyabur R M, Salauddin M and Park J Y, 2017 *Ceram. Int.* **43** S675-S681

IOP Conf. Series: Journal of Physics: Conf. Series **1052** (2018) 012052 doi:10.1088/1742-6596/1052/1/012052

Optimizing dimensions of unipolar Teflon-FEP piezoelectrets with micro-system-technology

F Emmerich and C Thielemann

Department of Engineering, University of Applied Sciences Aschaffenburg, Würzburger Straße 45, 63743 Aschaffenburg, Germany

E-Mail: florian.emmerich@h-ab.de

Abstract. In this work we present a new micro-system-technology based production process for unipolar ferroelectrets. Theoretical analysis of the influence of the air-gap size on the electric field distribution as well as on the induced charge on the electrodes shows superior performance of unipolar piezoelectrets with small air-gaps. For the production of these small air-gaps we developed a new design using a photoresist thermoforming master, an integrated micro-heater and shadow masks for metallization. Unipolar piezoelectrets produced with this technology exhibit increased d_{33}-coefficients compared to designs in previous publications. These piezoelectrets are highly preferable for energy-harvesting applications, as they promise high electric power output.

1. Introduction

Piezoelectricity in polymers can generally originate from three sources: the molecular structure of the polymer, embedded piezoelectric materials or tailored, macroscopic ferroelectric domains [1]. The latter are usually air-voids inside the polymer, showing piezoelectric behavior after appropriate electric charging. These so-called piezo- or ferroelectrets have been conducted for energy-harvesting applications, as they show good mechanical stability and high piezoelectric d_{33}-coefficients resulting in high electric power output [2]. Especially piezoelectrets with air-voids fabricated from structured Fluoroethylenepropylene (FEP) are promising examples of this class, with d_{33}-coefficients above 1,000 pC/N [3]. While these piezoelectrets are typically based on a bipolar charge distribution on opposite surfaces of the voids, only few research focuses on unipolar piezoelectrets. This may be explained by a complex production process allowing only for large air-gaps and thus comparatively small d_{33}-coefficients [4]. However, it should be pointed out that unipolar piezoelectrets storing only negative charges are clearly superior if thermal stability and longevity [4] are an issue.

2. Theoretical considerations on optimal dimensions in unipolar piezoelectrets

The performance of piezoelectrets is predominantly defined by their piezoelectric d_{xy}-coefficients. Equations for the calculation of the expected d_{33}-coefficient of piezoelectrets with artificial voids have already been derived for two-layer [5] as well as for multiple-layer designs [6]. Solutions of these equations are in good agreement with experimental findings, even if the real void geometry is mathematically simplified to parallel plates with an air-gap. However, these calculations rely on a bipolar charge distribution inside the voids. For calculating the d_{33}-coefficient of unipolar piezoelectrets, an adequate simplification of the geometry must be done, as shown in figure 1:

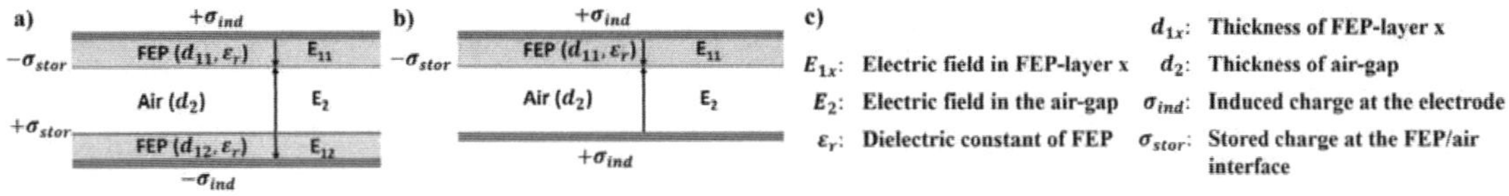

Figure 1. Schematics of a) bipolar and b) unipolar piezoelectret designs with c) explanation.

Here, evaluation of the electric field using Gauss' and Kirchhoff's law leads to a different electric field distribution and different values for the induced charge on the electrodes as compared to bipolar charge distribution [6]. The electric fields E_1 and E_2 are described as

$$E_{11} = \frac{d_2\sigma_{stor}}{\varepsilon_0(d_{11}+d_2\varepsilon_r)} \qquad E_2 = \frac{d_1\sigma_{stor}}{\varepsilon_0(d_{11}+d_2\varepsilon_r)}. \tag{1}$$

Assuming that thickness changes only occur in the air-gap d_2, the change of induced charge is calculated as function of air-gap size, to

$$\frac{\partial\sigma_{ind,\ top}}{\partial d_2} = \frac{\varepsilon_r d_{11}\sigma_{stor}}{(d_{11}+\varepsilon_r d_2)^2} \qquad \frac{\partial\sigma_{ind,\ bottom}}{\partial d_2} = \frac{-\varepsilon_r d_{11}\sigma_{stor}}{(d_{11}+\varepsilon_r d_2)^2}. \tag{2}$$

Combining these equations with a stress-strain relation, one derives a formula for the d_{33}-coefficient, which is independent of layer number or charge distribution type,

$$d_{33} = \frac{\sigma_{stor}\cdot\varepsilon_r}{Y}\cdot\frac{(1+d_2/d_1)}{(1+\varepsilon_r d_2/d_1)^2}. \tag{3}$$

However, values used for the unipolar case differ from that for the bipolar case: the charge density can achieve twice the value of a bipolar piezoelectret and the thickness of the solid layer can be reduced to half its value. Assuming a similar build-height as well as identical values for Young's modulus, this yields a higher d_{33}-coeffient, particularly for small air-gaps. The induced charge (figure 2 a), calculated from the electric field, as well as the change in induced charge (figure 2 b) is compared for unipolar and bipolar piezoelectrets built from 12.7 µm FEP. While high performance of FEP-based piezoelectrets at small air-gaps has generally been predicted by van Seggern [7], our calculations show a superior performance of unipolar piezoelectrets with thin FEP-foils for small air-gaps below 50 µm.

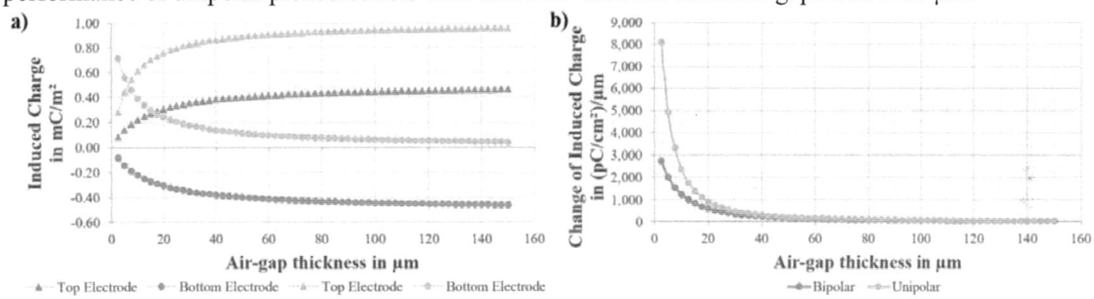

Figure 2. a) Comparison of the induced charge on the electrodes between unipolar (orange) and bipolar (blue) piezoelectrets for different air-gaps. Total charge-density in both systems is 1mC/m². b) Comparison of the change in the induced charge for the same systems.

3. Processing of unipolar piezoelectrets

A major challenge for the production of FEP-piezoelectrets is the lack of control over the dimensions in the micrometer-range, which is very difficult for conventional thermoforming molds [3] or laser-cut thermoforming substrates [4]. As stated above, especially these small dimensions show promising high d_{33}-coefficients. To achieve a reduction of dimensions with satisfying precision we developed a novel micro-system-based process for the production of thermoforming molds. We produced unipolar piezoelectrets in a thermoforming-process adapted from Zhang et al. [3], but using a thermoforming master created with a micro-system-technology-based approach. This allows to produce all dimensions inside the piezoelectret with micrometer-precision. With this technology we were further able to adjust the current design to optimize unipolar piezoelectrets.

Borofloat glass pieces (Schott, Germany) with dimensions of 34x34 mm² were used as substrate for the production of the thermoforming master. After dehydration of the substrate at 150 °C for 60 minutes,

IOP Conf. Series: Journal of Physics: Conf. Series **1052** (2018) 012052 doi:10.1088/1742-6596/1052/1/012052

the adhesion promoter TI Prime (MicroChemicals, Germany) was applied. In the next step, approx. 10 ml SU8-100 (MicroChem, USA) were applied on the substrate and spin coated at 3,000 rpm for 30 seconds. Subsequently, a softbake (SB) step at 65 °C for 30 minutes followed by 95 °C for 60 minutes was performed using a precision hotplate (Harry Gestigkeit, Germany). Heating and cooling was performed with a slope of approx. 3 °C/min. After cooling to room temperature, the SU8 was exposed with an intensity of 1,200 mJ/cm² using a Cloe UV-KUB 2 exposure system (Kloe, France). Following a resting phase of 20 minutes, a post exposure bake (PEB) with parameters identical to the softbake step was performed. Shortly after cooling down, the SU8 was developed for 4 minutes in mr-dev 600 developer (MicroChem, USA). In order to achieve steeper edges, the master was additionally developed for 1 minute in a separate beaker with fresh developer. The resulting SU8 height was determined with a Dektak XT (Bruker, USA) to around 65 µm. On the thermoforming masters aluminum micro-heaters were implemented, using a production process described elsewhere [8]. The thermoforming master was surface treated in an air-plasma (Diener, Germany) for 5 minutes to increase the adhesion of the micro-heater. Directly after the surface activation, we applied full-surface metallization with approx. 150 nm aluminium on the thermoforming master via a Nano 36 sputtering device (K. J. Lesker, UK). Subsequently, about 1.5 µm of the positive tone photoresist AZ MIR 701 (MicroChemicals) was spin coated at 3,000 rpm for 30 seconds on the metallized surface. After exposure and development, the necessary areas for the micro-heater remained masked. Then the remaining aluminium was etched using TechniEtch 80 (MicroChemicals), the remaining photoresist was removed with AZ 100 remover (MicroChemicals), and the sample was cleaned thoroughly.

In this work we produced thermoforming master for piezoelectrets with a void width of 1,000 µm, a space between voids of 500 µm and an approximate void-height of 65 µm. The implemented micro-heaters showed an average resistance of 115 Ohm. It should be noted, that the mentioned process is suitable for downsizing the width dimensions from centimeters to 5 µm and the height dimensions from 150 µm down to 5 µm, allowing for a broad spectrum of possible piezoelectret designs. A schematic of the production process for the thermoforming master with integrated micro-heater is depicted in figure 3.

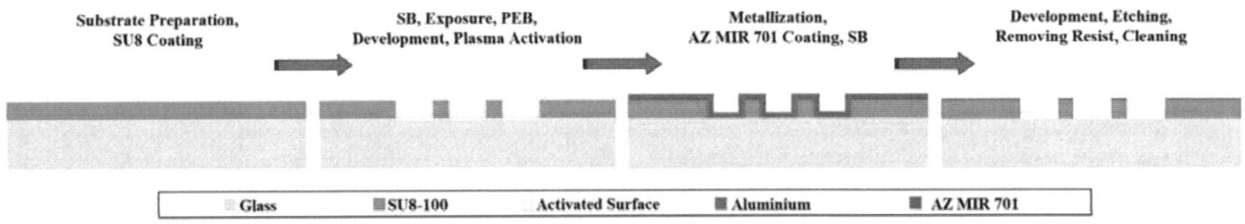

Figure 3. Schematic of the most important processing steps of the thermoforming master.

Piezoelectrets consisting of voids with tubular shape in 12.7 µm thick FEP-50A-foils (Wortmann, UK) with a width of 1,000 µm, spacing of 500 µm and height of about 65 µm were created during a thermoforming process at 100 °C on the SU-8-thermoforming master. During structuring, a load of 100 kPa was applied for 5 minutes. Parallel to the thermoforming process, a second 12.7 µm FEP-foil was metallized with approx. 150 nm aluminium applying shadow mask technology to achieve a metal layer corresponding to the channels in the structured foil. Subsequently, both foils were stacked according to the schematic in figure 4. With a static load of 50 kPa on the stack, local fusion-bonding was performed, using the integrated micro-heater design for 30 seconds. After bonding, the front of the stack was metallized with approx. 150 nm copper.

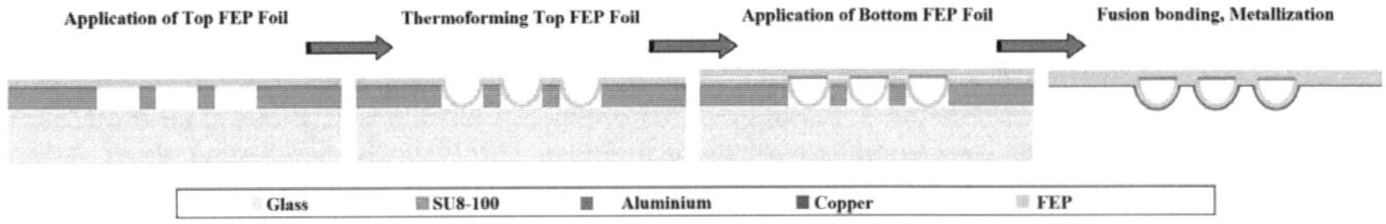

Figure 4. Schematic of the production process of optimized unipolar piezoelectrets.

Finally, the piezoelectrets were charged by applying a 2 kV saw-tooth-voltage in the right polarity between the electrodes for 10 s. A schematic of the finished piezoelectret is depicted in figure 5 a) and a cross section micrograph of a single void in the produced piezoelectret in figure 5 b).

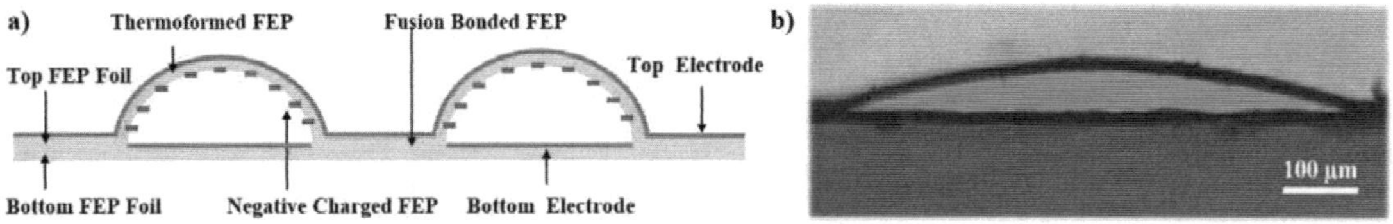

Figure 5. a) Schematic of the unipolar FEP-piezoelectret. b) Bright-field micrograph of a cross-section of a single void in a micro structured FEP-piezoelectret.

4. Characterization of unipolar piezoelectrets

Using a quasistatic approach, the produced piezoelectrets were characterized with respect to their piezoelectric properties. A load of 1.5 kPa was attached and removed for ten times on each piezoelectret. The generated charge was measured on a Spider 81B vibration-controller (Crystal Instruments, USA) with integrated charge amplifier. From these preliminary measurements piezoelectric coefficients have been determined between 500 and 1,850 pC/N. As these values stayed constant also for higher loads, we assume that the small weight totally compressed the samples. Compared to previously published results [4], the new design in combination with thinner FEP and minimized air-gaps resulted in a drastic increase of the d_{33}-coefficient. In addition, the durability of the fusion-bond is superior to glued structures.

5. Conclusion

Optimizing the production process of unipolar FEP-piezoelectrets allows for a noticeable increase of d_{33}-coefficients up to 1,850 pC/N for quasistatic measurements. Especially for applications demanding high thermal d_{33}-stability, this kind of piezoelectret is highly preferable. Future work will focus on miniaturizing width as well as the height of the artificial voids, for a further increase of the d_{33}-coefficient. Determination of the frequency response in dynamic measurements as well as calculations for the maximum value of the d_{33}-coefficient at distinct geometries are under evaluation. Optimizing the design of the integrated micro-heater will additionally reduce the thermal stress on the thermoformed FEP during fusion bonding.

This work was partially supported by Bayerisches Staatsministerium für Bildung und Kultus, Wissenschaft und Kunst in the frame of ZEWIS. We want to thank the German Academic Exchange Service (DAAD) for a travel grant.

References

[1] K. S. Ramadan, D. Sameoto and S. Evoy, *Smart Materials and Structures 2014*, vol. 23, 033001 (26pp).
[2] P. Pondrom, J. Hillenbrand, G. M. Sessler, J. Bös and T. Melz, *Applied Physics Letters 2014*, vol. 104, 172901 (5pp).
[3] X. Zhang, G. M. Sessler and Y. Wang, *Journal of Applied Physics 2014*, vol. 116, 074109 (8pp).
[4] D. Rychkov, W. Wirges, R. Gerhard and R. A. P. Altafilm, *Conference on Electrical Insulation and Dielectric Phenomena 2015*, 2015 (3pp).
[5] R. Kacprzyl, A. Dobrucki and J. B. Gajewski, *Journal of Electrostatics 1997*, vol. 39 (8pp).
[6] J. Hillenbrand and G. M. Sessler, *IEEE Transactions on Dielectrics and Electrical Insulation* 2000, vol. 7 (6pp).
[7] H. v. Seggern, S. Zhukov and S. Fedsov, *IEEE Transactions on Dielectrics and Electrical Insulation 2011*, vol. 18 (8pp).
[8] F. Emmerich and S. Ciba, *International Student Conference Prague POSTER 2017*, 2017 (5pp).

IOP Conf. Series: Journal of Physics: Conf. Series **1052** (2018) 012053 doi:10.1088/1742-6596/1052/1/012053

On the design guidelines for miniaturizing thermo-magnetically activated piezoelectric energy generator

Adrian Rendon-Hernandez[1,2] **and Skandar Basrour**[1,2]

[1] Université Grenoble Alpes, TIMA Laboratory, F38031 Grenoble, France
[2] CNRS, TIMA Laboratory, F-38031 Grenoble, France

E-mail: {adrian.rendon,skandar.basrour}@univ-grenoble-alpes.fr

Abstract. This article deals with experimental testing of a thermo-mangetically activated piezoelectric generator, we provide breakthrough in addressing the miniaturization issue of such power generators by reporting design rules. Three main design parameters were derived: the transducer's geometry, the gap distance, and the magnetic volume. Special attention was put into the design geometry of the transducer and the magnetic volume. Three different design were tested and compared. These tests revealed that correlated reduction of the transducer and magnets dimensions meant reduction of thermal operating cycles. An increase of 35% in power density is obtained by combining design strategies like to increase the transducer resonant frequency and to uniform distributed strain. An output power density during operating transition of 11.3mW/cm^3 is reached.

1. Introduction

Double step conversion methods for thermal energy harvesting such as thermal buckling [1], thermo-acoustics [2], and thermo-magnetic [3], can be considered an alternative solution for miniature thermal energy harvesting. Thermo-magnetic generators rely on the reversible change in magnetization of a suitable material such as the case of iron-nickel alloys, caused by exposing the material to a changing temperature. By utilizing the thermo-manetic field effect described earlier, a new device for generating electric power was previously presented in [4]. In previous work we reported a thermo-magnetic energy harvester able to harness the ambient thermal energy from a variation in environmental temperature and converts it into electrical energy. However, a key limitation of that power generator is that it is too bulky to be integrated into a wireless sensor node. With this goal, the present work seeks to offer design rules for miniaturization. The rest of the paper is organized as follows: the proposed power generator will be discussed in Section 2. The design parameters are presented in section 3. Section 4 shows the experimental measurements of test prototypes. Finally, Section 5 concludes with a summary.

2. Design

As shown in Figure 1, the power generator has two main sections according to their respective functions: the energy transducer and the triggering system. The energy transducer is composed

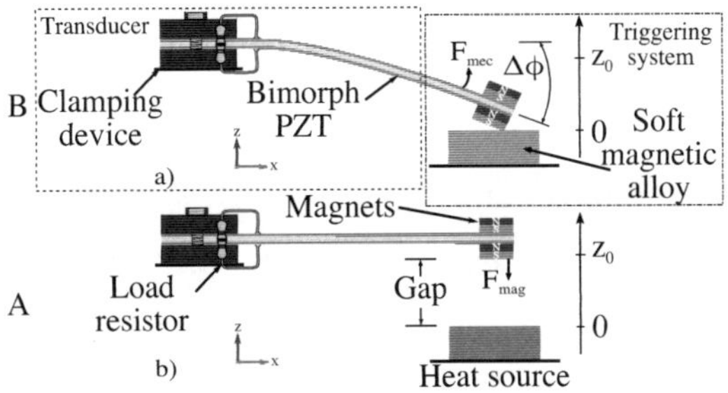

Figure 1. Schematic of the thermally activated piezoelectric generator: a) lateral view when closed position at low temperatures; b) lateral view when open position at high temperatures.

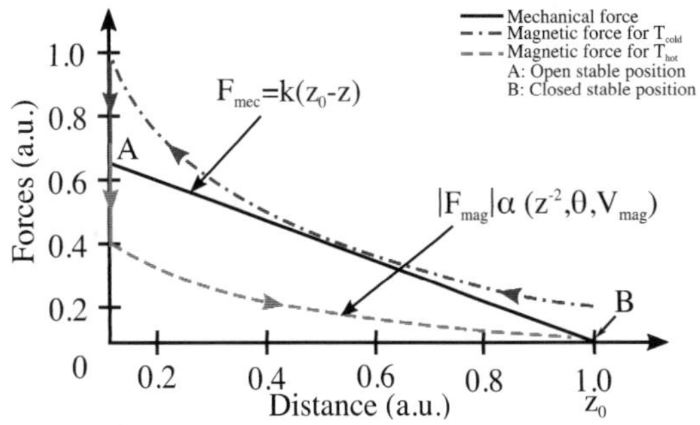

Figure 2. Schematic of actuation forces in dependence of the separation distance between soft magnetic alloy and permanent magnets. Temperature changes drive the stable positions.

of a piezoelectric bimorph-type cantilever beam, namely PSI-5H4E from Piezo Systems, Inc with a series-type polarization. Concerning the triggering system, it consists of two permanent magnets (NdFeB), that is the magnetic volume V_{mag}, that are attached to the free end of the beam, and a soft magnetic material (FeNi), fixed under the free end of the beam, forming an air gap.

2.1. Working principle

The generator has two stable positions: the closed position (B in Figure 2) and the open one (A in Figure 2). In the initial state, at a cold temperature, the soft magnetic material is magnetized so the cantilever beam is pulled-down due to magnetic force. After, when the environmental temperature increases, the magnetic properties of soft magnetic material change from ferromagnetic to paramagnetic, causing the magnetic force to decrease. The beam is pulled back to its initial state because of the spring-back force of the beam. This process of pulling-down and pulling-back of the beam is periodically repeated as long as the energy harvester is placed in an environment with cycled variations in temperature.

3. Design parameters

As can be seen from Figure 2, the cantilever beam has a linear restoring mechanical force F_{mec}, which depends on its geometry. The magnetic force, F_{mag}, is dependent on both the gap distance and temperature. It is trivial to verify that for complete deflection of cantilever, a positive resulting forces sum is expected. We derive three parameters design: the transducer's geometry, the gap distance, and the total magnetic volume.

3.1. Size reduction strategies

In Figure 3, we report the schematic representation of the overall system operation during 1 thermal cycle versus time. In this figure, a) corresponds to electric current input to change the temperature of FeNi alloy by Joule effect, b) is the temporal evolution of temperature at FeNi alloy, c) represents the actuation forces interaction during the thermal cycle, and d) is the tip displacement response of the transducer. Therefore, the size reduction strategy can be described as follows: the resonant frequency of the oscillation beam is drastically increased when its dimensions are reduced. This results in higher vibration energy of the transducer and better energy harvesting capability. The design geometry of the cantilever beam is geared towards the following goals: (1) Maximize the piezoelectric response for a given input. (2) Minimize the damping associated with the mechanical structure. This can be reached by

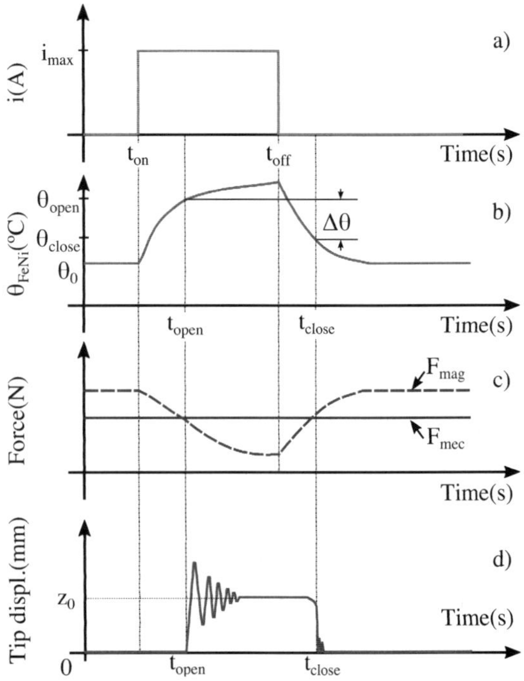

Figure 3. Transient response of the different parts of the micro-generator.

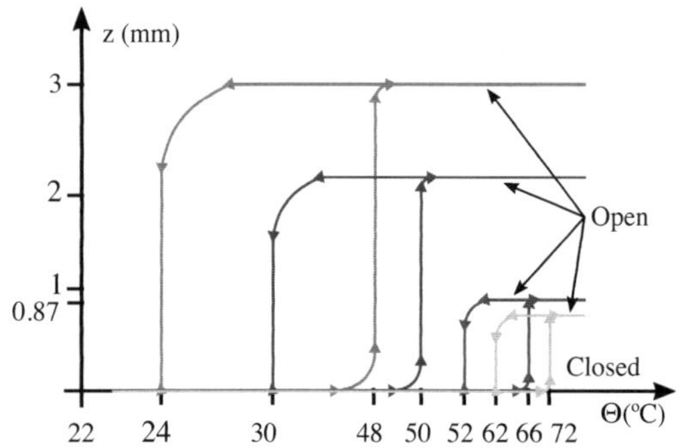

Figure 4. Experimental results for prototype A showing the thermal hysteresis, $\Delta\Theta$, and the temperature thresholds, Θ_{open} and Θ_{close}, as a function of the gap distance, z.

reducing the damping ration of the width-reduced beam; thus, a beam with smaller width vibrates with higher amplitudes and therefore has higher energy harvesting capability. (3) Improve scavenger robustness by producing an evenly distributed strain. This can be realized by having an increasingly triangular profile. In addition, we choose optimal load resistor matching the electrical impedance of the piezoelectric bimoprh to maximize power transfer.

4. Experimental testing

The first prototype shows that the gap distance is a key design parameter as reported in Figure 4. In fact this distance plays 4 roles: the first one concerns the thermal hysteresis, $\Delta\Theta$, the second one implies a shift in the temperature thresholds range, Θ_{open} and Θ_{close}, and the third one is associated to the magnetic damping and also the strain energy into the beam. Consequently, the design guidelines shall consider the influence of this parameter. For room temperature applications, having higher gaps is advantageous because it implies a thermal range shift towards these temperatures. Moreover, it increases the strain energy into the transducer. As all of these design parameters are correlated, a tradeoff has to be set for a correct operation of the generator. We constructed three different geometries of transducer, such cases are depicted in Figure 5. Firstly, we reduced the damping associated with mechanical structure by reducing the beam's width (from case A to case B). An increase of 35% in power density is obtained. Secondly, by reducing the V_{mag}, not only the generator's volume is reduced but also its resonant frequency is increased (from case B' to C'). This results in better power density per transition capabilities. Thirdly, using a triangular cantilever rather than a rectangle is to increase by an order of magnitude the power density per transition. This final design is the case C', which combines these earlier strategies (higher resonant frequency and uniformly distributed strain) to reach an output power density during transitions of 11.3mW/cm^3. Figure 6 depicts the typical responses of the design B' and C'. Table 1 summarizes experimental findings.

Table 1. Experimental results summary.

Item	Units	Transducers geometries				
		A	B	B'	C	C'
Gap	mm	2	2	1.17	2	1.15
V_{mag}	mm^3	39.2	39.2	6.28	39.2	6.28
V_{eff}‡	mm^3	140.6	38.1	57.84	23.7	13.96
Θ_{open}	°C	67	64	64	57	56
Θ_{close}	°C	34	46	31	32	28
$\Delta\Theta$	°C	33	18	33	25	28
f_r	Hz	90	85	100	120	250
R_{load}	kΩ	610	721	740	147	400
(P/V_{eff})†	mW/cm^3	0.147	0.227	0.178	7.468	11.3

‡Effective volume including the transducer's oscillations amplitude.
†Considering $40ms$ as the time period for closing and opening switching.

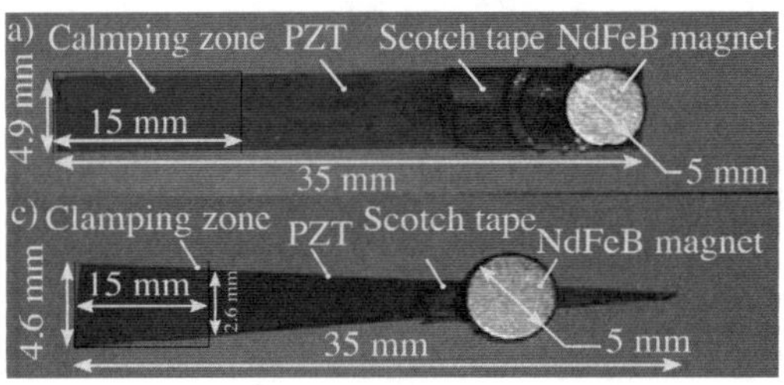

Figure 5. Photograph of two of the design of piezoelectric transducers: a) rectangular shaped, c) triangular shaped.

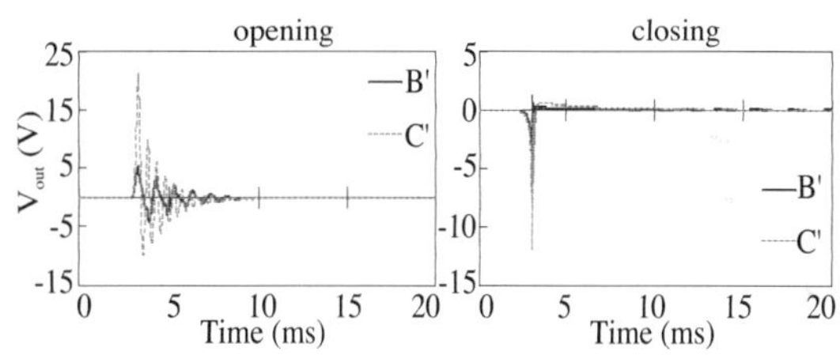

Figure 6. Experimental results regarding the output voltage at the opening and closing transitions for the prototypes B' and C'.

5. Conclusion and perspectives

Design guidelines for size reduction of a thermo-magnetically activated piezoelectric energy harvester were presented. Data collected shown that tuning of thermal hysteresis of such devices can be done by modulating the gap parameter. An increase in the power density during transitions was observed with a triangular shaped beam compared with a rectangular one. Current research includes reduction of soft magnetic alloy size to reduce the period time between transitions, thus increasing the generator's energy harvesting capability.

Acknowledgments

The authors wish to acknowledge the assistance of Mr.Gaëtan Debontride regarding the data acquisition stage. Adrian Rendon also thanks the CONACyT, for PhD Scholarship.

References

[1] E. Trioux, S. Monfray, and S. Basrour, "Butterfly micro bilayer thermal energy harvester geometry with improved performances," *Journal of Physics: Conference Series*, vol. 773, p. 012094, Nov. 2016.

[2] O. Puscasu, S. Monfray, J. Boughaleb, P. Cottinet, D. Rapisarda, E. Rouvire, G. Delepierre, G.Pitone, C. Matre, F. Boeuf, D. Guyomar, and T. Skotnicki, "Flexible bimetal and piezoelectric based thermal to electrical energy converters," *Sensors and Actuators A: Physical*, vol. 214, pp. 7–14, Aug. 2014.

[3] M. Gueltig, H. Ossmer, M. Ohtsuka, H. Miki, K. Tsuchiya, T. Takagi, and M. Kohl, "High Frequency Thermal Energy Harvesting Using Magnetic Shape Memory Films," *Advanced Energy Materials*, vol. 4, p. 1400751, Dec. 2014.

[4] A. Rendon-Hernandez and S. Basrour, "Coupled multiphysics finite element model and experimental testing of a thermo-magnetically triggered piezoelectric generator," *Journal of Physics: Conference Series*, vol. 773, p. 012024, Nov. 2016.

Paper-based water management system for microfabricated packageless fuel cell

Simon Hamel and Luc G Fréchette

Institut Interdisciplinaire d'Innovation Technologique (3IT), Université de Sherbrooke, 3000 Boulevard Université, Sherbrooke, J1K 0A5, Québec, Canada

Simon.hamel@usherbrooke.ca

Abstract. In this paper, we report the first successful implementation of a complete paper-based water management system (WMS) that collects the excess water at the cathode of a microfabricated packageless fuel cell (PLFC) and redistribute it to its anode and to an evaporation surface that simulate a hydride-based hydrogen generator. Five configurations of WMS were tested on a 1 cm^2 air-breathing PLFC and show that with the best configuration, the power output of the cell can be improved by 90% for at least 24h. We experimentally demonstrate that not only cathode flooding issues must be address to obtain an efficient WMS but also that anode drying plays a major role in poor performance of small air-breathing fuel cells. Finally, with the optimal configuration of WMS, the PLFC in air-breathing mode shows similar performances as with humidity controlled air forced-flow.

1. Introduction

Low battery life of portable electronics devices is currently a major concern for developers and consumers because of the slow growth in performances of lithium batteries [1]. Because of their theoretical high energy density and their instantaneous recharge, small air-breathing polymer electrolyte membrane fuel cell (PEMFCs) with hydride hydrogen generator are often seen as potential substitutes to power portable electronic devices. For the moment, bulky packaging issues combined with the absence of efficient water management systems decrease energy density of these systems to a level well below the performances of lithium batteries.

To reduce the volume of the packaging of small air-breathing PEMFC, a new approach to integrate all the necessary components of a fuel cell system on a single polyimide layer has been successfully developed in a previous work [2]. This packageless fuel cell (PLFC), reduced the thickness of the cell to 80μm while providing a 97% increase in volumetric power density.

Despite these encouraging results, small air-breathing fuel cell systems still suffer from low efficiency due to poor water management. Because of their size and low power output, they are not equipped with auxiliary components to stabilize their temperature and humidity level which lead to simultaneous flooding of the cathode and drying of the anode. Flooding of the cathode is well known and documented and happens when the cell produce more water than it can evaporates. Many systems using capillary channels integrated in the cathode current collector have been tested successfully on small air-breathing fuel cells and generated surface power density improvement from 10% to 30% [3–5]. Nevertheless, these systems are bulky and can hardly be optimized since they must be integrated in the current collector. On the anode side, strong electro-osmotic drag flow, combined with poor water diffusion through Nafion dries up the anode and increase its ionic resistance. This phenomenon is often neglected, but it has been showed that humidifying adequately the anode during operation can improve power density of the cell up to 30% [6,7].

Also, in a hydride based hydrogen generator, about half of the volume is occupied by a water reservoir required to complete the hydrolysis reaction. Therefore, the integration of a simple and passive water management system (WMS) which uses excess water produced at the cathode to humidify the anode and feed the hydrogen generator could increase maximal power output and energy density of the system (Figure 1). The integration of such system is significantly eased by the PLFC configuration where the current collectors are underneath the catalyst layers, which free the entire surface to collect and redistribute water.

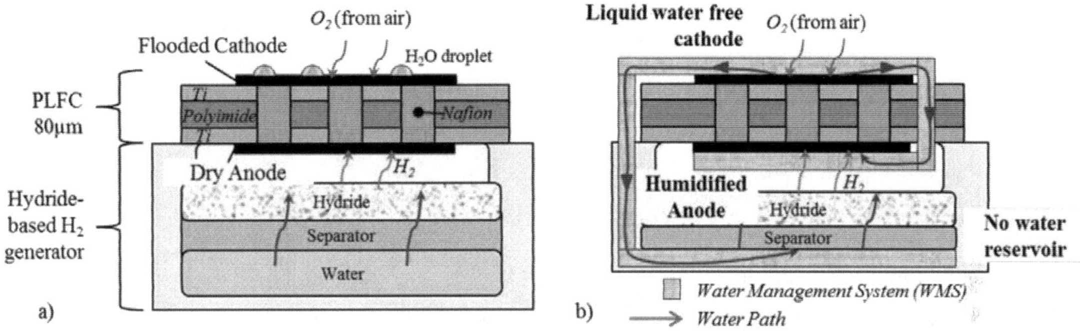

Figure 1. a) Integration of PLFC with hydride based H_2 generator showing poor water management and large water reservoir. b) Intented integration of paper-based water management system that collects liquid water at the cathode and redistribute it to anode and H_2 generator.

In this paper, we introduce for the first time a simple and efficient paper-based WMS that stabilize and increase significantly power output of the air-breathing PLFC without the injection of external energy. Five configurations of WMS are tested showing that power output improvement due to optimal humidification of the anode is as important as the one due to excess water removal at the cathode. Finally, we show that near to ideal working condition can be reached when the key dimensions of the WMS are adequately selected.

2. General design and test setup

Five water management system designs were prepared by patterning strips of 200µm thick hydrophilic paper (Berkshire® cleanroom wipes) with a commercial laser (LPKF protolaser U3). WMSs were divided in four regions; (1) Cathode draining, (2) Anode humidification, (3) Cathode to anode path and (4) Evaporation path (Figure 2a). At first, WMSs were used without evaporation path (4) to evaluate only the performance of the systems to transfer excess water of the cathode to the anode. The best configuration was then selected and an evaporation path added to test the whole system. WMSs were differentiated by the surface coverage of the hydrophilic paper on the catalyst layers, β, and the mean distance that a water droplet has to travel before reaching a hydrophilic strip, D_m (Figure 2b).

The WMSs were directly laid on the catalyst layers of a 1 cm^2 PLFC and hold in place with kapton tape. Fuels cell with water management system were finally inserted between two rigid metallic plates to prevent gas leakage between cathode and anode (Figure 2c).

a)
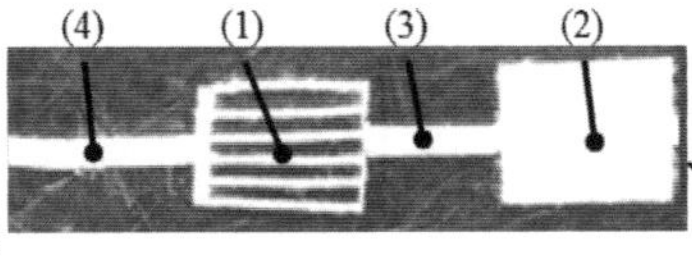

b)

	(1) Cathode		(2) Anode	
	β	D_m (mm)	β	D_m (mm)
WMS 1	0,2	1.6	0,2	1.6
WMS 2	0,2	1.6	0,84	0.05
WMS 3	0,2	1.6	1	0
WMS 4	0,2	0.67	1	0
WMS 5	0,3	0.45	1	0

c)
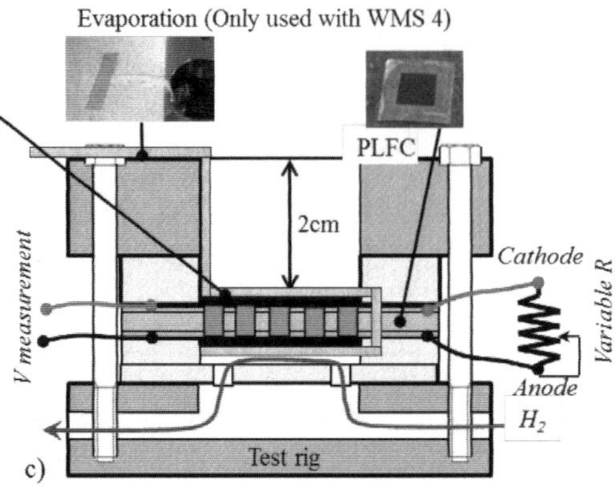

Figure 2. a) WMS 5 prior to installation on a PLFC. Cathode (1) : 0.5mm wide hydrophilic paper strips separated by 1mm, Anode (2) : fully covered. **b)** Hydrophilic coverage and mean water travel distance for the tested WMSs. **c)** Test setup to measure power improvement generate by the integration of a water management system on a PLFC.

3. Surface power density improvement of the PLFC with WMS introduction

3.1. Cathode and anode configuration optimisation

The WMSs were tested on a PLFC in air-breathing condition at ambient temperature until cathode flooding. Two PLFCs (PLFC 1 and 2) were used to conduct the tests to prevent unreliable results caused by their deterioration. VI curves were performed every 30 minutes to evaluate the performances of the PLFC and figure 3 illustrates the best result for each WMS.

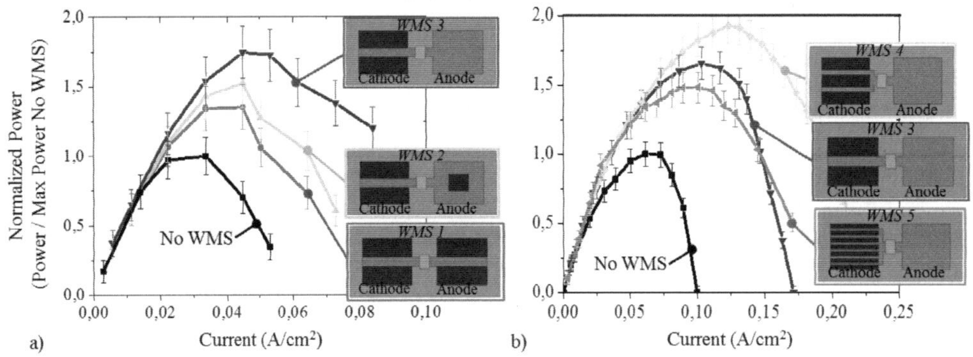

Figure 3. Power output increase caused by the water management system if a) cathode (PLFC 1) and b) anode (PLFC 2) configuration is optimized.

Figure 3a reveals that in the anode case, a higher coverage in hydrophilic material increase power output of the cell. The best result was obtained with a fully covered anode and generated about 60% more power than a cell without WMS. Consequently, the anode drying does have a significant impact on the power output of the cell and the catalyst layer alone does not provide enough water diffusion to ensure its proper humidification.

As presented at figure 3b, on the cathode side, minimal hydrophilic coverage with adequate reparation of the hydrophilic strips lead to an even greater increase of the power output. WMS 4 which has two 1mm wide strips separated by 2.6mm combined with a totally covered anode showed 90% surface power density improvement compared to a cell without WMS. Since hydrophilic strips on the cathode side tend to be saturated in water, oxygen can't reach the reaction sites located underneath. Therefore, a higher hydrophilic coverage doesn't result in better performances. At identical β, a better repartition of the hydrophilic strips (smaller D_m) will result in higher power output (Figure 3b).

3.2. Water management system vs forced flow conditions

A 1mm wide evaporation path (Figure 2c) was added to WMS 4 to test the long term performances of the system and compared the results with the same PLFC working in forced flow conditions. At figure 4a, it is shown that the addition of the evaporation path prevented the apparition of the cathode flooding that initially degraded dramatically the performances of the cell after 10h. Moreover, the evaporation path did not affect the maximal power output of the cell for 24h.

Finally, the PLFC without WMS was inserted in a test bench (*Fuel Cell Technologies inc.*) with controlled temperature, humidity and flows to extract its maximal performance under ideal conditions at 25°C. Figure 4c shows the comparison between the PLFC working in ideal conditions to the same cell operating in air-breathing mode with and without WMS. We can see that the cell with WMS 4 almost reach the performance of the cell in the ideal case. It confirms that the poor performances of small air-breathing fuel cell system are mainly due to poor water management and not oxygen diffusion at the cathode and that the simple paper-based WMS offers close-to-optimal performances.

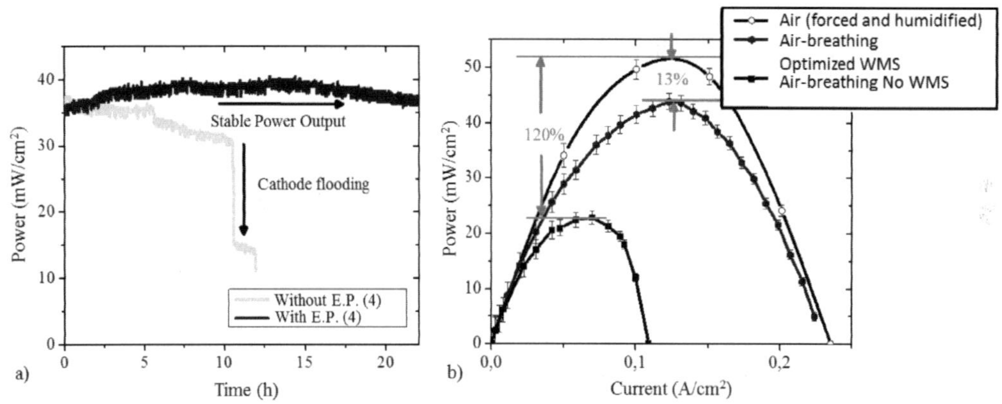

Figure 4. a) Power output stabilization by adding an evaporation path to WMS 4. **b)** Power comparison between a PLFC with and without an optimized WMS and a PLFC with optimal controlled forced flow at 25°C.

4. Conclusion

In this work, we showed that the PLFC configuration allow the integration of a simple and efficient WMS that improved its power output by 90% and stabilized it for at least 24h. The laser etched paper-based WMS allowed the air-breathing cell to reach comparable performances to its forced flow counterpart which proved that water management is the main issue to address to improve small air-breathing fuel cell. It also confirmed that even if this phenomenon is often disregarded in water management studies, anode drying is critical and solution must be integrated in future systems.

This work is an important step to solve water management issues in air-breathing fuel cell which delay the maturity for market readiness of this technology.

References

[1] S.B. Schaevitz, , International Society for Optics and Photonics, 2012: pp. 824802–824802.
[2] S. Hamel, T. Tsukamoto, T. Tanaka, L.G. Fréchette, , J. Phys. Conf. Ser. 557 (2014) 012007.
[3] Y. Zhou, X. Wang, X. Guo, X. Qiu, L. Liu, , Int. J. Hydrog. Energy. 37 (2012) 967–976.
[4] T. Metz, J. Viertel, C. Müller, S. Kerzenmacher, N. Paust, R. Zengerle, P. Koltay, , J. Micromechanics Microengineering. 18 (2008) 104007.
[5] D. Modroukas, S.C. Barton, V. Modi, L.G. Frechette, , in: ECS Trans., Los Angeles, CA, United states, 2005: pp. 419–428.
[6] U. Hojung, S. Ukjeong, K. Taepark, H. Meelee, K. Chun, D. Woongchoi, S. Kim, , Int. J. Hydrog. Energy. 32 (2007) 4459–4465.
[7] S. Hamel, L.G. Fréchette, , J. Power Sources. 196 (2011) 6242–6248.

Fabrication and Demonstration of Planar Micro-Reactors for Solar Steam Methane Reforming

Jean-Francois Dufault, Ines Esma Achouri, Nicolas Abatzouglou, Nadi Baridy, Luc G Fréchette and Mathieu Picard

Institut Interdisciplinaire d'Innovation Technologique (3IT), Université de Sherbrooke, 3000 Boulevard Université, Sherbrooke, J1K 0A5, Québec, Canada

Jean-Francois.Dufault @Usherbrooke.ca

Abstract. This paper presents a novel solar micro-reactor architecture for cheap hydrogen production by methane reforming. Using microfabrication techniques to create three-dimensional channel networks in diffusion-bonded plates, all the sub-processes required for methane reforming are integrated in a small unit, which is well suited for implementation on cheap and well-established solar parabolic troughs. Monolithic integration of the sub-processes allows heat recovery from the high temperature reaction to the lower temperature endothermic processes such as water vaporization and reagent preheating. To demonstrate the manufacturing feasibility as well as the sufficient catalytic activity, a 4.5W stainless steel demonstrator is built and tested in the laboratory. Results show a complete methane conversion for temperatures over 850°C at a space velocity of 35 000 ml/h*ml$_{cat}$.

1. Introduction

If produced from clean sources, hydrogen could become excellent energy carrier in the future. However, today's hydrogen mainly comes from fossil fuels. In particular, nearly 95% of the hydrogen produced currently comes from standard steam methane reforming (SMR) plants [5]. Within these plants, steam is added to methane and heated between 800 and 1000°C [6] to engage the catalytic SMR reaction:

$$CH_4 + H_2O \leftrightarrow CO + 3H_2 \qquad \Delta H = 206 \text{ kJ/mol} \qquad (1)$$

For hydrogen production purposes, the residual carbon monoxide is further oxidized in the mildly exothermic water gas shift reaction (WGS) to increase the hydrogen release per unit of CH_4:

$$CO + H_2O \leftrightarrow CO_2 + H_2 \qquad \Delta H = -41 \text{ kJ/mol} \qquad (2)$$

Within the SMR reactor, both reactions 1 and 2 can occur at the same time, leading to the creation of CO_2 even before the WGS reactor. The global reaction is endothermic and is represented by:

$$CH_4 + 2H_2O \leftrightarrow CO_2 + 4H_2 \quad \Delta H = 165 \text{ kJ/mol} \qquad (3)$$

The ratio between the released CO and CO_2 is stated by the thermodynamic equilibrium which mainly depends on temperature, pressure and steam to carbon ratio (S/C). However, even if it is the thermodynamic equilibrium that states the final concentrations, catalyst must be used to accelerate the reaction for commercial applications.

In conventional power plants, the energy necessary to complete the reaction comes from burning part of the natural gas feedstock. Consequently, a substantial feedstock saving can be achieved if the heat addition comes from a renewable source. That may results in costs reductions, as feedstock represents the biggest part of the hydrogen levelized cost of energy (LCOE) in conventional SMR plants [9].

Solar energy is the source of renewable energy typically considered for SMR as it is an abundant resource and can satisfy the high temperature required for the SMR reaction when concentrated [6]. Common solar concentrators considered for this application are parabolic dishes (PD) and solar towers (ST), which are well suited for high temperature applications since they can provide high solar concentration [6], [16]. On the other hand, conventional parabolic trough (PT) concentrators are cheap and commercially established, but come with lower concentration capability, which constitute a great challenge for solar-aided SMR. Consequently, there is no report of a successful implementation of an SMR reactor directly on the focal line of a parabolic trough. Moreover, no exhaustive studies on the feasibility, efficiency and interest to implement such a technology at high temperature seems to appear in the literature despite the several advantages that it could represent. Part of the challenge with this approach is in fitting the reactor along the narrow focal line of a parabolic trough.

This paper presents a new design of a fully-integrated solar micro-reactor capable of processing the entire SMR reaction at low solar concentration in an inch-wide unit. The micro-fabricated reactors can be stacked in thin arrays to cover the focal line of a parabolic trough. Taking advantage of the synergy between the endothermic and exothermic processes, the highly-integrated design redirects thermal losses from the high temperatures zones to satisfy the energy requirements of the lower temperature sub-processes, such as water vaporization and reagents preheating. A prototype is tested in the laboratory using a heating element to demonstrate chemical performance of the systems as well as its manufacturing feasibility. More precisely, the reactor shows a complete SMR reaction of methane at 850 °C and 900 °C for gas hourly space velocities of 35 000 and 70 000 SCCM/h*ml$_{cat}$ respectively.

2. Architecture

In this concept, elementary planar reactors are made of 2.5D metallic plates containing all the endothermic and exothermic sub-units required to complete the reforming process (Figure 1). The micro-reactors are then stacked and diffusion-bonded in parallel in the axis of the PT to form a monolithic unit at the focus line of the concentrator. Using this two-dimensional approach, layers can be manufactured by a standard batch process such as etching, punching and laser cutting. These manufacturing methods are compatible with a high volume production economy, instead of a scale economy typical to solar towers. The sub-units of the elementary reactors are placed so that the temperature increases linearly from the inlet on the cold side to the solar receiver on the hot side (Figure 1-a). First, water enters the central inlet and passes through nickel foam where it evaporates (Figure 1-b). This steam then flows to the heat exchanger entrance where it mixes with the methane flow. It then gradually heats up in the counter flow heat exchanger and flows into the SMR sub-reactor. The reformed gas then leaves the packed bed at a high temperature and releases its heat to the reactant mix and evaporating water, before it is exhausted. The remaining energy needed to vaporize water comes from the conduction inside the reactor's walls.

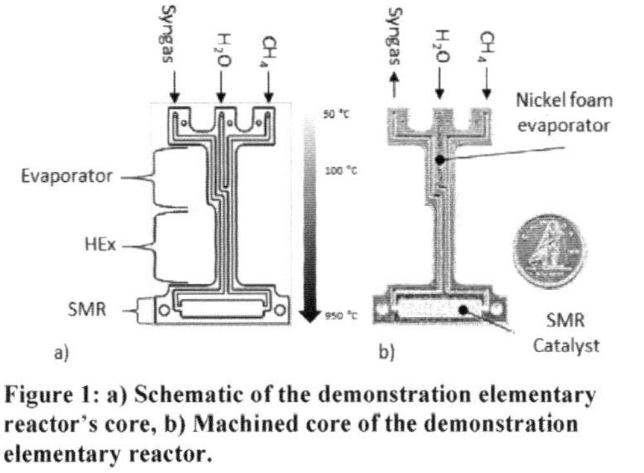

Figure 1: a) Schematic of the demonstration elementary reactor's core, b) Machined core of the demonstration elementary reactor.

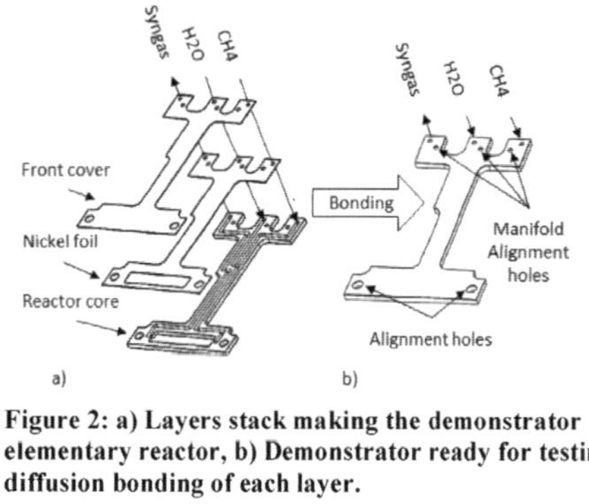

Figure 2: a) Layers stack making the demonstrator elementary reactor, b) Demonstrator ready for testing after diffusion bonding of each layer.

3. Prototype

The core of this elementary reactor prototype is micro-machined in a 1.2 x 25 x 50 mm stainless 316 sheet (Figure 1). For the demonstration purpose, stainless steel 316 is used due to its good resistance to

corrosion as well as its known properties and easy machining. The reactor is closed by a 0.004" thick SS-316 layer joined by diffusion bonding at 750°C for 90 minutes under a nitrogen atmosphere and an initial load of 50 MPa. The intermediate 0.005"-thick nickel foil shown on Figure 2 only acts as an interlayer to facilitate the diffusion bonding with no other structural function. All the faces were polished with a 1 μm suspension to facilitate bonding and increase tightness.
A Ni-Al spinel-based catalyst is used to perform the SMR reaction and is synthesized using the protocol developed by Abatzoglou [4]. First, a nickel 40 μm alumina powder (#43266 from Alfa Aesar) is mixed with water. Five percent of nickel (m/m) is then added to the alumina powder using an aqueous solution of $Ni(NO_3)_2*6H_2O$. After 90 min of stirring at 60°C, the mixture is heated up to 95°C under mixing until the water is evaporated. The impregnated powder is then dried at 105°C for 12 hours. Finally, the powder is calcined for 6 hours at 900 °C under atmospheric pressure to reach the desired spinel-phased catalyst. Since the catalyst can sustain high temperatures, it is inserted inside the reactor before the diffusion bonding.

4. Experiments

The test bench shown in Figure 3 is used to perform the tests. A first inlet provides the gas needed for the entire duration of the cycle and a second acts exclusively as a liquid water inlet. A third manifold collects the synthesis gas and brings it to the analysis system. A custom-made heating element is wrapped around the reactor base to provide the high temperature needed for the reaction to occur. A standard PID system using the feedback provided by a thermocouple controls the heating element's temperature. A CPU liquid cooling system chills the reactor's head and protects the manifold gaskets during the system ramp-up and the catalyst activation. The gas flows are controlled using three flow meters (Alicat) controlled with a LabVIEW interface. The first cylinder provides argon needed during the reactor's heating and cooling. The second cylinder contains a stoichiometric mixture of CO_2 and CH_4 needed to activate the catalyst and the third contains pure CH_4 for standard reactor operation. Once the synthesis gas escapes the reactor, it passes through a cooling coil submerged in an ice bucket and a separator to condense the unprocessed water. The gas then passes in a Drierite© packed bed for final drying. A mass spectrometer placed downstream continuously monitors the exhaust composition changes while gas is sampled and precisely analyzed using a Varian CP-4900 Micro GC.
During the tests, the reactor is first heated from room temperature to 950°C. The ramp up lasts about 8 min and only argon is injected in the reactor to avoid carbon deposition. The argon stream is then switched to a 1:1 CH_4-CO_2 stream to activate the catalyst under dry methane reforming conditions (DMR). When a complete conversion of methane is observed, water is injected in the reactor using a syringe pump (Harvard Apparatus 11 Pico Plus Elite) at a steam to methane ratio of 1:1. The change in CO_2 concentration observed by the mass spectrometer corresponds to water appearing in the reactor. At this point, the CH_4-CO_2 stream is changed for pure CH_4 and the water flow is increased to achieve a steam to carbon ratio of 2:1. The temperature is then reduced by increments of 50 °C until it reaches 650°C. For each temperature step, composition of the gas produced is analyzed with the gas chromatograph. Two sets of experiments corresponding to different mass flows are performed during the tests. The first one corresponds to a gas hourly space velocity (GHSV) of 35 000 ml/h*ml_{cat} at standard temperature and pressure, which represents a power conversion of 75 suns under a complete sun to chemical energy conversion. The second set of data is taken at a GHSV twice as large, i.e. 70 000 ml/h*ml_{cat}. A steam to carbon ratio of 2:1 is used for all the experiments.

5. Results

Figure 4 shows the methane conversion level achieved during the experiments. Complete methane conversion occurred for temperatures over 850° at low GHSV, and over 900°C under high GHSV. At low GHSV and temperatures higher than 750°C, the conversion is close to thermodynamic equilibrium. As the temperature decreases, the conversion departs from equilibrium with conversion 30% lower than the equilibrium at 650°C for low GHSV. At high GHSV the conversion level departs dramatically from equilibrium under 850°C and becomes as low as 26% at 650°C. The complete methane conversion speaks to the fact that the H_2O-CH_4 mixing length between the evaporator and the catalyst is sufficient to provide a homogenous mix to the SMR reactor.

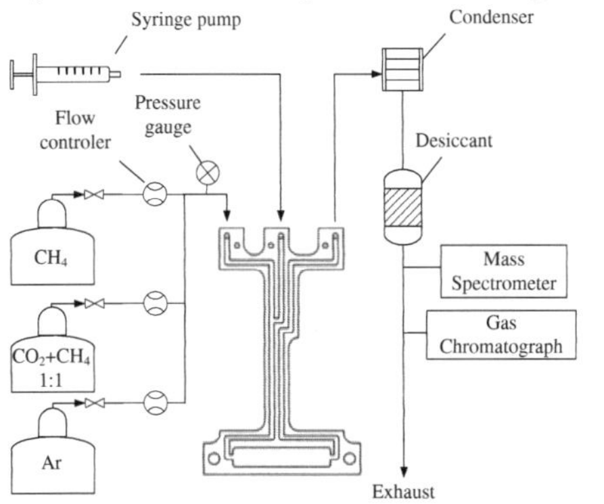

Figure 3: Schematic of the test bench used for experiments on the demonstrator reactor.

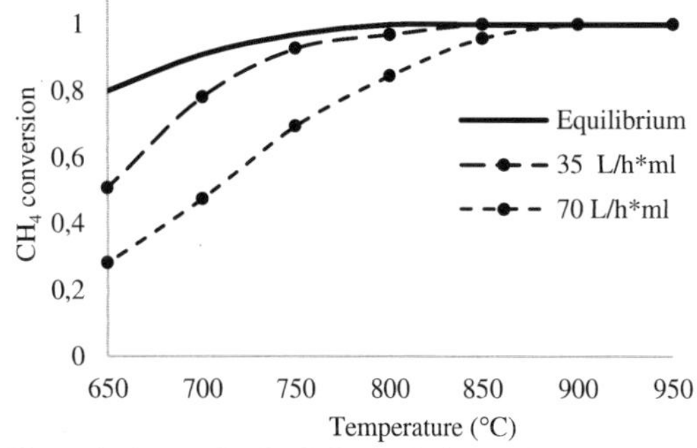

Figure 4 : Conversion level obtained with the demonstrator reactor at different temperatures and GHSV compared with the thermodynamic equilibrium at a Steam to carbon ratio of 2:1.

Three cycles were performed with this first demonstrator and no leak was detected during 20 h of operation. In addition to the high conversion level reached for SMR, these tests have also shown that dry methane reforming as well as mixed steam-CO_2 reforming with a complete methane conversion are achievable at 950°C and even colder.

It also shows the potential of the manufacturing process although further analysis must be done to determine the unit's resistance to degradation. Finally, it partly demonstrates the principle of integrating multiple sub-reactions and recuperating waste heat, underlying micro reactor concept and suggesting that the heat recovery is indeed implementable in such compact integrated units.

6. Conclusion

This paper presents a new design of solar micro-reactor for steam methane reforming. Using an integrated design, the system comprises all the necessary components to fulfil hydrogen generation with SMR. With each of these components being thermally coupled, this synergy allows reusing heat from the warmer processes to fulfil the energy requirement of the colder processes such as water vaporization and reagent preheating. A layered configuration was developed based on laminated metal sheets, compatible with eventual cost-effective batch manufacturing. A micro-reactor demonstrator was fabricated to demonstrate the reactor feasibility and its reforming capabilities, as well as validate its integrated layout and layered fabrication approach with catalyst integration. The tests made at 35 000 and 70 000 ml/h*mlcat shown a complete methane conversion level above 850 and 900°C, respectively. For the lowest GHSV, the methane conversion stays close to the equilibrium concentration until the temperature falls under 750°C. The reactor also resists to the strong temperature gradient and is able to complete water vaporization as well as reagent mixing inside the compact volume, which shows the feasibility of the laminated integrated elementary reactor for eventual operation in parallel.

References

[1] T. Ramsden and D. Steward, "Analyzing the levelized cost of centralized and distributed hydrogen production using the H2A Production Model," NREL, Golden, CO, Tech. Rep. TP-560-46267, 2009.
[2] C. Agrafiotis, H. von Storch, M. Roeb, and C. Sattler, "Solar thermal reforming of methane feedstocks for hydrogen and syngas production—A review," *Ren. Sustain. Energy Rev.*, vol. 29, pp. 656–682, Jan. 2014.
[3] M. Böhmer, U. Langnickel, and M. Sanchez, "Solar steam reforming of methane," *Sol. Energy Mater.*, vol. 24, no. 1, pp. 441–448, Dec. 1991.
[4] N. Abatzoglou, J. Blanchard, and I. E. Achouri, "A Novel Ni-Al Spinel-Based Catalyst for Dry Reforming of CH4-Rich Gasification Tail Gas Streams," presented at the 22nd European Biomass Conference and Exhibition, Hamburg, Germany, 2014, pp. 1004–1007.

Improved Omnidirectional 2D Photonic Crystal Selective Emitter for Thermophotovoltaics

Reyu Sakakibara[1,2], Veronika Stelmakh[1], Walker R. Chan[1], Michael Ghebrebrhan[3], John D. Joannopoulos[1,4], Marin Soljačić[4] and Ivan Čelanović[1*]

[1] Institute for Soldier Nanotechnologies, Massachusetts Institute of Technology, 77 Massachusetts Avenue, Cambridge, MA 02139, USA
[2] Department of Electrical Engineering and Computer Science, Massachusetts Institute of Technology, 77 Massachusetts Avenue, Cambridge, MA 02139, USA
[3] U.S. Army Natick Soldier Research, Development, and Engineering Center, 15 General Greene Avenue, Natick, Massachusetts, 01760, USA,
[4] Department of Physics, Massachusetts Institute of Technology, 77 Massachusetts Avenue, Cambridge, MA 02139, USA

E-mail: ivanc@mit.edu

Abstract. Hafnia-filled, two dimensional (2D) tantalum (Ta) photonic crystals (PhCs) are promising emitters for high performance thermophotovoltaic (TPV) systems because they enable, for a wide range of incidence angles, efficient spectral tailoring of thermal radiation. However, fabricating these PhCs to the required tolerances has proven to be a challenging task. In this paper, we use both focused ion beam (FIB) imaging and simulations to investigate the effects of fabrication imperfections on the emittance of a fabricated hafnia-filled PhC and to identify critical geometric features that drive the overall PhC performance. We demonstrate that, more so than uniform cavity filling, the key to the best filled PhC performance is the precise cavity period and radius values and thickness of the top hafnia layer.

1. Introduction

Thermophotovoltaic (TPV) systems are promising as small scale, portable generators to power small robotic platforms, sensors, and portable computational and communication equipment. In TPV systems, thermal radiation from an emitter at high temperature is converted to electricity by a low bandgap photovoltaic (PV) cell. A key factor for the system efficiency is the ratio of in-band emissivity—convertible by the PV cell—relative to the total emissivity. One approach to improve the conversion efficiency is to use two-dimensional (2D) tantalum (Ta) photonic crystals (PhCs) to spectrally tailor the thermal radiation to the PV cell bandgap. Indeed, using this approach we demonstrated a 4.3% fuel-to-electricity system efficiency [1] using PhCs coated with 20–40 nm hafnia as passivation layer.

In theory, dielectric-filled PhCs are a further improvement to coated PhCs. Filled PhCs have high in-band emissivity at a wide range of angles [2], as shown in Figure 1a. Because most thermal radiation is off normal, an omnidirectional filled PhC would increase total in-band radiated power by 55% at 1200 °C [3] compared to an unfilled PhC. Like the coated PhC, a hafnia-filled PhC would also have high-temperature stability and resistance to chemical contamination.

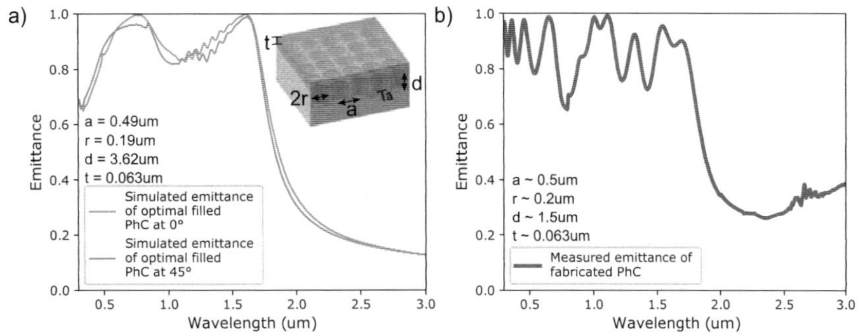

Figure 1. Filled PhCs are promising for enhancing thermal emission at all angles, but are difficult to fabricate. a) The optimal filled PhC for cutoff wavelength 1.8 μm, whose structure is shown in the inset, has high in-band emissivity independent of angle. b) Geometric imperfections in the fabricated PhC may explain the shape of the measured emittance spectrum.

However, filled PhCs are difficult to fabricate, largely because the cavity period a and radius r are reduced by approximately half (compared to the coated PhC) due to hafnia's high index of refraction (~2). The smaller sizes impact the fabrication in several ways: reduced cavity depths d (due to slower etch rates), more difficult cavity filling (due to higher cavity aspect ratios), and higher sensitivity to slight variations in PhC dimensions. The fabrication process is similar to that described previously [3]. We use atomic layer deposition (ALD) to fill the cavities.

The difficulty of fabrication is reflected in the mismatch between the measured emittance of our fabricated filled PhC and the simulated emittance, as shown in Figure 1b. In our fabricated PhC we have deliberately chosen a shallower depth of about 1.5 μm (aspect ratio ~4) in order to make the cavity easier to fill.

In this paper we use a combination of focused ion beam (FIB) imaging and simulations to quantify the impacts of fabrication imperfections and provide recommendations for improvement.

2. Simulation of Geometric Imperfections

The FIB image (Figure 2a) shows that the cavity filling is incomplete and there is a thick layer of hafnia covering the cavity.

Based on this, we construct a geometric model whose main features are a hollow core and

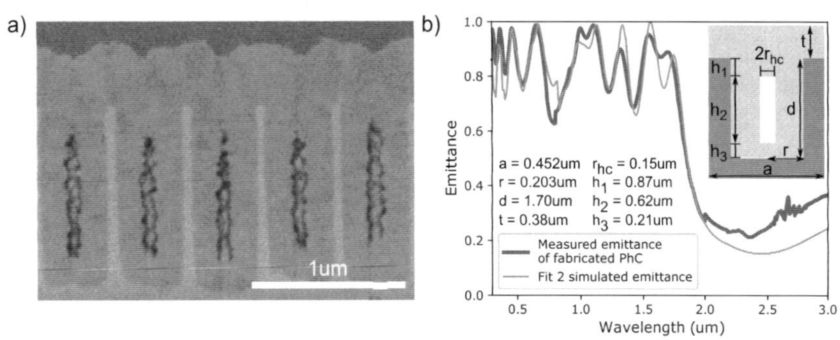

Figure 2. A focused ion beam (FIB) image of the fabricated PhC cavity cross section provides the basis for the geometric model for the fit. a) The FIB image clearly indicates incomplete filling of the cavity and a thick layer of hafnia above the cavity. b) Fit simulation with geometry shown in the inset has a close correspondence with the measured emittance.

thick top hafnia layer. We approximate the hollow core as a cylinder centered at the cavity center, and the hafnia layer as a simple slab (Figure 2b inset). We neglect secondary geometric effects such as scalloping of the top surface and the precise shape of the hollow core.

Our model is sufficient to capture the major features in the measured emittance spectrum: the position of the resonance peaks, cutoff, and shape of the long wavelength emittance, as shown in Figure 2b. The dimensions from the fit are reasonably close to those measured from the FIB. According to our fit, the volume of hollow core is about 21% that of the cavity. Also, the period a is about 40 nm shorter than that of the optimal PhC. Deviations of the fit from the measured emittance may be attributed to secondary effects such as scalloping of the hafnia layer, variations in cavity size across the sample, and assumptions about hafnia optical parameters.

3. Recommendations for improvement

Using the structure from the fit as a basis, we investigate what geometrical parameters can be changed to improve the emittance. For our simulations (not depicted) we vary a single parameter at a time while keeping all else equal. For our figure of merit (FOM), we calculate the ratio of the in band power to total radiated power [4], normalized to that of the optimal PhC, at 1200 °C.

However, we find that changing a single variable does not improve the emittance. In particular, increasing the cavity depth makes no difference. Reducing the thickness t of the top hafnia layer

Table 1. Dimensions (in µm) of simulated PhCs and their figures of merit (FOM) calculated at 1200 °C for $\lambda_{\text{cutoff}} = 1.8\,\mu\text{m}$. Bold indicates parameters changed from the fit.

PhC	a	r	d	t	Hollow core?	r_{hc}	h_1	h_2	h_3	FOM
Optimal	0.49	0.19	3.62	0.063	No	-	-	-	-	1
Fit	0.45	0.20	1.7	0.38	Yes	0.15	0.87	0.62	0.21	0.823
Increased d	0.45	0.20	**3.62**	0.38	Yes	0.15	0.87	0.62	0.21	0.823
Reduced t	0.45	0.20	1.7	**0.063**	Yes	0.15	0.87	0.62	0.21	0.821
No hollow core	0.45	0.20	1.7	0.38	**No**	-	-	-	-	0.756
Improved 1	**0.5**	0.20	1.7	**0.063**	Yes	0.15	0.87	0.62	0.21	0.897
Improved 2	0.452	**0.19**	1.7	**0.063**	Yes	0.15	0.87	0.62	0.21	0.905

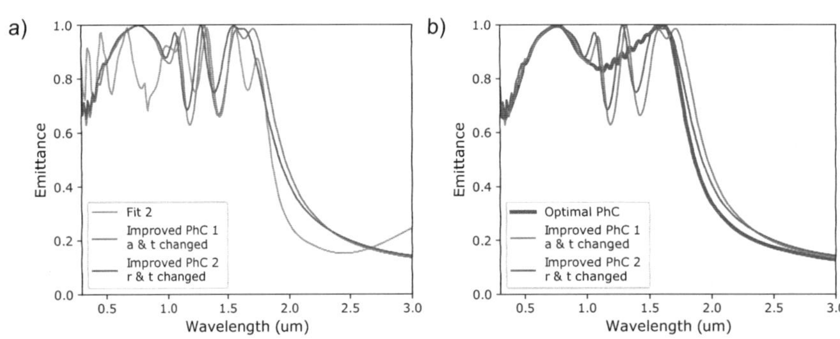

Figure 3. To improve the emittance, it is important to match the optimal a, r, and t values (see Table 1). a) Compared to the fit, Improved PhCs 1 and 2 show a higher in-band emissivity in the λ=0.5–1 µm and λ=1.5–1.7 µm ranges. b) Overall the emittances of Improved PhCs 1 and 2 are close that of the optimal PhC.

to 63 nm increases the in-band emissivity from 0.5 to 1 µm but shifts the cutoff towards a longer wavelength, which effectively increases the out-of-band emissivity. In addition, eliminating the hollow core appears to worsen our emittance according to our FOM (see Table 1).

Instead, it is simultaneously changing both t and either a or r that improves the emittance. As shown in Table 1, we change t to the optimal t, and either a to 0.5 µm or r to the optimal r. Compared to Fit 2 (see Figure 3a), both improved PhCs have improved in-band emissivity from 0.3 to 1.0 µm and 1.5 µm to the cutoff. The out-of-band emissivity from the cutoff to 2.7 µm becomes higher but improves from 2.7 to 3.0 µm.

The thickness t impacts the emittance both above and below the cutoff wavelength. Above the cutoff, the top layer creates Fabry-Perot resonances, whose peak locations can be estimated by considering reflection. Tuning t to roughly below $\lambda_{\text{cutoff}}/(4n)$ prevents destructive interference of reflected waves near λ_{cutoff} and eliminates high emittance above the cutoff. Below the cutoff, the higher emittance is likely due to the hybridization of Fabry-Perot modes and cavity resonances.

As Figure 3b shows, these emittances of the improved PhC actually better match that of the optimal PhC. This suggests that fabricating the correct a, r, and t to within ± 10 nm can make a PhC robust against a hollow core. The depth d appears to be less crucial than a, r, or t.

4. Conclusions and Future Work

We have found that the mismatch between the measured and the ideal emittance is due to the presence of a hollow core, a thick hafnia layer (t), and the deviation of a from the optimal a.

However, to improve the emittance it is more important to precisely fabricate the cavity period a and radius r, and to reduce the thickness t of the hafnia layer than to prevent the formation of the hollow core. With new and improved techniques for better geometrical control, such as stepper-based lithography and argon sputtering, it will be possible to achieve ~90% of the spectral selectivity of the optimal filled PhC.

Acknowledgments

This work was supported by the Army Research Office through the Institute for Soldier Nanotechnologies under Contract No. W911NF-13-D-0001 and the Micro Autonomous Systems and Technology Collaborative Technology Alliance under Contract No. 892730. We would like to thank Jim Daley and Tim Savas of MIT and Robert Geil of University of North Carolina at Chapel Hill for fabrication assistance, and Austin Akey and Stephan Krämer of Harvard University for focused ion beam imaging assistance. The fabrication was done in part at NSL at MIT and at CNS at Harvard University, a member of the National Nanotechnology Infrastructure Network (NNIN), supported by the National Science Foundation (NSF) under NSF Award No. ECS-0335765. The fabrication was supported as part of the Solid-State Solar-Thermal Energy Conversion Center (S3TEC), an Energy Frontier Research Center funded by the U.S. Department of Energy (DOE), Office of Science, Basic Energy Sciences (BES), under Award No. DE-SC0001299/ DE-FG02-09ER46577.

References

[1] Chan W, Stelmakh V, Ghebrebrhan M, Soljačić M, Joannopoulos J and Čelanović I 2017 *Energy Environ. Sci.* **10** 1367
[2] Yeng Y, Chou J, Rinnerbauer V, Shen Y, Kim S G, Joannopoulos J, Soljačić M and Čelanović I 2014 *Opt. Express* **22** 21711–8
[3] Stelmakh V, Chan W, Ghebrebrhan M, Soljačić M, Joannopoulos J and Čelanović I 2016 *J. Phys.: Conf. Series* **773** 012037
[4] Stelmakh V, Rinnerbauer V, Chan W, Senkevich J, Joannopoulos J, Soljačić M and Čelanović I 2014 *Proc. of SPIE* **9115** 911504–1–8

IOP Conf. Series: Journal of Physics: Conf. Series **1052** (2018) 012057 doi:10.1088/1742-6596/1052/1/012057

Urease enzyme as anodic catalyst in a microfluidic fuel cell

J. Galindo-de-la-Rosa[1], R. Balam-Vera[2], A. Álvarez[2], E. Ortiz-Ortega[1], N. Arjona[1], L.G. Arriaga[1] and J. Ledesma-García[2]*

[1] Centro de Investigación y Desarrollo Tecnológico en Electroquímica, Pedro Escobedo, Qro., C.P. 76703, México.
[2] Facultad de Ingeniería, División de Investigación y Posgrado, Universidad Autónoma de Querétaro, Centro Universitario Cerro de las Campanas, Querétaro, Qro., C.P. 76010, México.
E-mail: janet.ledesma@uaq.mx

Abstract. In this research, enzymatic bioanodes based in urease enzyme were developed using graphene oxide as support material on porous electrodes, for application in microfluidic devices that use urea as fuel. Urease is a metalloenzyme that due to its bimetallic site is able to carry out the oxidation of urea with the formation of ammonia and carbonic acid. Graphene oxide was used because of its characteristics that favour the process of enzymatic immobilization. The performance of the microfluidic fuel cell was carried in the absence and in the presence of 0.1 M urea in 0.1M phosphates buffer where can observed an open circuit voltage above 1 V and important power density value when urea is used as fuel (2.82 mW cm^{-2}). It was also evaluated by electrochemical techniques the effect of the concentration of urea on the power density generated by the microfluidic device, as well as stability, generating good energy conversion results for a first stage of research, where the objective is the creation of bioelectrodes with urease enzyme immobilized for application in Lab-on-a-Chip devices.

1. Introduction

Microfluidics fuel cells or membraneless fuel cells have attracted attention because of their potential applications because they exploit the co-laminar flow that occurs in the microscale to separate the fuel from the oxidizing species to generate electrical current from a variety of organic liquid fuels and oxidants under laminar flow regimen with low Reynolds number and high Peclet number, eliminating the need for a membrane performing this function, as in conventional fuel cells [1-3]. The anode has been the focus of research in this area, because it is where the oxidation reaction of the fuel occurs.

Among the types of electrodes used as anodes in several devices of energy conversion are those based in enzymes, which act as biochemical transducers that combine an electrochemical process with immobilized enzyme activity [4, 5]. The urease enzyme can be used for the construction of bioanodes due to its ability to carry out the oxidation of urea present in body fluids. Urease is a non-redox metalloenzyme involved in the overall nitrogen cycle in plants, algae, fungi and various microorganisms. Its bimetallic reaction site (Ni-II) catalyzes the rapid hydrolytic decomposition of urea to give ammonia and carbamate, this reaction continues hydrolyzing to form ammonia and carbonic acid as products [6, 7]. The urease activity is strongly dependent on the pH in the range of about 4.5-10.5 pH, with an increasingly slower reaction as the range increases, having an optimum pH between 7.4 pH. Urea use as fuel presents 2 Mcal / kg equivalent to 8.368MJ / kg of energy density, making it attractive for use in microfluidic fuel cells for biomedical applications, where obtaining a good open circuit potential close to the theoretical could be used for the feeding of different biomedical devices [8, 9].

$$\text{anode: } CO(NH_2)_2 + 6OH^- \rightarrow N_2 + 5H_2O + CO_2 + 6e^- \qquad E^0 = -0.746V$$
$$\text{cathode: } O_2 + 2H_2O + 4e^- \rightarrow 4OH^- \qquad E^0 = +0.401V$$
$$\text{complete reaction: } CO(NH_2)_2 + 3O_2 \rightarrow 2N_2 + 4H_2O + 2CO_2 \qquad E^0 = 1.147V$$

The nanostructured materials as graphene oxide are considering excellent candidates for enzymatic immobilization due to reduce the diffusion limitations and provide greater surface area improving the electronic transfer in the electrode [10, 11]. The low potency and stability of cells using enzyme electrodes is due to the complexity of enzymes and easy denaturation, since all enzymes have the characteristics of being sensitive to minimum changes in temperature and pH, their active sites can be reached to be disabled at the time of the enzyme-support immobilization, without forgetting that the shelf life and the poor power density of the microfluidic devices depend on the enzymatic stability [12]. Nanostructured supports favor enzymatic immobilization, mass transfer resistance and effective enzyme loading due to their higher surface area than macro-scale supports tend to have excellent biocompatibility, stability and reuse life of the enzyme [13]. Enzymatic immobilization can be performed without modifications on its surface and has a large surface area with abundant functional groups that facilitate the anchoring of the enzyme and the electronic transfer. In this work, graphene oxide was used as a support to immobilize the enzyme urease for the construction of enzyme electrodes for application in microfluidic fuel cells for urea oxidation.

2. Experimental

2.1 Bioelectrode construction.

A catalytic ink was prepared using 1mg urease (EC 3.5.1.5 Type III, powder, 15,000-50,000 units/g solid), 2mg of graphene oxide, 7 μL of Nafion solution (Sigma Aldrich), 500μL of 0.1M of buffer phosphate pH= 7.4, and 8 mg of tetrabutylammonium bromide (Sigma Aldrich). The catalytic ink was deposited by dipping on Toray carbon paper (EC-TP1-060T) whose dimensions were of 2.5x30 mm.

2.2 Microfluidic fuel cell evaluation.

The microfluidic device used was previously reported for application in enzymatic microfluidic fuel cells using the glucose oxidase [14], alcohol dehydrogenase [15], aldehyde dehydrogenase enzymes [16] immobilized at the bioanode. This device (Figure 1) consisted of two supporting plates made of Poly-(methyl methacrylate) (PMMA) and patterned using a computer numerical control CNC, a home-made silicone elastomer film (Silastic®, Dow Corning, prepared using an Elcometer® Film Applicator with a final thickness of 200 μm) used as both the gasket and cell channel structure.

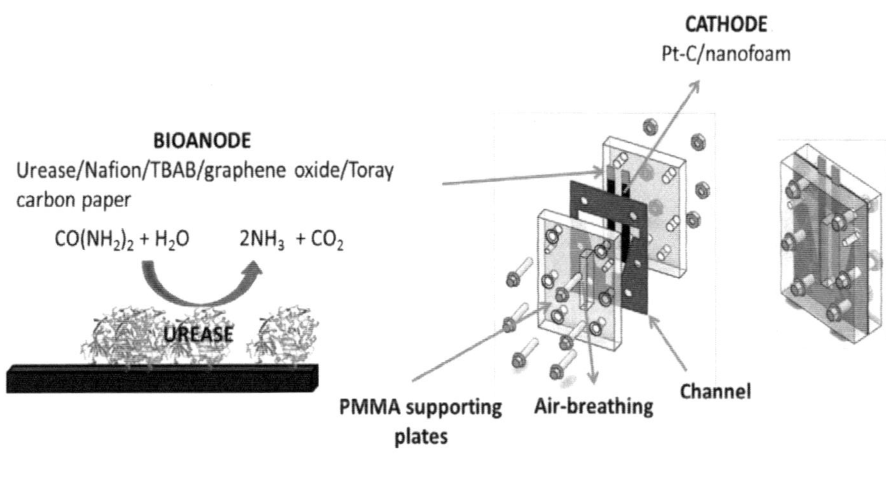

Figure 1. Scheme of air-breathing microfluidic fuel cell

The electrodes consisted of commercially available carbon nanofoam (Marketech Inc) for cathodes and 10 mm long × 2 mm wide Toray paper (100 µm in thickness) for bioanodes. The device was evaluated at different urea concentration N_2-saturated (4 U.P.D. Praxair) as the fuel and oxygen taken from the air as the oxidant within two aqueous streams that contained a phosphate buffer (pH 7.4) and 0.3 M KOH on the anodic and the cathodic side, respectively with 50 µL min-1 of flow rate for both streams.

3. Results

The performance of the microfluidic fuel cell was carried in the absence and in the presence of 0.1 M urea in 0.1M phosphates buffer (Figure 2) where can observed an open circuit voltage above 1 V and important power density value when urea is used as fuel (2.82 mW cm^{-2}). It can also be observed that in absence of urea the power density of the microfluidic device is very low, attributing it to the absence of substrate necessary to carry out the oxidation process at the anode. Further, the microfluidic device was evaluated used 10, 50 and 100 Mm of urea, increasing the current density linearly with the fuel concentration (Figure 3).

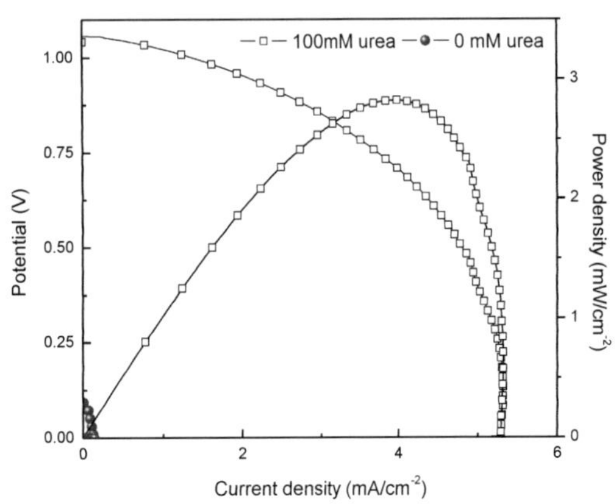

Figure 2. Polarisation and power density curves obtained from air-breathing microfluidic fuel cell using 0.1M urea in 0.1M phosphates buffer pH 7.4 as fuel and 0.3M KOH as oxidant, at 10 mV s^{-1}.

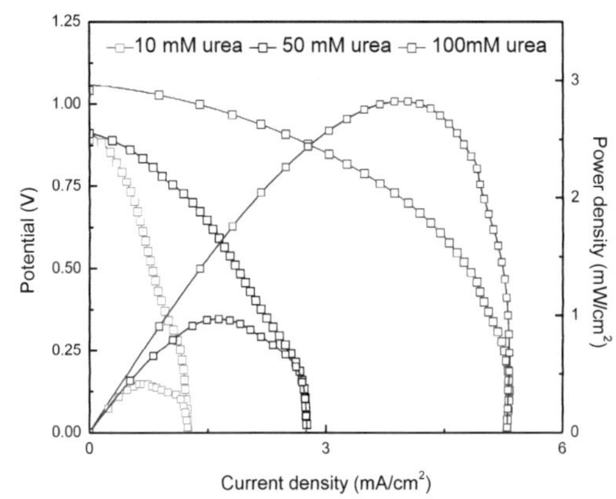

Figure 3. Polarisation and power density curves obtained from air breathing microfluidic fuel cell at different urea concentration in 0.1M buffer phosphates pH 7.4 at10 mV s^{-1} as fuel and 0.3M KOH as oxidant.

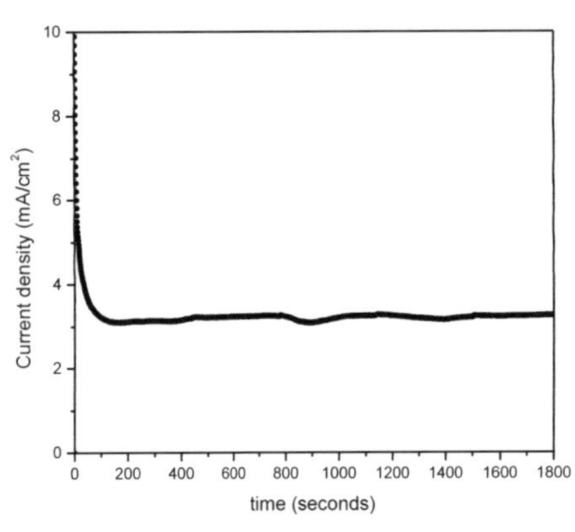

Figure 4. Chronoamperometry tests of air breathing microfluidic fuel cell evaluated with 0.1M urea in 0.1M phosphates buffer pH 7 as fuel and 0.3M KOH as oxidant at10 mV s^{-1}.

Finally, the stability of the microfluidic device was tested by chronoamperometry technique at 0.1M urea (Figure 4) showing excellent behaviour as function of the time. The results shows that the bioanode based in urease enzyme oxides successfully urea as fuel and could has potential possibilities of use in future biomedical applications.

4. Conclusions

Urea is an attractive fuel because it can be found in different biological fluids, such as blood and urine, having various interesting biomedical applications. The creation of enzymatic electrode based on the urease was important because using graphene oxide could develop a bioanode that can oxidize urea at different concentration that has selectivity and presented good stability when it was evaluated in microfluidic fuel cell obtaining a good open circuit potential higher of 1.0 V close to the theoretical open circuit potential reported.

Acknowledgements

Authors acknowledges to the Mexican Council for Science and Technology (CONACYT) for financial support through project Fronteras de la Ciencia Grant no. 611, Laboratorios Nacionales "Laboratorio Nacional de Micro y Nano Fluídica" Grant no. LN 271649, and "Sistema de almacenamiento de energía basado en la purificación y compresión no convencional de hidrógeno (Electroquímica)" Grant no. 246079 for the "Fondo Sectorial CONACYT-SENER Sustentabilidad Energética".

Reference

[1] Desmaële D, Nguyen-Boisse TT, Renaud L, Tingry S, 2016. *Microelectronic Eng.* **165** 23-26.
[2] Keang E, Djilali N and Sinton D, 2009, *J. Power Sources* **186** 353-369.
[3] Kundu A, Jang JH, Gil JH, Jung CR, Lee HR, Kim SH, Ku B, Oh YS, 2007, *J. Power Sources* **170** 67-78.
[4] Meredith MT and Minteer SD, 2011, *Sensor.Anal. Chem* **83** 5436–5441
[5] Zhou M, Dong S, 2011, *Acc. Chem. Res.* **44** 1232–1243.
[6] Mazzei L, Cianci M, Musiani F, Lente G, Palombo M and Ciurli S, 2017 *J. Inorg. Biochem.* **166** 182 -9.
[7] Wu Z, L. Guan, G. Shen, R. Yu. , 2002, Analyst **127** 391–395.
[8] Huffman HM, 1940, *J. Am. Chem. Soc.* 62 1009-1011.
[9] Schmidt-Rohr K, 2015, *J. Chem. Educ.* **92** 2094–2099.
[10] Zhang J, Zhang F, Yang X, Liu H, Zhang J and Gou S, 2010 *Langmuir* **26** 6083-6085.
[11] Zhou M, Zhai Y, Dong S. 20019, *Anal. Chem.* 81 5603-5613.
[12] Minteer SD, Liaw BY and Cooney MJ, 2007, *Biotecnology* **18** 228-234.
[13] Ahmed S, Ansari A, Husain Q, 2012, Biotechnology Advance **30** 512-523.
[14] Galindo-de-la-Rosa J, Moreno-Zuria A,Vallejo-Becerra V, Arjona N, Guerra-Balcázar M, Ledesma-García and Arriaga LG, 2016, *J. Phys. Conf. Ser.* **773** 012114.
[15] Galindo-de-la-Rosa J, Arjona N, Moreno-Zuria A, Ortiz-Ortega E, Guerra-Balcázar M, Ledesma-García J, Arriaga LG, 2017, *Biosens. Bioelectron.* **92** 117-124.
[16] Galindo-de-la-Rosa J, Arjona N, Arriaga LG, Ledesma-García J and Guerra-Balcázar M, 2015, *J. Phys. Conf. Ser.* **660** 012131.

Development of portable power unit with catalytic micro-combustor

K Higuchi[1], T Nakano[1] and S Takahashi[1]

[1]Department of Mechanical Engineering, Faculty of Engineering, Gifu University, 1-1 Yanagido, Gifu-shi, Gifu 501-1193, Japan

E-mail: shuhei@gifu-u.ac.jp

Abstract. A prototype of the portable power unit with coupling catalytic microcombustor, thermo-electric modules and air /fuel supply devices was developed. The developed power unit was completely self-standing; the air supply system was driven by a part of electricity generated at the generator unit. When the thermal input from the combustor was 12W, the total electricity is 330 mW and the net output electricity was 225 mW, which corresponded to the back work ratio of 32%. The final conversion efficiency of the total system reached 1.87%, which was order of world record.

1. Introduction

Hydrocarbon fuels have large energy density and therefore, their potentials for microscale power units have been attractive for decades [1, 2]. In the previous research, we have developed a very small combustor which has a monolithic porous catalyst layer on the inner wall of the ceramic tube whose inner diameter is 1.5mm or less [3]. (see Fig. 1) The developed microcombustor can continuously convert the fuel's chemical enthalpy to thermal energy for hours. If the thermoelectric (TE) modules are coupled with the combustor, we can obtain very small power unit that generate a certain amount of

Figure 1. Developed catalytic microcombustor working at 11.7W of thermal output. The inner diameter is 1.5mm, which is smaller than the quenching distance of butane-air mixture.

electricity and the rest heat as thermal energy.

In the previous PowerMEMS conference, we have discussed the feasibility of a micro-cogenerator which could self-stand with coupling our microcombustor, TE modules and micro-blower that supplied necessary air to the combustor [4]. In the present paper, we introduce a completely self-standing micro power unit with 12W class combustor which drives fuel and air supply system and generates 225mW as the net electricity.

2. Experimental setup

Figure 2 shows the components of the portable power unit and the microcombustor used as the heat source. The nominal thermal output of the microcombustor was 12.2 W at 7.0 sccm of fuel flow rate, and the fuel was mixture of propane and butane (propane/butane = 30/70 in vol%); the compressed liquid fuel for conventional gas lighter was stored in a small gas tank. The combustor was inserted the heat receiver made of copper, and 6 TE modules (1MD04-017-1, RMT Ltd., 3.8 × 3.8 × 2.3 mm) were set between the heat receiver and the heat sinks. The size of the heat receiver was 4 × 4 × 18 mm, and it has a hole whose diameter was 2.7 mm and the depth was 15mm. The microcombustor was inserted in to the hole and the temperature difference about 200K was achieved between the heat receiver and the heat sinks. The heat sink was cooled by ice blocks in order to set the low temperature side constant.

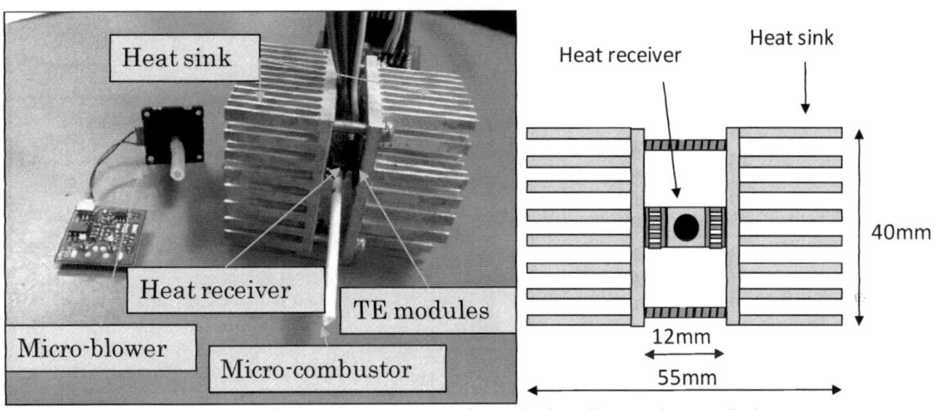

Figure 2. Components of micro-power unit and the dimensions of the power generator unit.

The schematic of the flow system power unit is shown in Fig. 3. The fuel is supplied by its vapor pressure and the flow rate was monitored by the mass flow meter. In order to stabilize the flow rate, the fuel tank was put into the water bath of 310K. The flow rate was adjusted by the small valve installed in the tank and the valve clearance was fixed though the experiment. The air is supplied by the micro-blower (Murata Manufacturing Co., Ltd., 20 × 20 × 1.85 mm) which is driven by a part of electricity generated at the TE modules and the rest was consumed at the variable resistor, which is accounted as the net output. The electricity consumed at the micro-blower including DC-DC converter was also recorded. In order to start the developed power unit, we first ran the micro-blower by an external power device. After the combustor attained steady state and generate sufficient electricity from the TE modules, then we switched the input of the blower to TE modules and removed the external power device.

The fuel flow rate was adjusted to be 7.0 sccm by the micro-valve of the fuel tank, and the air flow rate was controlled by the driving voltage of the blower. The relation between the driving voltage and the air flow rate was investigated in advance. In the present experiment, the driving voltage was set to be 12V, which can supply the stoichiometric air to the combustor. The variable resistor was set at 70Ω in order to maximize the net output. The net output was calculated by measuring the voltage between the resistor, and the consumed electricity at the blower with DC-DC converter was calculated by measuring the voltage and the current at the junctions of DC-DC converter.

3. Results and discussion

Figure 4 shows the histories of fuel flow rate. It is found that the flow rate slightly decreased from 7.5 to 5.8 sccm with time, but combustion was kept more than 20min and quenched finally. Because the air flow rate was constant, the equivalence ratio of the supplied mixture also gradually decreased in time, and it was the reason for the quenching.

The total output from the generator unit, the consumed power at the micro-blower and the net output consumed at the variable resistor were shown in Fig. 5. For the first 12min, the voltage at the TE modules was constant, which was 3.9V. The open-circuit voltage was about 7.8V, therefore it was thought that the variable resistor was adjusted properly. The net output was almost 225mW during the first 12min whereas the fuel flow rate varied from 7.5 to 6.9sccm. It was because the air flow rate was adjusted to be stoichiometry when the fuel flow rate was 7.0 sccm. Therefore, during the first 12 min, the mixture was slightly rich and the thermal input from the combustor was almost constant, which was 12.0W near the nominal condition. During this steady period, the ratio of the power consumed at the blower to the total output, which corresponds to the back work ratio, was 32 %.

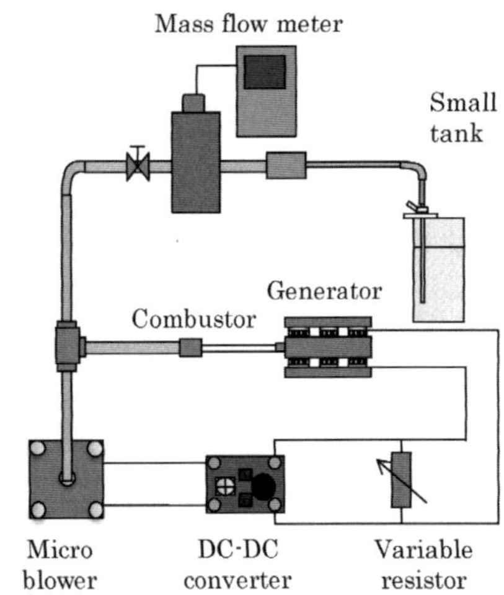

Figure 3. Schematic of the developed micro-power unit.

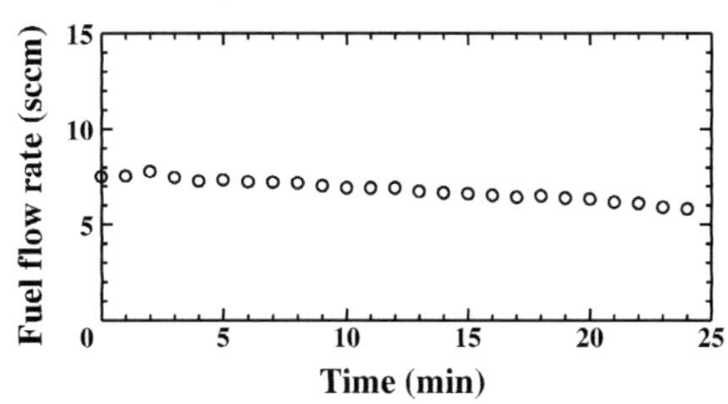

Figure 4. History of fuel flow rate.

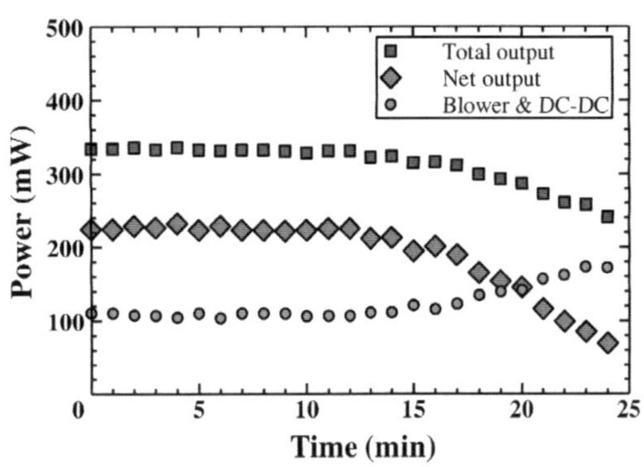

Figure 5. Histories of the total/net outputs and the consumed electricity at the blower.

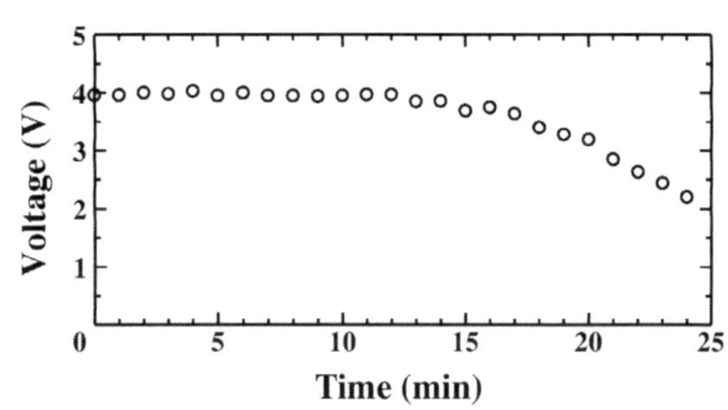

Figure 6. History of the voltage from the generator unit.

Figure 6 shows the history of voltage from the generator unit. After 12 min, it is found that the voltage from the TE modules decreased, which caused the increase of the consumed power at the blower, and eventually caused the large drop of the net output. The final conversion efficiency of the developed power unit is shown in Fig. 7. It is found that final efficiency of the system including fuel and air supply units reached 1.87%.

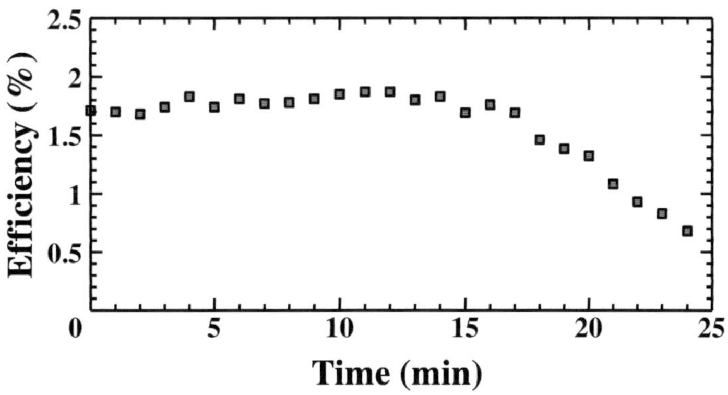

Figure 7. Final conversion efficiency from thermal input to the net output electricity.

4. Summary

We developed a self-standing micro-power unit whose final conversion efficiency was 1.87%. For the thermal input of 12W, the total output from the generator unit was 330mW and the net output was 225mW, which correspond to the back work ratio of 32%. Although there was the issue of stable fuel supply, the prototype recorded reasonable performance.

References

[1] Fernandez-Pello AC 2002 *Proceedings of Combust Institute* **29** 883

[2] Dunn-Rankin D, Martins Leal E, Walther DC 2005 *Progress in Energy and Combustion Science* **31** 422-465

[3] Takahashi S and Wakai K 2009 *45th AIAA/ ASME/ SAE/ ASEE Joint Propulsion Conference & Exhibit* AIAA 2009-5246

[4] Takahashi S, Tanaka M, Ieda N and Ihara T 2015 *Journal of Micromechanics and Microengineering* **25**(10)

Three channel high dynamic current measurement system for low power systems

S Heller, I Nematollahi, S Koeble, P Woias

Albert-Ludwigs-University of Freiburg, Department of Microsystems Engineering (IMTEK), Laboratory for Design of Microsystems, Georges-Koehler-Allee 102, 79110 Freiburg, Germany

Email: simon.heller@imtek.uni-freiburg.de

Abstract. The measurement of the dynamic power consumption is a very important task for the characterization and optimization of an energy-autonomous system. In this paper a measurement system is presented that allows the simultaneous measurement of three current ranges from 1 μA to 30 mA at a sample rate of 1 MHz with a selectable supply voltage in the range of 1.8 V to 3.3 V. With this system the full range of low to high power consumption is covered with a high resolution in time, as it is present in a wireless sensor node during data acquisition and wireless communication, respectively.

1. Introduction

In the field of energy harvesting there are two optimization goals in the design of a system: On the one hand the power consumption should be minimal, on the other hand functionality, reliability and availability should be maximized. To achieve this goal a good knowledge about the power consumption is mandatory.

In general, a precise multimeter is sufficient to measure low currents. However, in low power systems used for embedded wireless sensor nodes the supply current varies considerably over time. Energy is saved by turning off, respectively powering down units like wireless transceivers or sensors as soon as they are not needed. However, power consumption is increasing rapidly as soon as power-hungry units are turned on, e.g. the wireless transmitter. This results in a supply current ranging over several orders from a few 100 nA to a few 10 mA. The only system available on the market fulfilling the requirements of a highly dynamic current measurement system is the CX3324A "Device Current Waveform Analyzer" from *Keysight*. This system is able to measure currents in a range from 100 pA to 10 A with a maximum resolution of 16 bit and a maximum sample rate of 1 GS/s [1]. However, flashing a price tag of a five-figure amount, this system is too expensive to be accepted on a widespread basis.

Therefore, we present in this paper a low-cost current/power measurement system enabling the power analysis of many energy harvesting applications, by measuring currents in a range from 1 μA to 30 mA with a sample rate of 1 MHz and with a supply voltage range from 1.8 V to 3.3 V.

2. Concept and preliminary considerations

In general, a current through a load is measured by converting it into a voltage by a resistor. As the resistor is put in series to the load, a trade-off has to be found between an acceptable voltage drop and the resolution of the current measurement. Furthermore, a change in the supply voltage at the load is inevitable with a resistor in series and is often accompanied by a change in the load's supply current which introduces a measurement error.

To address this challenge the configuration shown in figure 1a can be used. The current sense resistor R_m is placed in the feedback loop of a non-inverting operational amplifier keeping the supply voltage U_{supply} of the load constant. The measurement range, respectively the resolution of the current sensing, is now limited to the supply voltage and the output swing of the amplifier. Replacing the sensing resistor by a current mirror (see figure 1b) allows for smaller supply voltages of the whole setup as the collector-emitter-voltage of a typical PNP transistor is in the range of a few 100 mV only. The output current I_{load} of the current mirror will reflect the supply current in the load at the adjusted supply voltage U_{supply}. To convert this current into a voltage, a sensing resistor or a transimpedance amplifier can be used. However, the problem of limited resolution is still present. This problem can be addressed by using a current mirror which mirrors the current several times (figure 1c). This modification enables a splitting of the measurement range while keeping the output voltage of the amplifier low. Due to the lack of integrated current mirrors with a defined ratio, the current consumption of the whole sensing system is a multiple of the load current.

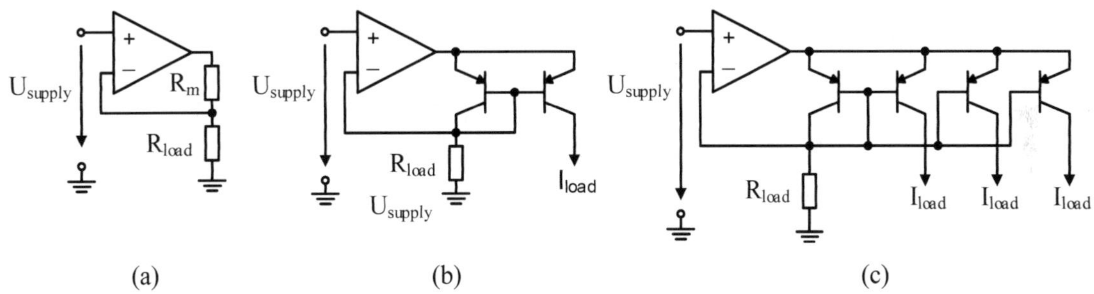

Figure 1. Concepts to measure the supply current through the load R_{load} while keeping the supply voltage U_{supply} for the device under test at a constant level.

3. Analog circuit design

The implementation of the previously presented concept is shown in figure 2. To minimize the gain error of the current mirror, a well-matched four PNP transistor array (THAT320) is used. An additional PNP transistor connected to the base of the current mirror reduces the error introduced by the base current. For temperature compensation, a small-tolerance 10 Ohm emitter resistor is used for each transistor in the current mirror. As the load current is mirrored three times, the total current through the current mirror can raise to values above 100 mA which cannot be delivered by a standard operational amplifier. Therefore, a NPN transistor controlled by the operational amplifier (OPA827) is used to handle higher currents. Due to oscillations, occurring for several load resistances placing the current mirror in the feedback loop, it is placed between the supply line and the collector of the NPN transistor.

The mirror concept presented here provides two main challenges. First, for the current-to-voltage conversion two approaches would be possible. On the one hand, transimpedance amplifiers could be used. The advantage of this approach would be the independence of the current on the collector-emitter-voltage as it is constant. However, a downside would be the increased complexity of the circuit: The output swing of the amplifier reduces with the ascending load current which would demand for a symmetric power supply. One the other hand, the conversion could be implemented by a simple resistor as it is done in this newly proposed system. A rail-to-rail operational amplifier (OPA4350) connected to the respective resistor is used as a buffer delivering U_{outx} for the ADC input. The error introduced by the transistor's Early effects is compensated in the digital post-processing by a look-up table.

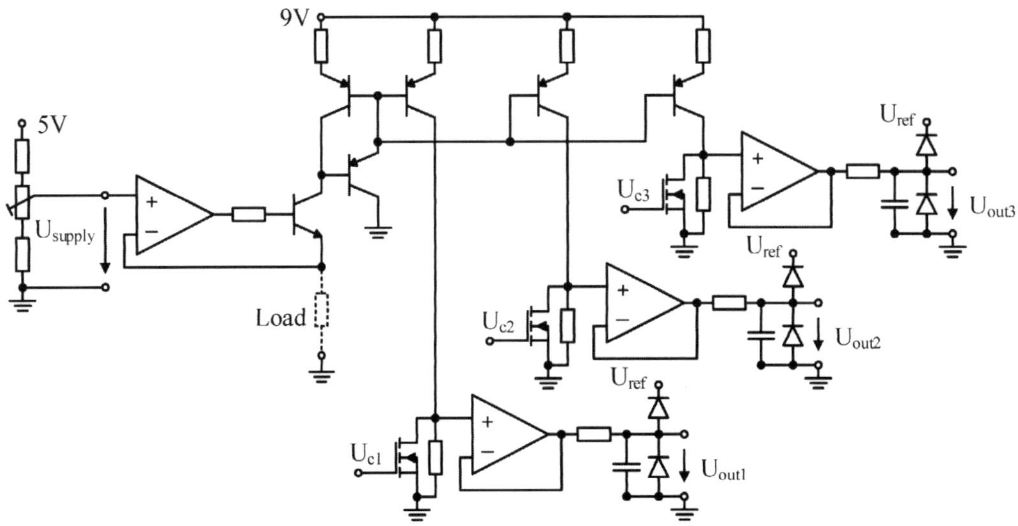

Figure 2. Schematic of the analog circuit.

A second challenge is the introduction of nonlinearities when one of the transistors enters the saturation region due to a high voltage drop on the respective current sense resistor. To avoid the saturation of a transistor an n-channel MOSFET, controlled by the digital circuit, is put in parallel to every current sense resistor to short it when its voltage exceeds the linear region of the bipolar junction transistor in the respective branch of the current mirror.
The system is powered from a unipolar 9 V supply enabling also a mobile use without any effort for split supply. Lower voltages for the buffer amplifiers and the digital circuit are generated by two voltage regulators. The supply voltage for the device under test is set in a range from 1.8 V to 3.3 V via a voltage divider with a digital 10 k resistor controlled by the digital circuit.

4. Digital interface and software design

The task of the digital circuit is the digitization and monitoring of the measured voltages as well as the data transfer to a computer for post-processing and data analyses. For the analog-to-digital conversion three single-channel ADCs (MAX11108) are used allowing a simultaneous sampling of all three channels. To prevent the transistor in the current mirror from entering the saturation region the ADC outputs are analyzed continuously. By evaluation the output value of the next-in-size current range an easy and fast control mechanism can be implemented. Introducing a hysteresis by setting an upper and a lower threshold minimizes the switching operations and enables an efficient post-processing of the data. Often, the data transfer to a PC is a bottleneck in a measurement system. To enable a continuous and simultaneous readout of all three channels, a combination of an FPGA (Spartan 6) and the FT232H USB UART/FIFO IC from FTDI is used (see figure 3). The FT232H chip runs in a parallel 245 FIFO mode allowing transfer rates up to 40 MB/s. Unfortunately its internal FIFO Buffer is limited to 1 kB in size resulting sometimes in a discontinuous data stream. Therefore, the FPGA acts as a second FIFO buffer in addition to the readout of the ADCs and the output monitoring. As the FTDI chip runs at a higher frequency than the FPGA respectively the sampling frequency, an overflow of the FPGAs buffer is impossible. On the PC side, a *LabVIEW* program is running that records and post-processes the data. The user can set the recording time as well as the supply voltage.

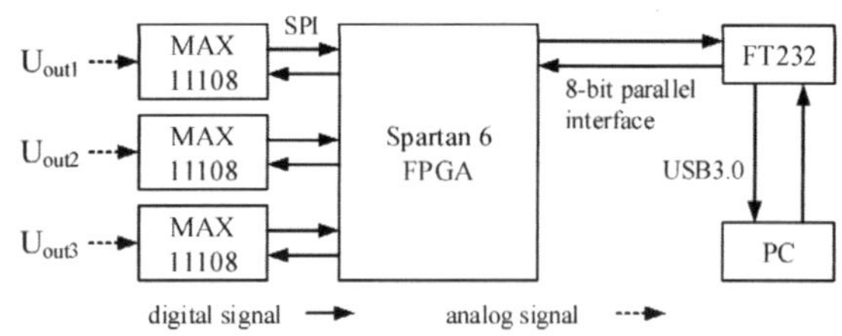

Figure 3. Schematic of the digital circuit.

5. Measurement results

In figure 4 the transfer characteristic of the measurement system is shown. The load was simulated with a resistor cascade. A strong linear relation with a $R^2 > 0.999$ was found between the load current and the output voltage. The offset error is related to the offset voltage of the buffer amplifier, the gain error is affected by the resistor tolerance as well as by the gain error in the current mirror. A look-up table in the post-processing step can easily compensate both.

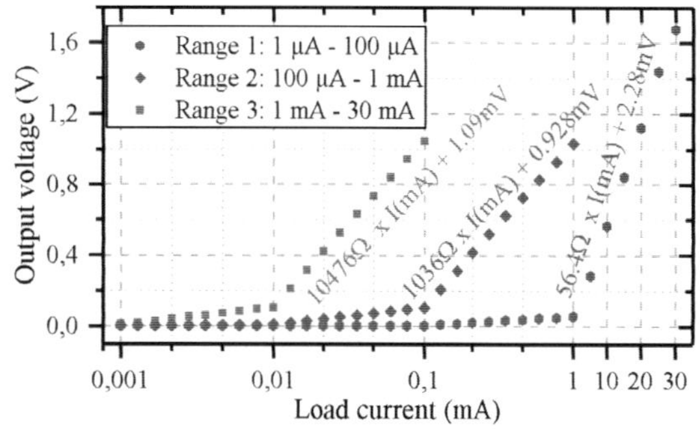

Figure 4. Output characteristic of the measurement system over a measurement range from 1 µA to 30 mA for all three measurement ranges.

The results of dynamic measurements are shown in figure 5 and 6. The power consumption of a MSP430 low-power microcontroller is depicted in figure 5. Short interrupts between the change of the low power mode (LPM) as well as the switching frequency of the internal voltage regulator can be clearly seen. When comparing the measured currents to the datasheet values [2] only small deviations can be found. In figure 6 the current consumption of a wireless data logger consisting of a wake-up receiver, a wireless data interface and two temperature sensors during the sensors readout and a wireless transmission is shown. All events can be clearly distinguished by their current consumption. Even the turn on of different modules during a task like the turn on of the amplifiers and the current sources as well as the start of the analog-to-digital conversion during the sensor readout are visible.

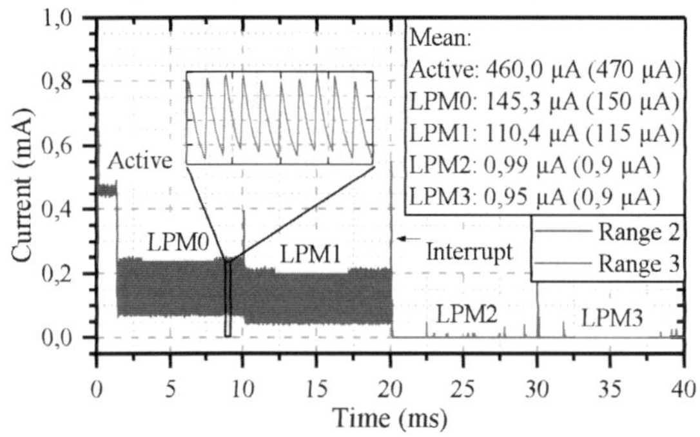

Figure 5. Current consumption of a MSP430FR5969 microcontroller in different power modes running at 8 MHz and 3 V supply voltage.

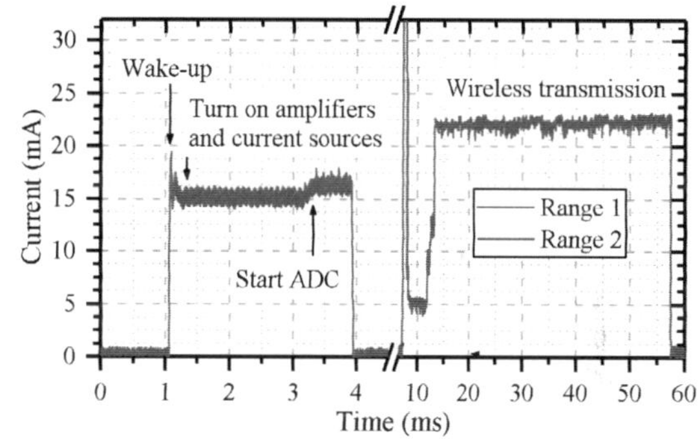

Figure 6. Current consumption of a data logger during a sensor readout and a wireless data transmission.

6. Conclusion and outlook

This paper reports the development of a low-cost measurement system able to measure the dynamic supply current of low power systems. With a sample rate of 1 MHz and a measurement range from 1 μA to 30 mA it is suitable for most energy-autonomous embedded systems involving microcontrollers, sensors and wireless transmitters. To set up the recording length and the supply voltage, a user interface is implemented in *LabVIEW*. This software also allows an initial post-processing and gives a first overview of the current consumption as well as derived quantities like the mean power, min/max power and the energy demand.

References

[1] CX3324A/CX3324 Device Current Waveform Analyzer, datasheet, Keysight, 2017.
[2] MSP430FR59xx Mixed-Signal Microcontrollers, datasheet, Texas Instruments, 2017.

DEVELOPMENT OF PIEZOELECTRIC VIBRATION ENERGY HARVESTERS FOR BATTERY-LESS SMART SHOES

H. Katsumura[1], T. Konishi[1], H. Okumura[1], T. Fukui[1], M. Katsu[2], T. Terada[3], T. Umegaki[3], and I. Kanno[3]
[1]Panasonic Corporation, 1006, Kadoma, Kadoma-city, Osaka, 571-8501, Japan
[2]ASICS Corporation, 6-2-1, Takatsukadai, Nishi-ku, Kobe, Hyogo, 651-2271, Japan
[3]Faculty of Engineering, Kobe University, 1-1, Rokkodai-cho, Nada-ku, Kobe, 657-8501, Japan
E-mail: tumegaki@aquamarine.kobe-u.ac.jp, kanno@mech.kobe-u.ac.jp

Abstract For the purpose of realising the battery-less smart shoes, we have developed piezoelectric vibration energy harvesters (PVEHs), and PVEH-driven sports shoes. The smart shoes are equipped with acceleration sensors in the sole and Bluetooth LE (BLE) modules, which were driven by PVEHs composed of piezoelectric bimorph cantilevers. We have confirmed the capability of the PVEHs to frequently transmit the sensing data from the beacons of BLE when the subjects practically run while wearing shoes.

1. Introduction

Many researchers have published presentations about "smart shoes" equipped with various sensors, radio communication modules and others for commercial uses [1]. However, the components in the smart shoes are usually driven by batteries, which are worried about giving out the power or frequent charging operation. On the other hand, shoe-mounted energy harvesting has attracted attention as a power source for mobile or wearable devices [2]. However, there have been few reports on the battery-less smart shoes in which sensors are driven by vibration energy harvesters. In this study, we have developed high-power piezoelectric vibration energy harvesters (PVEHs), which generate electric power enough to operate wireless modules and insole sensors by impulsive forces on the soles of shoes during walking and running, and test to transmit the sensing data. We demonstrated the capability of the PVEHs for the actual application to battery-free smart shoes.

2. Developments of piezoelectric vibration power device with piezoelectric films

2.1. Fabrications of piezoelectric vibration power device with piezoelectric thick films

In order to produce the PVEHs, we used the growth technique of piezoelectric thick films [3]. This is a revolutionary technique to form piezoelectric films by screen-printing on stainless steel substrates. We can form piezoelectric films with the thickness of 40 μm on both sides of the 100 μm-thick stainless steel by using this technique as shown in figure 1. We have already confirmed the dramatic improvement of the reliability of the fracture strength by the sintering process of the compressive residual stress over 500 MPa in the piezoelectric thick films. The experimental output spectrum of figure 2 has the resonant frequency 59.3 Hz and the maximum generation power 180 μW at the vibrational acceleration 0.98 m/s^2 and the load resistance 197.8kΩ.

2.2. A design method by the equivalent circuit model

Structure of the PVEH are optimized by theoretical models of equivalent circuit (figure 3) based on the lumped parameter model [4] and the piezoelectric constitutive equations [5]. The equivalent circuit model consists of mechanical and electrical systems, and a transformer connecting these two systems.

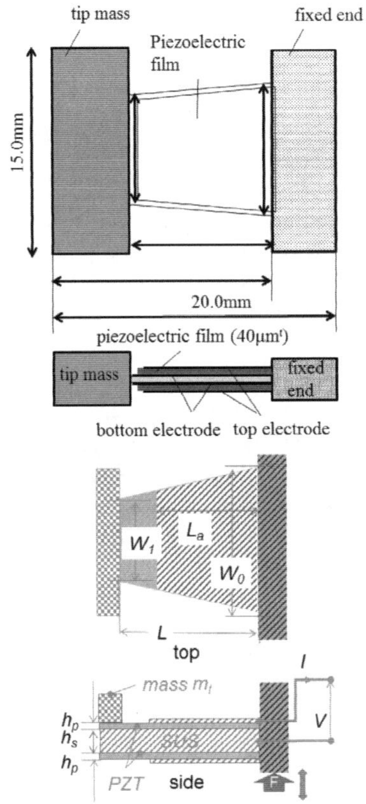

Figure 1. *A schematic illustration of PVEHs composed of a piezoelectric*

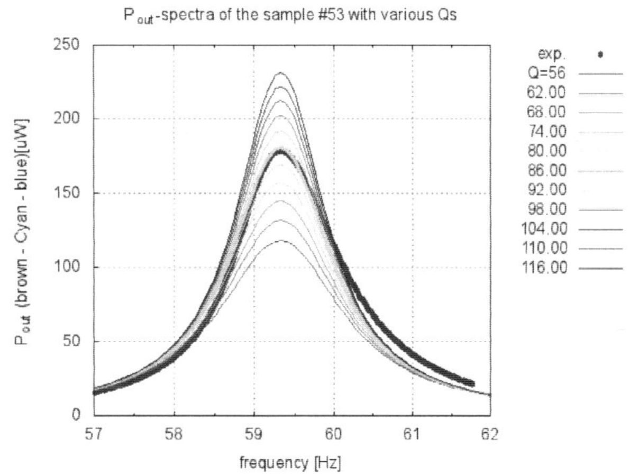

Figure 2. *Output power spectra with various Qs (lines) fitting to experimental values (dots).*

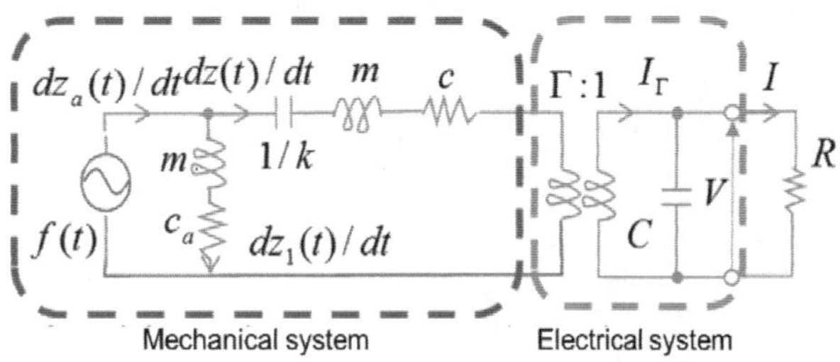

Figure 3. *An equivalent circuit of the cantilever type vibration energy harvester.*

The mechanical systems consist of a metallic cantilever beam with piezoelectric films, while the electrical ones are composed of electrode capacitances as shown in figure 1. In the case of sinusoidal excitation, i.e. $f(t)=F\sin\omega t$, we have formulated the output power P_{out} as follows.

$$P_{out} = 0.5R|V_G|^2\left[(R+R_G)^2 + X_G^2\right]^{-1}, \quad (1)$$

$$V_G = \left[K^2+1-(\omega/\omega_0)^2 + jD\right]^{-1}\Gamma^{-1}K^2F, \quad (2)$$

$$R_G = (\omega C)^{-1}DK^2\left\{D^2+\left[K^2+1-(\omega/\omega_0)^2\right]^2\right\}^{-1}, \quad (3)$$

$$X_G = -(\omega C)^{-1}\left\{\left[1-(\omega/\omega_0)^2\right]\left[K^2+1-(\omega/\omega_0)^2\right]+D^2\right\}\left\{D^2+\left[K^2+1-(\omega/\omega_0)^2\right]^2\right\}^{-1}, \quad (4)$$

$$K = \Gamma/\sqrt{Ck}, \quad (5)$$

$$D = (\omega/\omega_0)Q^{-1}, \quad (6)$$

$$Q = m\omega_0 c^{-1}, \quad (7)$$

where Γ is the turns ratio of the transformer, C is the capacitance of electrodes-sandwiched PZT, m is the tip-mass, and k is the spring coefficient of the trapezoidal cantilever formulated as follows.

$$\Gamma = 3(1-\nu_s)e_{31}(h_s+h_p)L_a(2L-L_a)k_d^{-1} \quad (8)$$

$$C = 0.5\varepsilon_{33}L_a\left[w_0(2L-L_a)+w_1L_a\right](Lh_p)^{-1} \quad (9)$$

$$\omega_0 = \sqrt{km^{-1}}, \quad (10)$$

$$k = k_nk_d^{-1}, \quad (11)$$

$$k_n = 8Y_p h_p^3 + 12Y_p h_p^2 h_s + 6Y_p h_p h_s^2 + Y_s h_s^3, \quad (12)$$

$$k_d = 6(w_0 - w_1)^{-3} L^3 w_0^2 \{r^2 - 2(1-r)[(1-r)\log_e|1-r| + r]\}, \quad (13)$$

$$r = (w_0 - w_1) w_0^{-1}, \quad (14)$$

where Y_p and Y_s are Young's modulus of the piezoelectric film and substrate, respectively. To accurately reproduce experimental values of the PVEH, we have performed fitting processes. More specifically, we have calculated the theoretical P_{out} spectra by the equation (1) with various Qs as shown in figure 2, where the Q means a mechanical quality factor defined by the equation (7). We can find the Q=86 spectrum is the best theoretical curve to fit the experimental ones in figure 2. Based on the extracted parameters, we can accurately reproduce output spectra by the theoretical calculations.

3. Harvesting characteristics of PVEHs in shoes

In order to estimate the generating power by PVEHs during running, we attached acceleration sensors in the soles of the prototype shoes as shown in figure 4. We measured the vibrational accelerations while 6 male subjects (21 to 27 years old students) jog on a pavement surface while wearing shoes. Figure 5 shows an example of the time dependent of the acceleration. The acceleration maxima are about 300 m/s^2 when the shoes land on the pavement surface.

Figure 4. *A shoe with acceleration sensors in the sole.*

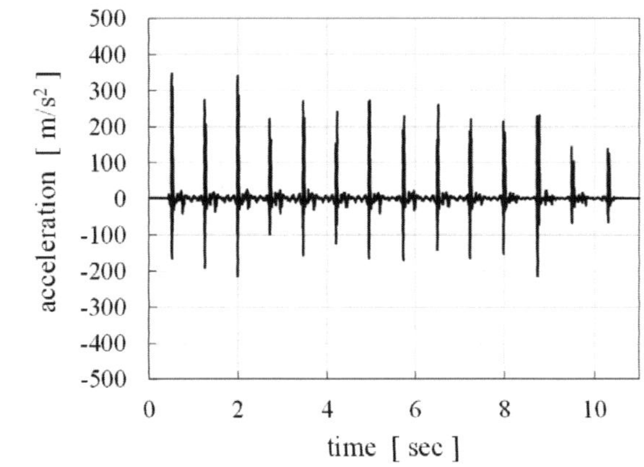

Figure 5. *An example of the time dependent of the acceleration at a sole.*

We measured the output generation characteristics of the above-mentioned PVEHs fixed on the vibrator in the excitation of impulses with acceleration maxima about 300 m/s^2. figure 6 shows an example of the time dependent of the output voltage. We can observe damped trains with voltage maxima about ±50V, which are appeared by impacts when the shoes land on the surface. Based on the measured voltages and the load resistance, we have estimated the ac power generation as 360 to 1200 μJ per one step, which is depend heavily on the subjects.

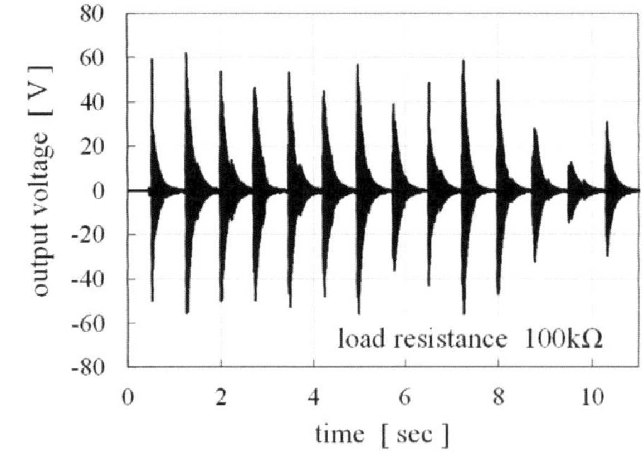

Figure 6. *A time dependent of the output voltage by the reproduced acceleration.*

4. Manufacturing prototype shoes

We have manufactured prototype sports shoes with a sole containing W20 mm × L60 mm × H14 mm resin housing, which contains two PVEHs with a power supply circuit, controller and Bluetooth LE module, as shown in figure 7. Further, we have incorporated the housing in the arch section of the shoe so as to avoid spoiling the design and the feeling. We have confirmed to be able to frequently transmit the sensing data from the beacons of Bluetooth LE when the subjects practically run while wearing shoes. So we can draw Bluetooth-transmitted foot pressures on a PC as shown in figure 7. Currently, we are applying this technique to the pose recognitions, feedback systems of voice, optical indicators of foot pressures and so on.

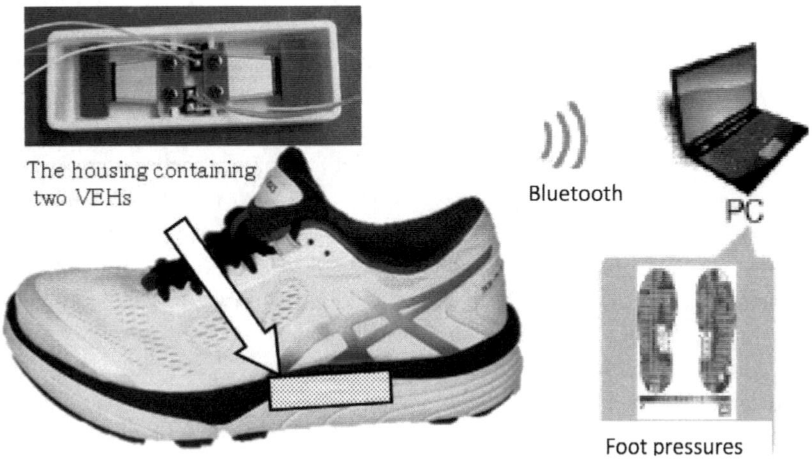

***Figure** 7. A prototype shoe with the PVEH module and Bluetooth-transmitted foot pressures drawn on a PC.*

5. Conclusions

We have developed PVEHs generated by impulsive forces on the soles of shoes during walking and running. We have also manufactured prototype sports shoes with a sole containing resin housing, which contains two PVEHs, a power supply circuit, controller, Bluetooth LE module, and others. We have confirmed to be able to frequently transmit the sensing data from the beacons of BLE when the subjects practically run while wearing shoes. Currently, we are developing smart shoes that can monitor runners' postures by using prototype shoes, and designing each device parameters using by equivalent circuit models.

References

[1] Pyeong-Gook Jung, et al., J. Dyn. Sys., Meas., Control 136(1), 011002 (2013).
[2] N. S. Shenck, J. A. Paradiso, IEEE Micro 21, 3, pp. 30-42 (2001).
[3] A. Oish, H. Okumura, H. Katsumura, H. Kagata, Journal of Physics: Conference Series, Power MEMS 2014, 557-1.
[4] A. Erturk and D. J. Inman, Journal of Intelligient Material Systems and Structures, vol. 19, pp.1311-1325, (2008).
[5] Ki B. Lee, Principles of Microelectromechanical systems, IEEE Press, 2011.

Heating and cooling the human body with wirelessly-powered devices

P.J. Goodrich [1], G. Fierro [2], V. Luu [1], H. Zhang [3], E. Arens [3]
(1) Department of Mechanical Engineering, University of California, Berkeley
(2) Department of Electrical Engineering and Computer Science, University of California, Berkeley
(3) Department of Architecture, University of California, Berkeley
November 14th, 2017

It's likely that you or someone you know is regularly dissatisfied with the thermal environment of the building that they work in. This is no coincidence - the statistics for occupant comfort in both the United States and abroad are miserable. In fact, current American Society of Heating, Refrigeration, and Air Conditioning Engineers (ASHRAE) standards only require 80% of occupants to report comfort. Numerous studies have come out in recent years that show occupant comfort leads to higher productivity [1] and happiness [2]. Rather than making expensive retrofits to buildings, PCS can provide thermal comfort that can be tuned to the needs of the user. If building designers and operators can find effective ways to allow building temperatures to vary over a wider range, while allowing occupants individual control of comfort, the potential for energy savings is enormous. However, many of these technologies require wired connections, don't communicate with each other, and don't seamlessly fit into daily life.

The goal of this work is to expand the use of PCS by reducing their need for wired connections. To do this, we developed a suite of devices that flexibly and efficiently transfer electrical power through wireless power coupling. The selection of thermal comfort devices was influenced by preliminary human studies, where we mapped the thermal sensitivity across the body. The data we collected from these studies are both quantitative (temperature of skin) and qualitative (user reported sensation). These results led us to devise a suite of four thermal comfort devices to be implemented into commercial buildings: a desk fan, heated/cooled wristpad, a heated/cooled chair, and a heated insole.

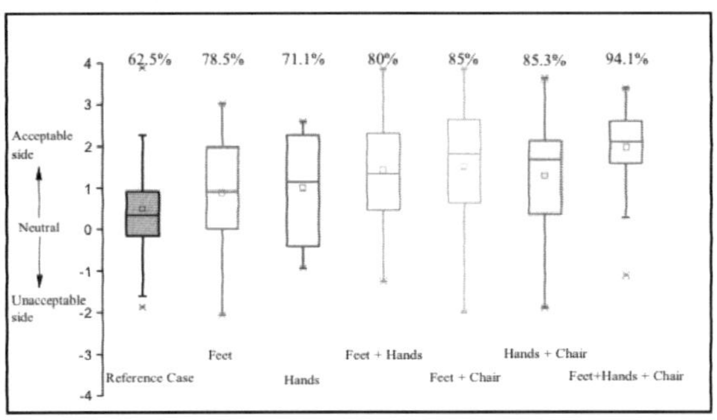

Figure 1: *User reported satisfaction of the initial user studies.*

All four of these devices operate through resonant inductive coupling. In this, a power source is fed through a Power Transer Unit (PTU) which converts the electrical energy into a magnetic field, propogating in the direction perpendicular to the coil. In application, the PTU is encased in a floor mat or a chassis mounted underneath a desk. On the device side, a Power Receiver Unit (PRU) is placed in the magnetic field where it is rectified and converted back into electrical energy. In this work, the PRU and PTU were developed by an industry partner - WiTricity Inc - specifically for these devices, and include security and safety protocols.

For our heated insole device - our main focus of work now - heat is delivered to six locations on the foot, which were selected based on our initial user studies. The target temperature for the bottom of the foot is set to 43C (109.4F) which we have previously found as the ideal temperature for providing comfortable warmth.

To undestand the power requirements to deliver the target temperature to the foot, a Finite Difference Model (FDM) as developed to approximate the solution of the 1D-Heat Equation that defines our thermal system. This model was validated and used to explore the design space of heat input, duty cycling, and their transient effects on the temperatures of the system. This parameter has guided our design, and allow us to work backwards to calculate the electrical requirements of our device.

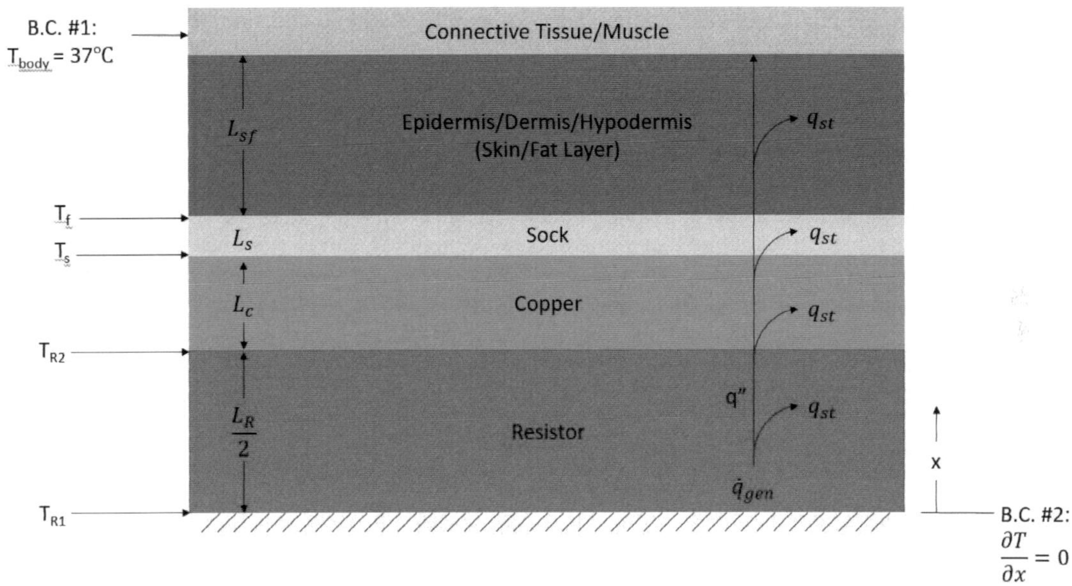

Figure 2: *Schematic of FDM Model*

Table 1: *Battery charge/discharge timing table of the heated insole under current prototype parameters of I_{in} = 1000mA, I_{out} = 210mA (corresponding to an individual heating duty cycle of 17.5%, which is the empirically determined duty cycle for maintaining our setpoint temperature)*

Design Variable	380mAh bat.	4.1Ah bat.
$TTC_{Idle,infield}$	24 min	4.3 hr
$TTC_{Active,infield}$	31 min	5.5 hr
$TTD_{Idle,notinfield}$	7.6 hr	82 hr
$TTD_{Active,notinfield}$	88 min	15.1 hr

References

1. Akimoto, T., Tanabe, S.I., Yanai, T. and Sasaki, M. 2010. *Build Environ* **45** 45-50
2. Steemers, K. and Manchanda, S. 2010. *Build Environ* **45** 270-8

IOP Conf. Series: Journal of Physics: Conf. Series **1052** (2018) 012062 doi:10.1088/1742-6596/1052/1/012062

Development of Rotational Electret Energy Harvester Using Print Circuit Board

M. Adachi[1], T. Miyoshi[1], K. Suzuki[1], Q. Fu[2], Q. Fang[2], and Y. Suzuki[1]

[1]Dept. of Mechanical Engineering, The University of Tokyo, Tokyo, Japan
[2]Huawei Technologies Co. Ltd., Shenzhen, China

Email: madachi@mesl.t.u-tokyo.ac.jp

Abstract. An improved 1-D electrostatic model is proposed for rotational electret energy harvester (EH). A test bench for rotational electret EH is developed, and the experimental data are compared with the prediction using the present model. A low-profile rotational electret EH is designed and prototyped using cost-effective PCBs. Ball bearings are used for the rotational support of a stainless steel rotor. In the rotational electret EH prototype with an air gap of 200 μm, the output power of 80 μW is obtained at 2 rps.

1. Introduction

Since the maximum output power of vibration energy harvesters (EHs) is defined by the velocity-damped resonant generator (VDRG) limit [1], rotational EHs [2] attract much attention for low-frequency vibration applications such as human motion. Among various transduction mechanisms, electret generators are advantageous in terms of higher output power at low frequencies and their low-profile structures. For electret EHs, keeping constant air gap between the rotor and stator is of crucial importance. Metal ball bearing is useful for reliable mechanical support of the rotational motion, but its integration with MEMS-friendly glass or silicon substrates is tricky. Nakano et al. [3] prototyped a rotational electret EH based on 1.5 mm-thick Si substrates, in which lengthy MEMS processes using seven masks are required. In addition, there is somewhat large discrepancy between the experimental data and the previous generator model [4]. The objectives of the present study are to design an electret rotational EH based on an improved generator model, and to develop the rotational EH prototype by using cost-effective PCB technologies.

2. Improved One-dimensional Electrostatic Model

In order to study the output power of rotational electret EHs, an 1-D electrostatic model [5] shown in Fig. 1 is assumed. In Fig. 1, d is the thickness of electret and g is the air gap. E_a, E_b, and E_c are the electric fields in between the electret surface and the counter electrode, inside the electret material, and in between the guard and the counter electrodes, respectively. σ_E is the surface charge density of electret, while σ_1 and σ_2 are the induced charge densities on electrode areas overlapped with the electret and the guard electrode, respectively. Based on the Gauss' law we get

$$\sigma_E = \varepsilon_0\varepsilon_1 E_a - \varepsilon_0\varepsilon_E E_b, \tag{1}$$

where ε_0, ε_1, and ε_E represent the vacuum permittivity, relative permittivities of the air and the electret, respectively. $V(t)$ and C_p are the voltage across the external load R_{load}, and the parasitic capacitance. Using the Kirchhoff's law, we get

$$V + E_a g + E_b d = 0, \text{ and } V + E_c(d + g) = 0, \tag{2}$$

where A_1 and A_2 are the effective surface areas overlapped with the electret and the guard electrode, respectively. In the present model, the gap between neighboring electrodes is excluded from the overlapping areas A_1 and A_2. Considering the charge conservation and the Ohm's law, we get

$$\frac{d}{dt}(\sigma_1 A_1 + \sigma_2 A_2) = -\left(\frac{V}{R} + C_p \frac{dV}{dt}\right), \tag{3}$$

where the induced charge densities σ_1 and σ_2 are given by

$$\sigma_1 = -\varepsilon_0 \varepsilon_1 E_a \text{ and } \sigma_2 = -\varepsilon_0 \varepsilon_1 E_c, \tag{4}$$

Figure 2 shows the power generation test bench using a rotational stage. A glass plate with patterned electret is fixed onto the rotational stage as a rotor, and another glass plate with counter electrodes is fixed onto a 5-axis stage with a vacuum chuck for keeping both plates in parallel. Alignment errors for the air gap and the rotational center are within 2 μm and 8 μm, respectively. Number of poles is chosen as 109. Figure 3 shows the rotor fabricated with MEMS technologies. For electret, 15μm-thick CYTOP EGG [6] is used as the high-performance electret. Surface potential of electret is -843 V. Parasitic capacitance is 128 pF.

Figure 4 shows the output voltage profile for the air gap of 65 μm at the rotational speed of 1 rps. Output voltage of 180 V_{p-p} is obtained. The experimental data are in good agreement with the present simulation results. On the other hand, the Bartsch et al's model [4] exhibits a sharp kink in the voltage waveform and overestimates the voltage amplitude as well as the output power. Output power is 200 μW at 1 rps, which is also in good agreement with the prediction with the present model (not shown).

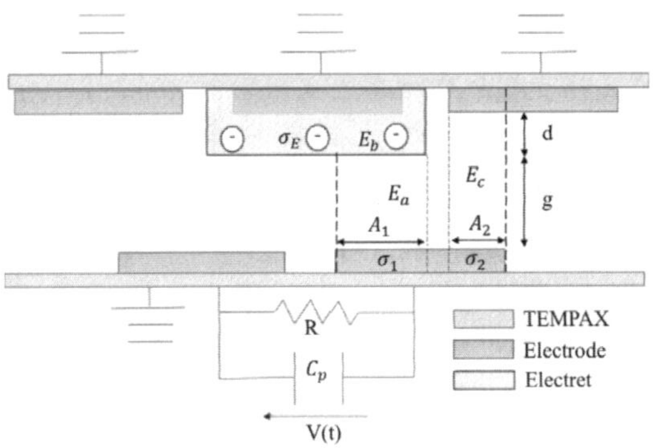

Figure1. One dimensional electrostatic model.

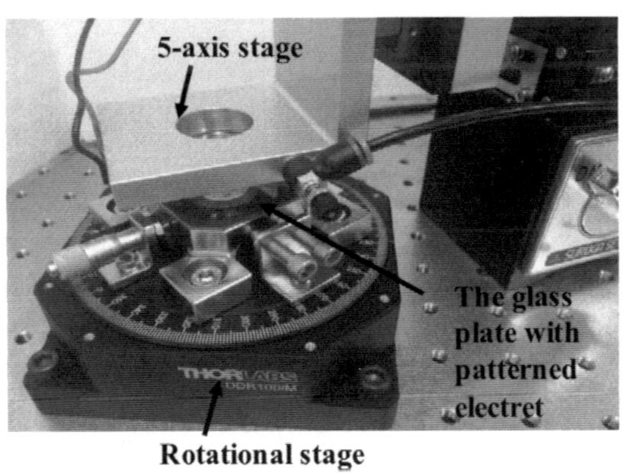

Figure 2. Power generation test bench using a rotational stage.

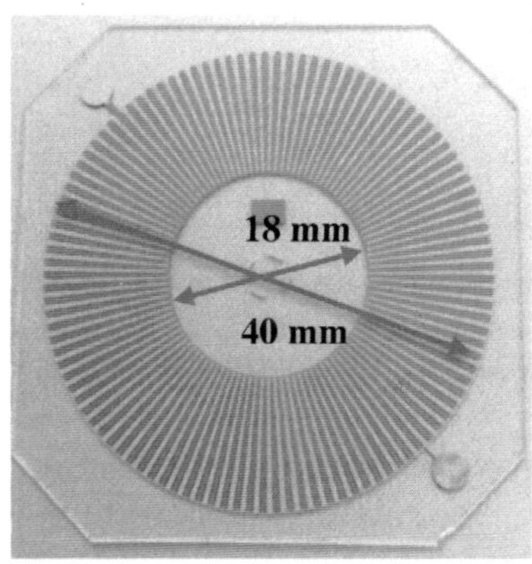

Figure3. Rotor with patterned electret on a glass substrate.

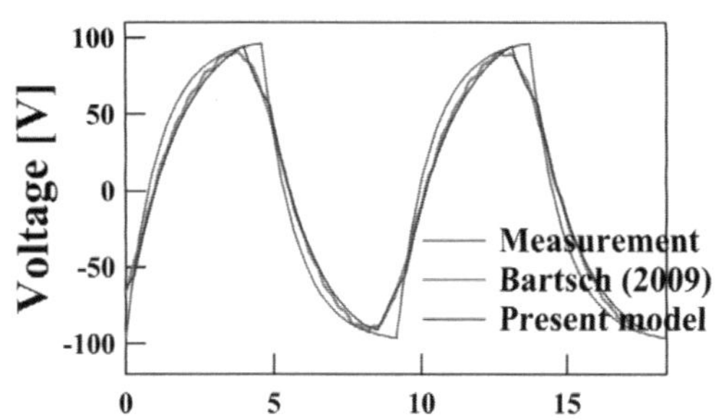

Figure 4. Output voltage waveform for the rotational electret generator at 1 rps in comparison with the present 1-D model and Bartsch's model[4].

3. Rotational Electret Energy Harvester with Eccentric Mass

Figure 5 shows the present rotational electret EH. The rotor and the stator are made of stainless steel, and the rotation is supported by preloaded miniature ball bearings at the center. Mass eccentricity is realized by perforation on the half of the rotor. Outer diameter is 39 mm, while the thickness of the rotational part is only 3.3 mm. Print circuit board (PCB) is used for the electret and electrode substrates. To reduce the parasitic capacitance between the adjacent electrodes and to enhance the output power, 170 μm-thick PCB with low permittivity (ε=2.4) is used as the stator substrate for the current collector electrodes (Fig. 6a). Outer/inner diameters are 38 mm and 18 mm, respectively. With 109 pole electrodes, the parasitic capacitance is 105 pF. On the other hand, 105 μm-thick heat-resistant PCB is employed for the electret substrate (Fig. 6c) in order to minimize thermal deformation during the curing process of the electret material, CYTOP EGG. The minimum electrode width and the minimum gap between comb electrodes are 210 μm and 50 μm, respectively. 15 μm-thick CYTOP EGG is spun-on the rotor PCB with patterned electrodes. Both PCBs are bonded onto the stainless steel rotor/stator by using spray-coated acrylic adhesive. After charging the CYTOP electret using soft X-ray [7], the rotor is assembled with the stator as shown in Fig. 6(d). For this particular device, the air gap is 200 μm. Surface voltage of electret on the PCB substrate is -876V. Note that the mechanical damping of the bearing is as low as 20 % of the estimated electrostatic damping.

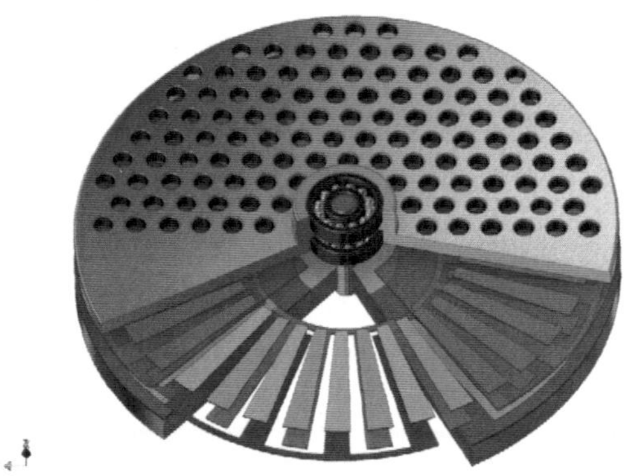

Figure 5. Schematic of the rotational electret EH.

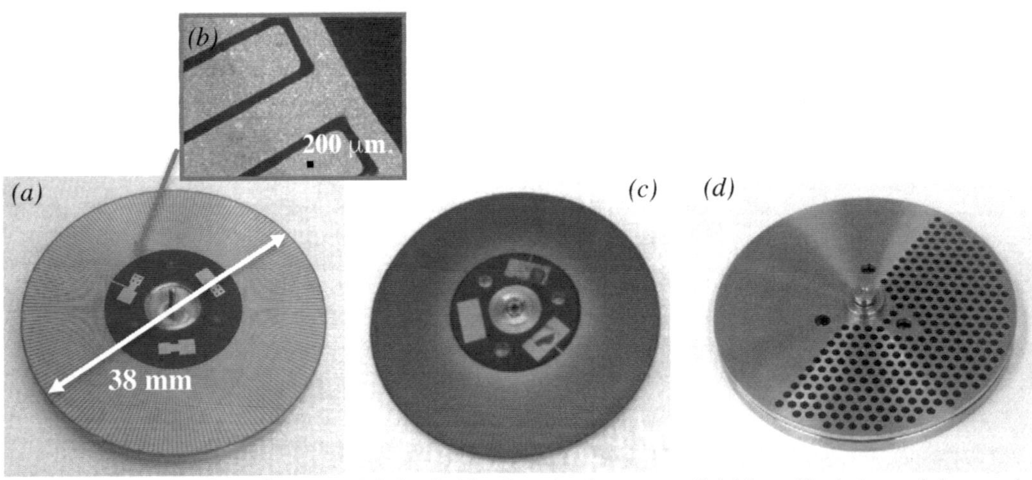

Figure 6. (a) PCB-based stator with backside electrical contact, (b) Magnified view of the comb electrodes, (c) PCB-based rotor with 15 μm-thick CYTOP EGG electret, (d) Assembled rotational electret EH prototype.

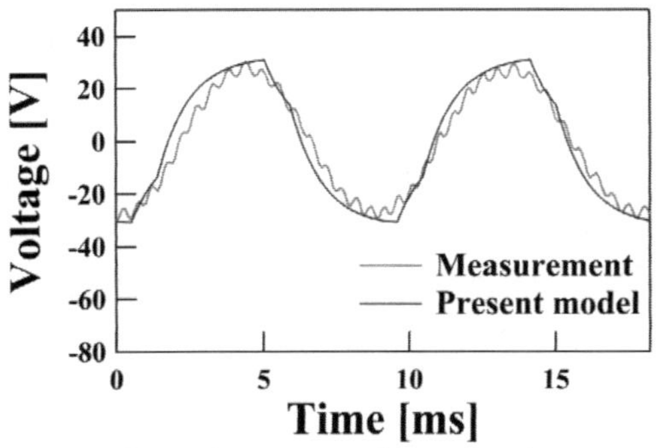

Figure 7. Voltage waveform at the constant rotational speed of 1 rps with external load of 9 MΩ.

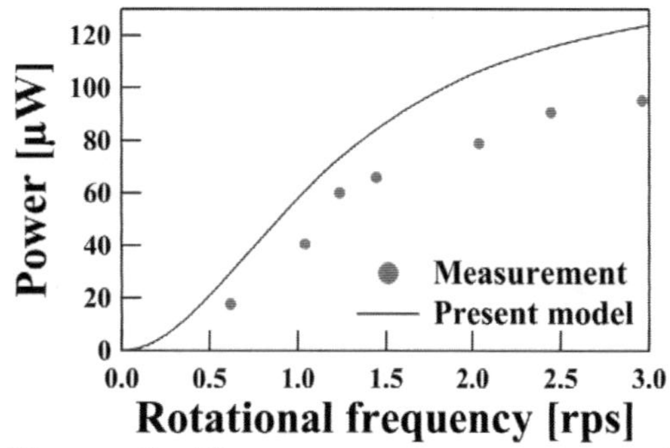

Figure 8. The output power versus the rotational frequency with external load of 10 MΩ.

4. Power Generation Experiment

Figure 7 shows the voltage waveform at the constant rotational speed of 1 rps with external load of 9 MΩ. Peak-to-peak voltage of 60 V corresponding to the output power of 44μW is obtained in the present PCB-based rotational EH. The output voltage is somewhat lower than the simulation results, but it trend is in accordance with the simulation. Figure 8 shows the output power versus the rotational speed with a 10 MΩ external load. The output power of 80 μW is obtained at 2 rps. The experimental data is about 20% smaller than the simulation results. It is probably due to insufficient control of the air gap and/or partial discharge of the electret during the assembling process. It is also demonstrated that the rotor rotates by applying acceleration/rotation to the device.

5. Conclusions

We have proposed an improved 1-D electrostatic model for rotational electret EH, and showed that its prediction is in good agreement with our experimental data. A low-profile rotational electret EH is designed and prototyped using two different kinds of PCBs for better performance. In the rotational electret EH with an air gap of 200 μm, the output power of 80 μW is obtained at 2 rps.

This work was partially supported by Huawei Technologies and JST CREST Grant Number JPMJCR15Q3, Japan.

References

[1] P. D. Micheson et al., *J. Microelectromech. Syst.*, 13 (2004), pp. 429-440.
[2] E. M. Yeatman, *Proc. IMechE*, 222, Part C (2008), pp. 28-36.
[3] J. Nakano et al., *J. Phys.: Conf. Ser.*, 660, (2015), 012052.
[4] U. Bartsch et al., *Proc. PowerMEMS 2009*, Washington D.C., (2009), pp. 332-335.
[5] Y. Tada, *Jpn. J. Appl. Phys.*, 31, (1992), pp. 846-851.
[6] K. Kashiwagi et al., *J. Micromech. Microeng.*, 21, (2011), 125016.
[7] K. Hagiwara et al., *Trans. IEEE, Dielectr. Electr. Insul.*, 19, (2012), pp. 1291-1298.

Characterization of glucose biofuel cell based on electrodes modified by carbon nano horns

Kenta Kuroishi, Takuma Ishida,Toshinari Doi, Yudai Fukushi, Satomitu imai, and Yasushiro Nishioka

Department of Precision Machinery Engineering, College of Science and Technology, Nihon University, Funabashi, Chiba 274-8501, Japan

E-mail: imai@eme.cst.nihon-u.ac.jp

Abstract. In this study, a film-like biofuel cell (BFC) using carbon nano-horn (CNH) as an electrode material was fabricated using micro-electro-mechanical systems (MEMS) technology. The cell was fabricated on a flexible polyimide (PI) substrate, and a porous carbon-coated aluminum (Al) electrode with a width of 3.0 mm and a length of 10 mm was formed using photolithography and screen-printing techniques. The BFC investigated consists of a porous carbon anode (area of 30 mm^2) modified by glucose oxidase and ferrocene, and a cathode (area of 30 mm^2) modified by bilirubin oxidase. The constructed BFC was able to produce a power density of 18.0 μW/cm^2 in phosphate buffer wherein 100 mM glucose was dissolved at room temperature, and this value was at least 10 times that of the BFC reported by Y. Fukushi et al. Furthermore, by examining the temperature-dependence of the BFC's power density, it was possible to attain a maximum power density of 39.9 μW/cm^2 in a phosphate buffer solution wherein 100 mM glucose was dissolved at 40 °C.

1. Introduction

Enzymatic biofuel cells (BFCs) that utilize enzymes as electrocatalysts instead of metals have attracted attention as ubiquitous sustainable power devices that can generate electricity directly from the blood or bodily fluids [1–3]. A characteristic feature of BFCs is their ability to generate power using various kinds of fuels, such as glucose, lipids, and hydrogen. In addition, they function under mild conditions, from room temperature to human body temperature under neutral pH conditions.

Y. Fukushi et al. reported a flexible biofuel cell equipped with a microchannel that mimicked the blood vessels in the human body [3]. Although a BFC fabricated by MEMS technology using ketjenblack (KB) as an electrode material has been reported by Y. Fukushi et al, its power density was only 1.45 μW/cm^2, and further improvement in the output was necessary for practical use. The use of KB as an electrode material presents some challenges: the enzyme used as the catalyst is easy to peel off, and the reaction area is small. Therefore, we endeavored to use a carbon nano-horn (CNH) with a specific surface area larger than that of the KB electrode and a structure that is thought to be less susceptible to enzyme peel-off.

CNH has been extensively studied as a catalyst support in hydrogen fuel cells, solar cells, capacitors, and so on, and it is thought to be highly suitable for use as an electrode material in BFCs, since it provides high conductivity and specific surface area, while simultaneously exhibiting high dispersibility. In a previous study of BFCs based on CNH, the CNH was modified by carbon-fiber microelectrodes

(CFMEs), and the performance was evaluated [4]. There has been no report on the creation of a film-like BFC using CNH to date. In this study, a film-like BFC using CNH as an electrode material was investigated to determine whether flexible glucose BFCs with a high power density could be realized.

2. Experimental

2.1. Power Generation Mechanism

The BFC investigated in this study consists of a porous carbon anode modified by glucose oxidase (GOD) and ferrocene, and a cathode modified by bilirubin oxidase (BOD). Figure 1 shows the power generation principle. Electrons removed from glucose by GOD are transferred to the anode through the ferrocene mediator, and protons are released from glucose at the same time. The electrons flow through the resistance and are delivered to the cathode, where oxygen is reduced by these electrons and protons. The power thus generated can be diverted through an external circuit connected between the anode and the cathode.

2.2. Design of glucose fuel cells with CNH electrode

Figure 2 shows the layout of the cathode and anode. These electrodes were fabricated by photolithography on a polyimide (PI) sheet, coated with aluminum (Al) and then modified by carbon black and respective enzymes. Both the anode and cathode had a width of 3 mm and a length of 10 mm. The gap between the anode and cathode was 1 mm. The anode and cathode were deposited on Al with a width of 3 mm and length of 30 mm. The areas of the anode and the cathode were 3 × 10 mm^2.

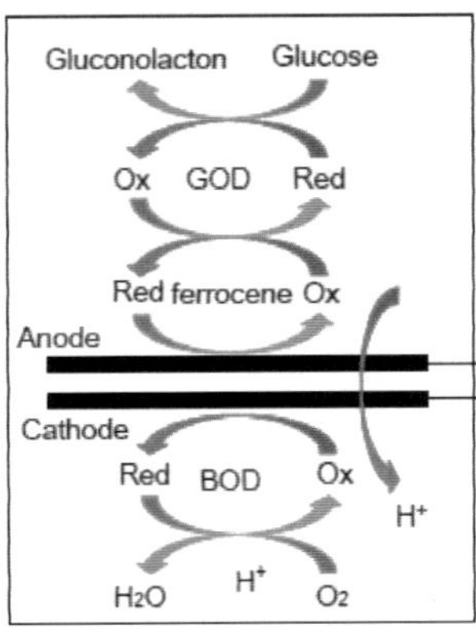

Figure 1. Possible mechanism of power generation.

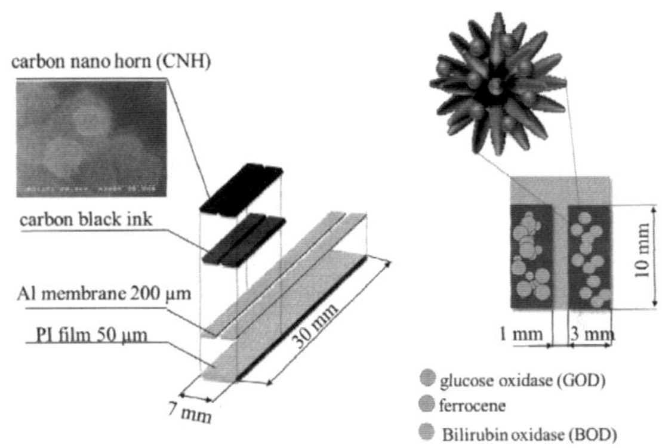

Figure 2. A porous carbon-coated aluminum (Al) electrode with a width of 3.0 mm and a length of 10 mm, prepared on a flexible polyimide substrate.

2.3. Fabrication processes of carbon-modified electrodes

Figure 3 (a) First, a layer of Al was deposited on the surface of a PI sheet by thermal vacuum deposition. (b) Then, the photoresist (S1818) was spin coated on the Al film. (c) The electrode pattern was optically defined by photolithography. (d) The Al film was partially removed by wet chemical etching using a solution mixture (dihydrogen phosphate (H_2PO_3) : nitric acid (HNO_3) : acetic acid (CH_3COOH) : H_2O = 2 : 1 : 1 : 1). (e) The photoresist was removed with acetone. After this, carbon-modified electrodes were formed on the surfaces of Al film by screen printing with carbon paste (MRX-713J-A, TAMURA Corporation), which was then cured at 150 °C for 20 min. (f) Finally, the surface of the carbon paste film was coated by screen printing with an ink mixture of carbon nano-horn (CNH, SIGMA-ALDRICH Inc.), and polyvinylidene fluoride (PVDF, KUREHA Inc.), which was subsequently cured at 100 °C for 20 min.

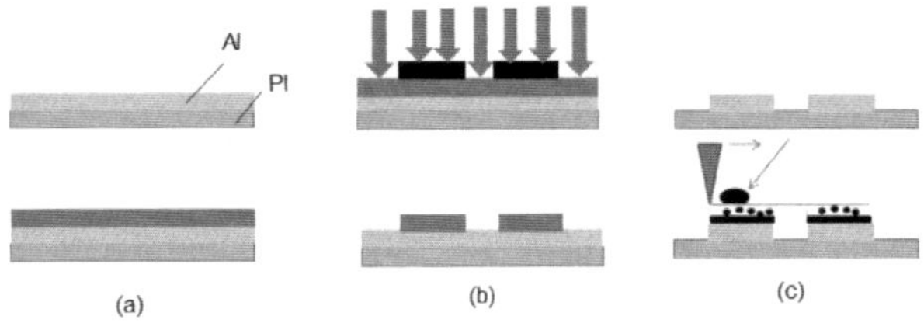

Figure 3. Fabrication process of carbon-modified electrodes: (a) Al deposition on PI, photoresist spin-coating, (b) exposure, Al wet etching, (c) photoresist removal, and screen printing with carbon paste.

2.4. Fabrication of the enzyme-modified electrodes

A small amount of GOD (0.05 mg) was mixed with 30 μl of phosphate buffer solution (PBS, 50 mM, pH 7.0). Then, 0.05 mg of the ferrocene mediator was mixed with 30 μl of PBS. Subsequently, 10 μl of the resulting solutions was dropped onto the porous carbon anode and left to dry for 90 min. Similarly, 0.05 mg of BOD was mixed with 30 μl of PBS. Then, 10 μl of the resulting solution was dropped onto the porous carbon cathode and left to dry for 90 min. As a reference, carbon electrodes were also fabricated using similar conditions to those previously reported.

2.5. Measurements

Electrochemical measurements were performed in an air-saturated, 0-mM, 50-mM, 100-mM and 200-mM PBS of glucose, at room temperature. Measurements were also performed over a range of temperatures, from 20 to 70 °C, in 100-mM PBSs of glucose. The power generated was evaluated by measuring the cell voltage while varying the external load resistance between 0 and 3 MΩ. W was calculated using the equation $W = V^2/R$, where R is the load resistance, and V is the generated voltage measured between the terminals of the BFC. The power density was calculated by dividing W by the area of the anode or cathode, which was 30 mm^2.

3. Results and discussion

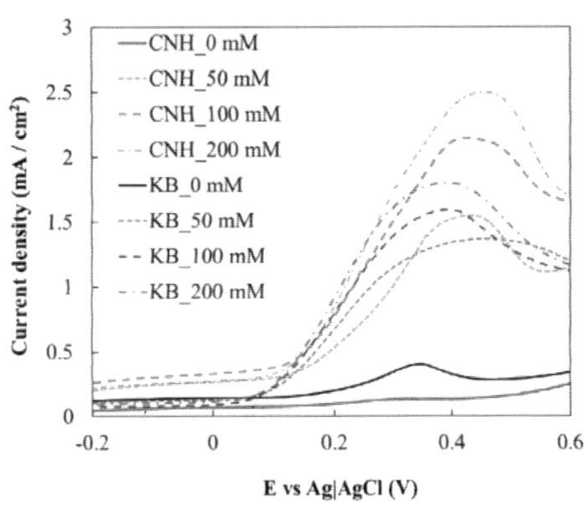

Figure 4. Cyclic voltammograms of the GOD/ferrocene complex modified anode in 0, 50, 100 and 200 mM glucose solutions for CNH or KB electrodes.

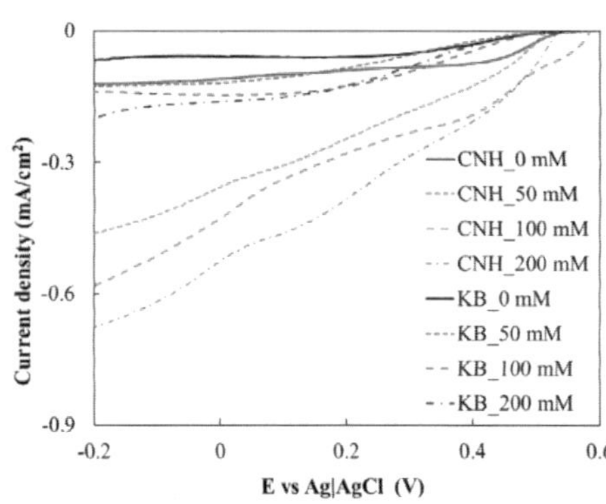

Figure 5. Cyclic voltammograms of the BOD-modified cathode in 0, 50, 100 and 200 mM glucose solutions at CNH or KB electrodes.

Figures 4 and 5 show the measured CV curves of the CNH and KB anodes and cathodes immersed in PBSs with glucose concentrations of 0, 50, 100 and 200 mM. The glucose oxidation current started to rise above 0.1 V for the CNH and KB anodes in 50, 100, and 200 mM glucose and showed the maximum values at 0.4 V. Depending on the concentration, the maximum current density increased and the maximum current density of the CNH anode was 2.50 μA/cm^2 with 200 mM glucose solutions.

Moreover, the maximum current density of the CNH cathode was 0.89 μA/cm^2 with 200 mM glucose solutions.

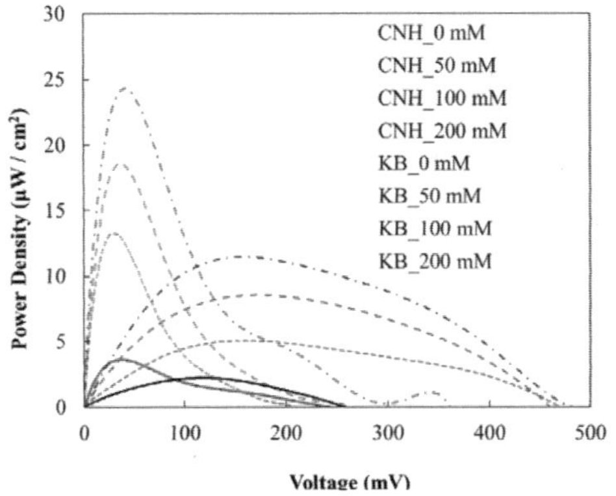

Figure 6. Relationships between the power density and output voltage of the BFCs with CNH and KB electrodes in 0, 50, 100 and 200 mM glucose solutions.

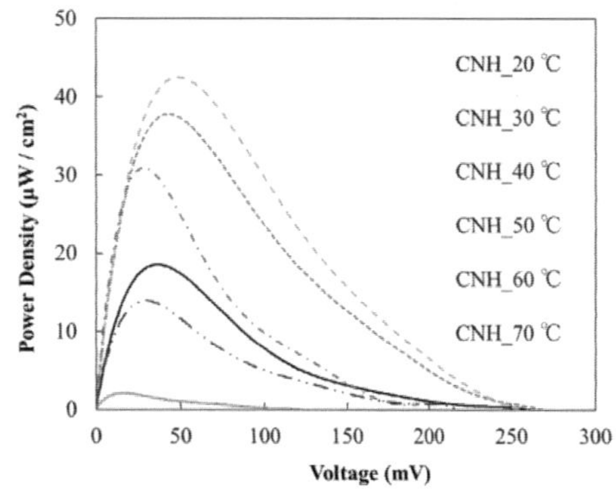

Figure 7. Relationships between the power density and output voltage of the BFCs with CNH and KB electrodes in 100 mM glucose solutions over the temperature range 20 – 70 °C.

Figure 6 shows a comparison of the relationships between the power density and voltage for the BFCs with CNH and KB electrodes immersed in fuels with different glucose concentrations (0, 50, 100 and 200 mM). The maximum power for the BFC with CNH electrodes was 18.0 μW/cm^2 at 37 mV for the 100-mM glucose solution, while that for the BFC with KB electrodes was 8.9 μW/cm^2 at 233 mV for the 100-mM glucose solution. Therefore, the power generation amount of the CNH electrode is twice or more than that of the KB electrode. This is probably due to the fact that the physical adsorption of the enzyme to CNH is superior than that to KB due to the three-dimensional structure.
Figure 7 shows the relationships between the power density and output voltage of the BFCs with CNH and KB electrodes in 100-mM glucose solutions over the temperature range of 20–70 °C. The maximum power for the BFC with the CNH electrodes was 39.9 μW/cm^2 at 55 mV in the 100-mM glucose solution at 40 °C. An output 2.2 times better than that obtained at room temperature (20 °C) could be confirmed. Below 60 °C, the output was lower than that at room temperature, and at 70 °C, almost no power generation could be confirmed. Since the enzyme is a protein, it is thought that thermal degeneration began at or above 60 °C, resulting in enzyme inactivation.

4. Conclusion

In this study, a film-like biofuel cell (BFC) using carbon nano-horn (CNH) as an electrode material was fabricated using MEMS technology. This BFC was able to produce a power density of 18.0 μW/cm^2 in a phosphate buffer, wherein 100 mM glucose was dissolved at room temperature. Furthermore, by examining the temperature-dependence of the BFC's power density, it was possible to attain a maximum power density of 39.9 μW/cm^2 in a phosphate buffer solution, wherein 100 mM glucose was dissolved at 40 °C.

References

[1] H. Becker and U. Heim, 2000 *Sensors Actuators: A Phys.*, **83** 130.
[2] E. R. Choban, J. S. Spendelow, L. Gancs, A. Wieckowski and P. J. Kenis, 2005 *Electrochim. Acta*, **50** 5390.
[3] Y. Fukushi,S. Koide,R.Ikoma, W. Akatsuka, S. Tsujimuraand Y. Nishioka, 2013 J. *Photopolym. Sci. Technol.*, **26** 303.
[4] Dan Wen, Xiaolong Xu and Shaojun Dong, 2011 *Energy Environ. Sci.*, **4**, 1358-1363

IOP Conf. Series: Journal of Physics: Conf. Series **1052** (2018) 012064 doi:10.1088/1742-6596/1052/1/012064

Alcohol dehydrogenase as bioanode for methanol and ethanol oxidation in a microfluidic fuel cell

J. Galindo-de-la-Rosa[1], D. Vite- González[2], J.A. Díaz-Real[2], N. Vázquez-Maya[2], A. Álvarez[2], L.G. Arriaga[1] and J. Ledesma-García[2*]

[1] Centro de Investigación y Desarrollo Tecnológico en Electroquímica, Pedro Escobedo, Qro., C.P. 76703, México.
[2] Facultad de Ingeniería, División de Investigación y Posgrado, Universidad Autónoma de Querétaro, Centro Universitario Cerro de las Campanas, Querétaro, Qro., C.P. 76010, México.
E-mail: janet.ledesma@uaq.mx

Abstract. In this work, an enzymatic bioanode was developed using functionalized carbon nanofibers for the immobilization of alcohol dehydrogenase enzyme, using Nafion, tetrabutylammonium bromide and glutaraldehyde onto Toray carbon paper. Two alcoholic fuels, methanol (MeOH) and ethanol (EtOH) were used to demonstrate the oxidation versatility of the electrode created as well as being able to convert energy from these fuels using an air-breathing microfluidic device. Commercial Pt/C (30% E-TEK) onto carbon nanofoam (Marketech Inc) was using as inorganic cathode. Carbon nanofibers were previously treated with nitric acid to produce oxygen-containing functional groups in order to facilitate the enzyme immobilization. A phosphates buffer solution pH 8.86 was used as electrolyte. The use of ethanol as fuel shows a better performance obtaining 8.643mAcm^{-2} and 0.847 V as current density and open circuit voltage respectively, compared to that use methanol as fuel (2.655 mAcm^{-2} and 0.567V).

1. Introduction

Alcohols are commonly used as fuels in microfluidic fuel cells due to easy transport, store and can be obtained from various sources, being the most used the methanol (MeOH) and ethanol (EtOH) [1]. Biological catalysts such as enzymes have been used for the alcohol oxidation reactions, since they have advantages over conventional catalysts, such as high specificity, selectivity and low cost [2]. Alcohol dehydrogenase enzyme carries out the alcohols oxidation to their corresponding aldehyde; ethanol is oxidized into acetaldehyde and methanol into formaldehyde, although with less efficiency. The cofactor of Alcohol dehydrogenase is NAD^+; both alcohol and NAD^+ may bind to the active center of the ADH, which contains zinc ions and during enzymatic alcohol oxidation, NAD+ is reduced to NADH [3]. Alcohol dehydrogenase enzyme structure is a homo tetramer of polypeptides, with 4 subunits each with its catalytic and co-enzymatic domain. To carry out the reaction requires addition of ethanol and ADH, the presence of NAD^+, and zinc as cofactor. Zinc fulfills structural and catalytic functions in the enzyme [4]. In the first instance, the zinc atom is bound to the oxygen of the hydroxyl group of alcohol, stabilizing the substrate. A serine unit presents an oxygen atom to the hydrogen of this hydroxyl group, and removes it by transfer of protons. In turn, NAD^+ removes hydrogen from the carbon attached to the hydroxyl group of ethanol, reducing to NADH [5].

Several supports have been reported, including different nanostructured materials have shown to increase the stability of immobilized enzymes and improve electrical conductivity for applications as biocatalysts in microfluidic fuel cells [6-8]. The creation of electrodes modified with this enzyme, which depends greatly on the enzymatic immobilization used for the bioelectrocatalytic process, where an efficient enzymatic confinement technique may serve to increase power density and stability of the microfluidic device when ethanol or methanol is used as fuel, being able in the future to have applications in the field of medicine and alcoholic beverage industry.

2. Experimental

2.1 Bioelectrode construction.

Carbon nanofibers were treated with HNO_3/H_2SO_4 (1:3) for 3 h at 40 ◦C in a water-bath sonicator, later was centrifuged and the precipitate was washed with deionized water were made to pH 7.0 and dried at 90°C for 8hrs to obtain carboxylic group functionalized. A catalytic ink was prepared using 1mg alcohol dehydrogenase (EC 1.1.1.1 initial activity ≥300 units/mg protein from Saccharomyces cerevisiae), 4mg of carbon nanofibers functionalized, 5 μL of Nafion solution (Sigma Aldrich), 500μL of 0.1M of buffer phosphate pH= 8.86, and 5 mg of tetrabutylammonium bromide (Sigma Aldrich). The catalytic ink was deposited by dipping on Toray carbon paper (EC-TP1-060T) whose dimensions were of 2.5x30 mm. The schematic representation of the bioandode developed as well as the reactions of alcohol dehydrogenase enzyme for the oxidation of ethanol and methanol is presented in figure 1.

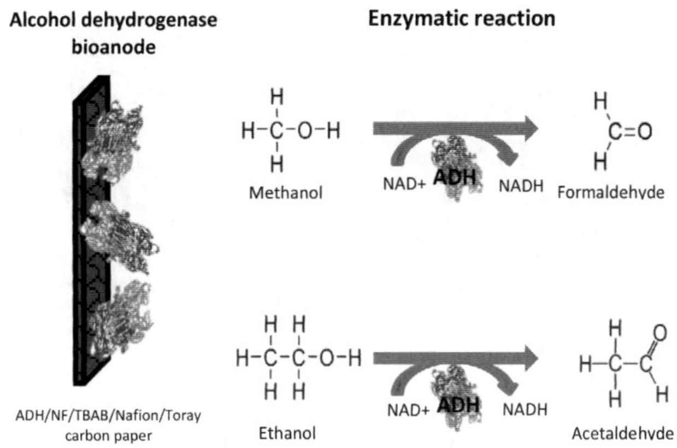

Figure 1. Scheme of bioelectrode arrangement and enzymatic reactions of the enzyme alcohol dehydrogenase.

2.2 Microfluidic fuel cell evaluation.

Schematic representations of the single microfluidic fuel cells used are illustrated in figure 2, whose design was previously reported [9]. The abiotic cathode Pt-based for the hybrid MMFC was prepared using commercial Pt/C (30% E-TEK) supported on carbon nanofoam (2×0.3 cm, Marketech®). Briefly, 120 μL of isopropyl alcohol and 14 μL of Nafion were added per mg of catalyst and mixed for 30 min. The catalytic ink was deposited on the carbon nanofoam surface using the spray technique.

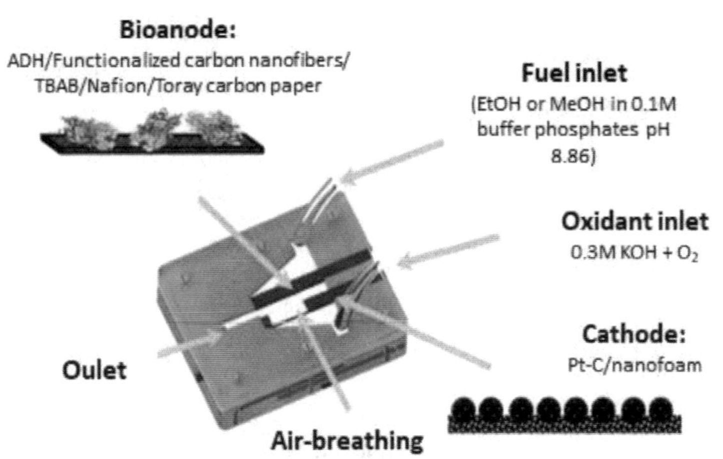

Figure 2. Scheme of air-breathing microfluidic fuel cell used for bioanode evaluation.

The microfluidic fuel cells were evaluated by injecting different fuel for two evaluation, 0.1M of methanol and 0.1M of ethanol in 0.1M buffer solution N_2-saturated (4 U. P. D. Praxair) and oxygen from the air as the oxidant within two aqueous streams consisting in 0.1M phosphate buffer (pH 7) and 0.3 M KOH injected on the anodic and the cathodic side, respectively, with a flow rate of 50 μL min^{-1} for both streams.

3. Results

Scanning electron microscopy technique was used to analyse the morphology of different electrode arrangements, as shown in figure 3, where a) is the naked carbon paper Toray and b) is when deposited on this paper the functionalized carbon nanofibres covering large part of the surface and where placing the catalytic ink q contains the enzyme alcohol dehydrogenase (Figure 3c).

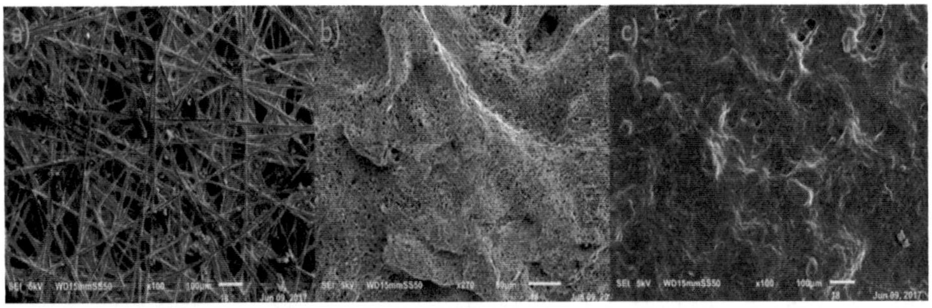

Figure 3. Scanning electron microscopy images of (a) Toray carbon electrode, (b) funcionalizated carbon nanofibers onto Toray carbon paper and (c) ADH bioanode on surface b.

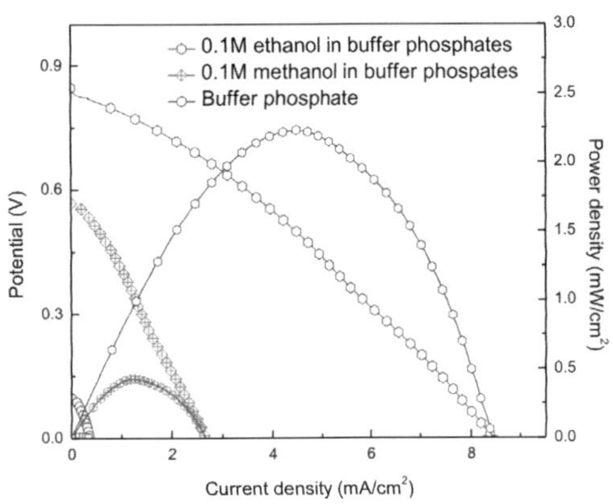

Figure 4. Polarisation and power density curves obtained from microfluidic fuel cell evaluated with ethanol or methanol as fuel in 0.1M phosphates buffer pH 8.86 and 0.3M KOH as oxidant. 10 mV s^{-1} scan rate.

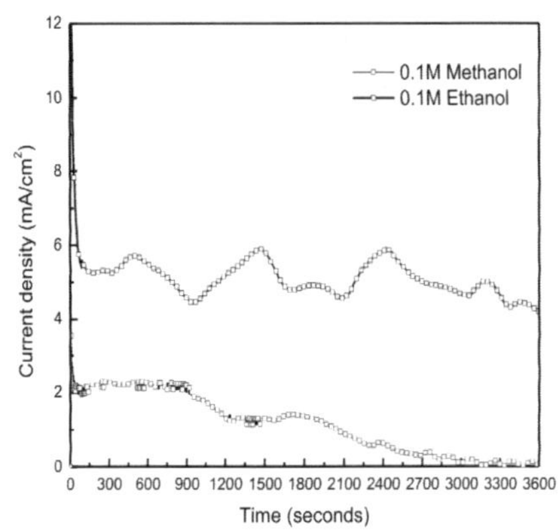

Figure 5. Chronoamperometry obtained from microfluidic fuel cell tested in the presence of 0.1M methanol or ethanol as fuel in 0.1M phosphates buffer pH 8.86 and 0.3M KOH as oxidant. 10 mV s^{-1} scan rate.

Table 1. Performance of microfluidic fuel cell using methanol and ethanol as fuel.

Fuel	Current density (mA/cm^2)	Potential(V)
Methanol	2.655	0.567
Ethanol	8.643	0.847

The use of ethanol as fuel shows a better performance obtaining 8.643 $mAcm^{-2}$ and 0.847 V as current density and open circuit voltage respectively, compared to that use methanol as fuel (2.655 $mAcm^{-2}$ and 0.567V) according to Figure 3 and table 1, it can be attributed that the enzyme has a greater affinity to the oxidation of ethanol than methanol, because methanol is a fuel difficult to handle in fuel cell technology, due in many aspects to its toxicity. It can also be observed that when there is no presence of fuel of an alcoholic nature, the performance of the cell is very low, indicating the oxidation selectivity of alcohol dehydrogenase enzyme.

Figure 5 shows the stability test by chronoamperometry technique of the air-breathing microfluidic fuel cell using ethanol or methanol as fuel. In the presence of methanol, the device shows a low stability and less current density is produced, decreasing the yield. In contrast, when ethanol is used as fuel, good stability is maintained during the tests. The results could be attributed to enzyme used as anode, whose affinity is greater when ethanol is used as fuel.

4. Conclusions

With the construction of an enzymatic bioanode based on the enzyme alcohol dehydrogenase and the use of functionalized carbon nanofibers, a microfluidic device was evaluated using ethanol and methanol as fuel, according to the developed analyses it was possible to determine that the electrode can oxidize these fuels, taking a greater affinity over ethanol, where a good power density could be obtained by using it. Due to the chemical nature of the methanol involved the immobilized enzyme activity and the stability of the microfluidic fuel cell, being a difficult fuel to handle.

Acknowledgements

Authors acknowledges to the Mexican Council for Science and Technology (CONACYT) for financial support through project Fronteras de la Ciencia Grant no. 611, Laboratorios Nacionales “Laboratorio Nacional de Micro y Nano Fluídica” Grant no. LN 271649, and Sistema de almacenamiento de energía basado en la purificación y compresión no convencional de hidrógeno (Electroquímica)" Grant no. 246079 for the "Fondo Sectorial CONACYT-SENER Sustentabilidad Energética".

Reference

[1] Galindo-de-la-Rosa J, Arjona N, Arriaga LG, Ledesma-García J and Guerra-Balcázar M, 2015, *J. Phys. Conf. Ser.* **660** 012131.
[2] Minteer SD, Liaw BY and Cooney MJ, 2007, *Biotecnology* **18** 228-234.
[3] Huntley A and Vallee BL, 1996, *Biochemistry* **5** 2026-2034.
[4] Ganzhorn AJ and Plapp BV, 1988, *J. Biol. Chem.*263 5446-5464.
[5] Plapp BV, Savarimunthu BR, Ferraro DJ, Rubach JK, Brown EN, Ramaswamy S, 2017, *Biochemistry* **56** 3632-3646.
[6] Arrocha AA, Cano-Castillo U, Aguila SA, Vazquez-Duhalt R, 2014, *Biosens. Bioelectron.* **61** 569-574.
[7] Aquino Neto S, Forti JC, Zucolotto V, Cieancaglini P and De Andrade AR, 2011, *Process Biochemistry* **46** 2347-2352.
[8] Galindo-de-la-Rosa J, Arjona N, Moreno-Zuria A, Ortiz-Ortega E, Guerra-Balcázar M, Ledesma-García J, Arriaga LG, 2017, *Biosens. Bioelectron.* **92** 117-124.
[9] Ortiz-Ortega E, Goulet M A, Lee J W, Guerra-Balcázar M, Arjona N, Kjeang E, Ledesma-García J and Arriaga L G 2014 Lab Chip **14** 4596.

Micro Methane-Oxygen Counterflow Diffusion Flames: Effects of Gravity on Flame Structures

Satoshi Kadowaki[1], Yusuke Hashimoto[1], Toshiyuki Katsumi[1], Thwe Thwe Aung[1], Tsuneyoshi Matsuoka[2], and Yuji Nakamura[2]

[1]Nagaoka University of Technology, Nagaoka 940-2188, Japan
[2]Toyohashi University of Technology, Toyohashi 441-8580, Japan

Corresponding author's e-mail address: kadowaki@mech.nagaokaut.ac.jp

Abstract We manipulated oxy-fuel combustion of methane in small scales to study the characteristics of micro counterflow diffusion flames under the gravity field. Methane-oxygen diffusion flames in counterflow burners were observed, where the burner distance was less than or equal to 1.0 mm and the apparent equivalence ratio was unity. The flame thickness and flame diameter were obtained as functions of the burner distance and gas flow rate. The minimum burner distance for the observation of flames became smaller as the gas flow rate increased. We obtained smaller values of the minimum burner distance when methane flowed out from the upper burner. The flame thickness and flame diameter decreased as the burner distance became smaller. In addition, the flame thickness and flame diameter had slightly larger values in the upper-methane case.

1. Introduction

Recently, we recognize that methane receives a lot of attention as the substitute of oil and coal, because of low emissions of carbon dioxide. Converting methane into power source of small-scale practical applications without nitrogenoxides emissions, we shall obtain the knowledge on micro methane-oxygen flames. Oxy-fuel combustion has been investigated in detail to clarify the fundamental phenomena [1-4]. In addition, small-scale combustion has been treated to elucidate the characteristics of micro flames [5-8]. Thus, we can study the characteristics of micro methane-oxygen flames, based on the previous information on oxy-fuel and small-scale combustion.

In the experiments, we manipulated oxy-fuel combustion of methane in small scales to study the characteristics of micro counterflow diffusion flames under the gravity field. Using the counterflow burner, we performed the experiments to obtain the flame thickness and flame diameter as functions of the distance between burners and the flow rate of gases. Moreover, we scrutinized the effects of gravity on flame structures.

2. Experimental procedures

Figure 1 shows the experimental apparatus. Stainless pipes (SUS316) with the inner diameter $\phi = 0.5$ mm were used, where the outer diameter was 1.59 mm. Methane was employed as fuel, and oxygen was adopted as oxidizing agent. The gas flow rates were controlled by digital mass flow controllers (Azbil, MQV9020, MQV9200). The flow rate of methane was set to Q_F = 4, 5, 6 and 7 mL/min, and that of oxygen was set to Q_O = 8, 10, 12 and 14 mL/min, where the apparent equivalence ratio was unity. Methane flowed out from the lower (upper) burner, and oxygen flowed out from the upper (lower) burner. Diffusion flames were observed between upper and lower burners whose distance d was set to be less than or equal to 1.0 mm.

Figure 2 shows the images of methane-oxygen counterflow diffusion flames at d = 1.0 mm. Methane-oxygen diffusion flames were observed by a digital microscope (Thanko, Dino-Lite Premier 500M). These types of micro diffusion flames appeared when oxygen was adopted as oxidizing agent. No micro flames appeared when air was adopted instead of oxygen.

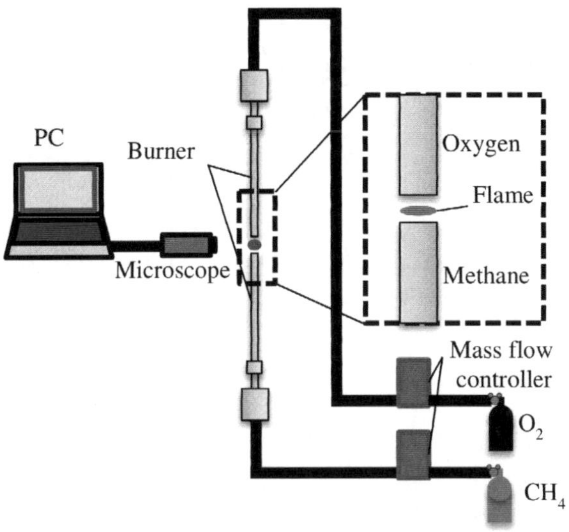

Fig. 1. Experimental apparatus in the lower-methane case

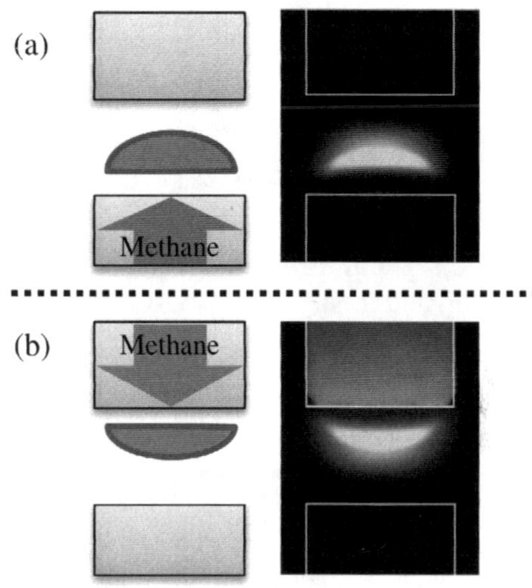

Fig. 2. Images of methane-oxygen counterflow diffusion flames at ϕ = 0.5 mm, d = 1.0 mm and Q_F = 7 mL/min; (a) Lower-methane case, (b) Upper-methane case

3. Experimental results

Figure 3 shows the images of methane-oxygen counterflow diffusion flames at Q_F = 4 & 7 mL/min and d = 0.1, 0.3, 0.5, 0.7 & 1.0 mL/min when methane flows out from the lower burner and oxygen flows out from the upper burner (lower-methane case). We observe micro diffusion flames in counterflow burners. It is found that the flame thickness and flame diameter decrease as the burner distance becomes smaller. In addition, the flame thickness and flame diameter at Q_F = 7 mL/min are larger than those at Q_F = 4 mL/min.

The images of methane-oxygen counterflow diffusion flames in the upper-methane case at Q_F = 4 & 7 mL/min and d = 0.1, 0.3, 0.5, 0.7 & 1.0 mL/min are shown in Fig. 4. We find that the flame thickness and flame diameter have slightly large values compared with the lower-methane case.

Figure 5 shows the minimum burner distance d_{min}, depending on the flow rate of methane. As the flow rate increases, the minimum burner distance becomes smaller. In addition, we obtain smaller values of the minimum burner distance when methane flows out from the upper burner.

Figure 6 shows the flame thickness δ_f and flame diameter D_f, depending on the burner distance. The flame thickness and flame diameter decrease as the burner distance becomes smaller. We obtain larger flame thickness and flame diameter when methane flows out from the upper burner. The obtained results denote that the gravity has an influence on the structure of micro flames.

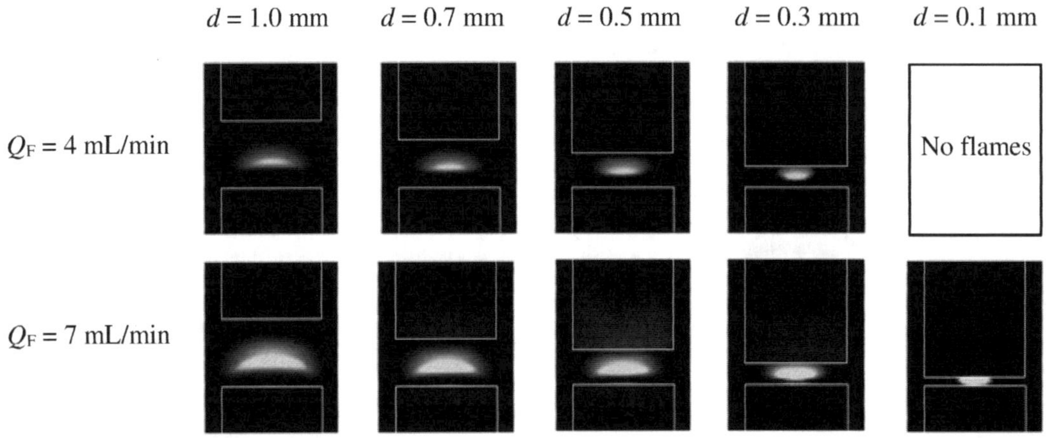

Fig. 3. Images of methane-oxygen counterflow diffusion flames in the lower-methane case at $Q_F = 4$ & 7 mL/min and $d = 0.1$, 0.3, 0.5, 0.7 & 1.0 mL/min

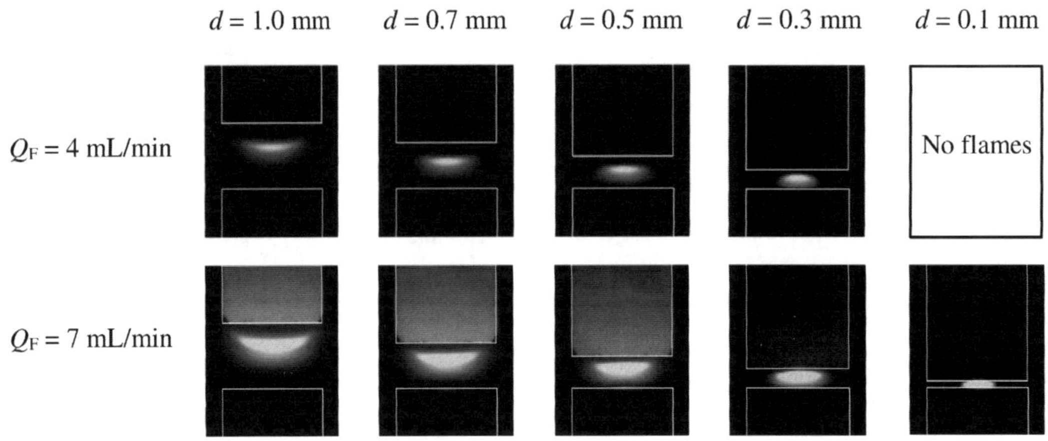

Fig. 4. Images of methane-oxygen counterflow diffusion flames in the upper-methane case at $Q_F = 4$ & 7 mL/min and $d = 0.1$, 0.3, 0.5, 0.7 & 1.0 mL/min

4. Conclusions

We have manipulated oxy-fuel combustion of methane in small scales to study the characteristics of micro counterflow diffusion flames under the gravity field. The obtained results are as follows:

We observe methane-oxygen diffusion flames in counterflow burners where the burner distance is less than or equal to 1.0 mm and the apparent equivalence ratio is unity. The minimum burner distance for the observation of flames becomes smaller as the gas flow rate increases. We find smaller values of the minimum burner distance when methane flows out from the upper burner. The flame thickness and flame diameter decrease as the burner distance becomes smaller. In addition, the flame thickness and flame diameter have slightly larger values in the upper-methane case.

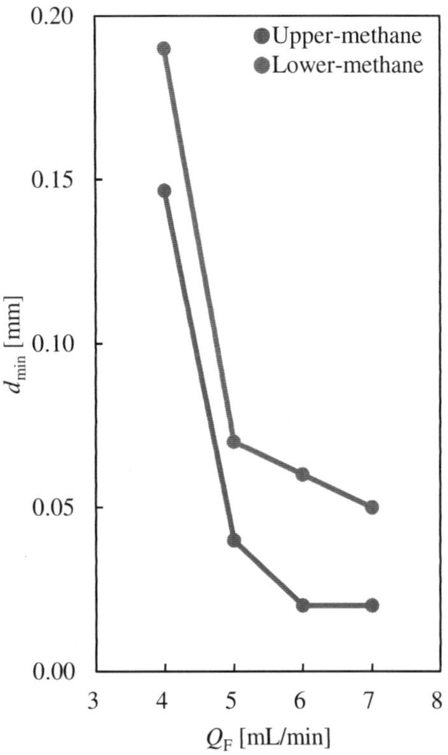

Fig. 5. Minimum burner distance, depending on the flow rate of methane

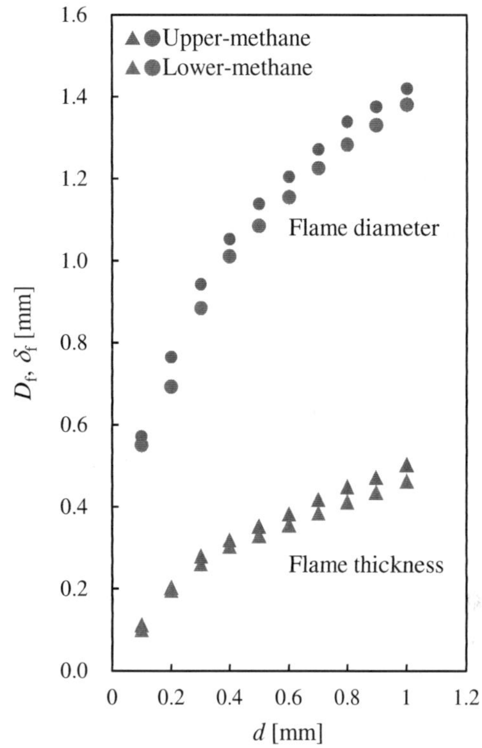

Fig. 6. Flame thickness and flame diameter in methane-oxygen counterflow flames at $\phi = 0.5$ mm and $Q_F = 7$ mL/min

In the near future, we will elucidate the characteristics of micro counterflow diffusion flames in detail, considering the heat transfer from flames to burners. We will clarify the mechanism of the dependence of flame size on the burner distance and gas flow rate to propose the new model on micro counterflow diffusion flames.

References

[1] K. Maruta, K. Abe, S. Hasegawa, S., Maruyama and J. Sato, *Proc. Combust. Inst.*, 31 (2007), pp. 1223-1230.
[2] H. Watanabe, J. Yamamoto and K. Okazaki, *Combust. Flame*, 158 (2011), pp. 1255-1263.
[3] H. Kobayashi, Y. Otawara, J. Wang, F. Matsuno, Y. Ogami, M. Okuyama, T. Kudo and S. Kadowaki, *Proc. Combust. Inst.*, 34 (2013), pp. 1437-1445.
[4] D. Shimokuri, S. Fukuba and S. Ishizuka, *Proc. Combust. Inst.*, 35 (2015), pp. 3573-3580.
[5] Y. Ju and K. Maruta, *Prog. Energy Combust. Sci.*, 37 (2011), pp. 669-715.
[6] Y. Saiki and Y. Suzuki, *Proc. Combust. Inst.*, 34 (2013), pp. 3395-3402.
[7] A. Hossain and Y. Nakamura, *Proc. Combust. Inst.*, 35 (2015), pp. 3413-3420.
[8] S. Kadowaki, R. Numata, Y. Hashimoto and T. Katsumi, *Mech. Engr. Lett.*, 2 (2016), 16-00170.

IOP Conf. Series: Journal of Physics: Conf. Series **1052** (2018) 012066 doi:10.1088/1742-6596/1052/1/012066

Preparation of conductive carbon paper based on carbon nanofibers and polypyrrole for biofuel cell application

Ricardo A Escalona-Villalpando[1], L G Arriaga[1], Shelley D Minteer[2] and J Ledesma-García[3*].

[1]Centro de Investigación y Desarrollo Tecnológico en Electroquímica, 76703, Santiago de Querétaro, México.
[2]Department of Chemistry, University of Utah, 315 South 1400 East, Salt Lake City, UT, 84112, United States.
[3]División de Investigación y Posgrado, Facultad de Ingeniería, Universidad Autónoma de Querétaro, 76010, Santiago de Querétaro, México.

Email: janet.ledesma@uaq.mx

Abstract. The development of novel conductive carbon paper has great potential application in biofuel cells (BFC) because it can increase the electrical conductivity. In this work, carbon nanofiber (CNF) was modified with polypyrrole (PPy) to obtain a conductive paper (CNF-PPy) for its application in BFC using lactate as fuel. The anode was prepared from lactate oxidase (LOx) with the $FcMe_2$-LPEI redox polymer, whereas the cathode was prepared from Laccase (Lc) with MWCNT-modified with anthracene, both enzymes were deposited on CNF-PPy. $FcMe_2$-LPEI/LOx show a pair of redox peaks at 0.27 and 0.16 V (vs. Ag/AgCl) by cyclic voltammetry (CV) in PBS 7.4 and presence of 20 mM lactate the current density increase to 68.12 μA cm^{-2}. The CV with AnMWCNT/Lc in PBS pH 5.7 presented an oxygen reduction potential of 0.61 V shown a current density of 200.6 μA cm^{-2}. The evaluation of BFC was carried out in a single compartment in pH 5.7 (0.1 M) obtaining a OCP of 0.59 V a current and power density of 432.5 μA cm^{-2} and 71.2 μW cm^{-2} respectively using 40 mM lactate. This result showed that CNF-PPy were essential in obtain of a high performance in the BFC that use lactate as fuel.

1. Introduction

Carbon conductive paper (CCP) have great importance for their application in various fields including the detection of elements, electronic components, biomedical systems, among others. CCP has an extensive utility for its versatility as mechanical flexibility, biodegradability, affordability, renewability, mainly [1]; its modification with conductive polymers such as polypyrrole (PPy) gives it a higher electrical conductivity, has a low toxicity, biocompatibility, chemically reactive and economically accessible [2]. CCP has a greater potential for application in the area of alternative sources of energy such as capacitors, fuel and biofuel cells (BFC), the latter being attractive because they can operate at room temperature, possess a high potential in biomedical application systems, such as micro pumps, pacemarkers, and so forth. Generally, one of its main challenges of the BFC is to increase its current and power density, so different strategies have been reported as using electrochemical mediators, metallic materials with a high electrical conductivity, carbon nanoparticles and a higher attention to use CCP that provide greater catalytic stability and electrical conductivity [3]. The importance of lactate as fuel is due to its high energy density and water solubility compared to other biofuels, such as glucose and alcohols. The secondary product of lactate oxidation is pyruvate generating a theoretical energy density of 73 Wh/L and a complete oxidation of pyruvate/air the theoretical maximum current density could be 367 Wh/L [4]. In this study, carbon nanofibers with polypyrrole (CNF-PPy) were modified by the *in situ* polymerization method to obtain a CCP, which was used for the immobilization of the lactate oxidase enzyme with a conducting polymer in the bioanode and laccase with carbon nanotubes modified with anthracene in the cathode and it was compared with a paper of commercial carbon (Toray carbon) in a BFC of lactate of a single compartment.

2. Experimental

2.1 Preparation of carbon nanofiber modified with polypyrrole in situ (CNF-PPy)

The preparation of CNF-PPy consisted of sonicating the 1 mol CNFs in an acid solution of EtOH bath with H_2SO_4 (0.5 M) at ratio 1:3 v/v during for 20 min, then 1 mol of pyrrole in H_2O_2 30 % was added dropwise in a range of 30 min with the help of a pump keeping the temperature below 25 °C and it was stirred for 45 min dried at 100 ° C after the reaction was stopped by adding 30 mL of deionized water (DI), the CNF-PPy were washed three times with DI and centrifuged at 3500 rpm and filtered using waltman paper, after it were dried at 100 ° C degrees for 24 h and compressed to 3700 lb (Scheme 1).

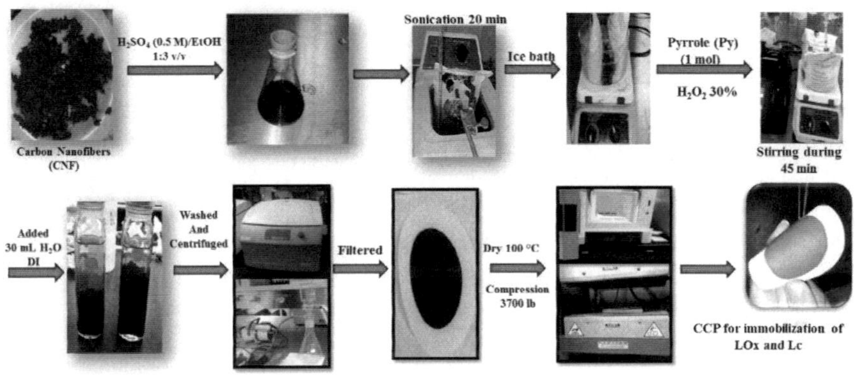

Scheme1. Preparation of CNF-PPy

2.2 Electrodes preparation

The immobilization of LOx was on CNF-PPy and carbon Toray (Ty) of area 0.25 cm^{-2}. The redox polymer was polyethylenimine-modified dimethylferrocene ($FcMe_2$-LPEI) which was prepared as previously reported [5]. The solution was prepared from 10 mg mL^{-1}, 100 U mL^{-1} LOx in PBS and 2-6% v/v EGDGE in a ratio of 56/24/3. Finally, 25 μL were deposited on CNF-PPy and Ty and allowed to dry for 24 h. For the immobilization of the biocathode, a solution of Lc 1.5 mg in 75 μL of PBS pH 6.5 with 7.5 mg of anthracene-modified carbon nanotubes AnMWCNT [6] was prepared by vortexing 4 times per 1 min, finally 25 μL of TBAB-nafion solution was added and mixed for 1 min. Finally, 9 mL of the solution was deposited on CNF-PPy and electrode using a brush and allowed to dry for 2 h.

2.3 Electrochemical methods

Electrochemical experiments were performed using a Biologic VSP Potenciostat/ Galvanostat. The $FcMe_2$-LPEI/LOx on Ty and CNF-PPy electrodes were evaluated in a PBS pH 7.4 (0.1 M) solution using an Ag/AgCl electrode and a platinum mesh as reference and counter electrode respectively. Cyclic voltammetry (CV) were performed at a scan rate at 2 mV s^{-1} and a potential range of 0.5 to -0.1 V, while the amperometry experiments were carried out applying a voltage of 0.3 V. While the electrodes with AnMWCNT/Lc were carried out by CV to a potential window of 0.8 to 0 V and a scan rate of 1 mV s^{-1}

in PBS pH 5.6 (0.1 M). The polarization curves for the BFC in a single compartment were linear polarization at 2 mV s^{-1}, using $FcMe_2$-LPEI/LOx as anode and AnMWCNT/Lc as cathode supported on Ty and CNF-PPy. The supported electrolyte consisted of a solution of PBS (0.1 M) at a pH of 5.6, using different concentrations of lactate 2, 10, 40 mM respectively.

3. Results and discussion

The characterization of $FcMe_2$-LPEI/LOx on CNF-Py by VC shows a pair of redox peaks at 0.27 V and 0.16 V (vs. Ag / AgCl) with a formal potential of 0.215 V belonging to the redox polymer $FcMe_2$ with a current density of 68.12 μA cm^{-2} measured at 0.26 V in the presence of 20 mM lactate. While $FcMe_2$-LPEI/LOx on Ty the pair of redox peaks were found at 0.27 V and 0.21 V (Ep' 0.24 V) respectively with a current density of 30.62 μA cm^{-2} (Figure 1A). On the other hand, the amperometric experiments were performed at 0.3 V and found that with the CNF-Py the current increase in the first addition of lactate was 6.03 times higher in comparison to Ty, to both electrodes the linearity range was 1-5 mM but obtaining a higher CNF-Py sensitivity of 29.8 μA mM^{-1} cm^{-2} and 5.97 8 μA mM^{-1} cm^{-2} with Ty (Figure 1B).

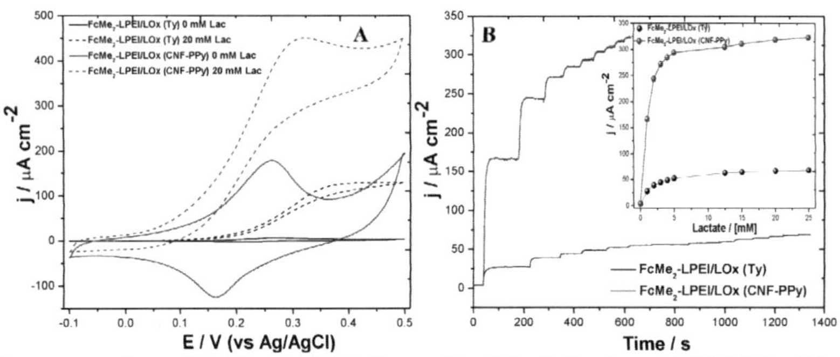

Figure 1. CV comparative of $FcMe_2$-LPEI/LOx on Ty (black line) and CNF-PPy (blue line) in the absence (solid line) and the presence (dashed line) of 20 mM lac in pH 7.4 PBS at a scan rate of 2 mV s^{-1}.

For AnMWCNT/LC the experiment was performed in PBS (0.1 M) at pH 5.6 saturate in N_2 and air by CV which are shown in Figure 2. The oxygen reduction reaction (ORR) potential starts at 0.62 V in both materials, which was due to the active site of the copper-containing enzyme T1 which allows the direct electron transfer between the enzyme and its surface [7]. The current density generated with Ty was 122.04 μA cm^{-2} while with CNF-Py it was 200.63 μA cm^{-2} 64.4% higher with the latter.

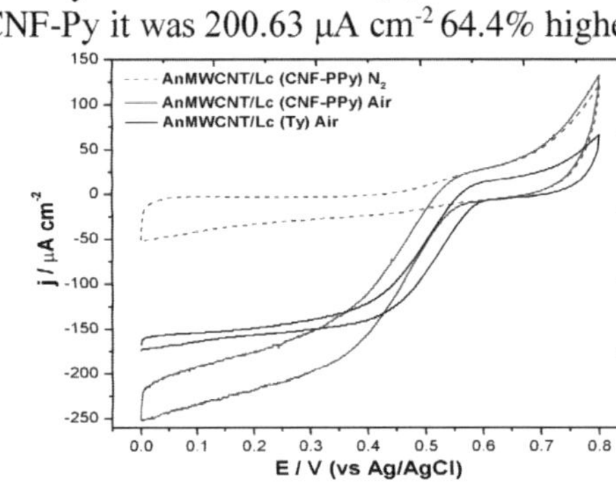

Figure 2. Comparative voltammograms of AnMWCNT/Lc on Ty (black line) and CNF-PPy (blue line) in 0.1 M PBS at pH 5.6 at a scan rate of 1 mV s^{-1}.

The BFC evaluation showed for the Ty a higher OCP of 0.47 V and a maximum current and power density of 9.5 μA cm^{-2} and 2.7 μW cm^{-2} respectively, using a concentration of 2 mM lactate (Figure 3A), while that at concentrations of 10 and 40 mM the performance decreases considerably (Figure 3 B). While with CNF-PPy the maximum OCP reached was 0.6 V and a maximum current and power density of 432.5 μA cm^{-2} and 71.5 μA cm^{-2} using 40 mM lactate, this result is equivalent to 45.5 and 26.5 times greater when compared to Ty and this result is comparable with others reported [8].

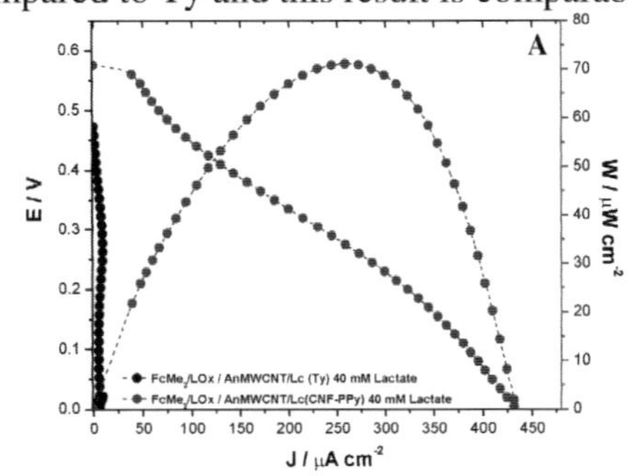

B

[Lac/mM]	OCP (V)	J/ μA cm^{-2}	W/ μW cm^{-2}
Ty			
2	0.47	9.5	2.7
10	0.45	7	1.9
40	0.44	5.3	1.4
CNF-PPy			
2	0.6	308.1	55.9
10	0.6	341.3	63.2
40	*0.58*	*432.5*	*71.2*

Figure 3. (A) Comparison of polarization and power density curves between Ty (black line) and CNF-PPy (blue line) in a BFC single compartment using 10 mM lactate characterized at 2 mV s^{-1} (B) comparative performance of BFC between TP and CNF-PPy using 2, 10 y 40 mM of lactate.

4. Conclusions

This work showed a higher enzymatic catalytic activity of $FcMe_2$-LPEI/LOx and AnMWCNT/Lc immobilized on CNF-PPy in the oxidation of lactate and ORR respectively, compared to commercial carbon Ty. The application of CNF-PPy in BFC the performance was 45.5 and 26.5 times higher with respect to Ty, exhibit the importance of the CNF polymerizing PPy to increase its electrical conductivity converting them into supports suitable for the immobilization of enzymes and its application in BFC of lactate.

Acknowledgments

The authors thank the Mexican Council for Science and Technology (CONACYT) for financial support through project Fronteras de la Ciencia 611, Ciencia básica 2015-Grant 256749, LN-280485 and SENER 246079.

References

[1] Li J, Qian X-r, Chen J-h, Ding C-y and An X-h 2010 *Carbohydr. Polym.* **82** 504–509.
[2] Paliotta L, De Bellis G, Tamburrano A, Marra F, Rinaldi A, Balijepalli S K, Kaciulis S and Sarto M S 2008 *Chem. Mater.* **20** 2574-2582.
[3] Wu K, Zhang Y, Wang Y, Ge S, Yan M, Yu J and Song X 2015 *ACS Appl. Mater. Interfaces* **7** 24330-24337.
[4] Sokic-Lazic D, Rodrigues A and Minteer S D 2011 *Electrochem. Acta* **56** 10772-10775.
[5] Milton R D, Lim K, Hickey D P, Minteer S D 2015 *Bioelectrochem.* **106** 56-63.
[6] Meredith M T, Minson M, Hickey D, Artyushkova K, Glatzhofer D T and Minteer S D 2011 *ACS Cat.* **1** 1683-1690.
[7] Jensen U B, Lörcher S, Vagin M, Chevallier J, Shipovskov S, Koroleva O, Besenbacher F and Ferapontov E 2012 *Electrochim. Acta* **62** 218-226.
[8] Escalona-Villalpando R A, Reid R C, Milton R D, Arriaga L G, Minteer S D and Ledesma García J 2017 *J. Power Sources* **342** 546-552.

Friction-induced fabrication of flexible supercapacitive microelectrodes

Shulan Jiang, Feng Wang, Hongbo Wang and Linmao Qian*

School of Mechanical Engineering, Southwest Jiaotong University, Chengdu, China

E-mail: linmao@swjtu.edu.cn

Abstract. With miniaturized electronic devices becoming increasingly pervasive, micro energy storage devices have attracted great attentions. Micro-supercapacitors have great potentials due to its long lifetime and high power densities. Here we demonstrate low-cost and lithography-free fabrication of flexible micro-supercapacitive electrodes over large areas by friction-induced nanofabrication of silicon nanostructures array and electrodeposition of MnO_2 nanostructures. The silicon nanostructures are fabricated by sliding a diamond tip on Si(100) surface under a given normal load combined with wet etching in potassium hydroxide (KOH) solution. Since the scratched pattern shows lower etching rate than that of non-scratched surface in KOH solution, such pattern can served as an etching mask to protect the silicon below from etching to some degree. Hence, the hillock silicon nanostructures can be realized on Si(100) surface. After reverse moulding, metal film sputtering and MnO_2 nanostructures integration, the flexible microelectrodes are fabricated and show good electrochemical performances. The proposed friction-induced fabrication method has potential in bio-implantable and wearable miniaturized devices.

1. Introduction

Nowadays, flexible micro energy storage devices have attracted great attentions because the miniaturized and wearable electronic devices become increasingly pervasive. Supercapacitors, also called ultracapacitors or electrochemical capacitors, have great potentials due to long lifetime and high power density compared with electrolytic capacitors and batteries [1]. Three-dimensional micro/nanostructures, which have the advantages of high mass loading and small ion transport lengths when used as microelectrodes, have the potentials of high energy and power densities over small footprint area. Photolithography and inductively coupled plasma etching are generally used for fabrication of micro/nanostructures. Photolithography is used to fabricate interdigital planar structures and high-aspect-ratio array structures. For example, Müllen et al. [2] fabricated graphene-based interdigital micro-supercapacitor by methane plasma-assisted reduction and photolithography processes, which shows good electrochemical performance. Jiang et al. [3] have reported the fabrication of carbon based supercapacitive microelectrodes by photolithography technique combined with pyrolysis and electrochemical deposition method. Wang et al. [4] have reported the fabrication of micro-supercapacitor using inductively coupled plasma etching technique to make the silicon template, then integrating the hydrous ruthenium oxide structures by cathodic deposition technique with the silicon template. In addition, manganese dioxide (MnO_2) is one of the most promising pseudocapacitive materials by virtue of its high theoretical specific capacitance, low-cost, environmental friendliness, and abundant resources, which is widely used in micro-supercapacitors [5].

While it still needs studying of cost-effective fabrication of micro/nanostructures for micro energy storage devices. Here we demonstrate low-cost and lithography-free fabrication of micro-supercapacitor electrodes over large areas by friction-induced nanofabrication of silicon nanostructures array and electrodeposition of MnO_2 nanostructures. The study of supercapacitive microelectrodes fabricaiton through friction-induced nanofabrication approach is carried out for the first time [6]. The fabrication parameters of the silicon nanostructures, and the influence of deposition time to the performance of microelectrodes will be investigated, respectively.

2. Experimental

The fabrication schematic is shown in Figure 1, which mainly consists of scratching, wet chemical etching, PDMS molding, film sputtering and electrodeposition. Firstly, after cleaned by deionized water and HF solution in advance, the silicon substrate (MEMC Electronic Materials, Inc., USA) was scratched line by line by a diamond tip on the surface under different loads through the homemade microfabrication apparatus. Then the samples were etched in the solutions of 20 wt.% KOH and isopropanol (v/v≈5 : 1) at 20±2 °C . After cleaning with deionized water, the silicon nanostructures array was fabricated and it will regarded as a template in the fabrication of mold. Poly(dimethylsiloxane) (PDMS, Sylgard® 184, Dow Corning Corporation) was used to be a soft template for pattern transfer from the surface of silicon micro/nanostructures. The base polymer and curing agent were mixed thoroughly, and the dissolved air bubbles were removed through evacuation for about 30 min. After careful solidifying at the hot plate with 80 °C for about 1 hour, the sample was taken down. When it was cooling to room temperature, the flexible nanostructures sample was peeling off from the friction-induced silicon template. It was deposited with platinum nanofilm by ion sputtering for 30 s. Then, the flexible nanostructures were integrated with MnO_2 nanostructures through galvanostatic electrodeposition method using electrochemical workstation (CHI 660E). As a comparison, platinum nanofilm was sputtered on plane PDMS substrate and it was then integrated with MnO_2 nanostructures.

The morphologies of the samples were characterized by an atomic force microscope (AFM, SPI3800N, Seiko, Japan) with silicon nitride probes (MLCT, Bruker Corp., USA) with a spring constant of 0.1 N/m. And the electrochemical performances of plane microelectrode and three-dimensional microelectrodes with different deposition time were also tested with cyclic voltammetry technique using the electrochemical workstation in 1 M Na_2SO_4 electrolyte.

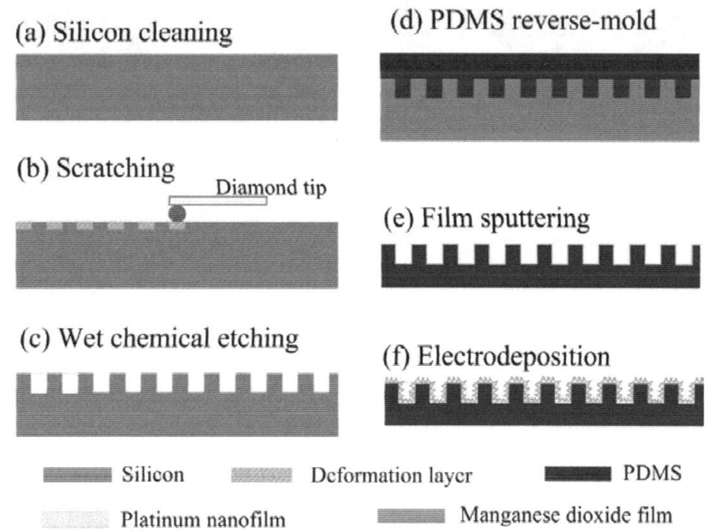

Figure 1. The schematic of friction-induced fabrication of supercapacitive microelectrodes.

3. Results and discussions

The scratched pattern is shown in Figure 2 a1, which is fabricated by the diamond tip under the given normal load of 7 mN. The cross-sectional profiles of the scratch pattern is shown in Figure 2 a2, the height of the scratch is about 50 nm. The scratched pattern is composed of amorphous layer and deformation layer, which shows different etching rate compared with non-scratched area when etching in KOH solutions [6,7].The remove of native oxide layer on the silicon substrate by HF solution before the scratch process is of great importance to the fabrication efficiency and surface quality of the nanostructures. Figure 2 b1~e1 are the morphologies of the typical nanostructures, and b2~e2 are the outlines of the nanostructures etched for 5, 10, 15 and 20 min, respectively. With the prolongation of etch time, the height of nanostructures increase from about 60 nm to 400 nm. Also the surface quality of the samples become better. Figure 2f and g are the typical morphologies of the silicon nanostructures array, which are scratched by diamond tip with the load of 7 mN and etched with 15 min and 20 min. The nanostructures array shows uniform height and width, and the effective surface area which can be used to load active materials is larger than that of the plane substrate. The etching of 20 min will be selected to fabricate microelectrodes in the following study.

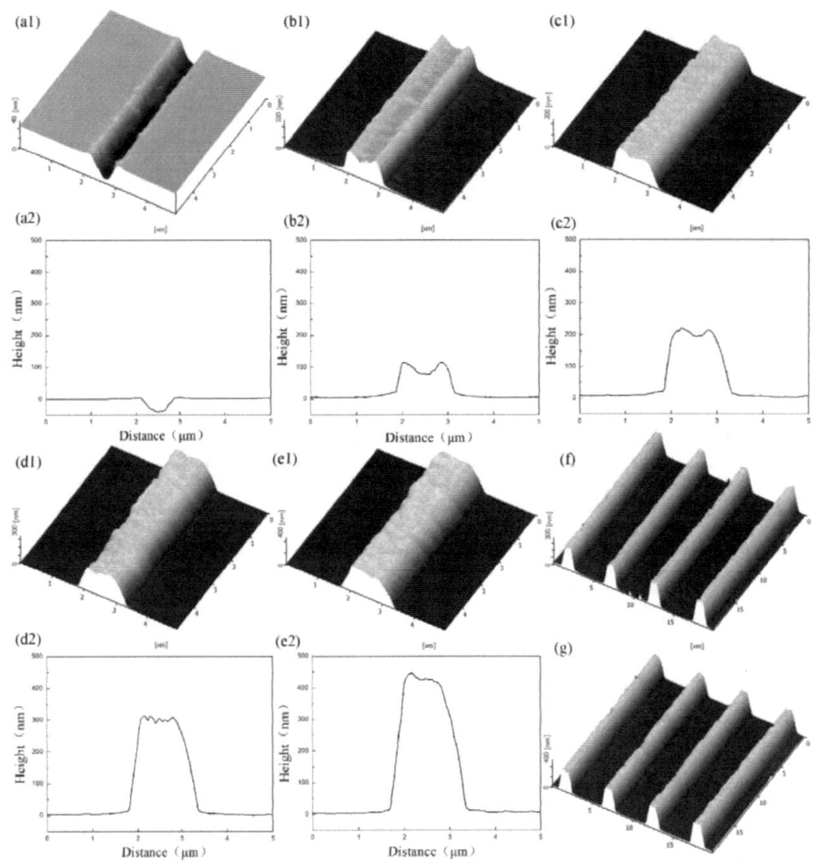

Figure 2.a1 and a2 are the AFM morphology and outline of the sample scratched. b1~e1 are the morphologies of the nanostructures and b2~e2 are the corresponding outlines of the nanostructures etched for 5, 10, 15 and 20 min respectively. f and g are the typical AFM morphologies of the silicon nanostructures pattern etched in KOH solution for 15 min and 20 min, respectively.

The MnO_2 nanostructures are deposited on the PDMS micro/nanostructures array, and the AFM images are shown in Figure 3a and 3b. Representative CV curves of the MnO_2/Pt/PDMS microelectrodes were shown in Figure 3c and 3d. The curves show near mirror-image current response on voltage reversal, which indicate that the structures have good capacitive behaviour. With the increase of scan rate, the voltammetric current increases. In addition, the currents of MnO_2/Pt/PDMS microelectrodes is much larger than that of the plane Pt/PDMS microelectrodes. And with the increment of deposition time, the current of the microelectrodes increase. The friction-induced microelectrodes have improved electrochemical performances, which show great potential in micro energy storage devices.

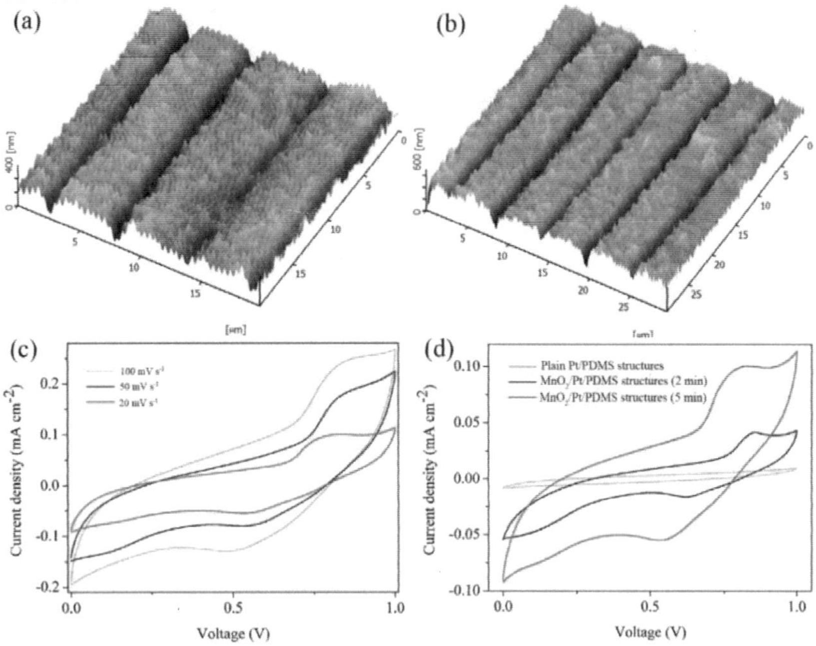

Figure 3 (a,b) The morphologies of MnO_2/Pt/PDMS structures. (c) CV curves of MnO_2/Pt/PDMS microelectrode at the deposition of 5 min. (d) CV curves of microelectrodes with different deposition time at the scan rate of 20 mV s^{-1}.

4. Conclusions

Flexible supercapacitive microelectrodes were fabricated through cost-effective and straightforward friction-induced fabrication of silicon nanostructures and PDMS reverse-molding. By sputtering metal film and integration MnO_2 nanostructures, the fabricated microelectrodes show good electrochemical performances. The proposed fabrication method exhibits great potential in bio-implantable and wearable miniaturized devices in the future.

References

[1] N.A. Kyeremateng, T. Brousse, and D. Pech. *Nature nanotechnology* 12 (2017) 7
[2] Z.S. Wu, K. Parvez, X. Feng, K. Müllen. *Journal of Materials Chemistry A* 2 (2014) 8288
[3] S.L. Jiang, T.L. Shi, D. Liu, H. Long, S. Xi, F.S. Wu, X.P. Li, Q. Xia and Z.R. Tang. *Journal of Power Sources* 262 (2014) 494
[4] X.F. Wang, Y. J. Yin, X. Y. Li and Z. You. *Journal of Power Sources* 252 (2014) 0378
[5] Z.J. Su, C. Yang, B.H. Xie, Z.X. Zhang, J.P. Liu, B.H. Li, F.Y. Kang and C.P. Wong. *Energy & Environmental Science* 7 (2014) 2652
[6] B.J. Yu, H.S. Dong, L.M. Qian, Y.F. Chen, J.X. Yu and Z.R. Zhou. *Nanotechnology* 20 (2009) 465303
[7] J. Guo, C.F. Song, X.Y. Li, B.J. Yu, H.S. Dong, L.M. Qian and Z.R. Zhou. *Nanoscale Research Letters* 7 (2012) 152

Acknowledgements

This work was supported by the National Science Foundation of China (No. 51605400), and the Fundamental Research Funds for the Central Universities (2682016CX027).

Solar cell efficiency improvement by photon absorption enhancement employing rare earth doped films

R Lopez-Delgado[1, 2], J C Melendres-Sanchez[1, 2], A J Cordova-Rubio[1, 2], M E Álvarez-Ramos[2] and Arturo Ayon[1].

[1]MEMS Research Lab, Department of Physics and Astronomy, University of Texas at San Antonio, San Antonio, TX, 78249, USA.
[2]Departamento de Física, Universidad de Sonora, Hermosillo, Son, 83000, México.

Corresponding author email: Rosendo.lopezdelgado@utsa.edu

Abstract. We report the synthesis and characterization of SiO_2:3-(trimethoxysilyl) propyl methacrylate (TMSPM) films doped with Eu^{3+} ions that exhibit down-shifting, photo-luminescent characteristics. The synthesized rare-earth films were observed to absorb UV photons and subsequently reemit photons in the visible range of the electromagnetic spectrum where there is a greater absorption by the silicon substrate. The incorporation of these films as a photon down-shifting layer on solar cells triggered improvements in the performance of the photovoltaic devices, especially in the short circuit current density (Jsc) leading to an increment in the power conversion efficiency (PCE) from 15.17 to 15.42%. This increment represents an improvement of the order of 1.6% on the power conversion efficiency on the silicon solar cells employed. The observed results could be conducive to promoting the proliferation of photovoltaic structures.

1. Introduction.

In recent years, silicon solar cells have captured a large portion of the total market of photovoltaic devices mostly due to its relatively high efficiency. However, even though the wavelength distribution of the solar spectrum extends from the infrared (2500 nm) to the ultraviolet (280 nm) one of the major problems in solar energy conversion is the relatively limited absorption range of the materials employed in solar cells [1-4]. The underlying reason is that in principle only photons with energy greater than the bandgap can be absorbed by the material leading to the loss of relatively low energy photons. But even photons with energies larger than the bandgap have a non-negligible probability of interacting with crystal phonons, in a process referred to as thermalization, rather than producing electron-hole pairs [3]. These fundamental losses can be ameliorated by the utilization of luminescent materials. To promote the absorption of low energy photons, near and below the bandgap, particles with up-conversion properties can be employed [5, 6]. On the other hand, in the case of energetic photons that contribute to thermalization, the alternative is to employ down-shifting materials, able to emit lower energy photons that fall within the range of absorption by the underlying solar cell[7, 8].

2. Experimental.

2.1. Doped SiO_2:TMSPM Film Synthesis.

A matrix host solution of SiO_2:TMSPM was prepared by mixing 62.3ml of Tetraethyl orthosilicate (TEOS), 69.9ml of Ethanol and 34.9 ml of 3-(trimethoxysilyl)propyl methacrylate (TMSPM) under stirring. In a second beaker, 21.6 ml of H_2O and 0.1 ml of Hydrochloric acid (HCl) were mixed. Then, the H_2O:HCl solution was added dropwise to the SiO_2:TMSPM solution under stirring. The resulting mixture was kept under stirring for another 30 minutes and then it was refrigerated for 24 hours.

A rare-earth ions solution (Eu^{3+}) was obtained by adding 0.0476 g of Europium(III) Nitrate Pentahydrate and 0.04 g of Triethylamine (TEA) to 10 ml of Ethanol (anhydrous). The rare-earth solution was cleaned by centrifugation at 4500 rpm for 10 minutes. The supernatant was removed and 10 ml of Ethanol was added. Finally, 1ml of Dimethyl sulfoxide (DMSO) was added to the Eu^{3+} solution. Three different concentrations were employed in order to study the influence of the concentration on the performance of the solar cell. To this end, 1.1, 3.3 and 5.6 ml of the Eu^{3+} solution were added to 10 ml of the SiO_2:TMSPM matrix. The final solution was deposited on the window side of the solar cell by spin coating technique with an angular velocity of 3000 rpm.

2.2. *Silicon Solar Cell Characterization.*

The Current – Voltage performance (I – V characteristics) of the solar cells was characterized using an Oriel Sol2A solar simulator under AM1.5G illumination at standard testing conditions (100 $mWcm^{-2}$) at room temperature. To this end, the photovoltaic device was irradiated with a simulated sunlight beam while connected to a current and voltage measurement system. The I and V values were collected simultaneously from open circuit to short circuit conditions; subsequently, the data was analyzed to obtain the power conversion efficiency of the solar cell. The external quantum efficiency characterization as a function of wavelength was performed with a Newport External Quantum Efficiency Measurement System (QEPVSI-B). The two characterization processes were performed before and after the deposition of the SiO_2:TMSPM films doped with Eu^{3+}. Figure 1 shows the flowchart of characterization and film deposition.

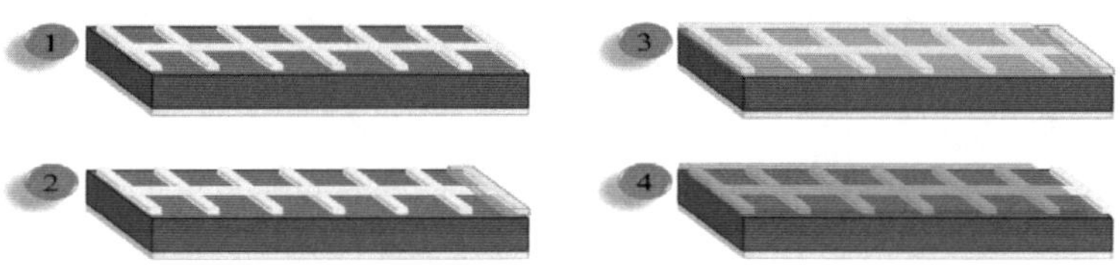

Figure 1. 1) I-V and EQE characterization, 2) Protection of the metallic bus bar, 3) Doped film deposition by spin coating technique and 4) Protection removal and I-V, EQE characterization to verify the influence of the luminescent film.

3. Results and Discussion.

3.1. *Characterization of SiO_2:TMSPM films doped with Eu^{3+} ions.*

The synthesised SiO_2:TMSPM films doped with Eu^{3+} ions absorb photons in the UV range of the solar spectrum and emit photons at wavelengths mostly in the red range, that are well suited for silicon absorption and electron-hole pair generation. The photoluminescent spectra of the doped film and the three most pronounced characteristic transitions of Eu^{3+} ions are shown in Figure 2[9].

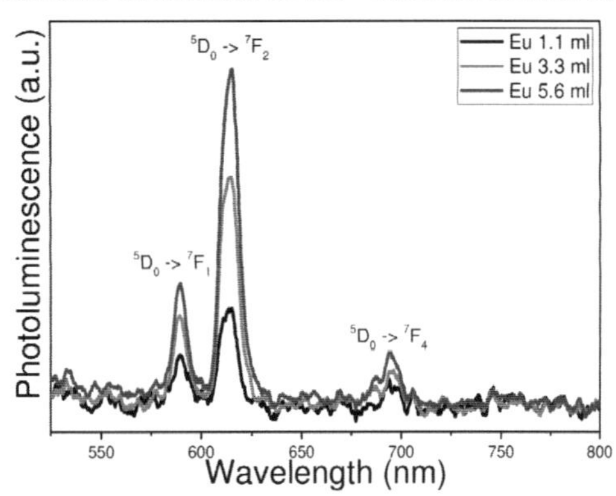

Figure 2. Photoluminescence spectrum of SiO_2:TMSPM films doped with Eu^{3+} ions under an excitation of 395nm.

3.2. Solar Cell Characterization.

The performance characterization of the silicon solar cells with and without the aforementioned photo luminescent, down-shifting Eu^{3+} doped SiO_2:TMSPM layer are shown in Figure 3. For characterization purposes, three sets of three silicon solar cells where measured before and after the deployment of the doped SiO_2:TMSPM layer. Figure 3 shows the I-V characteristics and the spectral response of the silicon solar cells before and after the application of SiO_2:TMSPM films doped with Eu^{3+} ions. EQE measurements indicate that the doped layer is improving the photocurrent generation specially for photons with a wavelength smaller than 550 nm. These EQE improvements are thought to be responsible for the current increase measured during the I-V analysis.

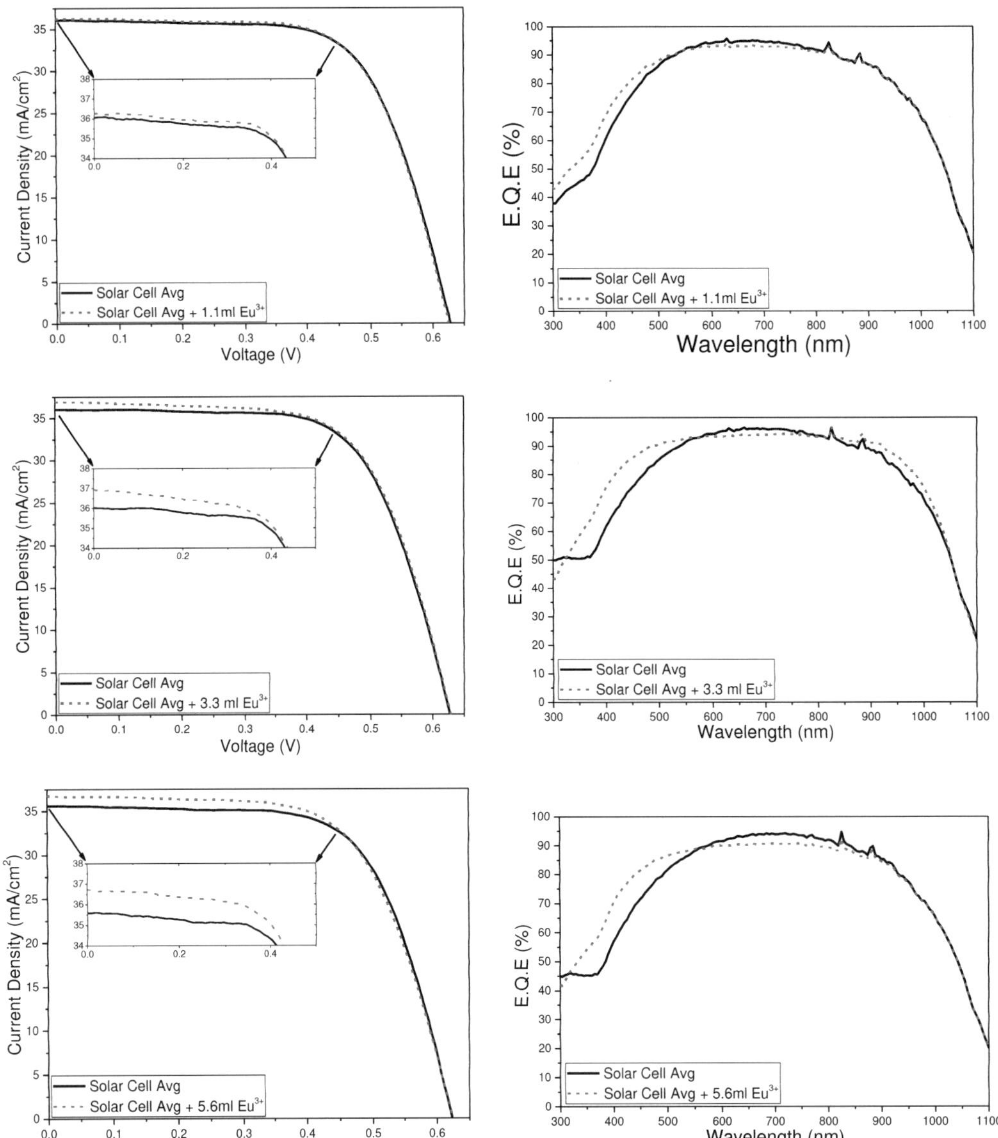

Figure 3. Current – Voltage characteristics and External Quantum Efficiency (EQE) of Silicon Solar Cells before (continuous lines) and after (dashed lines) application of SiO_2:TMSPM films doped with Eu^{3+} ions.

The incorporation of these films triggered improvements in the performance of the devices, particularly in the short circuit current density (Jsc) for which the measured values showed an increase from 35.94 to 36.20, from 35.94 to 36.76, and from 35.61 to 36.91 mA/cm2, for the samples with 1.1, 3.3 and 5.6 ml of Eu^{3+} solution respectively. The effect of the improved values of those variables led to an increment in the power conversion efficiency (PCE) from 15.48 to 15.63, from 15.34 to 15.53, and from 15.17 to 15.42%, respectively (See Table I). This increase represents improvements of approximately 1.0%, 1.2% and 1.6%, respectively, of the PCE of the fabricated devices. The effort described herein is considered a good fit to the generalized trend to improve the efficiency of solar cells with mass-compatible techniques that could serve to promote their widespread utilization. The averages of the solar cell performance parameters are listed in Table I.

Table I. Silicon solar cell performance before and after application of the doped SiO_2:TMSPM film.

Solar cell performance	Voc (mV)	Jsc (mA/cm^2)	FF (%)	PCE (%)
Solar cell set 1	628.7	35.94	68.52	15.48
Solar cell set 1 + 1.1 ml Eu^{3+}	628.8	36.20	68.70	15.63
Solar cell set 2	628.6	35.94	67.92	15.34
Solar cell set 2 + 3.3 ml Eu^{3+}	629.7	36.76	67.06	15.53
Solar cell set 3	624.5	35.61	68.26	15.17
Solar cell set 3 + 5.6 ml Eu^{3+}	625.9	36.91	66.76	15.42

4. Conclusions.

The influence of photo-luminescent (SiO_2:TMSPM) films doped with Eu^{3+} ions on the performance of solar cells has been characterized. Different concentrations were used in order to study their impact on solar cell performance. The experimental results indicate that the incorporation of Eu^{3+} doped films as down-shifting material on solar cells improves the photocurrent generation, and hence, the overall PCE was increased from 15.17% to 15.42% in the best scenario. This result represents an improvement of 1.6% in the photovoltaic performance. The experimental results evince that this strategy represents a promising way to improve the efficiency of existing solar cells which could promote the proliferation of photovoltaic energy harvesting.

Acknowledgements

The authors would like to acknowledge the U.S. Army Research Office (Grant W911NF-13-1-0110), National Council of Science and Technology (CONACYT), the Physics and Astronomy Department of the University of Texas at San Antonio and the Physics Department of University of Sonora, for the financial support provided for this research project.

References

[1] Hirst L C and Ekins- Daukes N J, 2011. *Progress in Photovoltaics: Research and Applications.* **19**(3): p. 286-293.
[2] Scholes G D, et al., 2011. *Nature chemistry.* **3**(10): p. 763-774.
[3] Huang X, et al., 2013. *Chemical Society Reviews.* **42**(1): p. 173-201.
[4] Richards B, 2006. *Solar energy materials and solar cells.* **90**(15): p. 2329-2337.
[5] Qu B, et al., 2016. *Journal of Alloys and Compounds.* **658**: p. 848-853.
[6] Van Der Ende B M, Aarts L, and Meijerink A, 2009. *Physical Chemistry Chemical Physics.* **11**(47): p. 11081-11095.
[7] Trupke T, Green M, and Würfel P, 2002. *Journal of Applied Physics.* **92**(3): p. 1668-1674.
[8] Zhang Q and Huang X, 2010. *Progress in Materials Science.* **55**(5): p. 353-427.
[9] Binnemans K, 2015. *Coordination Chemistry Reviews.* **295**: p. 1-45.

IOP Conf. Series: Journal of Physics: Conf. Series **1052** (2018) 012069 doi:10.1088/1742-6596/1052/1/012069

Energy Neutral Sensor System With Micro-scale Photovoltaic and Thermoelectric Energy Harvesting

Anand Savanth[1,3], Mathieu Bellanger[2], Alex Weddell[1], James Myers[3] and Mathias Kauer[2]

[1] University of Southampton, Southampton, UK
[2] Lightricity Ltd, Oxford, UK
[3] ARM Ltd, Cambridge, UK

E-mail: anand.savanth@arm.com

Abstract. Minimizing power conversion losses is critical for energy neutral operation of micro-scale energy harvested sensor nodes. These small form-factor sensor nodes rely on miniature harvesters with low output voltages that must be boosted with large conversion ratios to recharge batteries or super-capacitors. Selective Direct Operation (SDO), a technique to selectively avoid power conversion and thereby eliminate conversion loss in energy harvested systems has been demonstrated as an effective technique for light harvesters. This paper extends SDO to thermoelectric generators (TEGs). SDO exploits the ultra-low circuit functional voltages, enabling sensor systems to effectively harvest energy from cm-scale TEGs which output few 10's of mW but at low output voltages (100's of mV). PV cell construction from prior-work, TEG characterization and field measurements are presented in this paper to demonstrate the effectiveness of SDO and co-designing energy harvesters, power conversion circuits and digital sub-systems.

1. Introduction

Modern circuit design and silicon fabrication techniques have made possible ultra-low power miniature sensor systems with sufficient compute capability to enable the Internet of Things - specifically the leaf nodes that generate vast sensor data. A common theme of these techniques is aggressive and fast supply voltage scaling as lower voltage results in quadratic reduction in expended power. However, as demonstrated consistently by prior works [1][2][3][4], there exists a lower bound to the supply voltage below which operating energy increases despite an apparent reduction in power. This minimum-energy (MinE) point is determined by the static power. This is further exacerbated by heavily duty-cycled workloads in typical sensor applications.

Energy harvesting in such systems is critical to further enable untethered and extended operating periods as typical battery self-discharge is greater than the sleep power of such systems. However, this objective necessitates integrated power conversion which in practice is lossy. Typically, two stage conversion is used to 1) charge the 1.2~3.6V battery from the low voltages available from small form-factor energy harvesters and 2) to convert down from battery voltages to the low supply voltage at which MinE operation is possible. This two-stage conversion is lossy and state-of-the-art aims at high-efficiency conversion techniques at low input voltage

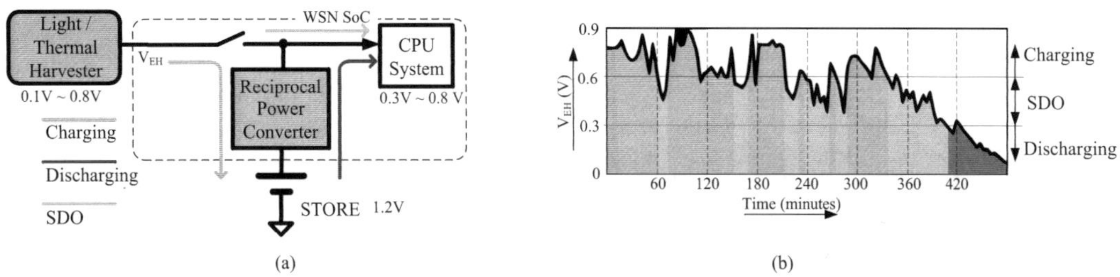

Figure 1: Selective Direct Operation [8] (a) block diagram and (b) conceptual operation

and power levels [5]. Note that conversion losses increase with conversion ratio and prior works aim to eliminate conversion stages under low input power conditions [6][7]. A related technique, Selective Direct Operation (SDO), has been demonstrated [8] with Photovoltaic (PV) cells to exploit the low operating voltage of MinE systems and minimize power conversion losses.

In this work, fabrication details of the high-efficiency PV cell are provided and the proposed SDO technique is extended to thermoelectric generators (TEGs) as an alternate source of energy harvesting. Measurements from real-world scenarios are presented supporting deployability of SDO for MinE sensor systems.

2. Power Conversion Losses and Selective Direct Operation

SDO is illustrated in Fig. 1 wherein a single reciprocal convertor is used for power conversion from harvester to battery and from battery to the CPU system. This minimizes losses associated with traditional two stage conversion. Further, at poor ambient energy levels, the system's ability to harvest energy is limited by power conversion overheads. For such conditions, SDO exploits the ultra-low operating voltage of subthreshold CPU systems to extend battery capacity by 1) direct operation of logic from harvester thereby eliminating conversion overheads and 2) avoiding battery discharge during these periods. SDO has been successfully demonstrated [8] with the MinE system executing an industry standard sensor benchmark software with ability to cold start at indoor conditions and charge a 1.2V NiMH battery when sufficient ambient energy is available.

2.1. Low-light high efficiency PV cell

The high efficiency PV cell for this work is fabricated by Lightricity. It is optimized for fluorescent and LED indoor lighting using high purity inorganic materials (Fig. 2). Measured results show >22 µW/cm^2 with 35% efficiency at 200 lux and can retain high efficiency down to very low light levels (10 lux). The low-light operating voltage (V_{MPP}) of ~1V at 200 lux significantly eases design constraints on maximum power point tracking (MPPT) and power conversion

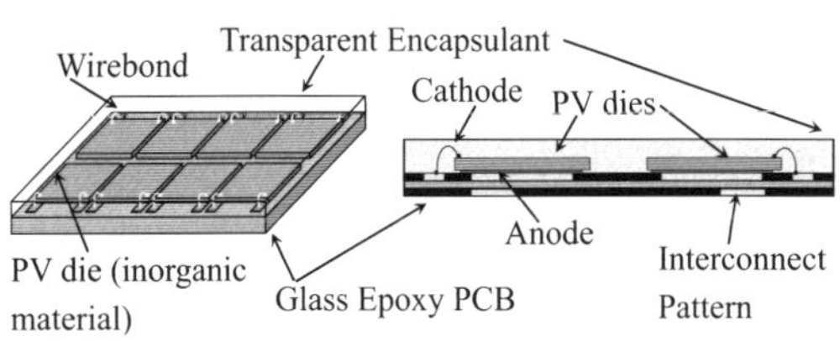

Figure 2: High efficiency PV cell for low-light conditions

Total size (mm)	:	23.8x10.2x1.5
Active area (cm^2)	:	2.15
V_{OC}, I_{SC}	:	1.15V, 49.8 µA
V_{MPP}, I_{MPP}	:	1V, 47.6 µA
P_{MPP}	:	47.6 µW
Power Density	:	22.1 µW/cm^2

Table 1: PV cell performance summary

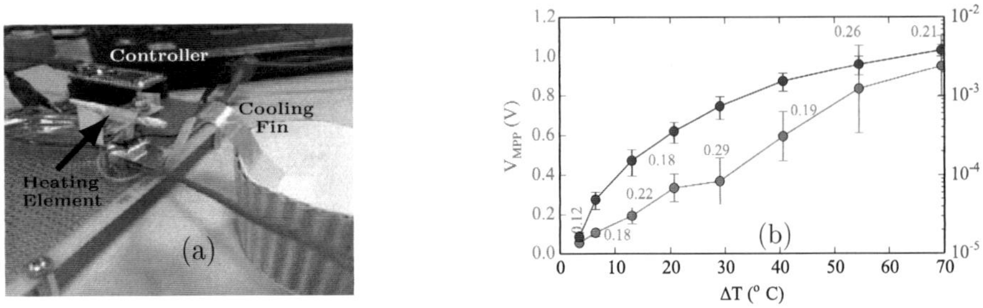

Figure 3: (a): Characterisation setup for TEG and (b): Measured power vs ΔT and variablity in V_{MPP}

circuits. The PV cells have been pre-characterized to estimate the impedance vs illumination characteristics and variation-aware models have been developed to aid with power convertor circuit simulation highlighting the effectiveness of harvester and circuit co-design.

2.2. TEG characterization and measurements

Similarly, commercially available cm^3 TEGs have been characterized by emulating differential temperature surfaces using a heater and cooling-fin setup as shown in Fig 3a. The ratio of V_{MPP}/V_{OC} is ≈0.5, however the ratio shows part-to-part variation and dependence on temperature (Fig 3b). The error-bars on the voltage and power trace highlight the variability observed when measuring 5 different samples for similar temperature gradients. However, the coefficient of variance (σ/μ) is 10x lower compared to commercial off-the-shelf micro-scale PV cells. Harvested power from ambient heat sources (hot-water dispenser and room radiator) is measured to be between 10-1000µW (Fig 4).

3. Measured Results and Energy Neutral Operation

Fig. 4 shows the system block diagram along with start-up comparator and V_{MPP} tracking scheme which modulates the convertor and CPU internal clock for MPPT during charging and direct operation (Fig 1b). The convertor is implemented in 175x210sq.µm area on a 1.94x1.94mm chip. When harvesting from PV cells V_{EH} in Fig 5a can vary between 0 to 1.1V while it is much lower with the TEG. ΔT as high as 40°C generates V_{MPP} = <0.6V. SDO is more effective for these applications where the low harvester output voltages impose large conversion overheads. Fig 5c shows the oscilloscope waveforms when the CPU is in sleep and active mode showing GPIO

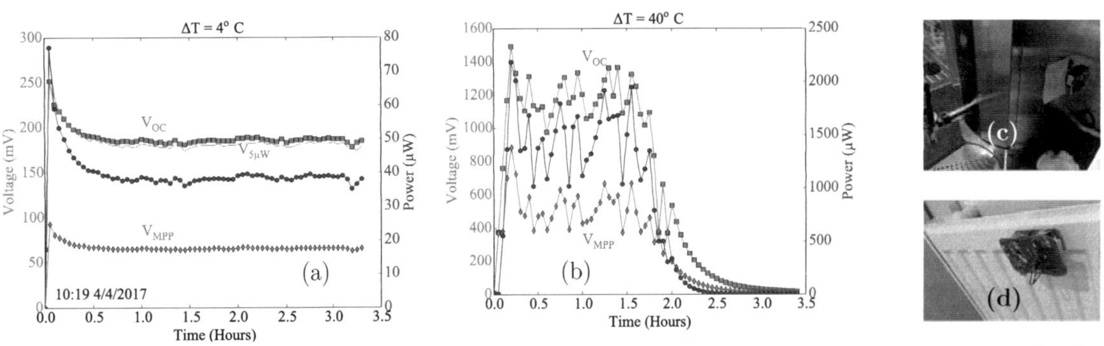

Figure 4: TEG measurements under field conditions (a,c): Low ΔT (4°C) conditions and (b,d): Large ΔT (40°C) conditions

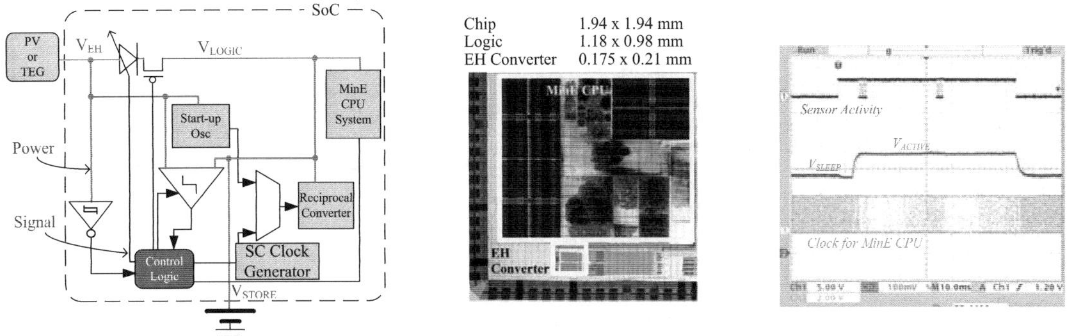

(a) (b) (c)

Figure 5: SoC block diagram and chip floorplan

activity in active mode. Also shown is the corresponding V_{LOGIC} and CPU clock frequency.

4. Conclusions

This paper presented experimental results demonstrating the effectiveness of co-designing energy harvesters, power conversion circuits and digital sub-systems in micro-scale energy harvesting sensor nodes. Prior work on SDO was extended by reporting harvester characterization methodology and field evaluation of SDO while harvesting from cm^3 form-factor PV cells and TEGs.

5. Acknowledgments

The authors extend thanks to Jedrzej Kufel for support with TEG measurements and characterization.

References

[1] J. Kwong et al., "*A 65nm Sub-Vt Microcontroller with Integrated SRAM and Switched-Capacitor DC-DC Converter*" in IEEE Int. Solid-State 580 Circuits Conf. (ISSCC) Dig. Tech. Papers, San Francisco, CA, USA, 581 Feb. 2008, pp. 318 - 616.

[2] M. Fojtik et al., "*A Millimeter-Scale Energy-Autonomous Sensor System with Stacked Battery and Solar Cells*" IEEE J. Solid-State Circuits, vol. 48, no. 3, pp. 801 - 813, Mar. 2013.

[3] D. Bol et al., "*Sleepwalker: A 25-Mhz 0.4-V Sub-mm² 7μW/Mhz Microcontroller in 65-Nm LP/GP CMOS For Low-Carbon Wireless Sensor Nodes*" in IEEE Int. Solid-State Circuits Conf. (ISSCC) Dig. Tech. Papers, San Francisco, CA, USA, Jan. 2013, pp. 20 - 32.

[4] J. Myers, A. Savanth, D. Howard, R. Gaddh, P. Prabhat, and D. Flynn, "*An 80nw Retention 11.7pJ/cycle Active Subthreshold ARM Cortex-M0+ Subsystem in 65nm CMOS For WSN Applications*" IEEE Int. Solid-State Circuits Conf. (ISSCC) Dig. Tech. Papers, San Francisco, CA, USA, Feb. 2015, pp. 1 - 3

[5] W. Jung, S. Oh, S. Bang, Y. Lee, D. Sylvester, and D. Blaauw, "*A 3nw Fully Integrated Energy Harvester Based on Self-Oscillating Switched-Capacitor DC-DC Converter*" in IEEE Int. Solid-State Circuits Conf. (ISSCC) Dig. Tech. Papers, San Francisco, CA, USA, Feb. 2014, pp. 398 - 399.

[6] S. Bandyopadhyay and A. P. Chandrakasan, "*Platform Architecture for Solar, Thermal, And Vibration Energy Combining with MPPT and Single Inductor*" IEEE J. Solid-State Circuits, vol. 47, no. 9, pp. 2199 - 2215, Sep. 2012.

[7] D. Bol, E. H. Boufouss, D. Flandre, and J. D. Vos, "*A 0.48mm² 5μW-10mw Indoor/Outdoor PV Energy-Harvesting Management Unit in a 65nm SoC Based on a Single Bidirectional Multi-Gain/Multi-Mode Switched- Cap Converter with Supercap Storage*" in Proc. 41st Eur. Solid- State Circuits Conf. ESSCIRC (ESSCIRC), Graz, Austria, Sep. 2015, pp. 241 - 244.

[8] Savanth, A. S. Weddell, J. Myers, D. Flynn and B. M. Al-Hashimi, "*Integrated Reciprocal Conversion With Selective Direct Operation for Energy Harvesting Systems*" in IEEE Transactions on Circuits and Systems I: Regular Papers, vol. 64, no. 9, pp. 2370-2379, Sept. 2017.

An Autonomous Power Management System with Event-driven Energy Harvester Switch

S. Yamada and H. Toshiyoshi

Research Center for Advanced Science and Technology (RCAST),
The University of Tokyo

e-mail: santa@iis.u-tokyo.ac.jp

Abstract. This paper presents a new use of piezoelectric PVDF film as an event-driven sensor switch that activates the power-management circuit for a subsequent wireless sensor ZigBee such that it would not consume extra power during the sleep mode. The circuit is designed to pick up a mechanical vibration with the PVDF film and control the self-holding circuit to keep flowing the drive current to the load until a reset pulse signal is received. The developed power management circuit can supply load current upwards of 33 mA at 3.3 Vdc for load resistance ranging between 100 Ω and 1 MΩ.

1. The first section in your paper

The concept of Internet-of-Things or IoT in coming years has presented an emerging vision of ultimate shapes of distributed wireless sensor nodes that would be distributed ubiquitously on the earth including artificial materials, human, and environments as schematically shown in **Figure 1**. Vast large amount of data collected by the IoT sensor nodes would be uploaded to the cloud and processed by the big-data analysis for the sake of prosperity of human life and company business activities. Such communication network of IoT data has similarity to biological neural sensor network or bionic network, because what wireless sensor nodes are to the big-data through the cloud is understood by what sensory receptors of living organism are to the brain through neural networks, as also schematically illustrated in **Figure 2**.

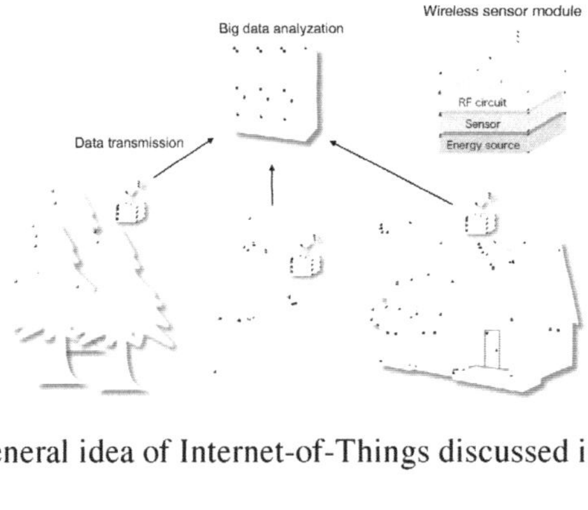

Figure 1 General idea of Internet-of-Things discussed in this work.

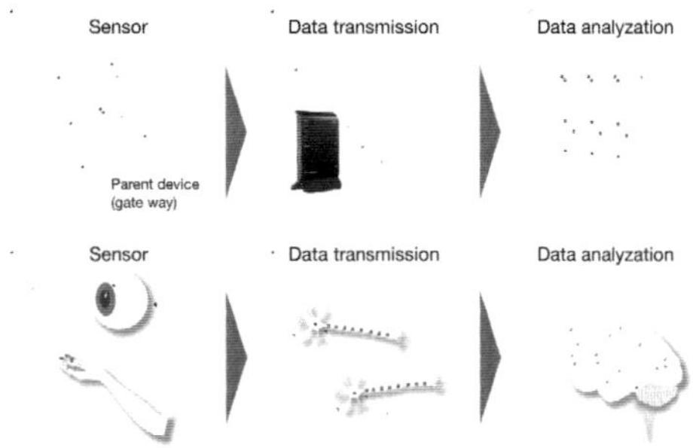

Figure 2 Structural similarities between IoT sensor network and biological sensory network.

In our recent publication, we have reported a prototype of such electronic sensory system called the e-Nerve that percepts physical input, transmits raw analog signals, and converts them into digital signals with a simple electromechanical circuitry [1]. In this system, a sensor works not only to detect the incoming signals but also to generate electric power to drive the rest of the signal processing units, such that the entire system can be autonomous without an explicit power source such as batteries. Collection of information through such e-Nerve system requires wireless communication capability because wire-communication cannot be used due to the overwhelming large amount of connections. Wireless is also required to make the entire e-Nerve network system to be maintenance-free from tiresome battery-replacement. This is where environmental energy harvesters [2,3] are used as an indispensable component to support the IoT applications. Nonetheless, most energy harvesters ever reported fail to supply enough power to keep the IoT wireless nodes operating all the time. Use of a timer IC would relax the power budget by firing the radio periodically rather than continuously but such operation might overlook those significant events during a sleep mode that should have been detected and reported. For this technical problem in this work, we have developed an event-drive type circuitry that does not miss an event to report by the wireless communication.

In this paper, we demonstrate a power management system based on the event-driven switch that activates circuits. A simple formation of switch is adopted by using a piezoelectric polymer PVDF (polyvinylidene fluoride) to detect mechanical excitation as an event.

2. Event-driven Circuit

Figure 3 shows the flowchart of the operation of event-driven signal processing. The system normally rests in sleep mode to save power consumption. When the energy-harvester switch is mechanically excited by the incoming impulse, for instance, the current holding module is activated to keep flowing the current to an external load until the reset pulse is provided to the RST port. The circuit diagram shown in **Figure 4(a)** is used to implement the event-driven function. At this moment of the research, we use an extra dc voltage source to supply 3.3 V_{dc} to the circuit and a PVDF cantilever (manufacture, product ID) as an energy-harvester switch. When the PVDF is mechanically excited to give a voltage to the hold-circuit, the pass-transistor FET is turned on to flow the current into the load. To stop the operation and turn the system into a sleep mode again, a pulse signal needs to be applied to the RESET port in order to discharge the voltage accumulated by the PVDF. For demonstration, we assembled a ZigBee wireless node (manufacturer, product ID) with a sensor on a breadboard as shown in **Figure 4(b)**. The sensor (manufacturer, product ID) is used to detect temperature and humidity, and the data is acquired by the ZigBee analog input and radio-transmitted to another ZigBee receiver to visualize the data on a PC screen.

IOP Conf. Series: Journal of Physics: Conf. Series **1052** (2018) 012070 doi:10.1088/1742-6596/1052/1/012070

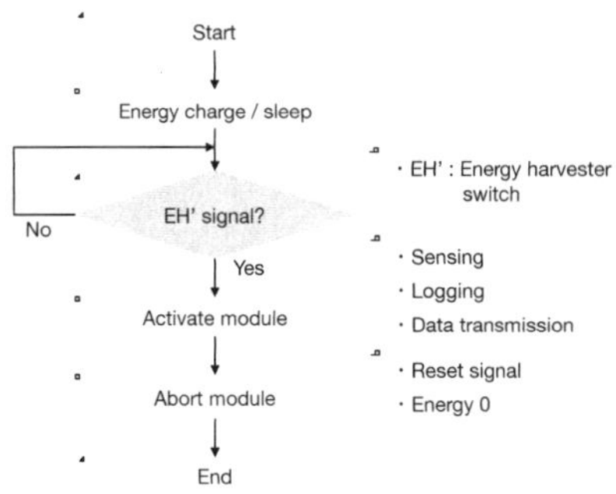

Figure 3 Block diagram of event-driven circuit for IoT sensor.

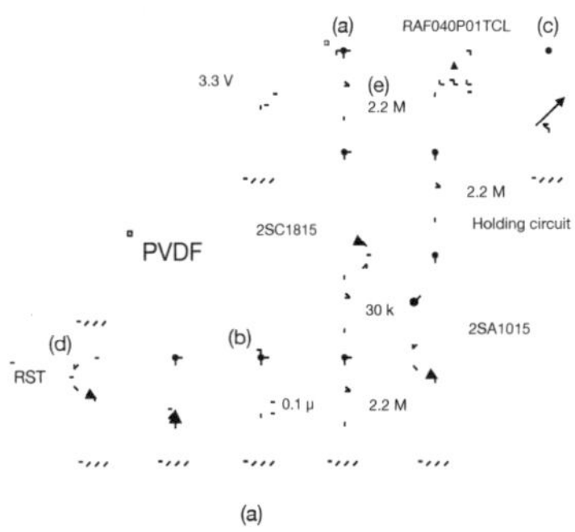

Figure 4 (a)Circuit diagram for the event-driven power management. (b) Event-driven ZigBee circuit implemented on a breadboard

3. Circuit Characterization

ZigBee operation was experimentally confirmed by observing the voltage waveforms of (a) the 3.3 Vdc power supply, (b) the NPN transistor base, (c) the load resistance 1 MΩ, and (d) the RST transistor; all of these points are marked in **Figure 4(a)**, and the waveforms measured at points (a) to (d) are compared as shown in **Figure 5.** During the sleep mode, the voltage at (c) remains to be 0 V, indicating that no current flows through the load resistance. When the PVDF event-driven switch is turned on, on the other hand, the PVDF switch gives a 0.6 V and maintains the level owing to the self-holding circuit, and then the voltage across the load resistance is flipped to 3.3 V. The voltage level remains high until the RESET pulse voltage is applied to (d). **Figure 6(a)** plots the source voltage level and the circuit output when the load resistance was varied from 100 Ω to 1 MΩ. These two voltage levels are seen superposed at the same value of 3.3 V, no matter what resistance value is used. It suggests that the circuit has a good current-drive capability upwards of 33 mA, which is large enough to operate a ZigBee. In contrast to the load current that is almost independent of the load resistance value, the self-hold circuit consumed power that slightly depends on the resistance value as shown in **Figure 6(b)**. Nonetheless, the internal loss of the circuit is as small as 20 nA or less, which is negligible compared with the load current, at (e) shown in **Figure 6(c)**.

4. Conclusions

In this work, we demonstrated an event-driven circuit to power-manage the wireless sensor node ZigBee by using a piezoelectric PVDF film as a trigger. The developed circuit is capable of supplying a current upwards of 33 mA to the load ranging between 100 Ω and 1 MΩ, which turned out to be powerful enough to drive a ZigBee node. The internal loss of the circuit was as small as 20 nA, which is negligible compared with the currents consumed by the sensor or the load.

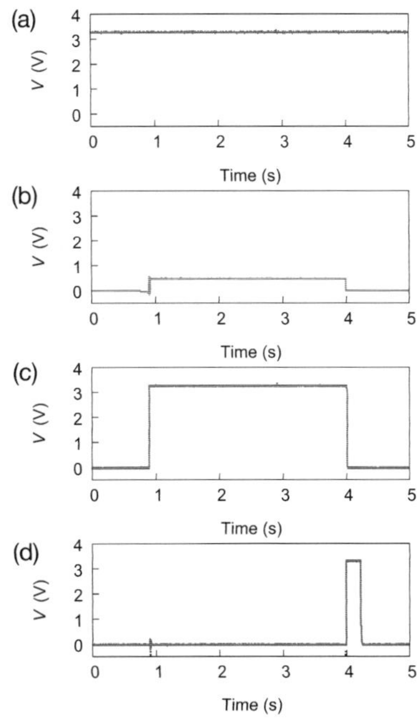

Figure 5 Waveforms measured on the event-driven circuit.

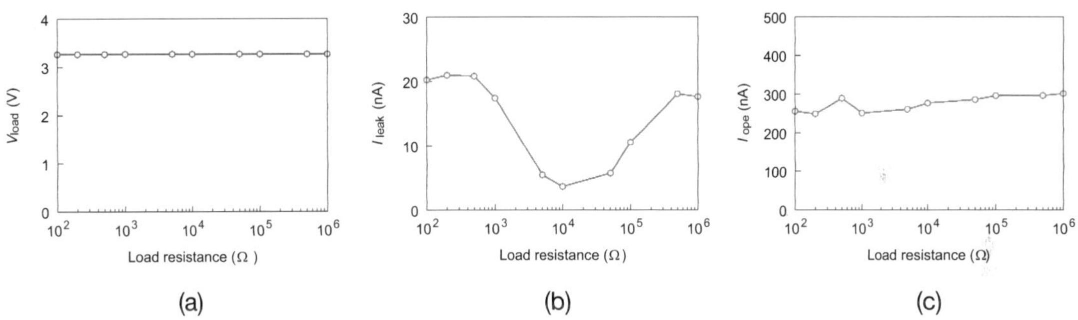

Figure 6(a) Load resistance influence on the output voltage. (b) Leak current in sleep mode with various load resistance. (c) Operational current of the holding circuit with various load resistance.

Acknowledgement
This work was supported by JST CREST Grant Number JPMJCR15Q4, Japan.

References

[1] S. Yamada and H. Toshiyoshi, "Self-Powered Artificial Sensory Nervous System Using Ring Oscillator for Pulse Density Modulation," IEEE Electron Device Letters, vol. 38, no. 10, pp. 1477-1480, 2017.
[2] J. Chun, B. U. Ye, J. W. Lee, D. Choi, C. Kang, S. Kim, Z. L. Wang, and J. M. Baik, "nanogenerator via electric double layer effect," Nat. Commun., vol. 7, no. May, pp. 1–9, 2016.
[3] S. Wang, L. Lin, and Z. L. Wang, "Nanoscale-triboelectric-effect enabled energy conversion for sustainable powering of portable electronics," Nano Lett., vol. 12, no. 12, pp. 6339–46, 2012.

IOP Conf. Series: Journal of Physics: Conf. Series **1052** (2018) 012071 doi:10.1088/1742-6596/1052/1/012071

Performance enhancement by an improved compact design for self-powered synchronous switching harvesting circuits

Weiqun Liu[1], Shuang Zhang[1], Adrien Badel[2], Fabien Formosa[2], Guangdi Hu[1]

[1]School of Mechanical Engineering, Southwest Jiaotong University, China
[2]Laboratoire SYMME, Université Savoie Mont Blanc, France
E-mail: weiqunliu@home.swjtu.edu.cn

Abstract. In the self-powered synchronous switching circuits, the electronic beakers for producing the desired switching control signal consumes a significant part of harvested energy. Moreover, the capacitors used in the electronic breakers brings the effect of decreasing the electromechanical coupling level thus hinders the available power. In this paper, a novel circuit design by combining two separate electronic breaker into a unified one in which only a single capacitor is required and the introduced parasitic capacitance is cut to half. Besides, the diodes in the envelope detectors are also removed. Consequently, both performance enhancement and cost reduction are achieved with this new circuit.

1. Introduction

Interface circuit is an important part of the piezoelectric generator and will significantly affect the performance. The synchronous switching circuits, such as SSHI (Synchronous Switching Harvesting on an Inductor) [1], SECE (Synchronous Electric Charge Extraction) [2], OSECE (Optimized SECE) [3] and so on, are considered as effective approaches for enhancing the generator's performance in the case of low electromechanical coupling level and providing better adaptability for load variations. A critical issue for these synchronous switching circuits is the self-powered realization. Various methods including electronic breaker [4-6], velocity control [7] , mechanical switches [8-10], integrated control circuits [11] etc. were proposed. Among them, the electronic breaker possesses the advantage of simplicity and reliability. Generally, a part of the extracted energy from the generator is consumed by the electronic breaker. Moreover, the parasitic capacitance in the electronic breaker, mainly referred to the envelope detector capacitance, imposes an equivalent effect of connecting a parallel capacitor of several times the parasitic capacitance to the generator, thus decreasing the coupling level, especially for the generators with small intrinsic capacitance. Consequently, the generator's performance degenerates in contrast with the ideal case.

Reducing the components of the self-powered circuit components while keeping the function could save the energy consumption and would be helpful to the performance. Recently, an alternative SSHI self-powered design has been proposed with fewer diodes used and better performance [12]. However, two capacitators are still used in the circuit for peak detectors. Compared with the diodes, the capacitance plays a more critical role in the by introducing the switching phase lag and decreasing the coupling level.

In this paper, we propose a simpler and more efficient circuit breaker for synchronous switch techniques dedicated to piezoelectric vibration energy harvesters. In comparison with the regular electronic breaker design for the SP-OSECE (Self-Powered OSECE) [6], this new CSP-OSECE (Compact SP-OSECE) circuit multiplexes the comparator in the envelope detector instead of the diode.

More important, only a single capacitor is used for both positive and negative peak detection. The parasitic capacitance of the electronic breaker is thus reduced to half so that the performance is improved.

2. Principle and analysis

Fig. 1 (a) presents the regular SP-OSECE design with the usual electronic breaker. In the circuit, the piezoelectric element is represented by an equivalent current source $i_{eq}=\alpha_0\dot{u}$ in parallel with an intrinsic capacitance C_0 and a leakage resistor R_0. Here, α_0 is the piezoelectric force factor and u is displacement of the piezoelectric generator. Two separate envelope detectors are used for determining the positive and negative piezoelectric voltage peak respectively. Each envelope detector is composed of an filtering RC network (R_{pi} an C_{pi}) and a diode D_{pi}. In addition, a PNP transistor T_{p1} is used as the comparator for the positive breaker and a NPN transistor T_{p2} is used as the negative comparator. At the beginning of the positive half-period, both switches S_1 and S_2 are opened which C_0 is charged by i_{eq} with the piezoelectric voltage V_p increasing. Meanwhile, C_{p1} is charged through R_{p1} and D_{pi} with V_{cp1} follows V_p while C_{p2} is charged through R_{b2} and T_{p2} as well. After V_p reaches the maximum value and starts to decline, V_{cp1} keeps this maximum value. As V_p decreases further, the comparator T_{p1} conducts and output the switching control signal to turn on S_1. Consequently, the voltage inversion is started in the form of LC oscillation (L_1 and C_0). When the induced voltage on L_3 at the secondary side of the transformer is higher than the load voltage V_{load} during the voltage inversion process, the energy stored on the transformer is transferred to the load. Afterwards, similar operations happens for the negative half-period. Clearly, in the regular electronic design, the current source i_{eq} has to charge C_0, C_{p1} and C_{p2} together. Consequently, the piezoelectric voltage V_p in the positive half can be approximated as

$$V_p \approx \frac{V_{load}}{m} + \int_{t_1}^{t_1+T/2} \frac{\alpha_0 \dot{u}}{C_0 + C_{p1} + C_{p2}} dt = \frac{V_{load}}{m} + \int_{t_1}^{t_1+T/2} \frac{\alpha_1 \dot{u}}{C_0} dt \qquad (1)$$

in which $\alpha_1=\alpha_0 C_0/(C_0+2C_p)$ is the modified piezoelectric force factor with $C_{p1}=C_{p2}=C_p$, t_1 is the start time of the positive half-period and T is the period. It can be inferred that the piezoelectric force factor α is decreased to α_1 by C_{p1} and C_{p2}.

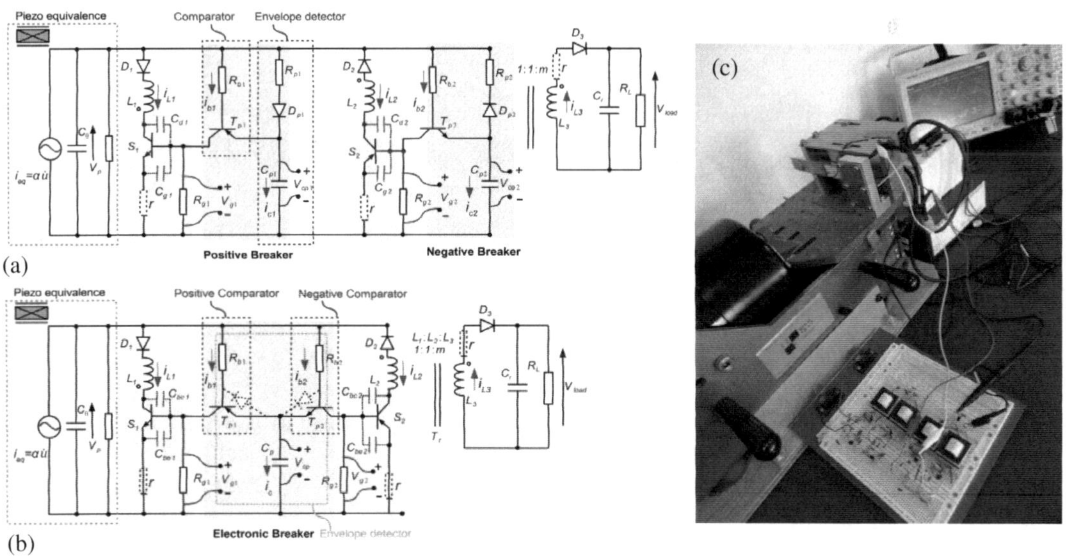

Figure 1 (a) SP-OSECE circuit with regular electronic breakers; (b) CSP-OSECE circuit.

In order to enhance the performance, the CSP-OSECE is proposed with a more compact design as seen in Fig. 1 (b). Instead of two separate electronic breakers, a unified one is constructed for both positive and negative half-periods with a single capacitor C_p. Moreover, the comparator transistor T_{p1} of the positive half-period is multiplexed for not only producing the switching signal but also rectifying the current instead of the diode D_{p2} in Fig. 1 (a) while T_{p2} covers the function of D_{p1} in the negative half-period as well. Two friction diodes are plotted with dotted lines in Fig. 1 (b) to indicate the multipurpose usage of the comparators. The current source i_{eq} charges C_p through T_{p2} and R_{b2} in the positive half-period and reversely charge C_p through T_{p1} and R_{b1} in the negative half-period. Both positive and negative voltage peak are reserved on the same capacitor while the comparators works similarly to detect the peak position as before. Fig. 2 presents the voltage waveforms and the switching signal of the two circuits. Easy to find, the CSP-OSECE circuit works the same as the regular SP-OSECE circuit in Fig. 1 (a) with fewer components. Moreover, due to the single capacitor configuration, we can write down the piezoelectric voltage in the positive half-period as

$$V_p \approx \frac{V_{load}}{m} + \int_{t_1}^{t_1+T/2} \frac{\alpha \dot{u}}{C_0 + C_p} dt = \frac{V_{load}}{m} + \int_{t_1}^{t_1+T/2} \frac{\alpha_2 \dot{u}}{C_0} dt \tag{2}$$

in which $\alpha_2 = \alpha C_0/(C_0+C_p)$.

By comparing eq. (1) and eq. (2), it is found that α_2 is larger than α_1, which means that better power performance is expected. It is also seen in Fig. 2 that the CSP-OSECE obtains higher load voltage than the regular SP-OSECE does.

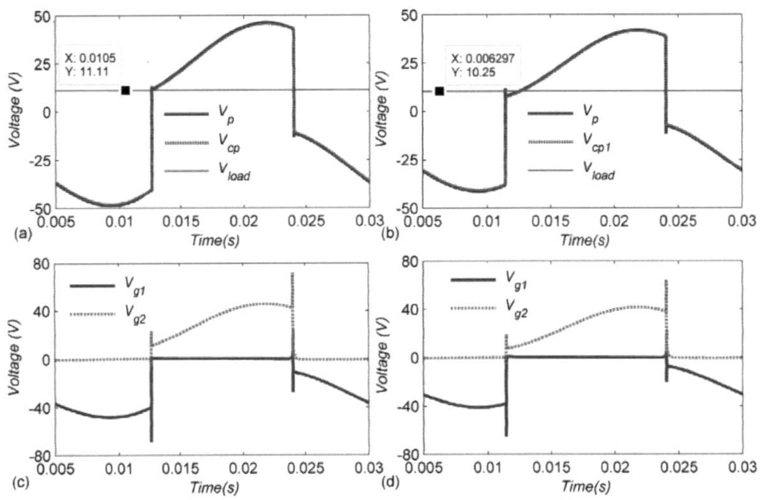

Figure 2. Operation waveforms of two self-powered circuits: (a),(c) CSP-OSECE; (b),(d) SP-OSECE.

3. Results

In order to verify the performance, a piezoelectric generator is subjected to both SP-OSECE and CSP-OSECE circuits as shown in Fig. 1 (c). The generator's parameters have been identified as follow: C_0=14nF, α=3.1e-3N V^{-1} and M=0.006kg. Here, M is the inertial mass. The resonant frequency is measured to be 41Hz in the short-circuit case. In addition, the envelope capacitance C_p is selected to be 1.75nF for both circuits and R_b is set to 3.3kΩ. The value of C_p is selected to obtain the optimal power performance of the circuits.

The constant displacement case is firstly investigated with the magnitude of u fixed as 1.25mm and the excitation frequency is selected to be 41Hz at the resonant frequency. Fig. 3 (a) shows the power results of the two circuits. Clearly, the load dependence properties of the two circuits is close to each other while the optimal power is obtained around 470kΩ in both cases. However, the CSP-OSECE obtains much better performance than the SP-OSECE except the small load case. A maximum power

IOP Conf. Series: Journal of Physics: Conf. Series **1052** (2018) 012071 doi:10.1088/1742-6596/1052/1/012071

of 0.301mW is obtained for CSP-OSECE with a power increase of 30% in consideration that the SP-OSECE circuit has a maximum power of 0.233mW.

The test with a constant acceleration amplitude of 0.5g was also performed with the optimal load of 470kΩ from 20Hz to 60Hz with the power plotted in Fig. 3 (b). Similar to the constant displacement case, the CSP-OSECE circuit always shows better performance. A power boost of 38% is found around the resonant frequency. Moreover, wide bandwidth is also obtained with the CSP-OSECE circuit.

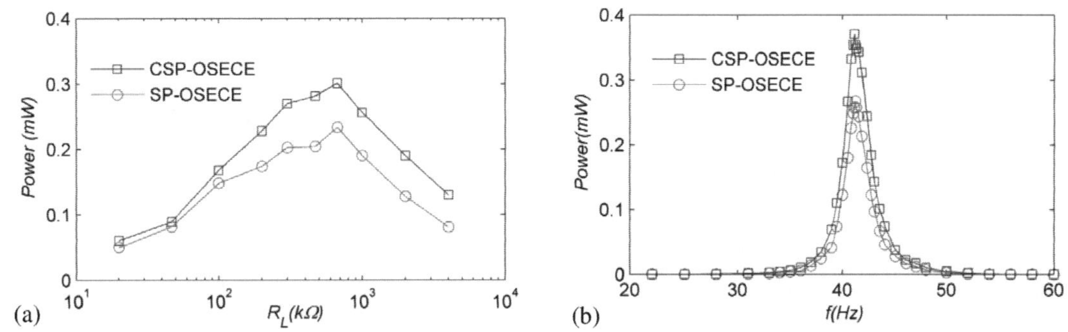

Fig. 3 Power performance of the generator using two self-powered circuits: (a) in the case of constant displacement amplitude; (b) in the case of constant acceleration.

4. Conclusion

A compact design of the self-powered synchronous switching circuit is proposed in this paper. Compared with the regular SP-OSECE design, the new CSP-OSECE circuit brings two improvements: First, the parasitic capacitance is reduced to half; Second, the circuit has fewer component by excluding the envelope detector diodes. The former improvement is especially useful about the enhancement of the power performance. It is validated by experiments in both constant displacement and constant acceleration cases. Meanwhile, the latter one saves the cost and also assembling space.

Acknowledgement

The work is supported by the NSFC (National Science Foundation of China) under grant No. 51505395, the Fundamental Research Funds for the Central Universities of China (No. 2682017CX26, 2682017CY05) and the Postdoctoral Science Foundation of China (2017T100709, 2016M600747).

References

[1] Guyomar D, Badel A, Lefeuvre E and Richard C 2005 *IEEE Trans. Ultrason. Ferroelectr. Freq. Control* 52(4) 584–595.
[2] Lefeuvre E, Badel A, Richard C and Guyomar D 2005 *J. Intell. Mater. Syst. Struct.* 16(10) 865–876.
[3] Wu Y, Badel A, Formosa F, Liu W and Agbossou A E 2013 *J. Intel. Mat. Syst. Str.* 24(12) 1445-1458.
[4] Liang J and Liao W 2012 *IEEE Trans. Ind. Electron.* 59(4) 1950–1960.
[5] Lallart M and Guyomar D 2008 *Smart Mater. Struct.* 17(3) 35030.
[6] Wu Y, Badel A, Formosa F, Liu W and Agbossou A E 2014 *J. Intell. Mater. Syst. Struct.* 25(17) 2165.
[7] Chen Y Y, Vasic D, Costa F, Wu W J and Lee C K. 2012 *Eur. Phys. J. Appl. Phys.* 57(3) 30903.
[8] Wu Y, Badel A, Formosa F, Liu W and Agbossou A E 2014 *J. Intell. Mater. Syst. Struct.* 25(14) 1658.
[9] Liu W, Formosa F, Badel A, Wu Y and Agbossou A 2014 *Sensors Actuators A Phys.* 216 106–115.
[10] Liu H, Liang J and Ge C 2015 *Appl. Phys. Lett.* 107(14) 141902.
[11] Boisseau S, Gasnier P, Gallardo M and Despesse G *Journal of Physics: Conference Series* 476 (Proc. PowerMEMS 2013) 012080.
[12] Eltamaly A and Addoweesh K 2017 *IEEE Trans. Power Electron.* 32(10) 7663.

Autoparametric Excitation and Self-powered SSHI for Power Enhancement in Piezoelectric Vibration Energy Harvester

H Asanuma, T Komatsuzaki, and Y Iwata

Faculty of Mechanical Engineering, Kanazawa university, Kakuma-machi, Kanazawa 920-1192, Japan

h-asanuma@se.kanazawa-u.ac.jp

Abstract. We proposed an autoparametric excitation harvester employing a microfabricated leaf spring for the base beam and a synchronized switch harvesting on inductor (SSHI) interface. Our harvester achieved miniaturization, low threshold acceleration of the autoparametric excitation, and increase in output power, compared with the previous work. The base beam for amplifying the excitation was microfabricated from a stainless steel film, through the photolithography followed by the wet-chemical etching. To trigger the autoparametric excitation, the main and the base beams are designed such that the resonance frequency for the base beam becomes twice higher than that for the main beam. The resonance frequencies obtained in experiment for the main and the base beams were 26.6 and 53.1 Hz, respectively. This study employed a self-powered parallel SSHI interface, which can increase the piezoelectric voltage and thus the output power, consuming only a small portion of the harvested energy. The harvester connected with the self-powered SSHI interface successfully displayed the autoparametric excitation at acceleration greater than 1.0 m/s^2, and the output power showed 1.12 mW at the frequency of 53.1 Hz under the acceleration of 2.0 m/s^2, which is 1.43-fold increase over the standard AC-DC interface.

1. Introduction

To increase output power for piezoelectric vibration energy harvesters (p-VEHs) significantly, the use of autoparametric excitation is one promising way [1, 2]. Previous studies proposed a harvester structure in which the main beam for a p-VEH is mounted perpendicular to the base beam for amplifying base excitation [1]. However, the structure has a problem of increasing harvesters' volume, because the base beam horizontally extends and requires a large area for clamping the end of the base beam. In addition, the output power reported previously is not so high for the size and for the applied acceleration. Specifically, the output power remains blow 0.2 mW at the applied acceleration of 19.6 m/s^2 [1]. To decrease the harvester's volume and lower the threshold acceleration, as well as to further increase the output power, we proposed a harvester employing a microfabricated leaf spring for the base beam and a nonlinear interface termed a synchronized switch harvesting on inductor (SSHI).

2. Harvester structure

Figure 1 shows a structure of the proposed harvester. The harvester consists of the main steel beam (AISI304, $41\times20\times0.3$ mm^3), copper tip mass (19.9 gram), piezoelectric ceramics (Fuji Ceramics, C-6, 37.7 nF), copper base mass (78.9 gram), copper clamp brackets (51.8 gram), and base steel beam

(AISI304, thickness 0.5 mm). The piezoelectric ceramics, which form a parallel bimorph, were bonded manually at the end of the main beam. The base beam was microfabricated from the stainless steel film through the photolithography followed by the wet-chemical etching [3, 4]. As shown in figure 1(c), compared with the previous work [1], the proposed structure for the base beam can reduce the harvester's volume significantly. The main beam vibrates perpendicular to the direction of the applied acceleration, whereas the base beam does parallel to the direction of the applied acceleration. To trigger the autoparametric excitation, the main and the base beams are FEM designed such that the resonance frequency for the base beam becomes twice higher than that for the main beam. The resonance frequencies obtained in experiment for the main and the base beams are 26.6 and 53.1 Hz, respectively.

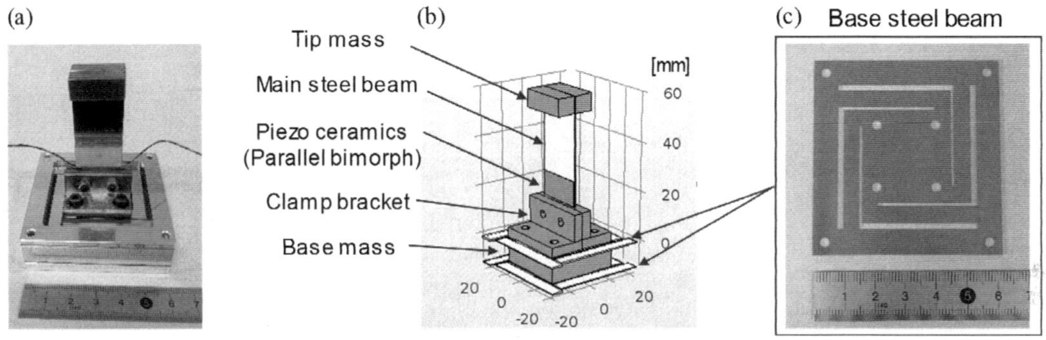

Figure 1. (a) Photograph, (b) schematic structure, and (c) photograph of base steel beam of the proposed autoparametric excitation harvester.

3. Results and discussion

Figure 2 shows the output power P_{out} vs the resistive load R_L, the standard AC-DC interface, the self-powered parallel SSHI interface, the typical transient response of the piezoelectric voltage V_p and the tip displacement for the harvester connected with the self-powered parallel SSHI interface, respectively. V_p is rectified through the full bridge diodes, and then the output DC voltage V_{DC} is obtained across the resistive load connected with the smoothing capacitor C_s. C_s is 10 μF, and L, R_{ed}, and C_{ed} used in the SSHI interface are 100 mH, 300 kΩ, and 2.2 nF, respectively. The output power is calculated from V_{DC}^2/R_L. As shown in figure 2(c), the self-powered parallel SSHI interface consists of three parts: the envelope detector, the comparator, and the digital switch, and it is the same as the standard interface adding the highlighted part drawn in red. The self-powered parallel SSHI interface can enhance V_p and thus the output power by the LC_p electrical oscillation between the inductor and the piezoelectric capacitance, consuming only a small portion of the harvested energy. The brief explanation on the workflow of the self-powered parallel SSHI is as follows. When the voltage $|V_{Ced}|$ across the envelope capacitor $|C_{ed}|$ is less than $|V_p|$, the bipolar transistor T_1 remains off and no switching action has occurred. After the displacement of the harvester reaches the maximum/minimum, $|V_p|$ starts to decrease and $|V_{Ced}|$ is greater than $|V_p|$. When the voltage difference of $|V_{Ced}-V_p|$ exceeds the threshold voltage of T_1, T_1 conducts. Subsequently, the bipolar transistor T_2 in the digital switch turns on as well, and then the voltage inversion of V_p caused by the LC_p electrical oscillation starts. After the inversion, T_2 and T_1 returns to off immediately. The detailed workflow is reported in [5]. As shown in figure 2(a), the impedance matched resistive load for both interfaces is 150 kΩ at the frequency of 53.1 Hz under the acceleration of 2.0 m/s^2, and the maximum output power for the self-powered SSHI interface showed 1.12 mW, which is 1.43-fold increase over the standard AC-DC interface. As shown in figure 2(d) and (e), the piezoelectric voltage and the tip displacement for the proposed harvester grew gradually, and then in 10 seconds, they rapidly enhanced by the autoparametric excitation. Therefore, our designed harvester successfully displayed the autoparametric excitation, and achieved both miniaturization and high output power.

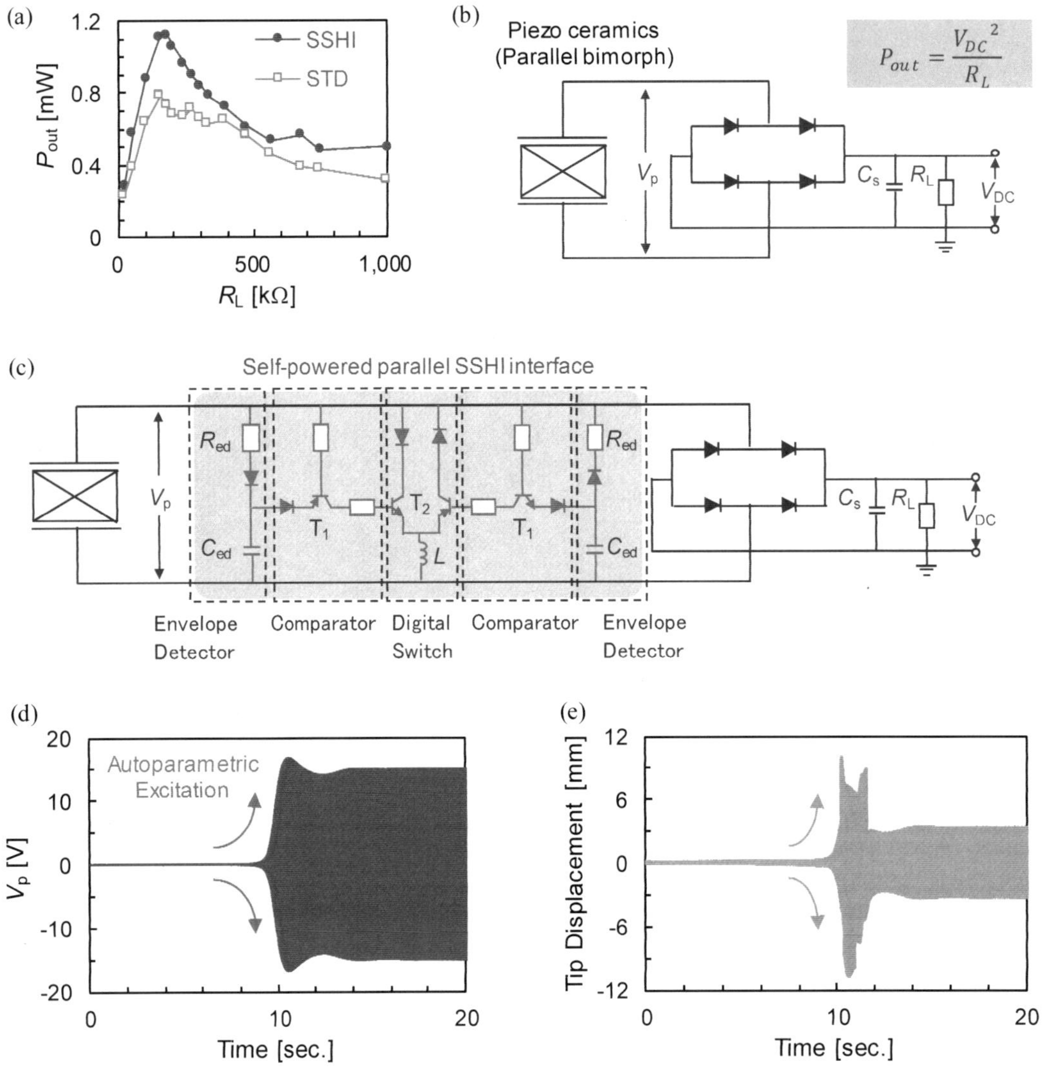

Figure 2. (a) Output power P_{out} vs resistive load R_L, (b) standard AC-DC interface, (c) self-powered parallel SSHI interface, typical transient response of (d) piezoelectric voltage V_p and (e) tip displacement for the harvester connected with the SSHI interface.

Figure 3 shows the output power and the peak-peak tip displacement vs the applied acceleration for the self-powered parallel SSHI and the standard interfaces under the autoparametric excitation. To prevent the breakage of the piezoelectric ceramics when applying high acceleration, the output power was evaluated at the frequency in which the autoparametric excitation started to display. Thanks to a large weight of the base mass and of the clamp brackets, the vibrational energy effectively transferred from the base to the main beam, and thus our harvester succeeded in lowering the threshold acceleration of the autoparametric excitation to 1.0 m/s^2. Furthermore, the harvester connected with the self-powered parallel SSHI interface showed 1.3-1.7 increase in output power over the standard interface, whereas the displacement showed the similar magnitude. Therefore, the proposed harvester achieved miniaturization, low threshold acceleration, and increase in output power, compared with the previous work.

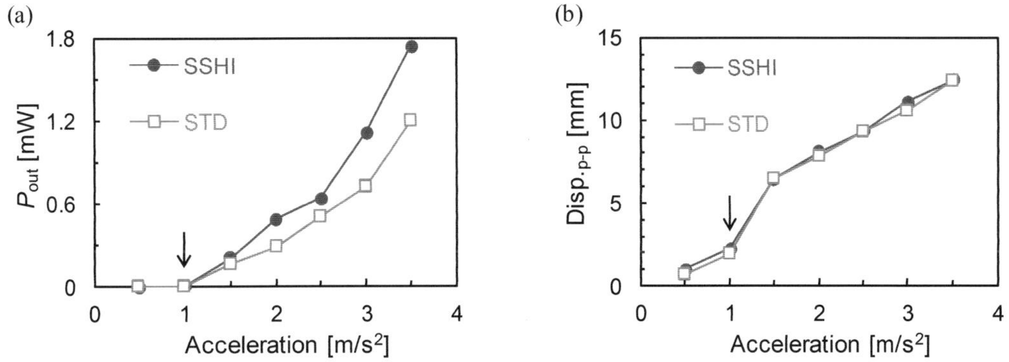

Figure 3. (a) Output power P_{out} and (b) peak-peak tip displacement Disp.$_{p-p}$ vs applied acceleration for the SSHI and the standard interfaces under the autoparametric excitation.

4. Conclusion

To achieve miniaturization, low threshold acceleration, and increase in output power, we proposed an autoparametric excitation harvester employing a microfabricated leaf spring for the base beam and a self-powered parallel SSHI interface. The base beam was microfabricated from the stainless steel film, through the photolithography followed by the wet-chemical etching. To trigger the autoparametric excitation, the main and the base beams are designed such that the resonance frequency for the base beam becomes twice higher than that for the main beam, and the resonance frequencies obtained in experiment for the main and for the base beams were 26.6 and 53.1 Hz, respectively. Thanks to a large weight of the base mass and of the clamp brackets, the vibrational energy effectively transferred from the base to the main beam, and thus our harvester succeeded in lowering the threshold acceleration of the autoparametric excitation to 1.0 m/s^2. The output power for the harvester connected with the self-powered SSHI interface showed 1.12 mW at the frequency of 53.1 Hz under the acceleration of 2.0 m/s^2, which is 1.43-fold increase over the standard AC-DC interface.

Acknowledgements

We would like to thank the Japan Society for the Promotion of Science (JSPS). This work was supported by Grant-in-Aid for Young Scientists (B) (No. 16K18039).

References

[1] Yu Jia and A. A. Seshia, *Sensors and Actuators A*, **220** (2014), pp. 69-75
[2] Z. Yan and M. R. Hajj, *Smart Mater. Struct.*, **24** (2015), 115012
[3] H. Asanuma et al., *Journal of Micromechanics and Microengineering*, **25** (2015), 104013
[4] H. Asanuma *et al.*, *AIP Advances*, **6** (2016), 075206
[5] Pang S, Kan J and Li W 2015 *Journal of Computational Information System* **11** 3203-3215

Numerical Investigation of Mechanically and Electrically Switching SSHI in Highly Coupled Piezoelectric Vibration Energy Harvester

K Sakamoto, H Asanuma, T Komatsuzaki, and Y Iwata

Faculty of Mechanical Engineering, Kanazawa university, Kakuma-machi, Kanazawa 920-1192, Japan

h-asanuma@se.kanazawa-u.ac.jp

Abstract. In aiming to increase output power for piezoelectric vibration energy harvesters, a self-powered synchronized switch harvesting on inductor (SSHI) using an electrical or mechanical switch has considerable attention. However, the advantages and disadvantages of the two switching technique for the self-powered SSHIs remains unclear. In addition, for a harvester with a high electromechanical coupling coefficient *k*, the piezoelectric damping force, which enhances by the SSHI's voltage increase, is likely to reduce the harvester's displacement and thus lower the output power. We developed simulation technique, and numerically investigated the performance for the electrical switch SSHI (ESS) and for the mechanical switch SSHI (MSS) harvester, considering the feedback of the piezoelectric damping force. The numerical investigation revealed that, for the ESS, the piezoelectric damping force reduces the displacement every switching on at the maximum/minimum displacement, and thus lowers the output power. In contrast, the MSS, in which the switch turns on only when the displacement exceeds the gap distance, achieved a higher output power, and exhibited intriguing phenomena that the output power continues to increase, whereas the displacement is held constant. Therefore, for a harvester with high *k*, the MSS can outweigh the ESS.

1. Introduction

To increase the output power for a piezoelectric VEH, Guyomar *et al* [1] developed a synchronized switch harvesting on inductor (SSHI) technique. The SSHI employs an additional set component of an active switch and an inductor, which are connected in parallel or series with the standard AC-DC circuit. When the switch turns on at every maximum/minimum of the harvester's displacement, the piezoelectric voltage was inversed by the *LC* electrical oscillation caused between the inductor and the capacitance of the piezoelectric material, thereby increasing the output voltage and thus the output power. However, the switch used in the early stage SSHI was controlled by an external power source.

Subsequently, a self-powered SSHI using an electrical or mechanical switch that requires no external power source was developed. The electrical switch, which consists of the envelope detector, the comparator, and the digital switch, successfully controlled on/off by consuming a small portion of the harvested energy [2,3]. The mechanical switch was designed by utilizing a main beam of the harvester [4,5]. The self-powered SSHI using the electrical and mechanical switches showed excellent performance, however, the advantages and disadvantages for the two switching technique are still not clear. In addition, for a harvester with a high electromechanical coupling coefficient *k*, the piezoelectric damping force, which enhances the SSHI voltage increase, suppresses vibration of the harvester and thus lowers output power [6].

This study investigates the performance of the electrical and mechanical switch SSHI numerically, considering the feedback of the piezoelectric damping force from.

2. Harvester Structure and Circuits

Figure 1 shows the harvester for the electrically-switching SSHI (ESS) and for the mechanically-switching SSHI (MSS). This study employs a cantilever-type VEH consisting of a stainless steel beam, copper mass, and piezoelectric bimorph connected in parallel. The squared electromechanical coupling coefficient k^2 is 0.038, which is several times higher than that for the previous studies [1].

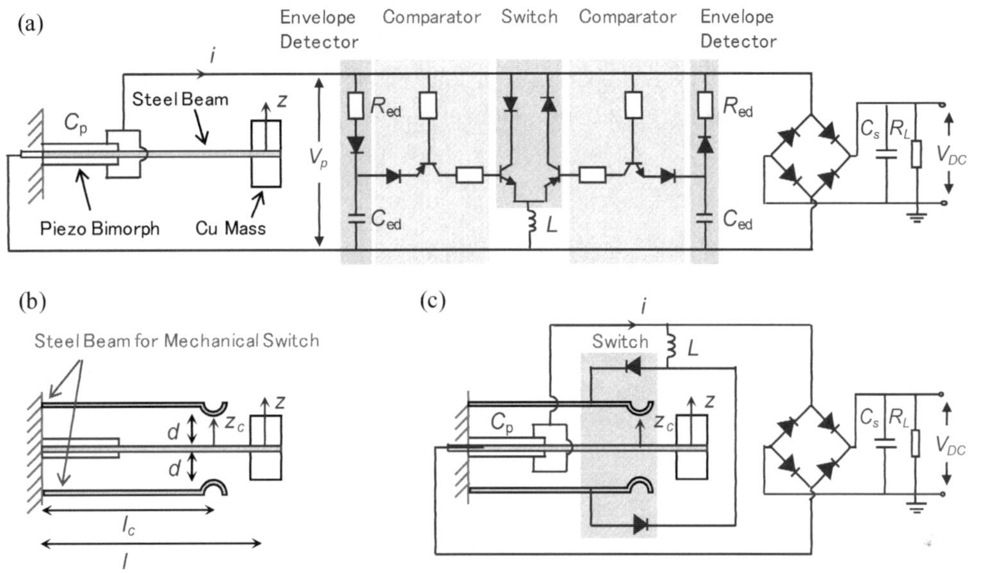

Figure 1. (a) ESS harvester, (b) structure of mechanical switch and (c) MSS harvester.

Based on the single degree of freedom (SDOF) model, the electromechanical equation for the piezoelectric vibration energy harvester are expressed as follows [1]:

$$M_{eff}\ddot{z} + D\dot{z} + K_{sc}z = F - \alpha V_p \tag{1}$$

$$i = \alpha\dot{z} - C_p\dot{V}_p, \tag{2}$$

where F is the external force, M_{eff} (9.4 gram) is the effective mass, D (0.036 Ns/m) is the mechanical damping coefficient, K_{sc} (344 N/m) is the short-circuited stiffness, z is the tip displacement of the harvester, α (7.73×10^{-4} N/V) is the force factor, V_{p} is the piezoelectric voltage, i is the outgoing current, C_{p} (44 nF) is the capacitance of the piezo bimorph, respectively. We stress that, unlike the SPICE-based simulation [3], our numerical model incorporates the feedback of the piezoelectric damping force αV_{p} from the electrical to the mechanical domain, which is enhanced by the SSHI's voltage increase. We used Matlab/Simulink and Simscape/Simeletronics to simulate the harvester's performance. R_{ed} and C_{ed} in the ESS are 300 kΩ and 2.2 nF, respectively.

As shown in figure 1(b) and (c), for the MSS harvester, the equation about mechanical vibration is changed as follows:

$$z_c = sz = \frac{3}{2}\left(\frac{l_c}{l}\right)^2\left(1-\frac{l_c}{3l}\right)z \tag{3}$$

$$M_{eff}\ddot{z} + D\dot{z} + K_{sc}z + F_{MS} = F - \alpha V_p \tag{4}$$

$$F_{MS} = \begin{cases} K_{MS}(z_c - d) + D_{MS}\dot{z}_c = sK_{MS}(z - d/s) + sD_{MS}\dot{z} & z_c > d \\ 0 & |z_c| \le d \\ K_{MS}(z_c + d) + D_{MS}\dot{z}_c = sK_{MS}(z + d/s) + sD_{MS}\dot{z} & z_c < -d \end{cases}, \tag{5}$$

where z_c is the displacement at the contact position l_c, d (0.41 mm) is the gap distance, F_{MS}, K_{MS} (66.3 N/m), and D_{MS} (0.3 Ns/m) are the total reaction force, the stiffness, and the mechanical damping coefficient of the mechanical switch, respectively. l and l_c in figure 1(b) are set to 53 and 30 mm. The impedance-matched R_L for the ESS and MSS are 150 and 180 kΩ, C_s is 10 μF, and the inductance L is 100 mH, respectively. All parameters used in the simulation are empirically attainable.

3. Results and discussion

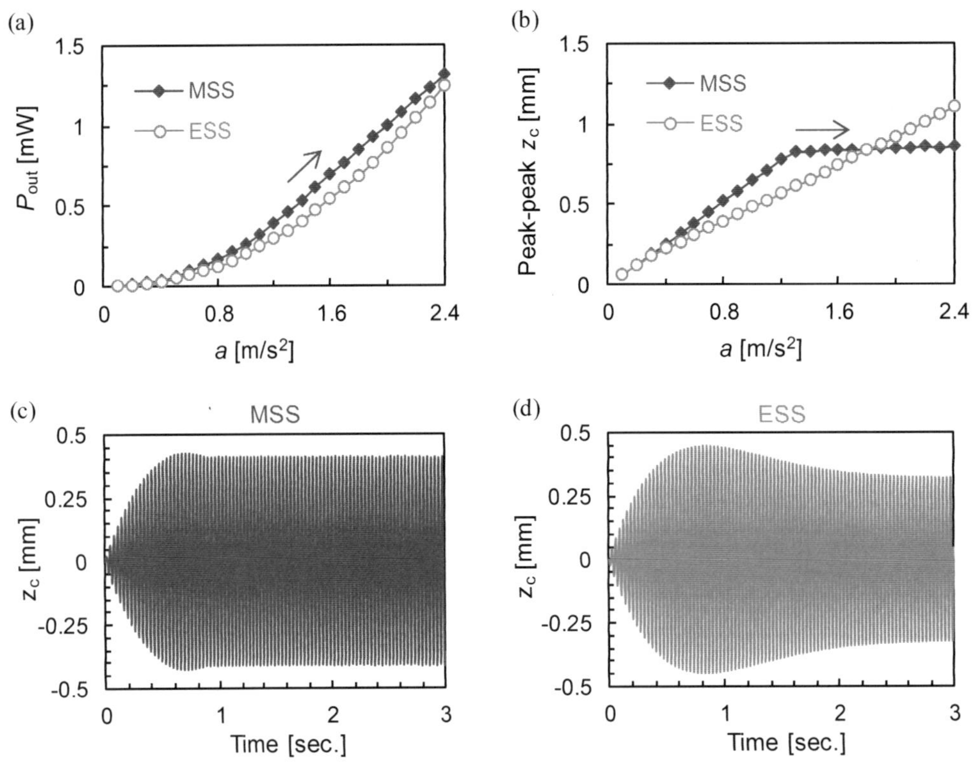

Figure 2. (a) Output power P_{out} and (b) peak-peak displacement z_c at the contact position l_c vs applied acceleration a, and transient response of z_c (c) for MSS and (d) for ESS.

Figure 2 shows output power P_{out} vs applied acceleration a at the resonance frequencies (30.4-30.8 Hz), and transient response of z_c for three seconds. For the ESS harvester, the piezoelectric damping force enhanced by the SSHI reduces the displacement and thus the resulting output power. Thus, for the harvester with high k, the ESS which turns on at every maximum/minimum displacement is not effective. In contrast, the MSS, which turns on only when z_c exceeds d, showed excellent performance. Intriguingly, the output power for the MSS continues to increase, whereas the displacement remains constant. The reason is explained as follows.

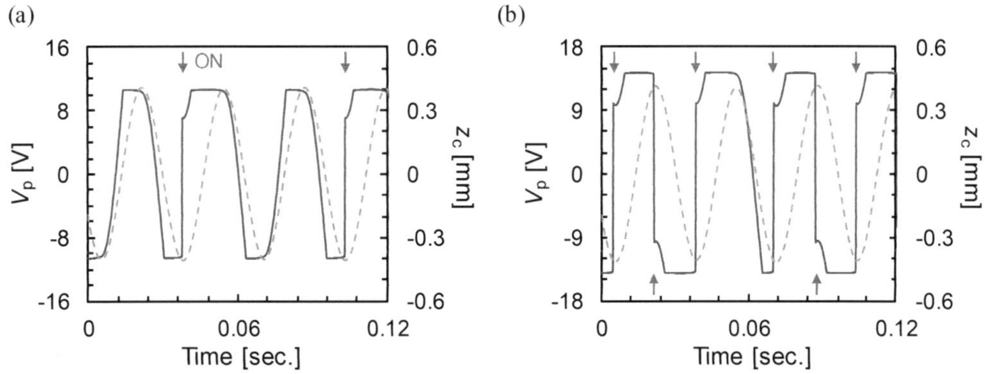

Figure 3. MSS's waveforms of the piezoelectric voltage V_p and of the displacement z_c at the contact position at acceleration of (a) 1.4 m/s^2 and (b) of 2.0 m/s^2.

Figure 3 shows waveforms of the piezoelectric voltage V_p and the displacement z_c at the contact position for the MSS at a of 1.4 and 2.0 m/s^2. As shown in figure 3(a), when the MSS turns on, the piezoelectric damping force reduces the displacement and skips the next switching on. However, the displacement gradually grows and MSS turns on again. Under higher excitation, the number of switching on increases, which leads to the increase in output power under the constant displacement.

4. Conclusion

We numerically investigated the performance of the ESS and the MSS harvester, considering the piezoelectric damping force enhanced by the SSHI's voltage increase in a harvester with high k. The displacement of the ESS decreases every switching on at the maximum/minimum displacement, which lowers the output power. In contrast, the MSS, which turns on only when the displacement exceeds the gap distance, showed the intriguing result that the output power continues to increase, whereas the displacement remains constant. Thus, for a harvester with high k, the MSS can outweigh the ESS.

Acknowledgements

We would like to thank the Japan Society for the Promotion of Science (JSPS). This work was supported by Grant-in-Aid for Young Scientists (B) (No. 16K18039).

References

[1] Guyomar D, Badel A, Lefeuvre E and Richard C 2005 *IEEE Transactions on Ultrasonics, Ferroelectrics, and Frequency Control* **52** 584-594
[2] Lallart M and Guyomar D 2008 *Smart Materials and Structures* **17** 035030
[3] Pang S, Kan J and Li W 2015 *Journal of Computational Information System* **11** 3203-3215
[4] Liu H, Liang J and Ge C 2015 *Applied Physics Letters* **107** 141902
[5] Asanuma H, Okubo H, Komatsuzaki T and Iwata Y 2016 *Proc. of PowerMEMS 2016, Journal of Physics: Conference Series* **773** 012082
[6] Makihara K, Yamamoto Y, Yoshimizu K, Horiguchi C, Sakaguchi H and Fujimoto K 2015 *Smart Materials and Structures* **24** 037005

Power-electronic-interface topology for MEMS energy harvesting with multiple transducers

Binh Duc Truong[1,2], Cuong Phu Le[1], Einar Halvorsen[1] and Shad Roundy[2]

[1] Department of Microsystems, University College of Southeast Norway, Campus Vestfold, Raveien 215, 3184 Borre, Norway.

[2] Department of Mechanical Engineering, University of Utah, 50 S Central Campus Drive, Salt Lake City, UT 84112, USA.

E-mail: `shad.roundy@utah.edu`

Abstract. Based on circuit simulations, this paper investigates a concept for a power electronic interface circuit for MEMS electrostatic energy harvesters. Two ordinary overlap-varying transducers are first electrically configured as a symmetric voltage doublers which enables the device to self-start from an initially low bias. The harvesting system is then reconfigured to couple with a buck-boost DC-DC converter in order to maximize the power delivered to an electronic load. The losses of electronic components due to diode voltage drop and parasitic resistance of inductors are taken into account for a feasibility investigation. Dependence of the maximum output power on inductance and switching frequency is explored.

1. Introduction

Development of standalone wireless sensors and implantable electronic devices over the past decades has been in a fast pace [1, 2]. Current portable electronics requires the use of batteries for supplying electrical energy [3]. They don't have sufficient energy density to last for a long time and may pose a toxic threat in implanted sensors [4]. Energy harvesting from vibration is one means for replacing batteries [5]. This energy conversion typically uses either piezoelectric, electromagnetic or electrostatic transduction even though some alternatives such as triboelectricity [6] and magnetostriction [7] also have been considered. In this paper, we focus on electrostatic harvesters due to their compatibility with microelectronics [8] and their suitability for microfabrication and miniaturization.

The power electronic interface forms an essential component in an energy harvesting system. Interface circuits ranging from simple passive diode rectification to complicated active-switch converters with smart control units have been reported in the literature. The primary aim is to maximize harvested power or harvester effectiveness [9–11]. In a recent work, an interesting configuration of Bennet's doubler composed of capacitors and diodes was proposed by de Queiroz [12]. When the output voltage across the storage capacitor is saturated [13], vibration energy is no longer scavenged. Therefore, for continued scavenging in a real system, control circuitry must reconfigure the system at this point. Several conceivable solutions were presented such as buck [14] and reversible buck-boost [15] converters. However, the former circuit is not able to synthesize an optimum load, which makes it hard to optimize the harvested power.

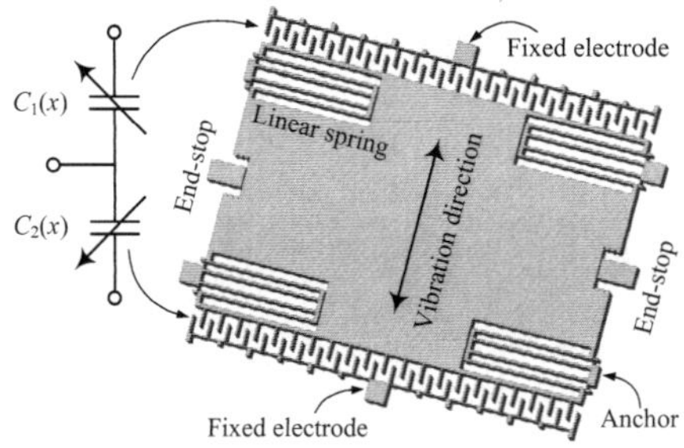

Figure 1: Schematic of the anti-phase overlap-varying electrostatic energy harvester.

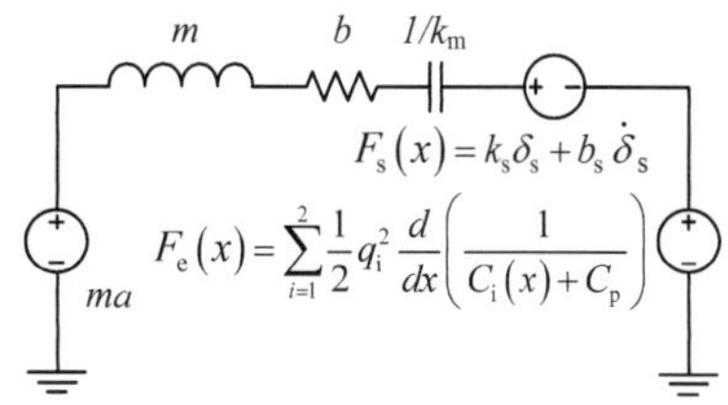

Figure 2: Equivalent circuit for mechanical domain.

Table 1: Model parameters

Parameters	Value
(a) *Device design*	
Proof mass, m	0.739 mg
Spring stiffness, k_m	16.41 N/m
Thin-film air damping, b	2.5e-5 Ns/m
Nominal overlap, x_0	110 μm
Nominal capacitance, C_0	15.16 pF
Parasitic capacitance, C_p	5.5 pF
Contact stiffness, k_s	3.361 MN/m
Impact damping, b_s	0.435 Ns/m
Max displacement, X_{max}	101.5 μm
(b) *Interface circuit*	
Biasing capacitor, C_b	1 nF
Storage capacitor, C_R	20 nF
ON resistance of SW, R_s	2 Ω
Inductance, L	150 μH
Inductor resistance, R_{sL}	45 kΩ/H
Output capacitance, C_S	1 μF
Load, R_L	5 MΩ

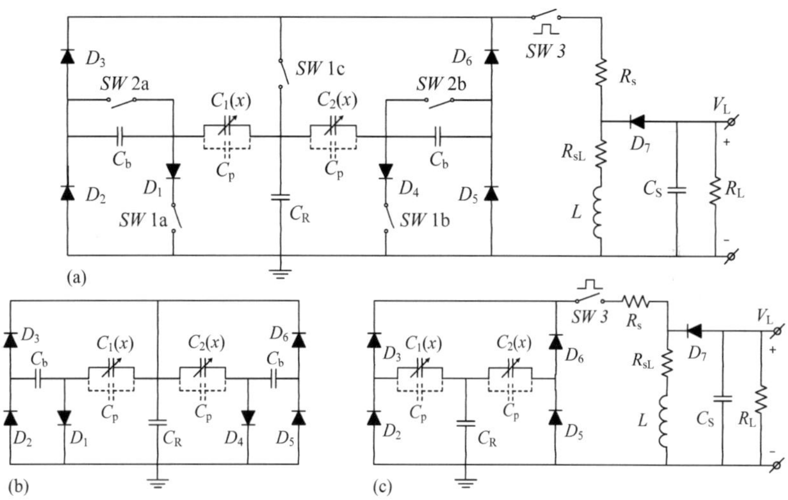

Figure 3: (a) Power electronic interface constructed by symmetrical voltage doubler and buck-boost converter and its corresponding operations when (b) SW 1abc [C], SW 2ab and SW 3 [O] and when (c) SW 1abc [O], SW 2ab [C] and SW 3 controlled by duty cycle. Notation: SW - switch, [C] - closed and [O] - open.

Many switches need to be continuously controlled in the latter topology which adds to the power consumption.

An advantage of the voltage doubler configuration is the potential to bootstrap itself from a low initial bias. In addition, a full-bridge rectifier and a buck-boost converter without an input filter capacitor may provide a resistive input impedance by tuning the duty cycle [16]. Based on circuit simulation, this contribution introduces an alternative interface circuit that combines those two interesting aspects. The question of how to optimize the ability of the processing circuit to harvest the maximum available output power is also an important issue in our study.

2. Energy harvester lumped-model

An anti-phase overlap-varying electrostatic energy harvester is considered for this investigation. Such a capacitor structure is shown in Figure 1. The proof mass m is suspended by four folded-beam springs with total linear stiffness k_m, while its motion is damped by a parasitic damping b and its maximum displacement X_{max} is defined by mechanical end-stops. The potential contact between the mass and the end-stops at sufficiently high accelerations are represented by the impact force F_s. The mass is also subject to the electrostatic force F_e generated by the transducers. This system is governed by the following differential equation

$$m\ddot{x} + b\dot{x} + k_{\mathrm{m}}x + F_{\mathrm{e}}\left(x\right) + F_{\mathrm{s}}\left(x\right) = ma \tag{1}$$

where $ma = mA\cos\left(\omega t\right)$ is the fictitious force with angular frequency ω. The two forces F_{e} and F_{s} are

$$F_{\mathrm{e}}\left(x\right) = \frac{1}{2}q_1^2\frac{\mathrm{d}}{\mathrm{d}x}\frac{1}{C_1\left(x\right)+C_{\mathrm{p}}} + \frac{1}{2}q_2^2\frac{\mathrm{d}}{\mathrm{d}x}\frac{1}{C_2\left(x\right)+C_{\mathrm{p}}}, \tag{2}$$

and

$$F_{\mathrm{s}}\left(x\right) = k_{\mathrm{s}}\delta + b_{\mathrm{s}}\dot{\delta} \tag{3}$$

where q_1 and q_2 are the charges on the transducers 1 and 2 respectively, $C_{1/2} = C_0\left(1 \pm \frac{x}{x_0}\right)$, C_{p} is stray capacitance in parallel with $C_{1/2}$, k_{s} is the end-stop stiffness, b_{s} is the impact damping and $\delta = |x| - X_{\max}$ is the relative displacement between the proof mass and the end-stops during impact. The equivalent circuit for the mechanical domain is depicted in Figure 2. Parameters of the design are listed in Table 1a.

3. Power electronic interface

Figure 3a shows how the harvester is coupled to the proposed circuit. Operation of the harvesting system can be divided into two schemes as presented in Figure 3b and 3c. At the beginning, *SW* 1*a*, 1*b* and 1*c* closed, *SW* 2*a*, 2*b* and 3 open, the two transducers are initially configured as a symmetric voltage doubler. The storage capacitor C_{R} is pre-charged to a voltage of V_0. For sufficiently high V_0, the voltage accumulated on C_{R}, namely $V_{C_{\mathrm{R}}}$, initially increases. After a certain number of cycles, steady state is achieved and $V_{C_{\mathrm{R}}}$ is maintained constant at $V_{C_{\mathrm{R}}} = V_{\mathrm{sat}}$. The states of switches 1*a*, 1*b*, 1*c*, 2*a* and 2*b* are then reversed. The energy harvesters are now connected to a buck-boost DC-DC converter without input filter capacitor and with diodes D_2, D_3, D_5 and D_6 functioning as a rectifier. Simultaneously, the transducers are primed by the "bias" voltage V_{sat} while C_{R} is acting as an energy storage. The electronic load is represented by the resistor R_{L} in parallel with the output capacitor C_{S}. When a buck-boost converter operates in discontinuous current mode, its input impedance R_{in} is driven through *SW* 3 by a pulse-width modulated square wave signal of duty cycle δ and frequency f_{s}. When f_{s} is much higher than the input vibration frequency f [16]

$$R_{\mathrm{in}} = \frac{2Lf_{\mathrm{s}}}{\delta^2}. \tag{4}$$

The time evolutions of voltages across C_{R} and R_{L} at the same input acceleration amplitude are shown in Figure 4 with different values of δ when the lossless inductor and the mathematically idealized diodes (i.e., zero voltage drop, near-zero leakage current and zero resistance when conducting) are utilized. After a transient time, V_{L} remains constant at V_{dc} which depends on δ. The power processing circuit is now supplying a DC voltage to the load. $V_{C_{\mathrm{R}}}$ is almost unchanged during operation of the buck-boost converter. Here f is chosen as $f = f_0 = 1/2\pi\sqrt{k_{\mathrm{m}}/m}$. The optimization toward low power consumption of control components is beyond the scope of this paper. However, it should be noted that R_{in} is independent of V_{L}, therefore feedback sensing is not essentially required and the control unit can be further simplified offering power savings. Parameters of the interface circuit are listed in Table 1b, adapted from [16].

To investigate the feasibility of such an introduced topology, diode and inductor losses need to be included. In this paper, we use a piecewise linear diode model which yields a computationally light weight albeit somewhat idealized representation of a diode. The diode losses are thus described by the forward voltage drop V_{D}, the resistance in forward conduction R_{on} and the resistance when off R_{off}. One of the crucial requirements for the energy harvesting system is to be able to operate with low input mechanical power. Hence, diodes with lowest possible losses are preferable. However, the important impact of forward voltage drop on efficiency of

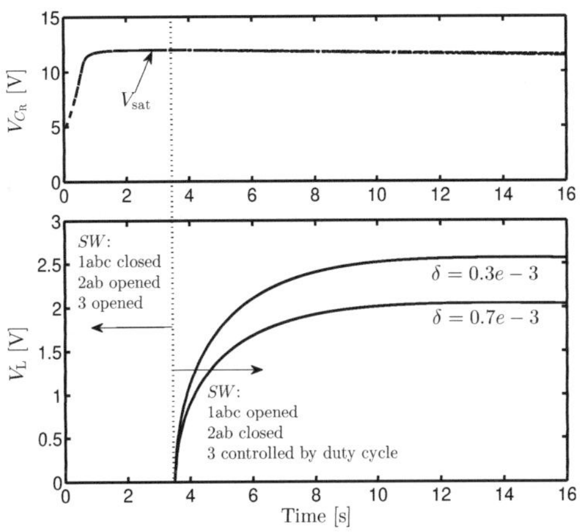

Figure 4: Operation waveforms of voltages across the storage capacitor $C_{\rm R}$ and the resistive load $R_{\rm L}$.

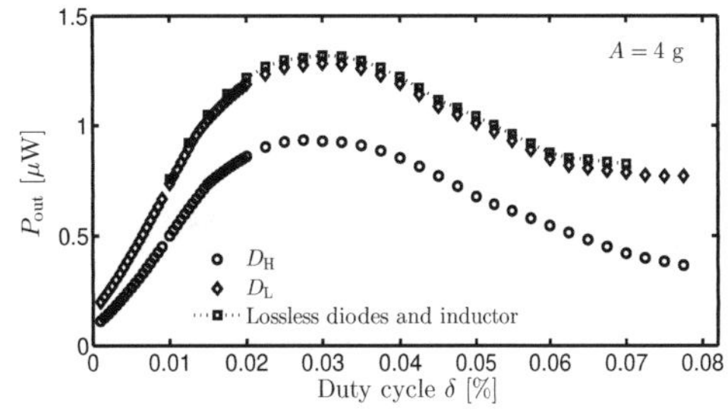

Figure 5: Average power versus duty cycles for different diodes in comparison with ideal case when losses are neglected.

Table 2: Diode parameters.

Parameters	Values
Diode voltage drop, $V_{\rm D}$	$D_{\rm H}$: 0.4 V
	$D_{\rm L}$: 26 mV
Conduction resistance, $R_{\rm on}$	0.5 Ω
Off resistance, $R_{\rm off}$	10 MΩ

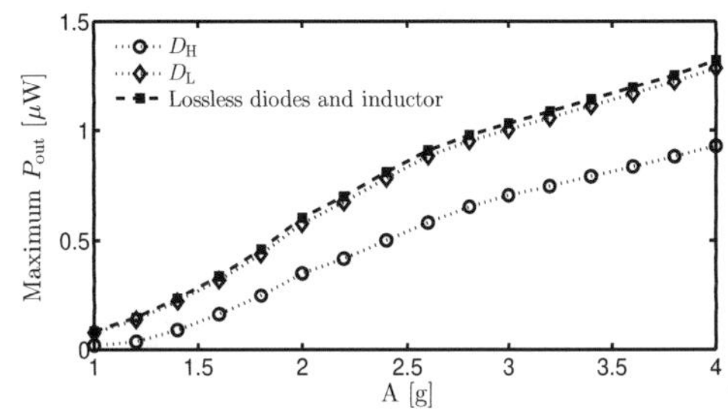

Figure 6: Maximum power versus input acceleration amplitudes.

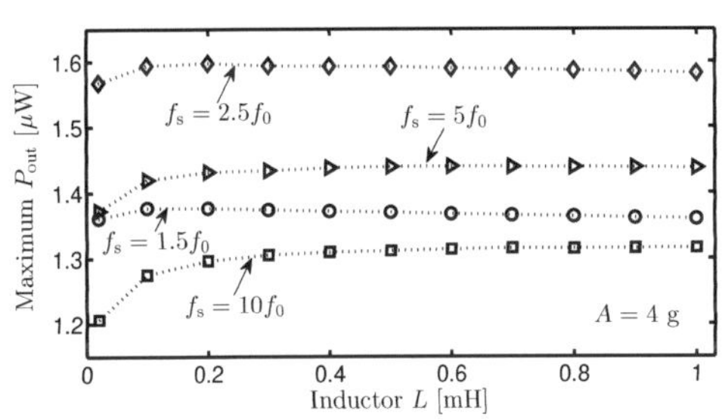

Figure 7: Maximum power versus inductance with different operating frequencies $f_{\rm s}$.

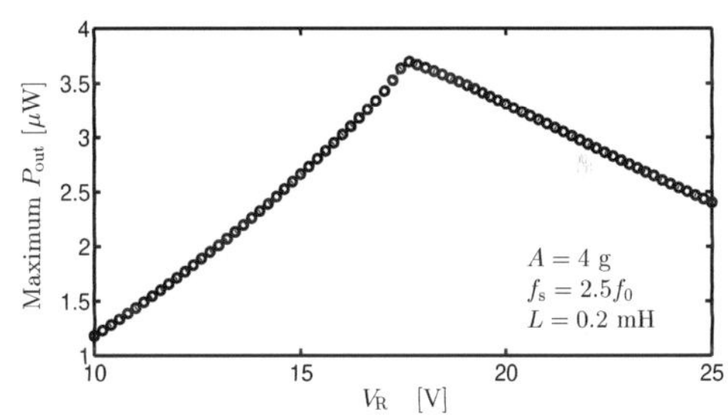

Figure 8: Dependence of maximum power on voltage $V_{\rm R}$.

the diodes is not clearly addressed in the literature and is worth further analysis. We choose to consider two different diodes called $D_{\rm H}$ and $D_{\rm L}$ as examples for this investigation, in which $V_{\rm D}$ is the specified value of the real diodes *PAD* (Pico-Amp Diode) and *SM*74611 respectively. Detail parameters are listed in Table 2.

Figure 5 presents corresponding output power $P_{\rm out}$ when $D_{\rm H}$ and $D_{\rm L}$ are used, comparing to the ideal case in that the losses of electronic components are neglected. The simulations show that the power obtained from using $D_{\rm L}$ is very close to that of the ideal case and both are higher than the other case as expected. The effect of the inductor parasitic resistance is insignificant in this circumstance. The results of maximum power delivered to the load are then expanded to different acceleration amplitudes in Figure 6. In addition, our simulations also predict that the minimum required initial voltage with the use of $D_{\rm L}/D_{\rm H}$ are 52 mV/0.79 V, which are about twice $V_{\rm D}$ of each diode. The maximum harvested power can be further improved with respect to the switching frequency f_s and the inductance L. Figure 7 shows that about 1.60 μW is achieved with $f_s = 2.5f_0$ and $L = 0.2$ mH. This is an increase of 25.2% in comparison with the case $f_s = 10f_0$ and $L = 150$ μH. These analyses open a room for maximizing the harvested power in future studies, which can be considered as function of f_s, L and even diode parameters.

To investigate the effect of V_{C_R} on the output power, we are now considering a particular circumstance when the transducers are directly coupled to the buck-boost converter (i.e., similar to Figure 3c) and C_R is replaced by a voltage source V_R. The dependence of maximum harvested power on V_R is shown in Figure 8, where $P_{max} \approx 3.7\,\mu$W is obtained with $V_R = 17.5$ V. Here $f_s = 2.5f_0$, $L = 0.2$ mH and $A = 4$ g. These results indicate that, in order to optimize the harvesting system performance, the saturation voltage of the doubler configuration at least needs to reach the optimal bias of the buck-boost converter. In case of $V_{sat} > V_R^{optimal}$, there is an optimal point before saturation at which switches 1a, 1b, 1c, 2a and 2b should be reverted their states to change the doubler configuration to the buck-boost converter.

4. Conclusion

In this paper, a new power electronic interface for MEMS capacitive energy harvesting was introduced and investigated. This circuit topology enabled to reconfigure the harvesting system from a symmetric doubler circuit to a buck-boost DC-DC converter when the voltage across the storage capacitor reaches an appropriate value. Simulation results showed that the minimum initial bias for efficient operation of the voltage doubler is about twice the voltage drop of diode, while the maximum harvested power strongly depends on the choice of diode, the switching frequency, the duty cycle and the inductance. Essential influence of the voltage V_R for the buck-boost converter configuration was highlighted, raising the requirement to carefully design the harvester so that the saturation voltage of the doubler circuit is at least equal to optimal value of V_R.

Acknowledgment

This work was supported by the Research Council of Norway through Grant no. 229716/E20.

References

[1] Patel S, Park H, Bonato P, Chan L and Rodgers M 2012 *Journal of NeuroEngineering and Rehabilitation* **9** 21 ISSN 1743-0003
[2] Fitzpatrick D 2014 *Implantable Electronic Medical Devices* (Elsevier Science) ISBN 9780124165779
[3] Baronti P, Pillai P, Chook V W, Chessa S, Gotta A and Hu Y F 2007 *Computer Communications* **30** 1655 – 1695 ISSN 0140-3664 Wired/Wireless Internet Communications
[4] Amar A B, Kouki A B and Cao H 2015 *Sensors* **15** 28889–28914 ISSN 1424-8220
[5] Wei X and Liu J 2008 *Frontiers of Energy and Power Engineering in China* **2** 1–13 ISSN 1673-7504
[6] Zheng Q, Shi B, Li Z and Wang Z L 2017 *Advanced Science* **4** 1700029 ISSN 2198-3844 1700029
[7] Zhou Y, Apo D J and Priya S 2013 *Applied Physics Letters* **103** 192909
[8] Fischer A C, Forsberg F, Lapisa M, Bleiker S J, Stemme G, Roxhed N and Niklaus F 2015 *Microsystems & Nanoengineering* **1**
[9] Roundy S, Wright P K and Rabaey J 2003 *Computer Communications* **26** 1131–1144
[10] Yen B C and Lang J H 2006 *IEEE Transactions on Circuits and Systems—Part I: Regular Papers* **53** 288–295
[11] Mitcheson P D, Yeatman E M and Green T C 2006 *Symposium on Design, Test, Integration and Packaging of MEMS/MOEMS* (Stresa, Lago Maggiore, Italy)
[12] de Queiroz A C M and Domingues M 2011 *Circuits and Systems (MWSCAS), IEEE 54th International Midwest Symposium* pp 1–4 ISSN 1548-3746
[13] Dorzhiev V, Karami A, Basset P, Marty F, Dragunov V and Galayko D 2015 *IEEE Electron Device Letters* **36** 183–185 ISSN 0741-3106
[14] Domingues M and de Queiroz A C M 2013 *IEEE International Symposium on Circuits and Systems (ISCAS2013)* pp 2960–2963 ISSN 0271-4302
[15] Wei J, Lefeuvre E, Mathias H and Costa F 2016 *Journal of Physics: Conference Series* **773** 012045
[16] D'hulst R, Sterken T, Puers R, Deconinck G and Driesen J 2010 *IEEE Transactions on Industrial Electronics* **57** 4170 –4177 ISSN 0278-0046

IOP Conf. Series: Journal of Physics: Conf. Series **1052** (2018) 012075 doi:10.1088/1742-6596/1052/1/012075

Magnetostrictive low-cost high-performance vibration power generator

T. Ueno

Kanazawa University, Kakuma-machi, Kanazawa-city, Ishikawa, 920-1192, Japan
ueno@ec.t.kanazawa-u.ac.jp

Abstract. A high-sensitivity, high-durability, low-cost vibration power-generating device using a magnetostrictive element (Fe-Ga alloy) is proposed. The device comprises a unimorph layer having a magnetostrictive element attached to a U-shaped frame with a permanent magnet for magnetic bias wound about by a coil. By magnetically saturating the part to which the frame element is affixed, an electromotive force is generated in the coil by vibration according to the change in the magnetic flux generated by the inverse magnetostrictive effect of the element. An evaluation of a prototype device (4 g) using an element of 4 × 0.5 × 16 mm is performed. With a weight of 1.7 g attached, an open-circuit voltage of 4 V at an oscillation of 88.7 Hz and 6 m/s^2 yields the effective power of 2.0 mW. With a weight of 10.2 g, an effective power of 0.39 mW is generated at a frequency of 28.4 Hz and an acceleration of 0.075 G. After repeated oscillation at 420 Hz and 8 G for 100 million times, the resonant frequency and voltage remain unchanged.

1. Introduction

The main problem with Internet-of-Things (IoT) and wireless sensor modules is battery power. Modules are said to work for several years with button batteries, but when the number of modules reaches 100 or 1,000, battery exchange becomes a very troublesome task. In addition, batteries lose charge capacity without using them, and their disposal is also troublesome. Batteries could be replaced with vibrational power generation. By replacing batteries with non-electrochemical power sources, the use of IoT devices and sensor modules could become far easier. At present, a vibration power-generating device using a Fe-based magnetostrictive material is under investigation. General environmental vibration occurs at frequencies of 20 to 100 Hz and accelerations of 0.05 G to 1 G. The device is required to generate power from mild to high-intensity vibrations. In this study, a power-generation device satisfying the criteria of high sensitivity, high durability, and mass productivity is proposed and tested to demonstrate these features.

2. Structure and principle of device

The structure of the device is shown in Fig. 1. The device has a unimorph base with a plate-like magnetostrictive element (Fe–Ga alloy [1]) adhered to a U-shaped magnetic frame (Fig. 1 top). A coil is wound around the unimorph and permanent magnet is arranged in the space within the frame. The frame is designed so that the part where the magnetostrictive element is bonded becomes magnetically saturated by magnetic bias; the other parts remain unsaturated, and a uniform stress is applied in the longitudinal direction of the element when the unimorph is bent [2].

The principle of power generation is shown in Fig. 2. The device is fixed to a vibration source, weight is attached to the tip of the frame, and the device is shaken. At this time, the device deforms such that the inertial force of the weight opens and closes the opening of the frame, uniform tensile and compressive stresses are generated in the longitudinal direction of the element, and the magnetic flux is increased or decreased by the inverse magnetostrictive effect. Figure 2 shows the cases for which the inertial force of the weight is downward (top) and upward (bottom). The change in the magnetic flux generated by the inverse magnetostrictive effect circulates in the closed magnetic circuit in series with the element, frame, magnet, and gap. The magnetic portion is magnetically saturated and, therefore, high in magnetic resistance, and the change in the magnetic flux of the device does not backflow through the magnetic portion. Therefore, an electromotive force proportional to the temporal change of the magnetic flux of the element is generated in the coil.

The critical design feature of this structure is the easily molded frame, which can be cut out and bent from a Fe plate, thereby significantly reducing the cost. The element can be fixed to the frame and the air-core coil can be fitted later, suitable for mass production. The generated power can be significantly increased by manipulating the layer thickness of the coil, while the generated voltage and resistance can be adjusted by manipulating the wire diameter. Furthermore, the unimorph is more easily vibrated by uniform curving compared to the conventional parallel-beam type [3], and also highly durable. Thus, the proposed structure satisfies both mass productivity and high performance.

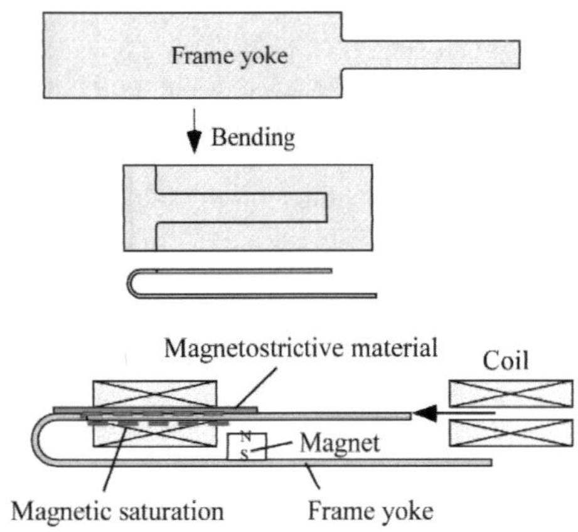

Figure 1. Frame (top) and device configuration (bottom).

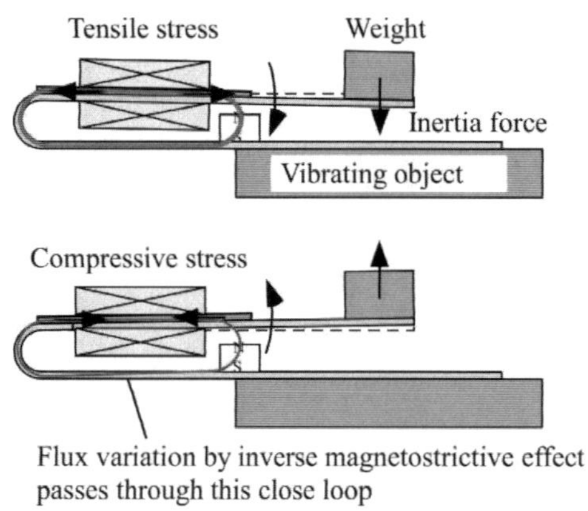

Figure 2. Principle of power generation: Inertial force downward (top) and upward (bottom).

3. Prototype evaluation

3.1. Prototype

In the prototype device, the magnetostrictive element was a 4 × 0.5 × 16-mm plate cut from a single-crystalline bulk $Fe_{81.4}Ga_{18.6}$ material by wire discharge machining. The frame was wire discharge machined from a 0.5-mm plate of cold-rolled Fe and then bent. The element was adhered to the frame with an epoxy-type structural adhesive. The coil had a wire diameter of 0.05 mm and 3500 turns (520 Ω), and the magnet was Nd–B–Fe measuring 4 × 3 × 2 mm. The coil was fixed in the unimorph section with silicone rubber. The weight of the device was 4 g.

3.2. Voltage characteristics

The power generation characteristics were measured with the system shown in Fig. 3. The device was fixed to the exciter with screws and vibrated. The vibration acceleration was measured by an acceleration pickup, while the displacement of the frame was measured with a laser displacement sensor. Weights of 1.7 g, 3.2 g, and 10.2 g were attached to the tip of the frame. Figure 4 shows the frequency response results of open-voltage/acceleration measured with white noise excited with a fast Fourier transform (FFT) analyzer. When the weights are 1.7, 3.2, and 10.2 g, the resonance frequencies are 89.8, 57.9, and 28.9 Hz, respectively, and the Q values are 185, 196, and 195, respectively.

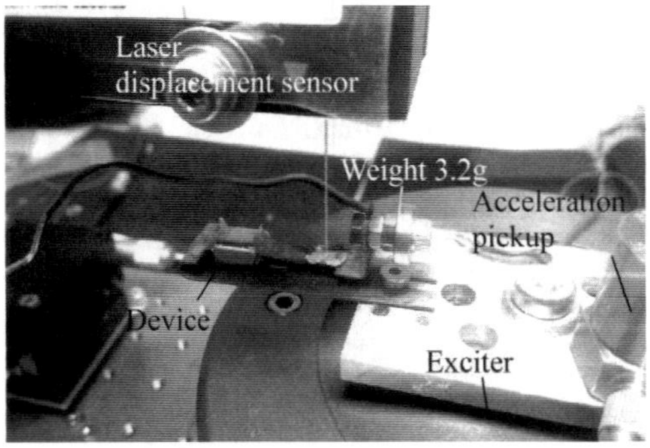

Figure 3. Prototype device and measurement system.

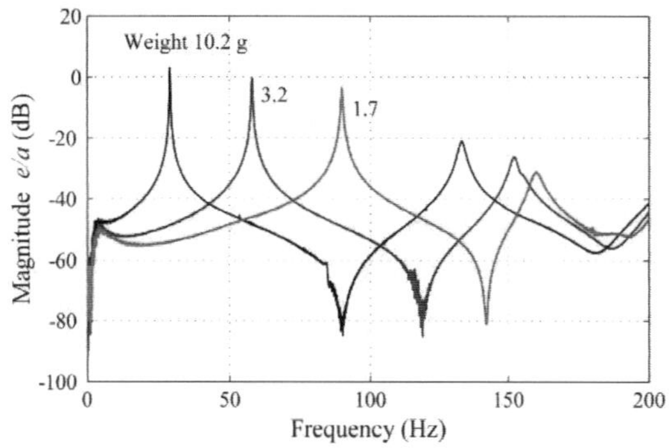

Figure 4. Frequency response of voltage/acceleration with varied weight.

The voltage waveform is shown in Fig. 5 under the weight of 1.7 g and sinusoidal wave oscillation of 88.7 Hz frequency and accelerations of 3.0, 6.0, and 9.0 m/s^2. The maximum voltage of 5.4 V is generated at the acceleration of 9.0 m/s^2. The relationship between the displacement of the tip of the frame and the change in the magnetic flux density of the element, as calculated by the time integration of the voltage waveform, is shown in Fig. 6. A magnetic flux density change of 1.3 T, equivalent to that of the conventional parallel-beam type [3], occurs. Under 6.0 m/s^2 acceleration, a resistance is connected as a load and the relationship between resistance and power (Joule loss) is measured and shown in Fig. 7. A maximum power of 5.1 mW and an effective power of 2.0 mW are obtained at approximately 1000 Ω. Figure 8 shows the relationship between the resistance and the displacement amplitude of the frame, and reduced vibration of the device by power consumption occurs.

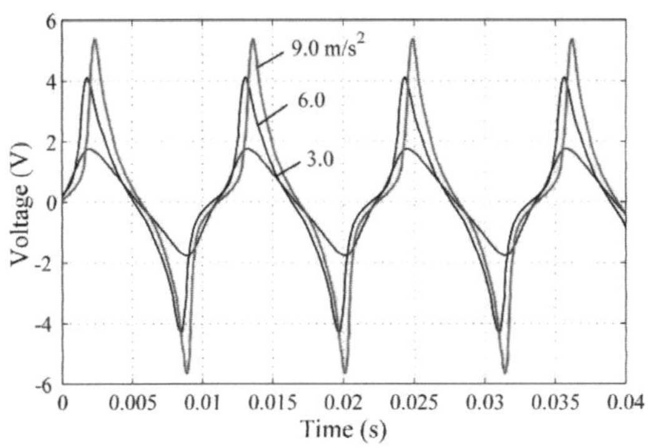

Figure 5. Open voltages (88.7 Hz).

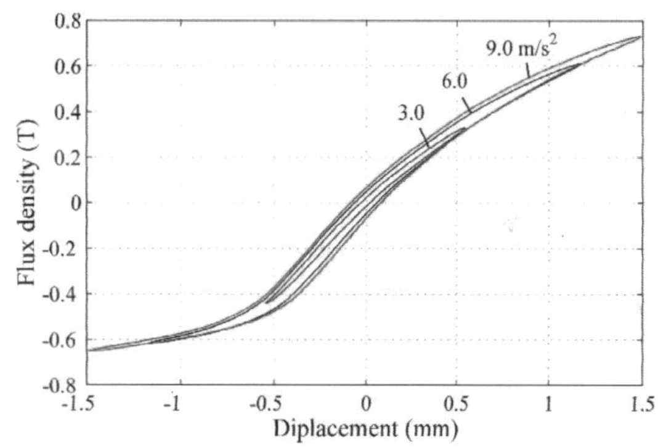

Figure 6. Flux density variation vs. displacement.

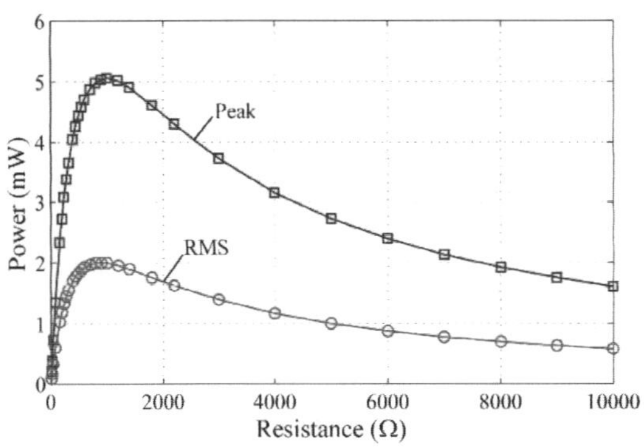

Figure 7. Power vs. resistance (0.6 G).

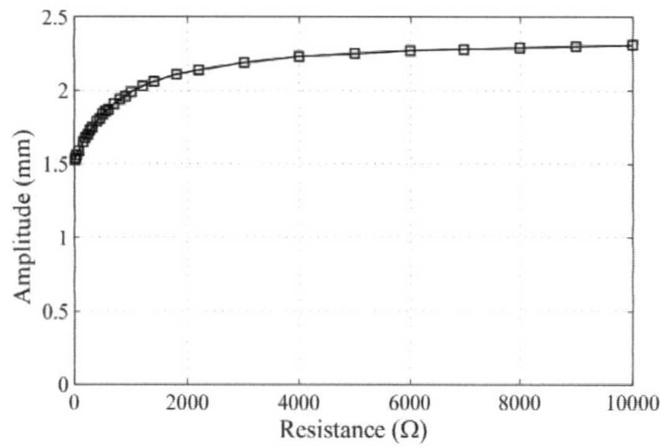

Figure 8. Amplitude vs. resistance (0.6 G).

Next, the voltage waveforms at the weight of 10.2 g, the frequency of 28.4 Hz, and accelerations of 0.34, 0.49, and 0.61 m/s^2 are shown in Fig. 9. A voltage of 1 V is generated at an acceleration of ~0.05 G and a practical voltage could be extracted with minimal vibration. Figure 10 shows the relation between the resistance and electric power at 0.75 m/s^2 acceleration. The maximum electric power of 1.05 mW and the effective electric power of 0.39 mW are obtained at a resistance of approximately 520 Ω. The difference in the matching resistance arises from the degree of coupling between electricity and machinery.

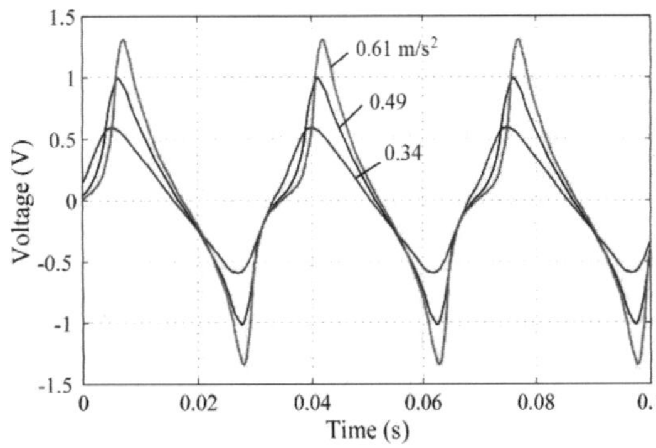

Figure 9. Open voltages (28.4 Hz).

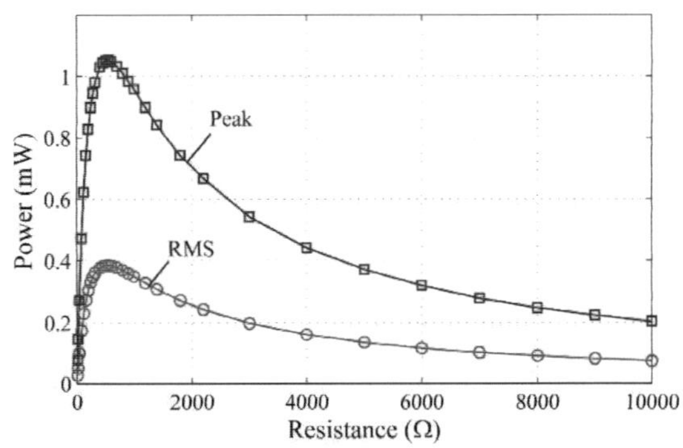

Figure 10. Power vs. resistance (0.075 G).

3.3. Fatigue test

Fatigue tests were performed with the apparatus shown in Fig. 11. Without weight on the tip, the device was shaken 0.1, 1, 10, and 100 million times at a resonance frequency of 420 Hz and an acceleration of 8 G at room temperature (25 °C). The resonance frequency was finely adjusted at the appropriate times, and the frequency characteristic of the voltage (1-G and 8-G frequency sweeps) was measured after each given number of cycles. Figure 12 shows the results. The frequency characteristics remain almost unchanged from the first cycle, even after 100 million cycles. After the test, neither the separation of elements nor metal fatigue is observed.

Figure 11. Apparatus for fatigue test.

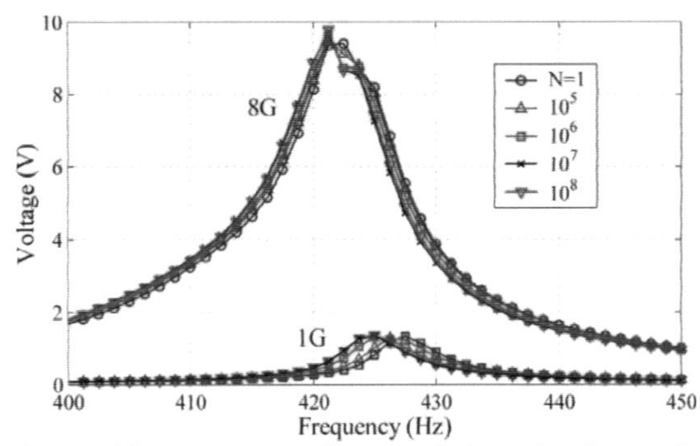

Figure 12. Frequency characteristics of voltage after cyclic vibration.

Acknowledgments
This work was supported by JST CREST Grant Number JPMJCR15Q1, Japan. The fatigue tests were conducted by Toyo Tire & Rubber Co.,LTD.

References

[1] A. E. Clark, M. Wun-Fogle, and J. B. Restorff, Magnetostrictive Properties of Body-Centered Cubic Fe-Ga and Fe-Ga-Al Alloy, *IEEE Trans. Magn.* 37, pp. 3238–3240, 2000.
[2] T. Ueno, PCT-JP2017-8437, 2017.
[3] T. Ueno, Performance of improved magnetostrictive vibrational power generator, simple and high power output for practical applications, *J. Appl. Phys.* 117, 17A740, 2015.

Energy Harvesting Devices for Condition Monitoring Applications of Pneumatic Combined Clutch-Brakes

D Hoffmann[1,5], K Ylli[1], A Willmann[1], D Stojakov[1], Y Manoli[1,2], M Tijero[3], M Mondragon[4]
[1] Hahn-Schickard, Wilhelm-Schickard-Str. 10, 78052 Villingen-Schwenningen, Germany
[2] Fritz Huettinger Chair of Microelectronics, Department of Microsystems Engineering – IMTEK, University of Freiburg, Germany
[3] IK4-IKERLAN, Arrasate-Mondragón, Spain
[4] Goizper, Antzuola, Spain

E-Mail: Daniel.Hoffmann@hahn-schickard.de

Abstract. In order to reduce unscheduled downtimes and maintenance costs, industrial machinery is being fitted with condition-monitoring systems (CMS) that measure important environmental factors and wear parameters of the machine. The electrical power required by these (often retrofitted) sensor systems can be supplied by energy harvesting devices. Pneumatic clutch-brakes provide an excellent application for both CMS and energy harvesting. These industrial machines transmit mechanical power through Kevlar brake pads which wear out due to the large frictional forces at play. The machine accelerates and decelerates from 0 to 1000 rpm and vice versa within 200 ms. This paper reports on energy harvesting devices that can use the large accelerations of the clutch-brake as an energy source to supply the CMS that measures the temperature and wear of the brake pads. A rotary system generates an average power output of 1.5 mW with the machine performing one clutch-brake action every four seconds.

1. Introduction

Clutches and combined clutch-brakes transfer the rotating movement of an electric motor to accelerate or decelerate a mechanical structure within a very short time. These devices may be pneumatic, hydraulic or electromagnetic. They are the key solution in machinery in different sectors, mainly where power transmission is needed: servo presses, lifting equipment, turbines, high-precision and planetary gears.

A combined clutch-brake as shown in Figure 1 (a) is a system with a dual braking and clutching function, where the functions may not both be activated at the same time [1]. The torque transmission is realized through friction between brake pads made of Kevlar. In this works' pneumatic device, the steel wheels of the system are forced onto each other when air pressure is applied and a clutch-brake action is performed. These brake pads are thus subjected to enormous forces and wear out over time. Traditionally, wear is monitored manually by qualified technical staff, which in turn requires machine downtime and thus increases costs. Condition monitoring of the temperature and wear of the brake

[5] To whom any correspondence should be addressed.

pads allows optimum use of friction material and avoids unscheduled downtime and other malfunctions in the clutch-brake. Sensors for condition monitoring can be integrated on the clutch-brake. For the energy supply and the maintenance free operation of the sensor system, energy harvesting is required.

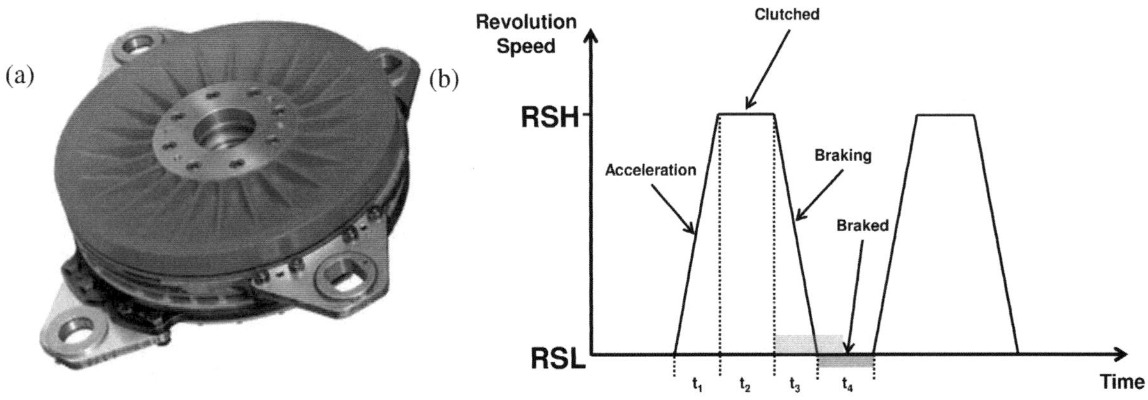

Figure 1. (a) Exemplary clutch-brake from Goizper [1]. (b) Schematic showing the different phases during the operation of a clutch-brake. "RSL" and "RSH" stand for "revolution speed low" and "high".

2. Operating Principle

Clutch-brakes come in a wide range of sizes and power classes. The boundary conditions for a typical clutch-brake include rotation speeds up to 1000 rpm and ambient temperatures up to 60°C. The clutch-brake action itself lasts less than 200 ms (acceleration to target value, deceleration to standstill, see Figure 2). If the application requires it, specific clutch-brakes can reach standstill within a rotation angle of 20°.

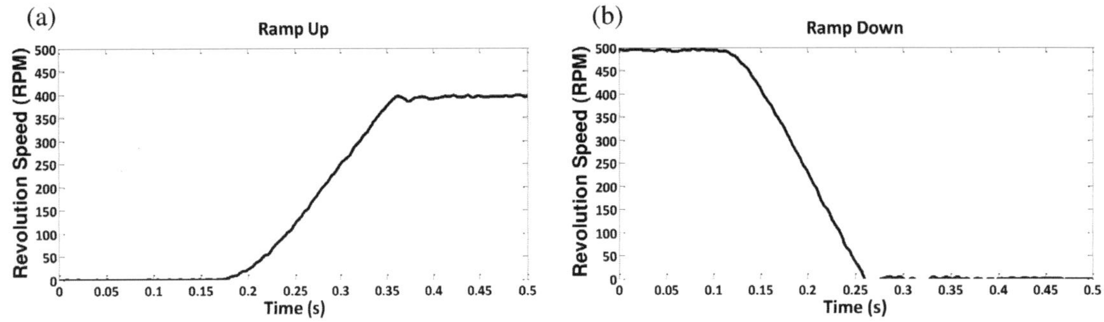

Figure 2. (a) Graph showing the rapid acceleration of a clutch-brake from 0 to 400 rpm. (b) Graph showing the rapid deceleration of a clutch-brake from 500 to 0 rpm.

In the measurements presented in this work, the harvesting device is placed at a distance of 192 mm from the center of the clutch-brake. During standard operation according to the schematic in Figure 1 (b), the device performs one clutch-brake action every four seconds, i.e. the "clutched" and "braked" states are $t_2 = t_4 = 4\ s$. The areas marked in blue and green represent the phases where temperature (blue) and wear (green) need to be measured by the CMS. The aim is for the CMS to sample the temperature at a rate of 2 ms for a total of two seconds and the distance to the brake pad, i.e. the wear, at a rate of 2 ms for a total of 500 ms. The accelerating states t_1 and t_3 for an exemplary device are shown in Figure 2.

These large accelerations offer an excitation source for a kinetic harvester, but require a mechanically extremely robust design.

3. Experimental Data

3.1. Rotary energy harvesting approach

The first approach builds on the harvester device presented in [2]. The proposed rotational inductive energy harvester was heavily reworked to withstand the harsh excitation conditions (see Figure 3). The device consists of an excentric mass mounted on a freely rotating wheel which houses 8 magnet segments of 45° angle and the back flux guide. The magnetic circuit thus forms a pendulum which can swing and rotate around the 8 coils that are placed at fixed positions within the housing.

During constant rotation of the clutch-brake, the pendulum of the harvester does not rotate against the coils due to centrifugal forces that hold it in position. Energy is mainly generated during the acceleration and deceleration phase. Table 1 shows the generated energy and the corresponding average power output using a 330 mF supercapacitor as energy storage for different number of clutch-brake actions and different motor speeds. It can be seen, that an average power of up to 1.7 mW can be achieved.

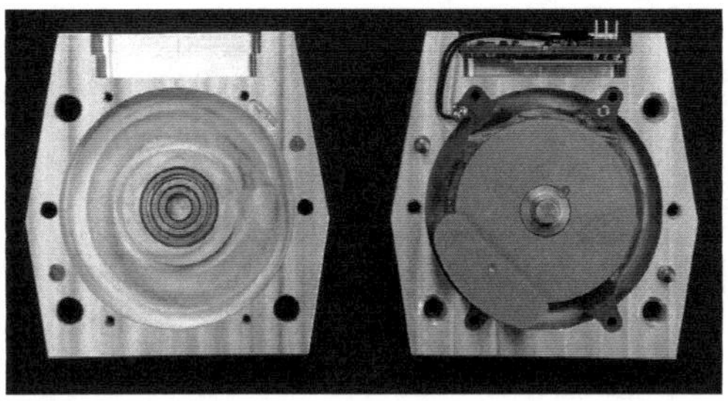

Figure 3. Rotational inductive energy harvester for mounting at the perimeter of the grey steel wheel in Figure 1.

Table 1. Energy stored in a 330 mF super-capacitor and average power output [mJ] / [mW] for time periods corresponding to the number of clutch-brake actions N.

RPM	N = 5	N = 10	N = 20
100	9.4 / 0.47	28.4 / 0.71	97.3 / 1.2
200	13.8 / 0.69	40.7 / 1.02	96.6 / 1.2
300	17.1 / 0.86	35.8 / 0.89	134 / 1.7
400	14.3 / 0.72	37.4 / 0.93	117 / 1.5
500	13.9 / 0.69	38.9 / 0.97	133 / 1.7

3.2. Sensor System

Apart from the energy harvester, the CMS consists of a Cypress PSoC BLE chip with integrated microprocessor and low-power Bluetooth transmitter, a thermocouple and an amplifier for the thermocouple (AD8495), a distance sensor (BAW M12ME-UAD50B-BP01) with corresponding DC/DC-converter (ADP1612) and an accelerometer (BMA280).

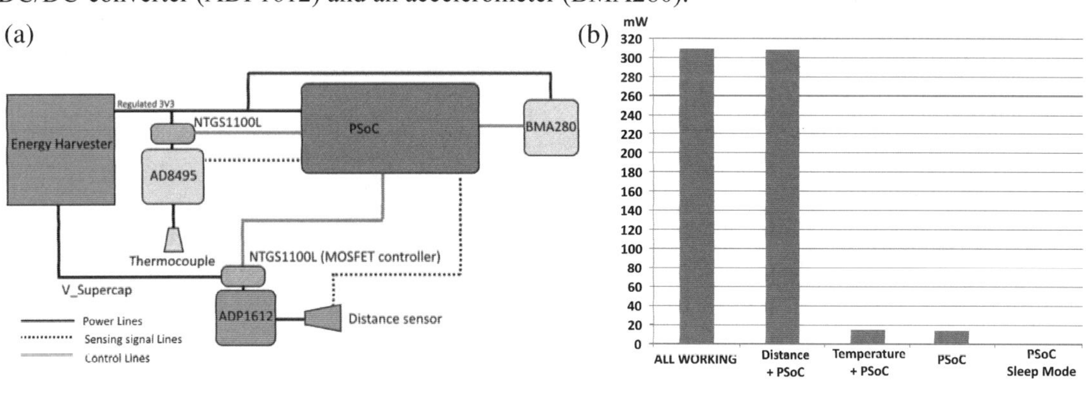

Figure 4. (a) Schematic of the CMS showing all components. (b) Power consumption of the CPS for different operation modes from 300 mW when all sensors are working to 0.45 mW in sleep mode.

The schematic of the CMS is given in Figure 4 (a), while the power consumption for different operating modes is shown in Figure 4 (b). When all electronics are operating, i.e. during the distance measurement, approximately 150 mJ of energy are needed for a duration of 500 ms (equivalent to

300 mW). The temperature measurement and data transmission requires 40 mJ for a duration of 2 s (equivalent to 20 mW). In sleep mode, the power consumption of the system is 0.45 mW.

Measurements and data transmission are performed discontinuously. Table 2 lists the required amount of power that results for different cycle times. As an example, one measurement per hour results in an average power consumption of 0.5 mW during that hour. The aforementioned rotary harvesting device is thus able to power the CMS for several measurements per hour. This is a very satisfactory result, as the wear of the brake pads is minimal for such short time frames.

Table 2. Average power consumption of the CMS depending on the cycle time, i.e. the interval between measurements. At one measurement per hour, the required power is an average of 0.5 mW.

Cycle Time (s)	Sleep Time (s)	Total Energy (mJ)	Average Power (mW)
60	57.5	216	3.6
600	597.5	459	0.77
3600	3597.5	1809	0.5

4. Conclusion

In this work we reported on energy harvesting on pneumatic clutch-brakes. These devices require very robust harvesters because of their large accelerations. At the same this provides a powerful energy source for kinetic harvesters. In addition to a rotary harvester, a condition-monitoring system was developed, which monitors the temperature and wear of the Kevlar brake pads of the clutch-brake.

Depending on the operating mode, the power consumption of the CMS ranges from 0.45 mW in sleep mode up to 300 mW for the brief time span when all sensors are powered on. The rotary harvester generates an average power output of up to 1.7 mW and is thus able to supply the CMS for more than ten measurement phases per hour (see Table 2). This represent a sufficient measurement rate as the brake pads do not noticeably wear out in such short time spans.

An alternative to the rotational system is in development and is based on the linear harvesters in [3], in order to use the large available accelerations. First simulations indicate that up to 100 mJ of energy could be generated in a single cycle.

5. References

[1] *Goizper Clutch-Brakes.* Available: http://www.goizper.com/en/industrial/brakes-and-clutches (2017, Sep. 17).

[2] K. Ylli, D. Hoffmann, A. Willmann, B. Folkmer, and Y. Manoli, "Investigation of Pendulum Structures for Rotational Energy Harvesting from Human Motion," *J. Phys.: Conf. Ser,* vol. 660, p. 012053, 2015.

[3] K. Ylli *et al,* "Energy harvesting from human motion: exploiting swing and shock excitations," *Smart Mater. Struct,* vol. 24, no. 2, p. 025029, 2015.

6. Acknowledgements

This research has received funding from the European Community's programme HORIZON 2020 under GA No. 644596 from the project SMARTER-SI. The H2020 project SMARTER-SI is part of the SmartAnythingEverywhere Initiative of the European Commission.

IOP Conf. Series: Journal of Physics: Conf. Series **1052** (2018) 012077 doi:10.1088/1742-6596/1052/1/012077

Flexible wireless power transfer system based on closed-loop magnetoinductive waveguides: solution to misaligned and rotational systems

Fralett Suárez Sandoval[1], Saraí M. Torres Delgado[2], Ali Moazenzadeh[1], and Ulrike Wallrabe[1]

[1] Laboratory for Microactuators, Department of Microsystems Engineering - IMTEK, University of Freiburg, Georges-Koehler-Allee 102, 79110, Freiburg, Germany

[2] Laboratory for Simulation, Department of Microsystems Engineering - IMTEK, University of Freiburg, Georges-Koehler-Allee 103, 79110, Freiburg, Germany

E-mail: wallrabe@imtek.uni-freiburg.de

Abstract. We present the characterization of a circular, 1D wireless power transfer system based on magnetoinductive waves travelling through an array of coupled LC resonators. This work builds upon the linear array that we have introduced in our former publications. The use of high quality double-spiral coils combined with an arrangement of 50% of overlapping between consecutive resonating cells renders low attenuation of the travelling wave. Thus, a receiver device located in the near field of the array is able to be powered from the energy travelling through the array. The newly-introduced flexibility of the array provides an additional degree of freedom for its end-application. We measured unoptimized system efficiencies of up to 60% for 1 mm of radial separation at 13.56 MHz. We present the measurement results of the transfer efficiency between the array and a receiver device for two terminating impedances of the array. We also discuss the feasibility of not implementing an impedance modulation scheme when the receiver device revolves inside the array. It was found that one can obtain small output voltage ripples for rotational frequencies as low as 207 revolutions per minute without a modulation scheme. The system is envisaged to recharge in- and on-body devices whose position cannot be directly controlled. The rotational wireless power transfer system could present an alternative to slip rings which are subjected to mechanical wear.

1. Introduction

Wireless power transfer (WPT) via magnetoinductive waves travelling through resonating arrays of one dimension have been demonstrated [1, 2, 3, 4, 5] before as a method to supply to a receiver device in the near-field of the array. However, standing waves in the arrays shift the transferred power and consequently the transfer efficiency to high and low values [2]. As a way to counteract the effects of standing waves one can either adjust the length of the array or change the operating frequency of the array according to the receiver position. An investigation on the dependence of the transmission parameter and transfer efficiency while varying these two parameters according to the receiver position has been reported in [3]. In former publications [4, 5], we formulated a control method based on inexpensive circuitry which permitted to shift in real time the standing waves along a magnetoinductive wave device to prevent the receiver to stand over a resulting

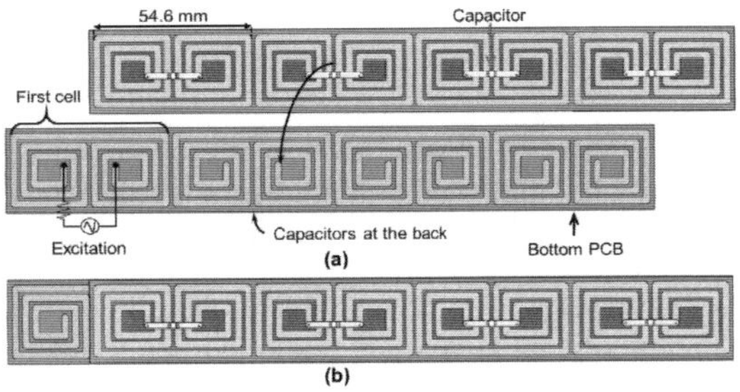

Figure 1. A 1D WPT system based on magnetoinductive waves. The array is composed by a brick wall configuration of eight cells of double-spiral coils. (a) Two boards before assembly. (b) The assembled array. More details of this system can be found in [4].

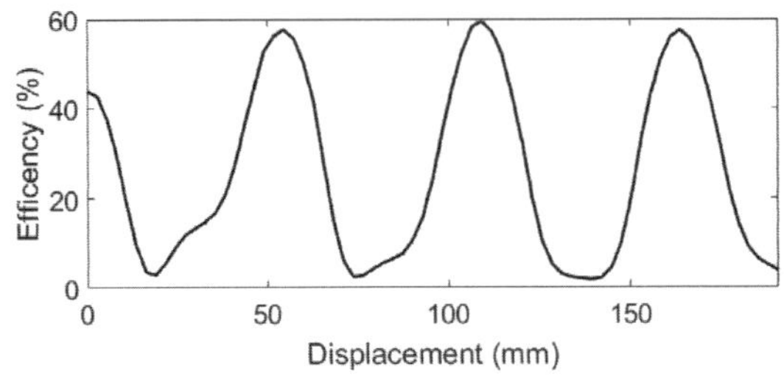

Figure 2. Efficiency between the system shown in figure 1 and a receiver scanned above the array at an axial separation of 6.15 mm. Note that the efficiency drops every second cell. The measurement was done with an excitation frequency of 13.56 MHz.

power null. This control method was implemented either at the resonance frequency of the transmitter array [4] or at a different operating frequency [5] inside the bandwidth of the array.

The present work experimentally investigates the efficiency characteristics of an 1D resonating array whose last cell is overlapped with the first, making a circular, closed-loop array. A flexible, closed-loop wireless charger tolerable to misalignment could, potentially, be applied to recharge batteries of implanted medical devices, such as a microstimulator for the vagus nerve [6] or equip every day use items like a backpack [7] with the possibility to recharge electronic devices that are just thrown in it. We also present a case study in terms of a rectified output voltage measured at the receiver device when the device is revolving inside the flexible wireless power transmitter array.

2. 1D resonating array for WPT

2.1. Efficiency of an open-loop array

This work builds upon the open-loop array of LC resonators that we have proposed in [4] as exemplified in figure 1. There we demonstrated how the array based on double-spiral coils renders a low attenuation of the travelling wave along the array due to the optimized quality factor of the coils. In [5], we compared the effect the double-spiral geometry has over the coupling coefficient between two consecutive cells for a lateral displacement sweep between these cells. The array of figure 1 is formed by two rigid double-sided printed circuit boards, both boards have 4 resonating cells, as shown in figure 1(a). The double-spiral coil of each cell is connected in series to a capacitor, thus driving them to resonate at 13.56 MHz. The boards were arranged by placing them on top of each other as shown in figure 1(b), making neighboring cells to have 50% of overlap.

To characterize the system we connected port one of a vector network analyzer (ZVL, Rohde & Schwarz GmbH & Co KG, Germany) in series with the first cell of the array, as shown in figure 1(a) and measure the two-port scattering parameters. The measurement was done for a logarithmic frequency sweep of 801 points between 1 MHz and 100 MHz. Port two of the analyzer is acting as the load impedance (50 Ω) in series with the receiver resonator, which has the same geometry as the cells of the transmitter array. The receiver is moved above the array at an axial separation of 6.5 mm. The system efficiency was calculated using equation (7) of [5]. Figure 2 shows the efficiency with respect to the receiver displacement. Note that the transferred

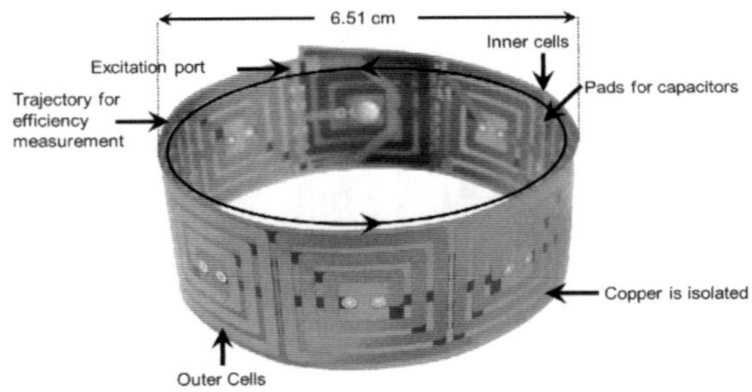

Figure 3. The closed-loop 1D array is demonstrated with a three-layer flexible printed circuit board on a polyimide substrate. Copper of 35 μm in thickness acts as the conducting material. Resonant behavior is achieved with surface-mount capacitors. The array is excited at the first cell and the last cell is overlapped 50% with the first.

power is position-dependent due to the interaction between the forward and backward waves.

2.2. Efficiency of a closed-loop array

A closed-loop system can be made by overlapping the last cell of a planar array with the first while maintaining the same 50% of overlapping between the last and the first cell. To investigate the behavior of the system with this closed-loop configuration we fabricated the array shown in figure 3. The array has three layers of conducting material (one layer less than the array of figure. 1) on a polyimide substrate. The most inner conductor layer is reserved for the capacitors (not present in figure 3) of each cell. Then one layer with four double-spiral coils is stacked over the conducting traces for the capacitors. The layer stack-up extends radially outwards with a second layer of four double-spiral coils displaced by 50% with respect to the first layer. There exist an isolation layer of 39 μm-thick polyimide between consecutive copper layers. The most inner and outer conducting layers are isolated as well, therefore, only the contact pads are exposed.

There is a solid-state relay connected in series with the last cell of the array, the relay can be controlled to set an impedance state for this cell. When the relay is activated, the last cell of

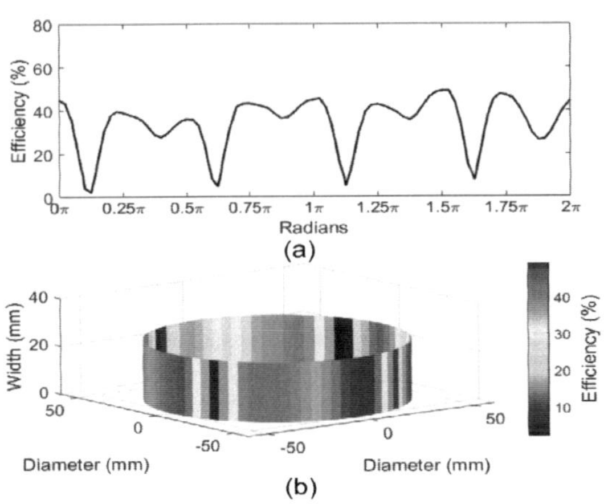

Figure 4. (a) Efficiency between the array of figure 3 with a closed circuit termination at its last cell and a receiver revolved inside the array. The radial separation is 1 mm and the measurement frequency is 13.56 MHz. (b) Efficiency on the surface of the closed-loop array.

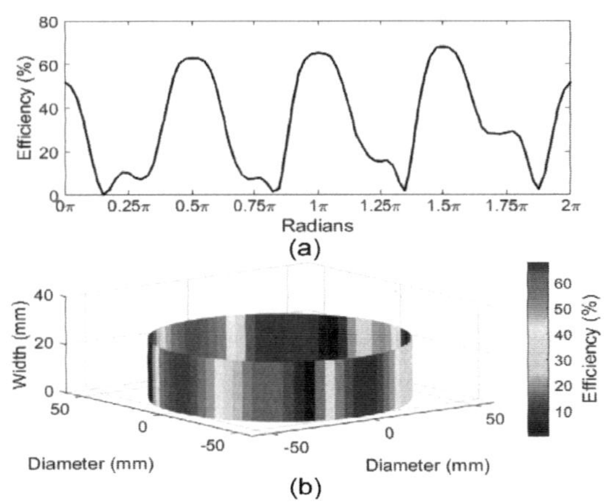

Figure 5. (a) Efficiency between the array of figure 3 with an open circuit termination at its last cell and a receiver revolved inside the array. The radial separation is 1 mm and the measurement frequency is 13.56 MHz. (b) Efficiency on the surface of the closed-loop array.

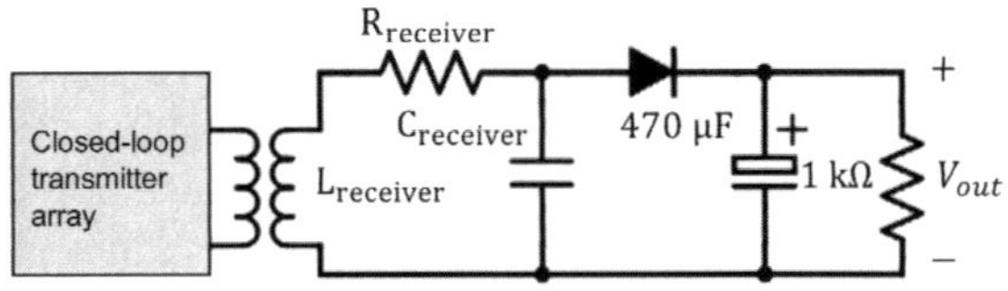

Figure 6. Simple rectifier circuit used to record the output voltage (V_{out} which is input to an analogue input of an Arduino Micro microcontroller [8].

the array is in operation (a closed circuit). When the solid-state relay is deactivated, the series circuit of the last cell is opened, therefore, the active diameter of the array is reduced by one cell. We characterized the system efficiency similarly to the planar array using the two-port scattering parameters. A receiver resonator was revolved inside the array with a radial separation of 1 mm following the trajectory depicted in figure 3. The result of this measurement is shown in figure 4(a) in a linear representation and in figure 4(b) as a surface plot showing the efficiency variation around the surface of the flexible array. Figure 4 depicts the efficiency when the array has all its cells in operation (closed circuit in the last cell). When the last cell is deactivated the efficiency varies as shown in figure 5. Both figures correspond to an excitation frequency of 13.56 MHz.

The efficiency behavior differs between figures 4 and 5 because when the solid-state relay is in the active mode, the last cell of the array is coupled to the first cell with the same coupling factor as the rest of pairs made with consecutive cells. Hence, the wave can continue travelling in the same direction once it reaches the end of the array. Since the coupling between the penultimate cell and the first cell in figure 3 is considerably lower than that between the first and the last cell of the same figure, once the wave reaches the last cell, most of it reflects through the same path it was travelling. We presented in [4, 5] how to sense the actual position of the receiver device and, from that position, adjust the length of the array so that the zero-efficiency spots are avoided. Note that although there exist the same number of zones with low efficiency in figures 4 and 5, the width of these zones is considerably smaller in figure 4 than those of figure 5, at the expense of reaching lower efficiency peaks in comparison to figure 5.

3. Delivering wireless power to a rotatory device

A method to modulate the active diameter of the closed-loop array is useful when the potential receiver device could be displaced from a fixed position but it is not foreseen for it to be in constant revolution. If power needs to be transferred to a revolving device a modulation scheme might not be even needed.

To investigate this hypothesis, we measured the output voltage after rectification as shown in figure 6. The output voltage is depicted in figures 7 and 8 for two different termination impedances. To record these figures the transmitter array was kept still while the receiver was revolved inside the array with a radial separation of 3 mm by mounting it to the axle of a stepper motor. The insets of each figure denote the waveform of the output voltage for different rotational frequencies. As expected, for a fixed RC output filter, in this case 470 μF and 1 kΩ, the larger the rotation frequency, the less the ripple of the output voltage, regardless of the termination impedance state.

4. Conclusion

We presented a flexible, circular, 1D wireless power transfer system based on magnetoinductive waves propagating through LC resonating arrays. It can transmit wireless power in misaligned-presumable configurations like body-worn devices. We measured unoptimized system efficiencies of up to 60% for a receiver device separated 1 mm in the radial direction at an excitation frequency of 13.56 MHz. The termination impedance modulation schemes that we have presented before could be applied as well to the closed-loop transmitter array. Nevertheless,

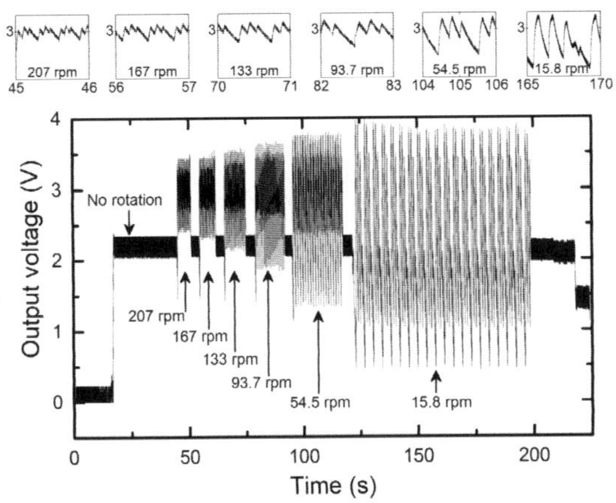

Figure 7. Recorded V_{out} of figure 6 with a termination impedance of the transmitter array shown in figure 3 set to an open circuit. The receiver device is revolved inside the transmitter array with varying rotation speeds. A smaller output voltage ripple is observed with increasing rotation speed.

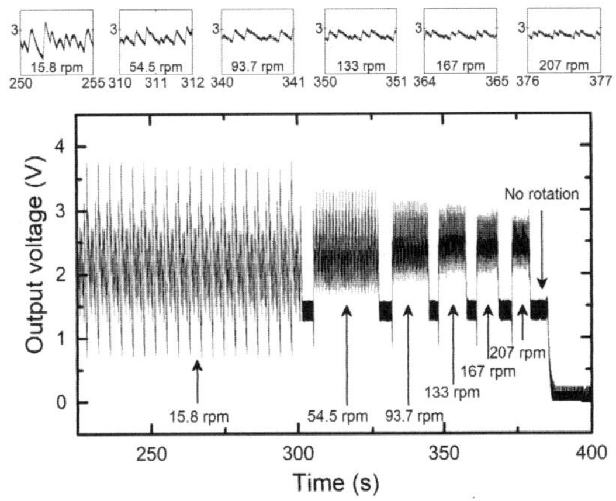

Figure 8. Recorded V_{out} of figure 6 with a termination impedance of the transmitter array shown in figure 3 set to closed circuit. The receiver device is revolved inside the transmitter array with varying rotation speeds. A smaller output voltage ripple is observed with increasing rotation speed.

modulation might not even be needed if the receiver is revolved inside the array with a sufficient rotational frequency and some output voltage ripple is admissible.

5. Acknowledgements

We would like to thank the laboratory for Electrical Instrumentation at the Department of Microsystems Engineering (IMTEK) of the University of Freiburg for providing us with access to measurement equipment.

References

[1] B Wang, W Yerazunis, and K H Teo. Wireless power transfer: Metamaterials and array of coupled resonators. Proceedings of the IEEE, vol. 101, no. 6, pp. 1359-1368, June 2013.

[2] C J Stevens. Magnetoinductive Waves and Wireless Power Transfer. IEEE Transactions on Power Electronics, vol. 30, no. 11, pp. 6182-6190, Nov. 2015.

[3] G Puccetti, C J Stevens, U Reggiani, and L Sandrolini. Experimental and Numerical Investigation of Termination Impedance Effects in Wireless Power Transfer via Metamaterial. Energies 2015, 8, 1882-1895.

[4] F S Sandoval, A Moazenzadeh, S M T Delgado, and U Wallrabe. Double-spiral coils and live impedance modulation for efficient wireless power transfer via magnetoinductive waves. 2016 IEEE Wireless Power Transfer Conference (WPTC), May 2016, pp. 14.

[5] F S Sandoval, S M T Delgado, A Moazenzadeh, and U Wallrabe. Nulls-free wireless power transfer with straightforward control of magnetoinductive waves. IEEE Transactions on Microwave Theory and Techniques, vol. 65, no. 4, pp. 10871093, April 2017.

[6] SetPoint Medical. Bioelectronic medicine for inflammatory diseases: the SetPoint platform. Viewed 26 September 2017, https://setpointmedical.com/technology/.

[7] T Deyle and M Reynolds. PowerPACK: A wireless power distribution system for wearable devices. 2008 12th IEEE International Symposium on Wearable Computers, Pittsburgh, PA, 2008, pp. 91-98.

[8] Arduino Micro. Viewed 30 September 2017, https://store.arduino.cc/arduino-micro.

Electromagnetic Energy Harvester with Embedded Ferrofluid In PCB Technology

Yi Chiu and Hao-Chiao Hong

Department of Electrical and Computer Engineering
National Chiao Tung University, Hsin Chu, Taiwan, ROC

Email: yichiu@mail.nctu.edu.tw

Abstract. This paper reports an electromagnetic energy harvester fabricated with Rigid-Flex Printed Circuit Board (PCB) technology by a commercial PCB supplier. Rigid FR-4 boards were used for transduction coil routing and mechanical support; flexible polyimide (PI) films were used to design elastic springs and magnet mounting platforms. Ferrofluid (FF) was injected into the embedded cavities inside the rigid PCB to increase the effective permeability of the medium around the magnet and thus increase the induced electromotive force (emf). Experiments showed up to 70% increase of output emf and 190% increase of output power compared with harvesters without embedded ferrofluid.

1. Introduction

Electromagnetic (EM) energy harvesters are attractive due to their simple structures and ease of operation. Among various fabrication technologies for EM harvesters, PCB technology offers the advantages of low cost, fast turn-around, high reliability, and compatibility with current electronic manufacturing technology [1-4]. In the past, rigid FR-4 PCB [1, 2] had been used to fabricate electromagnetic energy harvesters and flexible PI films [4] had been used to fabricate electret-based electrostatic energy harvesters. FR-4 boards and PI films can be laminated to form compound Rigid-Flex PCB which is also a very mature and common substrate in modern electronic manufacturing industry. In [5], such Rigid-Flex PCB was used to integrate piezoelectric and electromagnetic harvesters. In [6, 7], a wide-band electromagnetic energy harvester based on Rigid-Flex PCB was demonstrated where the nonlinear effect of the PI springs was used to increase the harvesting bandwidth.

PCB technology is well-suited for fabricating electromagnetic harvesters since sub-mm accuracy is good enough for device design and operation. However, the materials used in standard PCB industry, such as fiberglass cloth, epoxy binder, and copper, are non-magnetic. Therefore PCB-based electromagnetic harvesters are less efficient in energy conversion because the equivalent magnetic circuit of the device is not optimized to confine magnetic flux. In this paper, we demonstrate an electromagnetic energy harvester with embedded cavities fabricated by Rigid-Flex PCB technology. To increase the total magnetic flux in the device and guide the flux through the transducing coils to increase the output emf, ferrofluid is injected into the embedded cavities in the rigid PCB stacks. This fabrication process minimizes the number of components and assembly steps. Since no extra ferromagnetic components are used that may increase the volume or weight of the device, the device performance can thus be enhanced while maintaining low fabrication cost.

2. Principle and Design

The proposed device is based on our work presented at PowerMEMS 2015 [6, 7], as shown in Figure 1(a). The device is fabricated with Rigid-Flex PCB technology where flexible PI films are used for suspension springs and rigid FR-4 boards are used for mechanical support and coil routing. Two sets of transduction coils, each with three layers of routing, are designed on the top and bottom surfaces of the device where the magnetic flux has maximum gradient [7], as shown in Figure 1(b). Permanent magnets are attached to the central PI platform after the PCB is manufactured. The motion of the suspended magnets caused by external vibration induces an emf output in the coils in the rigid PCB. In the current device, embedded cavities were milled in the rigid boards during the PCB manufacturing. After the Rigid-Flex PCB device was manufactured, ferrofluid was injected into the cavities, as shown in Figure 1(b).

Magnetic circuit design is important in magnetic actuators such as the voice coil motors (VCM) in optical pickup heads [8]. Proper design of the high-permeability yokes in the magnetic circuits can increase and concentrate magnetic flux and enhance actuation sensitivity. The yoke structure in the current device is implemented by using the ferrofluid in the embedded cavities, as shown in Figure 1(b). Figure 2 shows the simulation of magnetic field distribution in devices without and with embedded ferrofluid. The ferrofluid has a relative permeability μ_r of 12. It can be seen that the cavity with ferrofluid has higher magnetic B field magnitude compared with that without ferrofluid. Futhermore, ferrofluid causes the magnetic flux lines to bend on the air/ferrofluid interface and results in a higher vertical flux component. These effects increase the total magnetic flux enclosed by the coils and thus the emf output, as shown in Figure 2(c).

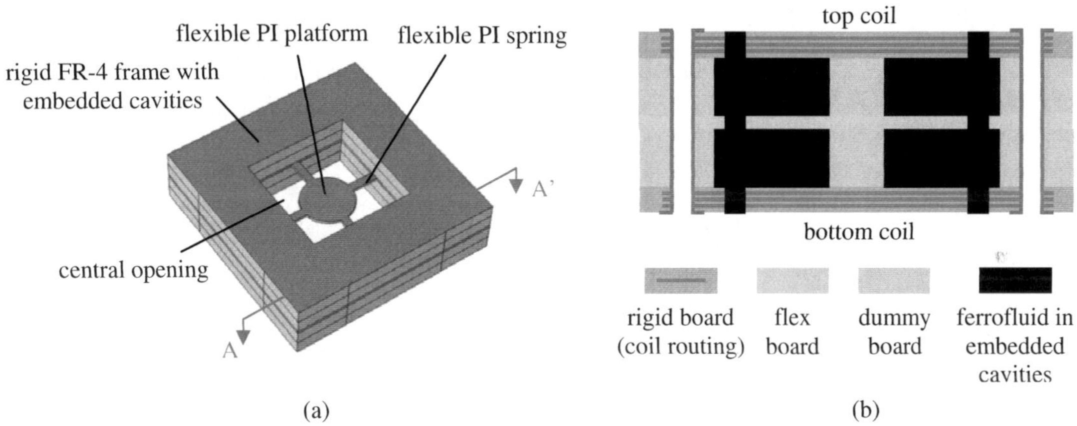

Figure 1. (a) Schematic of harvester fabricated by Rigid-Flex PCB technology, (b) AA' cross section.

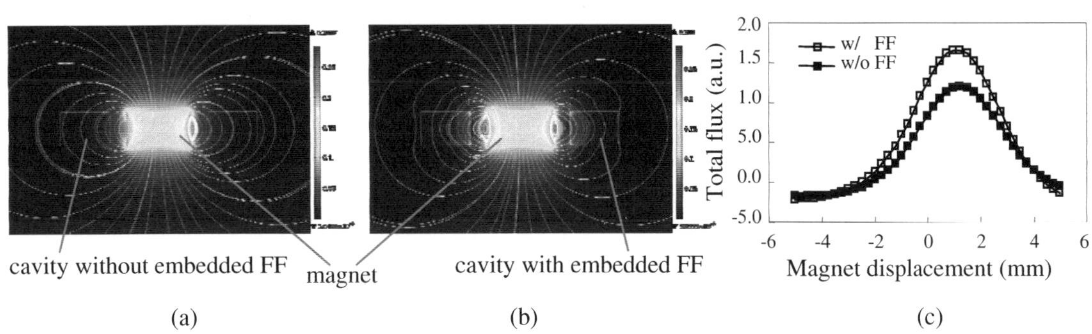

Figure 2. Magnetic field simulation showing B field magnitude and flux lines, (a) without FF, (b) with FF, (c) total magnetic flux vs. magnet displacement.

3. Device Fabrication and Measurement

The device was fabricated by a commercial PCB process, as shown in Figure 3. First, three kinds of circuit boards were prepared separately by standard processes, including (1) FR-4 boards containing three layers of coil routing , (2) thick dummy boards with milled cavities, and (3) PI films with punched geometry for springs and central platforms, as shown in Figure 3(a). The FR-4 coil routing and the dummy boards were then laminated and central opening were milled. Subsequently all boards were laminated. Via holes were then drilled and electroplated to connect routings on different layers, as shown in Figure 3(b). Next, vent/injection holes were drilled through the PCB stack and ferrofluid was injected into the cavities, as shown in Figure 3(c). To inject ferrofluid, the PCB assembly was immersed in ferrofluid and placed in a closed chamber. When the chamber was pumped and then vented, ferrofluid was pushed into and filled the cavities automatically due to the atmospheric pressure, as shown in Figure 3(c). Finally, two NdFeB magnets were attached to the central PI platform.

Figure 4(a) shows a fabricated and assembled device with two 2-mm-thick NdFeB magnets with 0.3-T surface flux density. The total dimension of the device is $20 \times 20 \times 4$ mm^3 and the central opening in the rigid boards is 10×10 mm^2. Both top and bottom coils have a measured internal resistance of about 70 Ω. In Figure 4(b), the top FR-4 PCB layer was removed to reveal the embedded cavities and the vent/injection holes drilled in the sandwiched PI film. Figure 4(c) shows the BB' cross-sectional view of the embedded cavities and the electroplated through via holes in the FR-4 PCB.

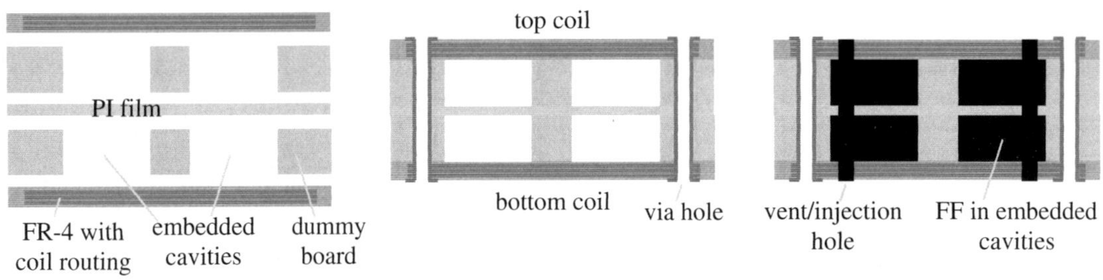

Figure 3. PCB manufacturing and FF injection processes

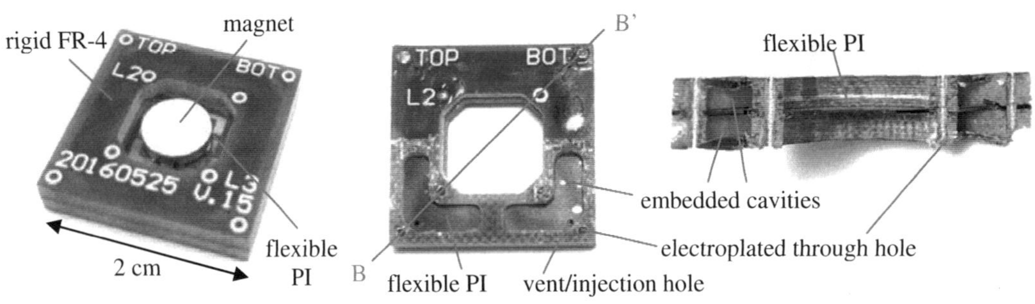

Figure 4. Fabricated harvester, (a) overview, (b) top view of delaminated device, (c) BB' cross-sectional view showing embedded cavities.

In vibration tests, two kinds of different cylindrical magnets (ϕ6×3mm and ϕ6×2mm) were used and the harvesters were tested before and after ferrofluid was injected. The measurement setup is shown in Figure 5 [7]. An Agilent 35670A Dynamic Signal Analyzer (DSA) was used to control the vibration and record data. The vibration amplitude was calibrated and fixed at 1 g_{rms} by using a commercial accelerometer. Both up-sweep and down-sweep were performed to characterize the frequency response of the harvester. The measured open-circuited output voltage V_{oc} when the top and bottom coils were connected in series is shown in Figure 6. The output voltage is increased by 70%

and 40% and the output power is expected to increase by 190% and 100%, respectively, for the two magnets after ferrofluid is injected. Therefore, ferrofluid can effectively increase the output voltage and power. With ferrofluid, the maximum open-circuited output voltage is 35.9 mV and 26.5 mV, respectively, for the two magnets. For matched load, the expected maximum output power $P_{opt,\,max}$ is 2.3 μW and 1.3 μW, respectively. The performance is summarized in Table 1.

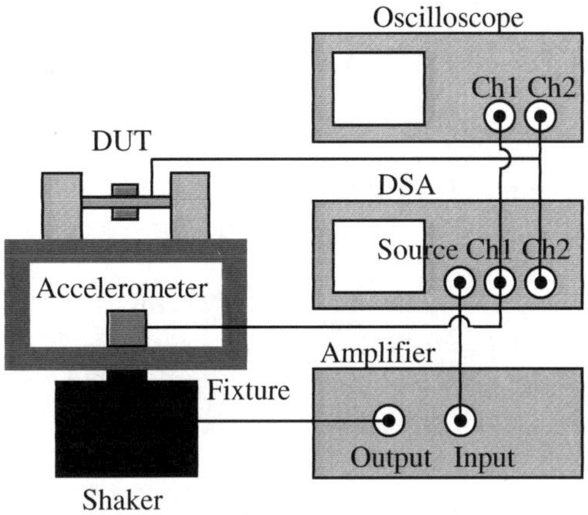

Figure 5. Test setup [7].

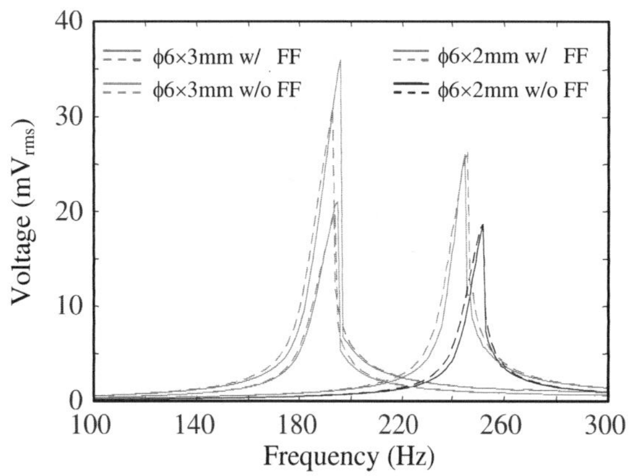

Figure 6. Measured open-circuit output voltage.

Table 1 Performance summary of proposed harvester

Magnet	$V_{oc,\,max}$ (mV$_{rms}$)		$P_{opt,\,max}$ (μW$_{rms}$)		Max power density (μW$_{rms}$/cm^2)	
	w/o FF	w/ FF	w/o FF	w/ FF	w/o FF	w/ FF
ϕ6×3mm	20.9	35.9	0.78	2.30	0.20	0.58
ϕ6×2mm	18.6	26.5	0.62	1.25	0.15	0.31

4. Conclusion

This paper reports an electromagnetic energy harvester fabricated with Rigid-Flex PCB technology. Ferrofluid was injected into the embedded cavities inside the PCB to increase the output emf. Experiments showed up to 70% increase of output emf and 190% increase of output power with embedded ferrofluid. Maximum expected output power at 1 g_{rms} is 2.3 μW$_{rms}$. The project was supported in part by the Ministry of Science and Technology, Taiwan, ROC (MOST 104-2221-E-009-168-MY2, MOST 106-2221-E-009-091-MY2, MOST 104-2221-E-009-156-MY3, MOST 106-2633-E-009-001). The authors are grateful for the support of the Chip Implementation Center, the National Center for High-performance Computing, the National Nano Device Laboratory, and the National Chiao Tung University Nano Facility Center, Taiwan, R.O.C.

References

[1] G. Hapitoglu, et al., Smart Mater. Struct., 19 015022, 2010
[2] D. Mallick, et al., Smart Mater. Struct., 24 015013, 2015
[3] C. Serre, et al., Microsyst. Techno., 15 1217–23, 2009
[4] Y. Chiu, et al., Proc. Power MEMS 2012, 149–52, 2012
[5] M. C. Wurz, et al., Proc. SPIE, Vol. 8688, 86881L, 2013
[6] Y. Chiu, et al., J. Phys.: Conf. Ser., 660 012083, 2015
[7] Y. Chiu, et al., Smart Mater. Struct., 25 125014, 2016
[8] K.-T. Lee, et al., Microsyst. Technol., 9, 232–242, 2003

IOP Conf. Series: Journal of Physics: Conf. Series **1052** (2018) 012079 doi:10.1088/1742-6596/1052/1/012079

Energy Harvesting Flex-Coil System for Pneumatic Pistons

J Esch[1,3], K Ylli[1], D. Stojakov[1], A Willmann[1], D Hoffmann[1], Y Manoli[1,2]
[1] Hahn-Schickard-Gesellschaft für angewandte Forschung e.V., Wilhelm-Schickard-Str. 10, 78052 Villingen-Schwenningen, Germany
[2] Fritz Huettinger Chair of Microelectronics, Department of Microsystems Engineering – IMTEK, University of Freiburg, Germany

E-Mail: Jonas.Esch@hahn-schickard.de

Abstract. Condition monitoring systems are being implemented in many branches of industry as they help reduce machine downtime and enable the application of predictive maintenance. For easy and maintenance-free retrofitting an autonomous power supply is necessary. This work presents results on a foil-based inductive energy harvester as part of a condition monitoring system (system-on-a-foil) which can be retrofitted to pneumatic pistons. A maximum of 180 µJ of energy was generated with a single coil of the best layout for one motion cycle (back-and-forth motion) of the piston with a speed of 1 m/s. The total setup with five coils and an additional flux guide generates 1 mJ per motion cycle. This setup allows measuring the capacitor voltage and environmental data and transmitting them via bluetooth low energy to a portable device.

1. Introduction

Condition monitoring systems are very important for modern industrial facilities. They help to reduce machine downtime and to increase productivity. Many machines use pneumatic pistons for linear motions. At the moment pneumatic pistons have only a proximity switch to detect the end position after a forward or backward motion. The proximity switch is triggered by means of a magnet, which is attached at the piston rod inside the piston case. To get more information a sensor system is being developed, which measures the position during the motion. Two different settings have been implemented. In the first setting hall sensors are placed outside at the case of the piston. In the second setting a combination of flat coils is used. Due to the moving magnet inside the piston, a voltage is induced in each of the coils. Based on the voltage characteristic the position of the piston rod can be calculated during the motion. For predictive maintenance additional sensors are necessary to measure vibrations, temperature, humidity and pressure. To generate the power for the microcontroller and the different sensors the coils are also used as an energy harvesting system.

To get a practicable condition monitoring systems for pneumatic pistons the system is designed as a system-on-a-foil. Such a label with thinned chips and flexible tracks can be easily applied to pneumatic pistons. Figure 1 shows a pneumatic piston with a schematic of the system-on-a-foil and an

[3] To whom any correspondence should be addressed.

inside view of the magnet including flux lines. The final energy harvesting setup with the maximum power generation is shown in Figure 2. This setup includes a power management, an environmental sensor and bluetooth low energy for wireless data transmission.

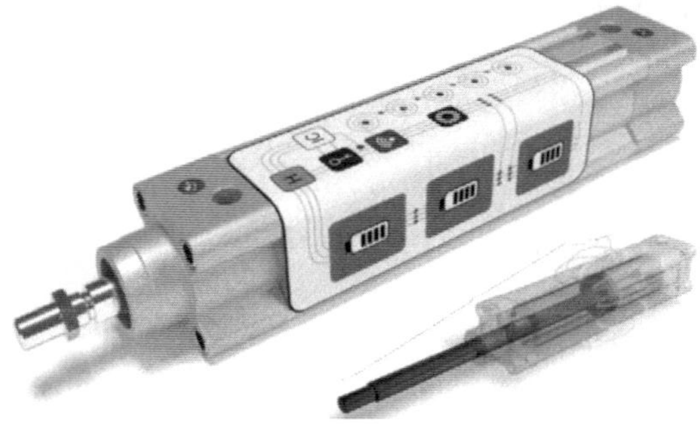

Figure 1. Schematic of the pneumatic piston with the system-on-a-foil including the energy harvesting coils and a small inside view to show the magnet and the flux lines [1].

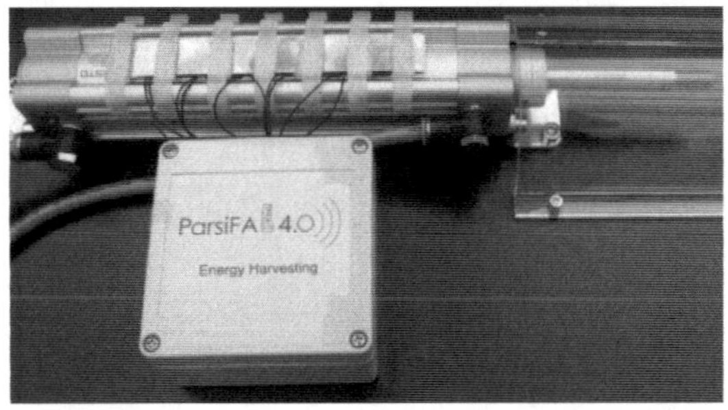

Figure 2. Photograph of the pneumatic piston including coils and flux guiding stripes connected with a power management.

2. Concept and Setup

A system-on-a-foil including an energy harvester means that the energy harvester coils also have to be implemented in the foil. In other energy harvesting systems, planar coils have been developed on PCB boards [2] or polyimide layers [3]. In this system the coils are designed on flexible circuit boards, with a copper layer height of 28 μm and various line widths and spacings (e.g. 120/80 means a copper line width of 120 μm and a spacing of 80 μm between copper lines). With the first manufacturing process a stripe (Figure 3) with two layers is generated. In a second step the stripe is folded to a multi-layer coil where 7 foldings imply a total of 14 coil layers (Figure 4). Each coil layer contains approximately 40-50 windings, depending on the line widths. Quadratic and circular layouts were designed and manufactured. The coil diameter is 18 mm and 10 coils can be placed next to each other along the test piston. The energy harvester uses the magnet from the proximity switch which moves with the piston rod. During piston motion a voltage is generated in the coils.

In order to get a higher voltage and power output, the original magnet (NdFeB N35) in the pneumatic piston was replaced by a bigger magnet with a higher magnetization (NdFeB N50).

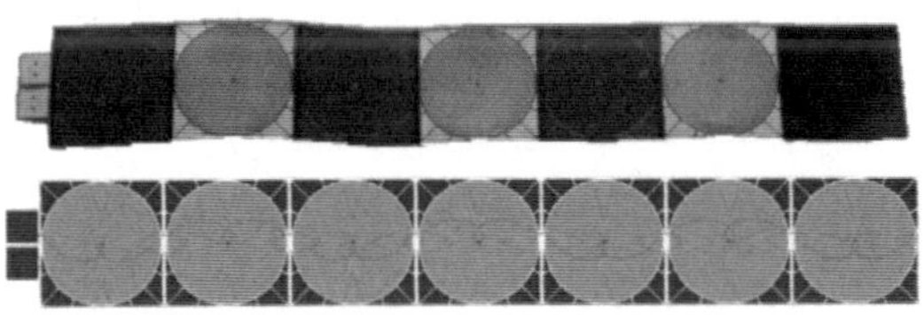

Figure 3. Photograph of the unfolded seven layer coil (top) and the layout (bottom).

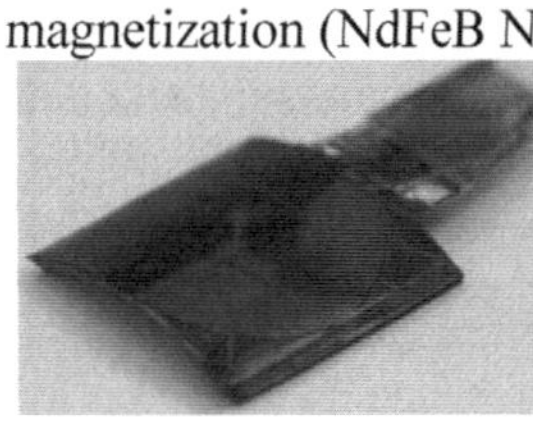

Figure 4. Photograph of the folded coil.

3. Experimental Data

3.1. Experiment setup

Pneumatic Pistons are used for linear motions in many industrial machines. Depending on the task such systems can move slow and fast. In the laboratory setup the velocity is measured by a distance measuring system and controlled by pneumatic throttle. For comparison all experiments were carried out with the same moving speed of approximately 1 m/s. The schema for the coil positions is given in

IOP Conf. Series: Journal of Physics: Conf. Series **1052** (2018) 012079 doi:10.1088/1742-6596/1052/1/012079

Figure 5. To each coil a load resistor was connected, which is equal to the respective coil resistance. The voltage at the load resistors was measured simultaneously and is shown in Figure 6. For the modified piston (Figure 6 b) the generated voltage is 10 times higher than for the original piston (Figure 6 a). The results are shown for the circular 7 folded layer coil setup. For the circular 10 folded layer setup the voltage is 45% higher.

The induced voltage in the first and the last coil is very low because the magnet starts and stops moving under this coils and the moving speed is very slow. When comparing the data of voltage and speed measurement there is a direct correlation between the voltage peaks and the moving speed.

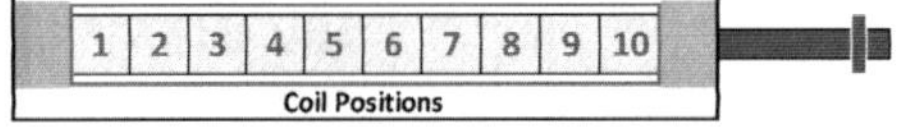

Figure 5. Schematic of the piston including the coil positions

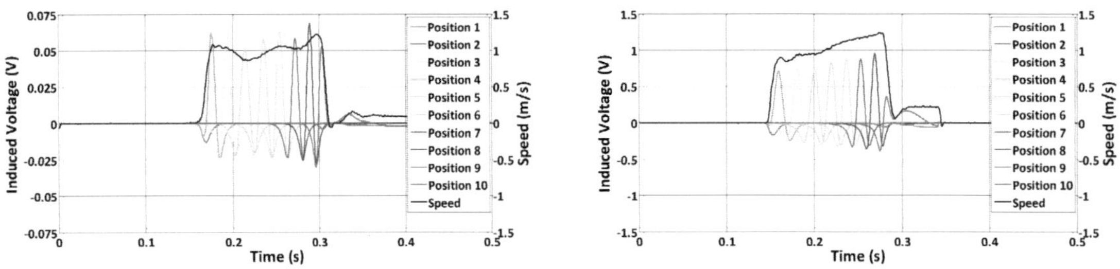

Figure 6. (a) Induced voltage for every coil for one piston motion with the original magnet including the speed measurement. (b) Induced voltage for every coil for one piston motion with the modified magnet including the speed measurement.

The generated energy for each coil and one motion cycle (one back-and-forth motion) is shown in Figure 7 and Figure 8. The energy is very low for coils at the end positions (Coils 1, 2, 9 and 10). This is because of the damping at the end which reduces the motion speed. Figure 7 shows a comparison between quadratic and circular coil designs and several line widths. The quadratic designs show extremely large resistances due to the fabrication process, where the copper line became unexpectedly narrow in the corners of each winding, thus creating a bottle-neck for the electrical current. When comparing circular coils of different line widths, experimental data indicates, that a comparatively small coil resistance leads to improved power outputs. An increased number of stacked coil layers also lead to an improved power output, although a stack of 10 folded layers loses its flexibility and is practically as rigid as a copper wire coil.

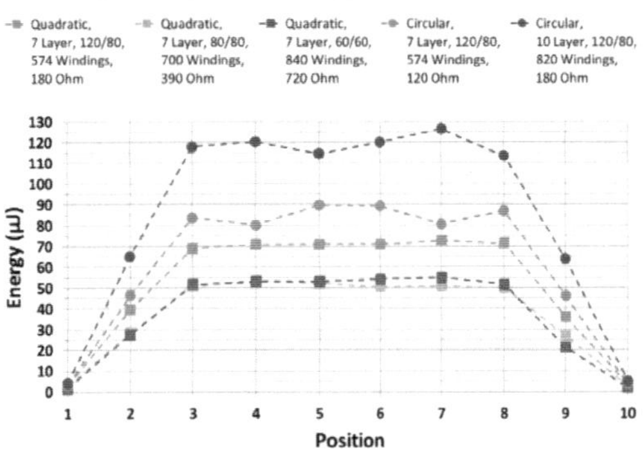

Figure 7. Energy comparison for different coil layouts and foldings.

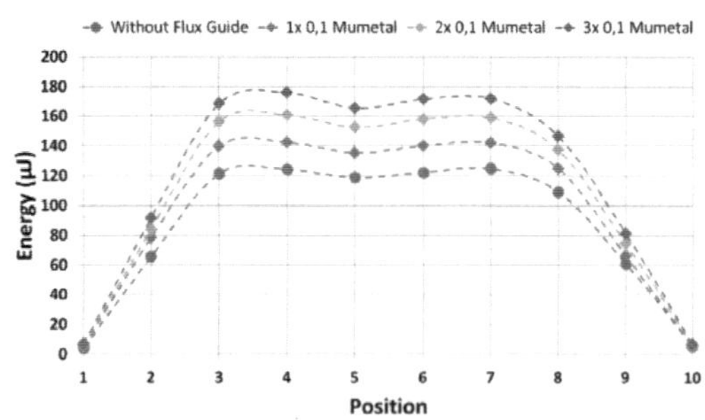

Figure 8. Energy comparison for varying number of flux guiding stripes.

The placement of thin layers of flux guiding material (mumetal) on top of the coils further improves the power output as shown in Figure 8. For the setting with the circular 10 folded layer coils adding three mumetal foils with a total height of 0.3 mm the generated power is 50% higher. The preceding simulations deviate by less than 5% from the experiments.

3.2. Demonstrator setup

To show the functionality of the energy harvester with a sensor system a new setup (Figure 2) was build. A diode rectifier was used to get a DC voltage of each coil. After the rectifier a power management with a capacitor and regulated voltage output was used to power a microcontroller with a bluetooth low energy communication interface. An environmental sensor was connected to the microcontroller in order to measure temperature, humidity and pressure. For generating usable voltage levels, two coils were used at each position and connected in series. In total, five coil pairs were positioned in the middle part of the piston and six mumetal foils were added above.

If the system starts the first time (cold-start) it needs about 80 seconds (one complete motion per second) to get enough energy in the capacitor for starting the sensor system and data transmission. Once the sensor system is started, the capacitor voltage is captured and transmitted every second whereas the environmental data is measured and transmitted every five seconds. In this configuration, the capacitor is still further charged as long as the motion of the piston continued.

4. Conclusion

In this work we presented an energy harvester for pneumatic pistons with no additional mechanical components. Only a replacement of the already existing magnet within the piston and mounting of coils were required to generate enough power for a sensor system with radio transmission. The functionality of the concept was demonstrated for an original pneumatic piston and a modified version. One motion cycle of the piston generates 1 mJ of energy. This is equivalent to a continuous output power of 1 mW when the piston does one motion cycle per second. For the final sensor system it is possible to use the first and last coil as a trigger signal to wake up the system. With such a configuration the power consumption can be reduced.

After the comparison of the simulation and experimental data an optimization algorithm was used to find the best coil layout. To get a coil layout with good manufacturability the feasible line width to space ratio is deposited as a boundary condition in the simulations. In addition to the line width to space ratio the line high was changed to 60 µm to reduce the coil resistant. Ongoing work includes a new fabrication run of the flex-coils and further experiments. An increase of 80% in power output is expected.

5. References

[1] Festo AG & Co. KG, *Forschungsprojekt ParsiFAl 4.0: Festo Unternehmen.* Available: https://www.festo.com/group/de/cms/12002.htm (2017, Sep. 21).

[2] Y. Li, W. Zhang, Y. Zhang, and J. Cao, "A batch-fabricated electromagnetic energy harvester based on flex-rigid structures," *Sensors and Actuators A: Physical*, vol. 263, pp. 571–579, 2017.

[3] D. Hoffmann, C. Kallenbach, M. Dobmaier, B. Folkmer, and Y. Manoli, "Flexible Polyimide Film Technology for Vibration Energy Harvesting," in *Proceedings Power MEMS*, pp. 344–347.

6. Acknowledgements

This research was supported by the German Federal Ministry of Education and Research (funding code: 16ES0435).

IOP Conf. Series: Journal of Physics: Conf. Series **1052** (2018) 012080 doi:10.1088/1742-6596/1052/1/012080

An electromagnetic energy harvester capable of frequency up-conversion and amplitude amplification under pulse excitation

D Zhu and L Evans

College of Mathematics, Engineering and Physical Sciences, University of Exeter, Exeter UK, EX4 4QF

E-mail: d.zhu@exeter.ac.uk

Abstract. This paper reports an electromagnetic energy harvester that is capable of frequency up-conversion and oscillation amplitude amplification to maximize its output power under pulse excitation. The harvester consists of a tube with a shuttle magnet travelling inside. The external force moves a complementary magnet set that reverses the direction of the magnetic force applied to the shuttle magnet. Such alternating magnetic force enables the shuttle magnet to move back and forth inside the tube and electric current can be induced in the coil wound around the tube. A prototype was fabricated and tested. Experimental results showed that when a 1 Hz footstep was applied to the complementary magnet set, the shuttle magnet oscillated at a frequency of 10 Hz and oscillation amplitude was amplified by 6 times. Peak power of over 70 mW was recorded and an average energy of 2.8 mJ can be generated from every footstep.

1. Introduction

Energy harvesting from footsteps has drawn a lot of attention in recent years. It enables electrical energy to be generated while a human is walking or running. Applications of such technology ranges widely from powering wearable electronic devices [1] to powering street lights [2]. One of the major challenges of harvesting energy from footsteps is that input force in this application normally has low frequency, e.g. ~1 Hz for walking and < 5Hz for running. Fundamental analysis of kinetic energy harvesting suggests that more mechanical power is available to be harvested if the operating frequency is higher and the resonant frequency of the kinetic energy harvester matches it [3]. This makes it difficult to harvest energy harvest from such low frequency movement. However, energy harvesting from footsteps does benefit from large input forces, which makes it still attractive as an alternative power source for various applications.

Two transducers are normally used to harvest energy from footsteps, i.e. piezoelectric and electromagnetic. In the piezoelectric method, piezoelectric materials can be applied where stress and strain are produced by the footsteps, such as shoe insole. The change of strain on the piezoelectric materials result in electric charges being generated. By collecting generated charges using electrodes, electric current can be generated. Traditional piezoelectric materials, e.g. PZT, are rigid and brittle, which makes it not ideal for such applications. Therefore, flexible materials that have piezoelectric properties have been investigated to tackle this problem. Luo *et al* [1] reported energy harvesting from footsteps by mounting porous polypropylene, a lightweight ferrelectret material on shoe insole. An average energy of over 100 μJ was generated by each footstep and this is sufficient to power a commercial ZigBee transmitter. In the electromagnetic method, the footstep causes a magnet to move with respect to a coil where current is induced. Ylli *et al* [4] presented two electromagnetic energy harvesters for human foot movement. They targeted swing movement and impact during human walking and average power of 0.84 mW and 4.13 mW were achieved by these two harvesters respectively.

This paper investigates a mechanism that can up-convert the operating frequency of an electromagnetic energy harvester and increase the oscillation amplitude of its inertial mass to maximize output power under pulse excitation.

2. Principle

The structure of the proposed energy harvester is illustrated in Figure 1(a). It consists of a cylinder tube with a coil wound around it. A shuttle magnet can travel back and forth inside the tube so that electric current is induced in the coil. One magnet is fixed to one end of the tube to provide repulsive force when the shuttle magnet approaches (the magnet on the left of the tube as in Figure 1(a)). A complementary magnet set (CMS) is suspended by a mechanical spring on the other side of the tube. The CMS consists of two bar magnets with opposite poles facing the end of the tube.

Figure 1(a) shows the original position of the CMS without external force. In this case, the magnet in the CMS that is aligned with the tube has opposite magnetic pole facing the shuttle magnet. Thus, the shuttle magnet moves towards the CMS due to the attractive force between them. When an external force is applied to the CMS as shown in Figure 1(b), the CMS moves downwards and the other magnet in the CMS is aligned with the tube. In this case, there is a repulsive force between the shuttle magnet and the CMS. This causes the shuttle magnet to move away from the CMS. When the shuttle magnet approaches the fixed magnet on the left, the repulsive force pushes it back towards the CMS where it once again encounters the repulsive force from the CMS and thus a higher frequency oscillation is produced. When the external force is released, the CMS returned to its original position thanks to the spring and the shuttle magnet moves back to its original position.

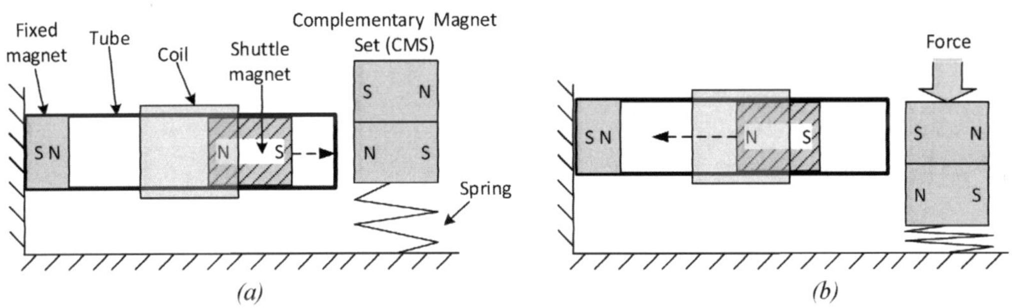

Figure 1. Schematic of the electromagnetic frequency up-converting energy harvester. (a) no external force (b) external force is applied.

3. Experimental

A prototype was manufactured for testing as shown in Figure 2. The plastic tube is 80 mm long. All magnets used here are made of NdFeB-38. The shuttle magnet is a 20 mm long cylinder magnet with a diameter of 15 mm. The fixed magnet on the left of the tube is a 15 mm × 15 mm × 5 mm (m) rectangular. The two bar magnets in the CMS have dimensions of 20 mm × 10 mm × 8 mm (m). (m) notates the magnetization of the magnets. The coil is wound using 75 μm thick copper enamel wire. It has around 500 turns and a resistance of 180Ω. All the supporting structures are made of 3D printed PLA.

In all experiment, the prototype was tested under human footsteps at around 1 Hz. The harvester was first connected to resistive loads and output voltage across various resistive loads was measured to find out the maximum output power. It was then connected to a bridge rectifier to charge a 2.2 mF capacitor and the charging rate was recorded.

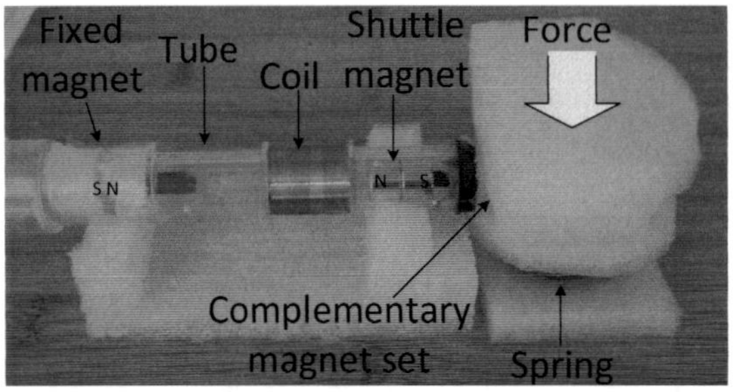

Figure 2. Photo of the testing prototype.

4. Results and discussion

When the footstep is applied to the CMS, the force moved the CMS downwards by 10mm. As explained in Section 2, the shuttle magnet is pushed to the left of the tube due to the repelling magnetic force by a maximum of 60 mm. This means the oscillation amplitude of the shuttle magnet is amplified by 6 times from the input displacement.

Figure 3 shows a typical waveform of the open circuit voltage of the energy harvester when an external pulse force is applied and released. When the external force was applied, the output voltage demonstrated a 10Hz oscillating waveform. This indicates that the shuttle magnet was oscillating at 10Hz inside the tube although this oscillation attenuated quickly due to friction between the shuttle magnet and the tube. When the force was released, the shuttle magnet was attracted towards the CMS and little oscillation occurred. Thus, only one peak was observed.

Then, peak output voltage of the energy harvester when connected to various resistive loads was recorded and peak output power was calculated using V^2/R, where V is the output voltage of the harvester when connected to a resistor, R. Ten peak voltage readings were taken for each resistive load and the average peak power was calculated. Figure 4 shows average peak output power when the harvester is connected to various load resistances. It is found that the maximum peak power was generated when connected to the optimum load resistance of 180 Ω.

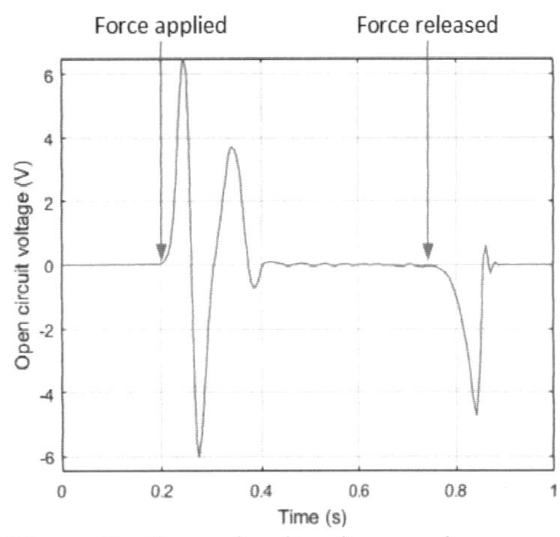

Figure 3. Open circuit voltage under one footstep.

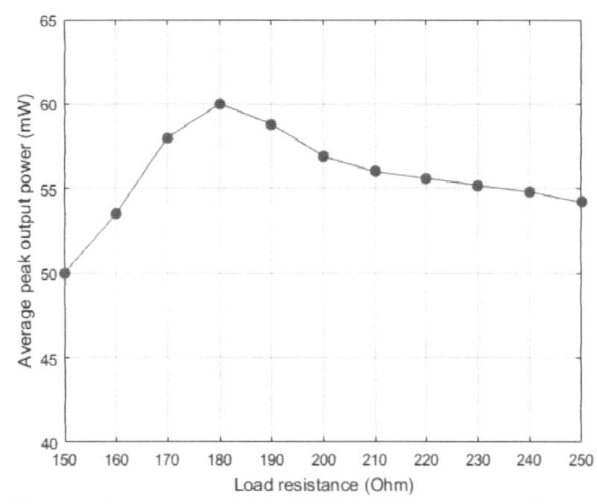

Figure 4. Average peak output power with variance of load resistances.

Figure 5 shows instantaneous output power of the energy harvester when it's connected to the optimum resistive load of 180Ω. Peak power of over 50mW and 20mW can be generated when the force is applied and released, respectively. Figure 6 shows accumulated energy produced by the harvester when connected to the optimum resistive load. An average energy of 2.8mJ can be generated from every step.

The energy harvester was also connected to a bridge rectifier to charge a 2.2 mF capacitor. Figure 7 shows the charging curve when the energy harvester was operated under a series of footsteps. It is found that the 2.2 mF capacitor was charged from 0 to 3.67 V within 20 seconds (19 steps). A total energy of 14.82 mJ was stored in the capacitor. In other words, an average power of 741 μW was delivered to the capacitor during this period.

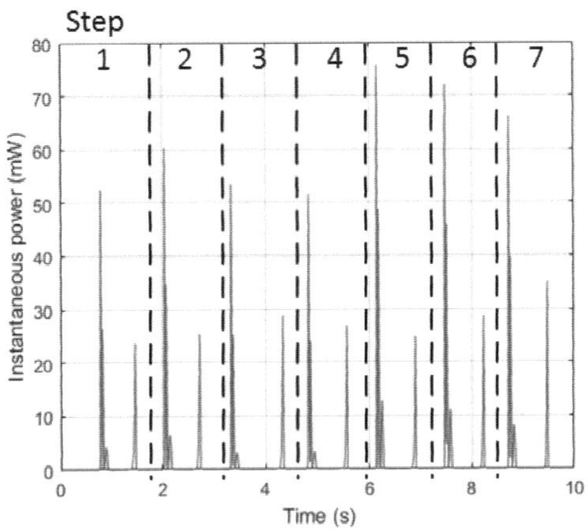

Figure 5. Instantaneous output power under a series of footsteps.

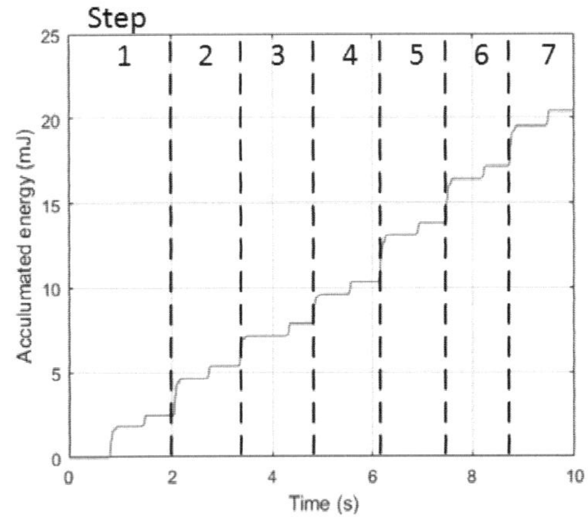

Figure 6. Accumulated output energy under a series of footsteps.

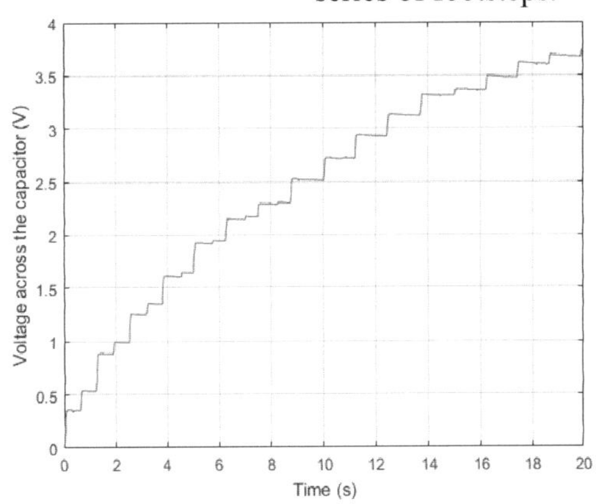

Figure 7. Voltage of the capacitor that was charged by the energy harvester.

5. Conclusion and future work

It has been demonstrated that the proposed mechanism is capable of up-converting the operation frequency by 10 times and amplifying input amplitude by 6 times when excited under pulse input. Such mechanism is suitable for energy harvesting from footsteps. It can be adapted for either integration with shoes or being embedded into floor. Future work will focus on further increasing the operation frequency by reducing the length of the tube and minimizing friction between the shuttle magnet and the tube.

6. Reference

[1] Luo Z, Zhu D and Beeby S P 2015 Multilayer ferroelectret-based energy harvesting insole, Journal *of Physics: Conference Series*, **660 (1)**.

[2] Pavegen Ltd, http://www.pavegen.com (accessed on 26 September 2017).

[3] Beeby S P, Tudor M J and White N M 2006 Energy harvesting vibration sources for microsystems applications *Meas. Sci. Technol.* **17** pp 175–195.

[4] Ylli K, Hoffmann D, Willmann A, Becker P, Folkmer B and Manoli Y 2015 Energy harvesting from human motion: exploiting swing and shock excitations *Smart Materials and Structures* **24 (2)**.

Electromagnetic energy harvester for atmospheric sensors on overhead power distribution lines

Z Wu[1], D S Nguyen[2], R M White[2], P K Wright[1], G O'Toole[3] and J R Stetter[3]

[1]Department of Mechanical Engineering, University of California, Berkeley, CA, USA
[2]Department of Electrical Engineering and Computer Science, University of California, Berkeley, CA, USA
[3]SPEC Sensors, LLC, Newark, CA, USA

E-mail: zway@berkeley.edu

Abstract. In this paper we present a small, credit-card sized, and low-cost electromagnetic energy harvester that couples energy from one conductor of an overhead power distribution line. We use a rectangular coil, placed near the power line without encircling it, to extract electrical power from the time-varying magnetic field. The addition of a U-shaped electrical steel sheet (a high-permeability material typically used in transformers) near the coil significantly increases the magnetic flux through the coil and consequently raises the output power. Experiments have shown that the harvester with an optimal load resistance obtains 230 mW (AC) when the power-line conductor current is 50 A_{rms}. The energy harvester is able to power a wireless gas sensor board having a carbon monoxide and an ozone sensors at 10 A_{rms}.

1. Introduction

Energy harvesters installed on electrical power lines (Fig. 1) have an ability to power sensors and wireless communication devices [1]. They can be used by utility companies to power different sensors that monitor the conditions of the power lines [2, 3]. Since this condition monitoring would require uninterrupted operations and minimal maintenance, the energy harvesters are more feasible than batteries which need regular replacement from time to time. Another possible use of the energy harvesters is to provide energy for a wireless sensor network built on the power distribution grid. For example, the air quality report provided by present technology is not localized [4], if many atmospheric sensors could be installed on the power distribution grid to form a dense sensing network, that enables people to obtain much more localized information about the present air quality. Energy harvesters that extract power from the electrical wires in a building were reported in [5, 6]. We envision that the energy harvesters installed on the overhead power distribution lines could power sensors for various interests:

A. Measurements useful to the power company: current; voltage; power and its flow direction; conductor temperature; AC frequency; phasing; outages; power theft data; *etc.*

B. Environmental sensors: particulate matter concentrations ($PM_{2.5}$, PM_{10}, *etc.*); local air temperature, relative humidity, pressure; gases such as NO_X, CO, CO_2, O_3, *etc.*

C. Other sensors: acoustic, optical, biological, vibration, acceleration, *etc.*

In this paper, we describe the design, modeling, and experimental study of the electromagnetic energy harvester with an output power range between several milliwatts to hundreds of milliwatts from an ac current up to 50 A_{rms}. We anticipate that the energy harvesters could be installed inexpensively on the overhead power distribution lines by a utility operator using a hot stick.

Figure 1. Hypothetical energy harvester (to scale) attached to an overhead power distribution line.

Figure 2. (a)1250-turn coil; (b)Coil with a laminated electrical steel core; (c) and (d)Coil with a laminated electrical steel core and a U-shaped flux guide.

2. Energy Harvester Design and Modeling

The energy harvester consists of a rectangular coil with a laminated electrical steel core is shown in Fig. 2b. To increase the output power from the harvester placed near the conductor without significantly adding volume to it, we made a U-shaped electrical steel sheet (flux guide) to direct more magnetic flux into the coil (Fig. 2d). The thin flux guide could eventually be embedded in the enclosure that contains the energy harvester. Energy harvesters that encircle the conductors [5, 7] could be challenging to install on the power distribution lines. The harvester in Fig. 2d is technically very similar to a transformer, and we use an equivalent circuit of a current transformer to model it (Fig. 3). The power delivered to the load resistance, R_{Load}, assuming the inductor is a linear component, is

$$P_{Load} = \frac{(|I_p|\omega L_\mu)^2 * R_{Load}}{N^2[(R_{Wire} + R_{Load})^2 + (\omega L_\mu)^2]} \tag{1}$$

where L_μ, R_{Wire}, and N are the inductance, resistance, and number of turns from the secondary side of the current transformer, respectively (Table 1). The frequency is set to 60 Hz, and I_p is the current (rms) in the power line conductor.

Table 1. Parameters for the secondary side

Number of turns, N	Inductance, L_μ	Resistance, R_{Wire}
1250	2.78 H	107.6 Ω

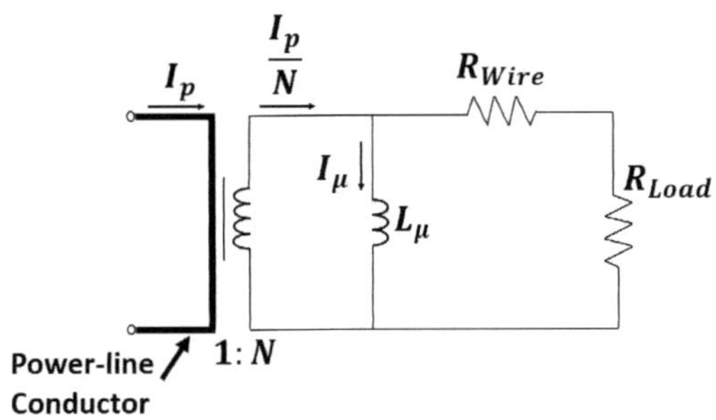

Figure 3. Equivalent circuit for the energy harvester using a current transformer model.

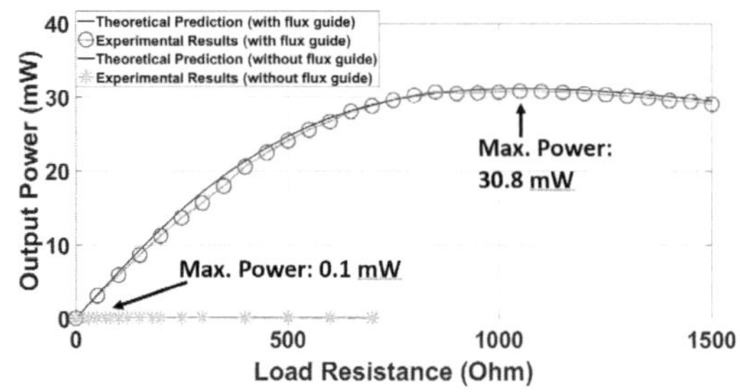

Figure 4. AC output power from the energy harvester versus load resistances. (I_p=10A$_{rms}$)

3. Experimental Results and Discussion

The energy harvester (Fig. 2d) was tested with varying load resistances when the current in the power-line conductor is 10 A$_{rms}$, in which the inductor still behaves linearly. Figure 4 shows both the model predicted by Eq. (1) and the experimental results, and they are very close to each other. The power output of the harvester without the flux guide is also shown here. We found that the maximum output power increases 300 times by adding the flux guide. In Figure 5, we show the output power of the harvester as we change the primary current in the conductor from 0 to 50 A$_{rms}$. The load resistor at each measurement point was adjusted to obtain the maximum output power for each power-line current. It was found that saturation occurred in the flux guide as the primary current reached about 15 A$_{rms}$ and persisted till 50 A$_{rms}$. With saturation in the flux guide, the harvester was still able to deliver more power when increasing the current, but at a lower increasing rate comparing to the theoretical prediction.

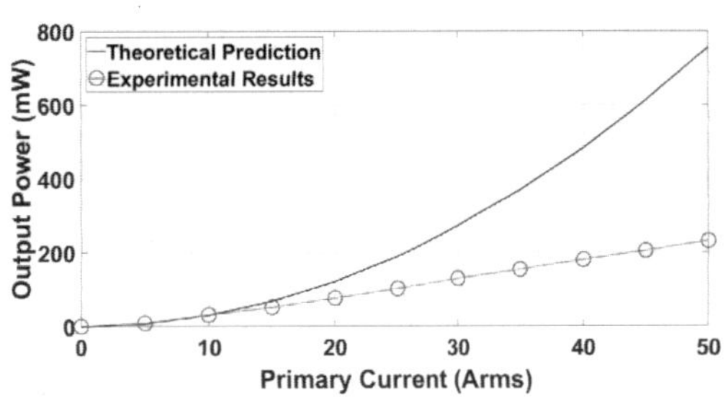

Figure 5. AC output power from the optimally loaded energy harvester versus primary current.

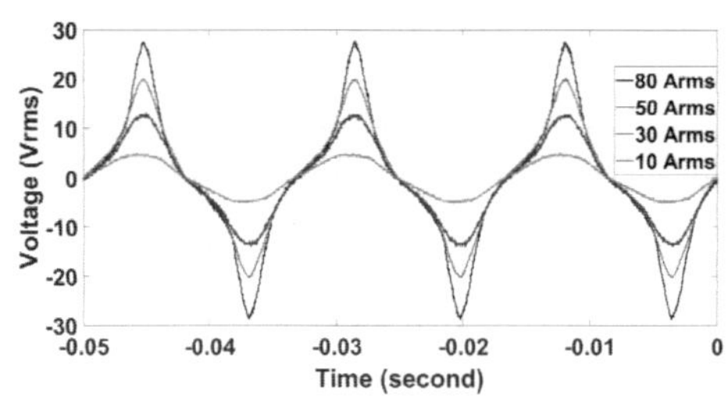

Figure 6. Waveforms of the voltage across the load resistance with respect to four different primary currents.

We find that the secondary-side inductor, L_μ shown in Fig. 3 actually behaves non-linearly as the current transformer couples to the power-line conductor. When the current, I_μ, flowing through the inductor rises due to either the high primary current or high load resistance, it increases the magnetic flux, and the flux guide becomes saturated. Hence the inductor in the circuit is no longer a linear component, and this is why there is a discrepancy between the theoretical model and experimental results (Fig. 5).

Four waveforms of the voltage on a fixed load resistance are shown in Figure 6, which correspond to current in the conductor of 10 A$_{rms}$, 30 A$_{rms}$, 50 A$_{rms}$, and 80 A$_{rms}$. This demonstrates that the voltage waveform is sinusoidal when there is no saturation (at 10 A$_{rms}$), and it is gradually distorted as the flux guide is saturated by the increased magnetic flux due to the rising current in the conductor. When the primary current reaches 80 A_{rms}, the flux guide would become heavily saturated. Moreover, the saturation would act as a protection for the load since it could prevent too much power from entering the circuit due to an abrupt increase in the primary current (i. e. lightning strikes on the power line.).

The energy harvester was connected to a power management circuit using an LTC 3108 step-up converter to operate a commercial gas sensor board from SPEC Sensors (Fig. 7), and the sensor board broadcast (with a Bluetooth Low Energy radio having a typical range of 132 ft) and stored data to an Android^{TM} phone. Fig. 8 shows how the output from an ozone sensor exposed to a brief burst of ozone. The sensor board can function with the energy harvester when the conductor current is larger than 10 A_{rms}.

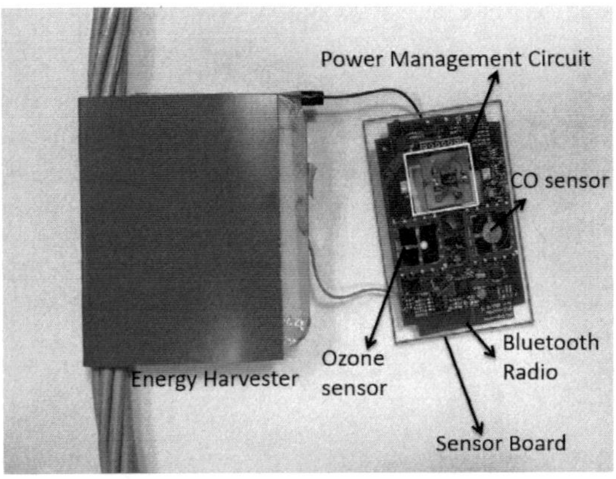

Figure 7. Energy harvester, power management circuit, and sensor board attached to an overhead distribution conductor.

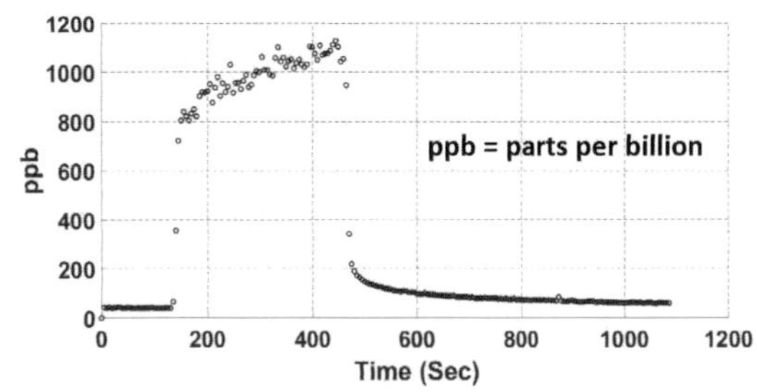

Figure 8. Ozone concentration from sensor response to brief ozone exposure

4. Conclusions and Future Work

In summary, we have shown a small, long-life, and low-cost electromagnetic energy harvester that can provide an AC output power of 230 milliwatts when placed near a power-line conductor carrying 50 A_{rms}. The harvester could power atmospheric sensors, as well as sensors that collect information of interest to power companies. Saturation was observed in the flux guide, and we plan future work to focus on modeling the nonlinear behavior of the magnetic material and mitigation of the saturation. We also plan to optimize the harvester's shape for maximum power output and test various magnetic materials for the core and flux guide.

References

[1] Paprotny I, Xu Q, Chan W W, White R M and Wright P K 2013 *IEEE Sensors Journal* **10** 190-201
[2] Roscoe N M and Judd M D 2013 *IEEE Sensors Journal* **10** 2263-70
[3] Yuan S, Huang Y, Zhou J, Xu Q, Song C and Thompson P 2015 *IEEE Trans. on Power Electron.* **30** 6191-6202
[4] Apte J S, Messier K P, Gani S, Brauer M, Kirchstetter T W, Lunden M M, Marshall J D, Portier C J, Vermeulen R C H and Hamburg S P 2017 *Environ. Sci. Technol.* **51** 6999-7008
[5] DeBruin S, Campbell B and Dutta P 2013 *Proc. of the 11th ACM Conf. on Embedded Networked Sensor Systems (Roma, Italy)*, pp 18:1-14
[6] Thorin O 2016 Power line induction energy harvesting powering small sensor nodes *Master's Thesis (KTH Royal Institute of Technology, Stockholm, Sweden)*
[7] Moon J, Donnal J, Paris J and Leeb S B 2013 *App. Power Electron. Conf. and Expo. (APEC) (Long Beach, CA, USA)* pp 3151-59

IOP Conf. Series: Journal of Physics: Conf. Series **1052** (2018) 012082 doi:10.1088/1742-6596/1052/1/012082

A MEMS Magnetic-Based Vibration Energy Harvester

A. Shin[1], U. Radhakrishna[1], Yuechen Yang[1], Q. Zhang[2], L. Gu[2], P. Riehl[2], A. P. Chandrakasan[1], J. H. Lang[1]

[1]MIT, Cambridge, MA, 02139, USA. [2]Analog Devices, Inc., Wilmington, MA, 01887, USA.

E-mail: `ujwal@mit.edu`

Abstract. This paper presents the design, analysis and integrated fabrication of a MEMS magnetic-based vibration energy harvester targeted for machine health monitoring. The design consists of Si-springs, permanent magnets as mass, and coils wound on the top and bottom side of the harvester package for mechanical-to-electrical energy conversion based on the Lorentz-force principle. The harvester is optimized to have its translational resonant-mode match external vibrations while separating higher-order modes. Mechanical and magnetic optimization of the harvester is carried out together with optimization of its power and control electronics in order to provide maximum output power from a vibration input that can vary its frequency by ±5%. The harvester achieves an open-circuit voltage amplitude of 145 mV and delivers 165 μW to a matched load at the resonance frequency of 45.7 Hz.

1. Introduction

The increasing importance of autonomous sensing and portable energy harvesting for the IoT domain is leading to miniaturization of energy-transducers that can power those applications [1][2]. Among several types of transducers that are available, MEMS (electro)magnetic-based (EM) vibration energy harvesters are attractive due to their compact form factor and high output-power density. This paper presents such an EM-energy harvester comprising a DRIE-etched silicon suspension, and pick-and-place N42 NdBFe magnets, copper coils, and back-iron. The target application is to provide over 100 μW to autonomous machine-health sensors from near-50-Hz vibrations with amplitudes less than 0.5 g.

2. Design and Optimization

Two high-level optimizations guide the design of the harvester. The first optimization concerns the harvester mechanics shown in Figure 1. The harvester comprises a spring-mass-damper system (K-M-B) that fits within a rectangular volume having dimensions L_1, L_2 and L_3. Here,

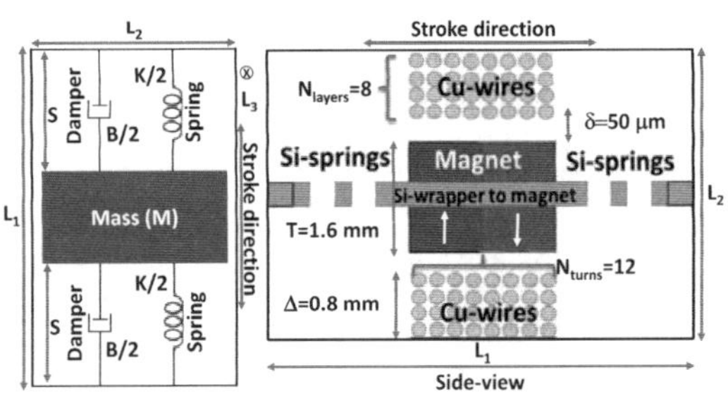

Figure 1. Harvester mechanics and magnetics for optimization of overall dimensions.

Design Variable	Value (mm)
L_1	14
L_2	10
S	3.5
l	3.5

Figure 2. Parameter optimization based on mechanics.

Design Variable	Value (mm)
L_3	3.3
T	1.6
Δ	0.8
δ	0.05
N_{layers}	8
N_{turns}	12

Figure 3. Parameter optimization based on magnetics.

B is a proxy for useful energy conversion. The mass moves in the L_1 direction, and at rest is centered within the harvester frame. The maximum bidirectional vibration stroke of the mass is S. If the harvester frame vibrates in the L_1 direction in the sinusoidal steady state with acceleration frequency ω and amplitude A, then the mass-motion amplitude X is given by (1) and the time-average mechanical power P_{M} harvested through B is given by (2). Solving (1) for B yields (3). Substitution of (3) into (2) shows that P_{M} is maximized at resonance, characterized by (4). In this case, P_{M}, obtained by combining (2), (3) and (4), becomes (5). Equation (5) expresses P_{M} in terms of the exogenous inputs ω and A, and the competing on-chip resources X and M. This power can only be achieved with: the B-loading design of (3), the spring design of (4), and $X \leq S$. From Figure 1 it is apparent that M is given by (6) in which case (7) holds. Finally, (7) is maximized with $X = S = L_1/4$ to yield (8). Several important conclusions can be drawn from (8). First, the competing resources of M and X should be given roughly the same space in the L_1 direction, thereby necessitating a large-travel spring. Second, L_1^2 in (8) indicates that the mass should move along the longest length. Thus a harvester designed to operate with out-of-plane motion is not optimal. Third, the mass should be designed with the largest possible density ρ. Fourth, the harvester should be loaded via B so as to exhibit maximum stroke. Fifth, power per volume ($L_1L_2L_3$) favors long thin designs. Due to these conclusions, magnetic energy conversion is chosen due to its compatibility with long-stroke motion. Permanent magnets are chosen as the mass since they offer the highest net ρ. Figure 2 describes the aforementioned variables designed from the mechanics optimization. The primary disadvantage of this design is that mass could be attracted to magnetic material in a transverse direction thereby locking up the harvester or damaging its springs.

The second optimization concerns magnetics. Shown in Figure 1 is a cross section of the magnetic energy converter through which B is implemented. The converter comprises of two anti-parallel magnets, and two stationary coils and backirons connected to the harvester frame. Assuming that the magnetic thickness of $T + 2\Delta + 2\delta$ is small compared to the magnet widths and lengths, and assuming that the backiron is highly permeable, the maximum magnetic flux density, and hence the maximum induced coil voltage V at a given ω and X behaves as in (9). Additionally, the coil resistance R_{coil} behaves as in (10). Therefore, the maximum time-average electrical output power $P_{\text{E}} = V^2/8R_{\text{coil}}$, achieved with a matched load, behaves as in (11), assuming sinusoidal steady state operation. For a given T, P_{E} is maximized with the smallest permissible δ and $\Delta = T/2 + \delta$. Thus, a thick coil and a thin gap is desired. A thin gap is also desired to prevent the magnets from moving too far in the transverse direction and locking up if they are shocked towards the backiron. Figure 3 lists the values of the aforementioned design variables resulting out of this magnetics-optimization. Finally, note that the optimal choice of B will not dictate maximum P_{E} and hence matched electrical loading under all vibration scenarios. Nonetheless designing for maximum P_{E} permits the greatest operating range.

3. Simulation and Analysis

Once the high-level optimizations yield the overall harvester dimensions for a targeted power, a spring design is chosen to provide the desired stroke S and resonant frequency. Accordion and 4-bar-linkage designs are considered for this purpose. While accordion springs provide the largest S with minimum underutilized area, their higher-order resonant modes (in-plane rotation and off-axis translation) are close to the desired fundamental translational mode. The 4-bar linkage has good modal separation but uses volume for anchors that could potentially be used for the mass. Therefore a combination of the two designs in the form of linked-accordion springs is chosen to provide good modal separation and the desired S. The beam-width w ($w = 30$ μm must be larger than the minimum width of 25 μm imposed by process limitation.) and beam length are optimized using simulations performed in ANSYS which show the lowest resonant modes are: translational motion along L_1 at 47 Hz; rotational motion about the L_3 direction (magnetic pole direction) at 283 Hz; and translational motion along L_3 at 575 Hz.

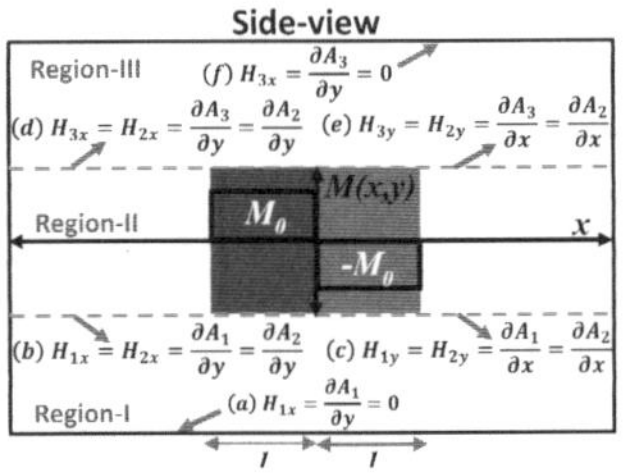

Figure 4. Magnetization and boundary conditions for voltage computation.

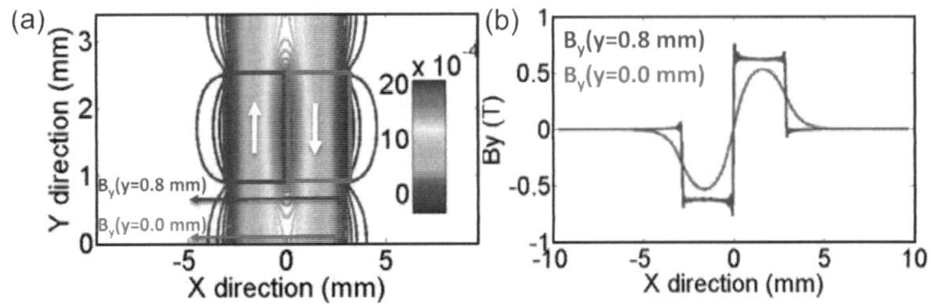

Figure 5. (a) Magnetic field distribution in the harvester and the magnetic flux densities along y-direction (B_y) that determine voltage across each turn in the top and bottom coil-layer.

Next, the coil-winding configuration is designed to maximize the mechanical-to-electrical transduction coefficient G while minimizing the coil resistance R_{coil}. This is done by computing the magnetic vector potential A by solving (12) in the space above and below the magnets where the coils are placed. Here M_K is the K-th Fourier component of magnetization given by (13) in region II, where $M_0 = 1.3$ T is the magnetization of the permanent magnets, l is the length of each magnetic pole in the stroke direction and $M_K = 0$ in regions I and III. The closed-form solutions for A_K are given by (14) in regions I and III and are given by (15) in region-II. The field distribution is computed numerically in Matlab using the boundary conditions highlighted in Figure 4 and is shown for the equilibrium position in Figure 5. Finally, the time-dependent voltage across each turn of the coil can be computed numerically using (16), where $v(t)$ is the velocity, and ω_0 is the radial resonant frequency. The choices for the total number of layers N_{layers} of 100 μm-diameter copper coils within T and the number of turns in each such layer N_{turns} are determined based on this magnetic-simulation and listed in Figure 3. The optimization-goal is to maximize the metric G^2/R_{coil} for efficient power extraction.

4. Fabrication and Packaging

The spring suspension is fabricated in silicon using a single cookie-cutter deep-reactive-ion through etch. To do so, 3 μm of CVD oxide is deposited on a 525-μm-thick wafer, and densified. The oxide serves as a hard mask for the through etch. Next, resist is spun on and patterned using direct-write technology and chemical development. Using the resist as a mask, the oxide is patterned using reactive-ion etching, and the resist is then removed. Finally, the through etch is executed. Importantly, the mask for the through etch is haloed [4] using 40-μm-wide trenches to reduce etch loading, resulting in essentially vertical etch side walls. Additionally, the through-etch mask is biased to accommodate a 5-μm blow-out per side wall. With this fabrication process, a minimum feature size of about 25 μm is possible. A spring photograph is shown in Figure 6 along with images of the spring-mass system during operation. 3D-printed plastic parts form the assembly that holds the MEMS spring and the two coils that are wound from 39 AWG wire on a mandrel using a lathe. Each coil has 100 turns resulting in a total $R_{\text{coil}} = 16\ \Omega$. The full assembly with the plan- and side-views is shown in Figure 7.

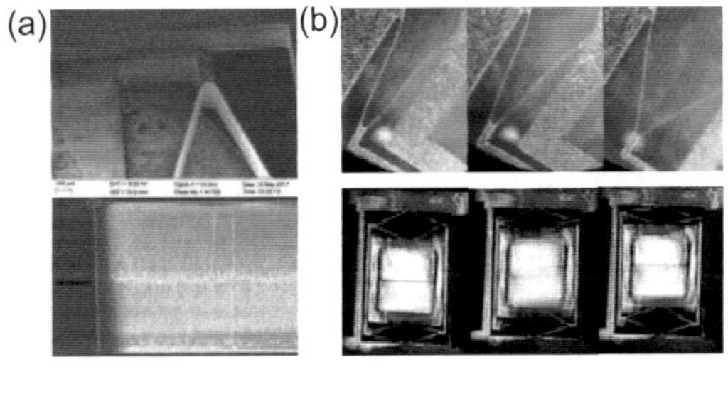

Figure 6. (a) SEM images of a spring and (b) images of beam-bending at various deflection-points during its motion.

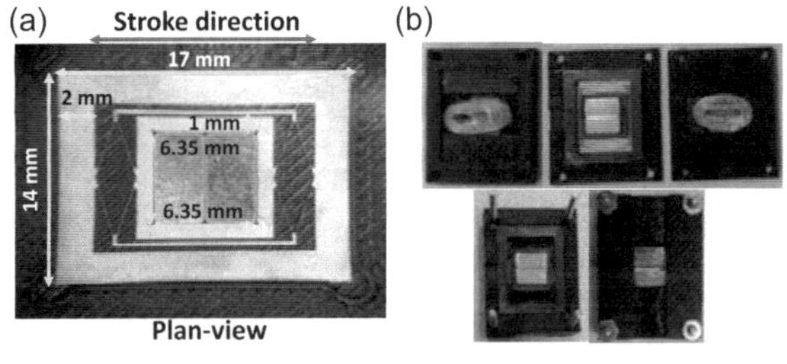

Figure 7. (a) Top-view of the fabricated suspension and magnets along with (b) individual 3D-parts and full harvester assembly.

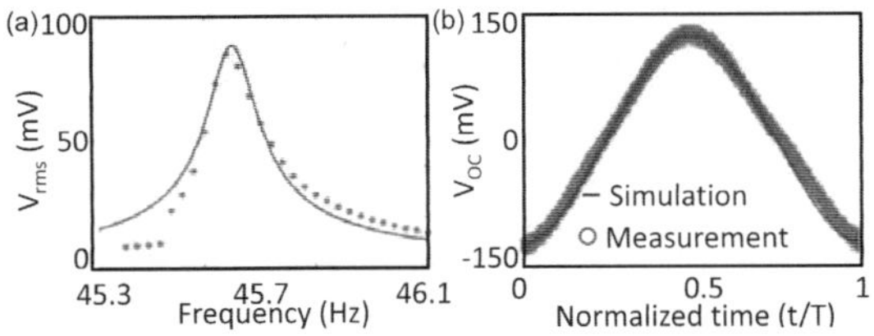

Figure 8. Measured and simulated (a) V_{rms} vs. frequency and (b) time-domain-V_{OC}.

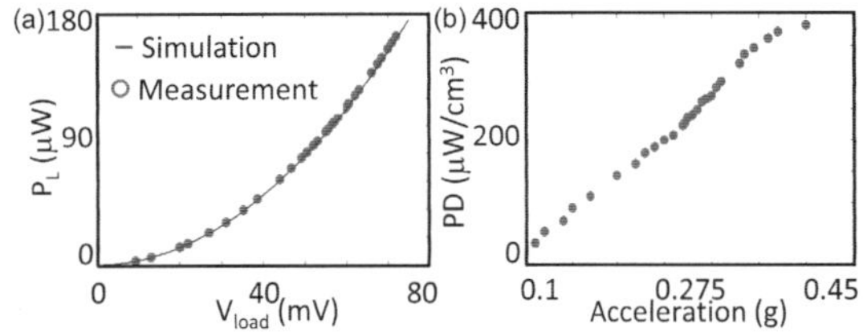

Figure 9. (a) P_L vs. V_{load} and (b) PD vs. a showing saturation due to stroke-limitation.

5. Experimental Performance

The measured frequency-sweep of open-circuit rms-voltage (V_{rms}) is shown in Figure 8 along with time-domain waveform of V_{OC}, indicating a peak $V_{\mathrm{OC}} = 145$ mV at a high quality factor of 133, which is matched well by simulation. At low input acceleration (a), the optimum load resistance R_{L} at resonance is matched to $R_{\mathrm{coil}} + R_{\mathrm{B}}$, the sum of coil and mechanical-damping resistance of harvester, which amounts to 1100 Ω. However since the maximum spring-stroke is limited to 1 mm before fracture, the maximum load-power P_{L} is delivered when this stroke is achieved at high $a = 0.4$ g, at which point the load-matching condition is $R_{\mathrm{L}} = R_{\mathrm{coil}} = 16\ \Omega$. Figure 9 shows P_{L} versus load-voltage V_{load} measured by increasing a, and demonstrates P_L=165 μW at $a = 0.4$ g. Given the active volume of 0.43 cm^3 (1 cm x 1.3 cm x 0.33 cm), the harvester achieves a power density (PD) of 382 μW/cm^3 (also shown in Figure 9), the highest among MEMS-based harvesters reported to date and summarized in Table 3 of [2]. The normalized power density (NPD) is 2400 μWcm^{-3}g^{-2} at $a = 0.4$ g with the peak NPD=4062 μWcm^{-3}g^{-2} at $a = 0.15$ g. A better figure-of-merit (FoM) to compare EM-harvesters is to normalize NPD to the resonant frequency, since the reported harvesters have different natural frequencies. This FoM is 98 μWcm^{-3}g^{-2}Hz^{-1} which is highest among EM-harvesters [2][3]. The harvester design can be further optimized with the inclusion of backiron and a larger spring travel, for which simulations show the power density increasing to 1146 μW/cm^3 and NPD=7162 μWcm^{-3}g^{-2}.

6. Summary & Conclusions

This paper presents a compact MEMS-based EM-harvester designed for near 50-Hz vibrational energy harvesting. Optimizing design guidelines are provided along with the fabrication details. The harvester demonstrates an open-circuit voltage of 145 mV and a load-power of 165 μW with a record power density of 382 μW/cm^3.

Table of Equations

(1) $X^2 = \frac{M^2A^2}{(K-M\omega^2)^2+B^2\omega^2}$ (2) $P_{\mathrm{M}} = B\omega^2X^2/2$ (3) $B = \sqrt{\frac{M^2A^2}{\omega^2X^2} - \frac{(K-M\omega^2)^2}{\omega^2}}$
(4) $K = M\omega^2$ (5) $P_{\mathrm{M}} = \omega XMA/2$ (6) $M = \rho(L_1 - 2S)L_2L_3$
(7) $P_{\mathrm{M}} = \rho\omega XA(L_1 - 2S)L_2L_3/2$ (8) $P_{\mathrm{M}} = \rho\omega AL_1^2L_2L_3/16$ (9) $V \propto T/(T + 2\Delta + 2\delta)$
(10) $R_{\mathrm{coil}} \propto 1/\Delta$ (11) $P_{\mathrm{E}} \propto \Delta T^2/(T + 2\Delta + 2\delta)^2$ (12) $\nabla^2 A = -\mu_0 \sum_{k=0}^{k=\infty}(\nabla X M_K)$
(13) $M_K = \frac{4M_0}{\pi k} sin^2(kl/2)$ (14) $A_K(x,y) = \frac{\mu_0 M_K}{k} \frac{sinh((\Delta+T)k)-sinh(\Delta k)}{sinh((2\Delta+T)k)} cosh(ky)cos(kx)$
(15) $A_K(x,y) = \frac{\mu_0 M_K}{k}[1 - sinh(\Delta k)\frac{cosh(ky)+cosh(k(y-T-2\Delta))}{sinh((2\Delta+T)k)}]cos(kx)$
(16) $V(t) = \omega_0 L_3(A(x_1(t)) - A(x_2(t)) = v(t)L_3(B_y(x_1(t)) - B_y(x_2(t)) = v(t)G(t)$

Acknowledgments

The authors would like to thank Mr. Dennis Ward for fabricating the MEMS suspensions and Ruize Xu in Prof. Kim group at MIT for providing test equipment.

References

[1] D. P. Arnold, in IEEE Transactions on Magnetics, vol. 43, no. 11, pp. 3940-3951, Nov. 2007.
[2] Y. Tan et. al., in Journal of Microelectromechanical Systems, vol. 26, no. 1, pp. 1-16, Feb. 2017.
[3] S. P. Beeby et al., in J. Micromech. Microeng., vol. 17, no. 7, p. 1257, 2007.
[4] R. Ghodssi et. al., "MEMS materials and processes handbook", Springer, 2011.

A broadband energy harvester using leaf springs and stoppers with response stabilization control

S. Kato[1], S. Ushiki[1] and A. Masuda[1]

[1]Department of Mechanical and System Engineering, Kyoto Institute of Technology, Matsugasaki, Sakyo-ku, Kyoto, Japan

E-mail: m66231105@edu.kit.ac.jp, masuda@kit.ac.jp

Abstract. This paper presents the design, fabrication and experimental verification of a broadband vibration energy harvester having a nonlinear oscillator using leaf springs and stoppers. In the conventional vibration energy harvester having a linear oscillator, there is a trade-off relationship between the magnitude of the resonance peak and the resonance band. Since the resonance band is expanded by using the nonlinear oscillator, the harvester can more efficiently generate electric energy in a wider frequency band. In this study, we designed and fabricated a nonlinear vibration energy harvester with leaf springs and stoppers, and verified the frequency response characteristics and the power generation performance of the proposed harvester by sinusoidal sweep tests with the input acceleration amplitude of 0.2 G_{rms}. The overall resonance bandwidth was approximately 33 Hz, which was over 75 % of the linear natural frequency of 42 Hz. The response stabilization control successfully destabilized the coexisting low-energy solution, and activated only the high-energy response in the resonance band.

1. Introduction

Vibration energy harvesting is a technology for generating electricity from ambient vibration. Since the power generation is effectively performed by resonating a mechanical oscillator with the source vibration, there is a tradeoff between the maximum power performance and the bandwidth. When the resonance band is expanded by introducing a nonlinear oscillator, the harvester can more efficiently generate electricity in a wider frequency band [1, 2]. However, since such nonlinear oscillators may have high and low-energy stable solutions coexisting in the resonance band, a mechanism to keep the oscillator responding in the high-energy solution is required to maintain the power generation performance. In the previous study done by Masuda et al. [3], a response stabilization control in order to globally stabilize the high-energy solution was proposed, in which a circuit switching between a positive and a negative load resistances depending on the response amplitude of the oscillator was introduced to give the harvester self-excitation capability. In the study done by Sato et al. [4], this response stabilization control was implemented to an energy harvester with a magnetically sprung oscillator. But the resonance band was not wide enough.

In this paper, an electromagnetic broadband vibration energy harvester with a nonlinear oscillator consisting of leaf springs and stoppers is designed and fabricated. The resonance band and the power generation performance in the frequency domain are verified by sinusoidal sweep tests. The response stabilization control is then applied to show the wideband operation of the proposed energy harvester.

2. Energy harvester design

2.1. Nonlinear oscillator

The mechanical design of the proposed energy harvester is shown in figure 1. An oscillator mass block sprung by a pair of leaf springs was arranged inside a casing so that the mass block could translationally move in the vertical direction. The leaf spring made of stainless steel thin plate with a thickness of 0.1 mm was designed in the Ω-like shape as shown in figure 2. Both ends of the spring were fixed to the casing with epoxy adhesive. The Ω-like shape of the spring was chosen to allow a large deformation of the spring without a hardening effect. Four stoppers made of PLA resin were placed close to the leaf springs. Since the effective length of the leaf springs was shortened after contacting the stopper, a hardening effect was imparted to the oscillator.

For the electromagnetic induction, a magnetic circuit consisting of four Neodymium magnets and two yoke plate was implemented in the mass block as shown in figure 3. A stator coil fixed in a coil holder was inserted in the airgap of the magnetic circuit, so that the magnetic flux passing through the coil moved along with the mass block motion. In order to allow as much flux as possible to pass through the coil, and to keep the leakage flux as little as possible, the yoke plates were used as a part of the structural frame of the mass block as an improved design, which made the mass block frame stiff enough to make the airgap as small as possible. As a result, the electromechanical coupling factor (the ratio of the induced voltage to the mass velocity) was improved from the previous design [5] by almost 50%.

A prototype with the improved magnetic circuit was fabricated as shown in figure 4. The specifications of the prototype harvester are listed in table 1.

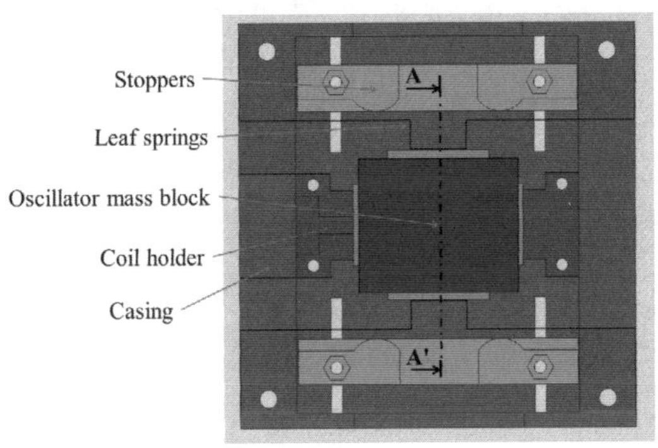

Figure 1. Design of the proposed harvester.

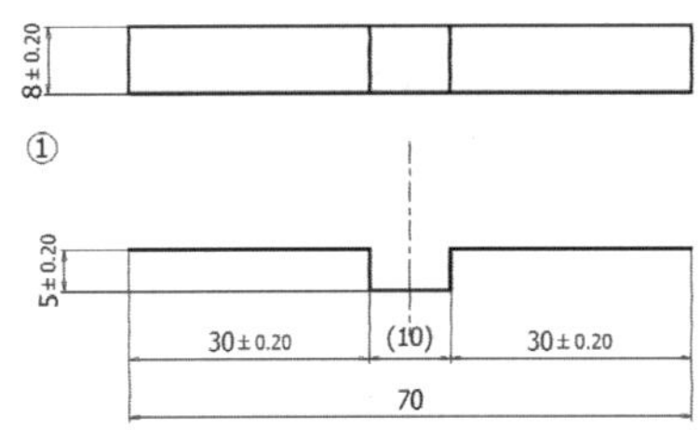

Figure 2. Leaf spring (unit: mm).

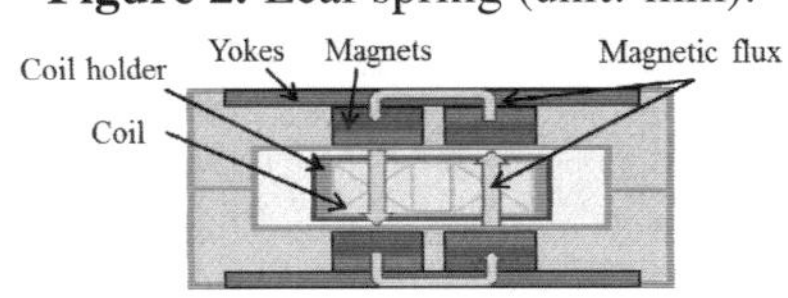

Figure 3. A-A' cross section.

2.2. Response stabilization control

In general, a nonlinear oscillator can have multiple stable solutions in the resonance band. The response stabilization control proposed by Masuda et al. [3] is a control of switching the load resistance between positive and negative values according to the response magnitude in order to globally stabilize the highest-energy solution by destabilizing the lower-energy solutions. The equation of motion of the harvester with the response stabilization control is given by

$$m\ddot{x}(t) + \left(c + \frac{\Phi^2(x(t))}{R(a) + R_G}\right)\dot{x}(t) + F(x(t)) = -mu_a \cos \omega t \tag{1}$$

where m is the mass and c is the mechanical damping coefficient, x is the relative displacement of the mass and u_a is the amplitude of the input acceleration. The functions $\Phi(x)$ and $F(x)$ are the

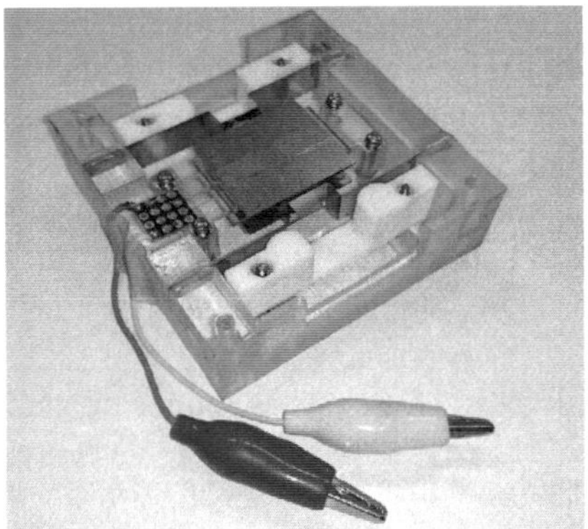

Figure 4. Photograph of the prototype.

Table 1. Parameter values.

Description	Value [unit]
Oscillator mass	8.6 [g]
Stiffness of leaf spring	365.4 [N/m]
Linear natural frequency	42 [Hz]
Electromechanical coupling factor	9.92 [Vs/m]
Internal resistance of coil	320 [Ω]

electromechanical coupling factor, and the restoring force of the nonlinear spring, respectively. $R(a)$ and R_G are the load resistance and the internal resistance of the induction coil, respectively, where a denotes the response amplitude.

When the response amplitude of the oscillator is larger than a predetermined threshold, the induction coil is connected to the positive load resistance (generation mode). If the amplitude drops below the threshold, then the circuit switches to the negative load resistance which is realized by a negative impedance converter (NIC) circuit. When connected to the negative resistance, the total damping coefficient $c+\Phi^2(x)/(R(a)+R_G)$ becomes negative, so that the oscillator is excited and entrained into the highest-energy solution (excitation mode). For the better performance of the stabilization, the threshold value should be determined depending on the excitation level and frequency [6].

For the implementation of the response stabilization control, the displacements of the mass block and the harvester casing, and the acceleration of the casing were measured by laser displacement sensors and an accelerometer, respectively, and all the signals were fed to the microprocessor (mbed LPC1768) to perform the load resistance switching. In this study, the power to drive the NIC circuit, the sensors, and the microprocessor was supplied by an external power source, which is to be self-supplied in the future study.

3. Experimental verification

An experimental verification of the resonance band and the power generation performance of the proposed energy harvester was carried out. The casing of the harvester was fixed to a vertical shaker and sinusoidal sweep tests in a frequency range of [35 Hz, 85 Hz] were conducted with an input acceleration amplitude of 0.2 G_{rms}. The stoppers were placed with a distance of 0.5 mm from the leaf spring. The load resistance and the negative resistance were set to 6.5 kΩ and -470 Ω, respectively. The threshold value was predetermined depending on the input frequency so that the response could be entrained to the high-energy solution favorably.

Figures 5 and 6 show frequency responses of the displacement amplitude and the generated power obtained by the sweep test with and without the response stabilization control. From the upper figures, it is found that the linear natural frequency was 42 Hz, and the tip of the resonance peak was extended to 75 Hz. Consequently, the proposed energy harvester showed significantly large bandwidth of 33 Hz, which was 78.5 % of the linear natural frequency. The maximum generation power at the peak frequency was 1.2 mW. In the lower figures, the responses with the proposed control were plotted. The red circles in figure 5 (b) shows the threshold values. It is obvious that the response stabilization control successfully destabilized the low-energy solutions, and the response of the harvester was always maintained in the high-energy solution.

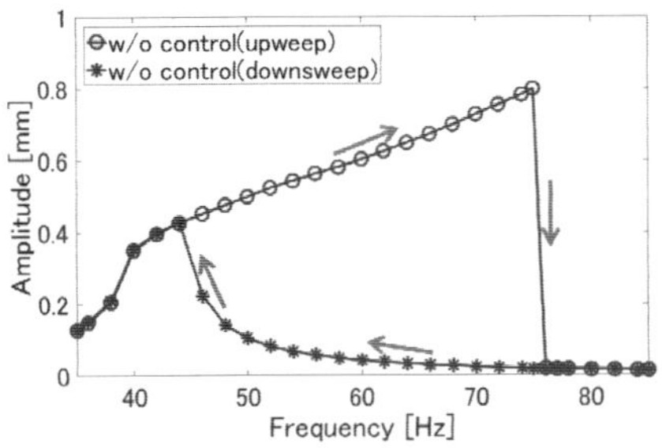

(a) Without control.

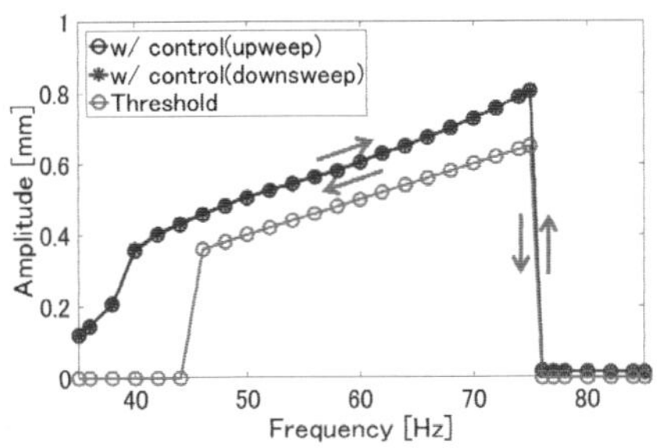

(b) With control.

Figure 5. Displacement response.

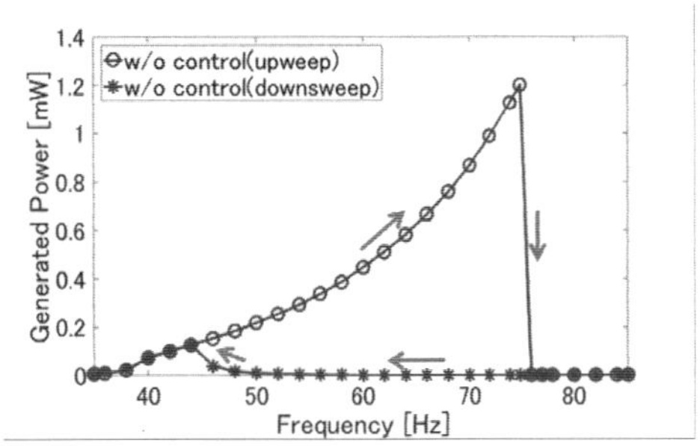

(a) Without control.

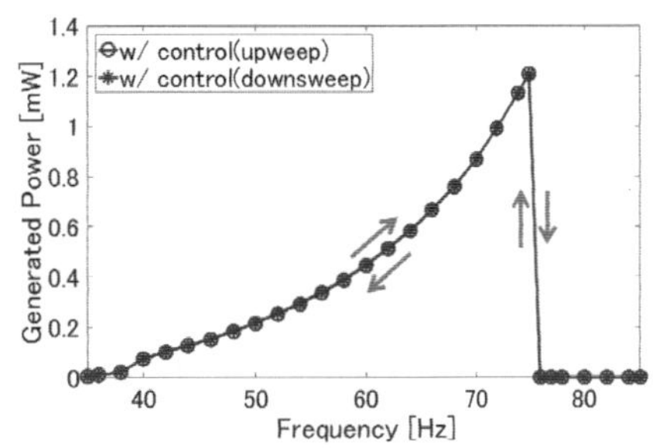

(b) With control.

Figure 6. Averaged power response.

4. Conclusion

In this paper, a broadband vibration energy harvester with a nonlinear oscillator using leaf springs and stoppers was presented, and sinusoidal sweep tests were carried out to examine its frequency response characteristics and power generation performance. It was concluded that, in the sweep tests at 0.2 G_{rms} excitation, the overall resonance bandwidth was approximately 33 Hz, which was over 75 % of the linear natural frequency. In addition, the response stabilization control successfully destabilized the coexisting low-energy solution, and activated only the high-energy response in the resonance band.

5. Acknowledgement

This paper is partially based on the results obtained from a project commissioned by the New Energy and Industrial Technology Development Organization (NEDO).

References

[1] A. Hajati, S. P. Bathurst, H. J. Lee and S. G. Kim 2011 Design and Fabrication of A Nonlinaer Resonator for Ultra Wide-bandwidth Energy Harvesting Applications *IEEE MEMS 2011* pp.1301-1304

[2] M. A. Karami and D. J. Inman 2010 Nonlinear Hybrid Energy Harvesting Utilizing a Piezo-Magneto-Elastic Spring *Proceedings of SPIE* **7643** 76430U pp.1-11

[3] A. Masuda, A. Senda, T. Sanada and A. Sone 2013 Global stabilization of high-energy response for a Duffing-type wideband nonlinear energy harvester via self-excitation and entrainment *Journal of Intelligent Material Systems and Structures* **24** 13 pp.1589-1612

[4] T. Sato, A. Masuda and T. Sanada 2015 Experimental verificatio of load resistance switching for global stabilization of high-energy response of a nonlinear wideband electromagnetic vibration energy harvester *Journal of Physics Conference Series* **660** 012109 pp.1-5

[5] S. Kato, A. Masuda and S. Ushiki 2017 A broadband nonlinear vibration energy harvester using leaf springs and stoppers *Proceeding of JSME Dynamics and Design Conference 2017* No. 508 pp. 1-11

[6] T. Sato, S. Kato, and A. Masuda 2016 Global stabilization of high-energy response of a nonlinear wideband electromagnetic energy harvester *Journal of Physics Conference Series* **744** 012085 pp.1-8

IOP Conf. Series: Journal of Physics: Conf. Series **1052** (2018) 012084 doi:10.1088/1742-6596/1052/1/012084

Sound power generation using magnetostrictive power generator

M. Aoki, and T. Ueno
Kanazawa University, Kakuma-machi, Kanazawa-city, Ishikawa, 920-1192, Japan
ueno@ec.t.kanazawa-u.ac.jp

Abstract. We propose a device that generates power from sound using a magnetostrictive material. The device includes a plate for receiving sound. The tendency of the change in the conversion efficiency of the device was confirmed by changing three parameters: its material, size, and fixation position. We also succeeded in generating 1.21 mW of power under a 117 dB environment. We aim to improve the output by enlarging the device.

1. Introduction

Generating electric power from sound is difficult because the physical energy is quite small. For example, the sound pressure level under a girder bridge with running trains is approximately 100 dB. In short, the sound intensity is only 10-2 W/m^2. Additionally, the conversion efficiency of a power generator that changes sound energy into electric energy is generally low. Eventually, we can only obtain small energy; thus, this technology is not practical. On the other hand, if we use a high conversion-efficiency device, we can obtain energy from microwatt to milliwatt level for applications in light-emitting diodes, sensors, and wireless modules. Additionally, we expect that it can simultaneously decrease noise. In the present study, we compared several efficiency properties of magnetostrictive power-generator conversion to improve performance.

2. Device principle

The magnetostrictive vibration power generator consists of a magnetostrictive material plate (iron–gallium alloy $Fe_{81.4}Ga_{18.6}$, 4 × 1 3 × 0.5 mm^3), a frame (0.5 mm thick), a coil (485 Ω; 3456 turns), and a magnet (Fig. 1). It can convert vibration energy into electrical energy using inverse magnetostrictive effect. Inverse magnetostrictive effect is a phenomenon in which the magnetic flux that passes through a magnetostrictive material changes because of mechanical stress. Fig. 2, shows that when a device receives vibration power from outside, the magnetostrictive material plate expands or is subjected to pressure because of the frame. Then, it generates electrical power via the coil

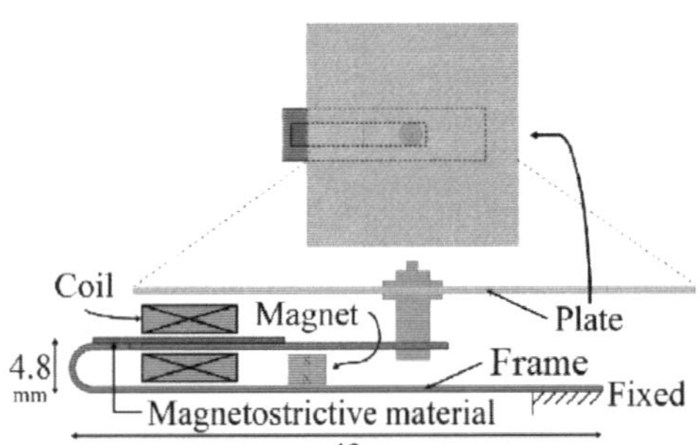

Figure 1. Device configuration

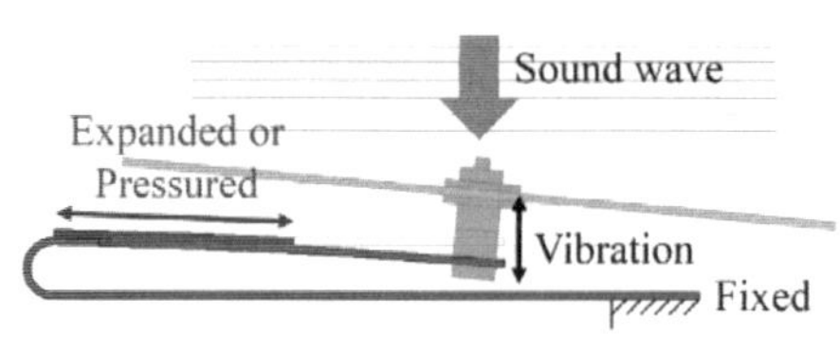

Figure 2. Expanding or pressing force from outside.

through electromagnetic induction [1]. Here, we place a plate at the tip of the device to receive the sound waves.

3. Experimental method

We conducted three experiments to compare the conversion efficiency characteristics of the plates by varying three parameters: materials, sizes, and fixation positions. In these experiments, input power was provided by a speaker and a function generator. Then, we measured the frequency characteristic of the sound pressure level using a sound-level meter. We assumed that the sound waves from the speaker were plane waves to simplify the input-power calculation [2]. The output was the power consumed by a matching resistance (670 Ω) directly connected to the device. Then, we measured the frequency characteristic of its electric potential difference using a fast Fourier transform analyzer when the device was excited by the speaker (Fig. 3). The frequency range was between 50 and 150 Hz.

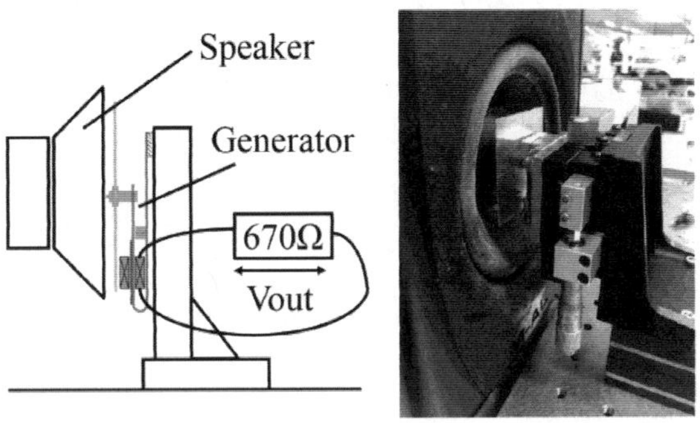

Figure 3. Measurement of the output

4. Experimental results

4.1. Different materials

In the first experiment, we compared four plates made of different materials: thick paper (1.07 g), cardboard (1.06 g), acryl (10.5 g), and aluminium (3.37 g). Their sizes were the same (50 × 50 mm^2). The result is shown in Fig. 4. The maximum conversion efficiency values were 0.92% for the thick paper, 1.8% for the cardboard, 0.050% for the acryl, and 1.4% for the aluminium. Comparison of the thick paper and cardboard, which had the same weight but different rigidity properties, showed that the maximum conversion efficiency of the latter with high rigidity was 2.0 times that of the former. Furthermore, the comparison results between the acrylic and other plates confirmed that only the acryl with a large mass had extremely low conversion efficiency. Therefore, we verified that the plate

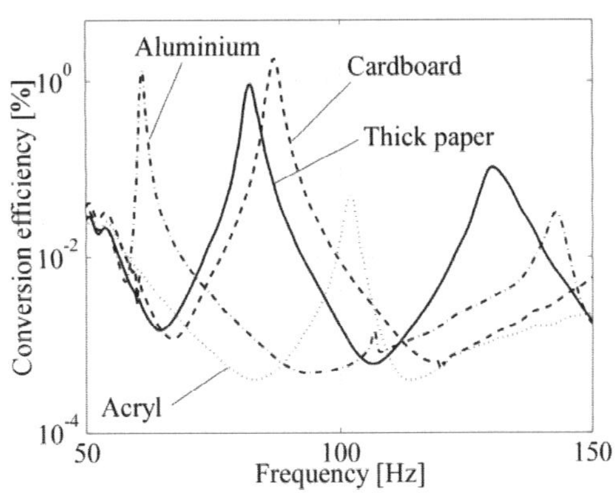

Figure 4. Conversion efficiency versus frequency in the comparison of plate materials

should be solid and not too heavy. In particular, the device with a cardboard plate generated 12.6 μW at 105 dB environment.

4.2. Different sizes

In the second experiment, we compared thick papers with three different sizes: small (35 × 35 mm^2), medium (50 × 50 mm^2), and large (71 × 71 mm^2). To maintain the same weight (2.15 g), we added weights to the small and medium papers. The result is shown in Fig. 5. The maximum conversion efficiency values were 0.94% for the small paper, 1.2% for the medium paper, and 0.014% for the large paper. We confirmed that a larger plate does not necessarily generate more energy than smaller ones. Fig. 6 shows each device open-circuit voltages at resonant frequency. Although they were excited by the same sound pressure level (105 dB), the device with a large plate received a lower voltage than the other two devices. We considered that the large plate has lower natural oscillation frequency than the other two types. The plate itself was deformed, and the stress was not efficiently transmitted to the magnetostrictive material plate. By using a material with higher rigidity, we believe that the deformation of the plate can be reduced and the output can be improved.

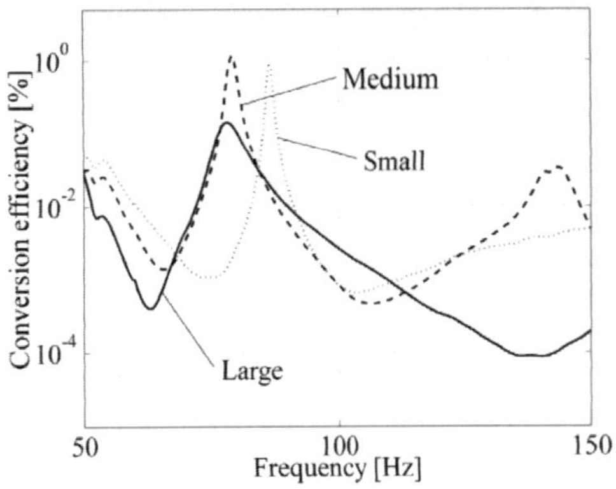

Figure 5. Conversion efficiency versus frequency

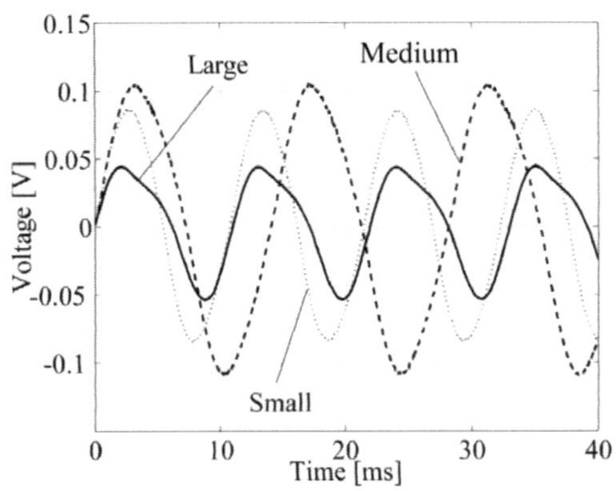

Figure 6. Time response of the open voltage in the comparison of the plate sizes (at 105 dB).

4.3. Different fixation positions

In the third experiment, we compared the plates at three different fixation positions: Positions 1, 2, and 3 (Fig. 7). We used an aluminium plate (3.37 g), and the size was 50 × 50 mm^2. The result is shown in Fig. 8. The maximum conversion efficiency was values were 0.20% at Position 1, 1.35% at

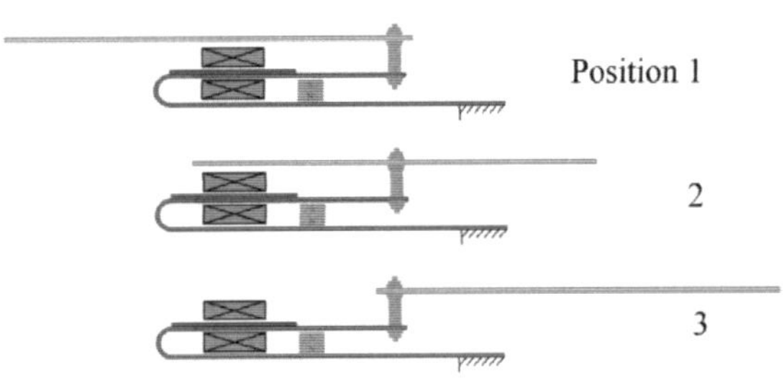

Figure 7. Definition of fixation positions

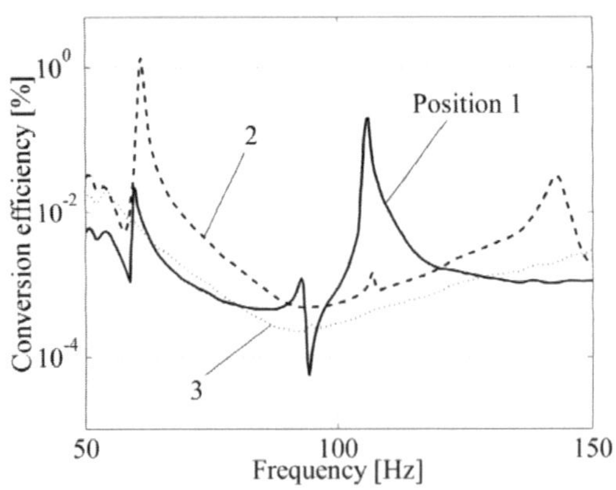

Figure 8. Conversion efficiency versus frequency, comparison of the plate fixation positions

2, and 0.012% at 3. We confirmed that the conversion efficiency strongly depends on the fixation position.

4.4. Use of a larger device

To drive sensors and wireless modules, power of several milliwatts or more is required, and the output was insufficient for the devices used in these experiments. We increased the size of the device to improve the output. The new device consisted of a magnetostrictive material plate (iron-gallium alloy, 8 × 25 × 1 mm^3), a frame (1 mm thick), a coil (333 Ω, 3910 turns), and a magnet. To prevent deformation of the plate itself, aluminium, which has the highest rigidity, was adopted as the plate. Three plates, namely, medium (50 x 50 mm^2), large (71 x 71 mm^2), and extra-large (90 x 90 mm^2) were prepared, and the characteristics of the conversion efficiency were measured between 50 and 200 Hz; the result is shown in Fig. 9. The maximum conversion efficiency values were 3.5% for the medium-size plate, 2.72% for the large plate, and 5.3% for the extra-large plate. Fig. 10 shows the open-circuit voltages of each device at resonant frequency. The use of aluminium with high rigidity reduced the deformation of the plates, and we can see that the outputs improved in proportion to the size of the plates. In particular, the device with an extra-large plate generated 1.21 mW at 117 dB environment.

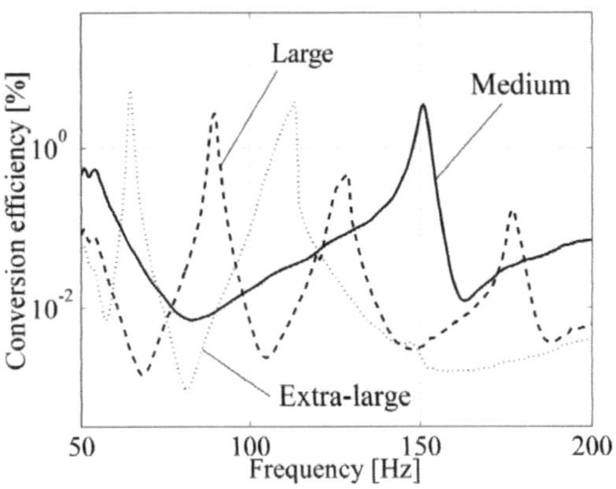

Figure 9. Conversion efficiency versus frequency with a large-size device

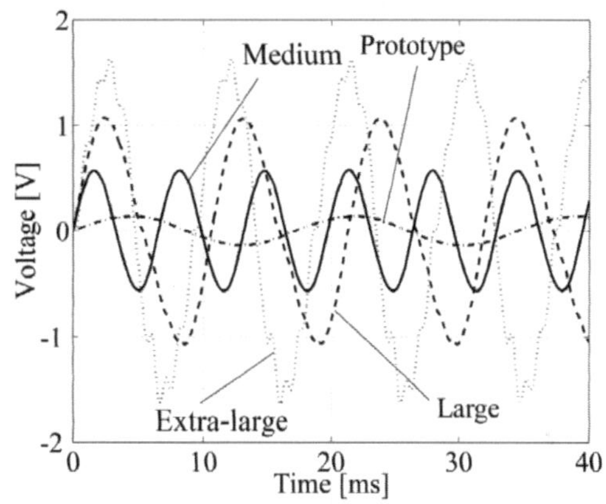

Figure 10. Time response of the open voltages in the comparison of the plate sizes with a large-size device (at 105 dB).

5. Summary

In this paper, we have proposed a magnetostrictive sound generator and experimentally verified the change in the conversion efficiency of the device due to the difference in the types of plate included in the device. As a result, the device that used a rigid and lightweight plate tended to have high conversion efficiency. In addition, in the case where a plate with insufficient rigidity was used, we believed that the conversion efficiency could be reduced by increasing the plate size. Furthermore, we confirmed that the fixation position of the plate greatly affected the conversion efficiency. As a result of enlarging the device to increase the output, we succeeded in obtaining 1.21 mW of electric power under a 117 dB environment.

Acknowledgments
This work was supported by JST CREST Grant Number JPMJCR15Q1, Japan.

References

[1] T. Ueno, PCT-JP2017-8437, 2017.
[2] Koyasu, Acoustic intensity level measurement method by sound intensity method, Journal of Japan Acoustical Society, Vol. 43, No. 12, p. 962, 1987.

IOP Conf. Series: Journal of Physics: Conf. Series **1052** (2018) 012085 doi:10.1088/1742-6596/1052/1/012085

Development of a Miniature Water Turbine Powered by Human Weight During Walking

K Ylli[1,3], D Hoffmann[1], A Willmann[1], Y Manoli[1,2]
[1] Hahn-Schickard, Wilhelm-Schickard-Str. 10, 78052 Villingen-Schwenningen, Germany
[2] Fritz Huettinger Chair of Microelectronics, Department of Microsystems Engineering – IMTEK, University of Freiburg, Germany

E-Mail: Klevis.Ylli@hahn-schickard.de

Abstract. Energy harvesting systems as an autonomous power supply for interconnected wearable devices have become increasingly important. The unobtrusive integration into clothing poses a great challenge to the feasible application of these harvesting devices. For systems worn on the lower body, a shoe-based energy harvester may offer the solution since the shoe offers a protected building space. This work presents the development of a water turbine system, which uses the human weight to cause water flows within the turbine. One reservoir (68 x 60 x 18 mm^3) is attached on either side of the turbine (58 x 54 x 16 mm^3). In a laboratory, the energy generated during one actuation of a reservoir was 3 mJ. During normal walking, with two separate systems, energies of 2.27 mJ and 2.26 mJ per step were achieved at a walking speed of roughly 4 km/h.

1. Introduction

The idea of powering body-worn sensor systems and devices with the energy expended by the human body is not new and is being investigated by several research groups around the world. Devices that harvest the kinetic energy of human motion have been presented in many varieties and different transduction mechanisms. Exemplary devices that use the swing motion of the leg as an energy source [1, 2] or the shock upon impact of the foot with the ground [3] can generate an average power of a few milli-Watts. Harvesting devices that use the force that acts on the shoe due to the weight of the user, either transduce the vertical motion directly, e.g. by using piezoelectric material [4], or translate the vertical motion into horizontal or rotational motion because of the limited height of the shoe sole.
One possibility of translating the downward motion of the foot into a different type of motion is to use fluidic reservoirs or "cushions". A reservoir can be placed underneath the heel and another under the forefoot. When the heel touches down and compresses the back reservoir, the fluid is pushed from the back of the shoe towards the front and vice versa as the weight shifts from the back to the front of the shoe. The fluid flow can then be used to drive the actual transduction mechanism. In [5] the fluid flow drives a differential piston which in turn drives gears and a generator to generate power outputs up to 800 mW. This work presents such a fluidic system. However, the present setup consists of a limited number of mechanical components to allow miniaturization, i.e. the fluid directly drives a turbine without gears.

[3] To whom any correspondence should be addressed.

2. Fabrication

The device consists mostly of 3D-printed components. The turbine wheel contains an axially polarized magnet with four magnet poles. A picture of the components as well as a schematic diagram are given in Figure 1. The coils were taken from a flat Faulhaber DC motor (type 2607T012SR). Three coils are placed at 120° phase shifts in a printed holder which also seals off the fluid-filled turbine compartment from the rest of the device. One of these coil holders is placed on either side of the main housing. A PCB board is added to the bottom holder to wire the three bottom coils in a star connection. The top coils are placed directly opposite the bottom coils and connected in opposite series to simultaneously create an induced voltage. The turbine body has a total size of 58 x 54 x 16 mm³.

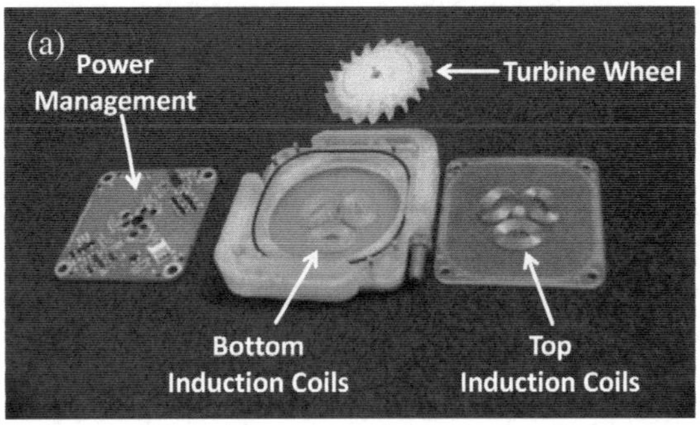

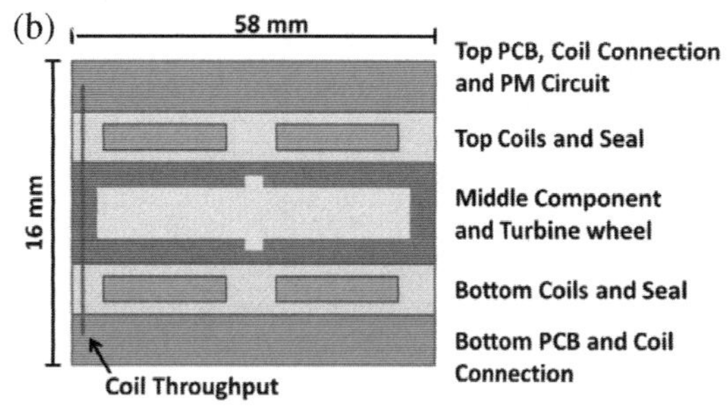

Figure 1. (a) Components of the water turbine. The main housing with the fluidic channels and the coil holders were 3D-printed. (b) Schematic showing the different layers of the turbine system.

Cylindrical pins are used to connect the two PCB boards. The fluid reservoirs are fabricated from shrinkage tube pieces of 50 mm length, which is very thin (0.4 mm walls) while being very robust as it has no seams. On either end, the tube is closed off with 3D-printed components resulting in a total size of 68 x 60 x 18 mm³. A bracket is placed over each reservoir in order to apply the weight of the test person uniformly on the shrinkage tube. Figure 2 shows the built device with fluidic channels embedded within the printed material in (a).

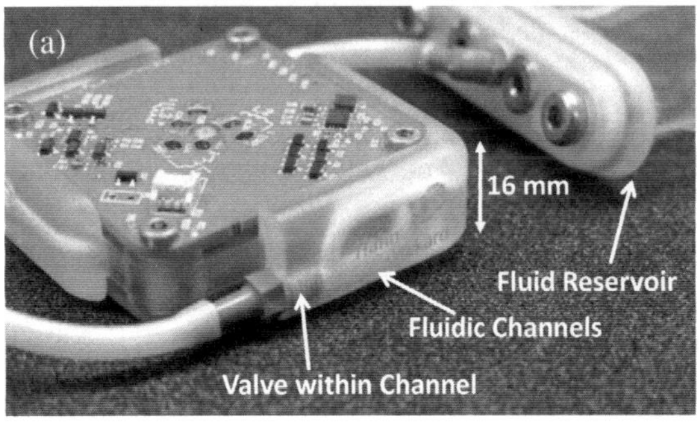

Figure 2. (a) Close-up of the built turbine and the fluid reservoir. The fluidic channels can be seen through the semi-transparent 3D-printed housing. (b) Laboratory test setup with a "Zwick-Roell" testing machine which can exert a force at a controlled speed.

The biggest problem during testing of this device was to properly seal the fluid reservoirs. The largest walked distance without leaking was about 200 m. As part of ongoing work, an improved seal is being developed. Also experiments are being conducted to find the optimal volume of water to be used in the system.

3. Experimental Data

3.1. Laboratory Experiments

In Figure 2 (b) the laboratory setup is shown, where a testing machine is used to compress the fluidic reservoir at defined speeds for a given distance. Because a single actuation can be performed, this can only mimic one part of a human step. The velocity of the testing machine was set to 10 mm/s for a total deformation of 5 mm. The recorded force generated under these actuation conditions reached up to ≈260 N. A varying ohmic load was attached to a three-phase rectifier based on the power-management in [6]. Table 1 shows the results. A maximum of 3 mJ could be generated for a single actuation. Variations were caused by the uncertainty in the exact amount of water in the reservoir at the start of each measurement.

Table 1. Generated energy for varying electrical loads and a single actuation of the fluid reservoir in a laboratory experiment.

Load [Ω]	40	45	50	55	60	65	70	75	80	85	90	95	100	105	110	115
Energy [mJ]	2.29	2.6	2.45	2.67	2.84	2.95	2.74	2.99	2.75	2.99	3.04	2.95	2.99	2.99	1.89	0.83

3.2. Real-Life Walking experiments

The complete device was placed within a sandal between the rubber sole and the cork insole for easy access as shown in Figure 3. Additional 3D-printed pieces were added to keep the turbine in place. In the future, a customized sole could contain dedicated compartments to fit the system.

(a)

(b)

Figure 3. (a) Full turbine system with two fluid reservoirs. (b) Placement of the turbine in a sandal (underneath the insole) for easier access during testing.

Testing was done on level ground at a normal walking speed of roughly 4 km/h, with a test person weighing approximately 70 kg. The rectified voltages measured across an ohmic load of 100 ohms for two separate turbine systems are shown in Figure 4. A 47 μF smoothing capacitor was used in both cases.

Each step contains two distinct voltage peaks that represent the two phases during stance where most of the body weighs on the heel or on the forefoot. In between the peaks, the voltage drops as the weight is balanced across the foot. The fluidic channels of the turbine are placed in such a way, that the turbine always turns in the same direction, hence the voltage only drops to zero very briefly during the swing phase of the foot, when no force acts on the fluid reservoirs.

The voltage peaks at 1 V (the resistance of each coil is 18 ohms). The corresponding instantaneous power output reaches up to 10 mW. For the two datasets of fifteen seconds each as shown in Figure 4, the generated energy is 24.94 mJ and 22.63 mJ for a total of 11 and 10 steps respectively. The energy per step thus amounts to approximately 2.3 mJ for either device. Considering the time span of 15 seconds, this translates into average power outputs of 1.66 mW and 1.51 mW.

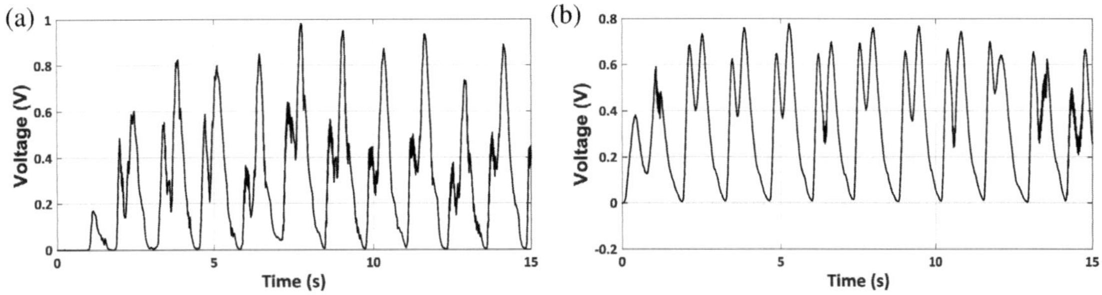

Figure 4. Rectified voltage output measured across an ohmic load of 100 ohms (using a 47 μF smoothing capacitor) for two separate turbine systems in (a) and (b).

4. Conclusion

The main issue with this device was the water sealing. After a few tens of meters, the reservoirs started leaking and when heavier test persons walked with the device, the shrinkage tube would slip out of the sealing mechanism due to the large forces exerted by the fluid. An improved seal is currently in development. Another solution may be to gradually reduce the amount of fluid in the system until the pressure build-up can be handled by the seal. The effects of this on the power output have to be investigated.

In terms of wearing comfort, there was no negative effect due to the turbine, other than the increased height of the shoe sole which had to accommodate the system and its maximum height of 18 mm. From a subjective point of view, the shoe felt very comfortable.

This device setup was able to generate average power outputs of roughly 1.6 mW or an energy of 2.3 mJ per step while walking. It contains only one moving part and thus is less noisy and prone to mechanical failure as other systems. Due to the lack of mechanical gears, the large forces cannot be translated into large speeds and thus the power output is significantly lower than for example in [5]. The presented device however is far smaller than the device in [5] and can actually be integrated in a shoe sole. Future work includes introducing a mechanical transmission while retaining the compactness of the device.

5. References

[1] K. Ylli *et al,* "Energy harvesting from human motion: exploiting swing and shock excitations," *Smart Mater. Struct,* vol. 24, no. 2, p. 025029, 2015.

[2] D. Carroll and M. Duffy, "Modelling, design, and testing of an electromagnetic power generator optimized for integration into shoes," *Proceedings of the Institution of Mechanical Engineers, Part I: Journal of Systems and Control Engineering*, vol. 226, no. 2, pp. 256–270, 2012.

[3] D. Hoffmann, B. Folkmer, and Y. Manoli, "Human Motion Energy Harvester for Biometric Data Monitoring," *J. Phys.: Conf. Ser,* vol. 476, p. 012103, 2013.

[4] J. Zhao and Z. You, "A shoe-embedded piezoelectric energy harvester for wearable sensors," (eng), *Sensors (Basel, Switzerland)*, vol. 14, no. 7, pp. 12497–12510, 25019634, 2014.

[5] C. Pylatiuk, F. Metzger, R. Wiegand, and G. Bretthauer, "Kinetic energy scavenging in a prosthetic foot using a fluidic system," (eng), *Biomedizinische Technik. Biomedical engineering*, vol. 58, no. 4, pp. 353–358, 2013.

[6] D. Hoffmann *et al,* "Energy Harvesting from Fluid Flow in Water Pipelines for Smart Metering Applications," *J. Phys.: Conf. Ser,* vol. 476, p. 012104, 2013.

6. Acknowledgements

This research (IGF-Project: 18984N) was supported by the "Programm zur Förderung der industriellen Gemeinschaftsforschung und -entwicklung (IGF)", Germany.

3D PRINTED MINIATURE WATER TURBINE WITH INTEGRATED DISCRETE ELECTRONIC ELEMENTS FOR ENERGY HARVESTING AND WATER FLOW MEASUREMENT

K T Adamski, J W Adamski, L Urbaniak, J A Dziuban, R D Walczak

Faculty of Microsystem Electronics and Photonics, Wrocław University of Science and Technology, 11/17 Janiszewski St., 50-372 Wrocław, Poland

krzysztof.adamski@pwr.edu.pl

Abstract An energy harvester fabricated by inkjet 3D printing and utilizing miniature water flow turbine with integrated mechanical and electrical components is shown. Flexibility of 3D printing design enables integration of electronic component with mechanical turbine and building of smaller and less complicated harvester than reported in literature. Turbine (ED 25.4 mm (1") and 14 mm long) is able to generate maximum 4 mW for 14 L/min of water flow and gives 0.7 V output voltage for optimal loading resistance (RL=55Ω). The harvester itself may be used as a self-suppling water-flow, zero-energetic sensor in the distributed nets of water consumption measurement.

1. Water turbine energy harvester

The proposed here harvester is based on a concept of multiphase electrical rotating generator. Almost similar concept have been presented in positions [1, 2] but our solution is based on the process of 3D printing of parts of the harvester (fig. 1). In our previous work we showed that 3D printing is enable to produce fine microstructures as a microfluidics chips with details dimensions down to 200 um [3] and with embedded actuation microstructure as in case of check microvalve [4]. Also utilized techniq can be used for preparing a biological structures as a reactors for carbon dioxide generation [5] or electrophoretic DNA separation [6].

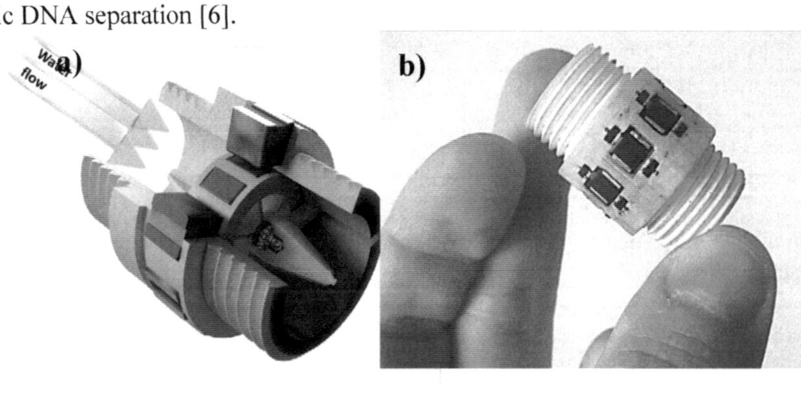

Figure 1. 3D printed harvester: a) schematic view, b) assembled device.

The device is 8-times smaller than presented in [1] and three times smaller in comparison to [2], but density of power generated by the harvester is almost at the same level. Harvester has also two times lower "starting" water flow, 1.4 L/min in comparison to 3 L/min as reported in [1]. The harvester (fig. 1) is designed to fit geometry and dimensions of typical ½" water pipe connectors (fig. 5b). It is printed by Projet 3510 printer (3D System, USA) with Visijet M3 Crystal build material and S300 support material. Electric part of the presented device contains nine small neodymium magnets (1x3x5 mm^3, magnets.eu, Poland) and 9 SMD 10 mH small (5x5x4 mm^3) inductor coils (Viking NL20JT103, Taiwan).

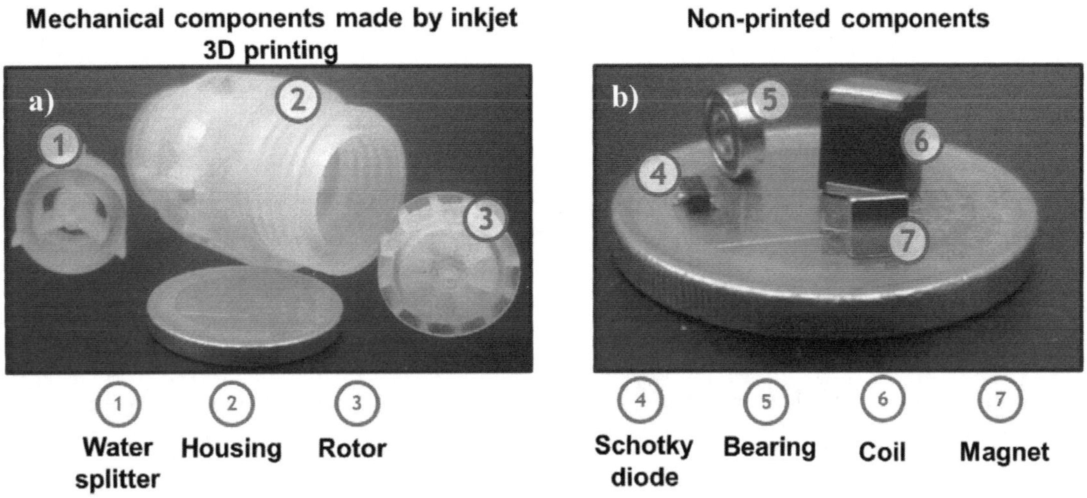

Figure 2. Components of 3D printed harvester divided in two group of parts - printed (a) and non-printed (b) parts.

2. Integration of electronic components

In the first version of the harvester outputs of each SMD inductors are connected with rectifiers (CS20D Graetz bridge, Diotec semiconductors, Germany). Development of application-specific mechanical components by 3D printing enables integration of inductors with smaller straight electronics elements (BAT54A, double Schotyky diode, NEXPERIA, Taiwan). Conductive paths are made by magnetron sputtering of cuprum (1.2 um thick) with resistance of 0.5 Ω/cm (fig. 3). Integration of electronic components and conductive paths lead to decrease of ED from 42 mm to 25.4 mm (1").

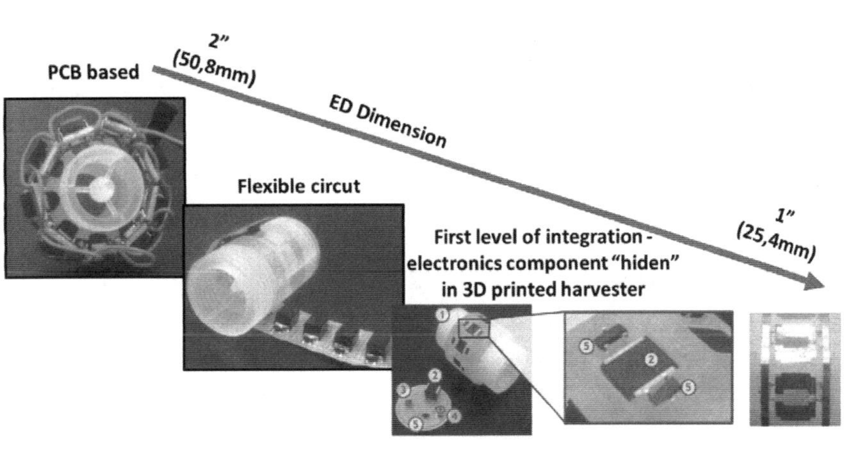

Figure 3. Electronics integrations and external dimension decreasing.

3. Electrical characterisation

The harvester has good electrical parameters (fig. 4). Maximal generated power of almost 4 mW for 14 L/min of water flow is achieved for 55 Ω loading resistor (fig. 4d). Output signal waveform can be combined in 3-phase power circuit (fig. 4 a, b, c) for more effective electrical power generation. The level of the harvested energy (~4 mW) seems to be high enough to supply low-power electronic components, including wireless communication modules.

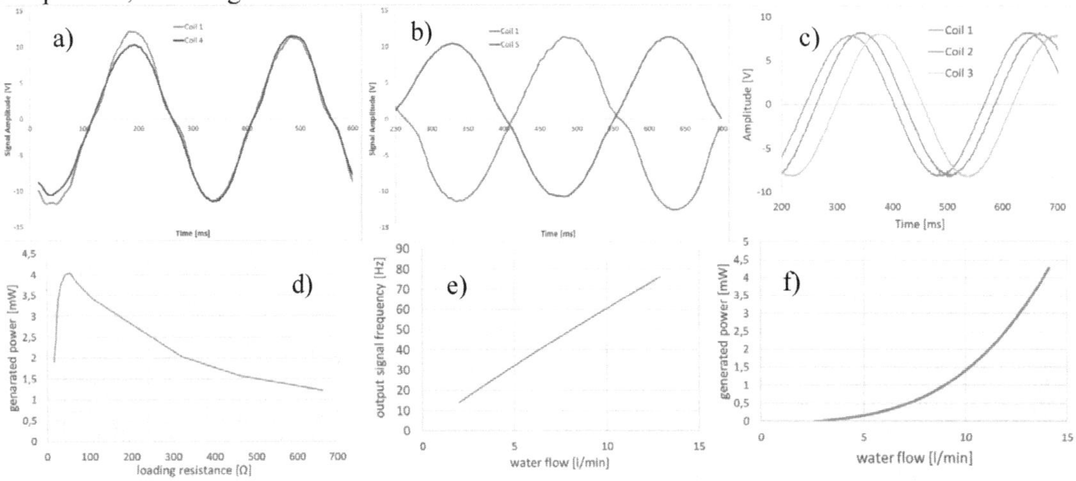

Figure 4. Electrical characteristics of the harvester: a) in phase inductors circuit, b) antiphase inductors circuit, c) signal for next 3 coils, d) generated power as function of loading resistance, e) frequency vs flow output signal, f) generated power as function of water flow.

4. Modular system

Thanks to 3D printing technique it was possible to produce modular and easy to adapt systems. It is possible to design and print modular turbines that are connected each to other to generated more power (fig. 5a). Also other components including energy management circuit, wireless communication or sensors (e. g. temperature sensors, conductivity sensors, pH sensor, MEMS microphone) can be easily mounted in 3D printing housing designed to specific sensor requirements. In developed by us system steering electronics with microprocessor (MKL05Z32VFK4), power managements circuit (BQ22504RGTT) and radio communication (NRF24L01) was placed in one of the modules (fig. 5d) and connected to turbine(s).

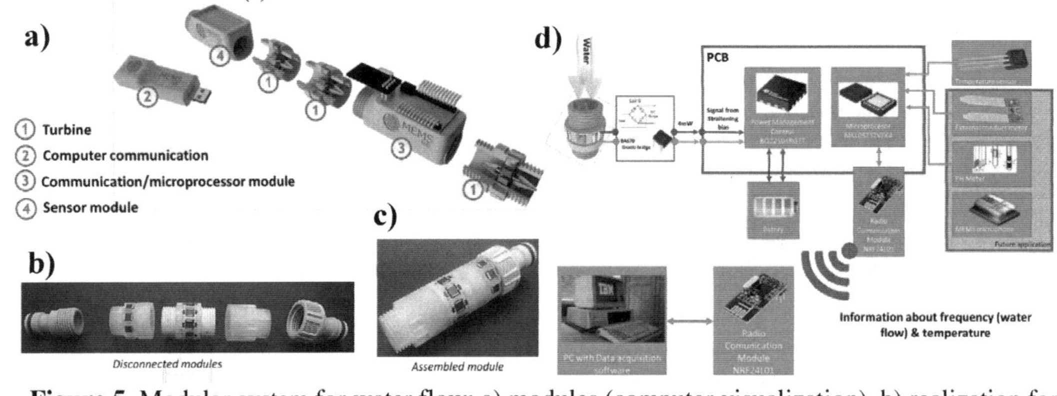

Figure 5. Modular system for water flow: a) modules (computer visualization), b) realization for ½" standard connectors – disconnected components, c) three turbines module, c) scheme of electronics system architecture.

5. Conclusion

Low-cost power suppling of distributed network system for water control and consumption measure is an important problem. Traditional powering from on-grid network is very expensive and trouble full. One of the solution is generation of energy directly from water flow with rotating water turbine. Generated in this way power can be used for system powering and also it is possible to correlate frequency of generated electricity with water flow rate. In this paper, design, system configuration and fabrication of 3D printed harvester (with embedded stainless steel bearings and electrical components) was presented.3D Printing is recognized as technology than can change the world [7] and it was confirmed that inkjet 3D printing can be use as simple method for fabrication a complex self-powering device. Results of energy harvesting were showed. It is finally concluded that after optimisation of the mechanical and electrical construction the proposed energy harvester it can be used for flow measurement and power supplying of low-power distributed net of sensors (temperature, humidity, conductometer etc.) for water consumption and pollution recognition in the near future. Thus a first generation of 3D printed milimechatronic device is here proposed.

Acknowledgments: The works are financed by Polish National Science Centre (NCN) under frame of the project no 2013/10/E/ST8/0342 and Wroclaw University of Science and Technology by young scientists statutory grant no 0402/00051/17.

References

[1] Boisseau, S.; Duret, A.-B.; Perez, M.; Jallas, E.; Jallas, E. Water flow energy harvesters for autonomous flowmeters. J. Phys.: Conf. Ser. 2016, 773, doi:10.1088/1742-6596/773/1/012019.

[2] Hoffmann, D.; Willmann, A.; Göpfert, R.; Becker, P.; Folkmer, B.; Manoli, Y. Energy harvesting from fluid flow in water pipelines for smart metering applications. J. Phys.: Conf. Ser. 2013, 476, doi:10.1088/1742-6596/476/1/012104.

[3] Walczak, R. Adamski, K. Inkjet 3D printing of microfluidic structures—On the selection of the printer towards printing your own microfluidic chips. J. Micromech. Microeng. 2015, 28, doi:10.1088/0960-1317/25/8/085013.

[4] Walczak, R.; Adamski, K.; Lizanets, D. Inkjet 3D printed check microvalve. J. Micromech. Microeng. 2017, 24, doi:10.1088/1361-6439/aa6152

[5] A Podwin et al., A step towards on-chip biochemical energy cascade of microorganisms: carbon dioxide generation induced by ethanol fermentation in 3D printed modular lab-on-a-chip, 2016. Phys.: Conf. Ser. 773 012052

[6] K. Adamski, W.Kubicki, R.Walczak, 3D Printed electrophoretic lab-on-chip for DNA separation, Procedia Engineering 168 (2016) 1454 – 1457, doi: 10.1016/j.proeng.2016.11.416

[7] Waldbaur, A.; Rapp, H.; Länge, K.; Rapp, B.E. Let there be chip—Towards rapid prototyping of microfluidic devices: One-step manufacturing processes. Anal. Methods 2011, 3, 2681e.

Power-generating shoes using a magnetostrictive vibration power generator

T.Minamitani, T.Ueno

Kanazawa University, Kakuma-machi, Kanazawa, Ishikawa 920-1192, Japan

E-mail:minamitani@staff.kanazawa-u.ac.jp

Abstract. Luminescent shoes that flash synchronously with walking motion can be used to enhance the safety of night walking. As a means of reducing reliance on batteries, which are typically used as power supplies, the potential use of the energy generated by walking has been investigated, although to date no practical power-generating shoe has been produced. Here, we examine the use of a magnetostrictive vibration generator developed in our laboratory as a power source for power-generating shoes. This generator has advantages over conventional piezoelectric or movable magnet type generators in terms of improved efficiency, higher robustness, and reduced electrical impedance. In this paper, we report on the fabrication method and experimental results obtained from testing the proposed magnetostrictive vibration generator for power-generating shoes.

1. Introduction

Typically, luminescent shoes use a sensor to detect interior vibrations caused by walking to induce LED flashes in synchronicity with footfalls. As a power supply for the sensors and LEDs, batteries are generally used, although power generation using the vibrations created in walking is being investigated as a method to reduce or eliminate the need for batteries [1]. The primary techniques for harvesting vibrational energy include piezoelectric and electromagnetic induction methods. As piezoelectric harvesters employ ceramic materials, they tend toward high internal impedance and brittleness. Electromagnetic induction harvesters require significant magnetic body motion to obtain sufficient changes in flux density change. For these reasons, conventional piezoelectric and electromagnetic induction harvesters are difficult to miniaturize and have not been put to practical use. A magnetostrictive vibration power generator developed at our laboratory is composed of magnetostrictive (Fe-Ga alloy, $Fe_{81.6}$ $Ga_{18.4}$) [2] and magnetic materials [3, 4]. Both the Fe-Ga alloy and the magnetic material are iron-based, enabling a simplified generator structure and, correspondingly, a highly durable device. Furthermore, the large reverse magnetostrictive effect of the Fe-Ga alloy results in a large flux density change of 1 T in the Fe-Ga alloy; as this is converted into voltage by the coil, the power generation device has a low output impedance, allowing for high levels of power generation amount from a small device. The magnetostrictive vibration power generator therefore has high power efficiency and is robust, making it suitable as a vibration power generator for use in shoes.

In our laboratory, we developed a U-shaped vibration power generation device; a description of the device and its suitability as a shoe-based power generator are provided on our homepage [5]. Because the generator has a high degree of springiness and operates with a body weight-generated force of ~ 100 N, it requires a strong structure and thus a large size. To improve on this design, we developed a device with a unimorph structure, making it possible to extract voltage using the impact acceleration occurring when the shoe's sole contacts the floor. In this report, we analyze acceleration as a function of shoe location, describe the method for fabricating a shoe-based power generator, and assess the output of the generator in action.

2. Structure of power generator and operation principle of power generation shoes

Figure 1 shows the structure of the unimorph-type generator. A magnetostrictive plate composed of Fe-Ga alloy is laminated onto a U-shaped magnetic frame on which a coil is wound. In addition, a permanent magnet is placed in the gap of the U-shaped frame to provide a bias magnetic field. The resonance frequency can be adjusted by changing the mass at the tip of the vibrating component. Because this device has a simple structure, it is easy to assemble and the number of turns of the coil can be increased to ramp up the power.

The principle of power generation is illustrated in figure 2. The generator is affixed to the sole of the shoe and, when the shoe's sole contacts the floor, the inertial force arising from the impact acceleration acts on a mass at the tip of the generator. This causes the magnetostrictive plate to expand and contract in a bending motion, causing the magnetic flux inside the plate to change as a result of the inverse magnetostrictive effect. As the magnetic flux varies with time, a voltage is generated on the coil by electromagnetic induction. The generator has a resonance frequency at which large amounts of power can be extracted.

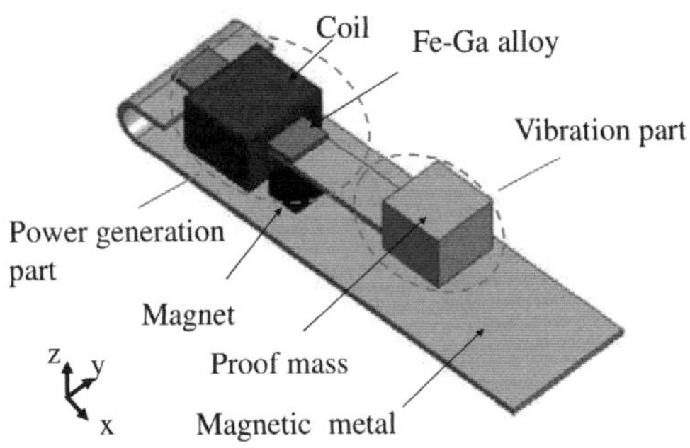

Figure 1. Configuration of magnetostrictive vibration power

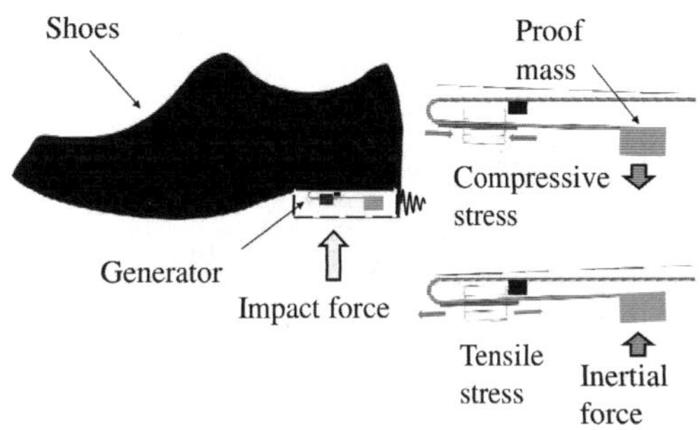

Figure 2. Principle of power generation.

3. Acceleration at shoe location and performance of generator

3.1. Acceleration at shoe location

To assess the suitability of the generator installation location on the shoe, acceleration as a result of walking motion was detected using an acceleration sensor attached to different locations om the shoe. The acceleration sensor was variously glued to the sole and in front of, next to, and behind the shoe. Figure 4 shows the average power spectrum of acceleration taken from walking 50 steps on corridor tile. Figure 5 shows examples of time response waveforms of acceleration taken over 5 steps. From these results, it is seen that the power spectrum of the acceleration of the sole is nearly flat with an impulse

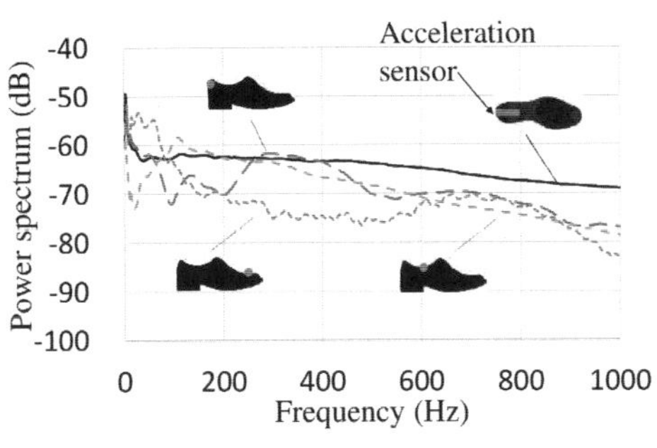

Figure 3. Power spectrum of acceleration response with respect to location of shoe.

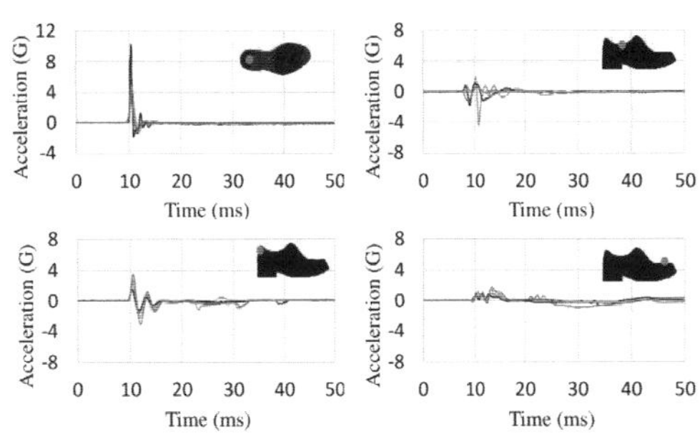

Figure 4. Time response of acceleration with respect to location of shoe.

spike of about 10 G. For the other shoe parts, the peak and frequency of the power spectrum differ by installation location, with the back of the shoe producing 7 G at 300 Hz, the side producing 6 G at 110 Hz, and the front producing 2 G at 50 Hz. These results show that the location of the installation site on the sole is suitable and that the resonance frequency of the device should be set in the target design from 100 to 400 Hz.

3.2. Performance of generator

The relation between the resonance frequency and the output voltage with respect to a test mass at the tip of the generator was then assessed. Figure 5 shows the dimensions of the assembly. The overall generator has the dimension 37 × 12 × 6.5 mm^3, the Fe-Ga component (manufactured by Fukuda Crystal Laboratory [6]) is 16 × 4 × 0.5 mm^3, the SPCC magnetic frame has a thickness of 0.5 mm, the coil has dimensions 10 × 8 × 5 mm^3, and the wire diameter is 0.05 mm. The 3 × 6 × 1.5 mm^3 magnet has 5,275 turns, and the generator has an internal resistance of 775 Ω.

In the experiment, the generator was fixed to the vibrator, a frequency sweep was performed under 0.1 G acceleration, and an oscilloscope was used to measure the resonance frequency and output voltage. The test mass was varied from 0, 0.26, 0.57, 0.97, to 1.7 g. Figure 6 shows the relationship between the resonance frequency and the output voltage of the generator with respect to the test mass at an exciter acceleration of 0.1 G. As the test mass increases, the resonance frequency decreases while the output voltage increases.

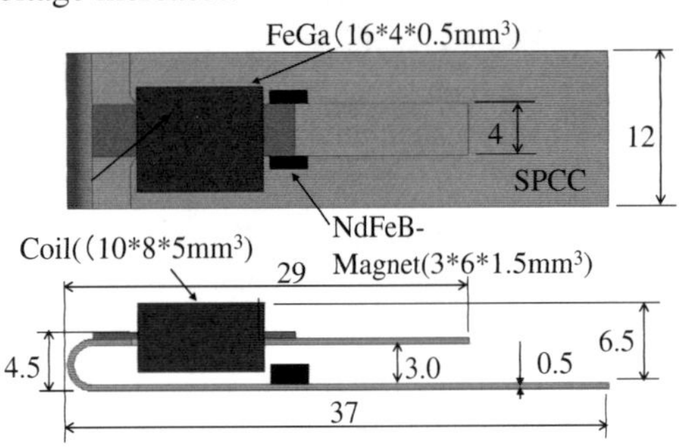

Figure 5. Dimension of generator.

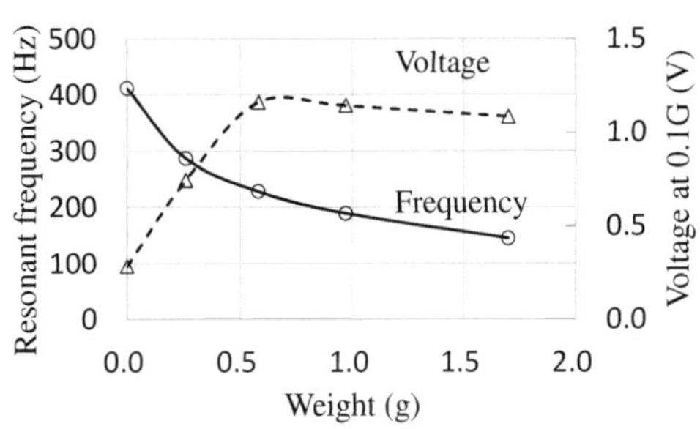

Figure 6. Relation between voltage, frequency and weight of proof mass.

4. Characteristics of generator attached to shoe

4.1. Output voltage depending on the location of the generator

Figure 7 shows the output voltage, Vp-p, with respect to generator installation location for varying test masses (10 steps). In the case in which the generator is attached to the shoe's sole, the output voltage is the highest and increases with the test mass. When the generator is attached behind the shoe, the output voltage is the maximum for the test mass for which the generator resonance frequency coincides with the resonance frequency of the shoe. When the generator is attached on the side or front of the shoe, the output voltage is small.

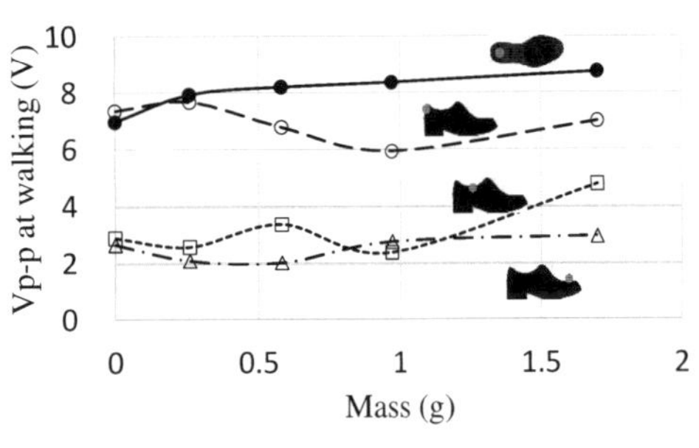

Figure 7. Output voltage with respect to location of shoe of generator.

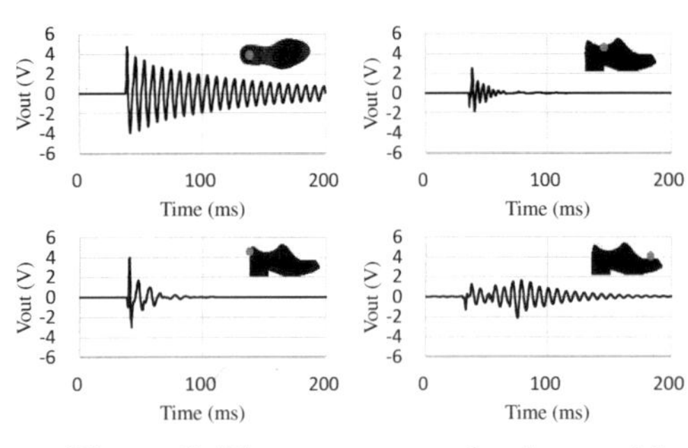

Figure 8. Time response of voltage with respect to location of shoe of generator.

Figure 8 shows examples of output voltage waveform for the 1.7 g test mass. Vibration is sustained when the generator is attached to the sole of the shoe. Several pulses are generated when the generator is attached to the back or side of the shoe. When the generator is mounted to the front of the shoe, pulse peaks occur at two points. The duration of vibration increases with the test mass.

4.2. Confirmation of energy and LEDs operation by walking motion

We then investigated the effect of the weight of the proof mass and the load resistance on energy when the generator was attached to the shoe's sole. Figure 9 shows the relationship between load resistance and energy calculated from Joule heating. As the test mass increases, the energy increases and the load resistance value producing maximum energy decreases. In this sample, the test mass was 1.7 g and the resistance was 150 Ω, producing a maximum energy of 55 μJ. For a 0 g test mass, the energy was 10 μJ.

Figure 10 shows a photograph of the generator and LEDs attached to the shoe sole with epoxy resin. The total weight of the generator is 10 g and it is connected directly to six LEDs. We confirmed that the LEDs could be caused to flash as a result of normal walking motion.

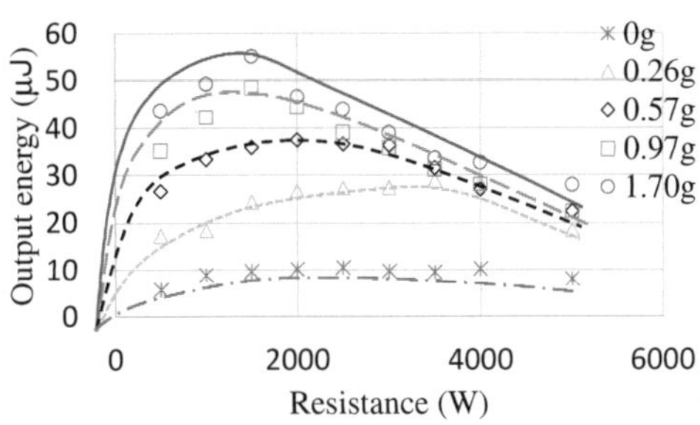

Figure 9. Relation between output energy and load resistance.

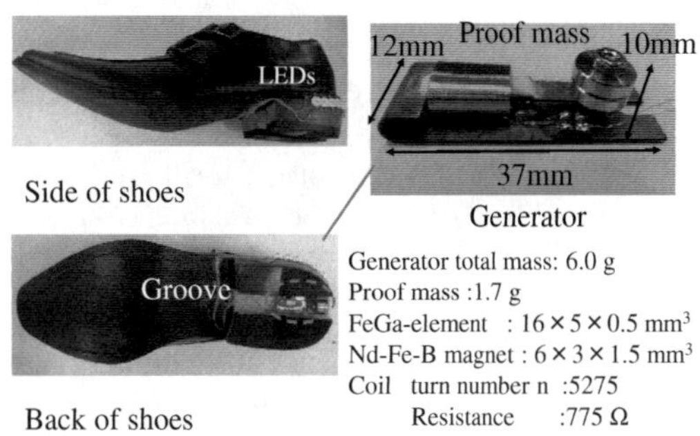

Figure 10. Photograph of shoe with generator and LEDs.

5. Conclusion

The acceleration of a generator as a function of position on a shoe was assessed and the characteristics of the generator with respect to mass loading were evaluated. It was determined that the shoe sole was a good place for attaching the generator and that a 1.7 g test mass produced optimal output characteristics. In actual walking motion, the generator produced an output voltage of 8 V and an energy of 55 μJ, and it was confirmed that this configuration could enable an array of six LEDs to successfully flash. Wireless modules require several hundred μJ of energy to operate; in future studies, we will assess the efficiency of the proposed generator with the goal of increasing generated energy through downsizing and using its output to power a wireless module.

Acknowledgments

This research was supported by JST CREST "Creating Innovative Environmental Power Generation Technology Using Micro Energy" (Issue No. JPMJCR 15 Q 1).

Reference

[1] Ylli K, Hoffmann D, Folkmer B and Manoli Y 2013 Journal of Physics: Conference Series 476
[2] Jayasimha A and Flatau A 2011 A review of magnetostrictive iron–gallium alloys Smart Materials and Structures 20(4) 1-15
[3] Ueno T 2015 J. Appl. Phys. 117 17A740
[4] Ueno T 2016 Science of Machines 16(9) 741-746
[5] Ueno laboratory (Kanazawa Univ.) HP http://vibpower.w3.kanazawa-u.ac.jp/power_shoes.html
[6] Fukuda Crystal Laboratory HP http://www.fxtal.co.jp

High performances low frequency vibration energy harvester with HSLD stiffness

Cyril Drezet[1,2], Najib Kacem[1], Noureddine Bouhaddi[1], Emmanuel Foltete[1] and Ziad Jabbour[2]

[1]University of Bourgogne Franche-Comté, FEMTO-ST Institute, CNRS/UFC/ENSMM/UTBM, Department of Applied Mechanics, 25000 Besançon, France
[2]TRAXENS, 13013 Marseille, France

E-mail: cyril.drezet@femto-st.fr, noureddine.bouhaddi@univ-fcomte.fr

Abstract. The contributions put forth by this paper first include the design and testing of a new high performance electromagnetic vibration energy harvester based on an optimized electromagnetic structure and a friction-less compliant folded-beam suspension. Then, High Static Low Dynamic (HSLD) stiffness principle is used to reduce the resonance frequency of the harvester while maintaining small static displacement.

1. Introduction

The Internet of Things (IoT) is the term used to describe a cluster of technologies enabling machine to machine (M2M) communication and machine to human interactions through the Internet [1]. Among those technologies, Wireless Sensor Networks (WSN) is a fundamental element. It is composed of wireless units able to sense a given physical phenomenon and cooperate to route the gathered data to a processing center. In some cases, these units can not be subject to maintenance on regular basis and there is a need to guarantee the energetic autonomy without battery replacement nor access to the power grid. Energy harvesting aims to benefit from the energy that can be found in the direct surrounding. Considering the various available sources (solar radiation, thermal gradient, mechanical movement, electromagnetic waves, *etc.*), vibration energy harvesting is a promising one [2].

Low frequency, vertical vibrations can profusely be found during road/rail transportation [3] or human motion [4]. In fact, those vibration sources' most energetic frequencies usually lie from 1 to 5 Hz. Scavenging energy in that frequency range implies large displacements (static and dynamic) which tends to increase the difficulty to build a vibration energy harvester which is both compact and efficient. Different solutions have been proposed to overcome this issue, among them we can cite frequency-up conversion [5], parametrically excited pendulum [6] or magnetic suspension [3]. However, each of these solutions has its own drawback which makes difficult an industrial application, respectively wear, amplitude threshold and high static displacement.

In this study, we propose a new electromagnetic harvester based on High Static Low Dynamic (HSLD) stiffness solution enabling efficient, space-limited and very low frequency vibration energy harvesting. Our approach relies on two steps; the first one is the design and testing of a high performances linear vibration energy harvester and the second one is the modification of the prototype to obtain HSLD stiffness behavior.

2. Linear vibration energy harvester

2.1. Theory

Vertically oriented, the behaviour of a single degree of freedom linear electromagnetic vibration harvester loaded with a resistive impedance R_{load} can be described by equations (6) and (2).

$$m\ddot{x} + c_m\dot{x} + kx + \alpha \cdot i + mg = -m\ddot{y} \tag{1}$$

$$L\frac{di}{dt} + (R_{int} + R_{load}) \cdot i = \alpha\dot{x} \tag{2}$$

Where m is the moving mass and x its relative position to the base, c_m the mechanical damping coefficient of the oscillator, k the suspension stiffness, g the gravity constant, y the base excitation, α the electromechanical coupling term, L and R_{int} the electromagnetic coil inductance and impedance.

A widely accepted assumption is to neglect the coil inductance L. Moreover, we decide to introduce the number of turns in the coil N: the coil resistance R_{int} is then equal to N^2R_0 and the coupling coefficient α to $N\gamma$. R_0 and γ are respectively called intrinsic resistance and coupling and only depend on the geometry and materials used in the magnetic structure (coil and magnets).

Considering that the harvester is subjected to a sinusoidal excitation $Asin(\omega_0 t)$ tuned around its resonance frequency $\omega_0 = \sqrt{k/m}$ and that impedance matching is satisfied (equation (3)), the mean P_{mean} over time of the instantaneous harvested power $P = R_{load}i^2$ is given by equation (4).

$$R_{load} = R_{int} + \frac{\alpha^2}{c_m} \tag{3}$$

$$P_{mean} = \frac{m^2A^2}{8c_m} \cdot \frac{\gamma^2/_{R_0c_m}}{1 + \gamma^2/_{R_0c_m}} \tag{4}$$

First, we observe that the harvested power is theoretically independent of N, in fact N does the balance between current and tension: if N is big, the harvester tends to ouput higher voltage and lower current.

Also, P_{mean} is limited by $m^2A^2 \cdot (8c_m)^{-1}$, to approach this limit, the harvester must be designed to maximize the term $\gamma^2 \cdot (R_0c_m)^{-1}$.

Finally, it is worth noting that when excited, the moving mass oscillates around a non-zero equilibrium position x_0 such as:

$$x_0 = -\frac{g}{\omega_0^2} \tag{5}$$

When harvesting high frequency vibration, this behavior can be neglected. However, dealing with very low frequency vibration, high x_0 values can lead to a large increase of the harvester volume and must be considered. For example, x_0 equals to 2.8 cm at 3 Hz.

2.2. Design

To assess the design rule extracted from theory (mainly maximizing $\gamma^2 \cdot (R_0c_m)^{-1}$), we decide to design a linear vibration energy harvester not accouting for the constraint of harvesting the very low frequencies.

The optimized design process is based on two steps, the first one is to choose and size the suspension and guiding of proof mass. This part not only has an influence on the mechanical damping c_m but also on the intrinsic coupling γ: it will be high if the airgap between magnet and coil is small and, this airgap can be minimized if the guiding has sufficient lateral stiffness to allow rectilinear motion. Among the numerous configurations available, we chose to use a

folded beam suspension which allowing highly rectilinear motion together with linear stiffness; furthermore, to allow high displacement and low damping, it has been machined out of a single piece of Beryllium Copper.
The second test is to design the electromagnetic coupling structure. We chose to base our optimization on the work done by Spreeman and Manoli [7] and their "magnet across coil with back iron" structure shown on Figure 1. Within a given volume of 35 mm×40 mm×50 mm, the objective is to maximize $\gamma^2 \cdot (R_0)^{-1}$ on a sufficient stroke of the moving mass while avoiding magnetic flux leakage. The main challenge being to find the right division between back iron, coil and magnet. Using the finite element sofware called FEMM, the final structure is made out of soft iron, Neodymium Iron Bore magnets and a copper coil counting 1000 turns (to obtain sufficient voltage). The equivalent scheme of the final prototype is given on Figure 2.

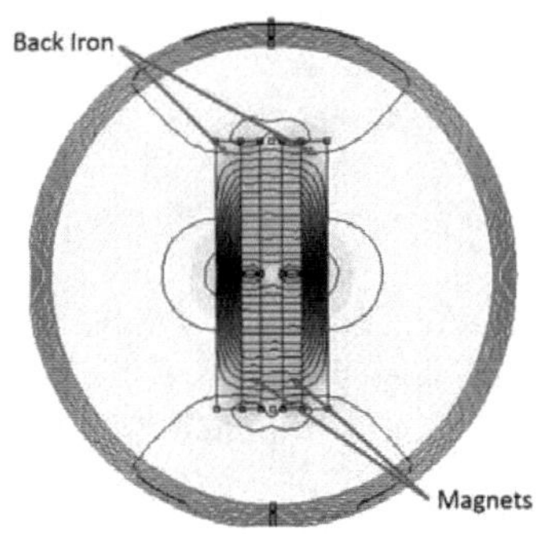

Figure 1. Screenshot of the FEMM software during electromagnetic structure optimization.

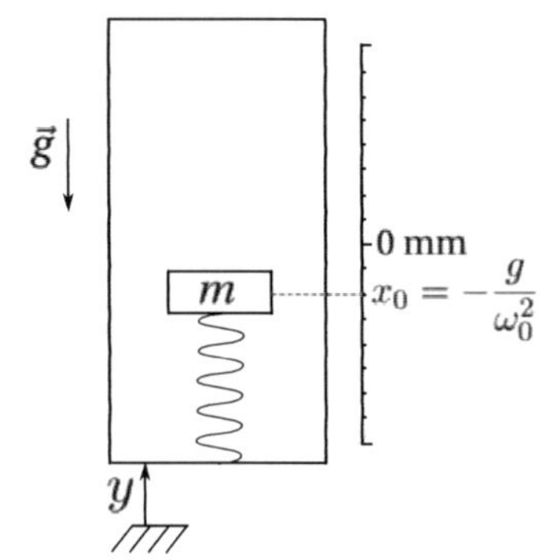

Figure 2. High performances linear vibration energy harvester prototype.

2.3. Experimental testing

The linear prototype has been tested on a high force shaker and several parameters have been identified. It has a resonance frequency of 15.3 Hz, the friction-less suspension gives a mechanical damping as low as 0.06 %. The coupling coefficient α is equal to 11.15 N.A^{-1} and the internal resistance R_{int} to 28 Ω. These characteristics give great energy harvesting performances, mostly when the excitation amplitude is low: for an excitation of 5 mg$_{RMS}$, the output power is 2.7 mW which correspond to 169 mW.g^{-2}.cm^{-3}. Comparing it to some harvesters of the literature [8] (Figure 3), the great performances of our prototype can be assessed.

3. HSLD stiffness vibration energy harvester

Our linear prototype shows good performances but does not comply to our objective consisting in harvesting the very low frequency. Lowering the resonant frequency (*e.g.* by decreasing the stiffness) would lead to a high static displacement not withstanded by the folded beam suspension. The solution that we introduce here is to use a HSLD stiffness.

3.1. Theory

Oscillators with HSLD stiffness is a technique used by vibration isolator designers [9]; it consists in adding a negative stiffness component to the linear oscillator around its equilibrium position to lower the dynamic stiffness while maintaning the static stiffness. Equation (6) becomes:

$$m\ddot{x} + c_m\dot{x} + kx - k_1(x - x_0) + k_3(x - x_0)^3 + \alpha \cdot i + mg = -m\ddot{y} \quad (6)$$

Hence, the harvester keeps its static stiffness k and its equilibrium position x_0. The dynamic stiffness is given by $k - k_1$ and the resonant frequency is $\sqrt{(k - k_1)/m}$. By correctly sizing the negative stiffness, the resonant frequency can therefore be tuned to the desired frequency.

3.2. Design

To add the negative stiffness to our linear prototype, we chose to use two parallel bistable buckled beams. The buckled beams machined out of Beryllium Copper sheets using EDM technology, they have been sized using a model by Cazottes *et al.* [10] based on the minimization of the internal energy of the beams. The equivalent scheme of the obtained prototype is given on Figure 4.

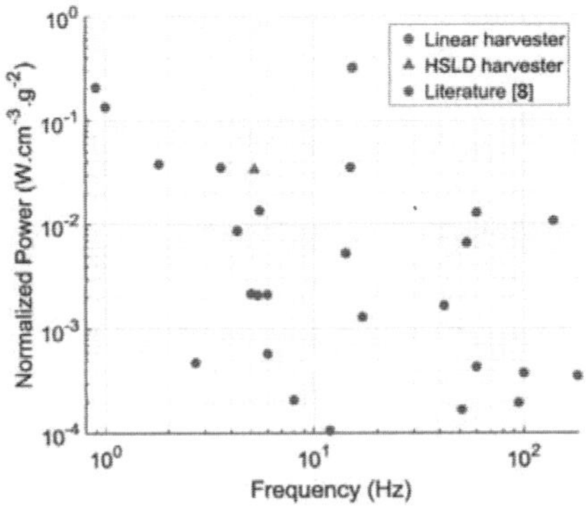

Figure 3. Normalized power versus frequency fo our prototypes versus litterature [8].

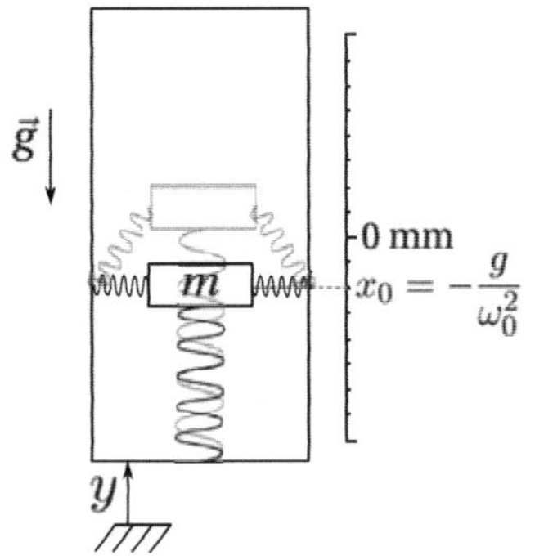

Figure 4. Vibration energy harvester with HSLD stiffness.

3.3. Experimental testing

Like the linear harvester, the HSLD harvester has been tested; the resonant frequency has been successfully decreased down to 5.2 Hz while maintaining the equilibrium position of the linear harvester. However, the mechanical damping is increased to 1.15 % (multiplied by 19). As could be expected, for the same excitation amplitude 5 mg_{RMS}, the harvested power falls to 252 μW (15.6 $mW.g^{-2}.cm^{-3}$). Refering on Figure 3, the HSLD prototype still gives good performances but in the same range as litterature.

4. Conclusion

While the designed linear harvester prototype shows high performances above its counterparts from the literature, it could still be improved by decreasing its volume. Moreover, tests using electronic circuitry to store the scavenged energy may be done. About the HSLD harvester, even if it shows good performances at low frequency and the concept has been proven, work is still to be done to further decrease the resonant frequency while limiting the impact on mechanical damping.

References

[1] Dastbaz M, Pattinson C and Akhgar B 2015 *Green Information Technology* 151–160 (Elsevier)
[2] Mitcheson P D, Yeatman E M, Rao G K, Holmes A S and Green TC 2008 *Proc. of the IEEE* **96** 9 1457–1486
[3] De Pasquale G, Soma A and Zampieri N 2012 *J. of Computational and Nonlinear Dynamics* **7** 4
[4] Saha C, O'Donnell T, Wang N and McCloskey P 2008 *Sensors and Actuators A: Physical* **147** 1 248–253
[5] Rahimi A, Zorlu O, Muhtaroglu A and Kulah H 2012 *Sensors and Actuators A: Physical* **188** 158–166
[6] Dai X 2016 *Sensors and Actuators A: Physical* **241** 161–168
[7] Spreeman D anf Manoli Y 2012 *Electromagnetic Vibration Energy Harvesting Devices* (Springer)
[8] Moss S, Payne O, Hart G and Ung C 2015 *Smart Materials and Structures* **24** 2
[9] Carella A, Brennan M, Waters S and Shin K 2008 *J. of Sound and Vibration* **315** 712–720
[10] Cazottes P, Fernandes A, Pouget J and Hafez M 2009 *J. of Mech. Design* **131** 10

A LOW-FREQUENCY THREE-DIMENSIONAL HYBRID ENERGY HARVESTER

J Zhang, T Chen, H Liu* and L Sun

School of Mechanical and Electric Engineering, Jiangsu Provincial Key Laboratory of Advanced Robotics, Collaborative Innovation Center of Suzhou Nano Science and Technology, Soochow University, Suzhou, 215123, China

Email: hcliu078@suda.edu.cn

Abstract. This paper reports a new design of a three-dimensional vibration energy harvester with high energy output and low working frequency. In a three-dimensional vibrating environment, this electromagnet harvesting structure will have voltage outputs from any directions under a low driving frequency. The volume is about 30mm*30mm*27mm. For an excitation acceleration of 1g, the peak voltage and power output of the electromagnetic part under the vertical vibration can reach to 1.3 V and 0.31mW/cm^3, respectively. When the excitation comes from horizontal direction, the peak voltage and power output of electromagnetic part can reach to 361 mV and 0.88mW at 30 Hz. The maximum output of triboelectric part under the vertical vibration can reach to 20V and 17.32μW/cm^2 at the excitation frequency of 35 Hz.

1. Introduction

Energy harvesting has become an active field of research in past few years and has showed tremendous applications in many areas [1-2]. Vibration based energy harvesting is an alternative to solar and thermoelectric energy harvesting for scenarios where sun and thermal gradients are not readily present [3]. Many vibration energy harvesting devices based on different conversion mechanisms have been proposed [4-5]. Research shows that the vibration in ambient environment are usually below 200 Hz and directionless, while the existing harvesters with high energy output always work at high frequency and single direction [6-8]. Wang's group [9-10] has developed a variety of triboelectric nanogenerators based on vertical contact-separation model, sliding model, and single-electrode model. The highest energy density is up to 45mW/cm^3. Liu et al [11] also proposed a hybrid generator with triboelectric and electromagnetic mechanisms based on vertical contact-separation model. The peak power density of triboelectric part and electromagnetic part can reach to 31μW/cm^2 and 28μW/cm^3 at a low operating frequency of 23.5 Hz.

In this study, we demonstrated a three-dimensional vibration energy harvester with high energy output and low working frequency. In a three-dimensional vibrating environment, the electromagnetic part and triboelectric part will both have electric voltage outputs.

2. Device and working principle

The schematic diagram of the fabricated harvesting device is shown in Fig. 1. The volume is about 30mm*30mm*27mm. It consists of top winding coil, bottom winding coil, permanent magnet, spring structure beam, sponge gasket, FEP (Fluoro-Ethylene Polymer) and two pieces of copper electrode. The permanent magnets are attached on the bottom of the spring structure beam and the sponge gasket with one piece of copper electrode and FEP is on the bottom of the magnet(Fig.1(a)). In order to save space, top winding coil of 0.2 mm in diameter and 500 turns is cohered into a hollow cylinder. A bottom winding coil consists of varnished wire with 0.2 mm in diameter assembled to the pedestal. On the surface of the bottom winding coil, the copper electrode of triboelectric part is attached on it(Fig.1(c)).

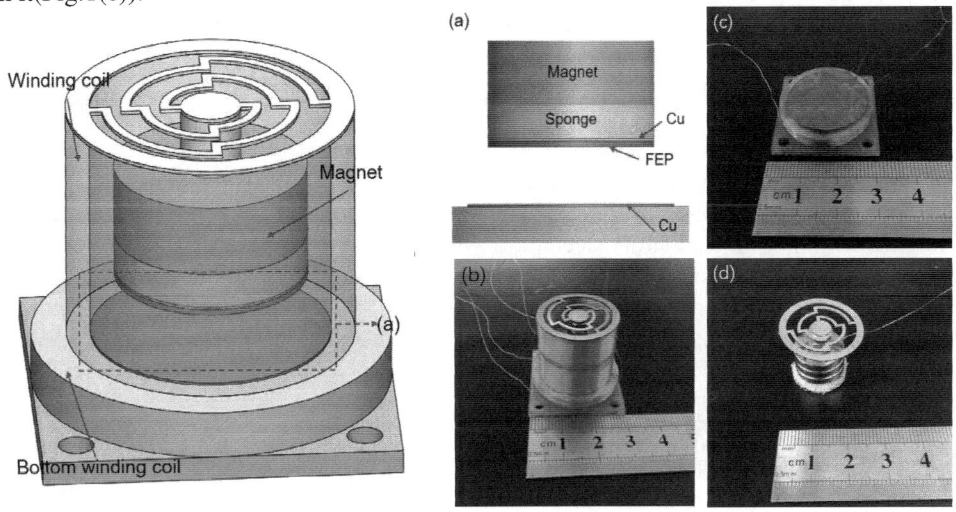

Figure 1. Schematic drawing of the hybrid energy harvester

The principle for electromagnetic and triboelectric parts are based on Faraday's law of electromagnetic induction and coupling effect of triboelectrification and electrostatic induction, respectively. As shown in Fig.2(a), as the harvesting device is driven by an external excitation, the spring structure will drive the permanent magnet to vibrate, which will change the magnetic flux across the winding coil. Hence, two parts of the electric voltage across the coil are generated from the electromagnetic part. At the same time, the periodic movement of the permanent magnet causes the periodic contact and separation between the copper foil and FEP.

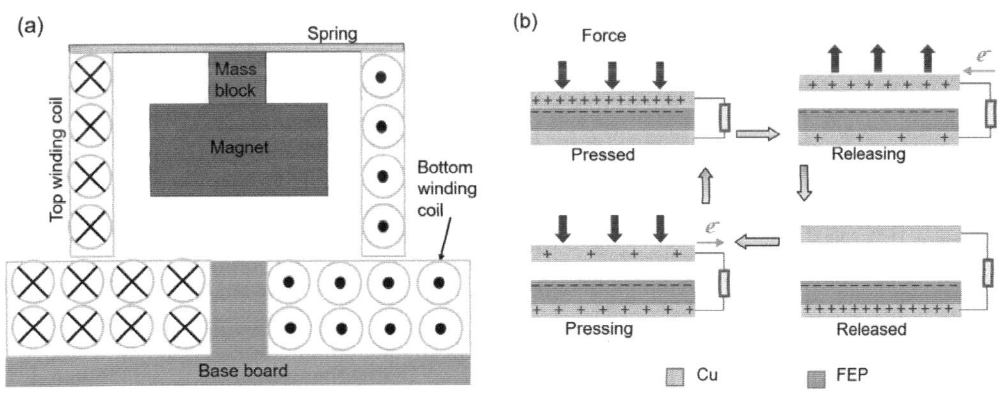

Figure 2. Illustration of the operating principle of the hybrid energy harvester

The contact materials Cu and FEP have large differences in the ability to attract and retain electrons according to triboelectric series. In Fig.2(b), as the Cu and FEP layers get in contact with each other by a pressure force, the positive and negative charges will be gathered on the contact surfaces of Cu and FEP layers, respectively, due to triboelectrification effect. But there is no electric potential difference (EPD) between the two contact materials, because the charges with opposite signs coincide at almost the same plan. When the Cu and FEP are separated, the top and bottom electrodes are connected. The established EPD drives the electrons to flow from the top electrode to the bottom electrode. As a result, the positive triboelectric charges on the top electrode will be gradually neutralized until the gap distance reaches to the maximum. Thus the current is generated from the triboelectric part during the releasing process. After that, the gap distance starts to decrease due to the periodic vibration. The top electrode gets close to the FEP layer gathering with negative charges, the electrons will flow back from the top electrode to the bottom electrode because of the electrostatic induction and thus the positive charges on the top electrode surface will increase again. Finally, as the two contact materials get in contact, the charge distribution is returned to the initial state again.

3. Experiment and results

The assembled device is tested by a vibration control system. The experimental results in Fig.3 shows the output performance of the top winding coil under vertical vibration in frequency domain and time domain. For an excitation acceleration of 1g, the peak voltage output across the top winding coil can reach to 369 mV at resonant frequency of 32.7 Hz. Fig.4 shows the output performance of the total winding coil in frequency domain and time domain. For an excitation acceleration of 1g, the peak voltage output across the total winding coil can reach 1.32V at resonant frequency of 32.7 Hz.

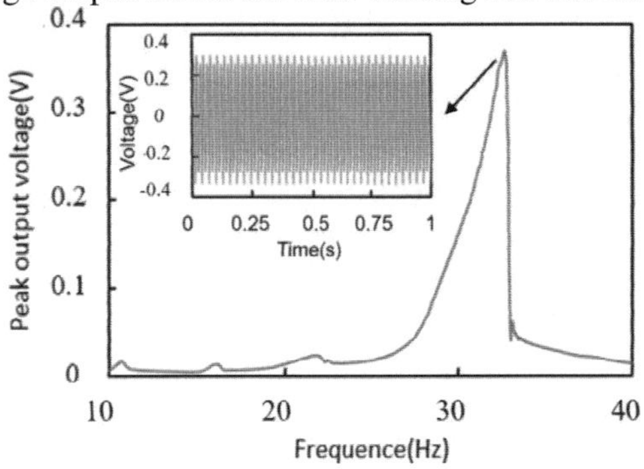

Figure 3. The voltage output of the top winding coil in frequency and time domain at 1g

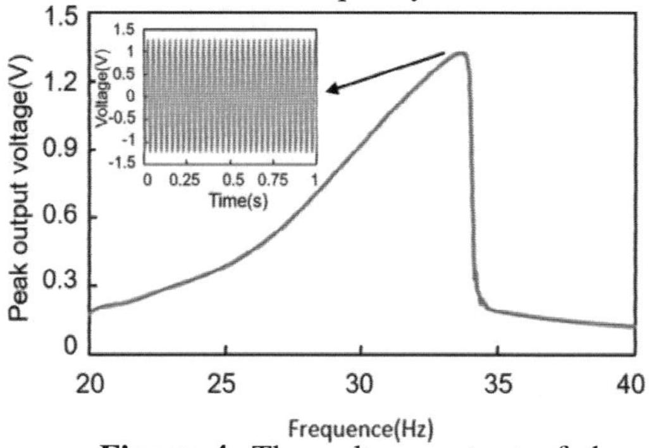

Figure 4. The voltage output of the total winding coil in frequency and time domain at 1g

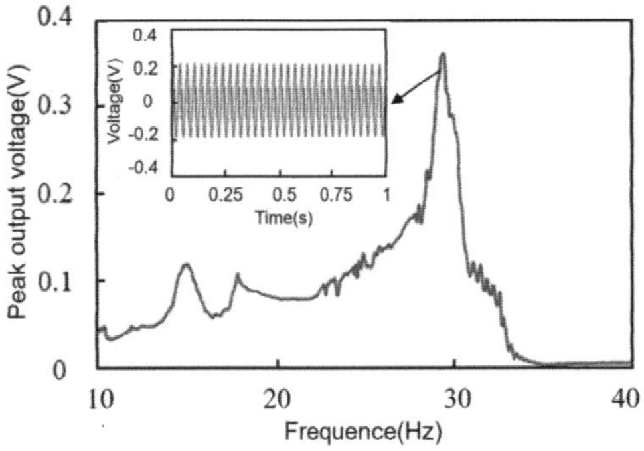

Figure 5. The voltage output of the bottom winding coil in frequency and time domain at 1g

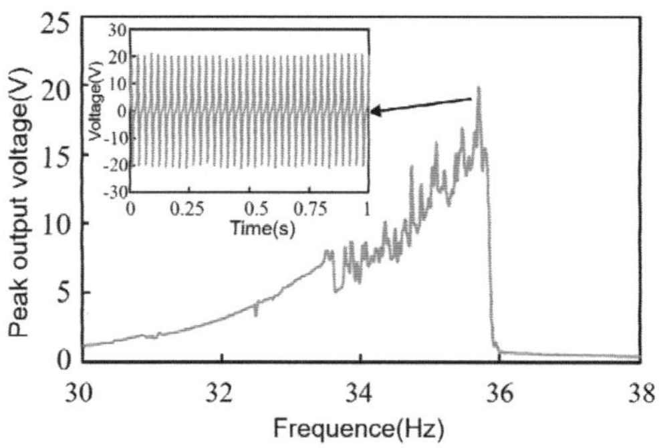

Figure 6. The voltage output of the triboelectric part in frequency and time domain at 1g

When the excitation comes from horizontal direction, the bottom winding coil's peak voltage output can reach to 361 mV at 30 Hz. Fig.5 shows the voltage waveform of the bottom winding coil in frequency domain and time domain at 30Hz. The output performance of the triboelectric part in frequency domain and time domain is shown in Fig.6. The maximum output voltage can reach to 20V at the excitation frequency of 35 Hz. For an excitation acceleration of 1g, the power output of the triboelectric part under vertical vibration at 35Hz is shown in Fig.7. When the top and bottom winding coil are connected into the total winding coil, the total power output under vertical vibration and the bottom winding coil under horizontal vibration is shown in Fig.8 at 32.7Hz.

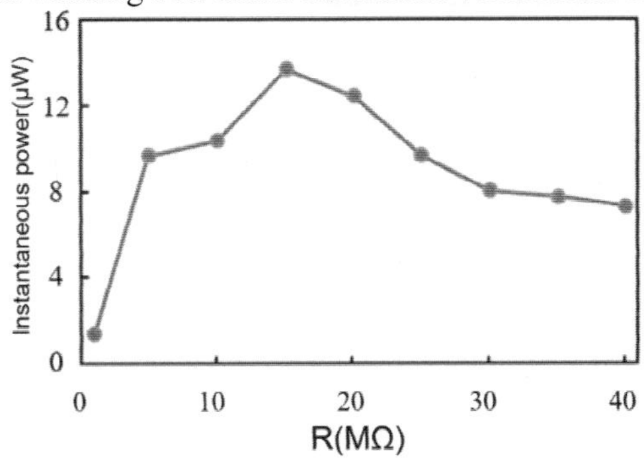

Figure 7. The power output of the triboelectric part at 35Hz

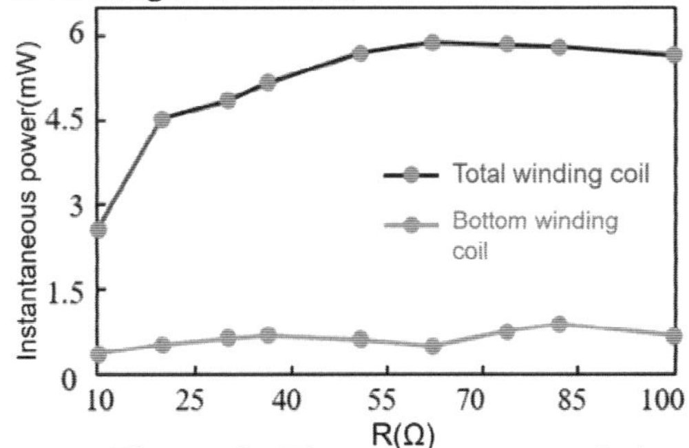

Figure 8. The power output of the total winding coil and bottom winding coil at 32.7Hz

4. Conclusion

In this paper, we proposed a three-dimensional hybrid vibration energy harvester, which can achieve high output performance from any directions under low excitation frequency. For an excitation acceleration of 1g, the power output under vertical and horizontal vibration can reach to 5.87mW at load resistance of 62Ω and 0.88mW at load resistance of 82Ω at 32.7Hz, respectively.

5. Acknowledgement

This work is funded by the National High Technology Research and Development Program of China (863 Program) (Grant No. 2015AA043502) and the National Natural Science Foundation of China (Grant No. 41527901).

References

[1] Beeby S P, Tudor M J and White N M, 2006, *Meas. Sci. Technol.* **17** R175-R195.
[2] Chen T, Xia Y, Liu H, 2016, J. Journal of Microelectromechanical Systems, **25** 845-847.
[3] Wang X, Niu S, Fang Y, 2017, J. Acs Nano. **11** 1728.
[4] Khaligh A, Zeng P, and Zheng C, 2010, *IEEE Trans. Ind. Electron.* **57** 850-860.
[5] Zhang X, Han M, Wang R, Zhu F, Li Z, and Zhang H. 2013, *Nano Lett.* **13** 1168-72.
[6] Roundy S, Wright P K, and Rabaey, 2003, J. *Comput. Commun.* **26** 1131-44.
[7] Sari I, Balkan T, Külah H, 2010, J. Journal of Microelectromechanical Systems,**19** 14-27.
[8] L. Dhakar, F. Tay, C. Lee, 2015, J. Microelectromech.S.,**24** 91-99.
[9] Bai P, Zhu G, Zhou Y S, W S, Ma J, Zhang G and Wang Z L, 2014, *Nano Res.* **7** 990-997.
[10] Tang W, Zhang C, Han C B and Wang Z L, 2014, *Adv. Funct. Mater.* **24** 6684-90.
[11] Liu H, Xia Y, Chen T, 2017, J. IEEE Sensors Journal, **99** pp.1-1.

Development of MEMS Air Turbine Micro Generator with Ball Bearing Mechanism and Magnetic Material

K. Kudo[1], K. Ebisawa[2], K. Mishima[1], M. Takato[2], K. Saito[2] and F. Uchikoba[2]

[1] Graduate School of Science and Technology, Nihon University, Japan

[2] College of Science and Technology, Nihon University, Japan

E-mail: uchikoba@eme.cst.nihon-u.ac.jp

Abstract. This paper proposes an electromagnetic type MEMS air turbine generator combining an air turbine fabricated by MEMS process and a magnetic circuit by multilayer ceramic technology. Introducing ball baring system for the turbine, two types of turbine rotor, a planar type and a rim type, were evaluated. The rim type achieved 290,135 rpm at the inlet flow of 2.4 L/min and the pressure of 0.3 MPa. The magnetic circuit was fabricated by multilayer ceramic technology. The circuit consist of Ni Cu Zn ferrite and silver 100 turn helical coil. The internal resistance was 2 Ω. The dimensions of the combined generator were 7.40 mm, 8.47 mm and 5.82 mm, respectively. When the load resistance was 8 Ω, the generator showed the maximum output. The maximum power was 2.41 mVA.

1. Introduction

Miniaturization of a power supply is important issue. A lithium ion secondary battery has been mainly used. However, the energy density has reached its theoretical limit. Therefore, miniature electric generators have been studied as the candidate to replace the battery. MEMS (Micro Electro Mechanical Systems) process have been used to achieve the miniaturization. MEMS process is based on the IC production process, and it produces a micro structure. The research of MIT UMGT (Ultra Micro Gas Turbine) which used the MEMS process was reported [1]. There are two methods for the MEMS power generation; an electrostatic type and an electromagnetic induction type. Since the structure is easy to fabricate by MEMS process, most of them are the electrostatic type [2]. However, its output power is small due to the charge saturation and the high internal impedance. On the contrary, the electromagnetic induction type is expected to achieve the high output power [3]. This type requires a complex magnetic circuit, and also, it does not contain a magnetic core usually. Therefore, the magnetic flux divers. To collect the magnetic flux, the area of the coil becomes wide, and then, the conductor loss increases.

We focused on the multilayer ceramic technology to form the miniature three-dimensional magnetic circuit. The multilayer ceramic technology provides the monolithic structure that has the helical coil and magnetic core. We previously developed the electromagnetic induction type MEMS air turbine generator combining with the multilayer ceramic magnetic circuit. The fluid-dynamic bearing system was adopted because of the low friction. However, its shaky motion reduced the rotational speed. Also, the gap between the magnet and the magnetic circuit was long because the flow path layer for the bearing should have been set between them. The maximum output power was 1.41 μVA [4].

In this paper, a miniature ball bearing is employed for the MEMS air turbine. It keeps the stable rotation. Moreover, by introducing a shaft, the permanent magnet can be separated from the turbine rotor. And then, the gap between the magnet and the magnetic circuit can be narrow. Two type rotors are fabricated with regard to the rotor structure, a planar type and a rim type. The rotational motion of these air turbines are evaluated, and then, the output power of the rim type is measured. The maximum power and the power loss are discussed.

2. MEMS Air Turbine

Developed MEMS turbines are shown in Figure 1 and Figure 2. The ball bearing of the planar type was arranged at the bottom of the rotor. A Compressed air flows from side inlet and to top out let. The bleads is formed on the disk. The dimensions of the fabricated turbine were 5.23, 5.20, 4.51 mm length, width and height, respectively.

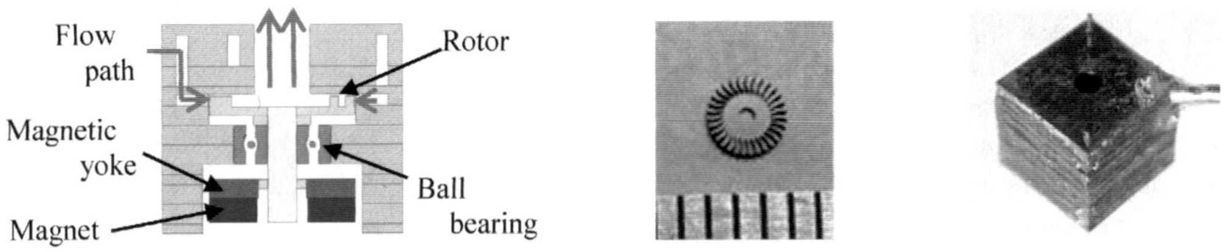

(a)Sectional view (b)Rotor view (c) Fabrication view

Figure 1. The schematic diagrams and the photograph of the planar type air turbine.

Two ball bearings were used for the rim type. The bearings are placed above and below the rotor. The compressed air flows from side inlet to side outlet. The blade are formed on the side wall of the rotor. The dimensions of the rim type were 5.24, 5.37, 4.64 mm length, width and height, respectively.

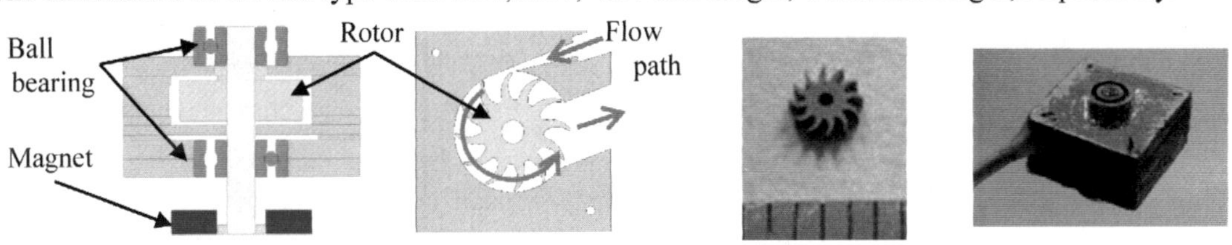

(a)Sectional view (b)Flow rate view (c)Rotor view (d) Fabrication view

Figure 2. The schematic diagrams and the photograph of the rim type air turbine.

Rotation measurements were performed using hall elements. A comparison of the rotational speed is shown in Figure 3. The flow rate varied from 0 to 1.0 L/min. at pressure of 0.3 MPa. Experimental result revealed that the rim type system had a superior potential to the planar type.

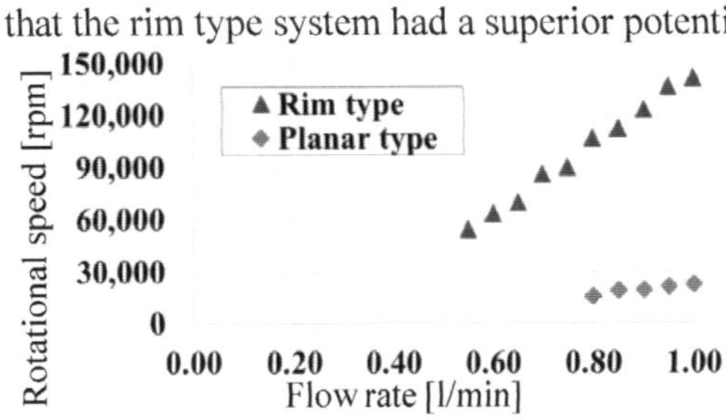

Figure 3. A comparison of the rotational speed of turbines.

3. Multilayer Ceramic Magnetic Circuit

The schematic diagrams of the multilayer ceramic magnetic circuit is shown in Figure 4(a) (b). The magnetic material was Ni Cu Zn ferrite with the permeability of 900. After the firing at 900 degrees Celsius, the four pieces were bonded and the closed magnetic circuit was obtained. The total number of coil turns were 100. The dimensions of the fabricated circuit were 7.40, 8.47, 2.36 mm, length, width and height, respectively. The measured DC resistance were 2 Ω.

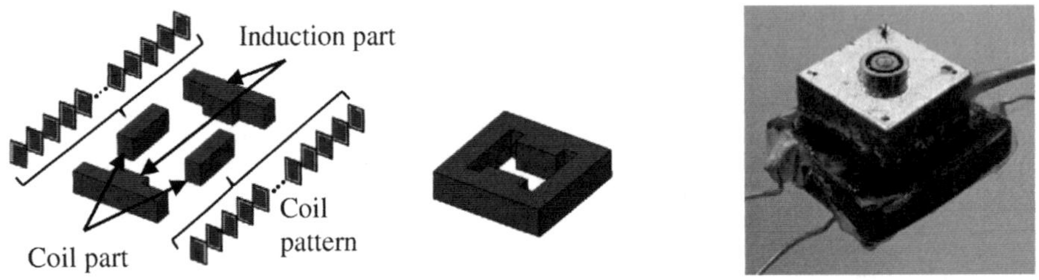

(a)Part view (b)Assembly view (c)Assembled generator

Figure 4. The schematic diagrams of the multilayer ceramic magnetic circuit and the photograph of the assembled generator.

4. Results and Discussions

The fabricated the MEMS air turbine generator is shown in Figure 4(c). The dimensions of the generator combined with the micro air turbine and the multilayer ceramic magnetic circuit were 7.40, 8.47, 5.82 mm, length, width and height, respectively. Compressed air was introduced into the MEMS air turbine, and the output waveform was observed. The maximum rotational speed was 290,135 rpm when the inlet flow was 2.4 L/min, and the pressure was 0.3 MPa. The load resistance dependence of the output voltage and the output power is shown in Figure 5. When the load resistance was 8 Ω, the voltage and the maximum power was 139 mV, 2.41 mVA, respectively. The output waveform is shown in Figure 6 at the load resistance is 8 and 1000 Ω.

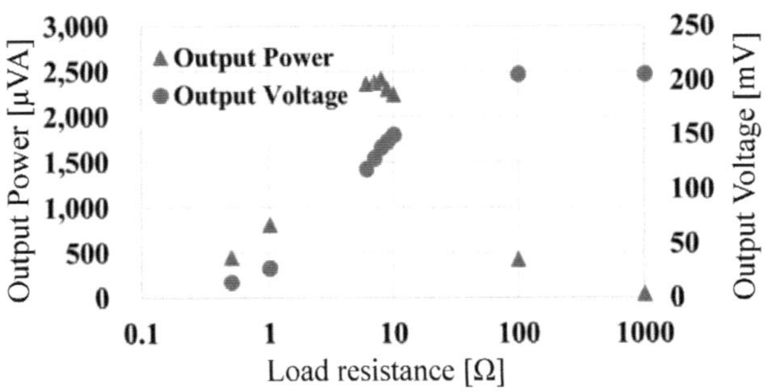

Figure 5. Relations between load resistance and output voltage and output power.2.4 L/min, 0.3MPa

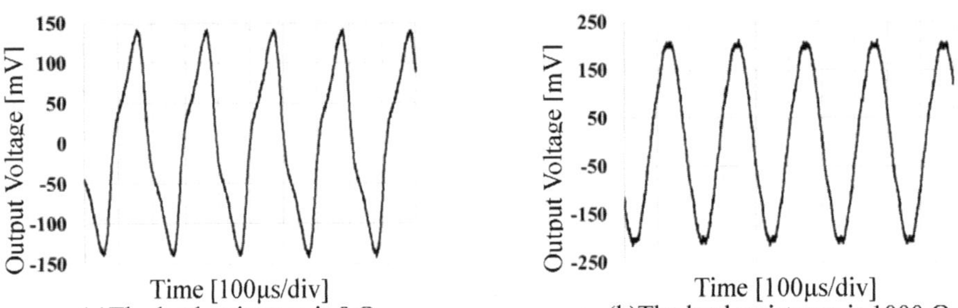

(a)The load resistance is 8 Ω (b)The load resistance is 1000 Ω

Figure 6. The output waveform at the load resistance is 8 and 1000 Ω.

When the load resistance R_L is connected to the generator and the current I flows through the circuit, the output voltage $V=R_LI$ is given by the following equation. At this time, the influence of the voltage drop due to the self-inductance L and the internal resistance r of the magnetic circuit are consider.

$$V = R_L I = N\frac{d\phi}{dt} - i\omega LI - rI \tag{1}$$

ϕ means the magnetic flux passing through the magnetic circuit of the generator. N is turn number of the coil. The value of L measured at the equivalent frequency of 290,135 rpm, with an impedance analyser Agilent 4294A was 241 μH. When 1 kΩ of R_L is connected, the current flowing the circuit is sufficiently small and the voltage drop due to self-inductance and internal resistance is small. Therefore, the output voltage is approximated by $d\phi/dt$. The surface magnetic flux density of the permanent magnet is 0.159T, and the magnitude of the magnetic flux on the magnet surface is 0.375μWb. If the all flux enter in the circuit, the amplitude of the output voltage is calculated to be 570 mV. In comparison with Figure 6 (b), 36% of the flux contributes to the actual power generation. It is considered that other magnetic flux leaks without entering the magnetic circuit. On the other hand, when R_L is 8 Ω, the reactance of ωL is calculated 7.3 Ω. The values of ωL and r are close to R_L. Therefore, their influence appears. The absolute value of the output voltage is estimated as follows.

$$|V| = \frac{N\,d\phi/dt\,R_L}{[(R_L + r)^2 + \omega^2 L^2]^{1/2}} \tag{2}$$

When calculated from equation (2), |V| becomes 367mV. Considering the contribution rate 36% of the magnetic flux to this result, it almost coincides with Figure 6(a).

When the current flows, the magnetic circuit occurs braking torque to the permanent magnet. In Figures 6(a) and 6(b), the rotational speed is almost equal. Therefore, it is considered that the braking torque by current is enough small in this system. The magnetic flux density from the magnetic circuit is estimated as follows.

$$B = \mu_0 \mu_r N I \tag{3}$$

μ_0 is the vacuum permeability, μ_r is the ferrite relative permeability , 900. B is calculated 1.96mT. When the magnetization of the permanent magnet and the magnetic flux density occurred by the magnetic circuit are perpendicular, the braking torque becomes maximum. If all of the magnetic flux contributes, the braking torque is estimated to about 42.9 pNm. This value is considered to be a sufficiently small value.

5. Conclusion

Two types of turbines with the ball barring were manufactured and the rotational speed was compared in this study. As a result, the rim type rotates faster than the flat planar type. Then, the electromagnetic type MEMS air turbine generator using the rim type was fabricated. The appearance size of the generator was 7.40, 8.47, 5.82 mm, length, width and height, respectively. The DC conductor resistance of multilayer ceramic magnetic circuit was 2 Ω. The load resistance dependence of the output voltage and the output power was measured. The maximum power of the generator was 2.41 mVA. Moreover, the measured voltage and power were compared to the theoretical value that was calculated taking account of the internal resistance and the self-inductance.

Acknowledgments

The part of this study was supported by Nihon University College of Science and Technology Project Research and JSPS KAKENHI Grants 16K18055. We appreciate the support. Specimen fabrication was supported by the Research Center for Micro Functional Devices of Nihon University.

References

[1] Epstein H A and Senturia D S 1997 *Science* **276** 1211
[2] Genda T, Tanaka S and Esashi M 2003 *TRANSDUCERS '03, The 121th Int.Conf. a Solid Stale Sensors, Actuators and Microsystems* p 492
[3] Herrault F, Ji C and Allen G M 2008 *Microelectromechanical Systems* **17** No.6
[4] Yokozeki Y, et.al. 2014 *Proc. Int. Conf. on Electronic Pachaging (Toyama)*p384

IOP Conf. Series: Journal of Physics: Conf. Series **1052** (2018) 012091 doi:10.1088/1742-6596/1052/1/012091

EVALUATION OF AN IMPACT SPRING-COIL-MAGNET SYSTEM WITH 3D-PRINTED SETUP

P. Mehne[1], P.Scholl[2], A.Rudmann[2], M. Kröner[1,] K. Van Laerhoven[3] and P. Woias[1]
[1]University of Freiburg - IMTEK, Germany
[2]University of Freiburg - ESE, Germany
[3]University of Siegen - Ubiquitous Computing, Germany

E-mail: philipp.mehne@imtek.uni-freiburg.de

Abstract. For several energy harvesting applications impact generators are one approach of generating energy to perform wireless transmissions to notify about an event. The goal is to harvest enough energy for a notification of a BLE-Chip to notify about one impact. Therefore magnet coil systems are one common approach. There are many variances assemble different arrangements of magnet, coils and springs. With this report we give an overview to behavior of coil behaviors by transmission of magnets. Therefore a setup is built with a 3D-Printer to reduce constructional effort. This transmission behavior of several 4 mm N45 neodymium magnets with different coils (height from 1-10mm, 100-400 windings) and springs setups are giving an idea about the practical approach and some comparison about simulation and measurement is done.

1. Introduction

It is obvious from the physical principles of mechanical energy harvesting that a trade-off between design complexity, size and generated energy has to be found when realizing a mechanoelectric generator for a dedicated application [1]. It is therefore helpful to have a modular system available for characterizing different elements of such a generator - and even for building prototype generators - to find a best combination of features with respect to application-specific requirements. This report presents an easy way of building generators by characterizing different set ups of magnet-coil systems in an electromagnetic energy harvester.
As a boundary condition, the outer dimensions of this generator were fixed to the approximate size of a pocket lighter. This was done with a look at the potential application of such generators as body-worn systems: They could be used, for instance, as autonomous impact sensors for pedometers or as energy-autonomous switches in medical applications, to be operated by a patient for signaling a dedicated request to a patient monitoring system. For all these applications the overall goal is thereby to find a generator setup which is capable to deliver enough energy from a single impact to send a wireless notification to a base station nearby. For such an application usually a Bluetooth data link is used, as has been done here.

2. Realized test set-up for a modular induction generator

A schematic setup of the generators studied is shown in figure 1. It consists of a tube, which holds permanent magnets located on top of a spring. A spacer tube is slid over this inner tube, which can be varied in its height to adjust the resting distance between magnet and coil. On top of the tube is a cover with a hole for a pushing piece which moves the permanent magnet through the coil, thus generating electrical energy via magnetic induction. The AC voltage generated in the coil is fed to a full-wave rectifier connected to a storage capacitor. With a given capacitance the energy extracted from a single movement of the magnet can be directly calculated from the change in the capacitor voltage. It is evident that already this simple set-up offers a large number of design parameters.

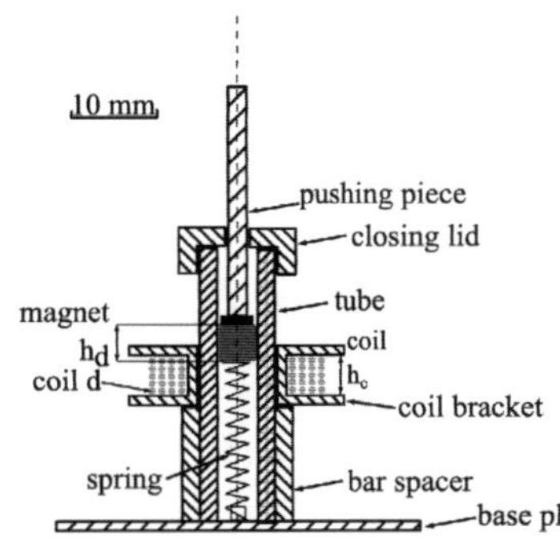

Figure 1. Basic design of the test setup, cross-sectional view of the 3D-printed realization.

Coils with variable height from h_c = (1, 2, 4, 6, 8, 10 mm) number of windings n = (100, 200, 300, 400), and wire diameters d = (0.055, 0.1, 0.22 mm) were evaluated.
Also, cylindrical N45 neodymium magnets with a diameter of 4 mm and different heights of 1, 2, 4 and 10 mm were used. In addition multiple magnets were stacked with both opposite and aligned poles. The AC voltage generated in the coil is fed to a full-wave rectifier connected to a storage capacitor. With a given capacitance the energy extracted from a single movement of the magnet can be directly calculated from the change in the capacitor voltage.
Figure 1 does also show a cross-sectional view of the realized test device, with a more detailed explanation of all parts. Building was done from acrylic photoresin with a Form Labs 2 stereolithographic printer: A base plate is printed together with a central tube for a co-axial alignment of the spring, the magnets and the coils, and for guiding the magnet movement with respect to the coil. It has an inner diameter of 4.5 mm and a wall thickness of 1.5 mm. The mechanical spring and the moving magnets are placed inside the tube, together with a pushing piece for initiating a movement of the magnet. On top of the tube is a threaded lid which aligns and holds the pushing piece.
To test different coil sizes, suitable bobbins were made via 3D printing and put over the tube using a second set of coaxial tubes as spacer, to fix their defined location. Taking the results of all experiments a final design of the generator was realized as shown in figure 4.
The main design improvement to be tested here is related to the distance of magnet and coil. Generally, to calculate the flux density B of a cylindrical magnet, the following equations can be used [2].

With: $dB = \frac{U_{ind} \cdot dt}{N \cdot A}$ (1) $B = \frac{B_r}{2}\left(\frac{h_d+z}{\sqrt{\left(\frac{d}{2}\right)^2+(h_d+z)^2}} - \frac{z}{\sqrt{\left(\frac{d}{2}\right)^2+z^2}}\right)$ (2)

where B_r is the remanence field flux of the permanent magnet, h_d and d its height and diameter, respectively, and z the distance from one pole.

3. Measurements with single coils

The spring used in these experiments is 17.5 mm long, with a diameter of 2.33 mm and a spring constant of 0.536 N/mm. For all measurements, the spring is compressed for one centimeter by exerting vertical force onto the magnet via the pushing rod. After a sudden release of this force, the magnet is pushed upwards, thus traveling though the coil. The resulting induced voltage at the coil is traced with a PicoScope 3206 MSO. Figure 2 shows the voltage characteristic of such an event. The minimum voltage and the maximum voltage values of the measurements are measured, as well as the reaction time. The measurements are repeated five times and the average and standard deviation is calculated. Overall, 241 combinations are measured with the first set up.
The results of the most promising are displayed in figure 3. It is remarkable that the 10 mm magnet induces less voltage than the 4 mm magnet. This is presumably related to the smaller change of the magnetic field. Also, a magnet of a higher mass must have a lower velocity with the same given input energy, when passing the coil, hence, dB/dt will be smaller.

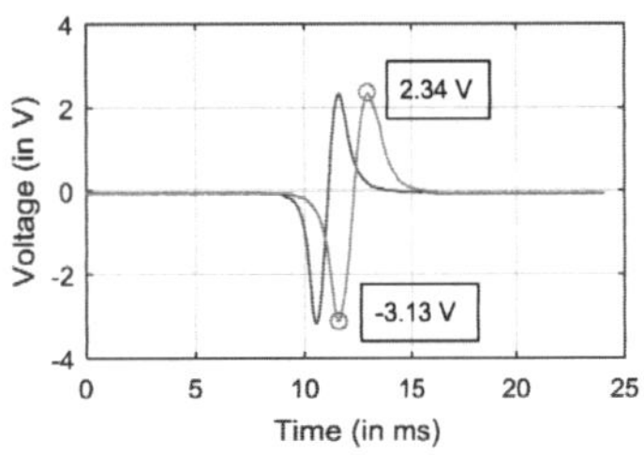

Figure 2. Voltage peak induced sequentially at two coils, with n=400, wire diameter d=0.1mm, h_c = 4mm, 4 mm magnet.

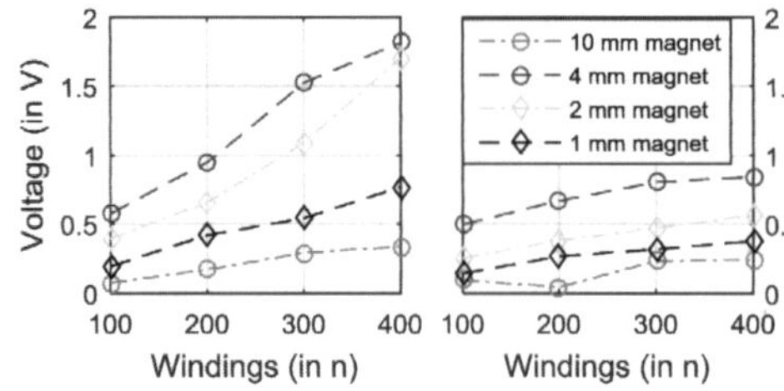

Figure 3. Maximum induced voltage with h_c = 4 mm and d = 0.1 mm on the left and N45 magnet with d = 0.22 mm as function of magnet size and turns.

In table 1 a collection of the most important measurements is shown. By assessing the basic parameters and combinations with the mean peak voltage of five measurements, all configurations were characterized regarding the maximum power to be harvested. The most promising configuration is used to evaluate further assemblies with parallel and serial coils as well as with stacked magnets.

Table 1. Table of most important measured values, sorted by peak power downwards.

Magnet height	Coil height	Wire diameter	Windings n	Resistance	$U_{peak\ mean}$	Standard deviation	Peak Power
4 mm	4 mm	0.22 mm	200	4.6 Ω	700 mV	40.5 mV	106.52 mW
4 mm	4 mm	0.1 mm	400	32.8 Ω	1814 mV	35.6 mV	100.32 mW
4 mm	4 mm	0.22 mm	300	6.9 Ω	802 mV	26.4 mV	93.22 mW
4 mm	4 mm	0.1 mm	300	27.7 Ω	1524mV	31.4 mV	83.85 mW
2 mm	4 mm	0.1 mm	400	36 Ω	1710 mV	66.3 mV	81.23 mW
4 mm	2 mm	0.1 mm	300	29 Ω	1375 mV	141.7 mV	65.19 mW
4 mm	4 mm	0.22 mm	400	11 Ω	840 mV	29.0 mV	64.15 mW
4 mm	4 mm	0.1 mm	200	15.5 Ω	950 mV	80.5 mV	58.23 mW
4 mm	2 mm	0.055 mm	400	124 Ω	2666 mV	98.9 mV	57.32mW

4. Improved Setup

Based on the results of the experimental study, an improved design of a mechanoelectric generator was realized, figure 4. As one central measure, the wall thickness of the central tube was reduced to only 260 µm and the coil is directly winded on this. The setup has a coil height of h_c = 4mm, n=700 windings, a h_d = 4mm N45 magnet giving the best performance in respect to peak voltage and power .This gives the big advantage that the coil can be mounted much closer to the magnet, which as a result of equation (2) leads to an increase in the induced voltage. Apart from that, it is possible to measure coils in serial and in parallel connection. The obvious disadvantage is that the tube material is very sensitive due to the strongly reduced stiffness of the thin walls. The mechanical spring with a spring constant of 0.54 N/mm (peak voltage 5.5 V, 700 windings, 4 mm coil height) is changed to a stronger spring with a spring constant of 0.854 N/mm. This leads to an increase of the peak voltage to 6.3 V and a faster movement of the magnet, whereas the transit time is decreased from 3.143 ± 0.09 ms to 2.671 ± 0.02 ms. This, however, is reducing the charging time of the capacitors as well.

Figure 4. Improved setup with two coils, with n=700, wire diameter d=0.1mm, h_c = 4mm, 4 mm magnet.

After evaluation this improvement does increase the peak voltage, but not the maximum power to be harvested. Also, the characteristics of the full wave rectifier play a significant role. The PMEG1201AESF is used. It has a very low forward voltage of 160 mV [3] is showing the best performance compared two different models (BAT 41 & BAT 43). Disadvantageous is the small package, which is very difficult to handle compared to the wired alternatives.

4.1. Results with multiple coils, magnets, diodes and capacitors

To increase the harvested energy, alternative configurations were experimentally evaluated and also studied by simulations. First, different magnet setups are investigated. Figure 5 shows a simulation of the magnetic density flux of three variants. The measurements indicated a slightly higher voltage peak when using 4 x 1 mm magnets instead of one single 4 mm magnet. With this simulation the reason is made visible: At the edges of the magnets there is still a slightly higher magnetic flux density than for a single magnet. The configuration using oppositely directed poles induces a higher single voltage peak, but the reaction time is slower. This is also understandable by evaluation of the simulation on the right hand side of figure 5, where a short and fast change is visible, but the flux density in the working area around the magnet is lower. Equation (1) is used to compare the measured data with the simulations. Assuming a constant magnetic field, the voltage would increase linear. With lower winding numbers, the theoretical values are matching, and of course the induced voltage will drop higher distance of the coil from the magnet. This is an expected behavior which can be seen in simulation, measurement and calculation.

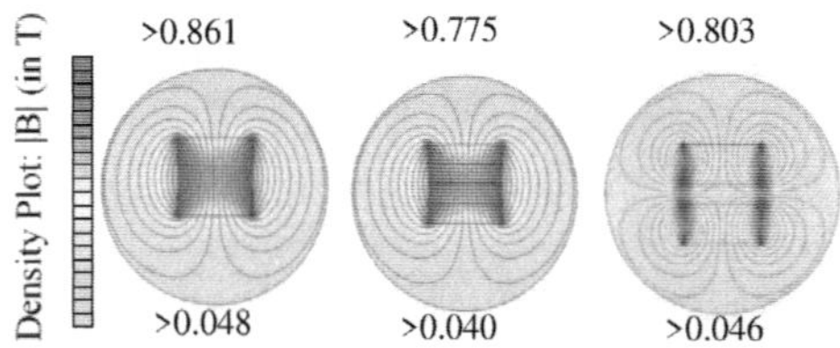

Figure 5. Finite Element Method Magnetics simulation of N45 magnets with 1 x 4 mm (left) 4 x 1mm (middle) and 2 x 2 mm (right)

4.2. Final Design and data transmission

The following design has seven coils. Every coil has an individual full bridge diode rectifier and a capacitor to store and evaluate the transformed energy. Two aligned magnets are used with a spacer of 4 mm in between. This leads to a higher number of trespassing magnets in respect to the coils, when using more magnets and additionally increasing the weight, which has some impact on the spring power Assuming the available spring tension (0.54 N/mm) when pushing it one cm down, there are 26.8 mJ available. Integrating the power over time, we reach with 2 magnets and seven coils 6.37 mJ in the setup in table 2. This is 23.46 % in the construction volume. Summing up the available energy in table 2, on the single capacitors, 360µJ are stored. This is only 5 % of the possible energy to transform. Nevertheless 360 µJ could be enough for some transmission protocols, like enOcean is able to transmit with 50 µJ [4]. Using Bluetooth low energy, the voltage levels are matched not to exceed 3.8 V. With a RF DUINO it is possible to send two notification if it is in sleep mode. If combining a battery to supply the sleep mode, and the coils, the chip needs 109 µJ.

Table 2. Resulting voltages on capacitors using setup with seven coils and two magnets.

Capacitor #	Voltage (V)	Capacitance (µF)	Energy (µJ)
1	2.73	9.63	35.89
2	2.09	43.5	95.01
3	1.89	63.3	113.06
4	2.04	43.5	90.51
5	2.64	28.8	100.36
6	3.18	9.63	48.69
7	2.58	9.63	32.05
Total	17.15	2.45	360.3

5. Conclusion

Some interesting effects are made visible, especially in respect to the magnet arrangements. It is more efficient to use aligned poles of the magnets with some spacers, that the coils are inducing voltage while trespassing. With 6.37 mJ is enough energy available in the system for radio transmissions, but the poorly energy extraction with the full bridge and multiple capacitors is some effort with low return of 360 µJ. In future, further works has to concentrate on the energy extraction and an oscillating system has to be compared. Further designs with other magnet types are interesting as well. To decrease energy demand, there seems to be some possibilities to improve the Bluetooth chips heavier. So even with the presented set up it could be possible to perform transmissions.

6. Acknowledgement

The authors would like to thank the German Science Foundation DFG for funding this project under the graduate research training program – GRK 1322 – Micro energy harvesting.

References

[1] C. Wei, X, Jing, A comprehensive review on vibration energy harvesting: Modelling and realization. Renewable and Sustainable Energy Reviews, Vol. 74, Jul. 2017 pp. 1-18.

[2] J. Leohold, How to calculate the magnetic flux density, 2017, *retrieved from https://www.supermagnete.de/faq/Wie-berechnet-man-die-magnetische-Flussdichte*

[3] PMEG1201AESF, 2 V, 0.1 A low VF MEGA Schottky barrier rectifier, Product data sheet, February 2015, NXP

[4] EnOcean – The World of Energy Harvesting Wireless Technology, White Paper, EnOecean, January 2016, p. 2

Self-powered wireless sensor node for flow and temperature sensing

Yushen Hu[1], Jingchi Yang[1], Ziyu Huang[1], Yulong Zhang[1,2] and Fei Wang[1,2,3]

1 Department of Electrical and Electronic Engineering, Southern University of Science and Technology, Shenzhen 518055, China;
2 Shenzhen Key Laboratory of 3rd Generation Semiconductor Devices, Shenzhen 518055, China;
3 State Key Lab of Transducer Technology, Shanghai Institute of Microsystem and Information Technology, Chinese Academy of Sciences, Shanghai 200050, China.
E-mail: wangf@sustc.edu.cn

Abstract. This paper presents the implementation and test of a self-powered wireless sensor node using enhanced triboelectric nanogenerator (E-TriG) as both energy harvester and wireless sensor node for flow rate and temperature detection. RF microcontroller with low power consumption is included for data conversion, signal processing and wireless signal transmission. The flow-rate detection is based on the maximum voltage of a capacitor, which is charged by E-TriGs during certain period. Experiment results show that this flow sensor is capable of detecting flow rate from 5.8 m/s to 17.3 m/s with a standard deviation of 0.3% and the temperature of 24.2 °C (±0.3 °C) in the actual test. Each operation requires less than 40 ms for data sampling, processing and transmission. With a Raspberry Pi, the wireless signal can be received and delivered to Internet and therefore, can be monitored easily from any portable terminal with internet-access.

1. Introduction

Energy harvesting devices have been recently developed to provide power supply for wireless sensors with low power consumption [1-2]. Power has been generated from thermal source [3], vibration source [4-5] and wind flow [6-8], at a range from nW to mW. Due to the fact that the energy generated from triboelectric nanogenerator is irregular and insufficient to directly power the embedded chip, a power management circuit combined with a large capacitor is typically needed [9].

Temperature and air flow monitoring are important for smart buildings, especially for a central air conditioning system. Herein, we proposed a self-powered wireless sensor node with compact size, short preparation time (including analog to digital conversion and signal processing) and low power consumption. A temperature sensor is embedded with the power management chip. Furthermore, triboelectric generator (TENG) is used not only as a power source, but also a flow rate sensor in this design. With this design, two different types of sensor can be integrated into the same node with self-powered wireless sensing function, which could be explored to more environmental information monitoring in the near future. The flow rate and temperature sensor node could be helpful to adjust the energy distribution and task allocation problems between different nodes in a more reasonable way.

2. Method

The power management circuit rectifies the output voltage from nanogenerator, stores that energy in

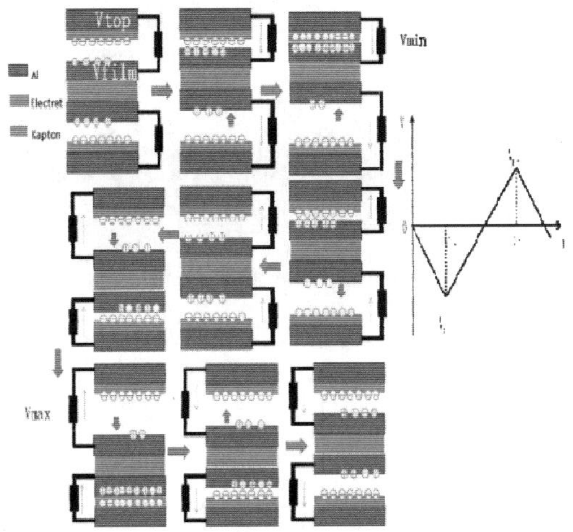

Figure 1. Principle of TENG devices. A soft film driven by wind periodically contacts the top/bottom layer. These layers and the soft film are fabricated with special material so that when they contact, the surface potential changes due to triboelectric effect. As a result, the voltage across the top/bottom layer and film serves as a voltage source.

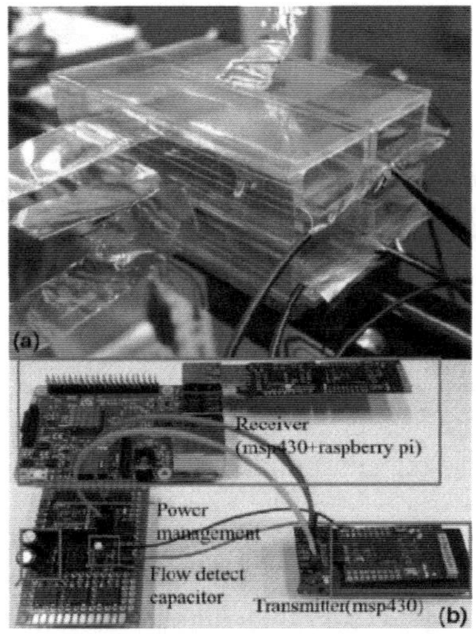

Figure 2. (a) Four TENGs are used in our experiment, (b) wireless sensor node with power management circuit, flow detection capacitor and MSP430 chip are shown in the bottom, server (raspberry pi) plus a receiver are shown on top.

the capacitor, and releases the stored energy when it is enough to power the RF chip for one cycle. The design and fabrication process of the enhanced triboelectric nanogenerator (E-TriG) was previously reported in [7]. Figure 1 shows the general principle of the E-TriG. Although various studies of using triboelectric nanogenerator as power supply and/or flow sensor have been conducted [8, 10-11], details about how flow speed is detected have not been revealed. Different from traditional frequency analysis [8], the flow rate detection method reported here is based on the maximum voltage detection of a capacitor that is charged by the flow-driven E-TriG. For temperature sensing, an embedded sensor is integrated in the power management chip (LTC3588-1). By using two ADC channels, therefore, it can read and send both the flow sensor signal and temperature sensor data simultaneously. Although four E-TriGs are used here to power the sensor node more efficiently, it is also possible to use fewer devices except that a longer charging time would be required.

Full experiment setup is shown in Figure 2. Four E-TriGs were used to charge for storage capacitor, power management accumulated the energy until it reached a certain voltage. Then power management would release part of the energy for sensing process. Two MSP430s with ultra-low power consumption were used as transmitter and receiver Micro Controller Units (MCUs), respectively. The transmitter can read and send both the flow signal and the temperature data at the same time. And the receiver would acquire the signal on the other end, then process the data and upload the information to a Raspberry Pi. Using the Raspberry Pi, we have developed a monitoring system, which can allow us to check the information of temperature and flow rate through the internet. The corresponding circuit diagram is shown in Figure 3. During our experiment, four E-TriGs were used which gave eight outputs in all. One of the outputs is used as a flow sensor which is connected with Ch1-flow capacitor (4.7 μF). Figure 3(a) takes one of the four generators as an example, where one output is linked to Ch1-flow capacitor before supplying energy to power management module. The remaining seven outputs are directly connected with the energy management chip LTC3588-1, which is shown in Figure 3(c). Through the power management, energy is saved into a storage capacitor (430 μF) until the voltage of the capacitor reaches a pre-defined threshold value. When this condition is fulfilled, a trigger signal is generated which will release the energy for MCU to start the sensing process described above. Figure 3(b) illustrates the simplified model of flow speed sensor. It

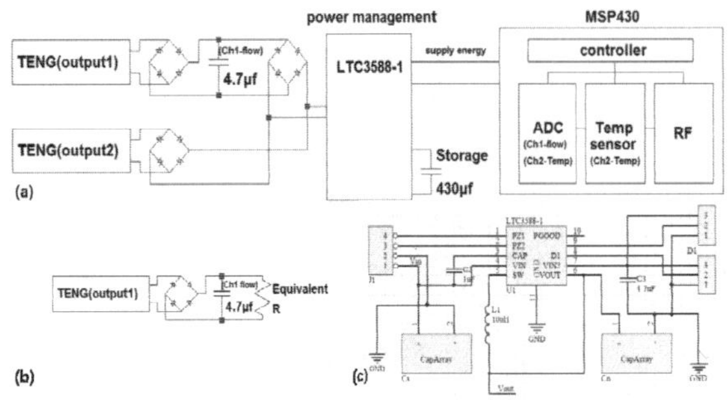

Figure 3. (a) Diagram of V-max method, (b) Equivalent resistance in the two ends of Ch1-flow capacitor, (c) Power management circuit.

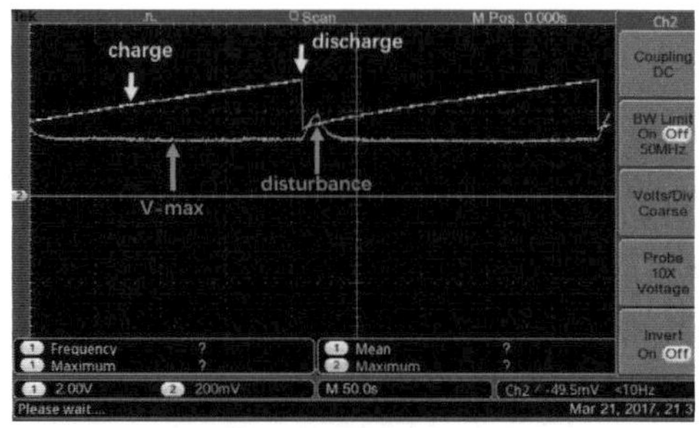

Figure 4. Voltage curve of storage capacitor (yellow line) and flow detection capacitor (blue line).

mainly consists of a rectifier bridge and a Ch1-flow capacitor, while all the rest part of the circuit can be seen as an equivalent load resistor R.

3. Discussion and experiment result

Figure 4 demonstrates the charge and discharge cycles during operation. When the storage capacitor (yellow line) starts charging, the voltage of flow detection capacitor (blue line) remains stable. This voltage is what we called V_{max}. Once the energy stored in that capacitor is enough to power MCU and data processing, MCU can read the voltage of flow detection capacitor (V_{max}). The ADC process only takes less than 35 milliseconds so that any disturbance, which might due to the variation of equivalent resistance between MCU input ports, can be ignored. Theoretical study about the relationship between air-flow rate and maximum power output of TENGs has been conducted in [12], which gave the formula below:

$$\frac{V^2}{R} = P_{max} = K \times U^3 \qquad (1)$$

Here, U is the flow rate, and K is a coefficient related to the material, structure and size of nanogenerator. P_{max} is maximum power output of TENG when TENG works in the steady state at a certain flow speed. The voltage V across the capacitor of equivalent circuit remains unchanged at steady state when the generated power equals to the consumed power.

According to the analysis above, we may use

$$V = A \times (U^{\frac{3}{2}}) + B \qquad (2)$$

to fit the test results. Moreover, due to the noise from measurement, a constant B is added to the data fitting model.

Quadratic relationship, therefore, can be achieved between maximum voltage and flow rate, which has been presented in Figure 5. According to the model for generated power, the test results from a few cycles were fit as the red line and the standard errors were also shown in the Figure 5. For faster calculation and data fitting from the MCU at lower power consumption, a linear model can also be used instead of Equation 2. The dash line in Fig.5 shows the fitting result with acceptable accuracy, especially for high flow rate range, as shown in the inset table bottom-right. The prototype application of this wireless sensor node is shown in Figure 6. This sensor node can send out 12 signals per an hour. According to the measurement, the temperature data fluctuates in a normal range at 24.2 °C (±0.3 °C). When we gradually moved the wireless sensor node towards the air blower, the flow rate was increased from 9.7 m/s to 13.2 m/s during half an hour, which agrees well with the previous results shown in Figure 5.

4. Conclusion

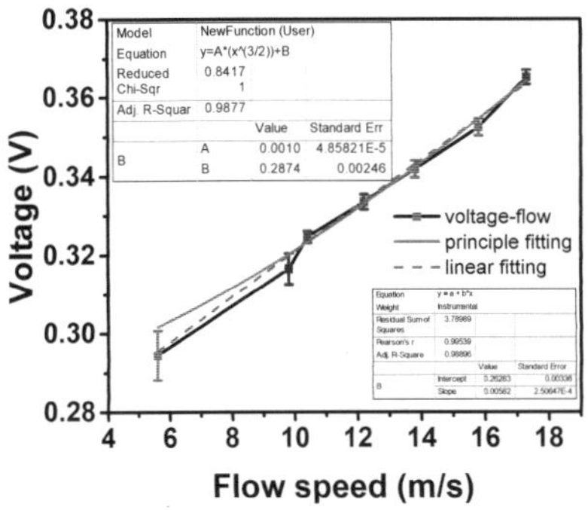

Figure 5. flow sensor Result of V-max method.

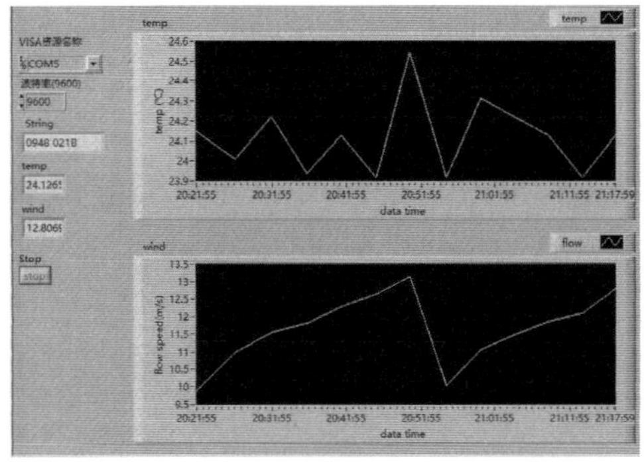

Figure 6. flow speed data and temperature data shown in upper computer monitor.

This paper presents a self-powered wireless sensor node with both temperature and flow rate detection. With electret enhanced triboelectric nanogenerator, the quantitative relationship between air-flow rate and maximum voltage of charged capacitor has been studied, with implementation details, power consumption, preparation time, and accuracy evaluation. With the self-power supply, this wireless sensor node could be promising for large scale network application in smart building and smart city.

Acknowledgement

This work was supported by National Natural Science Foundation of China (Project No.: 51505209) and Shenzhen Science and Technology Innovation Committee (Projects Nos.: JCYJ20170412154426330 and KQTD2015071710313656). Fei Wang is also supported by Guangdong Natural Science Funds (Project Nos.: 2015A030313812 and 2016A030306042).

References

[1] S. Sudevalayam and P. Kulkarni, IEEE Communications Surveys Tutorials, vol. 13, no. 3, pp. 443-461, 2011.
[2] C.Y. Sue, N.C. Tsai, Applied Energy, vol. 93, pp. 390-403, May 2012.
[3] M. Hyland, H. Hunter, J. Liu, E. Veety, and D. Vashaee, *Applied Energy*, vol. 182, pp. 518-524, 2016.
[4] P. D. Mitcheson, E. M. Yeatman, G. K. Rao, A. S. Holmes, and T. C. Green, Proc. IEEE 96(9), 1457–1486 (2008).
[5] F. Wang and O. Hansen, Sensors and Actuators A: Physical, vol. 211, pp. 131-137, March 2014.
[6] S. Wang, X. Mu, Y. Yang, C. Sun, A. Y. Gu, and Z. L. Wang, *Advanced Materials*, vol. 27, pp. 240-248, Jan 14 2015.
[7] Y. Wu, et.al, *in Transducers 2017*, Kaohsiung, June 18-22, 2017.
[8] Wang, Shuhua, et al. Advanced Materials 27.2 (2015): 240-248.
[9] Y. Hu, et.al, *in IEEE NEMS 2017*, Los Angeles, April 9-12, 2017.
[10] Wang, Shuhua, et al. "Elasto-aerodynamics-driven triboelectric nanogenerator for scavenging air-flow energy." ACS nano 9.10 (2015): 9554-9563.
[11] Su, Yuanjie, et al. Journal of Physics D: Applied Physics 49.21 (2016): 215601.
[12] Perez M, Boisseau S, Gasnier P, et al. An electret-based aeroelastic flutter energy harvester[J]. Smart materials and structures, 2015, 24(3): 035004.

A human locomotion driven hybrid energy harvester for wrist wearable applications

P Maharjan, J W Kim, J Y Kim, and J Y Park[1]
Micro/Nano Devices and Packaging Lab., Dept. of Electronic Engineering, Kwangwoon University, 447-1 Wolgye-dong, Nowon-gu, Seoul, 139-701 Korea.

E-mail: jaepark@kw.ac.kr (J Y Park)

Abstract. In this work, a human locomotion driven hybrid energy generator has been proposed and experimentally validated for scavenging significant biomechanical energy from natural motion of arm swinging, to be used for wrist-wearable smart device applications. A proposed hybrid electromagnetic (EM) and triboelectric (TE) generator can deliver much higher output power than that of the individual harvesting device unit due to the combinational operation of EM and TE mechanisms under the same applied mechanical movements. A novel curved structural design based energy harvester is proposed from the comparative analysis of human hand swinging motion with double pendulum motion, which helps to enhance the output performance and design it fit to wear on wrist. The swinging of human arm during walking and running initiates a freely movable magnetic metal ball to oscillate inside the curved structure through both EM and TE part. The EM part consists of two coils wrapped around the curved body, one at each end and micro-structured Polydimethylsiloxane (PDMS) film is attached on the inner wall of curved body along with copper electrode for TE part. A prototype of the hybrid harvester has been fabricated and tested. The EM and TE devices are capable of delivering maximum 84.37mW peak power at 49.2Ω matching load resistance and 1.86μW peak power at 10MΩ load resistance respectively which are considerably higher than the trending state-of-the-art devices.

1. Introduction

Wearable energy harvesting has been the subject of scientific and technological effort globally and has high demand in fitness and body health condition monitoring devices because of its prolonged life time and less maintenance requirement. For wrist-wearable devices, converting the biomechanical energy wasted during human locomotion is more efficient but challenging, as the device shape and size are matters in practical applications. Different harvesters have been developed to scavenge the ambient energy based on electromagnetic [1,2], triboelectric [3] and piezoelectric [4] mechanism. Recently, lots of reports have been demonstrated on harvesting energy from human body motion induced vibrations but most of them are not feasible to use as wearable device, which is our area of interest in this work [5-7]. Additionally, the human motion comprises very low frequencies (2-5Hz) based on the hand motion during normal walking, running and other daily activities. Human body spends less metabolic cost of stabilization in swinging arm during walking in comparison to walking without swinging arm [8]. Since, the act of swinging arm is natural act of human, it can be taken as a good source for converting this biomechanical energy into electrical power. An unintentionally driven

[1] To whom any correspondence should be addressed.

energy harvester is more challenging to develop as it's hard to maintain the same output power level for non-resonant vibrational harvester [7]. To address the above issues, a new structural geometry is proposed to harvest effectively from human locomotion, which works in non-resonant environment. The combination of two harvesting units enables to extract more electricity from single mechanical motion, which might fulfil the power requirement of wearable devices [1-2]. In this report, we mainly focused on designing the best structure for wearable harvester device. In the meantime, the hybrid mechanism on single applied impact has been studied to increase the overall power performance.

2. Hybrid energy harvester design and fabrication

In this work, we present a hybrid energy harvester driven by human locomotion, based on electromagnetic (EM) and triboelectric (TE) mechanism. During the natural motion activities like walking and running, the human arm swings to-and-fro which creates mechanical impact on harvester and hence the mechanical energy is converted into electrical power. The human gait analysis helps to understand the trajectory of human arm swing []. Figure 1 illustrates the human gait cycle for arm during running. This study shows that the curve made by the arm during walking and running is different from just a half circle curve similar by a simple pendulum.

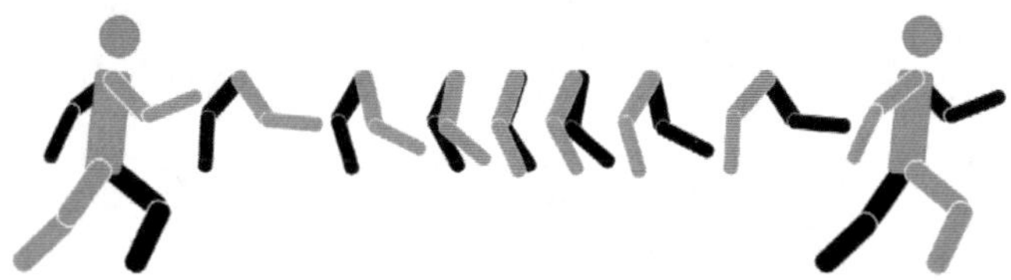

Figure 1. Human gait cycle showing arm swinging motion

During the structural design of the harvester, along with the consideration of this arm motion, a hollow curved structure has been fabricated via 3D printing with acrylonitrile butadiene styrene (ABS) polymer (Figure 2). It has a dimension of 100 mm length, 11 mm of diameter and 1mm thick wall. The EM generator consists of two coils (each 300 turns) connected in series wrapped around the curved body and a spherical NdFeB magnet of dimension Ø10 mm. Figure 3 shows the finite element simulation (FEM) of magnetic flux density and flux lines which helps to understand the variation of flux strength during the swinging as well as to determine the least possible distance between magnet and coil so that maximum flux linkage occurs.

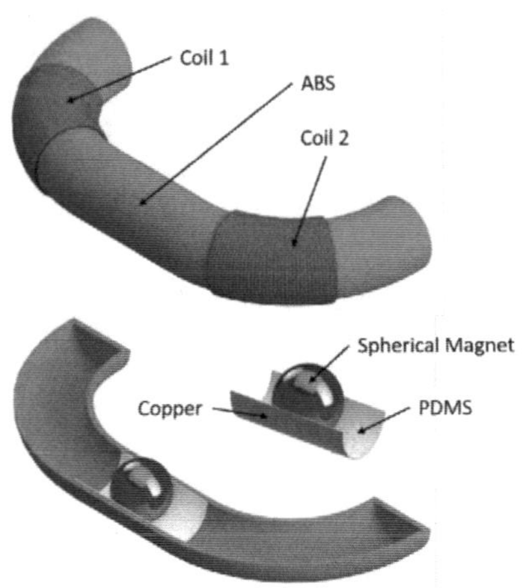

Figure 2. Schematic design of proposed hybrid energy harvester.

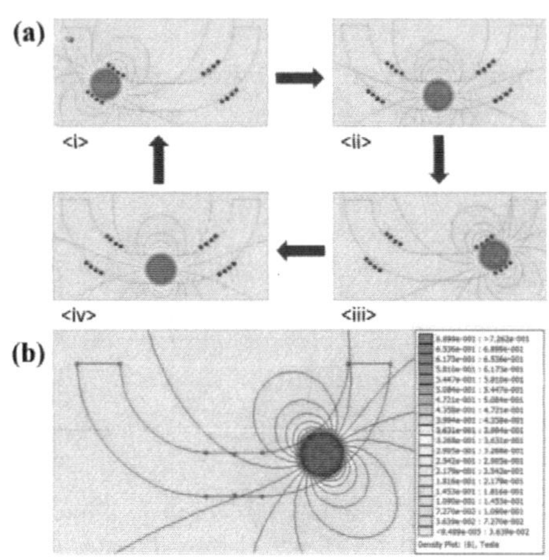

Figure 3 a-b. FEM analysis of magnetic flux density and stepwise illustration of working mechanism of harvester.

Similarly, TE generator consists of microstructure PDMS film as triboelectric material and copper film as electrode. The micro-structured PDMS film was fabricated by using sand paper as a template. First, a sandpaper with grit size of ISO/FEPA: P-240 was attached on Si wafer using double-side adhesive tape. The mixture of PDMS elastomer and curing agent in a ratio of 10:1 (w/w) was spin-coated over the sandpaper attached wafer and then cured at 70° C for 2 hours in an oven. The peel-offed micro-structured PDMS film was then attached on the inner wall of harvester tube using double-side conductive adhesive tape. Figure 4 shows the SEM image fabricated microstructure on PDMS. Here the metal magnetic ball and PDMS act as triboelectric material pair in order to generate power from rolling electrification and electrostatic induction. When a metal magnetic ball comes into contact with PDMS during rolling along it, there is flow of electrons between the Cu electrode and the ground. because of the swing motion of arm. As TE is designed as single electrode based structure, the contact element requires no electric contact which ultimately simplifies the device design and fabrication. Figure 5 illustrates the fabricated harvester device and the device being worn on wrist as wearable energy harvester.

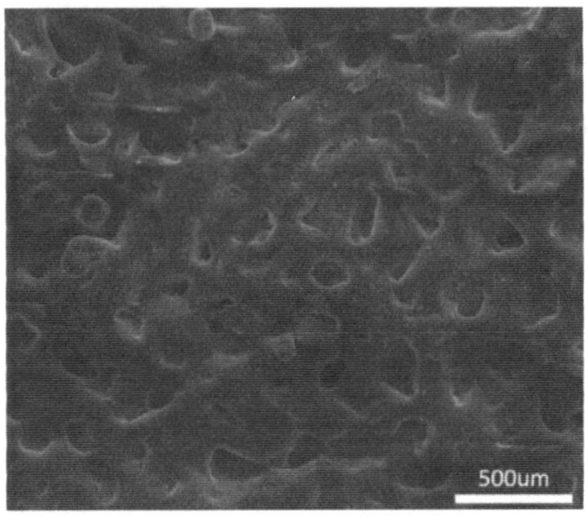

Figure 4. Scanning electron microscope image of micro-structured PDMS.

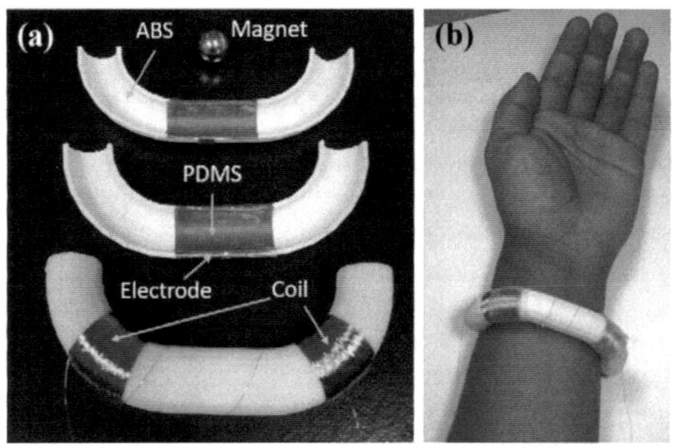

Figure 5. (a) Photograph of fabricated prototype harvester device along with pre-assembled parts and (b) harvester device being worn on wrist.

3. Experimental results and discussions

A prototype hybrid energy harvester device was fabricated which is capable of generating power from human arm-motion induced vibrations. All the components were assembled to build the prototype hybrid energy harvester and tested by wearing on wrist during walking and running. A low noise electrometer (Keithley 6514) was used to measure the output voltage signals of TE and oscilloscope was used to measure output voltage from EM generator. During the test, it shows that the applied human swing frequency varies from 1-2Hz which in turn impact on the magnetic ball to swing through the curve body in the up-converted frequency of 4-5Hz. This is because, the magnetic ball increases its acceleration due to gravity when it moves from high slope to lower and it also shows the similar phenomenon of pendulum oscillatory motion. Figure 6 shows the instantaneous voltage waveform of EM and TE generator across the optimum load and 4.82V and 5.3V are the peak voltages of EM and TE generator respectively. The hybrid harvester delivers 84.37mW peak power from EM at 49.2Ω matching load resistance and 1.86μW peak power from TE generator at 10MΩ load resistance. Figure 7 shows the peak voltages and peak power across the matching load resistance. Here, the EM has low output voltage and high current while TE has high voltage and low current. Thus, combining both mechanism helps to improve the overall output power.

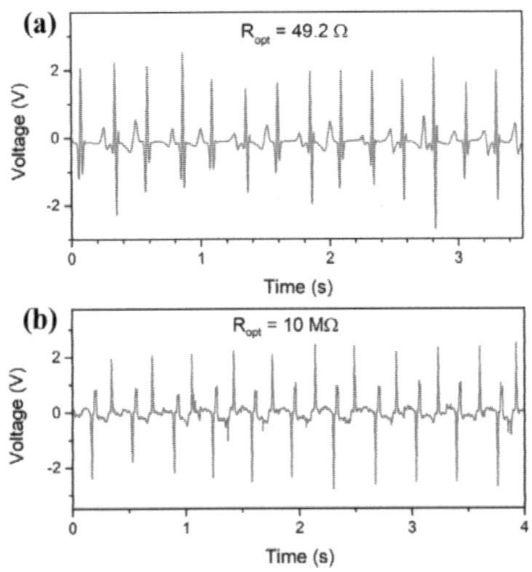

Figure 6. Instantaneous voltage waveform across optimum load resistance: (a) electromagnetic and (b) triboelectric generator.

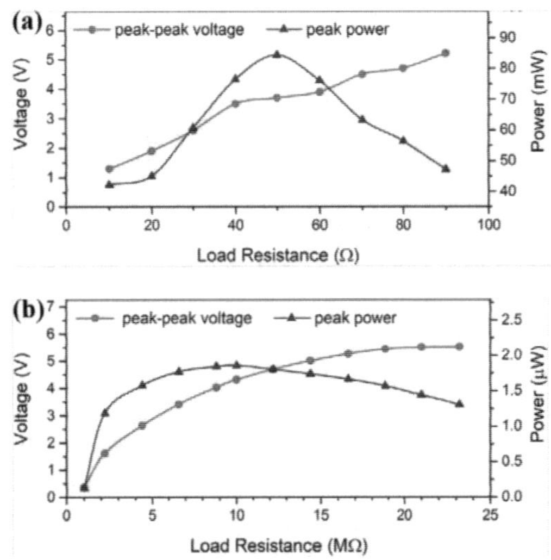

Figure 7. Peak-peak voltages and peak power across load resistance generated by (a) electromagnetic and (b) triboelectric generator.

4. Conclusions

A human locomotion based hybrid energy harvester was designed, fabricated and tested for harvesting electrical energy from biomechanical energy in order to power the wearable application such as fitness band, smart watch and other health monitoring sensors and electronic devices. The EM generates low voltage but high current while TE generates high voltage and low current. The combination of two mechanism to build a hybrid harvester helps to minimize the energy loss. Additionally, the implementation of microstructure on PDMS helps to increase the contact area between the magnetic ball and PDMS which helps to increase the output TE voltage. The EM generator is capable of delivering maximum 84.37mW of peak power across the optimum load resistance of 49.2Ω and the triboelectric generator delivers 1.86µW of peak power at 10MΩ matching load resistance. The main objective of the work was to design, develop and implement a feasible energy harvester to scavenge significant power from normal human activities like walking and running, specially from the swinging motion of arm, in order to power the wrist-wearable smart devices.

Acknowledgements

This research was supported by the Bio & Medical Technology Development Program of the NRF funded by the Korean government, MSIP(2017M3A9F1031270). The authors are grateful to MiNDaP (Micro/Nano Devices and Packaging Lab.) group members of Kwangwoon University.

References

[1] Salauddin M and Park J Y, 2017 *Smart Mater. Struct.* **26** 035011
[2] Halim M A, H. Cho and Park J Y, 2015 *Energ Convers Manage* **106** 393-404
[3] Rasel M S and Park J Y, 2017 *Appl. Energy* **206** 150-158
[4] Toyabur R M, Salauddin M and Park J Y, 2017 *Ceram. Int.* **43** 675-681
[5] Samad F A and Karim M F, 2016 *IEEE Sens. J.* **16** 1969-1974
[6] Quan T et al, 2015 *ACS Nano* **9** 12301-12310
[7] Hamid R and Yuce M R, 2017 *Sens. Actuator A-Phys.* **257** 198-207
[8] Ortega J D, Fehlman L A and Farley C T, 2008 *J. Biomech.* **41** 3303-3308.

THE PIEZOELECTRIC PZT THIN FILMS DEPOSITED ON METAL SUBSTRATES

T Ito, T Nishi, T Umegaki, H Hida and I Kanno

Department of Mechanical Engineering, Kobe University, 1-1 Rokkodai-cho, Nada, Kobe, Japan

E-mail: kanno@mech.kobe-u.ac.jp

Abstract. In this study, we deposited PZT thin films directly on various kinds of metal substrates by rf-magnetron sputtering, and measured their crystal structure, and ferroelectric and piezoelectric properties. We used metal substrates of austenitic stainless steel (ASS), ferritic stainless steel (FSS), Ti, Ni, Zr and W for the PZT deposition. From the x-ray diffraction (XRD) measurements, we confirmed the PZT thin film was directly grown on each substrate with a perovskite phase. The dielectric constant of the PZT thin films showed obvious dependence on the coefficient of thermal expansion (CTE) of substrates. This result suggests that the dielectric constant, which is one of the important parameters for the piezoelectric vibration energy harvesters (PVEHs), can be controlled by the CTE of the substrates. The PZT film deposited on the FSS showed the best piezoelectric coefficient $e_{31,f}$ of -2.4 [C/m^2], which was comparable to that of the PZT film formed on the platinum-coated metal substrates.

1. Introduction

Recently, many researches have reported piezoelectric thin-film devices to realize new functional micro-devices beyond the limitation of conventional MEMS. The PZT thin films have been the most popular piezoelectric thin films among piezoelectric MEMS including piezoelectric vibration energy harvesters (PVEHs). The PZT thin films are generally deposited on Pt-coated Si mainly because of the compatibility with MEMS fabrication process. On the other hand, from the viewpoint of the application to PVEHs, metal substrates are much more preferable than Si because of not only higher toughness for the large and long-term vibration, but also low cost. However, the deposition of Pt bottom electrodes on metal substrates causes the longer deposition process as well as increase the device cost. Therefore, direct deposition of the PZT thin films on metal substrates are preferable for the practical application of PVEHs. However, systematic study on the dependence of kinds of substrates has not been reported so far. In this study, we prepared the PZT thin films directly on a variety of metal substrates by rf-magnetron sputtering and evaluated their piezoelectric properties for the applications to PVEHs.

2. Deposition of PZT thin films

We prepared 2.5 μm-thick PZT thin films on the both sides of 50 μm-thick metal foils. We selected six different metal substrates; austenitic stainless steel (ASS), ferritic stainless steel (FSS), Ti, Ni, Zr and W for PZT depositions. The PZT thin films were deposited at the substrate temperature of around 600°C. In order to complete the crystallization to perovskite phase, post annealing was conducted for an hour at 650°C. Subsequently, the Pt top electrode was prepared. Crystalline structure of the PZT thin film on

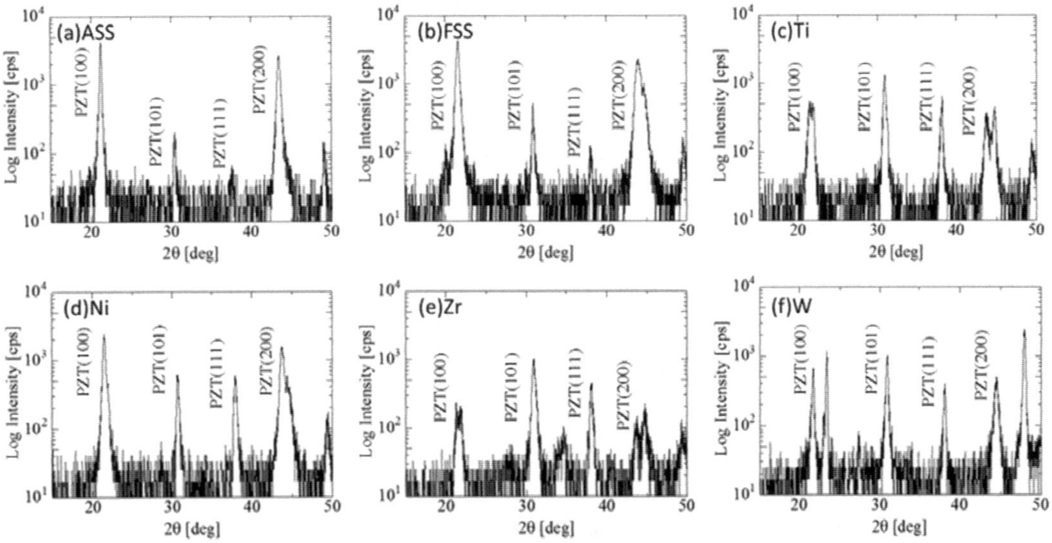

Figure 1. XRD patterns of PZT thin films on (a)ASS, (b)FSS, (c)Ti, (d)Ni, (e)Zr and (f)W substrates.

each metal substrate was examined by x-ray diffraction (XRD) measurement. Fig. 1 shows the XRD patterns of PZT thin films directly deposited on metal substrates. The measurement results indicated that polycrystalline PZT thin films with perovskite structure were successfully prepared on each metal substrate without Pt bottom electrode. Furthermore, the PZT thin films on ASS and FSS substrates were strongly oriented along c axis.

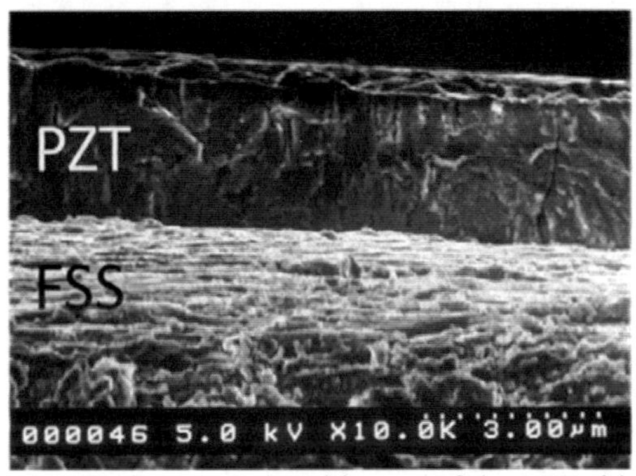

Figure 2. Cross-sectional image of the PZT thin film on ferritic stainless steel substrate.

Fig. 2 shows the cross-sectional image of the PZT thin film on FSS substrate. We confirmed the dense PZT thin films were grown on the FSS metal foils. Similar cross-sectional structures were observed in the PZT thin films on the other metal substrates.

3. Electric characteristics

3.1 Ferroelectric and dielectric properties

We measured ferroelectric properties of the PZT thin films using a Sawyer-Tower circuit and P-E hysteresis curves are shown in Fig. 3. The PZT thin films showed good ferroelectricity except the PZT on Zr substrates. We measured dielectric properties and results are listed in Table 1. Relative dielectric constant ranges from 180 to 421. Dielectric loss of the PZT thin films on the W substrate was as high as 34.8 %, while that of the PZT thin films on the other substrate was lower than 10 %.

The relationship between dielectric constant and coefficient of thermal expansion (CTE) is plotted in Fig. 4. The CTE of PZT thin film is 6.7 ppm/K [1], therefore the PZT thin films would have compressive stress except that on W substrates. Relative dielectric constant of the PZT thin film on the Ti substrates showed the highest values, probably because the CTE of Ti is close to that of PZT and compressive

Table 1. Properties of PZT thin films on metal substrates.

	ASS	FSS	Ti	Ni	Zr	W
Capacitance[pF]	147.8	244.5	294.5	156.7	unmeasurable	130.8
Dielectric loss	0.102	0.104	0.116	0.036		0.348
Area of electrode [mm^2]	0.218	0.198	0.197	0.224		0.206
Relative dielectric constant	192	348	421	197		180
CTE [ppm/K]	17.3	10.4	8.5	13.3	5.9	4.3

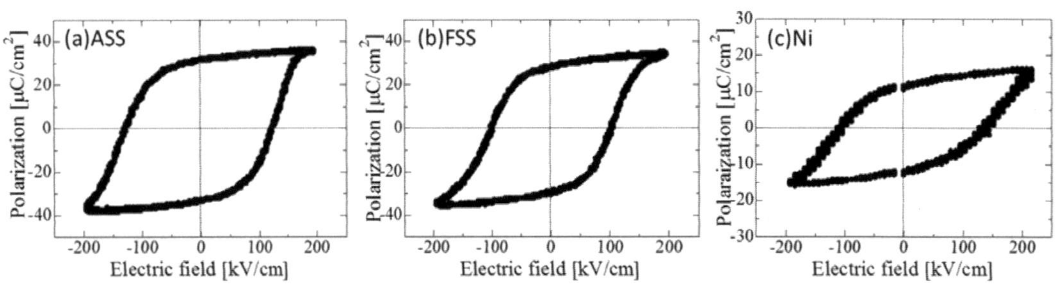

Figure 3. *P-E* hysteresis loops of PZT thin films on (a)ASS, (b)FSS and (c)Ni substrate.

stress of the PZT would be the lowest among them. These results suggest that the dielectric properties can be adjusted by the internal stress of the PZT thin films from the metal substrates.

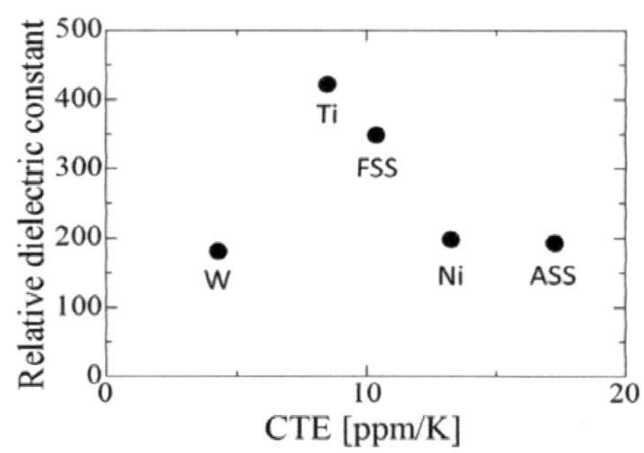

Figure 4. Relationship between dielectric constant and coefficient of thermal expansion.

3.2 Piezoelectric property

We fabricated 13 mm long bimorph cantilevers of PZT films on ASS, FSS and Ni metal substrates and evaluated the piezoelectric coefficient $e_{31,f}$ from the tip displacement by applying voltage to one side of the PZT layers as unimorph cantilevers. The tip displacement of the cantilevers as a function of applied voltage is shown in Fig. 5. The piezoelectric constant $e_{31,f}$ were calculated from the following equation [2].

$$e_{31,f} = -\left[\frac{h_s^3}{6}\left(\frac{Y_s}{1-\nu_s}-\frac{Y_p}{1-\nu_p}\right)\cdot\frac{\left(Y_p h_p + Y_s h_s\right)}{2Y_p h_p^{\ 3} + \left(Y_s+2Y_p\right)h_p^2 h_s + Y_s h_p h_s^{\ 2}} + \frac{1}{s_{11}^E + s_{12}^E}\left(\frac{h_s+h_p}{2}\right)\right]\cdot\left(\frac{h_p}{V}\right)\cdot\frac{2\delta}{x(2l-x)} \qquad (1)$$

where h_p, v_p and Y_p are the thickness, Poisson's ratio and Young's modulus of piezoelectric layers; h_s, v_s and Y_s are the thickness, Poisson's ratio and Young's modulus of a metal substrate; s_{11}^E and s_{12}^E are elastic compliance of piezoelectric layers; x, l, δ and V are the length of the top electrode, the length of cantilever, applied displacement to the cantilever and output electric voltage, respectively. The piezoelectric constant $e_{31,f}$ of PZT films on ASS, FSS and Ni substrates were calculated to be about -1.5, -2.4, -0.67 C/m^2, respectively. Those values are almost comparable with those of PZT thin films deposited on Pt-coated metal substrates [3].

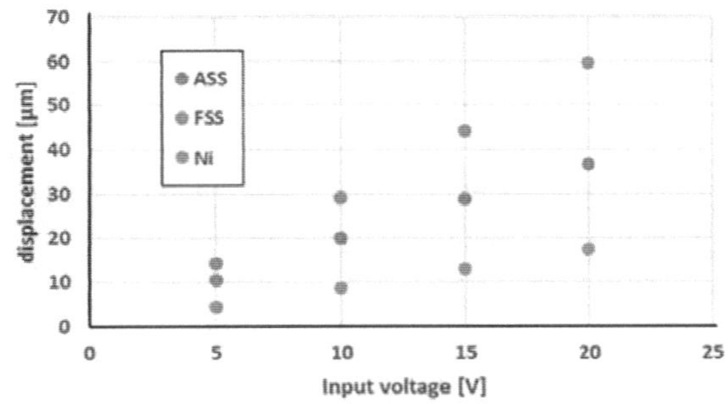

Figure 5. The output displacement plotted as a function of input voltage.

4. Conclusions

In this study, we measured crystalline orientation, ferroelectric and electrical properties of PZT thin films which were directly deposited on metal-foil substrates. We successfully prepared the high-quality PZT thin films directly deposited on a variety of metal foils. We selected six kind of metal substrates; austenitic stainless steel (ASS), ferritic stainless steel (FSS), Ti, Ni, Zr and W for PZT depositions. The PZT thin films showed good ferroelectricity except that on Zr substrates. The relative dielectric constant of PZT thin films was ranged from 180 to 421, and it strongly depended on the CTE of substrates. The PZT film deposited on the FSS showed the best piezoelectric coefficient $e_{31,f}$ of -2.4 C/m^2, which was comparable to that of the PZT film formed on the platinum-coated metal substrates.

References

[1] Hong E, Smith R, Krishnaswamy S V, Freidhoff C B. and Trolier-McKinstry S 2006 *Thin Solid Films* **510** 213
[2] Hida H, Hamamura T, Nishi T and Tan G 2017 *Jpn J Appl Phys* **56** 10PF08
[3] Nishi T, Ito T, Hida H and Kanno I 2016 *J Phys Conf. Ser.* **773** 012062

Acknowledgments

The authors acknowledge the financial support of Japan Science and Technology Agency (JST) A-STEP (AS272S002a) and CREST (JPMJCR16Q4).

IOP Conf. Series: Journal of Physics: Conf. Series **1052** (2018) 012095 doi:10.1088/1742-6596/1052/1/012095

10b piezoelectric energy harvesting devices and system
(Please choose category from list)

VERIFICATION OF SELF-TUNING 4DOF PIEZOELECTRIC ENERGY HARVESTER WITH ENHANCED BANDWIDTH

L.G.H Staaf, E. Köhler, A. D. Smith, P.D Folkow and P. Enoksson
Chalmers University of Technology, Sweden
E-mail: v96staaf@chalmers.se

Abstract
In this paper, we present an analytical model to predict enhanced bandwidth for a piezoelectric energy harvester with self-tuning, accomplished by a sliding mass. The model predicts that by implementing asymmetry of different piezoelectric cantilever lengths , the bandwidth can theoretically approach 60 Hz. Validation measurements demonstrate an increased 3dB bandwidth up to 21 Hz with 150 mW, by configuration 23/17 mm in open length – providing sufficient power for a ZigBee to continually transmit.

Introduction

Over the last decade, interest for autonomous intelligent wireless sensors (AIWS) has increased tremendously. Accompanying that expanding interest, investigation into powering these sensors has been the subject of increased attention [1-3]. Despite significant research effort, the biggest remaining challenge is substituting a conventional battery solution with a power source that does not need to be replaced. An alternative power source would be to use an energy harvester. To be able to replace a conventional battery solution, an energy harvester needs to be able to deliver sufficient power. Depending on the intended use of the AIWS and its corresponding surroundings, different energy harvesters can be used (such as photovoltaic, thermal and kinetic). In our case, we mainly have a vibration source and therefore a piezoelectric energy harvester solution (kinetic) is chosen. The main problem for a piezoelectric energy harvester is the bandwidth, which is usually quite narrow and makes the harvester ineffective in most real-life applications. To counter this problem, previous works implement an array of cantilevers, with the negative side of increase in size, which grow dramatically for each cantilever added to the harvester [4]. Another solution is to utilize coupled cantilevers in a 2DOF solution where the bandwidth is somewhat broadened [5-6].

In this paper we present a self-tuning piezoelectric energy harvester with enhanced bandwidth that is achieved by a sliding mass. The obtained enhanced bandwidth is based on analytical calculation. The concept for this 4DOF piezoelectric harvester was reported at PowerMEMS 2015 [7], where a maintained voltage output was achieved through a distributed stress over the piezoelectric cantilevers [8]. Further, the devices demonstrated that a sliding mass had a broader bandwidth than if the mass was fixed on the middle beam.

However, the harvester in this work differs from the former by being designed based on predictions of the analytical model. The model suggests that the bandwidth becomes wider by making the harvester asymmetrical via adjusting the stiffness of the piezoelectric beams. Presented is a bandwidth comparison, for symmetric and asymmetric setup, to verify the analytical prediction.

Harvester setup

The harvester contains a top piezoelectric cantilever (MIDE PPA-2014), connected to a back folded middle beam via a coupling. On the middle beam, a sliding mass is placed. The other end of the beam is connected to a bottom piezoelectric cantilever of the same type via a second coupling. The harvester is presented schematically in figure 1. The couplings and middle beam are made of aluminum. Specifications for the couplings, mass and middle beam are in table 1. The length given for the cantilevers and middle beam is the open length, which is the lengths between attachments and couplings.

Analytical model

An analytical harvester model was previously developed based on Euler-Bernoulli theory for homogeneous and isotropic beams with additional point masses [7]. The harvester model is transformed

10b piezoelectric energy harvesting devices and system
(Please choose category from list)

to a beam structure using a simple finite element (FE) model. In the present self-tuning system, the first two eigenfrequencies are found to be somewhat close to each other while the eigenfrequency for the third mode is much higher. It is thus imperative that the model adequately captures the two lowest eigenfrequencies. In figure 2, the first mode is presented. In our case, a system using only three beam elements with three-point masses was chosen, resulting in a four DOF problem. By adopting a Guyan reduction of the two nodal rotational degrees of freedom, u_d, the reduced 2×2 equations of motion system for the active nodal displacement DOF, u_a is

$$M\ddot{u}_a + Ku_a = 0. \quad (1)$$

The elements of M and K consist of polynomial expressions in terms of the geometrical and material parameters. Based on these expressions, the two lowest eigenfrequencies and the corresponding eigenmodes are expressed in a closed form involving only algebraic equations.

The analytical model gives a clear prediction that if an asymmetry for the harvester is developed, by altering the length of the piezoelectric cantilevers, the bandwidth for the harvester will be increased. The configuration of analytical model is as follows: top/bottom; 22/22, 23/21 and 24/20. . In figure 3, the difference in bandwidth between symmetric 20/20 (blue line) and 24/20 (green dashed) is presented and configuration the asymmetric 24/20 yields a theoretical bandwidth of 60 Hz

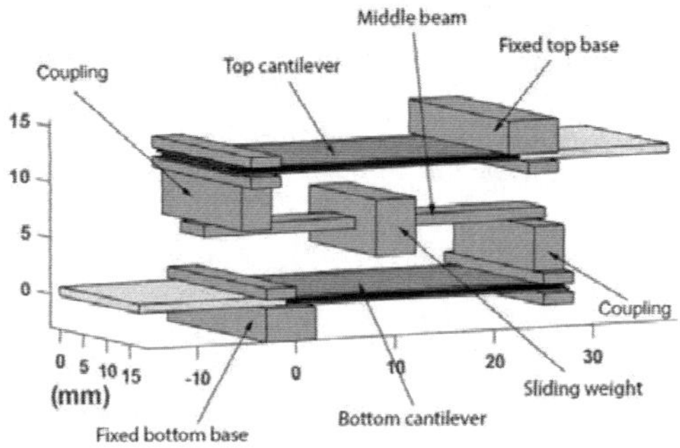

Figure 1, Schematic setup of the harvester

	Weight (g)	Thickness (mm)	Open length (mm)
Total	3.12	-	-
Coupling	1.04	-	-
Sliding mass	0.9	-	-
Middle beam	0.13	0.35	26.5

Table 1, Technical data for the aluminum parts, coupling, middle beam and sliding mass

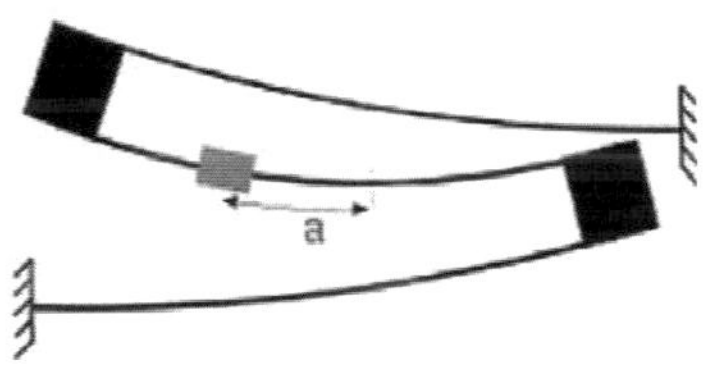

Figure 2, For mode 1, the masses move in the same direction and the position of the sliding mass can be related to the position with zero slope of the middle beam - the distance a, in our case a is close to zero

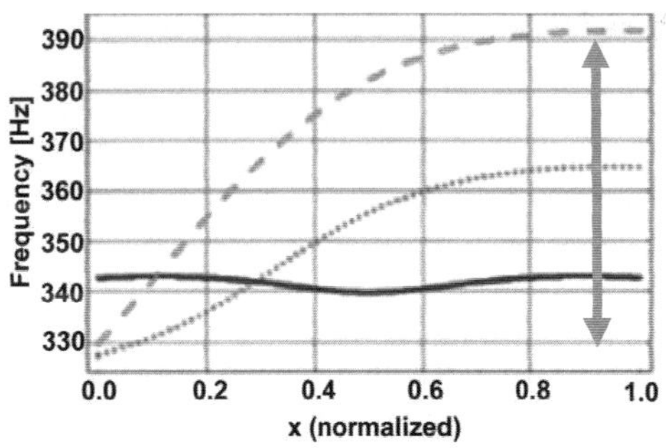

Figure 3, Eigenfrequencies for different positions x of a fixed mass. Upper/lower beam lengths are: 22/22 mm (solid blue), 23/21 mm (dotted red), 24/20 mm (dashed green). The theoretical bandwidth of 60 Hz is marked by the green arrow

Measurement setup

To validate the prediction of the analytical model, the harvester was measured with a symmetric setup and an asymmetric setup. The symmetry chosen is 20/20 and asymmetry 23/17. The length from the model 24/20 could not be achieved due to width restrictions of the coupling. When attached at the outer end, the open length was measured 23 mm with the used the piezoelectric cantilever. The total piezoelectric area is equal for both setups.

In figure 4, the measurement equipment is presented where A) the NI-USB 6210 is used to collect

10b piezoelectric energy harvesting devices and system
(Please choose category from list)

data, B) The harvester on shaker table, C) The tone generator which provide vibrations with an applied frequency between 100 – 300 Hz and amplitude of 1g.

Measurement

Before data collection, we swept the frequency several times to mitigate hysteresis in the measurement. While recording, the frequency was increased by 1 Hz at and was not changed until the mass was either stable or moved to a new position on the middle beam. Changes in the mass position occurred rapidly (within 1 to 2 seconds) when the frequency was changed. The RMS Voltage output is presented for the 20/20 and 23/17 case (figure 5) demonstrating that the 20/20 has a higher top output but the 23/17 has a much broader bandwidth. The 20/20 has one peak and the mass did not change position on the middle beam during the frequency sweep while the 23/17 case has two peaks. On the right side of the graph in figure 5, we can see that 20/20 and 23/17 has the same RMS voltage output. This output similarity depends on the mass position of the middle beam (which is the same for the 20/20 and 23/17). Apart from that similarity, we have a narrow frequency range (between 190 and 193 Hz) where the two configurations also have the same output. During this period, both configurations have the mass on same position on the middle beam. Between the two peaks for 23/17 we can see that the mass position affects the harvester in a way that the output is lower than for the 20/20 system. However, in the area left of the peak the mass position for 23/17 affects the system so that it yields a higher output than for 20/20 and gives the asymmetric system a wider bandwidth. The 3db bandwidth (RMS Voltage output) for 20/20 is 11 Hz and for 23/17 it is 21 Hz. The power output for the system above the 3dB bandwidth is over 150 mW (figure 6), which is more than sufficient to power our ZigBee wireless interface.

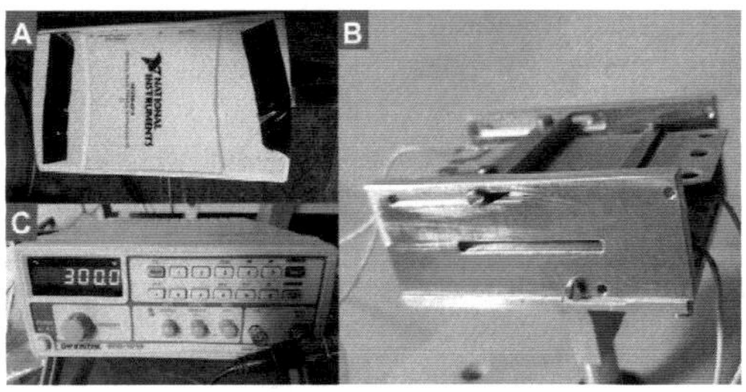

Figure 4, A) Texas Instrument USB-6210, B) The harvester on the shaker, C) Function generator SFG-1013

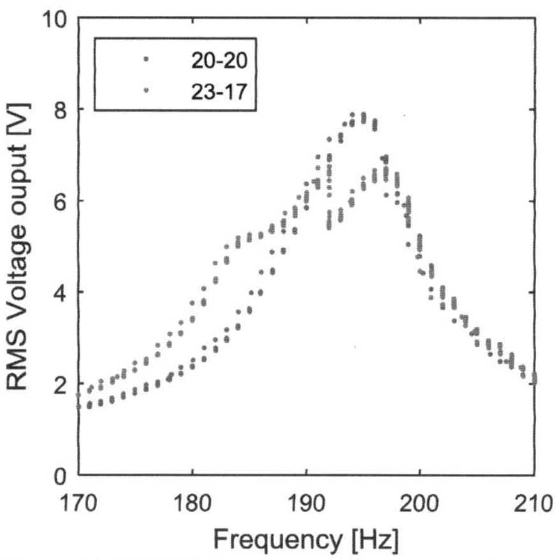

Figure 5, RMS Voltage output for symmetric 20/20 (blue) and asymmetric 23/17 (red)

Bandwidth and power output for real application

For one of our applications on a gas turbine, we provide power for a ZigBee (802.15.4). This particular ZigBee requires 8 mW during startup and 2 mW to be able to send and retrieve signals constantly. Previous measurements with this type of harvester (before the analytical model) demonstrated that we could not provide sufficient power for continuous broadcasting (as seen in figure 7). The power is one critical limit and voltage is the other. To be able to charge a supercapacitor, the voltage needs to be higher than 5.13 V

In our case both 3 dB bandwidth devices (configuration: 20/20, 23/17) are above the critical 5.13 V limit, which is required to maintain power for a ZigBee (802.15.4) equipped with a sensor and transmitting continuously [9]. In figure 6, we can see that the RMS power output has the same signal trend as the voltage output due to maintained total piezoelectric area, with 0.16 W above the critical limit. As seen in figure 5, a frequency step up converter could be used to convert current to voltage effectively achieving even wider bandwidth, with sufficient output for the ZigBee (802.15.4).

10b piezoelectric energy harvesting devices and system
(Please choose category from list)

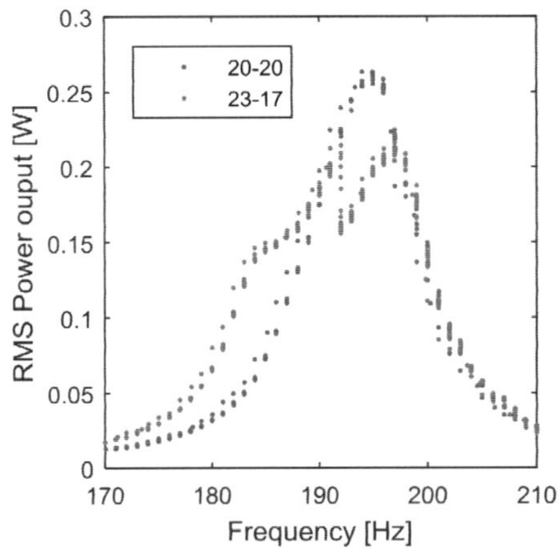

Figure 6, RMS power output for symmetric 20/20 (blue) and asymmetric 23/17 (red)

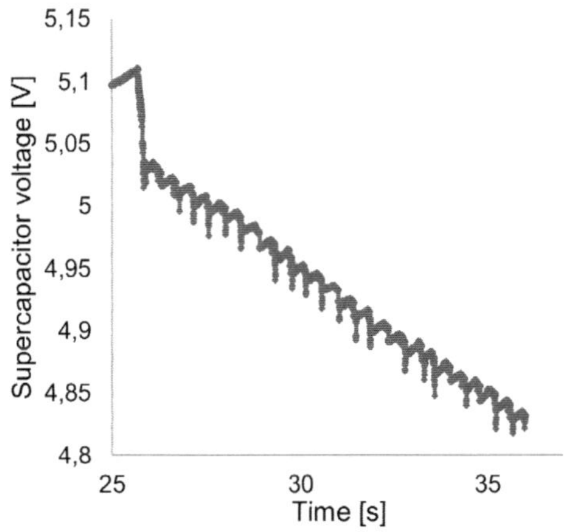

Figure 7, In the graph, measured voltage from a power supply (harvester + supercapacitor) where continuous transmitting performed by a ZigBee on an ex-service Rolls-Royce gas turbine is shown.

Conclusion

A harvester with broad bandwidth and sufficient output is one key to achieve successful autonomous intelligent wireless sensors. Our proposed enhanced harvester has a broad bandwidth of 21 Hz (3dB), which is achieved by an asymmetric setup. The output is slightly lower than for the symmetric setup, but is still above the critical limit (for the used circuitry) for both power (above 2 mW) and voltage (above 5.13 V) within its bandwidth. The prediction based on the analytical model proved to fit well with measurements, where the structure's asymmetric longitudinal change of the piezoelectric beams gives a greater bandwidth of 21 Hz, hencethe asymmetric setup demonstrates an effective method for extending the bandwidth of energy harvesters.

Acknowledgments

The Swedish Research Council, European project STARGATE, UDI energy harvesting toolkit and European project SMART Memphis is greatly acknowledged for their financial support

References

[1] Wong C, Dahari Z, Manaf A A and Miskam M A 2015 Harvesting raindrop energy with piezoelectrics: a review, *J. of Elect. Mater.* **1**, 13-21
[2] Zhen Y and Qing H 2010 A review of piezoelectric generator harvesting, *App. Mech. Mat*, **44-47**, 2945-2949
[3] Khan A, Abas Z, Kim H S and Oh I 2016 Piezoelectric thin films: an integrated review of transducers and energy harvesting, *Smart Mater. Struct.* **25**
[4] Li Y 2014 Modeling and tuning of energy harvesting device using piezoelectric cantilever array, *PhD Thesis*, West Virginia University, USA
[5] Zhou W, Reddy Penamalli G and Zuo L 2012 An efficient vibration energy harvester with a multi-mode dynamic magnifier, *Smart Mater. Struct.* **21**
[6] Wu H, Tang LH, Yang YW, Soh C K 2013 A novel two degrees-of-freedom piezoelectric energy harvester. *J. of Int. Mater. Syst. and Struct.* **24**, 357–368
[7] Staaf L G H, Köhler E, Soeiro M, Lundgren P and Enoksson P 2015 Smart design selftuning piezoelectric energy harvester intended for gas turbines *15th Int. Conf. on Micro and Nanotech. for Pow. Gen. and Energy Conv. Appl. (Boston) J. of Phys.:* Conf. Series **660**
[8] Staaf L G H, Köhler E, Parthasarathy D, Lundgren P and P Enoksson 2015 Simulation and experimental demonstration of improved efficiency in coupled piezoelectric cantilevers by extended strain distribution *Sensors and actuators A: Physical* **229**, 136-140
[9] Staaf L G H, Köhler E, Kemp J, Allen M, Zenkic S, Lindblom A, Christodoulou M, Roberts J, Lundgren P and Enoksson P 2016 Piezoelectric energy harvesting as energy source for autonomous intelligent wireless systems on gas turbines *EVI-GTI and PIWG Joint Conf. on Gas Turb. Instr. (Berlin) IEEE*

IOP Conf. Series: Journal of Physics: Conf. Series **1052** (2018) 012096 doi:10.1088/1742-6596/1052/1/012096

Using artificial gravity loaded nonlinear oscillators to harvest vibration within high g rotational systems

James Horne, Maisie M. Snowdon and Yu Jia

Department of Mechanical Engineering, University of Chester, Chester CH2 4NU, UK

E-mail: yu.jia.gb@ieee.org

Abstract. Energy harvesting within rotating environments can help to enable self-powered wireless sensing, which has been motivated in recent years by the advent of legislations mandating tyre pressure monitoring systems for automotive wheels. The centripetal acceleration ($a = \omega r^2$) within such rotational systems can attain 1,000's of g, which manifest as artificial gravity and can adversely suppresses the dynamic motion of oscillators. This paper investigates the possibility of using the high g conditions as a means of introducing nonlinear bi-stability, which can then allow an oscillator to benefit from a broadband response as well as mechanical amplification achieved from the bi-stable snap-through states. An experimental proof-of-concept prototype was designed, built and tested. By controlling the rotational speed ω of the apparatus, the masses of oscillators experienced a g-force of up to 90 g. Purely by increasing ω, an increase in transducer output was observed from the predicted amplification effect. However, beyond a certain threshold, output dropped to minimal as the potential barrier reached an insurmountable level. This work validates the proposed new mechanism that taps into the high g environment and opens a new avenue of design for vibration energy harvesting within rotational systems.

1. Introduction

Vibration energy harvesting have generally given more emphasis to translatory dynamics rather than rotational dynamics. While the core objectives in providing a self-sustaining power source for microelectronics and wireless systems are the same, the mechanism of operation are vastly different [1]. Rotational systems that do exist in the literature report on devices aimed at harvesting human movement, automotive wheels and rotor motion. In recent years, motivated by the mandated TPMS (tyre pressure monitoring system) in Europe and North America [2], there has bene a growing interest to develop energy harvesting within high speed rotational systems in order to power sensor instrumentation within these rotating environments.

Generally, linear oscillators have been applied to harvest shocks [3] or noise [4] excitation from these automotive wheel rotating systems. However, for systems with large centripetal acceleration due to either high speed or large radius ($a = \omega r^2$), a significant artificial gravity loading can manifest (1,000's of g). This can in turn statically suppress oscillatory motion and the harvester then becomes minimally responsive to ambient vibration. Therefore, the high g loading within rotational systems have previously been thought as something to avoid by positioning oscillators either parallel to the radial axis or near the centre of rotation [4]. This paper however, aims to deliberately utilise the high g artificial gravity in a specific configuration in order to introduce bi-stability and establish a design pathway towards high g induced mechanical amplification.

2. Theory

In rotational systems, centripetal acceleration results in artificial gravity loading radially outwards from the centre of rotation. A typical linear oscillator can either be stiffened or softened through axial loading [5, 6]. In the case where the oscillator behaves as a vertically upright cantilever, bi-stability can be achieved from the unstable equilibrium at the displacement $x = 0$ position [7]. With potential energy of the oscillator $U(x,t) = -0.5kx^2 + 0.25\mu x^4 - Ax\sin(\omega t)$, dynamic motion can be represented by equation 1 and axially loaded beam by equation 2.

$$m\frac{d^2x}{dt^2} + c\frac{dx}{dt} + \frac{dU(x,t)}{dx} = F(t) \tag{1}$$

$$EI\frac{\delta^4 x}{\delta y^4} + \rho B\frac{\delta^2 x}{\delta t^2} - \rho I\frac{\delta^4 x}{\delta y^2 \delta t^2} + \frac{\delta}{\delta y}[F(y,t)\frac{\delta x}{\delta y}] = 0 \tag{2}$$

where, $U(x,t)$ is the potential energy given by $-0.5kx^2 + 0.25\mu x^4 - Ax\sin(\omega t)$, x is the displacement, k is the spring stiffness, μ is the geometric nonlinearity, A is the input displacement amplitude, m is the proof mass, c is the damping, F is the driving force, E is the elastic modulus, I is the area moment of inertia, ρ is the mass density, B is the cross-sectional area of the beam, y is the axis along the beam length where $0 < y < L$, L is the beam length and t is time domain.

The terms ρB, EI and ρI denotes mass per unit length, flexural rigidity and rotatory inertia of cross-sectional area respectively. Taking into account the varying artificial gravity induced from centripetal acceleration, the forcing term $F(y,t)$ is given by $\pm\rho B(-\omega[r(x,t)]^2)(L-y)$; where, ω is the angular velocity, and r is the radial distance between the effective mass of the oscillator and the centre of rotation. This parameter itself modulates as a function of $x(t)$.

Figure 1 illustrates the modulation of the potential barrier as F varies. Instantaneous energy release can be achieved as the oscillator hops from one potential well to another. The modulation in the potential barrier helps to promote the potential barrier hopping.

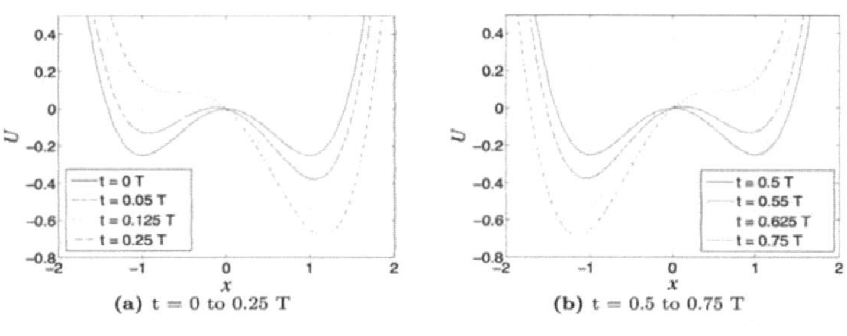

Figure 1: Periodic modulation of the bi-stability potential barrier in order to promotion intra-well hopping. U is the potential energy, x is the displacement of the oscillator, t is the time domain and T is the unit normalised time period.

3. Design and method

An embodiment of the proposed mechanism is shown in figure 2 with the cantilever tip mass radially directed towards the rotatory centre. A lab prototype is shown in figures 3 and 4.

Piezoelectric cantilevers made from PVDF, stainless steel substrate and mild steel end mass (~3 grams) were anchored onto a 3D printed frame, and attached to a rotor. The rotor system is both statically and dynamically balanced to minimise lateral vibration. Electrical connections from the piezoelectric transducer were routed out using a basic commutator mechanism, in order to electrically measure the oscillation of the cantilevers by a digital oscilloscope. The apparatus was mounted on a shaker to simulate ambient vibration, the shaker was programmed by a

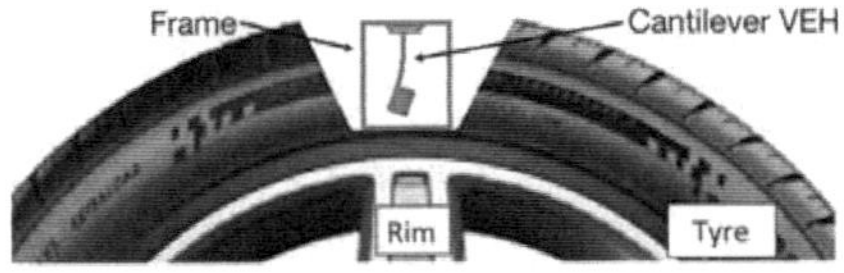

Figure 2: An embodiment of the proposed system: cantilever inversely upright (mass towards the centre) in the artificial gravitational field.

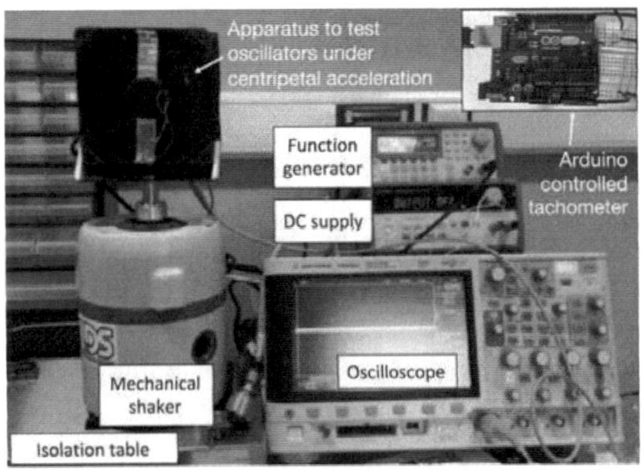

Figure 3: Experimental setup: the voltage controlled rotor with the cantilever oscillators, shaker apparatus controlled by function generator, measurement kit and an Arduino controlled tachometer.

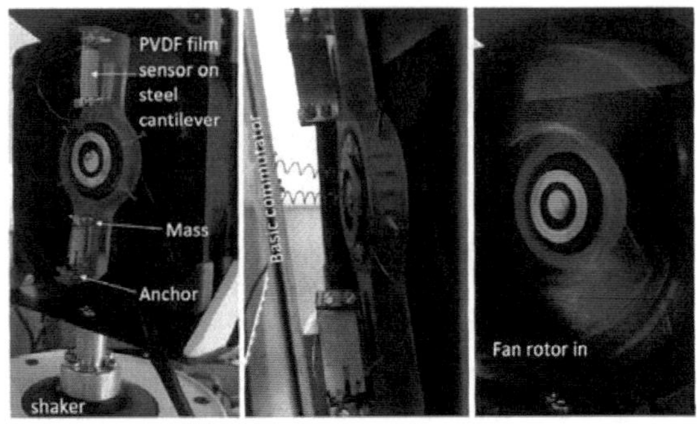

Figure 4: Experimental system in operation with the cantilever oscillators in the rotor. The piezoelectric cantilevers are wired to two metal rings at the centre of the rotor and the electrical signals are interrogated by two flexible springs serving as the commutator.

function generator and vibration monitored by an accelerometer, the rotor speed was controlled by a high precision DC supply, and the rotor speed of rotation was measured by an Arduino controlled tachometer.

4. Results and discussion

Figure 5 presents the time domain outputs and their respective FFT. The results illustrate examples of the system driven at different rotational speeds, which manifest as varying levels of artificial gravity loading on the proof masses. It can be seen that at a higher g level, response has fewer but higher frequency peaks. This is because the potential barrier becomes taller with higher g force, and the instantaneous energy release thus becomes greater once hopping takes place. However, after a certain threshold (>63 g), the response amplitude diminishes. This suggests that the potential barrier has attained an insurmountable level for the available ambient vibration.

Figure 6 summarises the average voltage output from the piezoelectric transducer at varying levels of g. An increase in output can be initially seen with larger g due to the greater energy release from crossing the higher bi-stable potential barrier. Beyond 63 g, the output returns to base levels as the bi-stable potential barrier becomes too high to cross over and the oscillator is trapped within the potential intra-well.

Conclusion and future work

This paper presents for the first time a centripetal acceleration induced bi-stable nonlinear oscillatory mechanism for vibration energy harvesting. A proof-of-concept prototype, with g-force up to 90 g, has experimentally validated the theoretical framework of utilising artificial

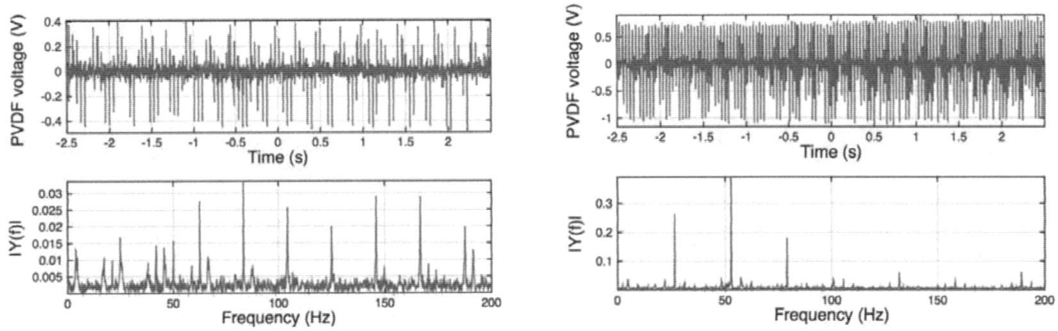

(a) 1230 RPM, 42 g (b) 1500 RPM, 63 g

Figure 5: Time domain and FFT response of the piezoelectric oscillator at various rotational speeds, resulting in varying levels of artificial gravity experienced by the proof mass.

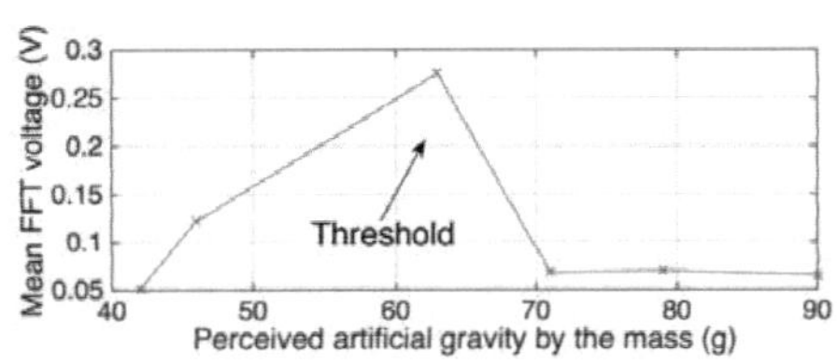

Figure 6: With higher gravity (g), the response increases from snap-through amplification. However, beyond a threshold, gravity loading traps the oscillator in one potential intra-well.

gravity as a means to introduce modulated bi-stability. Therefore, this study opens up new avenues of oscillator design to make use of the high g environments within rotational systems as a means of mechanical amplification for VEH. Future work will involve active and passive control of this nonlinear vibratory phenomenon, as well as experimentation with higher g scenarios.

Acknowledgments

Authors wish to thank the science and engineering technical team at the University of Chester; as well as sincere gratitude to Michael S. Horne for providing additional technical support.

References

[1] D. P. Arnold, Review of microscale magnetic power generation, *Magn., IEEE Trans.*, **43**(11), 2007, 3940?3951.

[2] Bartec Auto ID, Tyre pressure monitoring system legislation, 2016. Available: `http://www.bartecautoid.com/tpms-legislation.html`

[3] Z. Wang, R. Elfrink, R. J. M. Vullers, V. Van Acht, M. Tutelaers, S. Matova, J. Oudenhoven, and R. Van Schaijk, Large power amplification of a piezoelectric energy harvester excited by random vibrations, in 26th *IEEE MEMS*, 2013, 106-109.

[4] Y. Zhang, R. Zheng, R. Kaizuka, D. Su, K. Nakano and M.P. Cartmell, Broadband vibration energy harvesting by application of stochastic resonance from rotational environments, *Eur. Phys. J.*, **224**(14-15), 2687-2801.

[5] L.N. Virgin, S.T. Santillan and D.B. Holland, Effect of gravity on the vibration of vertical cantilevers, *Mechanics Research Comm.*, **34**(3), 2007, 312-317.

[6] X.F. Li, A.Y. Tang and L.Y. Xi, Vibration of a Rayleigh cantilever beam with axial force and tip mass, *J. Construct. Steel. Res.*, **80**, 2014, 15-22.

[7] M. Kawano, Y. Zhang, R. Zheng, K. Nakano and B. Kim, Design and manufacture of perpendicular bi-stable cantilever for vibrational energy harvesting on the basis of stochastic resonance, *J. Phys. Conf. Ser.*, **660**, 2015, 012104.

Interdigitated cantilever array topology for low frequency MEMS vibration energy harvesting

Yu Jia[1,2], Emmanuelle Arroyo[1], Sijun Du[1] and Ashwin Seshia[1]

[1]Nanoscience Centre, University of Cambridge, Cambridge, CB3 0FF, UK
[2]Department of Mechanical Engineering, University of Chester, Chester CH2 4NU, UK

E-mail: yu.jia.gb@ieee.org

Abstract. Micro-fabricated vibration energy harvesters enable merits such as miniaturisation, economies of scale for manufacturing, and ease of integration with semiconductor IC technologies. However, the frequency range of ambient vibration is generally low (10's Hz to 100's Hz). Existing MEMS vibration energy harvesters that target these frequencies typically are in the centimetre scale range. This sacrifices the miniaturisation aspect as well as introducing new challenges in packaging and integration for the unconventionally large MEMS devices. This paper proposes a new interdigitated fork cantilever array topology, which allows for up to about a third reduction in resonant frequency compared to the classical cantilever topology, for the same design area and without compromising on power optimisation. Further resonant frequency reduction is also possible, but at the expense of power optimisation. This opens up design flexibility to achieve low frequency MEMS resonators that are more suitable to practically target ambient vibration, without sacrificing the aforementioned merits of MEMS technology.

1. Introduction

Vibration energy harvesting (VEH) holds the promise to realise a self-sustaining on-board power source for autonomous sensors and microelectronics. MEMS fabrication technology brings several benefits to miniaturisation, ease of integration with IC and economies of scale for manufacturing. Micro-cantilever topology has been widely employed for MEMS VEH due to its good power responsiveness, ease of design, minimal fabrication process complexity, and relatively low frequency [1, 2].

However, the dominant active frequency of ambient vibration for most applications typically range from 10's Hz to 100's Hz [3], which presents a key design challenge for MEMS resonators. Therefore, the implemented micro-cantilevers tend to measure greater than several centimetres in dimension, which also result in atypical out-of-plane travel for the MEMS oscillators that can attain $\sim$1 mm in peak displacement [4]. The unconventionally large size of the required VEH device erodes the traditional scaling and integration benefits of MEMS technologies. Furthermore, large dies with substantial vertical travel also introduces a non-trivial and expensive packaging challenge.

To address the current undesirable volumetric sacrifice to accommodate low frequency devices, this paper presents a new interdigitated fork topology for micro-cantilever arrays. For a given design area (die size), the new design is capable of up to a third reduction in the fundamental resonant frequency without sacrificing power optimisation.

IOP Conf. Series: Journal of Physics: Conf. Series **1052** (2018) 012097 doi:10.1088/1742-6596/1052/1/012097

2. Design and simulation

The MEMS VEH devices were designed for a specific automotive application. Following FFT and STFT vibration analysis of the measured data, active frequencies were revealed to be in the vicinity of ~500 Hz. The design area considered here is a 25 mm^2 square die and the devices were fabricated as per the stack of material shown in figure 1a. The fabrication process employs AlN as the piezoelectric material, deposited on doped silicon as the device and bottom electrode layer. The proof mass was patterned from an un-etched silicon wafer. Due to etching tolerances, a minimum gap of 100 μm was observed between various proof mass entities.

A pair of classical cantilevers and a pair of fork cantilevers were separately fitted within the 25 mm^2 design space as shown in figure 1b. The new topology comprises a pair of interdigitated cantilevers: a fork and a reverse fork structure. The precise frequency of the fork cantilevers can be tuned during the design stage by adjusting the length of each mass arms that occupies the space of the accompanying reverse fork.

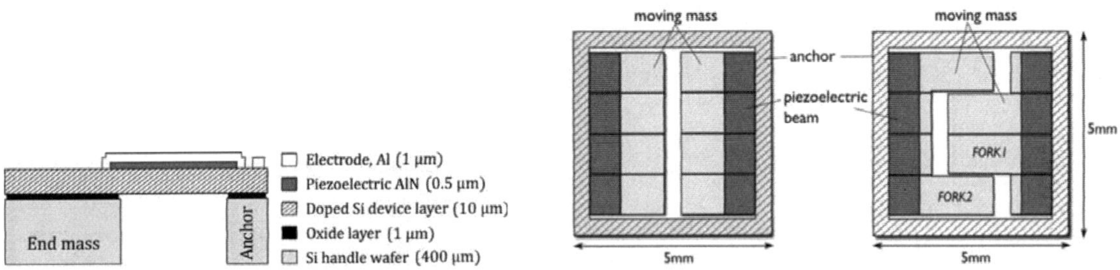

(a) Stack of materials for piezoelectric MEMS (b) Classical cantilever (left) and fork cantilever (right).

Figure 1: MEMS fabrication process and design of classical cantilever and the proposed interdigitated fork cantilevers. Cantilevers fitted within the same design area of 5 mm by 5 mm. Blue shaded region illustrate the active piezoelectric areas.

As shown in the simulation results summarised in table 1, the fork cantilevers can achieve ~30% lower resonant frequency than classical cantilevers, without compromising power output. All the devices have the same proof mass and active piezoelectric area to allow a fair comparison. By varying the mass arm length in the new fork design, the frequency can be varied. However, as the adjustment is in essence the reshaping of the same mass entity, the end mass does not quantitatively change. Instead, it is the effective length of the cantilever oscillator that is being altered. With a longer effective length, the resonant frequency is lowered and the maximum strain induced is also increased.

Table 1: Parameters for micro-cantilever designs within a fixed design area of 12.5 mm^2. Proof mass m, active piezoelectric area A, FEA simulated 1st mode resonant frequency f_0, average induced strain ε at acceleration of 1 g assuming Q of 100, and theoretical power P at resonance based on FEA and calculation.

	Classical cantilever	Fork cantilever I	Fork cantilever II
m (mg)	4.12	4.12	4.12
A (mm^2)	3.2	3.2	3.2
f_0 (Hz)	744	501	551
ε (1)	2.7×10^{-4}	4.0×10^{-4}	3.4×10^{-4}
P (μW)	1.71	2.64	2.12

3. Experimental results and discussion

A fabricated device is shown in figure 2, where an 11 mm square silicon die contains a pair of fork cantilevers and an equivalent classical cantilever for comparison. The die in the figure sits within a deep cavity leadless chip carrier (16 mm in length). A ~1 mm thick spacer (not shown) elevates the silicon die, in order to allow unrestricted travel of the shuttle during vibration. In both the classical and new designs, the cantilevers have four split electrode regions per device, which allow the possibility of connecting the separate piezoelectric regions in series to maximise voltage output [5].

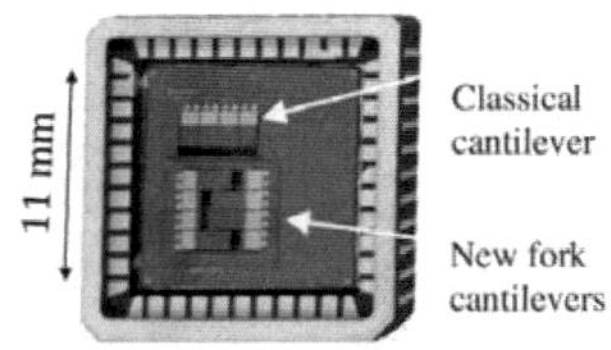

Figure 2: Photograph of a 11 mm square silicon die containing 1 classical cantilever device and two interdigitated fork cantilevers, in a chip carrier.

All devices were characterised on a mechanical shaker controlled by a waveform generator. The frequency domain power response is presented in figure 3. Measured resonant frequencies reveal 770 Hz for the classical cantilever, 525 Hz for Fork 1 and 586 Hz for Fork 2. This corresponds to 32% and 24% reduction in frequency for Fork 1 and Fork 2 respectively, over the classical cantilever. Apart from the anticipated offset in absolute values due to fabrication tolerance, the relative reduction is in line with the simulation.

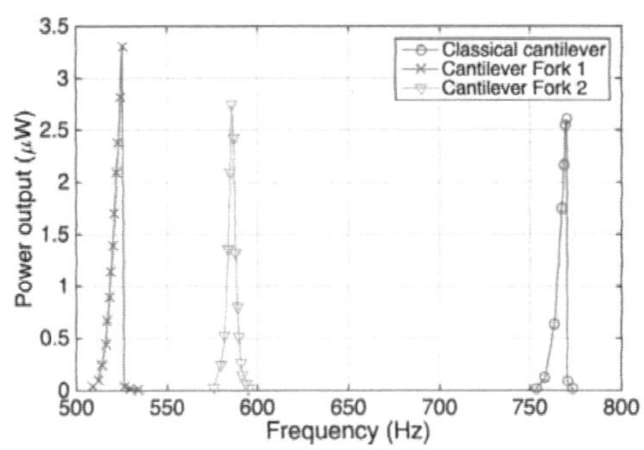

Figure 3: Experimentally measured power response in the frequency domain at 0.5 g of acceleration. Fork 1 peaks at 525 Hz, Fork 2 peaks at 586 Hz and Classical cantilever peaks at 770 Hz. Values are measured across matched load for each device.

Furthermore, at the 0.5 g of acceleration shown in figure 3, the resonant power output for the fork cantilevers performed slightly better than the classical cantilever. This is due to the larger induced average strain in the fork cantilevers. Figure 4 shows typical time domain responses of fork and classical cantilevers when subjected to band-limited (0 Hz to 800 Hz) noise excitation and the comparative results are presented in table 2.

The experimental results validate the design and theory in terms of demonstrating a new lower frequency cantilever topology within a given design space, without compromising on size or power output as compared to the classical design. This introduces new design flexibilities to accommodate the frequency ranges of varying applications, without having to scale up the overall size of the MEMS device.

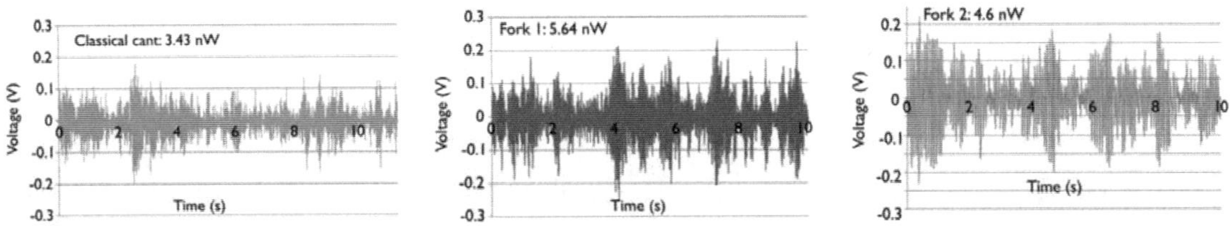

(a) Classic MC: 3.4 nW (b) Fork 1 MC: 5.6 nW (c) Fork 2 MC: 4.6 nW

Figure 4: Typical time domain average power responses when subjected to band-limited (0 Hz to 800 Hz) noise excitation at 0.00125 g^2/Hz.

Table 2: Experimental results when subjected to band-limited (800 Hz) noise excitation at 0.00125 g^2/Hz. The 1st mode resonant frequency f_0 was experimentally measured in air, power P is the average power record across the matched load resistance R for the first resonant peak.

	Classical cantilever	Fork cantilever I	Fork cantilever II
f_0 (Hz)	770	525	586
R (kΩ)	850	1,250	1,250
P (nW)	3.43	5.64	4.60

Within the 25 mm^2 die-level design area and the specific MEMS process explored within this paper, if the frequency of the MEMS devices were all optimised to the target frequency vicinity of ~500 Hz, 2 fork cantilevers can fit into the die area while only 1 classical cantilever can be accommodated. Therefore, the new fork cantilever topology will enable up to two times as many VEH devices to be manufactured per wafer.

Conclusion

A new interdigitated fork cantilever array topology has been numerically and experimentally investigated for MEMS piezoelectric vibration energy harvesting. The new topology allows for up to about a third reduction in resonant frequency for a given design area compared to a classical plain cantilever topology, without compromising on power optimisation. This opens up further flexibility in designing towards relatively lower frequency ranges needed for practical applications, while relaxing the tradeoffs on the miniaturisation and ease of integration benefits brought about by the MEMS technology.

Acknowledgments

This work was supported by EPSRC (Grant EP/L010917/1) and Innovate UK (Grant 102152).

References

[1] S.P. Beeby, M.J. Tudor and N.M. White, Energy harvesting vibration sources for microsystems applications, *Meas. Sci. Technol.*, 2006; **17**, R175-R195

[2] Y. Jia and A.A. Seshia, Power Optimization by mass tuning for MEMS piezoelectric cantilever vibration energy harvesting, *J. Microelectromech. Syst.*, 2015; **25**,1: 108-117.

[3] S.R. Anton and H.A. Sodano, A review of power harvesting using piezoelectric materials (2003-2006), *Smart Mater. Struct.*, 2007; **16**.

[4] R. Andosca, T.G. McDonald, V. Genova, S. Rosenberg, J. Keating, C. Benedixen and J. Wu, Experimental and theoretical studies on MEMS piezoelectric vibrational energy harvesters with mass loading, *Sens. Actuators A*, 2012; **178**: 76-87.

[5] S. Du, Y. Jia and A.A. Seshia, Maximizing Output Power in a Cantilevered Piezoelectric Vibration Energy Harvester by Electrode Design, *J. Phys. Confer. Ser.*, 2015; **660**:012114.

Effect of nonlinearities and objective function in optimization of an energy harvesting device

C D Gatti[1], J M Ramirez[1], M Febbo[2] and S P Machado[1]

[1]Grupo de Investigación en Multifísica Aplicada, Universidad Tecnológica Nacional FRBB and CONICET. 11 de abril 461, 8000 Bahía Blanca, Argentina.
[2]Instituto de Física del Sur, CONICET and Departamento de Física. Universidad Nacional del Sur (UNS). Av Alem 1253, 8000 Bahía Blanca, Argentina.

Abstract. This work presents a study of the impact of the linearity assumption of the mechanical model in the overall performance of an energy harvesting piezoelectric beam. Also, a brief assessment of geometrical optimization solutions using different objective functions is presented. The mechanical model of the harvester is based on both linear and nonlinear variants of the electrical and mechanical constitutive equations for the piezoelectric material. The nonlinear elastic, damping and electromechanical coupling parameters are obtained via least squares identification using physical experimentation; the experimental tests are performed at different ground excitation accelerations. The computational optimization of the harvester is done using the genetic algorithm implemented in Matlab. Different objective functions are tested, i.e. broadband maximum peak power, maximum power at a particular frequency and broadband mean power; the influence of the selection of each of them in the total recovery of the power of the device is analyzed. The most suitable function to recover the vibratory energy from conventional transport vehicles is found.

1. Introduction

A large number of works present devices based on piezoelectric materials that recover energy from vibrations taking advantage of material deformation [1-4]. This type of devices have a wide number of applications [5, 6], our main interest are the ones that operate excited by four stroke engine vibrations.

In the design and development of energy harvesting devices the most important variable is the efficiency of the energy conversion. Regardless of the device type and the power source used, a certain configuration of geometrical, mechanical and electrical parameters maximizes the total energy recovered. One the most effective methods to define this configuration is the use of optimization algorithms [7]. This necessarily implies the definition of variables and objective functions to be optimized. In most works [8-11], a linear mechanical model and a peak harvested power based objective function is used. As it will be shown in this paper, this choice of objective function does not always maximize the total energy recovered. Also, the linearization of the mechanical model entails a not negligible error in the system response.

This work presents two main developments; firstly a study of the impact of the linearity assumption of the mechanical model in the overall performance of an energy harvesting piezoelectric beam, lastly a brief assessment of geometrical optimization solutions using different objective functions. The most suitable function to recover the vibratory energy from conventional transport vehicles in the usual frequency range is found.

First, the mechanical model of the harvester based on linear and nonlinear variants of the electrical and mechanical constitutive equations for the piezoelectric material is presented. In the nonlinear model, the nonlinear parameters are identified using the least squares method fed with experiments performed in the laboratory subjecting the device to different amplitudes of ground acceleration. Then, the computational optimizations performed with genetic algorithms implemented numerically in Matlab are presented. Three different objective functions are tested: broadband maximum peak power, maximum peak power at the RMS frequency of usual use and broadband mean power in the range of frequency use. The influence of the objective function choice in the total energy recovery of the device is analyzed and the most suitable for the application at hand is determined. At last, conclusions regarding the inadequate use of a linear model to optimize are presented.

2. Harvester model and parameters identification

The device design takes advantage of the vibration of a piezoelectric beam subjected to the excitation of its base by the transport vehicle's engine. A potential difference between the electrodes of the piezoelectric due to beam deformation is generated.

The mathematical model is based on the scheme shown in Figure 1 which consists of a cantilevered stainless steel beam with a piezoelectric sheet (MFC M8507P2) attached to the upper surface. This beam is excited in the base by means of a shaker (Labworks ET-132), and a triaxial accelerometer (PCB Piezotronics ICP 356A32) is used to measure the accelerations.

The beam model is based on a Bernoulli-Euler formulation, considering a nonlinear piezoelectric constitutive equations [12], which is presented by the authors in [13]. A Lagrangian approach is used to derive the system of differential equations, which result:

$$M\ddot{q} + (Bq\,\mathrm{sgn}(q) + B_n q^2)\mathrm{sgn}(\dot{q}) + Kq + K_n q^2\,\mathrm{sgn}(q) - (\tilde{\theta} + \tilde{\theta}_n q\,\mathrm{sgn}(q))v = f, \quad (1)$$

$$C_p\dot{v} + \frac{v}{R_L} + (\tilde{\theta} + \tilde{\theta}_n q)\dot{q} = 0. \quad (2)$$

where M, B, B_n, K, K_n, $\tilde{\theta}$, $\tilde{\theta}_n$, Q, f, C_p and R_L are the modal mass, linear and nonlinear damping, linear and nonlinear stiffness, linear and nonlinear electromechanical coupling, displacement and modal force, internal capacitance of the piezoelectric unimorph sheet and load resistance. Table 1 presents the numerical values of the main parameters of the model for all study cases.

The nonlinear differential equations are solved analytically using harmonic balance method, so that the dynamic response of the system is obtained. From this, it is possible to find the voltage $v(t)$ to compare with the experiments and to identify the nonlinear parameters that are presented in Table 2 (which are related to those in equations (1,2) in [12]). The identification is performed using least squares method. Figure 2 shows the voltage curves at different levels of acceleration. As a first singular feature, it is possible to observe that the curves bend to the left (softening response) as the amplitude of the base acceleration increases.

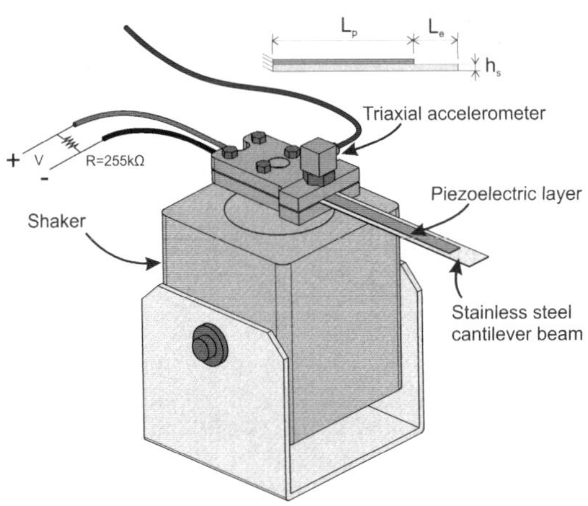

Figure 1. Harvester and shaker scheme.

Table 1. Numerical values of the main parameters of the model.

Parameter	Stainless steel	Piezoelectric
Width (mm)	12.7	8
Thickness (mm)	h_s	0.3
L_e (mm)	21.5	-
L_p (mm)	-	L_p
Density (kg/m^3)	7900	5440
Young's modulus (GPa)	193	15.85
Charge constant d_{31} (pm/V)	-	-170
Electrical permittivity (nF/m)	-	16.81
Electrical capacitance (nF)	-	38.11
Load resistance (kΩ)	255	

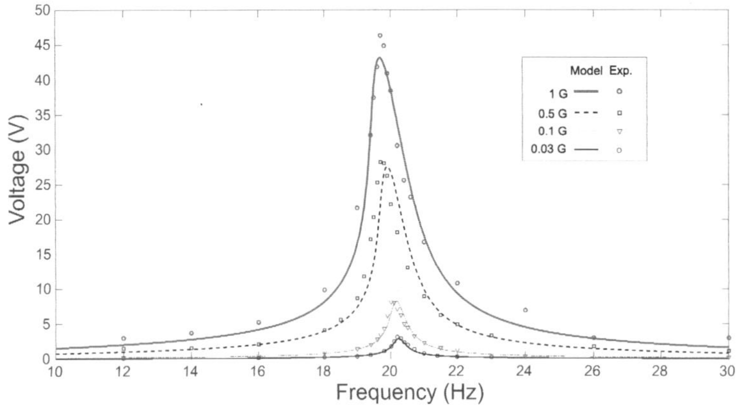

Figure 2. Voltage-frequency curves used for the identification of parameters.

Table 2. Nonlinear parameters identified.

Parameter	Value
b_{11} (N/m^2)	1.11 10^9
b_{111} (N/m^2)	6.38 10^{12}
c_{111} (TPa)	-79.42
e_{311} (kC/m^2)	-113.42

Finally, the following power equation used to perform the optimizations is obtained from the voltage:

$$P = \frac{V^2}{R_L}. \tag{3}$$

3. Optimizations and results

The geometric optimizations of the harvester are done using a genetic algorithm implemented in Matlab (*ga* command). The variables to be optimized are the piezoelectric length ($10\,mm < L_p < 85\,mm$) and the thickness of the stainless steel beam ($0.1\,mm < h_s < 1.5\,mm$). Three optimizations with the nonlinear model and three with the linear model ($B_n = K_n = \tilde{\theta}_n = 0$) are performed, proposing three different objective functions: broadband maximum peak power (A), maximum peak power at the RMS frequency (B) and broadband mean power (C). In order to adequately compare the performance of each optimized device, the energy recovered by the device along an urban way of a conventional transport vehicle is evaluated. This energy is obtained by integrating the temporal signal of the electric power (P) obtained by the model.

Table 3 shows the different study cases. As can be observed, the parameters optimized with the linear model differ markedly from those optimized with the nonlinear model. In addition, it is observed that the optimization C obtains the device that recovers more energy; this shows that choosing the broadband mean power as objective function is the best option for this type of applications. The recovered energy is shown in Figure 3, comparing the linear and nonlinear model.

Finally, to evaluate the impact of assuming a linear model to perform the optimization, we introduce the optimized variables by the linear model within the nonlinear model. Case C is used to perform the analysis because it is the one with the most energy recovered. The recovered energy is 0.44 J, approximately 34% less than the power that it would have recovered (see Table 3 and Figure 4, nonlinear model of case C).

Table 3. Optimization cases and results obtained.

Case	Objective function	Linear model			Nonlinear model		
		L_p (mm)	h_s (mm)	Recovered energy (J)	L_p (mm)	h_s (mm)	Recovered energy (J)
A	Broadband maximum peak power	37	0.3	0.48	85	1.24	0.39
B	Maximum peak power at RMS frequency	60	0.44	0.74	85	0.82	0.51
C	Broadband mean power	52	0.53	0.82	85	1.19	0.68

4. Conclusions

From the numerical results two important conclusions are obtained. First, the optimized parameters obtained with the linear and nonlinear models are not the same. The linear model overestimates the

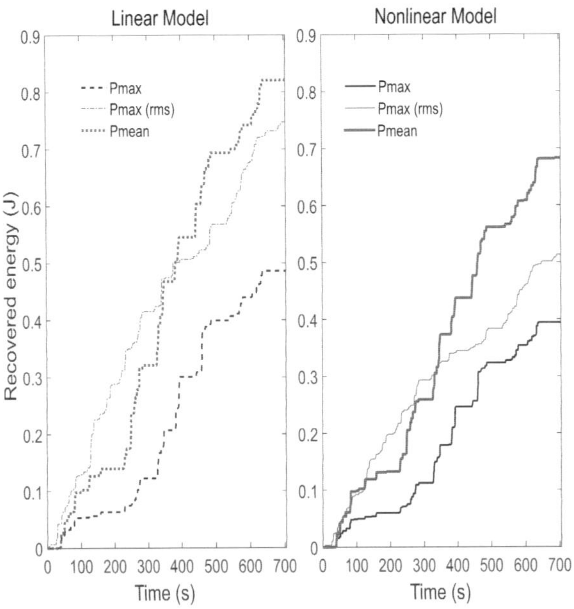

Figure 3. Recovered energy with different objective functions.

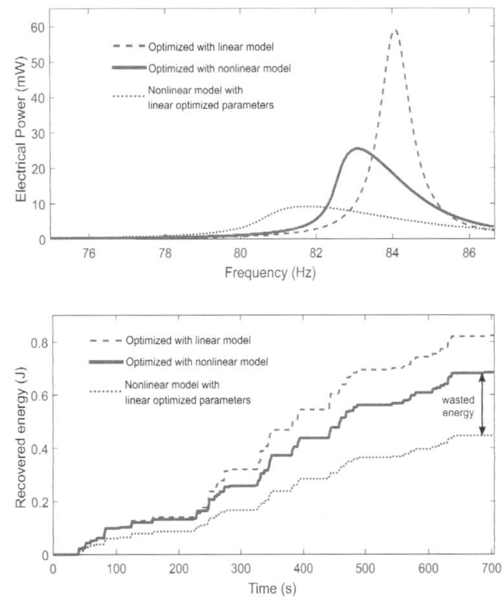

Figure 4. Recovered power and energy: comparison between optimization with linear and nonlinear models.

generated power in comparison with the nonlinear one. Additionally, a low power is obtained when the parameters optimized by means of the linear model are used in the nonlinear model.

Second, there is a great influence of the objective function selected in the optimization algorithm. It is clear that selecting the broadband mean power as objective function, instead of using the other two functions (most frequently used), the total recovered energy is considerably larger.

Acknowledgments

The authors wish to thank CONICET, UNS, UTN FRBB, ANPCyT under grant PICT 2013-2065 for their financial support.

References

[1] Roundy S, Wright P K and Rabaey J 2003 *Comput. Commun.* **26** 1131-44
[2] Priya S 2007 *J. Electroceram.* **19** 167-84
[3] Febbo M, Machado S, Ramirez J and Gatti C 2016 *J. Phys. Conf. Ser.* **773** 012058
[4] Erturk A and Inman D J 2008 *J. Intel. Mat. Syst. Str.* **19** 1311-5
[5] Xiao H and Wang X 2014 *Fuel Cells* **280** 28
[6] Febbo M, Machado S, Gatti C and Ramirez J 2017 *Energ. Convers. Manage.* **152** 166-75
[7] Goldberg D E 1989 *Reading: Addison-Wesley*
[8] Shafer M W, Bryant M and Garcia E 2012 *Smart Mater. Struct.* **21** 085008
[9] Park J, Lee S and Kwak B M 2012 *J. Mech. Sci. Technol* **26** 137-43
[10] Lefeuvre E, Badel A, Richard C and Guyomar D 2005 *J. Intel. Mat. Syst. Str.* **16** 865-76
[11] Shu Y-C and Lien I 2006 *J. Micromech. Microeng.* **16** 2429
[12] Leadenham S and Erturk A 2015 *Nonlinear Dynam.* **79** 1727-43
[13] Gatti C D, Ramirez J M, Machado S P and Febbo M 2016 *J. Phys. Conf. Ser.* **773**

Optimization analysis of a magnetic-piezoelectric current sensor

Po-Chen Yeh[1] **and Tien-Kan Chung**[1,2,*]

[1] Department of Mechanical Engineering, National Chiao Tung University, Hsinchu, Taiwan
[2] International College of Semiconductor Technology, National Chiao Tung University, Hsinchu, Taiwan
[*]Corresponding Author: tkchung@nctu.edu.tw

Abstract. In this paper, we reported an optimizing analysis for a magnetic-piezoelectric current sensor. The optimizing approach is conducted by investigating the magnetic force interaction between the sensor and the wire by changing the magnetization direction of the magnet and the relative orientations. Furthermore, we established a magnetic bending-moment based analysis and derive the governing equation to quantitatively correlate the magnetic force interaction to the mechanical bending moments. The analysis results show that the modelled results are highly consistent with the experimental results.

1. Introduction

To date, using wireless sensors network for electricity usage monitoring (i.e., wireless current sensing network) is general in smart grid applications. However, wireless current sensing network requires a large amount of current sensors with high sensitivity, linearity, and accuracy for precisely measuring the electricity use. Furthermore, for the need of long-term current monitoring, the current sensors have to own self-powered capabilities. To address this, some researchers recently proposed magnetic-piezoelectric self-powered current sensors [1]-[4] for monitoring electricity use. However, when the current carried by the wire is generally low in practical conditions, distinguishing the signals produced by the sensors from noises becomes difficult. Therefore, to increase the sensitivity of the sensors, an experimental-based optimizing approach was proposed (i.e., our previous work [4]). However, the experimental-based optimization in our previous work merely provides the optimizing concept in certain conditions, thus cannot be applied to more complete and complicated situations. Hence, to establish a complete optimizing approach, we present a magnetic bending-moment based optimizing approach for the piezoelectric current sensors.

2. Design

2.1. Piezoelectric current sensor

The design of the self-powered piezoelectric current sensor is shown in figure 1 (for design detail, please see [4]). The designs of the experiments are shown in figure 1(b)-(c). As shown in figure 1(b)-(c), we investigated magnetic force interaction by changing magnetization-direction of the magnet and the relative orientation between the sensor and the wire. Figure 1(d)-(e) is the illustration and photograph of the testing setup, respectively. Partial experimental results of three difference orientation cases of the sensor are shown in figure 1(f).

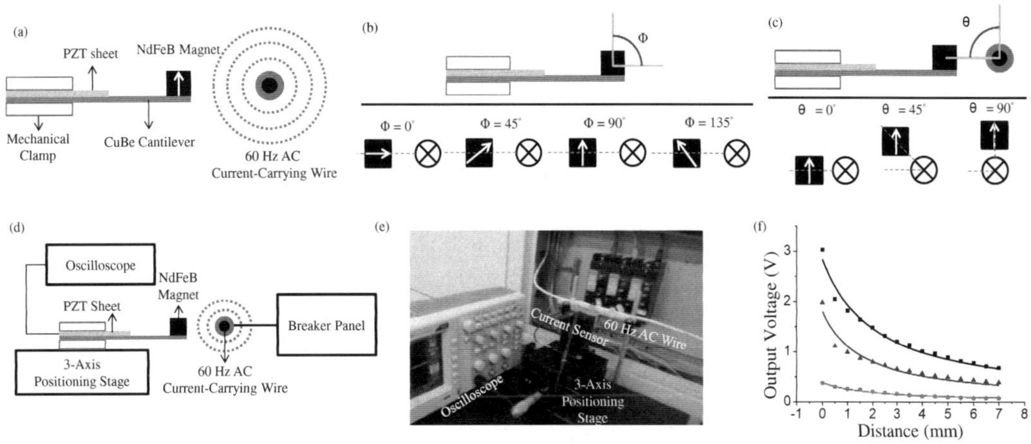

Figure 1. The illustration of the magnetic-piezoelectric current sensor: (a) The design of the sensor. The design of experiment for investigating (b) the magnetization and (c) the orientation induced magnetic force interaction. (d) The illustration and (e) the photograph of the testing setup. (f) The partial results of three difference orientation cases of the sensor. [4]

2.2. *Optimization analysis of current sensor*

To calculate the magnetic force interaction between the magnet and the wire, we establish the magnetic bending-moment based analysis and derive the corresponding governing equation to quantitatively correlate the magnetic force interaction to the mechanical bending moments. The general governing equation is shown as equation (1).

$$\vec{\mathbf{M}} = \vec{\mathbf{F}} \times \vec{\mathbf{r}} \propto \vec{\mathbf{B}} \times \vec{\mathbf{H}} \times \vec{\mathbf{r}} \quad (1)$$

In equation (1), a mechanical bending moment M subjecting to the beam is estimated through the magnetic interaction force. Where B is the magnetic flux density produced by the magnet, H is the magnetic field intensity produced by the wire, and r is the lever-arm distance from the subjecting location of the magnetic force to the pivot (note: the pivot is the fixing location of the sensor).

Based on the assumption of the 3-axial coordinate system shown in figure 2(a)-(c), equations (2)-(4) determine the 3-axial magnetic flux density at a location around a rectangular magnet in a 3-axial coordinate system [5]. In the 3-axial coordinate system, x_1, x_2, y_1, y_2, z_1, z_2 shown in figure 2(b) and 2(c) define the magnet's two reference planes which are used in the equation (2), (3), and (4). Where B_x, B_y and B_z is the x-axial, y-axial, and z-axial magnetic flux density, respectively, μ_0 is the vacuum permeability, M_s is the magnetization of the magnet, and (x_1, x_2), (y_1, y_2) and (z_1, z_2) denote the locations of the edges of the magnet respecting to the x-, y-, z-axes. For demonstration and comparison with the experimental results in our previous works [4], we set $x_1 = -2.5$, $x_2 = 2.5$, $y_1 = -5$, $y_2 = 5$, $z_1 = -2.5$, $z_2 = 2.5$ mm (which are the dimensions of the magnet) and $M_s = 9.95 \times 10^5$ A m^{-1}.

The illustration of the magnetic field intensity produced by the wire is shown in figure 3. The 3-axial magnetic field intensity is derived from Biot-Savarts Law [2] and H_x, H_y, H_z are shown in equations (5), (6), and (7). Where H is the magnetic field intensity (A/m), I is the current in the wire (A) which is set as 8 A in the modeling, x and z is the x-axial and z-axial distance from the point P to the center of the wire.

IOP Conf. Series: Journal of Physics: Conf. Series **1052** (2018) 012099 doi:10.1088/1742-6596/1052/1/012099

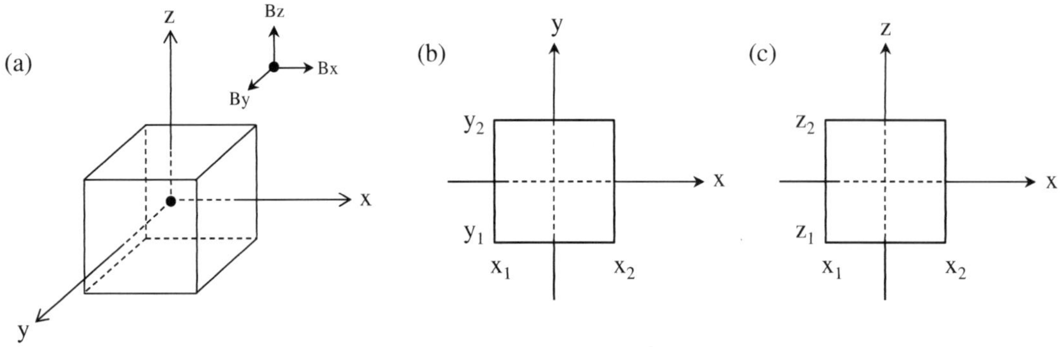

Figure 2. The 3-axial x-y-z coordinate system of the rectangular magnet: (a) The illustration of x-y-z coordinate system. (b) The referenced x-y plane and (c) the referenced x-z plane from cross-sectional view of the magnet.

$$\mathbf{B}_x(x,y,z)=\frac{\mu_0 M_s}{4\pi}\sum_{k=1}^{2}\sum_{m=1}^{2}(-1)^{k+m}\ln\left[\frac{(y-y_1)+[(x-x_m)^2+(y-y_1)^2+(z-z_k)^2]^{1/2}}{(y-y_2)+[(x-x_m)^2+(y-y_2)^2+(z-z_k)^2]^{1/2}}\right] \quad (2)$$

$$\mathbf{B}_y(x,y,z)=\frac{\mu_0 M_s}{4\pi}\sum_{k=1}^{2}\sum_{m=1}^{2}(-1)^{k+m}\ln\left[\frac{(x-x_1)+[(x-x_1)^2+(y-y_m)^2+(z-z_k)^2]^{1/2}}{(x-x_2)+[(x-x_2)^2+(y-y_m)^2+(z-z_k)^2]^{1/2}}\right] \quad (3)$$

$$\mathbf{B}_z(x,y,z)=\frac{\mu_0 M_s}{4\pi}\sum_{k=1}^{2}\sum_{n=1}^{2}\sum_{m=1}^{2}(-1)^{k+n+m}\tan^{-1}\left[\frac{(x-x_n)(y-y_m)}{(z-z_k)[(x-x_n)^2+(y-y_m)^2+(z-z_k)^2]^{1/2}}\right] \quad (4)$$

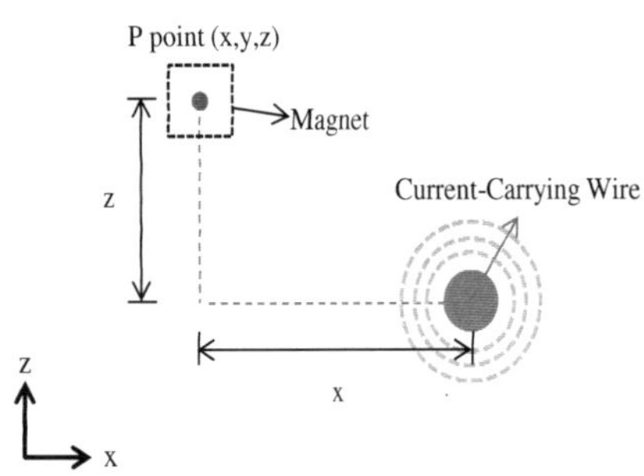

$$\mathbf{H_x}=\frac{I}{2\pi}\frac{z}{x^2+z^2} \quad (5)$$

$$\mathbf{H_y}=\mathbf{0} \quad (6)$$

$$\mathbf{H_z}=\frac{I}{2\pi}\frac{x}{x^2+z^2} \quad (7)$$

Figure 3. The illustration and the derived equations for estimating the magnetic field intensity at any point located outside of the wire.

3. Modeling results and discussion

Through using equations (2)-(4), the partial modeled results of the x-axial and z-axial magnetic-flux-density (B_x and B_z) over the x-y and y-z plane surfaces of the magnet are obtained, as shown in figure 4(a), 4(b), 4(c), and 4(d), respectively. Base on the results shown in figure 4(a)-(d), the average magnetic-flux-density is calculated by using weighted average method (i.e., add up the products of value times weight to get the total value, and consequently divide the total value by the total weight) and subsequently converted to vector-form expression. In addition, above same method is used to calculate the magnetic field intensity produced by the wire.

Finally, by substituting the vectors of magnetic-flux-density and magnetic-field-intensity to equation (1), we can obtain the bending moment subjecting to the beam. The comparison of modeled results and experimental results [4] of four critical magnetization-direction (Φ= 0°, Φ= 45°, Φ= 90° and Φ= 135°) at the relative-orientation θ= 0° are shown in figure 4(e). In figure 4(e), the modelled results are consistent with the experimental results.

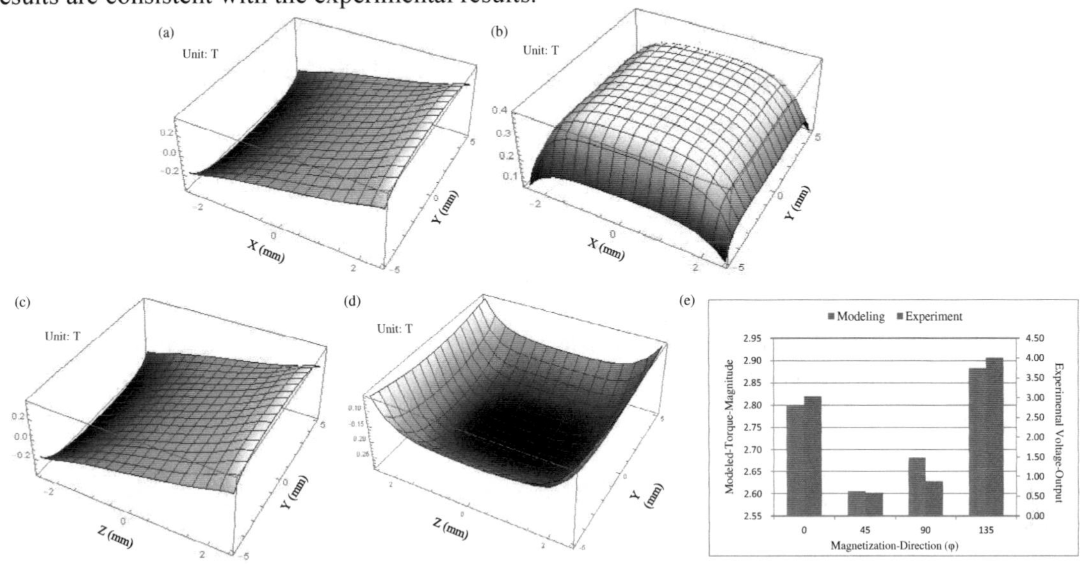

Figure 4. The partial modeled results of the x-axial and z-axial magnetic-flux-density (B_x and B_z) over the x-y and y-z plane surfaces of the magnet. (a) x-axial and (b) z-axial magnetic-flux-density over the x-y plane surface of the magnet. (c) x-axial and (d) z-axial magnetic-flux-density over the y-z plane surface of the magnet. (e) The comparison between the modeled and experimental results.

4. Conclusion

We proposed an optimizing approach for the magnetic-piezoelectric current sensor. The modelled results are consistent with the experimental results. By using the approach, the modeling-based approach can be used to further optimize the sensor performance.

Acknowledgment

Support for this work was provided by The Taiwan Ministry of Science and Technology (No. 105-2628-E-009 -001 -MY2)

References

[1] Xu Q R, Paprotny I, Seidel M, White R M, and Wright P K 2013 *IEEE Sens. J.* **13** 1055-1064

[2] Wang D F, Suzuki Y, Suwa Y, Kobayashi T, Itoh T, and Maeda R 2013 *Meas. Sci. Technol.* **24** 125109

[3] Lao S B, Chauhan S S, Pollock T E, Schroder T, Cho I S, and Salehian A 2014 *Actuators* **3** 162-181

[4] Yeh P C, Chung T K, Lai C H, and C M Wang 2016 *Appl. Phys. A Mater. Sci. Process*, **122** 1-10

[5] Furlani E P 2001 *Permanent Magnet & Electromechanical Devices: Materials, Analysis, and Applications* (San Diego: Academic Press)

[6] Wagner B and Benecke W 1991 4th *Proc. Inter. Conf. MEMS* (Nara, Japan) p 27-32

IOP Conf. Series: Journal of Physics: Conf. Series **1052** (2018) 012100 doi:10.1088/1742-6596/1052/1/012100

Simple method for quality factor estimation in resonating MEMS structures

S Larsson[1], P Johannisson[1], D Kolev[1], F Ohlsson[1], S Nik[2,3], J Liljeholm[2,4], T Ebefors[2,5] and C Rusu[1]

[1]RISE Acreo AB, Sweden
[2]SILEX Microsystems AB, Sweden
[3]IMEC, Belgium
[4]KTH-MST, Sweden
[5]Spinverse AB, Sweden

Email: cristina.rusu@ri.se

Abstract. The quality factor of a packaged MEMS resonating structure depends on both the packaging pressure and the structure's proximity to the walls. This type of mechanical constraints, which causes energy dissipation from the structure to the surrounding air, are applicable for oscillating energy harvesters and should be considered in the design process. However, the modelling of energy losses or the measurements of their direct influence inside a packaged chip is not trivial. In this paper, a simple experimental method to quantify the energy loss in an oscillating MEMS structures due to the surrounding air is described together with preliminary results. The main advantage of the method is the ability to characterize the damping contributions under different vacuum and packaging conditions without requiring any packaging of the harvester chip or fabrication of multiple devices with different cavity depths.

1. Introduction

Usually microelectromechanical system (MEMS) resonating structures need to be packaged for protection from external influences, as well as for reliability since it prevents the fabricated structures from excessive displacements. However, the package introduces additional parasitic damping mechanisms and thus energy losses, such as viscous drag losses (packaging pressure) and squeeze film damping (proximity to the walls). These effects must be considered in the design process when resonating structures are manufactured [1,2]. Usually, for the experimental characterization of these damping effects, extra processing steps are needed that can be of high complexity (hermetic packaging, various cavity depth, electrical connections, etc.) [1], be time consuming, and require optically transparent packaging for characterization of the mechanical behavior in addition to the electrical properties.

In this paper, a simple experimental method to quantify the energy loss of oscillating MEMS structures due to the surrounding air, without requiring any packaging or complex microtechnology fabrication is described together with preliminary results.

IOP Conf. Series: Journal of Physics: Conf. Series **1052** (2018) 012100 doi:10.1088/1742-6596/1052/1/012100

2. Experimental characterization

This section describes the experimental techniques that allows characterization of the Q-factor and thereby the energy loss of oscillating MEMS structures due to packaging.

2.1. Measurement setup

A laser Doppler vibrometer (LDV) measures the velocity of a moving object by measuring the frequency shift in the reflected light via interferometry leading to very accurate velocity measurements. The LDV used has a special objective that allows measurements of MEMS structures.

The motion is excited by a shaker which is driven by the LDV (Polytec Inc.) internal signal generator and an amplifier. The fact that the shaker excitation is phase locked to the detection implies that phase sensitive measurements can be carried out, showing how different parts of the structure moves relative to each other.

As the damping is typically dominated by viscous effects [2], it is essential to control the pressure during measurements as this allows the necessary sealing pressure for the MEMS component to be determined. This has been achieved by placing both the small-size shaker (The Modal Shop, Inc.) and the sample in a vacuum chamber (Figure 1). The air is evacuated using two pumps; a mechanical pump for pressures down to about 10 mbar, and a turbo pump for pressures down to about 10^{-5} mbar. Such low pressures are not possible to obtain in a sealed device but this allows the damping to reach the intrinsic damping regime and indicates the level of the internal damping mechanisms in the device.

For measurements of the squeeze film damping effect on the Q-factor, a distance structure is manufactured by fine mechanics (Figure 2). This squeeze-film structure can be easily attached to the sample holder to obtain a selected gap size to the oscillating structure.

2.2. MEMS sample

The measured MEMS structures are silicon cantilevers and meanders, as shown in Figure 3, with proof mass to decrease the resonance frequency as we target low-frequency energy harvesting. The MEMS structures are manufactured on SOI wafers with a 400 µm thick handle and a 15 µm thick device layer. On the device layer, a 2 µm SiO_2 is grown and a bottom electrode is deposited, on top of which the sol-gel PZT is spun. The PZT and the bottom electrode are patterned by wet and dry etch, respectively. The top electrode is sputtered TiW/Au layers and patterned by wet etch. The proof mass is released by DRIE Si etch. The chips have dimensions of about 6mm x 4mm.

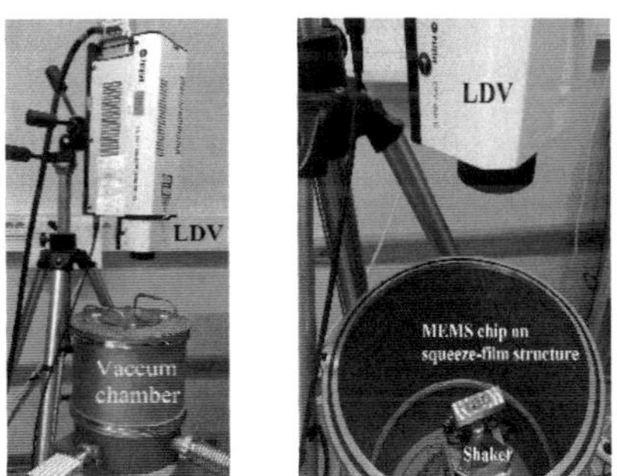

(a) (b)

Figure 1. (a) Image of the experimental setup to estimate the energy losses showing the LDV with a special objective that allows measurements of MEMS structures. Below the LDV, the vacuum chamber is seen. The measurement is made through the glass window. (b) View of the small-size shaker inside the vacuum chamber with a MEMS chip on the sample

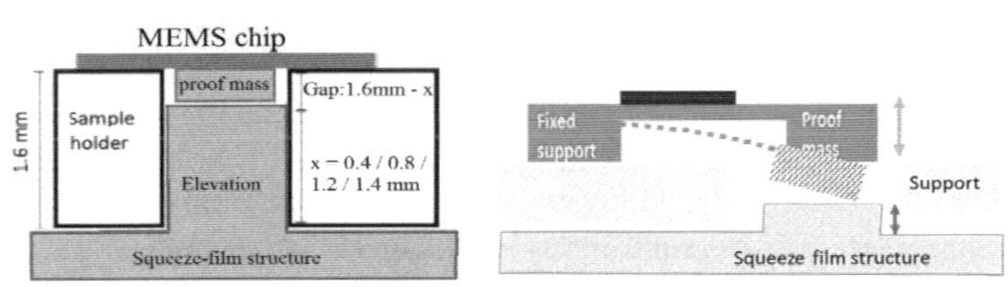

Figure 2. Sketch of the elevation module placed under sample for squeeze film damping experiments.

(a) 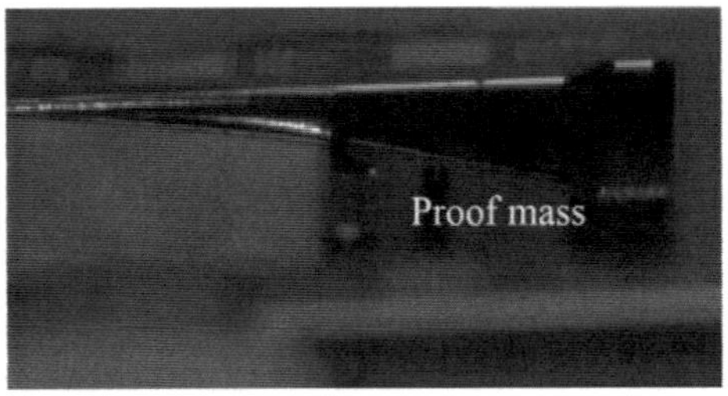

(b) 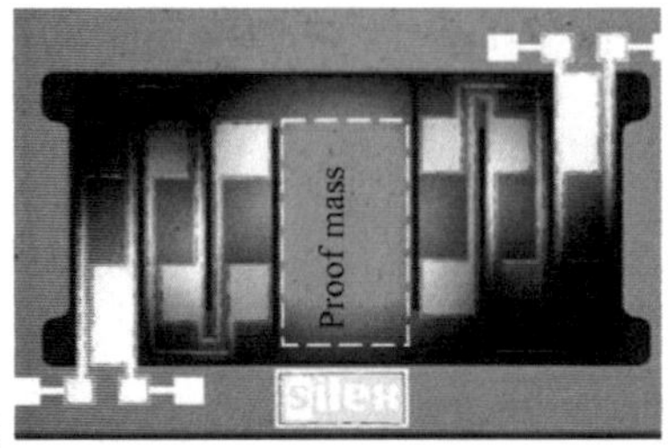

Figure 3. Image of MEMS (a) cantilever and (b) meander structure with proof mass.

3. Results and Discussions

Tests were done to define the optimal values (e.g. excitation amplitudes) for the LDV, mechanical shaker and vacuum chamber (Figure 4). Using these settings, testing of the measurement setup shows the expected behavior of the Q-factor, as shown in Figure 5.

The novelty of our measurement system is the possibility to have a multi-probe measurement setup by combining in one system the LDV, shaker, vacuum and squeeze-film structure. Thus, it is possible to simultaneously measure the Q-factor and the influence of damping in an easy way. The Q-factor for MEMS meanders and cantilevers as a function of pressure with and without wall is shown in Figure 6. Three different regimes can be observed in accordance with theoretical results [2,3]. At low pressure, the Q-factor saturates on the intrinsic damping level, then drops rapidly in the molecular regime and changes more slowly in the viscous regime above 1 mbar.

Q-factors as a function of gap distance for cantilever and meander samples at 1000 mbar are shown in Figure 7. It can be seen that squeeze film damping has little or no significance with a gap larger than 400 μm for the meander structure. The experimental results indicate that the proposed method can quantify both drag and squeeze film damping losses in resonating MEMS structures without packaging.

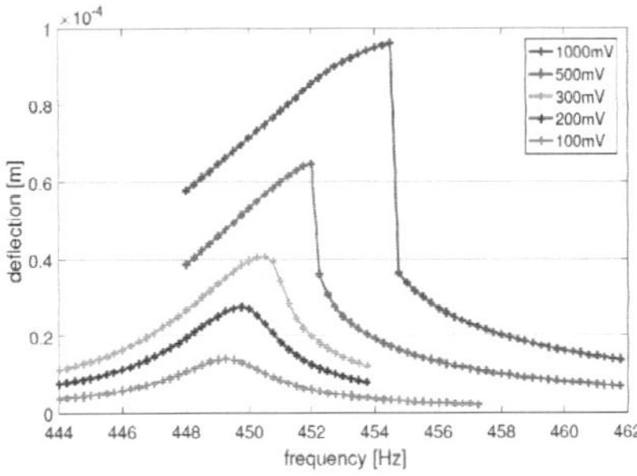

(a)

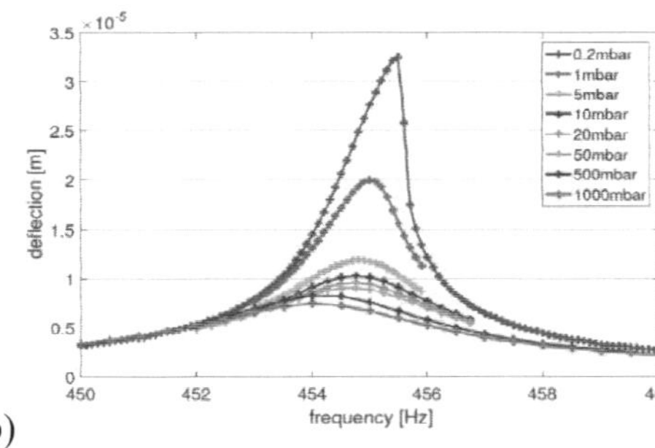

(b)

Figure 4. Spectra for the first resonance peak measured with the single frequency sine sweep method of MEMS cantilever as function of frequency and (a) excitation amplitudes in air, (b) pressure in the vacuum chamber with the excitation amplitude of 100mV.

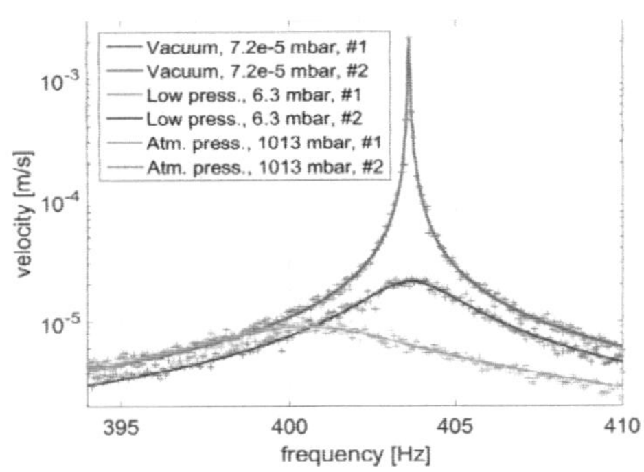

Figure 5. Measurement data (crosses) and fitted curves (solid lines) for the meander structure at three different pressures

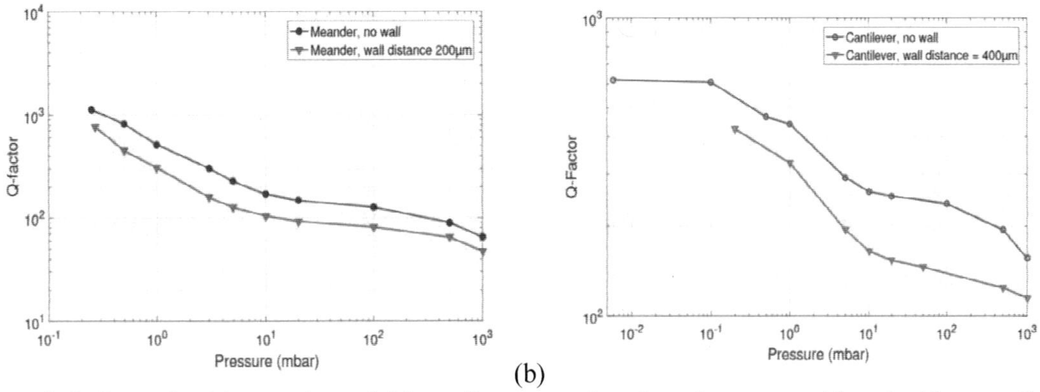

(a) (b)

Figure 6. Q-factor for (a) meander and (b) cantilever as a function of pressure with and without wall.

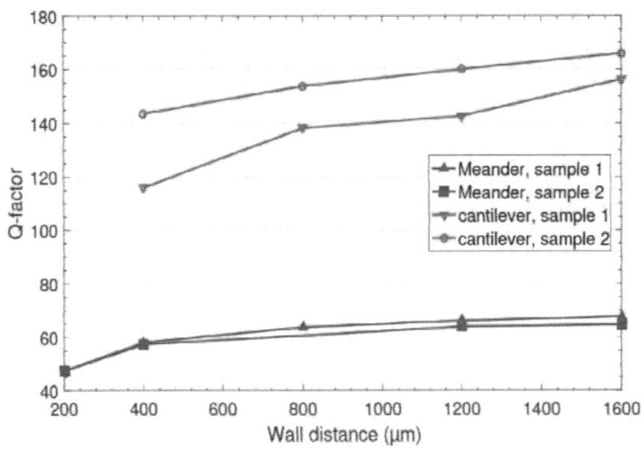

Figure 7. Q-factor as a function of the wall distance for cantilever and meander samples at 1000 mbar.

4. Conclusions

A simple experimental method to quantify the energy loss of oscillating MEMS structures due to the surrounding air is described. The combination of LDV measurements with the shaker, the squeeze-film structure and the placement of the sample in a vacuum chamber gives the possibility to study mechanical characteristics (e.g., resonance frequencies and the corresponding modes), piezo-mechanical coupling (by monitoring the output voltage with selected loads) and the damping mechanisms under varying pressure and excitation conditions. Thus, measurements with this combination setup can serve as a benchmark on the accuracy of numerical simulations, but also gives complementary information about the sample behavior that is difficult to obtain numerically.

Acknowledgements

This work was performed in the Smart-MEMPHIS project with funding from the European Union's Horizon 2020 research and innovation program, grant agreement No. 644378.

References

[1] R Elfrink, et al. "Vacuum-packaged piezoelectric vibration energy harvesters: damping contributions and autonomy for a wireless sensor system", J. Micromech. Microeng. 20 (2010) 104001.

[2] F.R. Blom, S. Bouwstra, M. Elwenspoek, J.H.J. Fluitman. 1992. "Dependence of the Quality Factor of Micromachined Silicon Beam Resonators on Pressure and Geometry." J. Vac. Sci. Technol. B 10 (1): 19–26.

[3] M. Bao, H. Yang. 2007. "Squeeze Film Air Damping in MEMS." Sensors and Actuators 136: 3–27.

IOP Conf. Series: Journal of Physics: Conf. Series **1052** (2018) 012101 doi:10.1088/1742-6596/1052/1/012101

Multibeams energy harvester for rotational low-frequencies

J M Ramírez[1], C D Gatti[1], S P Machado[1] and M Febbo[2]

[1]Grupo de Investigación en Multifísica aplicada, Universidad Tecnológica Nacional FRBB and CONICET. 11 de Abril 461, 8000 Bahía Blanca, Argentina.

[2]Instituto de Física del Sur, CONICET and Departamento de Física. Universidad Nacional del Sur (UNS). Av Alem 1253, 8000 Bahía Blanca, Argentina.

josemramire@gmail.com

Abstract. The mechanical vibration induced by rotation speed is gaining interest in the last years. Comparing to other mechanical vibration sources, the excitation by rotational speed from low power wind turbines is really a challenge since the vibration frequency is lower than 10 Hz and it generates accelerations higher than 1 g. In this paper, a rotating piezoelectric energy harvester is designed and analysed. In order to achieve the best electrical power performance, the prototype is designed to have low natural frequencies varying the hub positions and dimensions of the multi-beams, decreasing the crossing-frequency (natural frequency/rotation speed) and finding the optimal electrical load resistance. The dynamic behaviour of the harvester is simulated using a nonlinear one-dimensional finite element formulation. The rotating piezoelectric beam is formulated by means of a geometrically nonlinear finite element with six mechanical degrees of freedom and one electrical degree of freedom per node. The simulations are performed from 1 to 4.5 Hz (60 – 270 rpm) rotation speeds. A performance assessment is done investigating the influence of the hub positions, crossing-frequencies, electrical resistance and rotation speeds over the voltage and power generation. Regarding harvester's performance, promising results are obtained since the output power 3.72 mW @ 3.61 Hz. In this sense, this article provides a prototype at low rotational frequency capable to scavenge energy from low power wind turbines.

1. Introduction

Sensing electronic devices in wind turbines are known for their small size and low power consumption. The most commonly used sensors are ultrasonic wind sensors, weather, pressure, humidity temperature and temperature transmitters. Mostly, electrochemical batteries are the best choice today to powering these sensors or electronic devices. Despite the advantages of batteries such as low cost, high energy density and small size, they have disadvantages such as low durability, inconveniences to recharging and environmental pollution. On the other hand, the evolution of alternative energy sources to powering electronic devices with low power consumption is gaining interest in the science community. In the last decade, one source of alternative energy is the energy harvesting from mechanical vibrations under base acceleration excitation [1],[2]. Between them, mechanical vibration induced by rotation speed has drawn much attention in the last years [3],[4]. In comparison with to other mechanical vibration sources, rotational excitation from low power wind turbines is really a challenge since the vibration frequency is lower than 10 Hz and the accelerations are higher than 1 g.

In this paper, a rotational piezoelectric energy harvester comprising of multi-beams is designed and analyzed. In order to obtain the maximum power [5], the prototype is designed to have low natural frequencies near the rotation speeds, varying the hub positions (R) and the dimensions of the multi-beams. The dynamic behavior of the harvester is simulated using a nonlinear one-dimensional finite element formulation developed by Ramírez, et al [6]. The rotating piezoelectric beam is formulated by means of a geometrically nonlinear finite element with six mechanical degrees of freedom and one electrical degree of freedom per node. The simulations are performed from 1 to 4.5 Hz (60 – 270 rpm) rotation speeds at three hub positions (R). A performance assessment of the influence of the hub distance (R) on the crossing-frequencies is investigated. For the selected dimensions, the crossing-frequency phenomenon is induced by the softening effect [7], which yields a natural frequency near the rotational frequencies. The effect of rotation speeds and electrical load resistance over the voltage and power generation are assessed. Finally, the study of the results obtained concludes that the proposed design is a promising prototype to scavenge energy from low power wind turbines.

2. Design and modelling

The proposed harvester consists of multiple beams with attached masses and a MFC 2814 P2 piezoelectric sheet (fabricated by NASA and commercialized by Smart Material Corporation) mounted on one of the multiple beams. The material properties of the harvester are show in table 1. The prototype is mounted on a rigid plate that has the possibility to adjust the hub position (R) with respect to axis of rotation (see figure 1). Likewise, the rigid plate rotates at a constant angular velocity Ω.

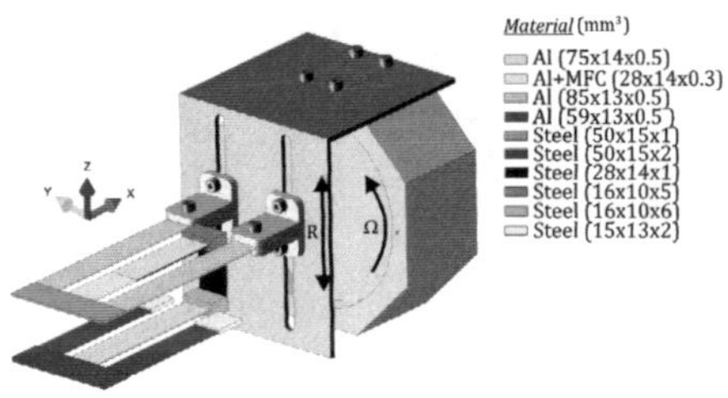

Figure 1. Schematic of the rotating energy harvester prototype.

Table 1. Material properties for aluminium, steel, and MFC 2814 P2.

Aluminum		MFC 2814 P2		Steel	
E_1	67 GPa	E_1	30.3 GPa	E_1	210 GPa
E_2	67 GPa	E_2	15.85 GPa	E_2	210 GPa
Density	2700 kg/m^3	Density	5440 kg/m^3	Density	7850 kg/m^3
		Piezoelectric constant d_{31}	-2.1 E+2 pm/V		
		Capacitance	30.78 nF		

2.1. Finite element formulation

A one-dimensional finite element is used for modeling three-dimensional rotational energy harvesters [6]. The rotating piezoelectric beam is formulated by means of a geometrically nonlinear finite element with six mechanical degrees of freedom and one electrical degree of freedom per node. The electromechanical equations of motion in matrix form are:

$$\begin{gathered} \boldsymbol{K_T}\,\widehat{\boldsymbol{U}} + \boldsymbol{D}\,\widehat{\boldsymbol{V}}_{\boldsymbol{G}} + \boldsymbol{C_c}\widehat{\boldsymbol{V}}_{\boldsymbol{G}} + \boldsymbol{M}\,\widehat{\boldsymbol{a}}_{\boldsymbol{G}} - \boldsymbol{\Theta}\,\widehat{\boldsymbol{V}} = \boldsymbol{F_E} \\ \boldsymbol{\Theta}^T\,\widehat{\boldsymbol{U}} + \boldsymbol{C_P}\,\widehat{\boldsymbol{V}} + \boldsymbol{Q_E} = \boldsymbol{0} \end{gathered} \tag{1}$$

where $\widehat{\boldsymbol{V}}$ and $\widehat{\boldsymbol{U}}$ are the nodal voltage and displacement vectors. $\widehat{\boldsymbol{V}}_{\boldsymbol{G}}$ and $\widehat{\boldsymbol{a}}_{\boldsymbol{G}}$ are the nodal velocity and acceleration vectors. $\boldsymbol{F_E}$ and $\boldsymbol{Q_E}$ are the mechanical and electrical loads. $\boldsymbol{K_T}$ is the total stiffness matrix which includes the material, geometric, and rotation stiffness, $\boldsymbol{D}$ is the damping matrix. $\boldsymbol{\Theta}, \boldsymbol{C_P}$ are the electromechanical and capacitance matrices, respectively.

3. Numerical results and discussion

3.1. Natural frequencies of vibration

One of the key points of the harvester design in this case is to obtain low natural frequencies to achieve the resonance phenomenon at low rotational speeds. All the simulations of the rotating system are performed at constant angular velocity from 1 – 4.5 Hz (60 to 270 rpm) for three hub positions, R = -0.03 m, R = 0 m and R = +0.03 m as shown in figure 2. The study of the harvester starts with the modal analysis; this requires designing specific dimensions of the beams and the tip masses of the harvester to obtain a low natural frequency. After this process, the first natural frequency of the non-rotating system is 4.7 Hz. Additionally, the influence of the hub distance (R) on the crossing-frequency phenomenon, which yields a natural frequency near the rotational speeds, is evaluated. The results show that the natural frequency decreases as the rotational speed increase for all three cases of R. The decrement of the natural frequencies is due to the softening effect of the rotating structure, i.e. the contribution of negative stiffness [7]. The values of the crossing-frequency are: 3.61 Hz at R = -0.03 m, 3.30 Hz at R = 0 m and 3.53 Hz at R=+0.03 m.

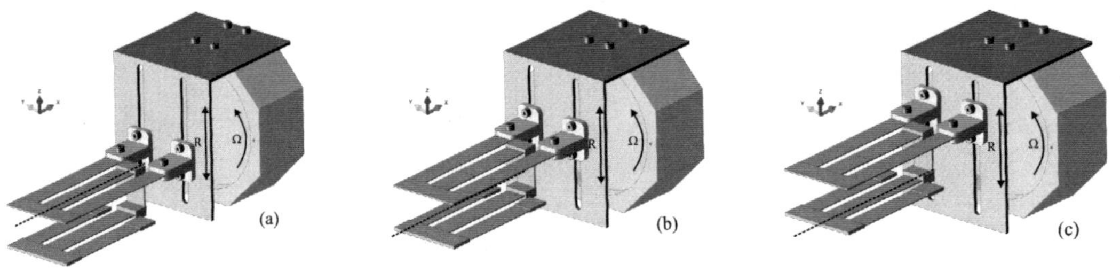

Figure 2. The schematic of three positions of the harvester (a) R=-0.03, (b) R=0 and (c) R=+0.03m).

3.2. Energy harvester performance

The harvester prototype is excited by rotational speed from 1 to 4.5 Hz (60 – 270 rpm) for all three R varying the load resistances values: 10, 100, 330 and 1000 kΩ. The voltage generated is shown in figures 3-5. Clearly, it can be seen that the maximum voltage is generated when the resonance phenomenon is reached and above this value, the voltage generation decreases. It is important to note that increasing the rotation speeds higher than 4.5 Hz (270 rpm), the voltage generation starts to increase due to the resonance of the second bending mode. However, going beyond this frequency, the stress limits of the whole structure of the harvester are exceeded.

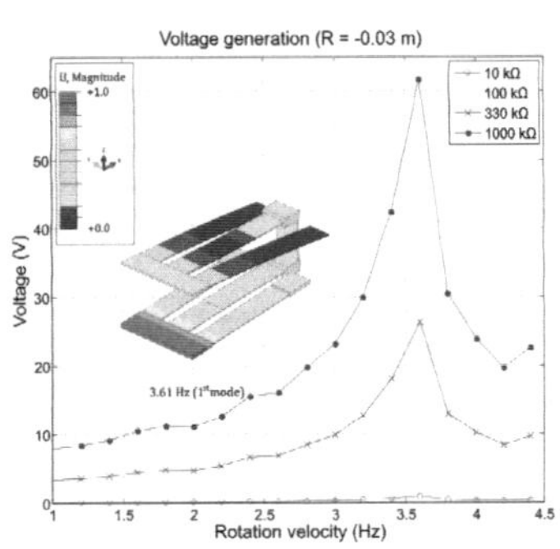

Figure 3. Voltage vs. rotation velocity (R = -0.03 m).

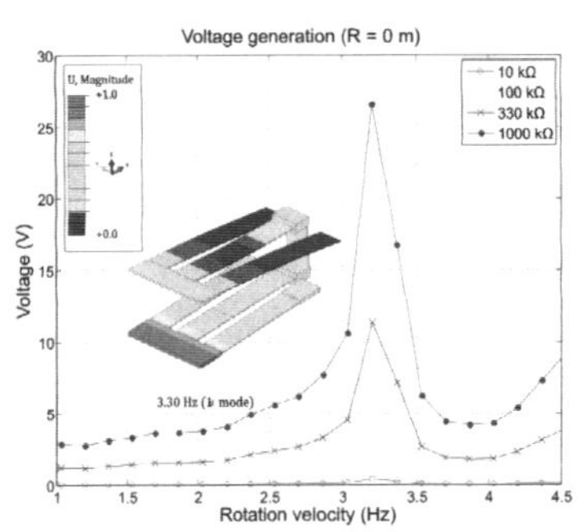

Figure 4. Voltage vs. rotation velocity (R = 0 m).

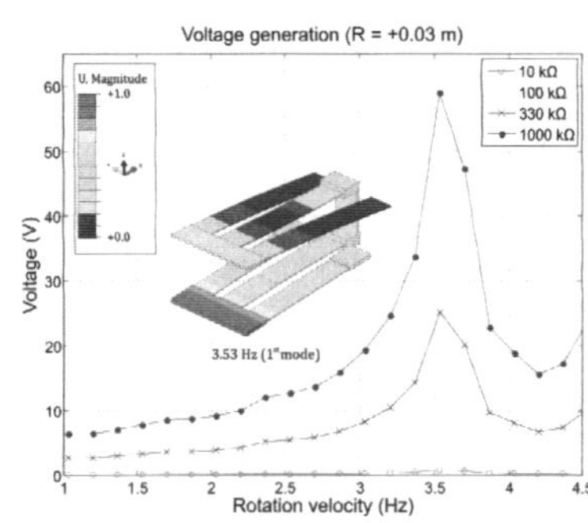

Figure 5. Voltage vs. rotation velocity (R = +0.03 m).

IOP Conf. Series: Journal of Physics: Conf. Series **1052** (2018) 012101 doi:10.1088/1742-6596/1052/1/012101

In figure 3, the maximum voltage is generated at 3.61 Hz, where the resonance phenomenon is reached. The maximum voltages for different load resistance values are: 0.92 V – 10 kΩ, 8.80 V – 100 kΩ, 26.36 V – 330 kΩ and 61.70 V – 1000 kΩ. Similarly, figures 4-5 show the maximum voltages generated as a function of rotation speed are: 0.40 V – 10 kΩ, 3.82 V – 100 kΩ, 11.35 V – 330 kΩ and 26.59 V – 1000 kΩ at R = 0 m and 0.88 V – 10 kΩ, 8.49 V – 100 kΩ, 25.20 V – 330 kΩ and 59.01 V – 1000 kΩ at R = +0.03 m. A summary of the maximum voltage and power generation as a function of rotational speed from 1 to 4.5 Hz (60 – 270 rpm) for all hub distances are illustrated in figures 6 and 7, respectively.

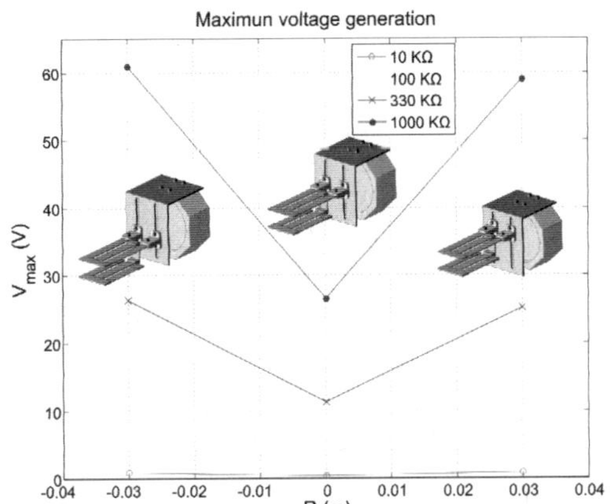

Figure 6. Maximum voltage generation as a function of R.

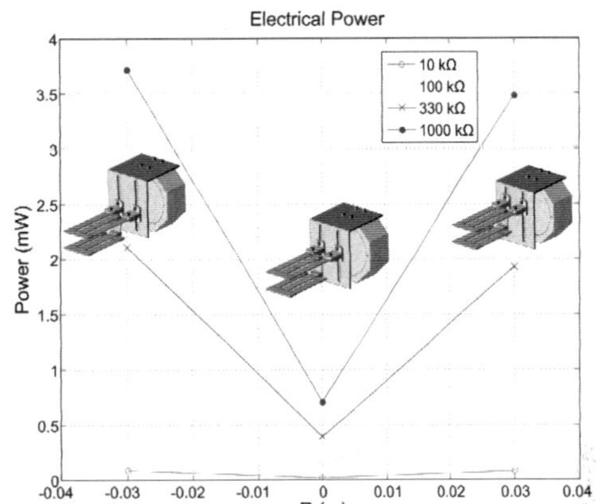

Figure 7. Maximum electrical power as a function of R.

As it can be observe in this last figure that the best configuration of the prototype is for R = -0.03 m (Ω = 3.61 Hz – 216 rpm, 61.7 V) which gives a value of electrical power $P = V^2/1\ M\Omega = 3.72$ mW.

4. Conclusions

This paper presents a novel low frequency rotating energy harvester. The rotating piezoelectric beam is formulated by means of a geometrically nonlinear finite element with six mechanical degrees of freedom and one electrical degree of freedom per node. In order to design the harvester, the simulations are performed from 1 to 4.5 Hz (60 – 270 rpm) rotation speeds. Several design features of the prototype such as the hub position (R), the tip masses and dimensions of the multi-beams are considered to achieve the crossing-frequency phenomenon as lower as possible. Regarding the harvester performance, promising results are obtained with a power of 3.72 mW at 3.61 Hz for R = -0.03 m. With this generation this prototype is good candidate to harvest energy from low power wind turbines.

Acknowledgments

The authors wish to acknowledge the supports financial from CONICET (PIP Nº:11220120100346), UNS, UTN FRBB and ANPCyT under grant PICT 2013-2065.

References

[1] Erturk A and Inman D J 2011 *Piezoelectric energy harvesting* (Wiley).
[2] Mitcheson P D *et al* 2008 *P. IEEE.* **96** 1457-1486.
[3] Gu L and Livernore C 2010 *Appl. Phys. Lett.* **97** 081904-1-3.
[4] Jin-Chen Hsu *et al* 2014 *Smart. Mater. Struct.* **23** 075013-1-13.
[5] Priya S and Inman D 2009 *Energy harvesting technologies* (Springer).
[6] Ramírez J M *et al* 2017 *Eng Struct* (Submitted).
[7] Saravia C M *et al* 2011 *Eur J Mech A-Solid* **30** 432-441.

T-shaped Piezoelectric Vibratory MEMS Harvester with Integration of Highly Efficient Power Management System

Seyedfakhreddin Nabavi, Ahmed Aljaroudi and Lihong Zhang

Department of Electrical and Computer Engineering, Faculty of Engineering and Applied Science, Memorial University of Newfoundland, St. John's, NL, Canada

Email: snabavi@mun.ca

Abstract. This paper introduces a T-shaped piezoelectric vibratory MEMS energy harvester. The prototype measurement results clearly demonstrate that the T-shaped harvester can achieve higher normalized power efficiency by a factor of 1.95 in comparison with the conventional piezoelectric harvester. Since the magnitude of harvested voltage in this type of energy harvesters is only a few millivolts, utilizing the traditional power electronic components is impractical. Thus, in this paper we also propose a highly efficient power management system without requiring external bias voltages. The numerical simulation illustrates that our proposed power management system can rectify and boost the harvested voltage up to 5V. In order to keep the output voltage on the load side at a stable level, a JFET transistor is utilized as a voltage limiter. Thus, the processed output voltage cannot exceed 4V even if a higher harvested voltage may be produced by the harvester.

1. Introduction

Energy harvesting from environmental resources has been recognized as a vivid solution for providing power supplies with no life time worries. Among the harvestable energy resources, energy harvesting from vibrations (including, wind flow, ocean wave, human motions and etc.) has gained higher attention, due to their ubiquity in environment and ease of trapping. Kinetic energy of vibration can be converted to electricity, by electrostatic-, electromagnetic- and piezoelectric-based techniques. Each technique above has its own advantages as well as drawbacks. Among them, the piezoelectric-based technique features sound advantages for MEMS or portable energy harvesters [1].

In the literature, various approaches, such as tuning proof mass size [2] or optimizing piezoelectric film and beam aspects [3][4], have been proposed to enhance energy conversion efficiency of the piezoelectric harvesters. Although those methods can enlarge efficiency to some extent, considerable efficiency enhancement is not possible due to the utilized conventional cantilevers with clamped-free boundary in the harvesters. On the other side, the magnitude of the harvested voltage by this type of energy harvesters is only a few millivolts. Therefore, utilizing the conventional rectifier circuitry (e.g., a silicon diode bridge) as a power management system for vibratory MEMS harvesters is impractical.

To overcome this aforementioned issue, Tan *et al.* [5] proposed an active rectifier. The proposed converter can rectify low input voltage (V_{in}=1.2V) with 70% efficiency. However, this converter includes two op-amps to magnify the harvested voltage, which are supplied with 3.3V external power supplies. Unfortunately this required external power supply makes this proposed method less useful for practical energy harvesters. To eliminate the required external power supply, Wahbah *et al.* [6] proposed a voltage doubler rectifier. Since this circuit can rectify minimum 1.8V AC signal with efficiency of 24%, most of the harvested voltage would be dissipated due to its low efficiency. In a recent study [7], efficiency of the rectifier interface circuit for piezoelectric MEMS harvesters has been largely improved up to 80% with a pre-charge battery required. However, this efficiency is valid only if the amplitude of the generated voltage is at least 2V. For the voltage at the millivolt level, the reported efficiency is zero.

Therefore, besides introducing a T-shaped MEMS energy harvester with the feature of high power density, we are motivated to propose a novel self-supplied power management circuit with extremely high efficiency in AC-to-DC signal conversion when the amplitude of the AC input signal is adequately low. This paper is organized as follows. A brief literature review on the MEMS efficiency improvement methods and different architecture of power management circuitry is surveyed in this section. In Section 2, the proposed structure of T-shaped piezoelectric MEMS harvesters is explained. The architecture of proposed power management system is discussed in Section 3. Section 4 reports the experimental results. Eventually, the conclusion is made in Section 5.

2. T-shaped piezoelectric MEMS harvester

The cantilevers with clamped-free boundary are recognized as the most common structure for piezoelectric vibratory MEMS harvesters. However, this type of energy harvesters have quite narrow operational bandwidth and in turn low efficiency in conversion of vibration to electricity [8]. Therefore, to eliminate these aforementioned issues, we are aimed to develop a novel piezoelectric MEMS harvester that can feature high power density. Our proposed harvester has a T-shaped structure with an integration of two proof masses at tip locations. This new structure allows our proposed harvester to vibrate in both torsion and bending modes, while the conventional harvesters just oscillate in the bending mode.

3. Power management system

The overall block diagram of the proposed power management system with the capability of working under extremely low input voltage is displayed in Figure 1. Thanks to low threshold-voltage MOSFET, the rectifier can rectify the harvested AC signal to DC with the least voltage difference. And this converted DC voltage is accumulated in the start-up capacitor (*C_startup*). Once its level reaches 1V, the voltage detector, which is a supervisory circuit, is activated so that the microcontroller is turned on. Thus, the microcontroller can generate constant PWM control signal for switching the boost-converter transistor. Consequently, the harvested voltage, whose amplitude can be boosted, is stored in load capacitor (i.e., *C_boost*). In order to provide a passive feedback control, the amount of the delivered voltage to load can be limited to a certain level by using a JFET transistor. By using numerical simulation, the performance of the system will be demonstrated in the following section.

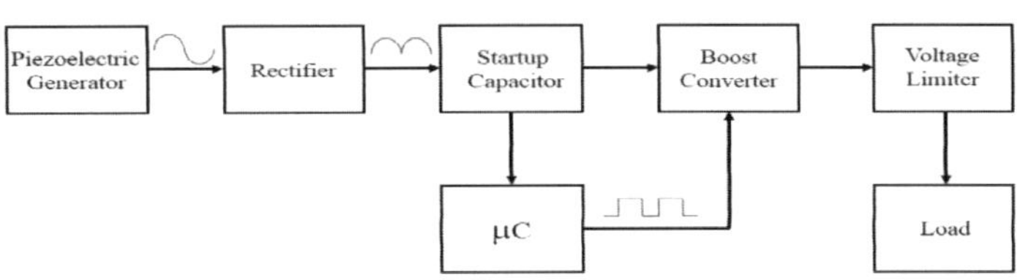

Figure 1. The overall block diagram of the proposed power management system.

4. Experimental results and discussion

In first step, the proposed T-shaped piezoelectric MEMS harvester, by using a commercial micro-fabrication process, i.e., *PiezoMUMPs*, was fabricated. In order to demonstrated and compare the proposed harvester superiority in terms of the energy conversion efficiency, we also fabricated the conventional piezoelectric harvester, which is a simple clamped-free cantilever. The SEM images of the fabricated piezoelectric harvesters with their sizes are illustrated in Figure 2.

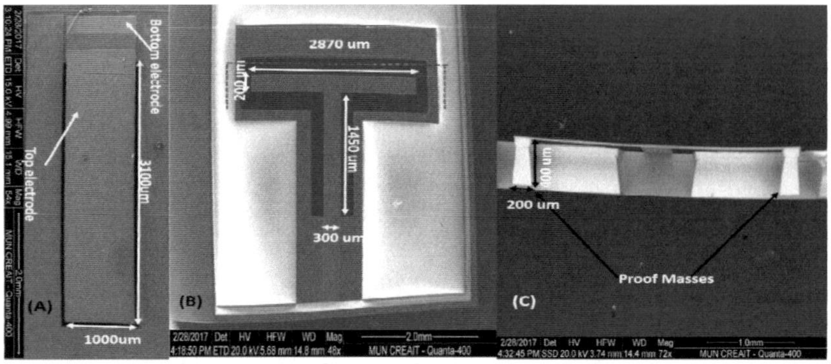

Figure 2. Fabricated piezoelectric MEMS energy harvesters (A) conventional unimorph cantilever, (B) top view and (C) side view of the proposed T-shaped harvester.

The prototyped harvesters were excited by a shaker (4809 manufactured by Brüel & Kjær), while the shaker excitation frequency is equal to resonant frequencies of the harvesters. Then, the Normalized Power Density (NDP) with various load resistances were measured for display in Figure 3. It is worth noting that the measured NDP for the T-shaped harvester is greater than the conventional piezoelectric one by a factor of 1.95, while the optimum load resistance for both harvesters is around 0.1MΩ.

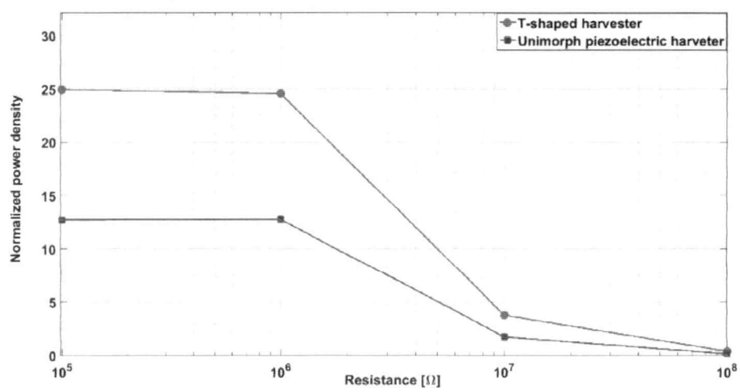

Figure 3. Measured normalized power density of the proposed T-shaped harvester (red) and conventional unimorph piezoelectric cantilever (blue).

The performance of the proposed power management system has been verified by numerical simulation. In this way, the schematic in Figure 4(A) was implemented in LTspice and it was assumed the piezoelectric generator can generate maximally 1.5V. It should be noted that all the utilized components are available in the market as commercial discrete devices. Therefore, their models were defined according to the provided information by the foundries. The performance of each individual part of the proposed power management system is demonstrated in Figure 4(B). It is obvious that generated 1.5V by the piezoelectric generator can be boosted up to 5V shortly. To limit the boosted voltage to a constant level, the JFET transistor with 4V threshold-voltage is utilized. As shown in Figure 4(B), the proposed power management system, without requiring external bias voltage, can enlarge the harvested voltage sufficiently as the power supply for low-voltage consumer electronics products.

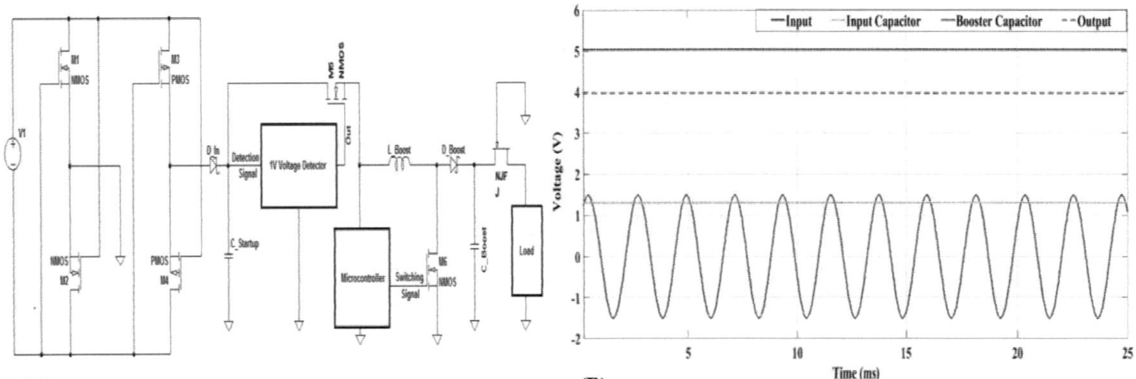

(A) **(B)**

Figure 4. (A) Schematic of the implemented power management system in LTspice, (B) The performance evaluation of the proposed power management system.

5. Conclusion

In this study, we presented a T-shaped piezoelectric MEMS energy harvester. The prototype measurement results show that it can achieve higher normalized power efficiency by a factor of 1.95 in comparison with the conventional piezoelectric harvester. Since the harvested voltage magnitude by this type of harvesters is only a few millivolts, we proposed a highly efficient power management system for converting AC signal to DC, and also enlarging the harvested voltage magnitude on the load side. The numerical simulation results clearly showed that the proposed power management system could boost the magnitude up to 5V when the harvested voltage was only 1.5V. To maintain a stable output voltage level, a JFET transistor as a voltage limiter was utilized. Thus, the delivered voltage at the load side cannot exceed 4V if a higher harvested voltage by the harvester is expected.

Acknowledgment

This work was supported in part by Natural Sciences and Engineering Research Council of Canada (NSERC), Canada Foundation for Innovation (CFI), Research and Development Corporation (RDC) of Newfoundland and Labrador (through its Industrial Research and Innovation Fund and ArcticTECH R&D Award), Memorial University of Newfoundland, and CMC Microsystems.

References

[1] S. Nabavi and L. Zhang, "Portable Wind Energy Harvesters for Low-Power Applications: A Survey," *Sensors*, vol. 16, no. 7, p. 1101, 2016.

[2] Y. Jia and A. A. Seshia, "Power optimization by mass tuning for MEMS piezoelectric cantilever vibration energy harvesting," *J. Microelectromechanical Syst.*, vol. 25, no. 1, pp. 108–117, 2016.

[3] S. Nabavi and L. Zhang, "MEMS piezoelectric energy harvester design and optimization based on Genetic Algorithm," in *IEEE International Ultrasonics Symposium (IUS),*2016, pp. 1–4.

[4] S. Nabavi and L. Zhang, "Design and Optimization of Piezoelectric MEMS Vibration Energy Harvesters Based on Genetic Algorithm," *IEEE Sensors Journal*, vol. 99, 2017

[5] Y. K. Tan and S. K. Panda, "Optimized wind energy harvesting system using resistance emulator and active rectifier for wireless sensor nodes," *Power Electron. IEEE Trans.*, vol. 26, no. 1, pp. 38–50, 2011.

[6] M. Wahbah, M. Alhawari, B. Mohammad, H. Saleh, and M. Ismail, "An AC--DC converter for human body-based vibration energy harvesting," *Microelectronics J.*, vol. 55, pp. 1–7, 2016.

[7] S. Du, Y. Jia, C. D. Do, and A. A. Seshia, "An Efficient SSHI Interface With Increased Input Range for Piezoelectric Energy Harvesting Under Variable Conditions," *IEEE J. Solid-State Circuits*, 2016.

[8] S. Nabavi and L. Zhang, "Design and Optimization of Wideband Multimode Piezoelectric MEMS Vibration Energy Harvesters," in *proceedings 31st Eurosesnosors, Paris, France*, 2017, vol. 1, no. 4, pp. 586: 1-4.

Development of vibration energy harvester with 2D mechanical metamaterial structure

Y. Umino[1], T. Tsukamoto[1], S. Shiomi[1], K. Yamada[1], and T. Suzuki[1,2]

[1]Graduate School of Science and Technology, Gunma University, 1-5-1 Tenjin cho, Kiryu-shi, Gunma, Japan
[2] JST PRESTO, Japan

E-mail: suzuki.taka@gunma-u.ac.jp

Abstract. In this paper, we propose a vibration energy harvester with high power generation efficiency in low-frequency wide band. The proposed device is a bimorph type composed of two piezoelectric layers and a middle elastic layer having a flexible mechanical metamaterial structure. By controlling the flexibility of the elastic layer by the microstructure, the strain of the piezoelectric layer and the power generation amount are increased. As a result, the proposed device achieves high power generation of 47.4 μW in the low-frequency band of 21.1 Hz. This power is 1.6 times larger than that of the conventional flat plate and satisfies the minimum electric power amount required as a sensor node for WANs.

1. Introduction

Wireless sensor networks (WSNs) are composed of huge amounts of sensor nodes, and energy harvester is expected to be used for power supply of the sensors [1]. In particular, many piezoelectric cantilever type energy harvesters have been proposed due to high power generation efficiency [2]. Main vibration source such as human motion and household appliance has the low-frequency band below 30 Hz [3][4]. In the vibration power generation, the power generation amount is closely related to the resonance frequency. So, it is necessary to reduce the resonance frequency in power generation under the low-frequency wide band. However, since the resonant frequency increases as the device become smaller, it is difficult to realize high power generation efficiency in the low-frequency wide band. Most of the reported piezoelectric energy harvester usually corresponds to a high-frequency band over 100 Hz [5]. In this research, we focus on the mechanical metamaterial structure [6] to control mechanical properties exceeding the material own properties. We propose a bimorph type harvester composed of two piezoelectric layers and a middle elastic layer having a flexible mechanical metamaterial structure for vibration energy harvesting from human motion and household appliance.

2. Design of mechanical metamaterial energy harvester

2.1. Principle of mechanical metamaterial energy harvester

The proposed harvester is a bimorph type as shown in figure 1, with the neutral axis at the center of the device. Since strain and output voltage generated in the piezoelectric layer are proportional to the distance from the neutral axis, increasing the distance between the piezoelectric layer and the neutral axis can increase the power generation amount. The proposed harvester has a flexible 2D metamaterial structure in the elastic layer and maintains low rigidity even in a thick elastic layer. Therefore, since the device can obtain large strain in the piezoelectric film while maintaining high flexibility, high power generation amount and low resonance frequency are simultaneously realized.

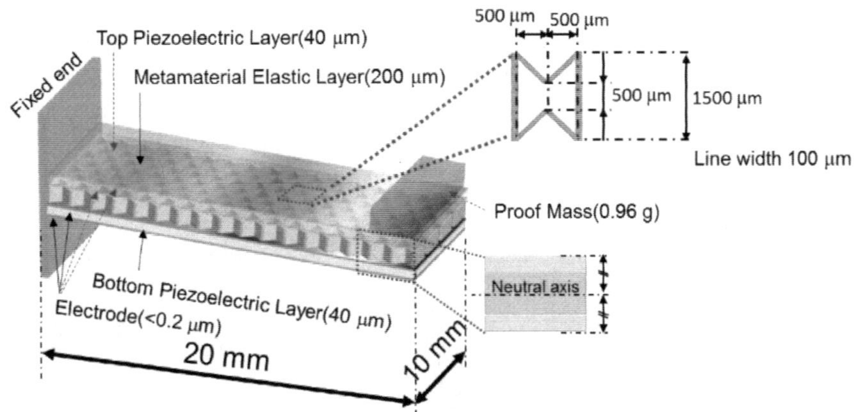

Figure 1. Schematic of the metamaterial energy harvester. The piezoelectric layer is a flexible organic material PVDF and the elastic layers are made of polymer material SU-8 which is a photoresist capable of thick film patterning. The device is fabricated by a simple process only by photolithography.

2.2. FEM analysis

Prior to device fabrication, we estimated the resonant frequency of the metamaterial harvester using modal analysis by FEM (Finite Element Method). Table 1 shows the resonant frequency obtained by using by FEM analysis software COMSOL Multiphysics 5.2a, and mode shape is shown in figure 2. Compared with the flat plate type, the metamaterial type has a resonant frequency of 80% and the flexural rigidity is reduced to 62%. From the results, it is found that the metamaterial elastic layer improves the flexibility under same device size.

Table 1. Results of FEM analysis. Each device has same thickness of the elastic layer and same device volume.

		Meta material	Flat plate
Resonant	1st	21.1	26.2
frequency	2nd	94.2	122.8
[Hz]	3rd	253.8	314.4
Flexural rigidity [Pa·m^4]		3.42×10^{-5}	5.54×10^{-5}

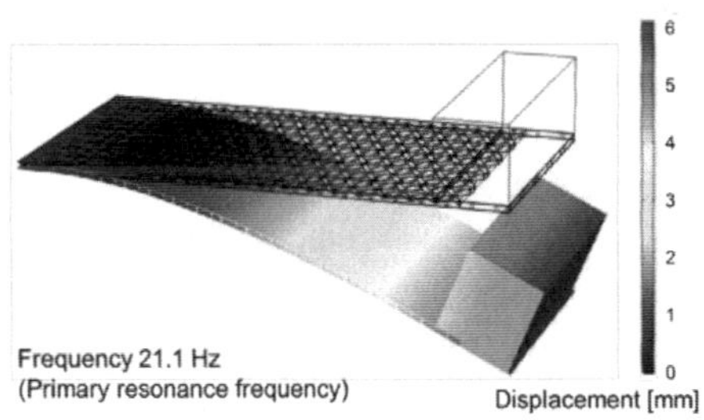

Figure 2. Mode shape of the metamaterial harvester at first resonant frequency. Three mode shapes of the proposed metamaterial harvester from first to third resonant frequency are as same as these of the conventional cantilever beam with the flat plate type elastic layer.

3. Experiments

3.1. Device fabrication

Figure 3 shows a photograph of the fabricated device and a SEM image of the elastic layer. As a piezoelectric layer of the device, a PVDF film (KF Piezo, Kureha Corp.) was used. The film has a film thickness of 40 μm and Al electrodes are deposited on both sides.

The fabrication process is divided into three steps by using two PVDF films. In the first step, a SU-8 elastic layer is made by photolithography on one PVDF film. Here, an elastic layer having a mechanical metamaterial structure is prepared. In the next step, the other PVDF film is separately cut out and a SU-8 thin film with a film thickness of 2 μm is formed on the film. In the final step, Using the formed SU-8 thin film as an adhesive layer, a PVDF film adheres to the metamaterial elastic layer made in the first step.

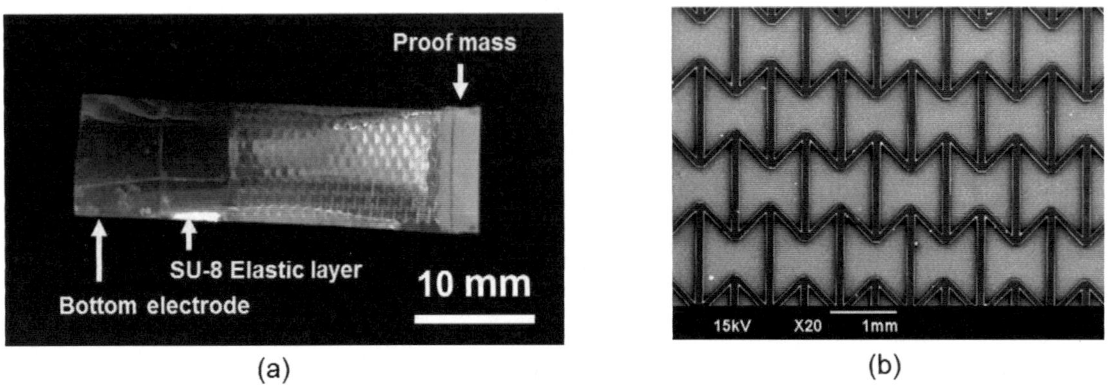

Figure 3. (a) Photograph of the fabricated metamaterial harvester, (b) SEM image of metamaterial elastic layer. The line width of the metamaterial structure is 100 μm and the aspect ratio is 2 with respect to the film thickness. The flexibility can be improved by optimizing the 2D structure by reducing the line width.

3.2. Vibration test

We evaluated the power generation and resonant frequency of the harvester by the vibration test. The experimental setup is shown in figure 4. To control acceleration under each vibration frequency, the excitation force was adjusted by an acceleration sensor fixed on the vibration table. The output voltage of the harvester was measured by an oscilloscope with an internal resistance of 10 MΩ. The power generation amount was calculated by measuring the power on one side. In the calculation, it is assumed that the same power generation amount generated because both piezoelectric films were at equal distances from the neutral axis.

The experimental frequency response curve of the fabricated device is shown in figure 5, and the power generation amount at resonant condition is shown in table 2. As a result, we achieved a high power generation amount of 47.4 μW at low-frequency of 21.1 Hz, which is 1.59 times higher than that of the flat plate type. Since the occupied volumes of the devices are equal in the two devices, the increase in the power generation amount directly corresponds to the increase in the energy density. The electric power generation by the piezoelectric cantilever is proportional to the square of the tip displacement. From table 2, the square of the tip displacement ratio of the metamaterial and the flat plate types is 1.56. This calculated value was in good agreement with 1.59 which is the ratio of actual power generation amount.

In addition, the response frequency capable of generating electric power is also widened. The bandwidth in the metamaterial harvester on the power generation amount of over 10 μW is 3 Hz, which is 1.5 times wider than one of the flat plate type harvester.

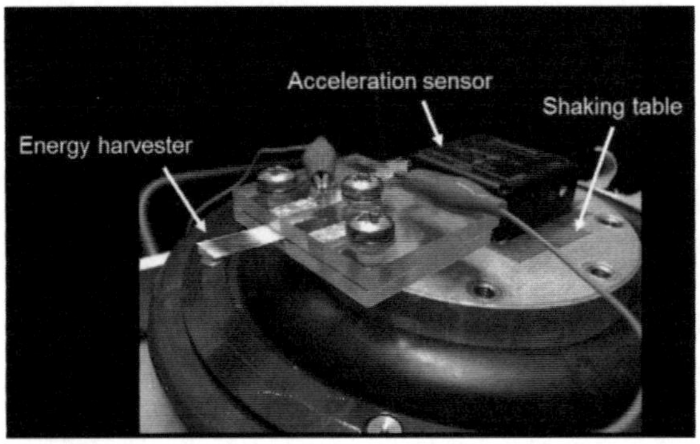

Figure 4. Photograph of experimental setup of the vibration test. The electrode of the energy harvester is directly connected to the oscilloscope.

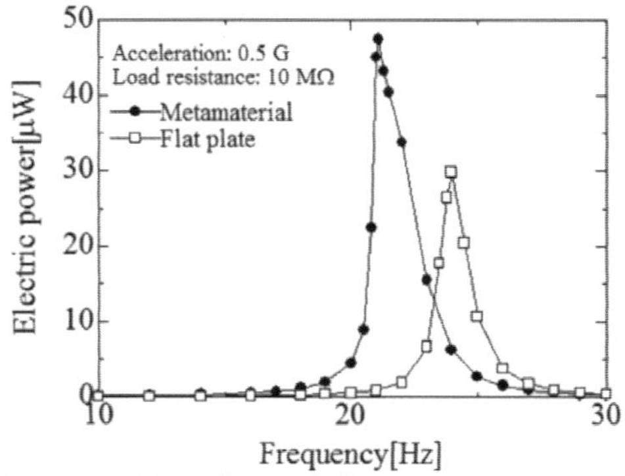

Figure 5. Experimental frequency response of power generation amount. In the proposed device, the resonance frequency is lowered and the power generation is increased as compared with the flat plate type.

Table 2. Power generation amount at resonant condition. Oscillating condition is same as figure 5. The energy density is a value of dividing the power generation amount by the device volume.

	Resonant frequency [Hz]	Output voltage [V]	Total electric power [μW]	Energy density [μW/cm^3]	Tip displacement [mm]
Meta material	21.1	15.4	47.4	55.4	7.5
Flat plate	24.0	11.7	29.8	34.8	6

4. Conclusion

In this study, we propose the piezoelectric vibration energy harvester for the low-frequency band. In the proposed harvester, the lower frequency and the higher power generation than the conventional flat plate type harvester are achieved by the metamaterial structure of the elastic layer. In the near future, we plan to optimize the device shape such as structure and film thickness for various applications.

Acknowledgement

This research was partially supported by JSPS Science Research Grant JP17H03196, JST PRESTO Grant Number JPMJPR15R3, and MEXT Nanotechnology Platform Project (The University of Tokyo Microfabrication Platform).

References

[1] S. Faisal and Z. Sherali, 2016 *Renewable and Sustainable Energy Reviews* **55** 1041–1054
[2] H.S. Kim, J.H. Kim and J. Kim 2011 *International Journal of Precision Engineering and Manufacturing* **12** 1129–1141
[3] E. Dechant, F. Fedulo, D.V. Chashin, L.Y. Fetisov, Y.K. Fetisov and M. Shamonin, 2017 *Smart Materials and Structures* **26** 065021
[4] K Ylli, D Hoffmann, A Willmann, P Becker, B Folkmer and Y Manoli 2015 *Smart Materials and Structures* **24** 025029
[5] C.F. Wei and X.J. Jing, 2017 *Renewable and Sustainable Energy Reviews* **74** 1–18
[6] B. Sukhwinder 2015 *e-Polymers* **15** pp 205–215

Three-axis MEMS DC magnetic sensor using magnetic force interaction with the piezoelectric effect

Po-Chen Yeh[1#], Hao Duan[1#], and Tien-Kan Chung[1,2*]

[1] Department of Mechanical Engineering, National Chiao Tung University, Hsinchu, Taiwan
[2] International College of Semiconductor Technology, National Chiao Tung University, Hsinchu, Taiwan
[#]Equal Contribution (Co-First Author).
*Corresponding Author: tkchung@nctu.edu.tw

Abstract. In this paper, we report a novel three-axis MEMS DC magnetic sensor using magnetic force interaction and piezoelectric effect. The sensor includes a back-side-etched silicon diaphragm, a sol-gel PZT thin film with electrodes, and two patterned electroplated Ni thick films. Each patterned Ni film has specific magnetization direction, defined as specific sensing area. As a DC magnetic fields is applied to the sensor, impulse magnetic force and torque are induced to deflect and twist specific Ni films, respectively. Through the mechanical coupling between the Ni films and the PZT film, the deflection and bending are transmitted to the PZT film to produce piezoelectric peak voltages. Experimental results show that, by analysing full duration and half maximum of voltage outputs produced in two sensing areas, the sensor is able to sense three-axial DC magnetic fields.

1. Introduction

Today, magnetic sensors are widely used in personal electronics, data storage systems, automotive and navigation systems, etc. [1]. Among these applications, Hall-effect sensors and magnetoresistive (MR) sensors are two major magnetic sensors used in the applications. However, the Hall-effect sensors need a magnetic concentrator to sense a magnetic field parallel to the sensors, while the MR sensors require an additional reset-coil to reset the magnetization direction of magnetic materials used in the sensors. In addition, an additional power source is required for both sensors (i.e. not "self-powered"). Recently, piezoelectric based magnetoelectric (ME) sensors [2] have been intensively studied. Due to the ME effect, the ME sensor can directly sense an AC magnetic field under a specifically applied DC magnetic bias field. However, regardless of AC or DC magnetic field sensing, these typical ME sensors can only achieve single axis magnetic field sensing due to the sensing mechanism. In addition, the ME sensors also require an additional power source to provide the DC magnetic bias field. This means the sensor is not fully self-powered.

Hence, to address these issues (i.e. only can achieve single axis magnetic field sensing and must require additional power source), we recently demonstrated a self-powered three-axis piezoelectric MEMS AC magnetic sensor [3]. However, although the AC magnetic field sensing has many applications [4], the DC magnetic field sensing is more practical and thus has more general applications [5]. Thus, based on our previous research work [6], we provide a novel piezoelectric three-axis MEMS DC magnetic sensor in this paper.

2. Design

Figure 1(a) is an illustration of top view of the magnetic sensor. Figure 1(b) and Figure 1(c) shows the cross-section view of the sensor along C-C' line in figure 1(a). As shown in figure 1(a) and 1(b), the sensor consists of a back-side-etched silicon diaphragm, a sol-gel lead-zirconate-titanate (PZT) piezoelectric thin film with top and bottom Pt/Ti electrodes, and two patterned electrodeposited Ni thick films. The two Ni films are magnetized orthogonally as two different sensing areas (i.e., area A and B has magnetization direction along +Z and +X axial direction, respectively).

When applying a DC magnetic field to the sensor, impulse magnetic force and torque are induced, and consequently corresponding deflection and torsion are produced in specific sensing areas (i.e., in specific magnetized Ni films). The mechanical coupling between the Ni films and the PZT film transmits the deflection and torsion to the PZT film. These further transform PZT film to produce peak voltage outputs due to the piezoelectric effect. The instant peak voltages are generated/detected while the DC magnetic field is switched "on" and "off". Based on this, sensing principle of three-axis magnetic fields is shown in figure 1(b)-(c). When applying a X- or Y- axis DC magnetic field to the sensor, an impulse magnetic torque is induced in sensing area A while an impulse magnetic force is induced in sensing area B. Similarly, when applying a Z-axis DC magnetic field, an impulse magnetic force is induced in area A while an impulse magnetic torque is induced in sensing B. These forces and torques produce corresponding deflections and torsions, consequently generate associate strains, and finally convert to different piezoelectric voltage outputs. Though voltage analysis, the applied three-axis magnetic fields are obtained.

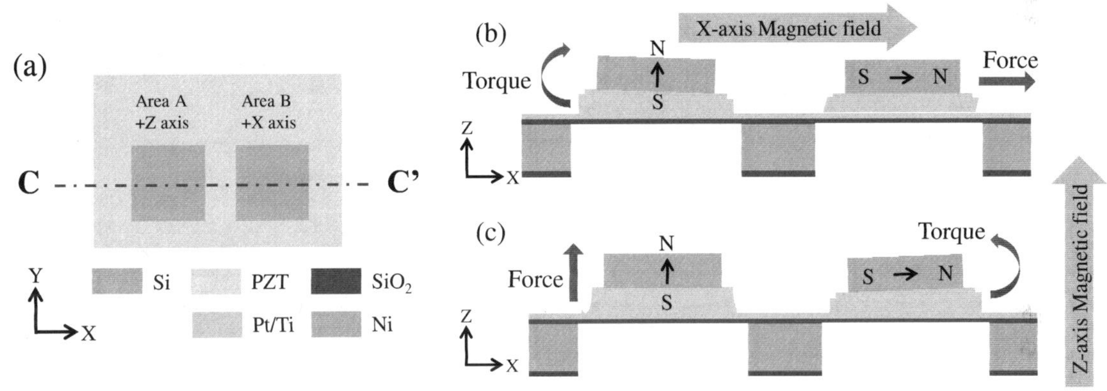

Figure 1. The illustration of (a) top view, (b) and (c) C-C' cross-section view of the sensor when applying in-plane (X-axis) and out-of-plane (Z-axis) magnetic fields, respectively. Magnetization directions of nickel films are indicated by black arrows [3].

3. Fabrication and testing

Fabrication of the sensor is reported in our previous works [3], [7]. After the sensor is fabricated, two sensing areas are magnetized to different but specific directions [4] by applying a DC magnetic field of 3000 Oe [8]. The photographs of the sensor are shown in figure 2.

Figure 3(a)-(b) are the illustration and photograph of the test setup, respectively. The setup consists of a function generator, C-shape electromagnet, charge amplifier, voltage amplifier, and oscilloscope. The sensor, electromagnet, and charge amplifier are placed in a magnetic field shielding box. By the shielding, ambient magnetic noises is eliminated during testing. The testing procedure is described below. The function generator is used to provide a square-wave signal (3 Hz, 50% duty cycle) to the electromagnet to produce a DC magnetic field ranging from 1 to 20 guess. The charge amplifier and voltage amplifier are used to amplify voltage outputs generated by the sensor. Finally, the oscilloscope is used to record the amplified voltage outputs.

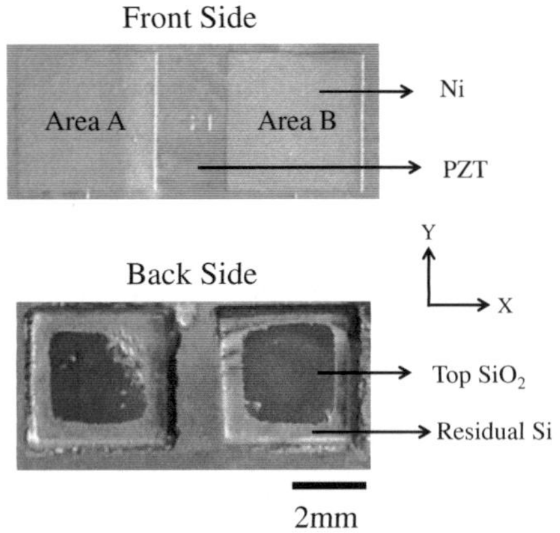

Figure 2. The photograph of the sensor [3].

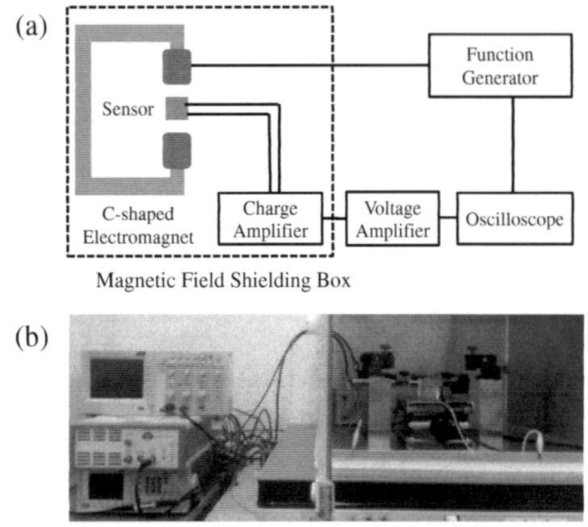

Figure 3. (a) Illustration (b) photo of test setup.

4. Result and Discussion

The voltage waveforms in sensing area A when applying X- and Z- axis magnetic fields of 20 gauss are shown in figure 4(a)-(b). The voltage waveforms in sensing area B when applying X-, Y-, and Z-axis magnetic fields of 20 gauss are shown in figure 4(c)-(e). The enlarged waveforms are shown in figure 4(f)-(j). When applying X- and Z- axis magnetic field to area A, the FDHM of the voltage outputs are 24.2, 1.0 millisecond. When applying X-, Y-, and Z- axis magnetic fields to area B, the FDHM of the voltage outputs are 22.2, 22.8, 0.9 millisecond.

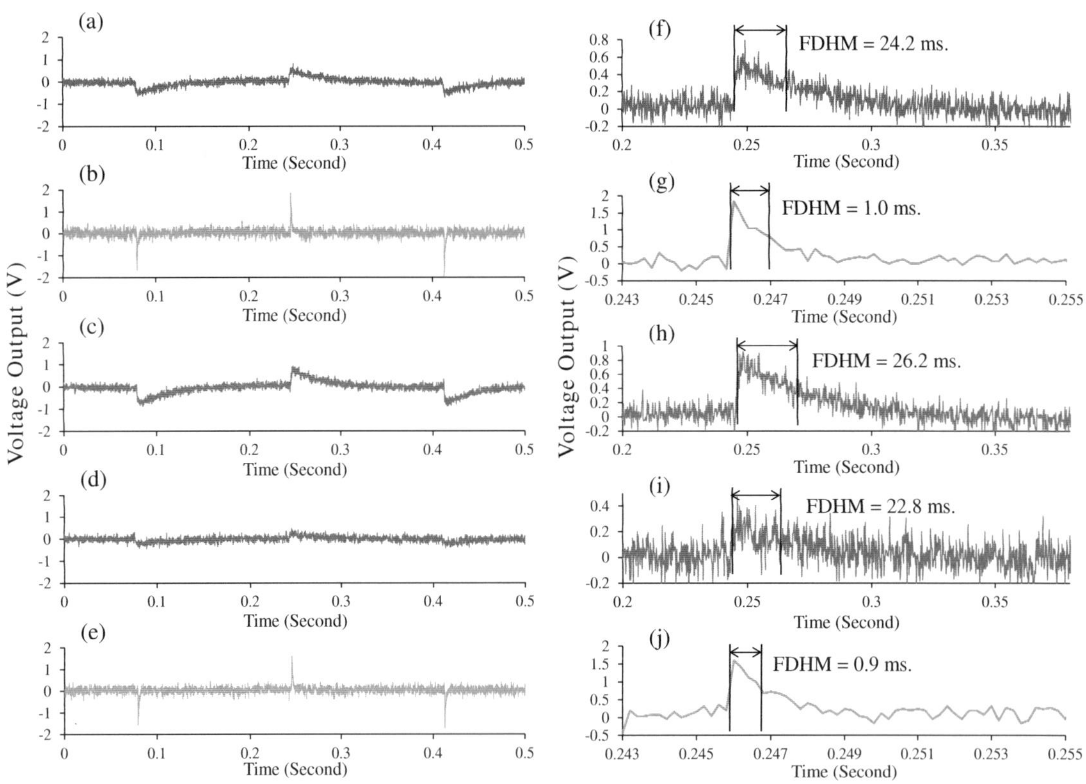

Figure 4. Voltage outputs in two sensing areas of the sensor under different axial magnetic fields: (a)-(b) voltage outputs in area A when applying X- and Z- axis magnetic fields. (c)-(e) voltage output in area B when applying X-, Y-, and Z- axis magnetic fields. (f)-(j) enlarged waveforms of (a)-(e).

Furthermore, we varied the applied magnetic field and obtain the results shown in figure 5(a)-(b). In figure 5(b), the sensitivities in linear regions of area A when applying X-axis and Z-axis magnetic fields are 0.03705 V/G in 2-20G and 0.08776 V/G in 3-20G, respectively. The sensitivities in linear regions of area B when applying X-, Y-, and Z -axis magnetic fields are 0.04097 V/G in 4-20G, 0.02613 V/G in 6-20 G, and 0.07525 V/G in 13-20 G, respectively. Finally, by comparing the FDHM and sensitivities, the direction and magnitude of these applied axial magnetic fields can be distinguished.

(a)

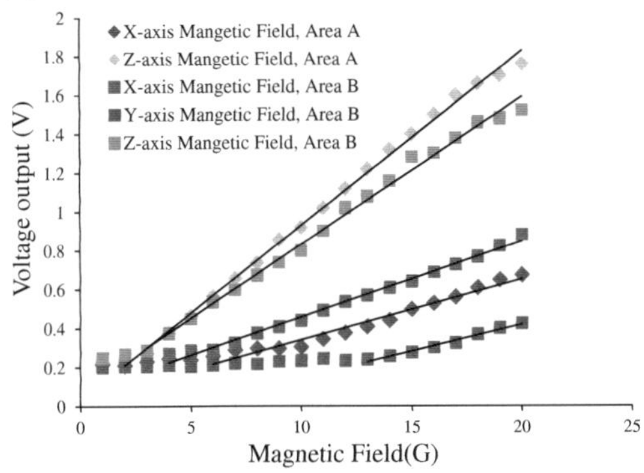

(b)

Magnetic Field Direction	Magnetization Direction of Nickel Thick Film: Area A Along +Z Direction	Magnetization Direction of Nickel Thick Film: Area B Along +X Direction
X	0.03705 V/G (2-20 G)	0.04 V/G (2-20 G)
Y		0.02613 V/G (4-20 G)
Z	0.08776 V/G (6-20 G)	0.07525 V/G (13-20 G)

Figure 5. Sensing performance and sensitivity: (a) Voltage outputs in different area of the sensor under different axial magnetic fields. (b) Sensing sensitivities in linear regions of the voltage outputs.

5. Conclusion

We demonstrated a self-powered piezoelectric three-axis MEMS DC magnetic sensor. Experimental results show that the sensor can successfully senses three-axial DC magnetic fields and thereby will create more magnetic-field-sensing applications in the future.

Acknowledgement

The work is supported by Taiwan Ministry of Science and Technology (No. 105-2628-E-009-001-MY2)

References

[1] Lenz J and Edelstein S A 2006 *IEEE Sens. J.* **6** 631–49
[2] Marauska S, Jahns R, Greve H, Quandt E, Knöchel R and Wagner B 2012 *J. Micromech. Microeng.* **22** 065024
[3] Yeh P C, Duan H, and Chung T K 2017 *Int. Conf. on Solid-State Sensors, Actuators and Microsystems.* (Kaohsiung, Taiwan)
[4] Yeh P C, Chung T K, Lai C H, and Wang C M 2016 *Appl. Phys. A. Mater. Sci. Process.* **122** 1–10
[5] Wang D C, Li P and Wen Y M 2016 *Rev. Sci. Instrum.* **87** 105001
[6] Hung C F, Yeh P C, and Chung T K 2017 *Sensors* **17** 308
[7] Chung T K, Keller S and Carman G P 2009 *Appl. Phys. Lett.* **94** 132501
[8] Chung T K, Wang H M, Chen Y J, Lin S H, Chu H J, Lin P J and Hung C F 2016 *Appl. Phys. Expres.* **9** 043003

Film stress dependence on deposition temperature in scandium aluminium nitride thin film

R Takei[1], N Makimoto[1], T Tabaru[2], M Akiyama[2], T Itoh[1,3] and T Kobayashi[1]

[1] Research Center for Ubiquitous MEMS and Micro Engineering, National Institute of Advanced Industrial Science and Technology, 1-2-1 Namiki, Tsukuba, Ibaraki, Japan
[2] Advanced Manufacturing Research Institute, National Institute of Advanced Science and Technology, 807-1 Syukumachi, Tosu, Saga, Japan
[3] Graduate School of Frontier Science, The University of Tokyo, 5-1-5 Kashiwanoha, Kashiwa, Chiba, Japan

r.takei@aist.go.jp

Abstract. This paper reports investigation of stress in ScAlN thin films deposited on 4-in silicon wafers at various substrate temperature. ScAlN has a 5 times higher piezoelectric constant, but simultaneously ScAlN thin film has a large film stress. To relax the stress, the substrate temperature was changed between 170 and 250°C. Film stress, crystal quality, piezoelectric constant, and Sc concentration was investigated for the deposited ScAlN films. The film stress was relaxed as the substrate temperature increases.

1. Introduction

A ScAlN thin film has a large piezoelectric effect and can be deposited using a standard RF magnetron sputtering method [1,2]. Therefore, the ScAlN film is promising to a piezoelectric silicon micro electro mechanical systems (MEMS) vibration energy harvester. However, the ScAlN thin film has a large compressive stress, making it challenging to process the ScAlN film on a silicon wafer. In this work, stress, crystal structure and piezoelectric constant were investigated for the ScAlN thin films deposited at various substrate temperature to relax the strong stress.

2. Experimental

Three types of ScAlN thin films were deposited on each 4-in. Si wafers at different substrate temperature. A thermal oxide film was formed to electrically isolate the ScAlN film from the substrate, and then a 0.1-µm-thick Pt thin film was deposited with a 0.03-µm-thick Ti adhesion layer. A ~2-µm-thick ScAlN film was deposited on the Pt film by RF magnetron sputtering using an AlSc alloy target at substrate temperatures of 170, 210, and 250°C, respectively. The deposition conditions are summarized in Table 1. A gas mixture of 4 sccm nitrogen and 6 sccm argon was introduced to the chamber, and the gas pressure was controlled at 0.3 or 0.4 Pa.
A film stress of the deposited ScAlN films was investigated. The stress was evaluated by Stoney's formula from measurement of a wafer bow. The crystal structure and orientation was also investigated by X-ray diffraction (XRD) measurements.

Table 1. Deposition conditions.

	RF power (W)	Ar/N_2 flow (sccm)	Pressure (Pa)	Temp. (°C)
#1	500	6/4	0.3	170
#2	400	6/4	0.4	210
#3	400	6/4	0.4	250

Capacitors of the ScAlN film sandwiched with two Ti/Pt electrodes were fabricated to investigate the piezoelectric constant of the ScAlN films deposited at the temperature of 170 and 250°C. On the deposited ScAlN films, Ti/Pt thin films were formed again, and then dot-capacitors with a diameter of 400 μm were fabricated using photolithography and argon ion beam milling. Voltage was applied to the capacitor via two probes, and displacement in the perpendicular to the film was measured using a double beam interferometer. The effective piezoelectric constant ($d_{33,f}$) was extracted from the measured displacement.
The concentration ratios of Sc were measured for the ScAlN films deposited at the temperature of 170 and 250°C by Rutherford backscattering spectrometry. For this measurements, the ScAlN films were deposited under the similar conditions to the deposition described above.

3. Results and Discussion

Figure 1 shows stress of the ScAlN films deposited at various substrate temperature, indicating that the stress is mitigated as the temperature increases. The stress at the temperature of 250°C is half of that at 170°C.

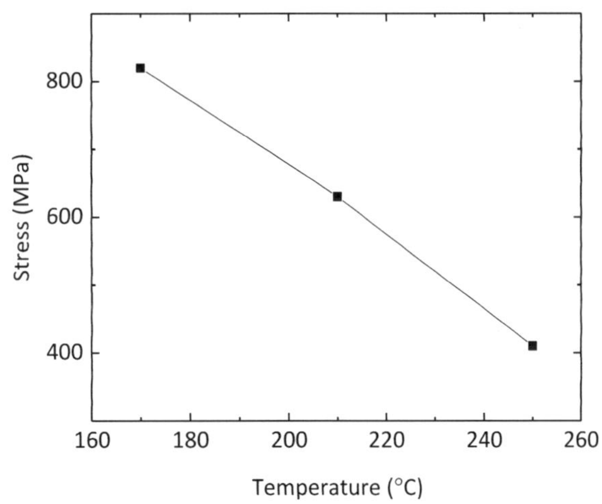

Figure 1. Stress of the ScAlN films.

Figure 2 shows 2D-XRD patterns of the ScAlN films. Diffraction peaks of Pt-(111) and ScAlN-(0002) are shown together in the patterns. Pt exhibits a c-axis crystal orientation, but the c-axes of the ScAlN deposited at 170, 210, and 250°C were tilted by 4.1, 16.0, and 16.5°, respectively. Figure 3 shows theta-2 theta scan around ScAlN-(0002) peak. The FWHM becomes wider as the substrate temperature increases, indicating that an increase in the substrate temperature slightly degrades the crystal quality. Also, the peak position slightly shifts to larger theta as the temperature increases. This is because the compressive stress is relaxed so that the lattice constant becomes larger as the temperature increases.

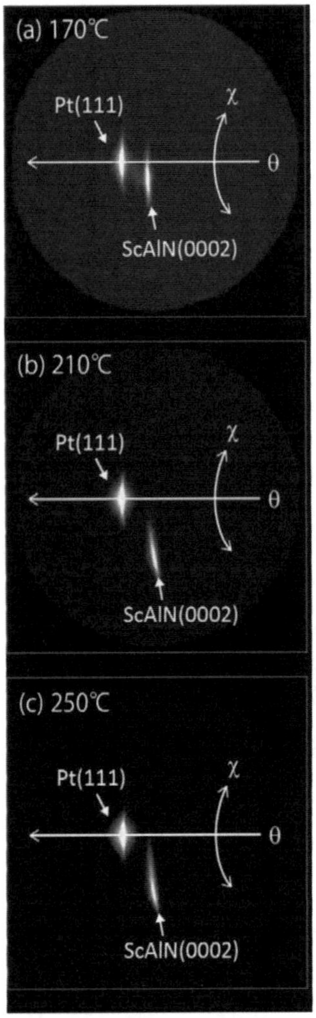

Figure 2. 2D-XRD patterns.

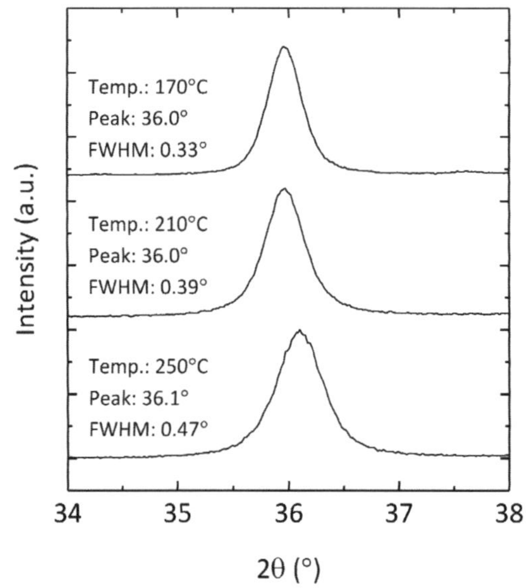

Figure 3. XRD patterns of a theta-2 theta scan.

Figure 4 shows displacement of the capacitors when voltage was applied to them. The effective piezoelectric constants ($d_{33,f}$) were measured to be 9.3 and 6.9 pm/V for the ScAlN films deposited at the temperature of 170 and 250°C, respectively. And the Sc concentrations were 39.3 and 37.0%, respectively. A decrease in the concentration from 39.3% to 37.0% decreases the piezoelectric constant by ~10% [2]. However, $d_{33,f}$ of the ScAlN film deposited at 250°C degraded by 25% as compared with the film deposited at 170°C. In addition, the degradation of $d_{33,f}$ due to the inclination of the c-axis is less than 5%. Therefore, an increase in the deposition temperature more degrades the piezoelectric effect. However, although the decrease in piezoelectric constant is 25%, the film stress can be dramatically relaxed to be half.

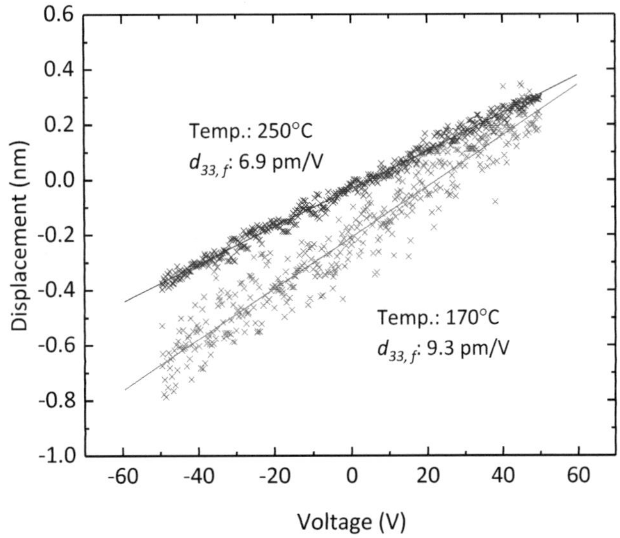

Figure 4. Displacement of the ScAlN capacitors.

4. Conclusions

In this work, we investigated the piezoelectric constant of the ScAlN films deposited at various deposition temperature. An increase in the temperature degrades the piezoelectric effect in the ScAlN films, but simultaneously the film stress can be dramatically relaxed. Therefore, this stress relaxation method may be effective to find an appropriate design of a ScAlN MEMS device.

[1] Akiyama M, Kamohara T, Kano K, Teshigahara A, Takeuchi Y and Kawahara N 2009 *Adv. Mater.* **21** 593
[2] Akiyama M, Kano K and Teshigahara A 2009 *Appl. Phys. Lett.* **95** 162107

Acknowledgments

This work was partly supported by the New Energy and Industrial Technology Development Organization (NEDO).

Impact-driven up-conversion in piezoelectric MEMS energy harvesters with pulsed excitation

Pontus Johannisson, Fredrik Ohlsson, Cristina Rusu
RISE Acreo AB, Arvid Hedvalls Backe 4, SE-411 33 Gothenburg, Sweden
E-mail: pontus.johannisson@ri.se

Abstract. The potential of impact-driven frequency up-conversion in a MEMS EH is evaluated using numerical simulations. The investigated design is compared to a conventional cantilever EH in terms of output power and loss rate. The upshifting can lead to significantly increased output power at a similar loss rate but as the time scale for the loss is long, the benefit is limited. This also requires an effective upshifting process. The design of the impact introduces a length scale that must be selected with excitation, gravity, and pre-stress taken into account. This makes this type of EH application-dependent as a non-optimal choice may result in low output power.

1. Introduction

One way to power sensors is with piezoelectric energy harvesters (PEHs), where PZT cantilevers is a common choice [1]. One alternative is a bridge configuration, which, due to its inherent nonlinearity, has been suggested as a way to adapt to an increasing excitation frequency [2]. Further benefits are the mechanical robustness and the uniform PZT strain, which is higher at a given deflection than in a comparable cantilever [3, 4]. Another concept is the use of impacts to obtain frequency upshifting.[1] One example is a scrape-through PEH [6], where a high-frequency PEH (PZT-H) was excited by impacts from a low-frequency PEH (PZT-L). The PZT-H, however, generated little power (30 nW) compared to the PZT-L (145 nW). This makes the approach questionable since the space needed for the PZT-H could be used to optimize the PZT-L. In a PEH by Gu and Livermore [7], power was mainly harvested *during* impact.

In this paper, impact-driven frequency up-conversion in a MEMS PEH is evaluated using numerical simulations. The discussion is kept as general as possible to assess the approach. Assuming a medical application, the use of magnets is not allowed and the excitation is pulsed.

2. Problem statement

For manufacturing simplicity, the structure is made of silicon with PZT deposited on top. With this method, the PZT is thin, 2 μm. The PZT material, modeled as PZT-5H, is operated in the 31 mode, i.e., the strain is orthogonal to the electric field. A tungsten proof mass is used to reach the energy target, which is > 1 μJ per excitation pulse. Large tungsten pieces can be attached to MEMS structures, but the drawback is mass production issues. We assume that the impact occurs between the device and the capping as this reduces the height and complexity

[1] Impacts can generate nonlinear effects somewhat comparable to those in a bridge, see, e.g., Ref. [5].

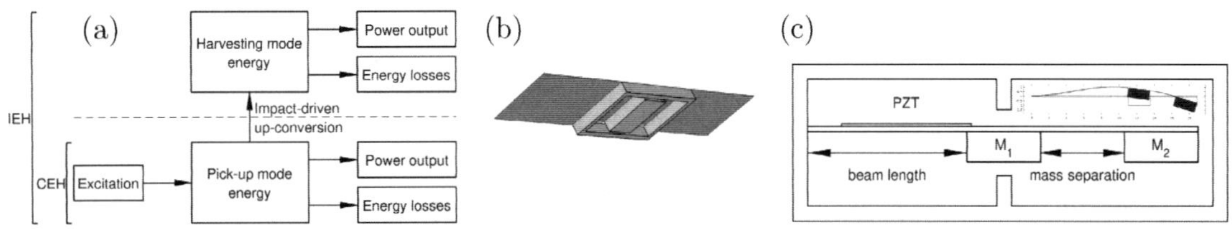

Figure 1. (a) The impact and cantilever EH approaches. (b) A conceptual double bridge IEH. (c) The double cantilever EH. The inset shows the harvester (second) mode.

compared to having multiple devices. Thus, the impact transfers energy between modes. We aim for a factor of 10 frequency up-conversion to clearly bring out the scaling effects.

A pulsed excitation is targeted. As the device velocity is bounded, the excitation pulse must have both positive and negative values and we model it as a single period of a sine wave. The maximum amplitude is $1g$. Thus, the EH is not operated in resonance, as is often assumed in the literature, and the output power is limited accordingly. A low-frequency application is targeted and the excitation single-period sine wave frequency is 40 Hz. As the excitation spectrum is wide, matching of the device to the excitation is facilitated.

The aim is to investigate if up-conversion can be effectively achieved and if the resulting EH can extract more energy than one without up-conversion. We compare with a traditional cantilever EH, and refer to this and the impact EH as the *CEH* and the *IEH*, respectively.

3. Design considerations

The impact approach is visualized in Fig. 1a. The excitation generates *pick-up mode* motion, which transfers to the *harvesting mode* via the impact(s) and results in a certain fraction of energy transfer.[2] For the CEH, these are the same mode. Due to the thin PZT, the mechanical and electrical processes are weakly coupled (as confirmed below) and there (approximately) is an *excitation phase*, with a certain energy input, and a *harvesting phase*, with power output and energy reduction due to losses. The IEH should obtain maximum output energy per pulse by reducing the energy output time scale and/or increasing the loss time scale, without reducing the energy input.

First, a double bridge as in Fig. 1b was considered. The PZT is on the inner bridge and power is output after the frame impacts the capping, which has an indentation for unimpaired proof mass motion. This, however, reduces the input energy dramatically. With reasonable dimensions, a 1 g proof mass, and a 40 Hz pick-up mode, the beam thickness is ~10 μm. This is incompatible with the excitation as it puts the bridge in the strongly nonlinear regime. (The nonlinear potential term becomes important at deflections comparable to the beam thickness [8, Eq. (1)].) In words, as the excitation increases the deflection, the spring force increases superlinearly. The proof mass velocity changes signs, and then a deceleration occurs although the pick-up mode is linearly matched to the excitation. Thus, in this application, a linear pick-up structure should be used and we instead consider the double cantilever EH in Fig. 1c, where M_1 impacts the capping and M_2 corresponds to the inner bridge proof mass.

4. Numerical simulations

To assess the IEH, the potential for output power increase, loss reduction, and efficient conversion, respectively, must be estimated. In all cases, the EH width is 10 mm and the beam thickness is 80 μm.

[2] There may also be "leakage" to other modes, which is neglected here.

4.1. Output power
For comparison, a traditional CEH with a 5 mm × 10 mm × 1 mm tungsten proof mass (weighing 0.97 g) and a 12.7 mm beam (cf. Fig. 1c) has the first resonance at 40 Hz. Directly after the excitation, the EH energy is $W = 7.1$ μJ. (With no output power, $W = 7.3$ μJ, confirming the weak coupling.) The PZT and the load resistance are optimized for maximum output power [9]. Thus, the PZT is at the CEH base (maximum strain) and is 10.4 mm long. With a load resistance of 4.5 kΩ, the optimal maximum output power is $P = 23$ μW. Based on average power, the time scale is $\tau_{\text{el}} = 2W/P = 620$ ms. (The excitation occurs over 25 ms.)

A 400 Hz CEH is obtained by shortening the beam to 1.0 mm. Using a 400 Hz excitation with $11g$ amplitude, the energy after excitation is the same as before.[3] An optimal maximum output power of 240 μW ($\tau_{\text{el}} = 58$ ms) is reached with a 1.0 mm PZT and a 4.6 kΩ load resistance. Thus, the 400 Hz CEH can have significantly increased output power.

To investigate if an IEH can achieve similar output power, the EH in Fig. 1c is simulated. The proof mass is now split in two equal pieces. The pick-up (first) mode (at 40 Hz) and the harvesting (second) mode (at 400 Hz) have been designed by optimizing the beam length (18.8 mm) and the mass separation (2.8 mm). Placing the PZT from 6.8 mm to 11.1 mm, using a load resistance of 1.1 kΩ, and exciting the harvesting mode with the same energy as before, the maximum output power is $P = 110$ μW ($\tau_{\text{el}} = 130$ ms). While lower than the power for the 400 Hz CEH, this is still almost five times higher than for the 40 Hz CEH.

4.2. Energy losses
It has been found for comparable CEHs [10] that squeeze film damping is the dominating loss mechanism. For the IEH pick-up and harvesting modes, the proof masses move normal to the capping. The deflection is not small compared to the capping distance, but we nevertheless use analytical estimates for this regime for a general discussion. The drag coefficient, c_d, is known [11, Eq. (3.17)][4], and this can be used to calculate the Q-value.[5] For a harmonic oscillator, the power decay with time is proportional to $\exp(-2\zeta\omega_0 t)$, i.e., the characteristic time $\tau_{\text{sf}} = 1/(2\zeta\omega_0) = Q/\omega_0$. Since the Q-value is proportional to ω_0, the time scale is independent of which mode is oscillating and nothing is gained from frequency upshifting in this respect.

Relevant to our study, Elfrink et al. [10] found that in vacuum-packaged devices, the squeeze film damping was comparable to the intrinsic damping. For the "Type 5" CEH with 5 mm×5 mm silicon proof mass and 600 Hz resonance frequency, the Q-value was 1000, cf. [10, Table 2]. Using this value, we find that $\tau_{\text{sf}} \gg \tau_{\text{el}}$ for both the CEHs discussed above.

4.3. Mode conversion efficiency
Simulation of the mode coupling requires impact modeling and a simple lumped IEH model, where the capping acts as a stiff linear spring, has been used. This causes an ideal impact on a very short time scale. Using the same masses, resonance frequencies, and excitation as above, the energy at the end of the excitation is 7.3 μJ at a maximum deflection of 490 μm. After placing the capping at a selected distance, the average energy in the two springs can be found.

One important complication is that the static deflection due to gravity is 160 μm and if the device orientation is not known, then the equilibrium position uncertainty is 33 % of the maximum deflection. With in-plane (or no) gravity, a capping distance of 300 μm leads to a high impact velocity, shortly before the end of the excitation. This distributes the energy effectively between the modes but the system energy is reduced by the impact to 5.4 μJ. However, when

[3] We here, temporarily, assume that the energy can be ideally transfered to the 400 Hz CEH.
[4] Note that Eq. (3.17) of [11] has incorrect dimensions as a factor $1/\omega$ is missing.
[5] For an assumed proof mass motion, the energy is known. Calculating the energy loss in one period using c_d, the Q-value can be obtained from its definition.

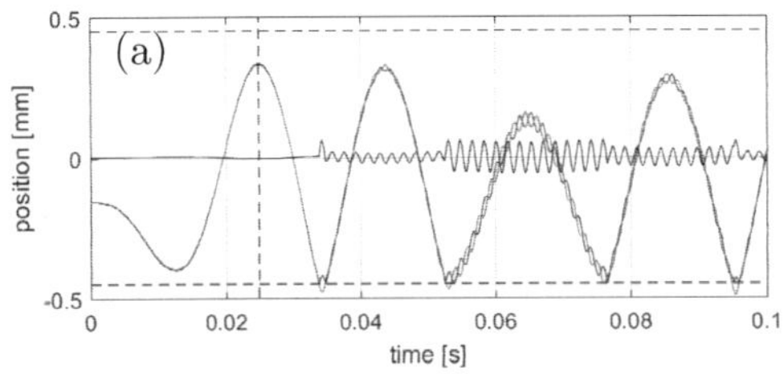

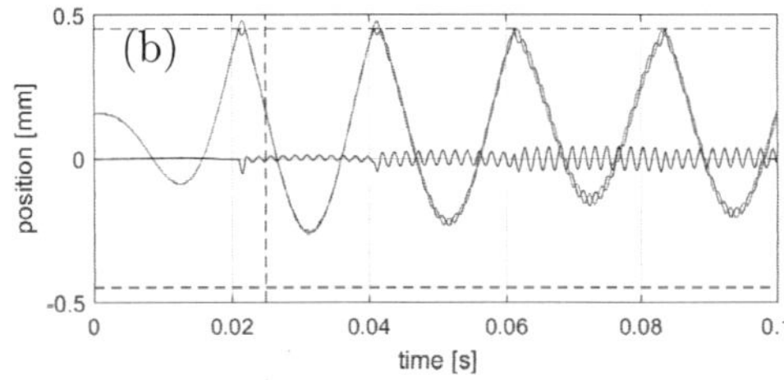

Figure 2. The position of M_1 (blue), M_2 (red), and their separation (black). The dashed lines indicate the position of the capping and the end of the excitation. (a) Gravity acts downwards and impact occurs after the end of the excitation. (b) Gravity acts upwards. The earlier impact reduces the system energy.

aligning gravity with the up or down directions, the system energy is significantly reduced to about 3 μJ due to the early impact. Increasing the capping distance to 450 μm gives the results in Fig. 2. In Fig. 2a, the first impact occurs after the excitation while in Fig. 2b, the energy is 5.3 μJ due to the impact. The impact velocity is here reduced and more impacts are needed to effectively distribute the energy between the modes. Thus, the optimal capping distance is both excitation- and orientation-dependent and it is expected that any device pre-stress changes the optimal capping distance.

5. Conclusion

The potential of impact-driven frequency up-conversion in a MEMS PEH has been studied. It was found that the upshifting can lead to significantly increased output power at a similar loss rate, implying that a higher fraction of the available energy is obtained as useful output. However, the time scale for the loss is long and the benefit is therefore limited. An IEH, however, requires an effective upshifting process. The capping distance introduces a length scale that must be selected with both the maximum deflection due to excitation and the static deflection due to gravity and pre-stress taken into account. If information about this is not present during the IEH design process, then the impact may reduce the total system energy considerably or the mode conversion may be ineffective.

Acknowledgments

The Smart-MEMPHIS project has received funding from the European Union's Horizon 2020 research and innovation program under grant agreement No. 644378.

References

[1] Roundy S and Wright P K 2004 *Smart Mater. Struct.* **13** 1131–1142
[2] Marzencki M, Defosseux M and Basrour S 2009 *J. Microelectromech. Syst.* **18** 1444–1453
[3] Hajati A and Kim S G 2011 *Appl. Phys. Lett.* **99** 083105
[4] Gafforelli G, Xu R, Corigliano A and Kim S G 2014 *Energy Harvesting and Systems* **1** 179–187
[5] Soliman M S M, Abdel-Rahman E M, El-Saadany E F and Mansour R R 2008 *J. Micromech. Microeng.* **18**
[6] Liu H, Tay C J, Quan C, Kobayashi T and Lee C 2011 *Microsyst. Technol.* **17** 1747–1754
[7] Gu L and Livermore C 2011 *Smart Mater. Struct.* **20**
[8] Hajati A, Xu R and Kim S G 2011 Wide bandwidth piezoelectric micro energy harvester based on nonlinear resonance *Power MEMS*
[9] Patel R, Popov A A and McWilliam S 2009 Parametric study of a cantilevered piezoelectric energy harvester *Power MEMS*
[10] Elfrink R, Renaud M, Kamel T M, de Nooijer C, Jambunathan M, Goedbloed M, Hohlfeld D, Matova S, Pop V, Caballero L and van Schaijk R 2010 *J. Micromech. Microeng.* **20**
[11] Bao M and Yang H 2007 *Sens. Actuators* **136** 3–27

Study and Modeling of a Traveling Wave Piezoelectric Transformer

T Martinez[1], **G Pillonnet**[2], **D Vasic**[1,3] **and F Costa**[1,4]

[1]Laboratoire SATIE, ENS Paris-Saclay, France
[2]Univ. Grenoble Alpes, CEA, LETI, DACLE, LGECA, Grenoble, France
[3]Université Cergy-Pontoise, Cergy-Pontoise, France
[4]Université Paris-Est, Créteil, France

E-mail: thomas.martinez@satie.ens-cachan.fr

Abstract. This paper reports the study and the modelling of a new topology of traveling wave piezoelectric transformer (TWPT). Unlike conventional piezoelectric transformers, the proposed transformer uses progressive waves to carry out the electromechanical conversion. Moreover, associated with an appropriate converter, it enables any kind of AC-DC or AC-AC power conversion. An analytical modelling of the system is carried allowing defining numerically the voltages, currents, displacements and stresses in the piezoelectric transformer over time as a function of the geometrical and physical parameters of the materials.

1. Introduction

In conventional piezoelectric transformers (PT), the coupling between primary and secondary sides is realized by mechanical coupling through a standing wave. In general, these transformers will only provide one or two output voltages with fixed phase shift compared to the input thus limiting their use to simple power converter topology. The generation of a traveling wave inside the PT instead of a stationary wave allows obtaining at the secondary side numerous output voltages with the same amplitude and different phase-shifts between them. The traveling wave piezoelectric transformer could thus be used in a universal power electronic converter able to realize different complex electrical conversion like DC/DC, DC/AC or DC/AC multi-phases. To the author's knowledge, only one paper [1] proposes the use of a traveling wave for electric to electric conversion but for a bulk wave and without considering the generation of a multiphase system at the output. In this paper, we will present a piezoelectric transformer converting energy by generating a flexural traveling wave leading to an octophase system at the outputs.

2. Structure and working principle of the TWPT

2.1. Geometry of the TWPT

The common geometry used to generate a traveling wave is a ring [2]. This architecture presents the assets of working at a resonant frequency of the element and allows to obtain the highest energy transfer between

IOP Conf. Series: Journal of Physics: Conf. Series **1052** (2018) 012107 doi:10.1088/1742-6596/1052/1/012107

the primary and secondary sides of the transformer. As stated above, two modes of propagation of the wave are possible: a flexural wave and a bulk wave. For a flexural wave; the TWPT must consist in two piezoelectric ceramics polarized in opposite directions and gathered together (figure 1). For a bulk wave, only one piezoelectric ceramic is necessary. In the following, we will focus on the flexural wave generation. The objective of this device is to generate a flexural wave in the thickness direction that will propagate along the circumference of the ring. This flexural wave appears near the resonance.

2.2. Generation of the traveling wave

The excitation of a resonance with only one signal will lead to a standing wave. The generation of a traveling wave can be realized by the generation of two standing waves with a phase shift of $\frac{\pi}{2}$ and a $\frac{\lambda}{4}$ displacement between them. The combination of these two standing waves leads to a progressive wave that will propagate alongside the circumference of the ring. The position of electrodes on the surface of the piezoelectric ceramic depends on the mode of resonance we want to excite. On figure 2 is represented the typical scheme used for the mode 3 of flexural mode. In this mode of resonance, the ring will present six antinodes and six nodes that will propagate along the circumference of the TWPT (figure 3). In order to generate the standing waves two main sets of electrode are used. The ones in light gray are driven by a sinusoidal voltage and on the ones in dark gray is applied a voltage with a phase shift of $\frac{\pi}{2}$ compared to the first one. These two sets of electrodes are responsible for each standing wave. The electrodes in white represent the output electrodes. In this scheme, we can see 8 output electrodes with a size twice smaller than the input electrodes. It means that we will obtain an octophase system at the output of the TWPT.

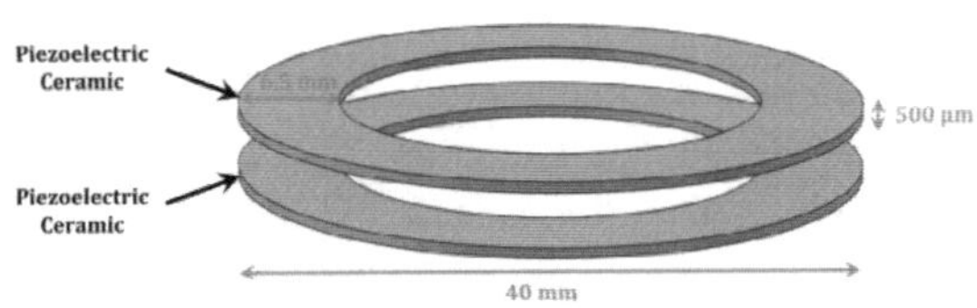

Figure 1. Ring Architecture of our TWPT

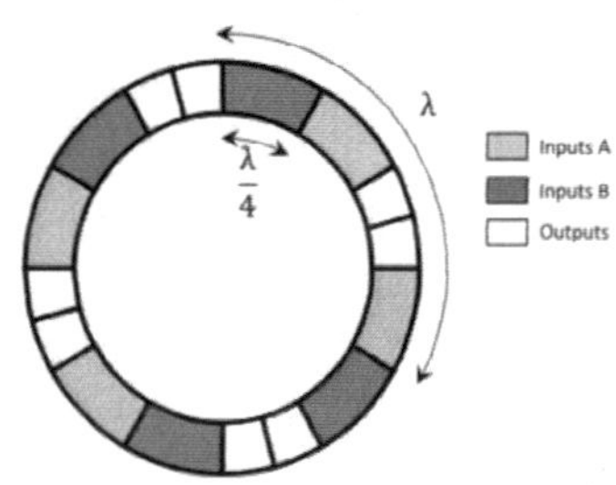

Figure 2. Electrode scheme on the surface of the TWPT

3. Analytical modelling

It is only useful to know the displacement of the flexural wave, in transformer applications, it is necessary to have a complete analysis of the electrical performances of the TWPT. In order to do that, we propose to obtain a model linking all the current and voltage values for the different electrodes.

The analysis is based on the Euler-Bernoulli beam theory. For this analysis, we separate the transformer into different sections corresponding each to one electrode. We make the hypothesis that these sections are treated as beams as represented in figure 4. Furthermore, we consider that the TWPT vibrates at a specific frequency and we thus realize a harmonic analysis at pulsation ω. The flexural wave appears in direction x_3 and propagates alongside direction x_1. We considered thus the flexural stresses and moments (resp. F1 and F2) and the linear and angular speeds (resp. U1 and U2).

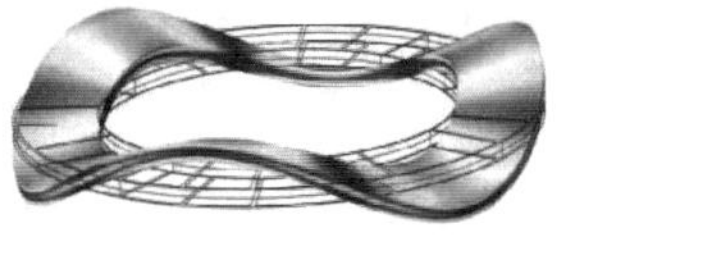

Figure 3. Flexural deformation of the ring

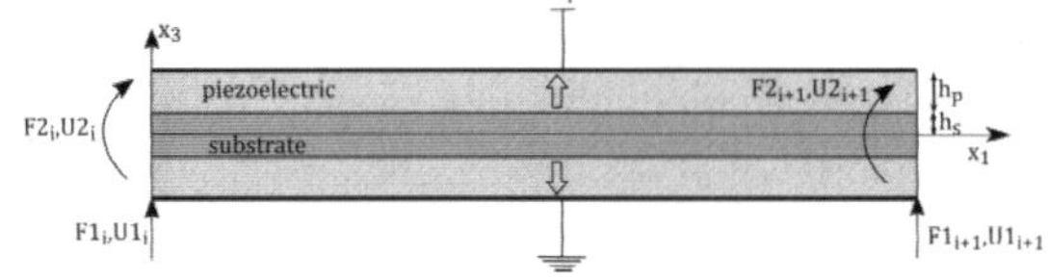

Figure 4 : Beam element representing a ring section under flexural stress.

From the Euler Bernoulli beam theory and piezoelectric equations, we find that the propagation of a flexural wave in the direction 3 alongside direction 1 is governed by equation (1)

$$K_b \frac{\partial^4 u_3}{\partial x_1^4} + \rho_b \frac{\partial^2 u_3}{\partial t^2} = 0 \tag{1}$$

with K_b the flexural rigidity and ρ_b the linear density of the TWPT depending on the material characteristics and geometry of the transformer. From this equation, we can determine the form of the deformation u_3.

$$u_3(x_1) = \alpha_1 \cos(\lambda x_1) + \alpha_2 \sin(\lambda x_1) + \alpha_3 \cosh(\lambda x_1) + \alpha_1 \sinh(\lambda x_1) \tag{2}$$

with α_i coefficients that depend on the boundary conditions and λ the wavelength of the wave. Furthermore, by writing these conditions, we can determine transfer matrixes G_L and T_L depending on the geometric parameters and linking the mechanical variables (stresses and speeds) at one side to the ones at the other sides and the voltage on the electrode considered.

$$B_i = \begin{pmatrix} F1_i \\ F2_i \\ U1_i \\ U2_i \end{pmatrix} = G_L \begin{pmatrix} F1_{i+1} \\ F2_{i+1} \\ U1_{i+1} \\ U2_{i+1} \end{pmatrix} + T_L V_i = G_L B_{i+1} + T_L V_i \tag{3}$$

As contrary to previous work [3] where are only considered fixed or free boundary conditions, here, the matrixes are associated one with each other's along the TWPT. This allows obtaining a matrix P_i linking the mechanical variables at one point as a function of all the voltages on the electrodes (V is the matrix column of all the voltages on the electrodes).

$$B_i = \begin{pmatrix} F1_i \\ F2_i \\ U1_i \\ U2_i \end{pmatrix} = P_i V \tag{4}$$

Thereafter, from the equations of piezoelectricity we have $I_i = j\omega C V_i - N_b(U2_i - U2_{i+1})$. The current flowing through an electrode is a function of the voltage V_i and the angular velocities U2 at the extremities of the section (C is the clamped capacitance and N_b is an electromechanical coupling coefficient). Thus from this equation and (4) we determine an admittance matrix Y_{tot} linking all the currents and voltages on the electrodes:

$$I = Y_{tot} V \tag{5}$$

This admittance matrix is valid without considering if an electrode is an input or an output.

4. Simulation results

To simulate the device, we considered a TWPT whose dimensions are presented in figure 1. The material considered for the simulation is a PZT with the following characteristics: a density $\rho = 7500$ kg/m^3, a piezoelectric coefficient $d_{31} = -123 \times 10^{-12}$ C/N, an elastic compliance $s_{11} = 12.3 \times 10^{-12}$ m^2/N and a permittivity $\varepsilon_{33} = 1475\ \varepsilon_0$. The system was driven by sinusoidal voltages of amplitude 15V and charged with load resistance of 10 kΩ. Losses were also taken into account by using complex properties of the material.

The first analysis realized is a frequency sweep to observe the resonant frequencies of the piezoelectric transformer. For example, the voltage gain of the converter as a function of frequency is represented in figure 5. We see on the graph two resonances ($f_1 = 3980$ Hz and $f_2 = 11040$ Hz) corresponding to the 3rd and 5th modes of resonance. Due to the scheme of electrodes the third resonance presents the highest voltage gain.

Furthermore, the admittance matrix is composed of nonlinear functions of the resonance frequency and thus it is difficult to replace the matrix by a simple equivalent circuit. Nonetheless, at a specific frequency we can completely determine the type of admittance linking two electrodes (capacitance or inductance). This process was realized in Simulink where the admittance matrix was represented by a set of inductances or capacitances depending on the admittance matrix value. The input electrodes were driven by two sinusoidal voltages of amplitude 15V, at frequency f_1 with a phase shift of $\frac{\pi}{2}$ and a load resistor of 10 kΩ was plugged at each output electrode. Figure 6 shows the voltages at the output of the TWPT, depicting the octophase system. The amplitude of the output voltage corresponds to the value found with the frequency sweep.

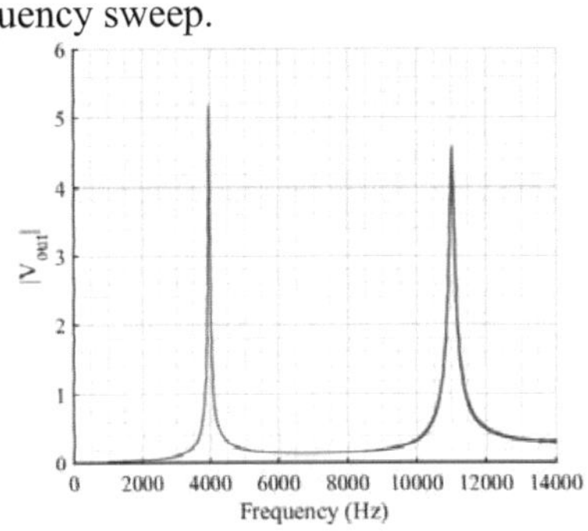

Figure 5. Voltage gain of the TWPT as a function of the excitation frequency

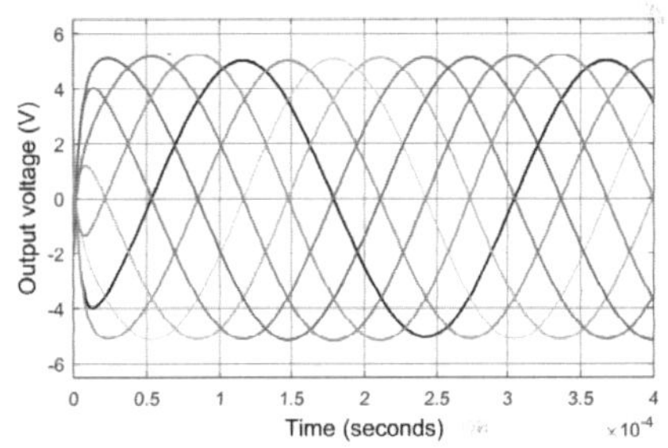

Figure 6. Output voltages at the output of the TWPT ($f_1 = 3980$ Hz)

5. Conclusion

This paper presents the architecture of a new traveling wave piezoelectric transformer. The analytical analysis of the transformer is presented and simulated. The results of the simulation of the model fit quite well with the finite element analysis. This new architecture of piezoelectric transformer allows obtaining a multi-phase system at the output thus opening the possibility of a universal power converter able to realize different kind of conversions.

References

[1] Kauczor, C., Schulte, T., & Grotstollen, H. “Piezoelectric Transformer of Travelling Wave Type” in Actuator 2004, pp. 395–398.

[2] Duan, W. H., Quek, S. T., & Wang, Q. (2009). A novel ring type ultrasonic motor with multiple wavenumbers: design, fabrication and characterization. Smart Mater. Struct, 18, 125025–13.

[3] D. Vasic, E. Sarraute and F. Costa, "Dynamic 4×4 chain matrix for analytical modeling of piezoelectric actuator," 2001 IEEE Ultrasonics Symposium. Proceedings. Atlanta GA, 2001, pp. 985-988 vol.2.

Rectified Output Power Analysis of Piezoelectric Energy Harvester Arrays under Noisy Excitation

Sijun Du[1], Yu Jia[1,2], Emmanuelle Arroyo[1] and Ashwin A. Seshia[1]
[1]Department of Engineering, University of Cambridge, Cambridge CB2 1PZ, UK
[2]Department of Mechanical Engineering, University of Chester, Cambridge CH1 4BJ, UK
E-mail: sd672@cam.ac.uk

Abstract.
In the past decade, vibration energy harvesting has emerged as a potential alternative solution to power wireless sensor nodes. In real-world implementations, external excitation can be very noisy and includes noise signals in a wide frequency band. In order to accommodate operation under noisy excitation, arrays of energy harvesters with different resonance frequencies are often employed to improve responsibility. Due to the nature of noisy excitation and the difference in resonance frequencies, the response voltage signals from each harvester can be very different in amplitude, frequency and phase. In this paper, an array with two cantilevered energy harvesters is studied to analyze the rectified output power with different configurations using full-bridge rectifiers (FBR). The experiments show that connecting the two harvesters in parallel or in series before connecting with a FBR results in significant voltage cancellation due to phase mismatch. The most efficient way to extract energy is to use two FBRs for the two cantilevered energy harvesters, individually, and charge to one single storage capacitor connected at the outputs of the two FBRs.

1. Introduction

Among different vibration energy harvesting topologies, piezoelectric materials are widely adopted due to their high power density in mm^2 scale miniaturized energy harvesting systems [1]. Although many different active rectification circuits were proposed in the past decade [2–10], full-bridge rectifiers (FBR) are widely employed in commercially available systems due to their simplicity and stability. In alignment with miniaturization trends for wireless sensor systems, MEMS piezoelectric harvesters (PH) are drawing much research interest due to their compatibility with conventional CMOS technology [11–14].

A miniaturized piezoelectric harvester (PH) in 10's mm^2 scale can typically generate raw AC power up to 100's μW varying by different piezoelectric materials, structures and ambient conditions. In order to achieve the maximum possible output power, the PH is excited at its resonance frequency with a sinusoidal excitation signal under the highest operational excitation level. Although the resulting 100's μW peak power is promising to fully sustain operations of most wireless sensors, the usable rectified DC power in real-world implementations can be much lower than this peak value of around 10's μW. This significant power decrease is mainly due to two reasons: the wide frequency bandwidth of the real-world vibration conditions where the PH is implemented and the power extraction efficiency of the rectifier. In order to increase the effective frequency bandwidth for real-world noisy vibration environments, an array of two or more cantilevers with different resonance frequencies aiming at different frequency bands was designed in [15–18]. Due to the wide bandwidth of the excitation signal and the difference in resonance frequencies, the voltage signals generated in each cantilever can be very different in frequency, amplitude and phase. As a result, by simply connecting the harvesters up in series

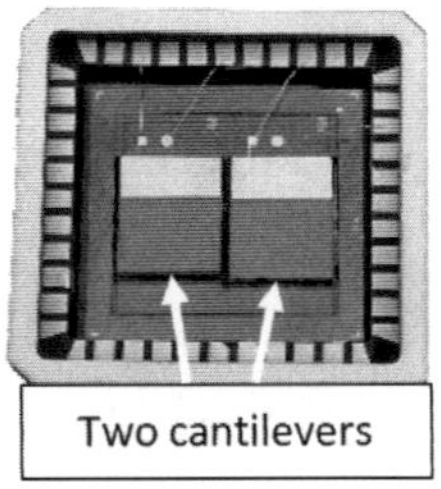

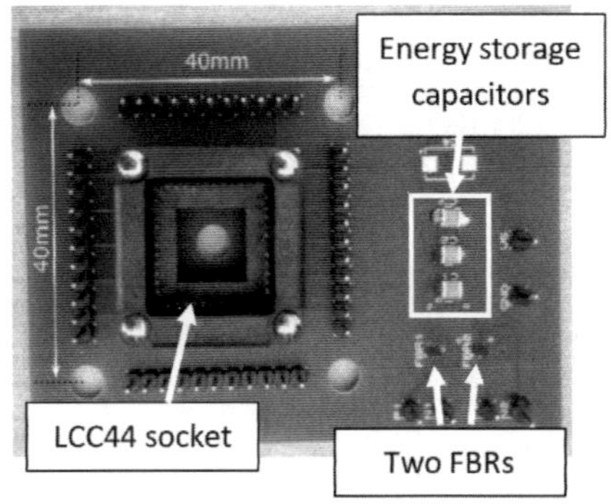

Figure 1: Photos of the MEMS device with an array of two micro-cantilevers (left) and the test board with one LCC44 socket and two FBRs (right).

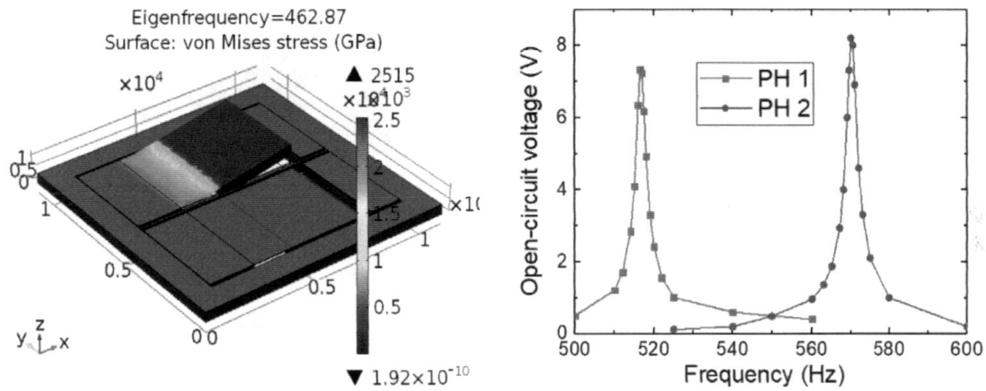

Figure 2: COMSOL model of the MEMS device with an array of two micro-cantilevers (left) and the measured resonance responses of the two cantilevers (right).

or parallel, it is possible to reduce the overall output power generated. Hence, power extraction methods from these cantilevers should be analyzed to increase the rectified DC power. In this paper, the rectified DC power of an array of two cantilevered piezoelectric harvesters (PH) using up to two FBRs under wide frequency-bandwidth excitation is analyzed and experimentally evaluated. The results show that each cantilever requires an individual FBR to extract and rectify the energy. By contrast, if only one FBR is employed while the two micro-cantilevers are connected in parallel or series before connecting with the FBR, the resulting DC power is extremely low due to the voltage cancellation cased by out-of-phase voltage combination.

2. Experiments

Figure 1 shows the fabricated MEMS device with an array of two cantilevers and the test board with two FBRs and a few energy storage capacitors. The size of the MEMS die is 11 mm × 11 mm and the two micro-cantilevers can be seen in the figure. During the measurements, the MEMS device is placed in a LCC44 leadless chip carrier, which is placed in the LCC44 socket on the test board fixed on a shaker.

Figure 2 shows the COMSOL model for the MEMS device comprising an array of two cantilevers. The device is designed in a MEMS process with AlN (Aluminum Nitride) as the piezoelectric material and the simulated model is with the same dimension as the device to be tested in the following experiments. The resonance frequencies of the two cantilevers are designed to be 463 Hz and 501 Hz, respectively. Due to fabrication tolerances, the fabricated MEMS cantilevers are measured to have resonance frequencies at 517 Hz and 570 Hz, respectively, as shown in the right sub-figure of Fig. 2

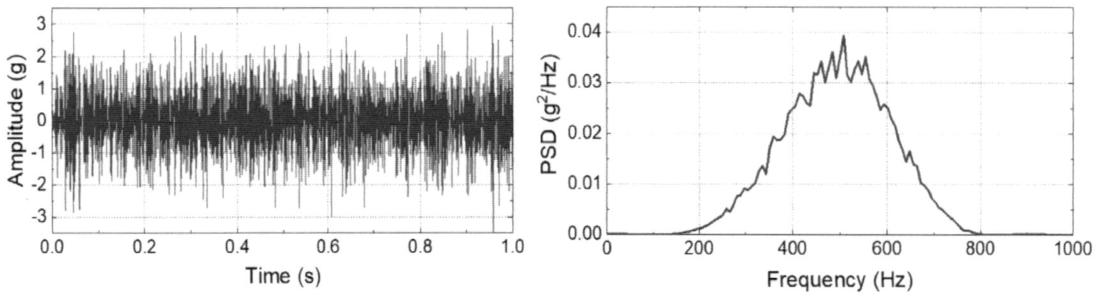

Figure 3: Noisy excitation profile signal centered at 500 Hz in time-domain (left) and frequency-domain (right).

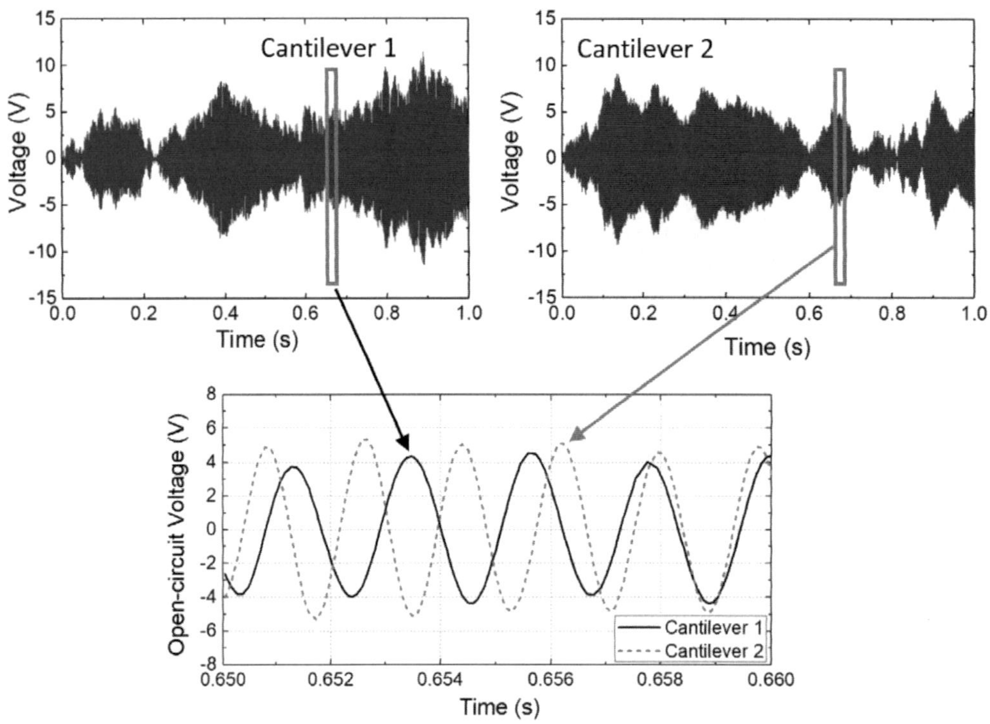

Figure 4: Simulated open-circuit voltage signals from the two micro-cantilevers under noisy excitation.

A vibration profile is used for following experiments instead of using a sine wave signal to mimic the real-world excitation. The time-domain and frequency-domain plots of the profile is shown in Fig. 3. This vibration profile has a peak frequency at 500 Hz with broadband noise. The MEMS device with two cantilevers is excited under this profile.

Fig. 4 shows the simulated open-circuit voltage from the two cantilevers excited with the profile data shown in Fig. 3. It can be seen that the output signals from the two cantilevers are significantly different in amplitude in the two upper sub-plots of Fig. 4. A zoomed-in figure shows the two signals in a 10-ms period and it shows that there is a significant phase mismatch for the two cantilevers due to the difference in the response characteristics. If the two cantilevers are connected in parallel or in series before connecting to a FBR, voltage cancellation due to phase mismatch can result in significant energy waste.

The following measurements are split into two parts: testing the two cantilevers separately and testing them together in different connection configurations. The left sub-figure in Fig. 5 shows the measured DC power while the two cantilevers are tested individually. In the plot,

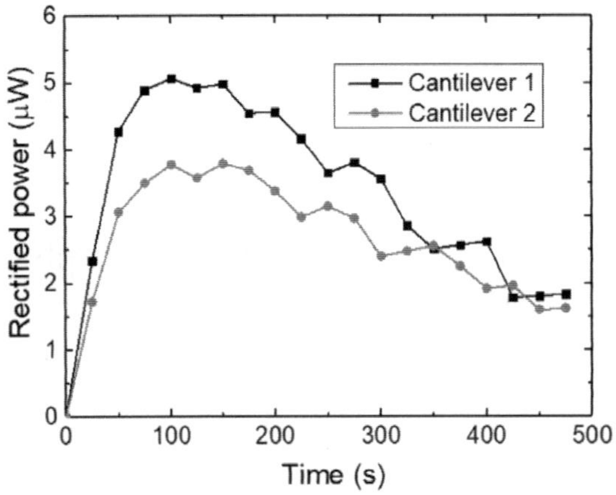

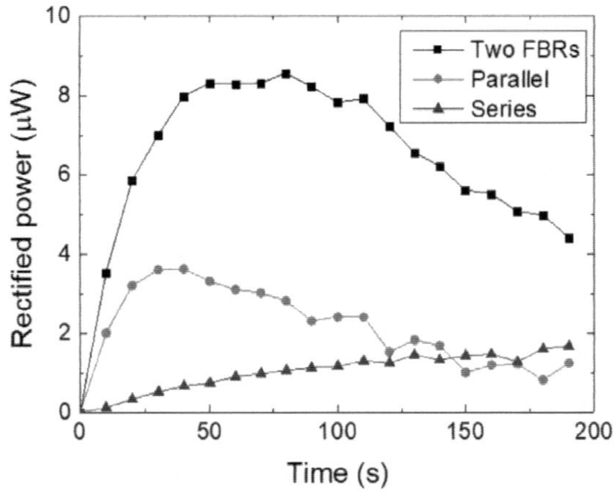

Figure 5: Measured rectified DC power. Left: DC power by two cantilevers while measured individually. Right: Two cantilevers are measured together using two FBRs for each one and while connecting them in parallel and series before connecting to a single FBR.

Table 1: Summary of measurement results

Individual tests		Two cantilevers together	
Cantilever	DC power	Type	DC power
No. 1	5 μW	Using two FBRs	8.4 μW
No. 2	3.8 μW	One FBR (parallel)	3.3 μW
Mathematical sum	8.8 μW	One FBR (series)	2.7 μW

the horizontal axis shows a representative measurement time axis plotted up to 500 s and the vertical axis shows the rectified DC power. The measurements were performed to charge energy storage capacitors via a FBR and the DC power is then calculated from the measured voltage across capacitors for every 25 seconds. These two measurements show that the two cantilevers can output peak power values of 5.0 μW and 3.8 μW, respectively, when they are measured individually.

The two cantilevers are then measured together to show the DC power under different configurations. The results are shown in the right sub-figure of Fig. 5. For the first configuration (labeled as "Two FBRs"), two FBRs are used (one for each cantilever) and the outputs of the two FBRs are connected together to charge sharing energy storage capacitors. In this case, measured peak DC power is around 8.4 μW. For the following two configurations, only one FBR is employed and the two cantilevers are connected in parallel/series prior to connecting to the FBR. In these two cases, the measured DC power levels are around 3.3 μW and 2.7 μW, respectively.

The measurement results are summarized in the table 1. It can be seen that, due to the phase mismatch between the two cantilevers under noisy excitation, the DC power values for parallel and series configurations are much lower than the mathematical sum of the power values individually measured from the two cantilevers. This is due to the voltage cancellation when excited under noisy excitation. When each cantilever is connected up to a separate FBR, the measured output power is almost the mathematical sum and there is very little energy loss in this case.

3. Conclusion

This paper analyzes the phase mismatch for multiple piezoelectric cantilevers connected up in either parallel or series excited under noisy excitation. A MEMS piezoelectric harvester with an array of two micro-cantilevers were fabricated to test the theory. During the measurements, a vibration profile with broadband noise is used to excite the MEMS device. Simulation and experimental results show a significant voltage cancellation if the micro-cantilevers are directly connected in parallel or in series prior to rectification. In order to extract and rectify energy generated from an array of cantilevers, each cantilever requires an individual FBR in order to eliminate voltage cancellation and to maximize the extracted DC power. If only one FBR is employed by connecting all the micro-cantilevers in parallel or in series, the resulting DC power can be extremely low due to the phase mismatch across the cantilevers.

References

[1] P. D. Mitcheson, E. M. Yeatman, G. K. Rao, A. S. Holmes, and T. C. Green, "Energy harvesting from human and machine motion for wireless electronic devices," *Proceedings of the IEEE*, vol. 96, no. 9, pp. 1457–1486, 2008.

[2] S. Du and A. A. Seshia, "An inductorless bias-flip rectifier for piezoelectric energy harvesting," *IEEE Journal of Solid-State Circuits*, vol. 52, no. 10, pp. 2746–2757, 2017.

[3] D. Kwon and G. A. Rincon-Mora, "A single-inductor 0.35 um cmos energy-investing piezoelectric harvester," *IEEE Journal of Solid-State Circuits*, vol. 49, no. 10, pp. 2277–2291, 2014.

[4] S. Du, Y. Jia, and A. A. Seshia, "An efficient inductorless dynamically configured interface circuit for piezoelectric vibration energy harvesting," *IEEE Transactions on Power Electronics*, vol. 32, no. 5, pp. 3595–3609, 2017.

[5] Y. K. Ramadass and A. P. Chandrakasan, "An efficient piezoelectric energy-harvesting interface circuit using a bias-flip rectifier and shared inductor," in *Solid-State Circuits Conference - Digest of Technical Papers, 2009. ISSCC 2009. IEEE International*, Conference Proceedings, pp. 296–297,297a.

[6] M. Dini, A. Romani, M. Filippi, and M. Tartagni, "A nanopower synchronous charge extractor ic for low-voltage piezoelectric energy harvesting with residual charge inversion," *IEEE Transactions on Power Electronics*, vol. 31, no. 2, pp. 1263–1274, 2016.

[7] S. Du, Y. Jia, C. D. Do, and A. A. Seshia, "An efficient sshi interface with increased input range for piezoelectric energy harvesting under variable conditions," *IEEE Journal of Solid-State Circuits*, vol. 51, no. 11, pp. 2729–2742, 2016.

[8] G. D. Szarka, B. H. Stark, and S. G. Burrow, "Review of power conditioning for kinetic energy harvesting systems," *Power Electronics, IEEE Transactions on*, vol. 27, no. 2, pp. 803–815, 2012.

[9] S. Du, Y. Jia, C. Zhao, S.-T. Chen, and A. A. Seshia, "Real-world evaluation of a self-startup sshi rectifier for piezoelectric vibration energy harvesting," *Sensors and Actuators A: Physical*, vol. 264, pp. 180–187, 2017.

[10] E. E. Aktakka and K. Najafi, "A micro inertial energy harvesting platform with self-supplied power management circuit for autonomous wireless sensor nodes," *IEEE Journal of Solid-State Circuits*, vol. 49, no. 9, pp. 2017–2029, 2014.

[11] A. Hajati and S.-G. Kim, "Ultra-wide bandwidth piezoelectric energy harvesting," *Applied Physics Letters*, vol. 99, no. 8, pp. –, 2011.

[12] H. Liu, C. J. Tay, C. Quan, T. Kobayashi, and C. Lee, "Piezoelectric mems energy harvester for low-frequency vibrations with wideband operation range and steadily increased output power," *Microelectromechanical Systems, Journal of*, vol. 20, no. 5, pp. 1131–1142, 2011.

[13] S. Du, Y. Jia, S.-T. Chen, C. Zhao, B. Sun, E. Arroyo, and A. A. Seshia, "A new electrode design method in piezoelectric vibration energy harvesters to maximize output power," *Sensors and Actuators A: Physical*, vol. 263, pp. 693–701, 2017.

[14] Y. Jia, S. Du, and A. A. Seshia, "Twenty-eight orders of parametric resonance in a microelectromechanical device for multi-band vibration energy harvesting," *Scientific Reports*, vol. 6, p. 30167, 2016.

[15] H. C. Lin, P. H. Wu, I. C. Lien, and Y. C. Shu, "Analysis of an array of piezoelectric energy harvesters connected in series," *Smart Materials and Structures*, vol. 22, no. 9, p. 094026, 2013. [Online]. Available: http://stacks.iop.org/0964-1726/22/i=9/a=094026

[16] X. Huan, H. Yuantai, and W. Qing-Ming, "Broadband piezoelectric energy harvesting devices using multiple bimorphs with different operating frequencies," *Ultrasonics, Ferroelectrics, and Frequency Control, IEEE Transactions on*, vol. 55, no. 9, pp. 2104–2108, 2008.

[17] Y. Jia, S. Du, and A. A. Seshia, "Micromachined cantilevers-on-membrane topology for broadband vibration energy harvesting," *Journal of Micromechanics and Microengineering*, vol. 26, no. 12, p. 124007, 2016.

[18] H. Zhang and K. Afzalul, "Design and analysis of a connected broadband multi-piezoelectric-bimorph- beam energy harvester," *Ultrasonics, Ferroelectrics, and Frequency Control, IEEE Transactions on*, vol. 61, no. 6, pp. 1016–1023, 2014.

IOP Conf. Series: Journal of Physics: Conf. Series **1052** (2018) 012109 doi:10.1088/1742-6596/1052/1/012109

Shape effects in doubly clamped bridge structures at large deflections

Fredrik Ohlsson, Pontus Johannisson and Cristina Rusu
RISE Acreo AB, Arvid Hedvalls Backe 4, SE-411 33 Gothenburg, Sweden
E-mail: fredrik.ohlsson@ri.se

Abstract. The shape of a doubly clamped bridge structure depends on its deflection. At large deflections, where the system exhibits nonlinear behaviour, the shape effect becomes significant. We present a general method, based on variational analysis, for computing corrections to the nominal linear regime shape function. The method is used to compute the first non-trivial correction and quantify the corresponding improvement in the large deflection regime. The model obtained is also validated using FEM simulations.

1. Introduction

The interest in nonlinear mechanical structures for vibrational energy harvesting applications has increased over the past decade. Advantages compared to linear resonator systems include increased bandwidth, limited deflection and improved structural durability [1–5]. In addition, bi-stable modes of operation can be used to target low frequency and amplitude excitations [6].

A model system featuring mechanical nonlinearities is the doubly clamped bridge with a central proof mass shown in Fig. 1. The boundary conditions induce stretching of the two beams which results in a contribution to the restoring force nonlinear in the proof mass displacement δ. In this paper we present a model for the bridge system which accounts for the dependence on δ of the shape of the beams, defined as the functional coordinate dependence of the displacement field. The model uses variational analysis to compute corrections to the nominal shape function, which is valid at infinitesimal deflections where the bridge is described by pure bending, by minimizing the strain energy. We explicitly compute the first non-trivial correction and the corresponding strain energy reduction. Furthermore, we consider an example of a bridge structure and validate the predictions of our model using FEM simulations.

Although the present paper is concerned with a purely mechanical model of the doubly clamped structure, an electromechanical model of a piezoelectric harvester could be constructed based on the derived shape function, extending the work by Gafforelli et al. [7].

2. Theoretical models

2.1. Large deflection model

At large displacements δ, the total strain in the beams is the sum of bending and stretching strains according to Green-Lagrange

$$S_{33} = -x_1 \frac{\partial^2 w}{\partial x_3^2} + \frac{1}{2}\left(\frac{\partial w}{\partial x_3}\right)^2, \tag{1}$$

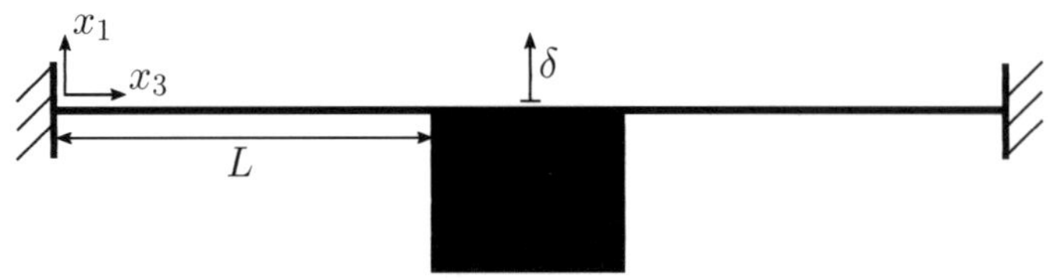

Figure 1. Illustration of the doubly clamped bridge structure with definitions of proof mass displacement δ, beam length L and coordinates x_1 and x_3.

where w is the x_1 component of the displacement field and S_{33} is the total strain in the beam.

Introducing the dimensionless coordinate $\xi = x_3/L$, the total strain energy $\mathcal{V}$ of the structure can be obtained by integrating the energy density $c_{33}S_{33}^2$, where c_{33} is the elastic modulus, over the total system volume and enforcing the doubly clamped boundary conditions

$$\mathcal{V} = A_b \int_0^1 \left(\frac{\partial^2 w}{\partial \xi^2}\right)^2 d\xi + A_s \int_0^1 \left(\frac{\partial w}{\partial \xi}\right)^4 d\xi \,. \tag{2}$$

The constants A_b and A_s, corresponding respectively to bending and stretching, are given in terms of moments of area for the beam cross-section A as

$$A_b = \frac{1}{L^3}\int_A c_{33}x_3^2 dA \;,\quad A_s = \frac{1}{4L^3}\int_A c_{33} dA \,. \tag{3}$$

The shape function $w = w(\xi, \delta)$ is the function that minimizes the strain energy $\mathcal{V}$. To describe the shape function we use a power series Ansatz for w

$$w(\xi, \delta) = \sum_{i=0}^{\infty} a_n(\delta)\xi^n \,, \tag{4}$$

where the coefficients $a_n(\delta)$ are determined by the doubly clamped boundary conditions, the symmetry with respect to $\xi = 1/2$ and the condition that w extremizes the functional $\mathcal{V}[w]$

$$\frac{\delta \mathcal{V}}{\delta w} = 0 \,. \tag{5}$$

In particular, the boundary conditions imply the vanishing of the coefficients a_0 and a_1.

Given the shape function, and corresponding strain energy, the dynamics of the bridge system is governed by the Euler-Lagrange equation of motion for δ.

2.2. Shape function approximations

Approximate expressions for the shape function are obtained by truncating the series at some power N, where the symmetry of the problem implies that it is sufficient to consider N odd. The lowest order for which all boundary conditions can be satisfied is $N = 3$, for which the unique solution is

$$a_2 = 3\delta \;,\quad a_3 = -2\delta \,. \tag{6}$$

The minimization requirement is trivial for $N = 3$, in the sense that all coefficients a_n are uniquely determined by the structural boundary conditions[1].

[1] We also note that for $N = 3$ the shape function is the exact solution for the doubly clamped structure in Euler-Bernoulli theory describing infinitesimal displacements δ.

The first non-trivial correction to the shape function is obtained for $N = 5$, where the boundary conditions and symmetry constraints can be used to solve for the first four coefficients

$$a_2 = 3\delta - \frac{1}{2}a_5 \ , \ a_3 = -2\delta + 2a_5 \ , \ a_4 = -\frac{5}{2}a_5 \, . \tag{7}$$

The remaining coefficient a_5 can then be determined by the variational condition (5), which amounts to solving the equation $\partial\mathcal{V}/\partial a_5 = 0$.

The method described above can be applied to compute corrections to arbitrary order N. Although the power series is not a perturbation series expansion in the ordinary sense, the strain energy $\mathcal{V}$ is a decreasing function of N since a solution of the same energy as for $N-2$ is obtained by setting the added coefficients to zero.

2.3. Validation model

In order to validate the model we consider a bridge structure with beams of length $L = 5$ mm, width $W = 3$ mm and thickness $H = 5\,\mu$m, chosen to comply with MEMS manufacturing constraints while being sufficiently weak. The proof mass is taken to be a $2 \times 3 \times 0.6$ mm^3 block of tungsten to obtain a low frequency system which can be expected to operate in the large deflection regime when subjected to moderate excitations. The theoretical approximations are evaluated and compared to the results obtained using a large deflection FEM simulation model accounting for geometric nonlinearities.

3. Results and simulations

3.1. Shape function approximations

The solution to the $N = 5$ system can be presented in a normalized way, which makes it independent of the mechanical and geometrical properties of the bridge structure, by introducing the length scale λ according to $\lambda^2 = A_b/A_s$. From the definition of λ it follows that the characteristic length scale of large deflection effects is the thickness of the beam [5]. The normalized coefficient a_5 and the reduction of the strain energy in the $N = 5$ model relative to the $N = 3$ model, as functions of the normalized proof mass displacement δ, are shown in Fig. 2. As expected, at $\delta \approx \lambda$ the coefficient a_5 and the strain energy reduction become significant.

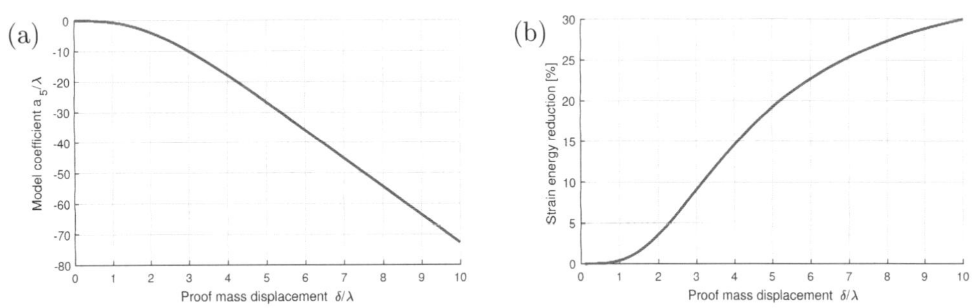

Figure 2. (a) Normalized model coefficient a_5 in the $N = 5$ model as a function of normalized proof mass displacement. (b) Relative strain energy reduction in the $N = 5$ model.

3.2. Validation model results

The validation structure is evaluated using the $N = 3$ and $N = 5$ approximations and the resulting shape functions are compared to the FEM simulation in Fig. 3. Similarly, the total strain energy of the structure at static proof mass displacement is shown in Fig. 4a. The results show that at large deflection the $N = 5$ model gives significant improvement over the nominal $N = 3$ case. For example, at $\delta = 2H$ the relative error in strain energy compared to simulation is reduced from 27% in the $N = 3$ model to 12% in the $N = 5$ model. Finally, in order to exemplify the impact on the dynamics of the structure, the frequency of unforced and undamped oscillations is computed from the potential strain energy and shown in Fig. 4b as a function of amplitude. Again, the results demonstrate improved agreement with the FEM simulation results in the $N = 5$ model.

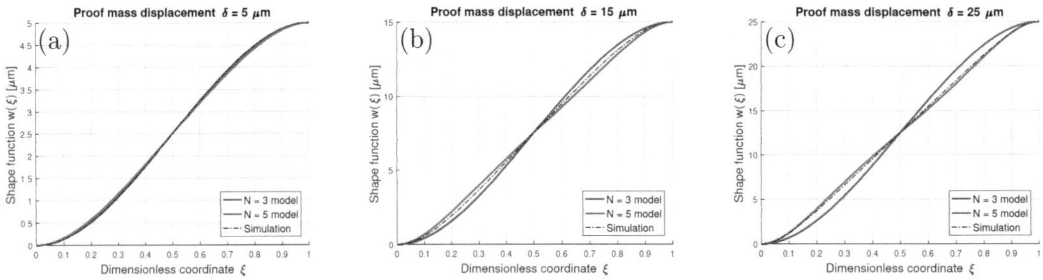

Figure 3. Shape function $w(\xi)$ for (a) $\delta = 5\,\mu$m, (b) $\delta = 15\,\mu$m and (c) $\delta = 25\,\mu$m.

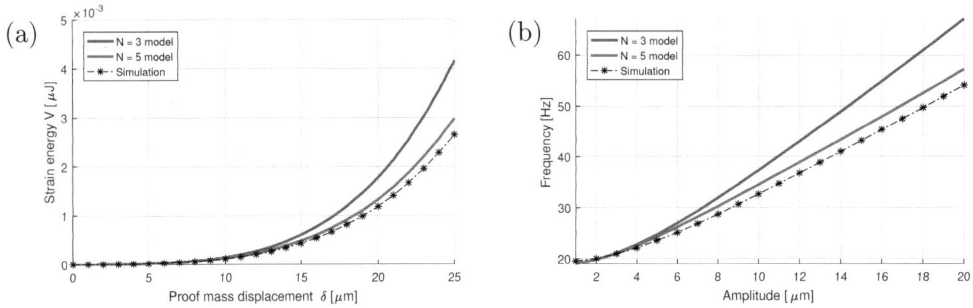

Figure 4. (a) Total strain energy of the structure as a function of proof mass displacement. (b) Frequency of unforced oscillations as a function of amplitude.

4. Conclusions

We have presented a method for computing corrections to the shape of a doubly clamped bridge structure at large deflections. The lowest order corrections have been computed and shown to provide significant improvements of both static and dynamic properties of the structure.

Acknowledgments

This work was performed in the Smart-MEMPHIS project with funding from the European Union's Horizon 2020 research and innovation program under grant agreement No. 644378.

References

[1] Cottone F, Vocca H and Gammaitoni L 2009 *Phys. Rev. Lett.* **102** 080601
[2] Marzencki M, Defosseux M and Basrour S 2009 *J. Microelectromech. Syst.* **18** 1444-1453
[3] Zhu D, Tudor M J and Beeby S P 2010 *Meas. Sci. Technol.* **21** 022001
[4] Tang L, Yang Y and Soh C K 2010 *J. Intel. Mater. Syst. Struct.* **21** 1867-1897
[5] Hajati A, Xu R and Kim S G 2011 *Proc. PowerMEMS* 3-6
[6] Xu R and Kim S G 2015 *J. Phys. Conf. Ser.* **660** 012013
[7] Gafforelli G, Xu R, Corigliano A and Kim S G 2014 *Energy Harvesting and Systems* **1** 179-187

IOP Conf. Series: Journal of Physics: Conf. Series **1052** (2018) 012110 doi:10.1088/1742-6596/1052/1/012110

A piezoelectric generator based on PVDF/GO nanofiber membrane

Kaidi Li [1, 2, 3], Xia Liu[1, 2, 3], Yifeng Liu[1, 2, 3] and Xiaohong Wang[1, 2, 3]

[1] Institute of Microelectronics, Tsinghua University, Beijing 100084, P.R. China
[2] Tsinghua National Laboratory for Information Science and Technology, Tsinghua University, Beijing 100084, P.R. China
[3] Department of Microelectronics and Nanoelectronics, Tsinghua University, Beijing 100084, P.R. China

E-mail: wxh-ime@tsinghua.edu.cn

Abstract. We report a vibration-powered micro-generator, based on electrospinning polyvinylidene fluoride (PVDF) / graphene oxide (GO) composite nanofiber film. With the help of the electrospinning process, the PVDF nanofibers chemically wrapped by GO lamellae, in which the polar β -phase nanocrystals are formed and uniaxially self-oriented by the synergistic effect of mechanical stretching, high-voltage alignment, and chemical interactions. The special core-shell structure formed during electrospinning process gives nanofiber better piezoelectric properties. And the high piezoelectric coefficients nanofiber constitutes a film generator with better performance than previous work with similar sandwich structure.

1. Introduction

Piezoelectric energy harvesting technology is the simplest way to collect energy without the need for complex geometries and additional units. Commonly, piezoelectric material includes piezoelectric ceramics, metallic oxide, and polymer. Compared with ceramics and metallic oxide, polymer is considered the most promising material in the field of biomedical care, wearable energy harvesting devices and so on[1], because of its structural flexibility, ease of processing with large sensitive areas, simplicity in device design and potential for low-cost implementation. Among them, polyvinylidene fluoride (PVDF) and its copolymers are very typical type of piezoelectric polymer. While its shortcomings are also obvious, the piezoelectric coefficient of PVDF is far from the piezoelectric ceramic.

As semicrystalline polymer, PVDF usually has four kinds of crystalline phase: α , β , γ and δ . The piezoelectricity of PVDF is mainly contributed by polar crystalline phases such as β phase (TTTT conformation) and γ phase (T_3GT_3G' conformation), rather than the nonpolar α phase (TGTG′ conformation). In these polar phases, the β phase imparts the largest spontaneous polarization of the polymer per unit cell, which leads to the highest piezoelectric response. So it is obvious that the content and orientation of the β phase and the degree of crystallinity is the key to improving the piezoelectric property of PVDF. The uniaxial or biaxial stretching of polymer films or fibers at a given temperature and electrical polarizing are the most common. [2-4]

method to enhance the β phase content. Electrospinning is a good way to combine mechanical stretching and electrical polarizing together, in the process of manufacturing fiber at the same time to complete the polarization. There is a problem that the fiber has the possibility of depolarizing after electrospinning. To solve this problem, Liu et.al [5] add graphene oxide (GO) into PVDF solution and then obtain core-shell fibers with a piezoelectric coefficient comparable to that of a piezoelectric ceramic. The functionalizing oxygen groups on the GO lamellae surface allow it to interact and chemically react with PVDF molecular line. [3]In the process of electrospinning, GO migrates toward the surface of the nanofiber with the evaporation of the solution and the action of electrostatic forces, which makes the PVDF nanofibers get higher content of the β -phase and more uniaxial orientation.

In this paper, we use electrospinning fabricating PVDF/GO nanofiber membrane. And the whole generator is sandwich structure with nanofiber membrane between two layers of Al electrode. Because the membrane is composed of nanofibers with high d_{33} piezoelectric coefficient, it has good piezoelectric properties, which makes the devices own better output performance.

2. Design and fabrication

The polarization direction of electrospinning PVDF / GO piezoelectric nanofibers is mainly along the out-of-plane direction, and the nanofibers are stacked in a certain thickness to a certain thickness of the piezoelectric fiber membrane. The device configuration is a structure consisting of a glass base frame, a single piezoelectric nanofiber film (layer sandwiched between a pair of metal (Al) electrodes). The use of PDMS elastic support around the fiber membrane helps torestore the fibrous membrane to its original and the device was sealed by tape, as illustrated in figure.1. The nanofiber membrane in this paper has a thickness of 200 microns. Before fabricated the whole device, the fiber membrane is subjected to a uniform pressure for a certain period

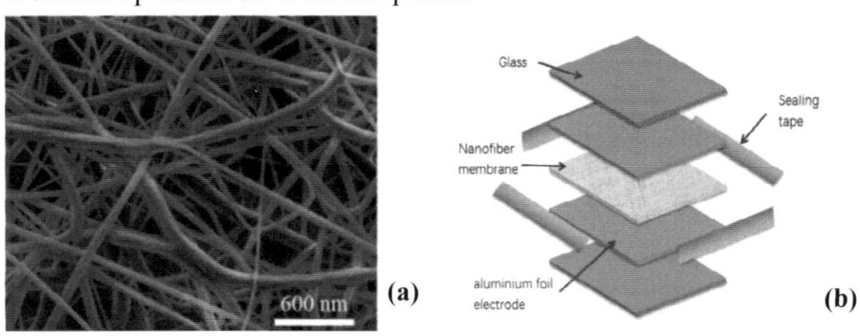

Figure 1. (a) SEM visualize the nanofiber in the film (b) Device structure diagram

of time to increase the density of the fiber membrane and to reduce the internal resistance of the fiber membrane.

The polymer fiber energy collector with counter-electrode structure has low energy conversion efficiency and poor power output, while the energy collector with interdigital structure and sandwich structure has energy output up to 20V or so. The parallel assembly of the devices can improve the electrical output of the device, such as the energy collector [6-7] of the interdigitated three-layer PVDF nanofiber fabric, which achieves a 20V voltage output. Sandwich structure of the energy collector can achieve a large area of energy conversion, thereby enhancing the power output.

IOP Conf. Series: Journal of Physics: Conf. Series **1052** (2018) 012110 doi:10.1088/1742-6596/1052/1/012110

A mixture of N,N-Dimethylacetamide (DMAC) and acetone (4:6 by volume/volume) was used as the solvent. GO lamellae is added to the solvent and mixed with ultrasound and then uniformly dispersed in the solvent to make solutions with GO mass ratios of 1.0wt %. Subsequently, PVDF (16 wt %) was added to the above solution and stirred at 60 ° C for 2 hours.

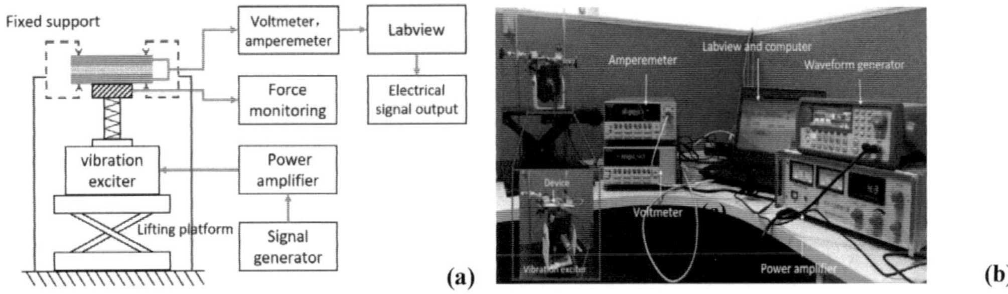

Figure 2. Schematic diagram components (a) and photo (b) of the test system for the electrical signal of the device

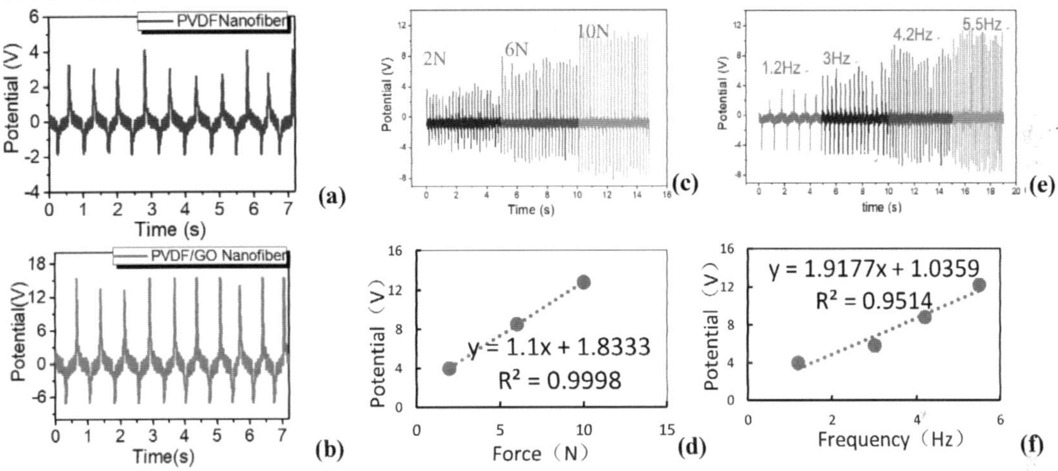

Figure 3. In the case where the operating frequency of the exciter is 1.4Hz and the output force is 20N, (a) the output of the PVDF / GO nanofiber membrane and (b) the output of pure PVDF nanofiber membrane with same condition. (c) The voltage output signal of the device under different force (d) The output signal is proportional to the force. (e) Voltage output signal of device under different vibration frequency (f)The output signal is proportional to the frequency.

The former solution into the 50mL syringe, connected to the stainless steel needle (diameter 0.25mm). The positive electrode of the high voltage source is then connected to the needle, the silicon substrate is grounded, and the distance of the needle from the substrate is 12 cm. The solution propulsion rate was 0.1 mL / h. The high pressure source was set to + 30kV high voltage to start the preparation of PVDF nanofibers. PVDF / GO nanofibers were collected using parallel electrodes. Finally, a membrane composed of nanofibers with core-shell structure is obtained.

3. Testing system and result

The power output test system consists of three parts: device, force device and electrical signal acquisition equipment. The fixture is used to fix the device to the electromagnetic exciter, and the height of the device is adjusted by the lifting platform. Since the magnitude of the electromagnetic exciter is proportional to the magnitude of the current, the magnitude of the force can be calculated by controlling the applied current signal. The biasing device utilizes a signal generator and a power amplifier to trigger an electromagnetic exciter so that the electromagnetic exciter generates a force of a certain magnitude and a certain frequency and applies it to the device. The electrical signal acquisition is measured by a voltmeter (nano voltmeter) and collected in real time using Labview software. The whole system as illustrated in figure.2.

The peak voltage (Vp-p) of the output voltage of the PVDF / GO nanofiber membrane is about 21V, and the operating frequency of the exciter is 1.4Hz, of which the output force is 20N. As the control group, the PVDF nanofiber membrane with same thickness composed energy collector has Vp-p about 5V, as shown in figure.3. When the output power of the exciter is 2N, 6N, 10N, the average peak-to-peak voltage (Vp-p) of the device is 6.5V, 12V, 21V, respectively, indicating that the voltage output of the device is proportional to the size of the force, as illustrated in figure.3. The force of the exciter is set to 10N. When the frequency of the excitation force of the exciter is 1.2Hz, 3Hz, 4.2Hz, 5.5Hz, the average peak-to-peak voltage (Vp-p) of the device is 8V, 10V, 14V, 20V respectively; indicating the device output voltage size and Frequency is proportional, shown in figure.3.

4. Conclusion

It can be seen that with the help of excellent piezoelectric performance nanowires, membrane generators have a very good voltage output performance. PVDF / GO nanofiber generator has voltage output peak-to-peak (Vp-p) about 21V, the output voltage is proportional to the size of the force and the operating frequency. The structure and manufacturing of the device also provide design ideas for similar energy conversion devices.

References

[1] Swallow, L. M., Luo, J. K., Siores, E., Patel, I., Dodds, D. 2008,A Piezoelectric Fibre Composite Based Energy Harvesting Device for Potential Wearable Applications. *Smart Mater. Struct.*, 17, 025017.

[2] Kanik, M., Aktas, O., Sen, H. S., Durgun, E., 2014, Bayindir, M. Spontaneous High Piezoelectricity in Poly(Vinylidene Fluoride) Nanoribbons Produced by Iterative Thermal Size Reduction Technique. *ACS Nano*, 8, 9311−9323.

[3] Jiang, Z. Y., Zheng, G. P., Zhan, K., Han, Z., Yang, J. H., 2015, Formation of Piezoelectric β-Phase Crystallites in Poly(Vinylidene Fluoride)-Graphene Oxide Nanocomposites under Uniaxial Tensions. J. Phys. D: *Appl. Phys.*, 48, 245303.

[4] Li, L.; Zhang, M.; Rong, M.; Ruan, W. Studies on theTransformation Process of PVDF from α to β Phase by Stretching.RSC Adv. 2014, 4, 3938−3943.

[5] Liu, X., Ma, J., Wu, X., Lin, L., Wang, X. , 2017. Polymeric Nanofibers with Ultrahigh Piezoelectricity via Self-Orientation of Nanocrystals. ACS nano, 11(2), 1901-1910.

[6] Lerf, A., He, H., Forster, M., and Klinowski, J. , 1998. Structure of graphite oxide revisited‖. *The Journal of Physical Chemistry B*, *102*(23), 4477-4482.

[7] Fuh, Y. K., Kuo, C. C., Huang, Z. M., Li, S. C., Liu, E. R. , 2016. A Transparent and Flexible Graphene - Piezoelectric Fiber Generator. *Small*, 12(14), 1875-1881.

[8] Garain, S., Jana, S., Sinha, T. K., Mandal, D. , 2016. Design of in situ poled Ce3+-Doped electrospun PVDF/Graphene composite nanofibers for fabrication of nanopressure sensor and ultrasensitive acoustic nanogenerator. *ACS applied materials & interfaces*, 8(7), 4532-4540.

IOP Conf. Series: Journal of Physics: Conf. Series **1052** (2018) 012111 doi:10.1088/1742-6596/1052/1/012111

Printed MEMS-based self–contained piezoelectric-based monitoring device for smart grids

Hélène Debéda[1], Isabel Rua-Taborda[1], Egon Fernandes[2], Sid Zarabi[2], David Nairn[2], Lan Wei[2], Armaghan Salehian[2]

[1] Université de Bordeaux, Laboratoire IMS, Talence
[2] University of Waterloo, Waterloo, Canada

E-mail: helene.debeda@ims-bordeaux.fr

Abstract. In this work, screen-printed piezoelectric PZT films are printed on a meandering-shaped 301SS stainless steel substrate to harvest power from current-carrying wires by piezo-electromagnetic transduction. Process optimization lead to good adhesion of the PZT on the stainless steel substrate and minimum bending of the multilayer {electrode/PZT/ electrode/stainless substrate}. A permanent magnet is integrated on the substrate tip for electromagnetic coupling. A power of 0.05 µW is harvested from 1 A of current flowing in a wire 12.5 mm from the tip mass. The normalized power is comparable to other available MEMS-based technologies in the literature. The development of a wireless sensor network for real time electricity monitoring application is also discussed.

1. Introduction

An emerging application of piezoelectric microstructures is in the field of vibration energy harvesting for the Internet of Things and wireless sensor networks. Our objective is to develop a low cost MEMS-based self –contained piezoelectric-based monitoring device for smart grids. The MEMS is based on a piezo-magnetic Energy Harvester (EH) operating through interactions of a tip magnet with the AC electromagnetic field around a wire carrying an AC current. Then, the induced magnetic force results in vibrations in the MEMS where a piezoelectric material is integrated and consequently in an induced voltage in the piezoelectric layers. This harvested power can be used to power a sensor in a self-contained system for electricity monitoring applications. For the maximization of the electromechanical coupling, thick piezoelectric layers of tens of micrometers are required. The low-cost screen-printing process is well adapted to this films scale. Screen-printed piezoelectric thick-films are usually implemented on silicon substrates suitable for further electronics integration [1-2]. Nevertheless, use of metallic substrates (stainless steel platforms, aluminum or copper) instead of silicon substrates has already confirmed their excellent performances such as flexibility, toughness and high-efficiency of power generation [3-5]. Zhu et *al* have obtained a power of 240 µW harvested at 66.2 Hz with an acceleration $g=2.8\ m.s^{-2}$. In this work, the energy harvester MEMS is made of a stainless substrate supporting the piezoelectric printed layers. This choice is justified by the low Young's modulus of the stainless steel substrate compared to alumina substrates classically used in thick film technology, or to silicon wafers. Finally, despite the RoHs restriction, the PZT ($PbZrTiO_3$) piezoelectric material is selected because of its outstanding electromechanical properties. This paper reports the fabrication and the optimization of the low-frequency cantilevered-zigzag energy harvester designed in a previous work [6] and the experimental testing of the piezoelectromagnetic energy harvester for current monitoring applications. Details of the wireless sensor network developed for real time electricity monitoring application is also presented.

2. Energy harvester processing

For the processing of the MEMS piezo-magnetic energy harvester, PZT layers sandwiched between two electrodes are integrated by screen-printing on the top of a stainless steel substrate. This substrate is first laser micromachined into a meandering shape which allows maximum strain under vibrations at the expected resonance frequency of 60 Hz [6].

2.1. Screen-printing, firing and polarization

Commercial pastes from Electroscience Laboratory or home-made pastes are respectively used for the deposition of the gold electrode layers and the piezoelectric PZT layers. The selected 301SS stainless substrate withstands the high temperature necessary for the PZT densification. The PZT piezoelectric paste is prepared with the PZT: LBCu powder blended with an organic binder. After manual mixing in a mortar, the particle dispersion is optimized using a three-roll mill to achieve good viscosity [7]. The EH fabrication then consists of subsequently printing of Au, PZT and Au layers on the 301SS substrate, with a drying step at 120 °C between each deposition (Figure 1 a and b). The layers are then isostatically pressed at 40 MPa before their co-firing at 900 °C in air to improve densification and thus piezoelectric properties [8]. The final PZT's thickness is ~60µm. In addition, compared to previous results [6], adhesion is clearly improved with the co-firing of all the printed layers. Slight bending of the multilayer is also observed because of an adapted PZT thickness leading to good compensation between the respective expansion's coefficients. Finally, to to achieve the desired piezoelectric properties, the fired PZT layers are wire-bonded and poled under 200V applied at 280°C, just below the Curie temperature (Figure 2). The beams named bi+ or bi- are polarized in opposite directions as in bimorph configurations according to their tension or compression under vibration.

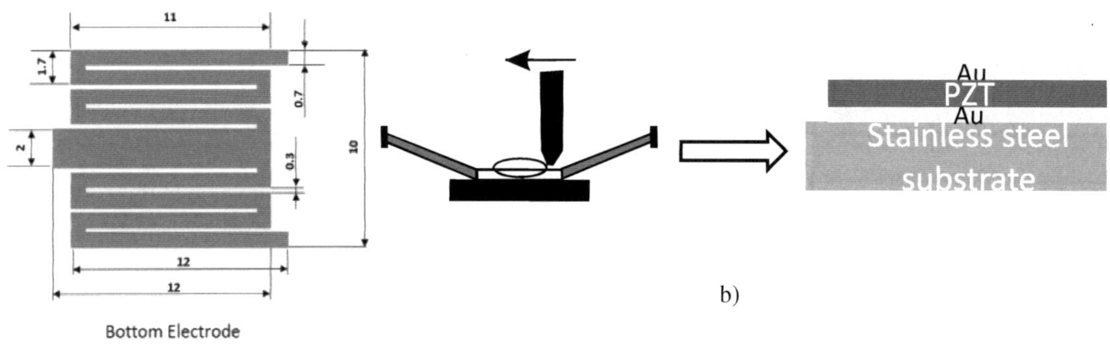

a) b)

Figure 1. Energy harvester process (a) dimensions in mm of the gold electrode (b) successive screen-printing and drying of the Au bottom electrode, PZT layer and Au top electrode

2.2. Characterization

The electromechanical properties are strongly influenced by the densification. The previous work showed a better densification and reproducibility of PZT thick films when using the PZT powder mixed with the eutectic LBCu instead of borosilicate glasses [8]. The µm particle size of the different powders (<10µm) also helps to improve densification. The porosity of the layers are analysed through SEM cross-sections by using ImageJ analysis software (Figure 3). The evaluated porosity of PZT on the stainless substrate (24%) is comparable with that on layers fired on alumina substrates. This porosity is nevertheless higher than that observed for PZT free-standing layers, because of the binding with the substrate reducing the lateral shrinkage. Electromechanical characterization, with conductance and susceptance measurements, confirm the success of the polarization step with a main resonance peak at 321 kHz for the polarized bi+ beams (Figure 4) .

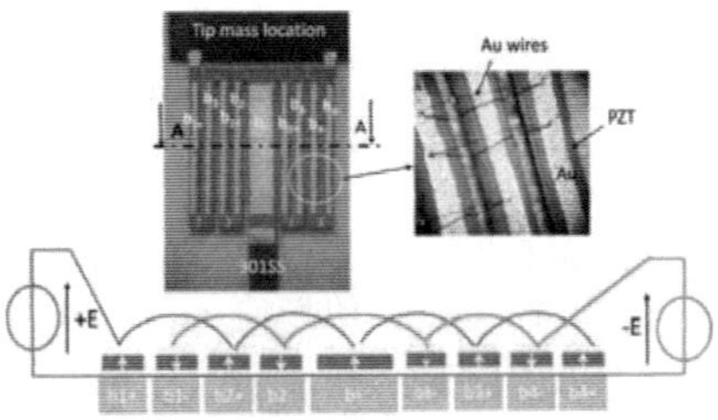

Figure 2. Wiring bonding of the beams bi+ and bi- for the PZT polarization of each adjacent beam

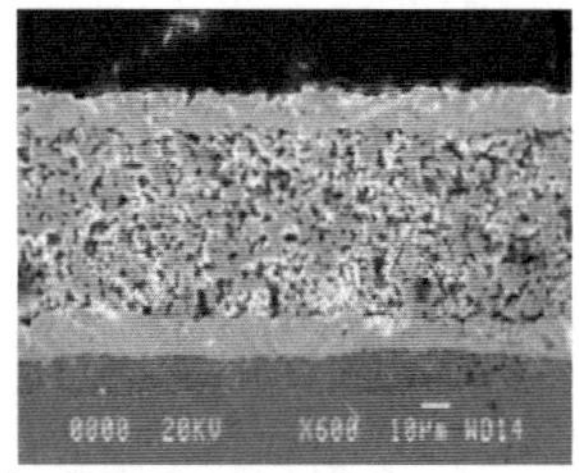

Figure 3. SEM image of the cross-section of the multilayer SS substrate Au/PZT/A

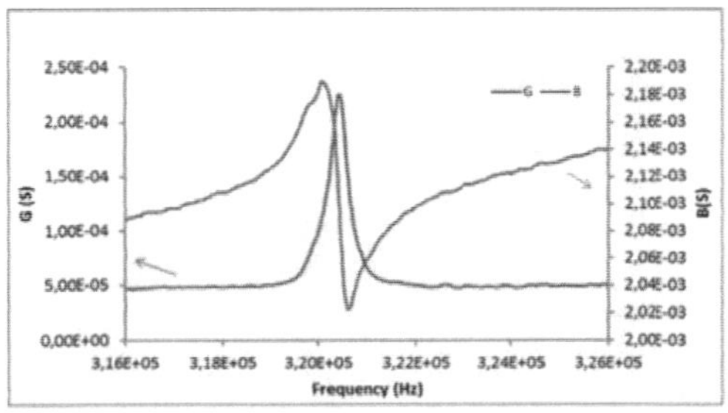

Figure 4. Conductance G and susceptance B measured for all the b+ beams connected

3. Device tests

3.1. Energy harvester tests

Once polarized, tests under vibration without tip-mass are carried out using a shaker to confirm the good electromechanical coupling. Voltage output in an open circuit state is taken across the top and bottom electrodes of the b+ beams (Figure 5a). The acceleration is fixed at 0.4g and the frequency is swept around the resonance vibration of the EH MEMS. With this configuration, the maximum RMS voltage produced by the harvester without tip mass is 242 mV with a natural frequency of 209 Hz. This value, similar to that obtained during previous tests [6], confirms the good reproducibility of the process. For further tests under magnetic field, all b+ and b-beams are connected in parallel and a neodymium magnet (dimension 12.7x3.2x3.2mm^3) is finally epoxied to the substrate's tip. With the addition of this magnet the resonance frequency is reduced down to 60 Hz. Then, a magnetic field is created by a 1A current flowing in an AWG10 wire (Figure 5b). The harvested power is finally measured with different load resistances. For a distance of 12.5 mm between the magnet and the wire's surface, a maximum power of 0.05 μW is measured for a 1.8 MΩ resistor. With the volume of 169,7mm^3 of the EH, the normalized power 0,053μW/mm^3 is comparable to other available MEMS-based technologies in the literature [9-10].

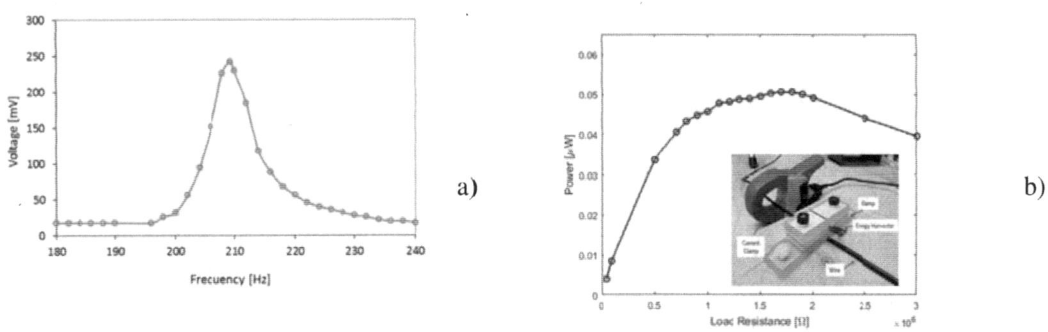

Figure 5. EH tests (a) Experimental open-circuit voltage of the EH without tip mass at 0.4g (b) Experimental set-up of the EH with magnetic tip-mass and harvested power for different load resistances

3.2. Electronics
Piezo-electromagnetic harvesters using simple cantilever beam structures have already been integrated with power management and wireless communication circuits to form an individual node in a wireless sensor network for electricity grid monitoring (Figure 6) [11]. With its innovative lower power circuit design and dynamic power management scheme, the system is fully self-contained and non-invasive. With the simple cantilever beam EH attached at 6mm away from an electric wire carrying currents in the range of 7.6A to 30A, the state-of-the-art electronic system can achieve a read-transmit duty cycle from <1min to 2.5min and capable of reading a 60 HZ AC sensor signal with a peak voltage in the range of 100 mV to 900 mV with an accuracy of 91.4% [11]. The integration of the electronic system is in progress.

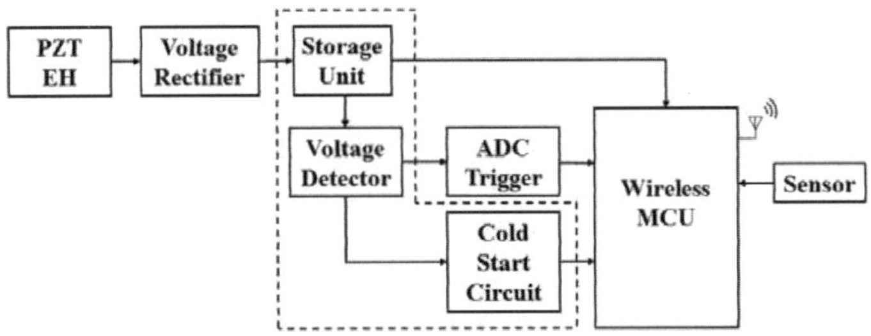

Figure 6. System diagram for the sensor node for electricity grid monitoring

4. Conclusion
In summary, a fully printed piezo-electric energy harvester was successfully fabricated, with a symmetric cantilevered zig-zag geometry. Adhesion of the thick printed layers (~60µm) on the selected 301SS stainless steel substrate is improved and bending of the multi-layer has been reduced. A power of 0.05 µW, harvested from 1 A of current flowing in a wire 12.5 mm from the tip mass, is comparable to other available MEMS-based technologies in the literature. Tests are under progress to integrate this EH in a system for electricity monitoring applications.

Acknowledgements
This work was supported Waterloo Institute of Sustainable Energy, Cisco Systems, UWUB grant, IRPG and AMADEUS LABEX. The authors also thank I. Favre for the wire-bonding process.

References
[1] Xu R, Lei A, Dahl-Petersen C, Hansen K, Guizzetti M, Birkelund K, Thomsen E.V, Hansen O 2012 *Sensors and Actuators A* 188 pp.383– 388
[2] Costache F, Pawlik B, Rieck A, 2017 *Eurosensors XXXI, Paris, France*
[3] Chen J J, Lien Y C, Kuo C L,Wu W J, 2015 *IEEE NMEMS , Xi'an, China*
[4] Zhu D, Beeby S P, Tudor MJ, Harris N R 2011 *Sens. Act. Phys.*, vol. 169, no 2, p. 317–325
[5] Chen C T, Lin S C, Lin T K, Wu W J. 2016, *PowerMEMS* IOP Publishing Journal of Physics: Conference Series 773
[6] Fernandes E, Zarabi S, Debéda H, Lucat C, Nairn D, Wei L, Salehian A 2016 *PowerMEMS, Paris*, Journal of Physics: Conference Series 773 012109
[7] Rua I, Fernandes E, Zarabi S, Nairn D, Wei L, Salehian A, Debéda H, 2017 *DTIP, Bordeaux*
[8] Debéda H, Clément P, Llobet E, Lucat C, 2015 *Smart Mater. Struct.* 24, 025020
[9] Chen W, Cao Y, Xie J 2015 *J. of Mechanical Science and Technology* 29 4313-4318
[10] Olszewski O Z, Houlihan R, Mathewson A, Jackson N, 2016 *J. of Physics: Conference Series* pp 012039
[11] Zarabi S, Fernandes E, Rua I, Debéda H, A. Salehian, Nairn D, Wei L, 2017 *IEEE ASICON, Guiyang, China*

IOP Conf. Series: Journal of Physics: Conf. Series **1052** (2018) 012112 doi:10.1088/1742-6596/1052/1/012112

Orientation Dependence of Power Generation on Piezoelectric Energy Harvesting Using Stretched Ferroelectric Polymer Films

A Kobayashi[1], Y Koshiba[1], Y Ueno[1], T Kajihara[1], Y Tsujiura[1], M Morimoto[1,2], S Horike[1], T Fukushima[1], I Kanno[1], and K Ishida[1]

[1]Kobe University, 1-1 Rokkodai-cho, Nada-ku, Kobe 657-8501, Japan
[2]University of Toyama, 3190 Gofuku, Toyama, 930-8555, Japan

kishida@crystal.kobe-u.ac.jp

Abstract. The piezoelectric vibration energy harvesters were fabricated by using uniaxially stretched poly (vinylidene difluoride/trifluoroethylene) copolymer (P(VDF/TrFE)) film, and the relationship between piezoelectric power generation and molecular orientation was investigated. The molecular orientation in the stretched P(VDF/TrFE) films was evaluated with polarized Fourier transfer infrared (FT-IR) spectra measurement. In stretched films, the main-chains of P(VDF/TrFE) were aligned along the stretching direction. The piezoelectric properties and the electric power generation of stretched P(VDF/TrFE) films were strongly depended on their molecular orientation, measuring by cantilever-type energy harvesters. The piezoelectric coefficient(e) and output power observed in the energy harvester with the film stretched in the longitudinal direction of cantilever were 16.9 mC/m^2 and 222 nW, respectively. These values were approximately 2.1 and 3.5 times these of the unstretched elements.

1. Introduction

Vibration energy harvesting using piezoelectric material have attracted much attention in the recent years [1-4]. Inorganic piezoelectric materials with high piezoelectric coefficient have been widely used for vibration power generation, however, the resonant frequencies of these harvesters are higher than that of environmental vibration. To effective use the environmental vibration, the resonant frequency should be adjusted to around that of environmental vibration widely distributed under 100 Hz. The soft organic ferroelectrics are promising candidates for the element of the vibration harvester because of their low resonance frequency, whereas, the low output power of them has been reported.

In our previous work, we used poly (vinylidene difluoride/trifluoroethylene) (P(VDF/TrFE)) (Fig.1 (a)) which is the typical organic ferroelectric polymer [5-6], and tried to improve the performance of harvesters with the multi-stacked P(VDF/TrFE) films. The energy harvester with the cantilever

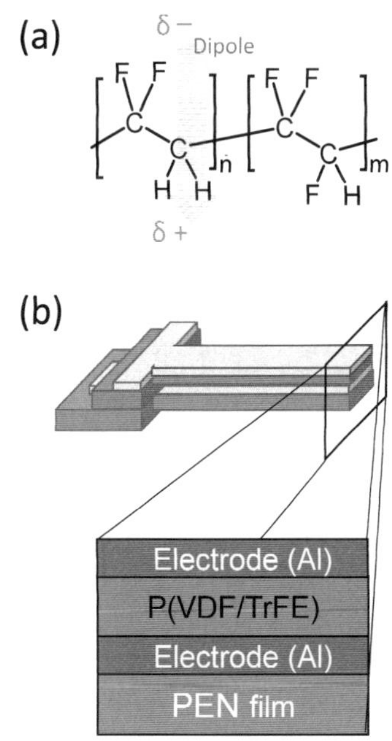

Figure 1. Chemical structure of P(VDF/TrFE) (a) and schematic structure of cantilever-type energy harvester (b).

structure generated electric power with a resonant frequency of approximately 25 Hz that is comparable to the environmental vibration. The power density of four-layered harvesters was larger than that of the single layer harvester [7]. Moreover we fabricated energy harvester with stretched P(VDF/TrFE) films and investigated the relationship between piezoelectric power generation and molecular orientation in P(VDF/TrFE) films [7]. Therefore, in this study, the fabrication of the energy harvesters using the uniaxially oriented P(VDF/TrFE) films and the piezoelectric power generation of these harvesters were reported in more detail.

2. Experimental methods

P(VDF/TrFE) was dissolved in methyl ethyl keton, and spin-coated onto a silicon substrate. The thickness of the P(VDF/TrFE) films was 2.0μm, approximately. The P(VDF/TrFE) layer was peeled off from the substrate. The free-standing P(VDF/TrFE) films was uniaxially stretched to ratio of 300 % in air, and aneled at 140 °C for 2h. The stretched film, onto which the 80 nm Al electrodes were deposited, was laminated to the 25-μm-thick poly(ethylene naphthalate) (PEN) substrate. The cantilever structure consisted of Al/P(VDF/TrFE)/Al was used as a design for power generation and piezoelectric measurements.This structure is illustrated in Fig.1 (b). For poling, a triangular external electric field was applied to these energy harvester at room temperature. The piezoelectric properties were evaluated using a laser Doppler vibrometer and optical microscope (Ono Sokki LV-1720). The displacement of the cantilever corresponding the piezoelectric vibration was measured with applying a sinusoidal voltage between the top and bottom electrodes. The electric power generation was measured by applying mechanical vibrations with an acceleration of 10 m/s^2.

3. Results and discussion

Figure 2 shows the polarized Fourier transfer infrared (FT-IR) spectra of stretched P(VDF/TrFE) films. In the spectrum which was measured with the incident light polarized perpendicular to the stretching direction, the high-intensity absorption at 1400 cm^{-1}, corresponding to the CH_2 wagging mode, was

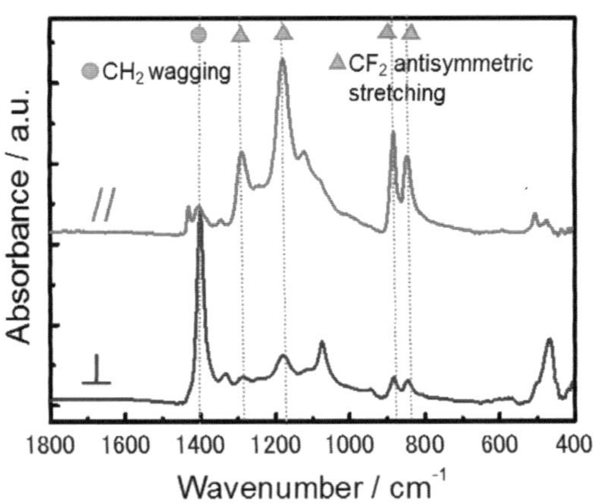

Figure 2. Polarized FT-IR spectra of stretched P(VDF/TrFE) films.
Stretched direction parallel to polarized IR : //, perpendicular to polarized IR : ⊥.

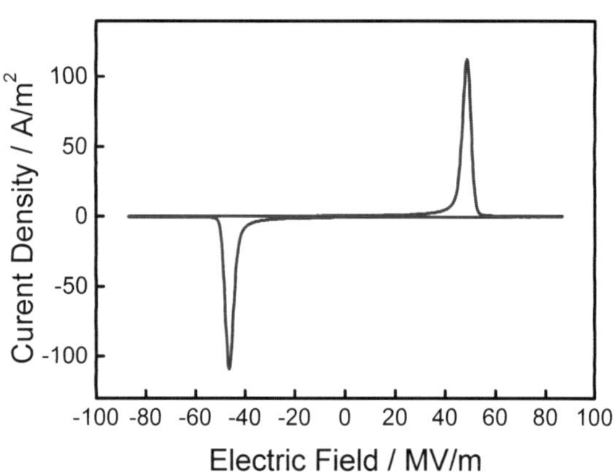

Figure 3. *J-E* curve of stretched P(VDF/TrFE) film.

observed; on the other hand, the intensities of absorption peaks at 1290, 1190, 880 and 840 cm^{-1} are much weaker than those in the spectrum with parallel polarization. These absorption peaks were assigned to CF_2 antisymmetric stretching, which is perpendicular to the polymer chains [8]. These results suggested that the main-chains of P(VDF/TrFE) were aligned along the stretching direction.

The current density electric field (*J–E*) curve of the stretched P(VDF/TrFE) film was shown in Fig.3. This measurement was carried out by applying a triangular electric field. There are two peaks in the negative and the positive electric field, respectively. These peaks indicated polarization switching. The remnant polarization was 70 mC/m^2 which was comparable to the value of the unstretched P(VDF/TrFE) film.

The properties of vibration power generators were measured around the resonant frequency approximately 25 Hz. The three type cantilever structure are illustrated in Fig.4, and the resonant frequency, effective voltage, output power generation, and piezoelectric constant *e* of these energy harvesters were shown in table 1. In unstretched P(VDF/TrFE) film (Type A), the piezoelectric constant *e* was 8.24 mC/m^2; and this value of P(VDF/TrFE) film stretched in the longitudinal direction (Type B) and widthwise direction (Type C) of cantilever was 16.9 and 2.19 mC/m^2, respectively. The highest *e* value was observed in Type B energy harvester, in which main-chains of P(VDF/TrFE) were aligned to the longitudinal direction of the cantilever. The output power of Type A, B and C energy harvesters were 61.2, 222 and 22.4 nW, respectively. The highest value also observed in Type B energy harvester. Here, this value of P(VDF/TrFE) film stretched in the longitudinal direction and widthwise direction of cantilever was approximately 3.5 and 0.4 times these of the unstretched elements, respectively [7].

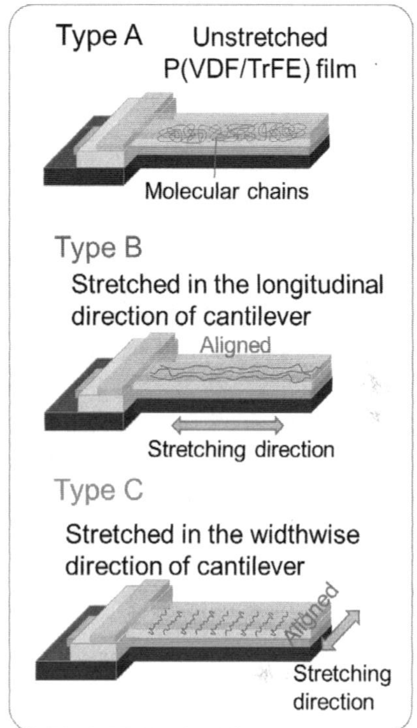

Figure 4. Schematic structures of cantilever-type energy harvesters.

Table 1. Maximum output power and piezoelectric constants of unstretcherd and stretched P(VDF/TrFE) films.

	Type A Unstretched P(VDF/TrFE) film 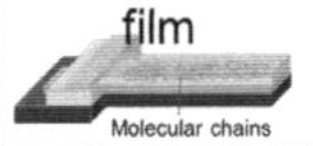	Type B Stretched longitudinal direction 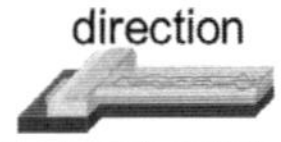	Type C Stretched widthwise direction 
Resonant frequency (Hz)	24.1	24.6	23.3
Effective voltage (mV)	423	735	330
Piezoelectric coefficient *e* (mC/m^2)	8.24	16.9	2.19
Output power (nW)	61.2	222	22.4

These results suggested that the molecular orientation in P(VDF/TrFE) films affected the piezoelectric constant and output power, and the optimal molecular orientation for vibration power generations was to be parallel their long chain axis to the longitudinal direction of the cantilever.

In conclusion, we fabricated the cantilever-type energy harvesters using stretched and unstretched ferroelectric polymer films. In the energy harvester in which P(VDF/TrFE) main-chains were aligned to the longitudinal direction of the cantilever, the highest *e* and output power values were observed. Thus, the uniaxially oriented P(VDF/TrFE) films showed potential for the element of the vibration harvester using in the environmental vibration.

References

[1] Morimoto K, Kanno I, Wasa K and Kotera H 2010 Sensor. Actuat. A-Phys. **163** 428
[2] Xua R, Lei A, Dahl-Petersena C, Hansenb K, Guizzettib M, Birkelunda K, Thomsena E V and Hansena O 2012 Sensor. Actuat. A-Phys. **188** 383
[3] Bowen C R, Kim H A, Weaverb P M and Dunnc S. 2014 Energy Environ. Sci. **7** 25
[4] Matsumoto A, Horie S, Yamada H and Matsushige K 2007 Appl. Phys. Lett. 90, 202906
[5] Fulukawa T, Date M and Fukuda E 1980 J. Appl. Phys. **151** 1135
[6] Kuroda Y, Koshiba Y, Misaki M, Ishida K and Ueda Y 2013 Applied Physics Expresss 6 021601
[7] Kajihara T, Ueno Y, Tsujiura Y, Koshiba Y, Morimoto, M. Kanno I, and Ishida K 2017 Jpn. J. Appl. Phys. **56** 04CL04
[8] Kim K J, Reynolds N M and Hsu S L 1989 Macromolecules **22** 4395

Acknowledgments
This work was partly supported by CREST, JST, and JSPS KAKENHI, JAPAN.

Investigation of piezoelectric energy harvesting from human walking

R Kakihara, K Kariya, Y Matsushita, T Yoshimura and N Fujimura

Graduate School of Engineering, Osaka Prefecture University, Sakai 599-8531, Japan

tyoshi@pe.osakafu-u.ac.jp

Abstract. Energy harvesting from human foot motion is attractive for the electrical power source of wearable sensors. For human foot motion, three types of energies are available: bending in the sole, heel strike and swing motion. In this study, the magnitudes of these energies were analyzed by dynamic measurements using various sensors. Then, appropriate piezoelectrics and device structure for the energy harvesting were discussed for each energy. From these results, it was concluded that utilizing bending in the sole and organic piezoelectrics such as PVDF is the best way for the energy harvesting from walking.

1. Introduction

Wearable devices have been rapidly developing, and there is an increasing demand for self-sustaining power source. Energy harvesting from human motion is one of the solutions. Since the energy derived from foot motion is the largest among the various human motions [1], there are various reports, which demonstrate prototypes of shoes with energy harvester [2, 3].

Conversions from mechanical energy to electrical energy can be realized through four basic mechanisms, including electromagnetic, electrostatic, inverse magnetostrictive and piezoelectric. Since the piezoelectrics have features of large power density and ease of miniaturization, various piezoelectric harvesters attached to shoes have been reported. For example, Shenck N et al. reported the harvester for shoe sole, which produced an output power of 8.4 mW [4].

In this study, the maximum electrical power that can be expected for piezoelectric harvesters equipped in shoes is discussed. Three types of energy can be obtained from human foot motion: bending in the sole, heel strike and swing motion. The magnitudes of these energies were analyzed by dynamic measurements using various sensors and maximum electrical power converted from these energies during walking was calculated. Moreover, the optimum structure for efficiently converting to electrical power by utilizing piezoelectric effect is investigated.

2. Experiments

Bending strain during walking was measured to investigate the input energy to shoes by bending in the sole. Strain gauges (KFEM-5-120-C1L3M2R, KYOWA, Japan) were used for the measurement. PET sheet with a thickness of 500 μm and Young's modulus of 2 GPa was cut into a foot shape. Four strain gauges were attached to the both sides of the PET sheet as shown in figure 1(a). The PET sheet was placed under the insole of a shoe. For unrestricted measurements, the portable strain amplifiers were fabricated, and the output voltage from the strain amplifiers was recorded by Arduino UNO during walking. The force by the heel strike was measured by pressure sensors (Flexiforce, Nitta, Japan) which were put on the insole. The swing motion was measured by a 9-degree sensor attached to the side of the shoe. The direction of acceleration and angular velocity are also shown in figure 1(a). These signals were also detected by Arduino UNO at the same time.

3. Results and discussion

The measurement results of bending strain in the sole, the force of the heel strike and swing motion are shown in figure 1(b). From the measurement of bending strain, it was found that the maximum strain was obtained at [C], which is around the boll of the foot. While similar strain was observed at [A] and [B], the strain at [D], which is around the arch of the foot is much lower than at other positions. The pressure of the heel strike reached about 0.9 MPa. From the measurement result of acceleration and angular velocity, walking state can be analyzed. It can be separated stance phase and swing phase.

The calculated bending stress in the PET sheet at the boll of the foot is shown in figure 2. The bending stress is about ten times larger than the pressure of the heel strike shown in figure 1(b). This indicates that larger electrical power can be expected for the bending in the sole. For the detailed analysis, the output power was estimated for the bending and the heel strike. The assumed structure of the harvester for the bending is the piezoelectrics attached on the PET sheet. The strain energy U, which is stored in the piezoelectrics, is calculated by

$$U = \iiint \frac{1}{2}\sigma\varepsilon \, dxdydz = \int \frac{1}{2}\frac{\sigma^2}{E}\, dV = \int \frac{1}{2}\varepsilon^2 E \, dV \qquad (1)$$

where ε is strain, σ is stress and E is Young's modulus. The bending strain in the piezoelectrics is proportional to the distance from the neutral plane. Therefore, the bending strain ε can be expressed by

$$\varepsilon = \varepsilon_{Measure}\frac{h_{piezo}}{h_{PET}} \qquad (2)$$

where $\varepsilon_{Measure}$ is measured strain, h_{piezo} and h_{PET} are thickness of piezoelectrics and the PET sheet, respectively. From equations (1) and (2), it is indicated that the stored strain energy for the bending is proportional to the cube of thickness of the piezoelectrics. For the energy harvesting from the heel strike, the piezoelectrics are placed simply under the heel. The stress in the piezoelectrics is independent of the thickness, so the stored strain energy is

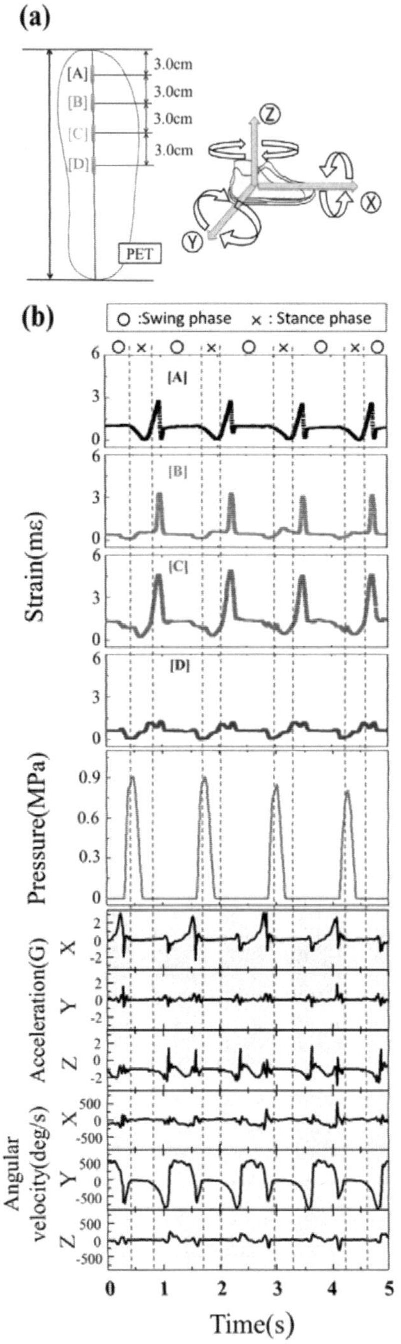

Figure 1. (a) Direction of acceleration and angular velocity and strain measurement points. (b) Measurement results from various sensors during walking.

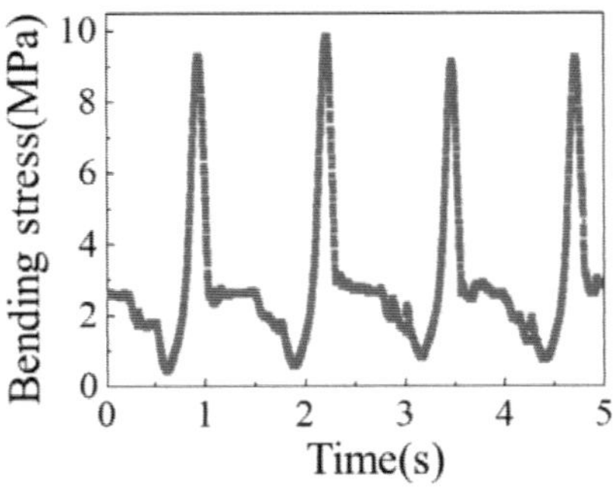

Figure 2. Bending stress on PET at the boll of foot.

Table 1. Properties of piezoelectrics and PET.

	PVDF	PZT	PET
Young' s modulus (GPa)	2.0	80	2.0
Bending strength (MPa)	75	100	73
Thickness (μm)	Various	Various	500
k_{31}	0.11	0.35	-
k_{33}		0.69	

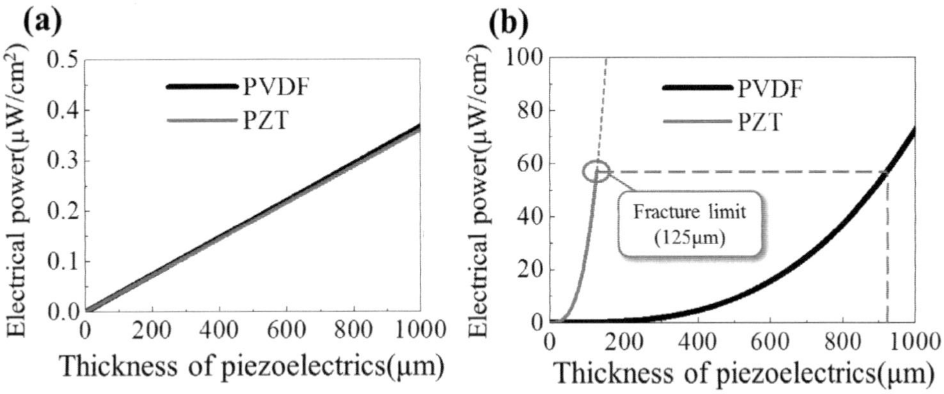

Figure 3. The dependence of the electrical power on the thickness of PVDF and PZT. (a)Heel (33mode) (b)Bending (31mode).

proportional to the thickness. For the non-resonant type piezoelectric energy harvester, the electrical power P converted from the strain energy is given by $P = k^2W = k^2\frac{dU}{dt}$, where k is electromechanical coupling coefficient of the piezoelectrics. In this study, $Pb(Zr,Ti)O_3$ (PZT) and Poly Vinylidene DiFluoride (PVDF) are selected as typical piezoelectrics. The properties of PZT, and PVDF as well as PET are summarized in table 1. While PZT has an advantage of high k, it has a disadvantage of brittleness. PVDF has completely opposite properties, which are low k and high flexibility.

Figure 3 shows the harvested power from the heel pressure and the bending strain around the ball of the foot as a function of the thickness of PZT and PVDF. The thickness of the PET sheet was assumed to be 500 μm. For the piezoelectric transduction, 33 mode and 31 mode are used for the heel pressure and the bending strain, respectively. While PZT has about 40 times higher k^2 than PVDF, for the heel pressure, the harvested electrical power for PZT and PVDF is almost same. This is because the deformation and the input energy for PZT by the heel pressure is one-fortieth of those for PVDF. As a result, the low k^2 of PVDF is compensated by the high input energy.

On the other hand, large differences are observed for the bending in the sole as shown in figure 3(b). PZT can harvest larger electrical power than PVDF at the same thickness. However, the electrical power is limited at 60 μW/cm^2 due to the fracture limit for PZT, which is about 1250 με at the thickness of 125 μm. PVDF can harvest larger electrical power than PZT, when the thickness is more than 900 μm. While the use of such thick PVDF sheets is not easy, it is not necessary to be a single sheet. Bimorph structure composed of non-piezoelectric plate and PVDF, and multilayer structure can harvest the almost same amount of the electrical power, when the total thickness of PVDF is same. It is also revealed that electrical power harvested by the bending is more than 100 times larger than that by heel strike, although k of 33 mode is higher than that of 31 mode. These results indicate that the increase of the input mechanical energy is quite effective to increase the harvested electrical power.

Figure 4 shows the electrical power as functions of the thickness of PVDF and PET. The electrical power increases with increasing the thickness of PET as well as PVDF. This is because the strain of PVDF increases with increasing the thickness of PET. The electrical power more than 10 mW can be expected when the actual device size becomes more than 10 cm^2.

The energy harvesting from the swing motion is also discussed. In this case, the kinetic energy is transferred to the mass attached to the piezoelectric oscillators. The kinetic energy of a 1 g mass in a shoe during walking is calculated as shown in figure 5. The average kinetic energy is about 500 μJ/s. From this result and the theoretical conversion efficiency of the piezoelectric oscillators, it can be expected that maximum electrical power is less than 50 μW. This calculation is consistent with the previous results, in which the obtained output powers from the swing motion are less than ~mW [5]. Hence, the direct application of the force to piezoelectrics such as the bending in the sole is superior to the use of the kinetic energy of the swing motion.

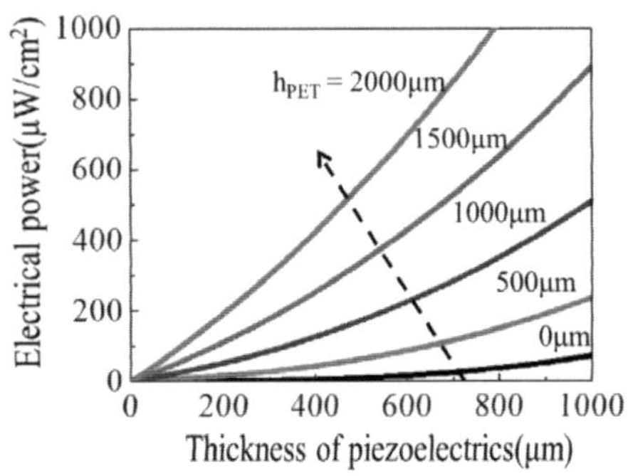

Figure 4. The dependence of the electrical power on the thickness of PVDF and the substrate.

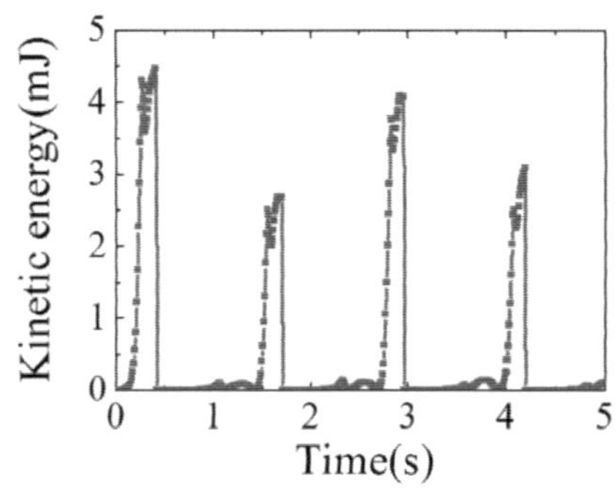

Figure 5. Kinetic energy of a 1 g mass in the shoe during walking.

4. Conclusion

In this study, three types of input energy to the shoes, which are bending in the sole, heel strike and swing motion, were analyzed by dynamic measurement using various sensors. Input mechanical energy by bending in the sole is largest among three energies. It is suggested that utilizing bending in the sole with organic piezoelectrics such as PVDF is the best way for energy harvesting from walking.

References

[1] Starner T 1996 *IBM Syst. J.* **35** 618
[2] Kymissis J, Kendell C, Paradiso J and Gershenfeld N 1998 *Second IEEE Int'l Conf. Wearable Computing*, **19** 132
[3] Zhao J and You Z 2014 *Sensors* **14** 12497
[4] Shenck N and Paradiso J 2001 *IEEE Micro* **21** 30
[5] Ylli K, Hoffmann D, Willmann A, Becker P, Folkmer B and Manoli Y 2015 *Smart Mater. Struct.* **24** 025029

Design and optimization of a flapping water flow energy harvester

Jorge Antonio Nieves Juárez[1,2], Ivo Neftali Ayala Garcia[1] and Dibin Zhu[1*]

[1]Collegue of Mathematics, Engineering and Physical Sciences, University of Exeter, Exeter, UK, EX4 4QF
[2]Technical University of Queretaro, Queretaro, Mexico

*E-mail: d.zhu@exeter.ac.uk

Abstract. This paper reports a flapping water flow energy harvester. It consists of a bluff body and a cantilever based resonator. The bluff body facing the water flow creates mechanical instability. The bluff body can control forces and pressures applied to the tip of the resonator, i.e. a wing, so a continuous and stable oscillations of the resonator can be achieved. Key parameters for initiating and maintaining such oscillation include position and size of the bluff body as well as the angle of attack. These parameters were simulated in ANSYS and simulation results were verified experimentally. In order to test the energy harvesting capability of this structure in water, a MFC patch was attached to the cantilever and output power was measured when the harvester was placed in laminar water flow with the speed of 1 m/s. A maximum power of 52 μW was recorded.

1. Introduction

Energy harvesting concerns methods to obtain energy from ambient resources such as vibrations, heat, water and wind. It is regarded as a promising energy source to power small electronic devices, e.g. wireless sensors. It is most suitable for applications where maintenance is difficult and costly. Monitoring using wireless sensors in water industry is challenging because the size of water networks is vast and a majority of water pipes are buried underground. There is an increasing demand to innovate the existing water networks to enable maintenance free wireless monitoring. A reliable power source is essential for these wireless sensors. The work reported in this paper focused on methods to harvest energy from water flow.

Some work on energy harvesting from water flow has been reported. Both piezoelectric and electromagnetic transducers have been considered to convert water flow into electrical energy. Wang *et al* [1] developed a piezoelectric energy harvester which generated electrical energy from oscillation caused by the Kármán vortex street behind a bluff body in a water flow. However, the output power was only 0.7 nW in water flow of 1 m/s. Water flow energy harvesters based on electromagnetic transducers have also been reported [2, 3]. However, existing devices are all propeller based and harvest energy from rotation. Such devices require regular maintenance to keep the bearing lubricated and extend its life time. The device presented in this paper operates based on the flapping oscillation of a cantilever structure flow and thus has longer lifespan. It is mitigated from a device initially designed for air flow application [4]. Some key parameters were optimized using ANSYS CFD and initial experimental results of its energy harvesting capability in water are reported.

2. Operation principle, simulation and device prototyping

2.1. Principle

The water flow energy harvester presented here evolves from a previously reported air flow energy harvester based on flapping operation principle, as shown in Figure 1, to convert flow into mechanical oscillation to harvest energy [4]. The device consists of a bluff body in charge to create responsible for creating mechanical instability and a cantilever based resonator. The bluff body can control forces and pressures applied to the wings so a continuous and stable oscillations of the resonator can be achieved. One of the ends of the of the cantilever beams is anchored while the other hold a wing. Previous study showed that the position and size of the bluff body as well as the angle of attack (α in Figure 2) have significant effect on the performance of this flapping structure.

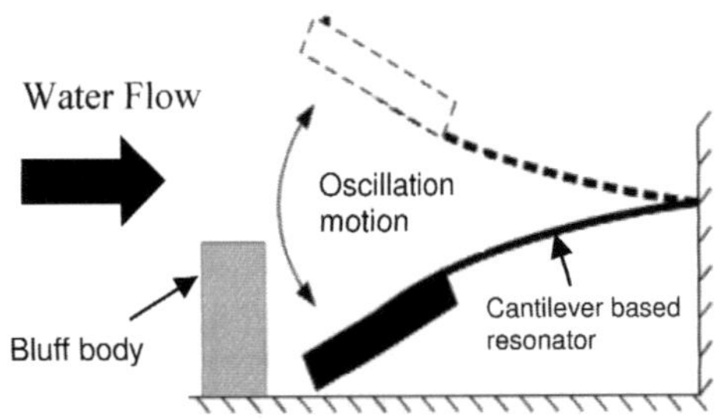

Figure 1. Basic operational Principle of the flapping energy harvester.

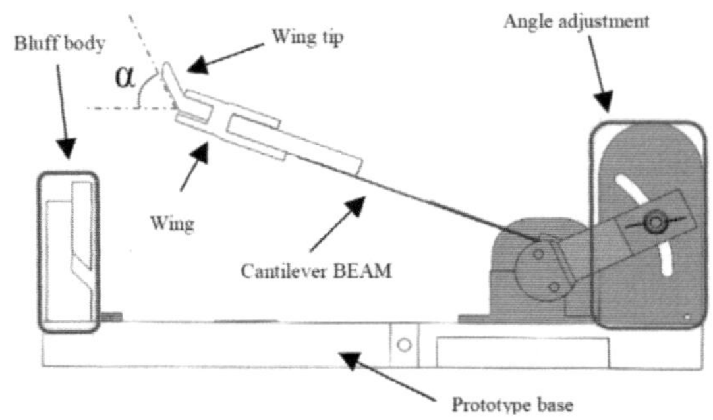

Figure 2. Schematic of the testing prototype of the flapping energy harvester.

2.2. Simulation

ANSYS simulations has been carried out to maximize the lifting force by varying positions and sizes of the bluff body as well as the angle of attack. Figures 3 and 4 show simulated pressure difference between top and bottom of the wing tip. This also gives an indication of the lift force applied to the wing. The higher the pressure difference there is, the larger lifting force is applied to the wing. Simulation results in Figure 3 showed that when α is 45°, the pressure difference between top and bottom of the wing tip became maximal which means the maximum lift force can be achieved at such degree. In addition, simulation results also show that when the bluff body position is placed 10 mm to the edge of the wing tip, the maximum pressure difference, thus, maximum lifting force, can be achieved as shown in Figure 4. As shown in Figure 5, such combination is able to provide a pressure difference of 1931 Pa. Given the wing tip area of 0.00385 m^2, the maximum lift force is 7.4 N. Such a force can initiate a constant oscillation with a maximum displacement of 50 mm.

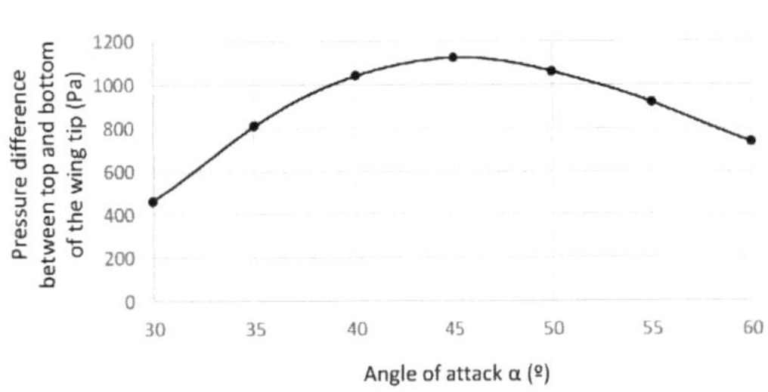

Figure 3. Various angles of attack shows provide different pressure values. 45° is shown as the optimum degree.

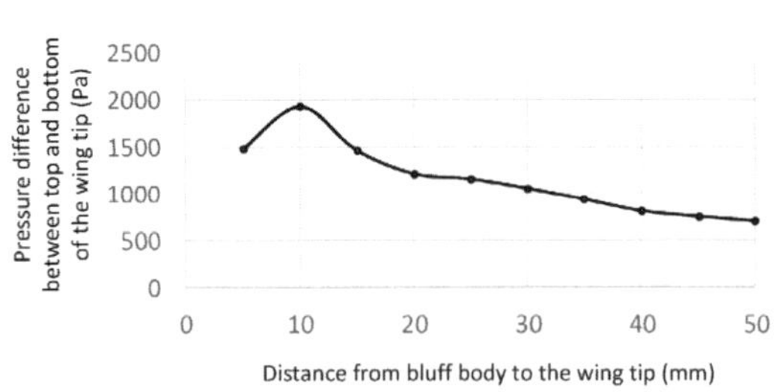

Figure 4. Distance from bluff body to the wing tip gives different pressure values.

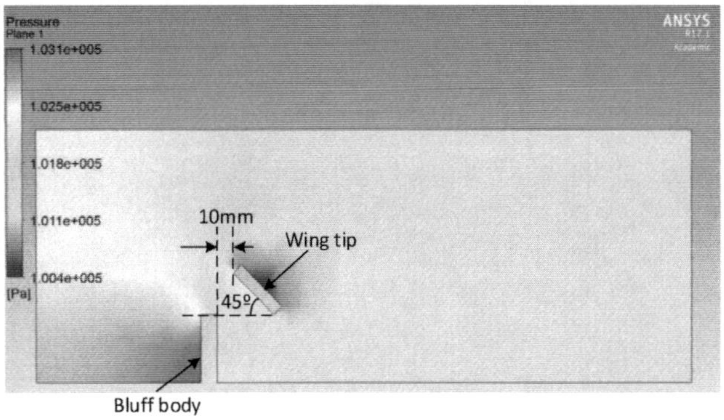

Figure 5. Simulation results of pressure in the tunnel when α is 45° and the distance from the bluff body to the wing tip is 10 mm.

2.3. Prototyping

In order to verify the simulation results in ANSYS, a prototype with adjustment capability on the bluff body and angle of attack was fabricated as shown in Figure 6. The base, bluff body and wing were all 3D-printed using PLA. The cantilever beam was made of 0.3 mm thick Beryllium Copper (BeCu) due to its excellent mechanical properties. In order to test the energy harvesting capability of this structure in water, a MFC [5] strip is attached to the cantilever beam as a transducer to convert mechanical oscillation into electrical energy as shown in Figure 7. The MFC strip generates an electrical signal when the mechanical stress is applied.

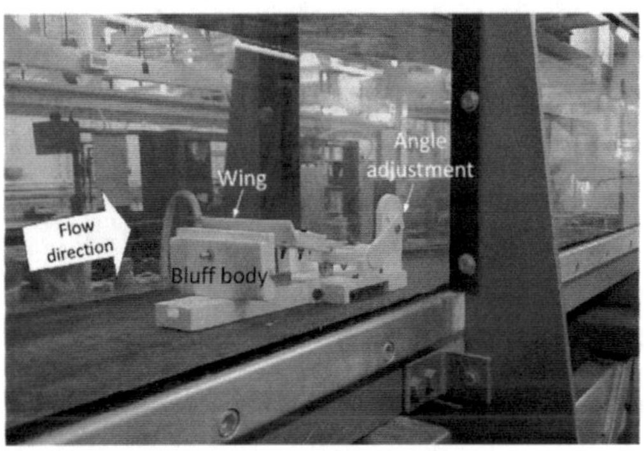

Figure 6. A prototype of the flapping water energy harvester placed in a hydraulic channel.

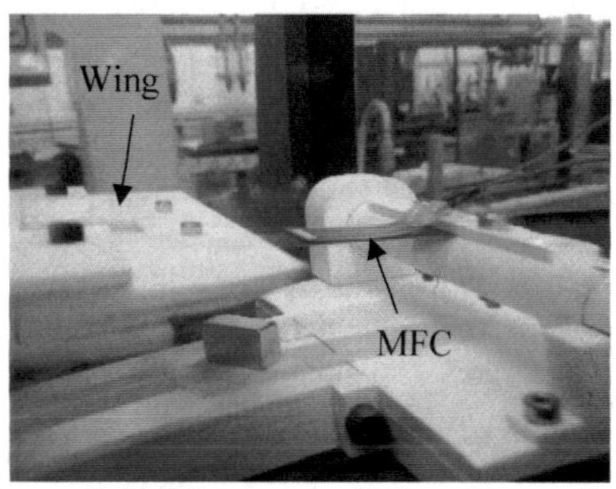

Figure 7. MFC strip attached to the cantilever beam.

3. Tests and Results

The prototype was tested in a hydraulic channel under the water flow with a speed of 1 m/s. It was measured that due to the existence of the bluff body, the speed of the water stream that meets the wing tip was doubled to 2 m/s. As the channel is made of steel, the prototype was anchored to the bottom of the channel using strong magnets. The device was located in the centre of the water flow path.

First test was carried out to find the optimum location of the bluff body and the angle of attack. It was found that when the angle of attack is around 45 and the bluff body is around 10 mm away from the wing tip, stable oscillation of 0.9 Hz with a maximum displacement of 5 cm occurred. This matches the simulation results presented earlier. This setting was kept for the remaining tests.

The MFC was connected to various resistive loads while its output voltage was measured. Thus, the output power can be calculated and optimum load resistance can be observed. Figure 8 shows the experimental results of output power when the harvester was connected to various load resistance. The optimum load resistor was found to be 3 MΩ while the maximum output power was measured as 52 μW.

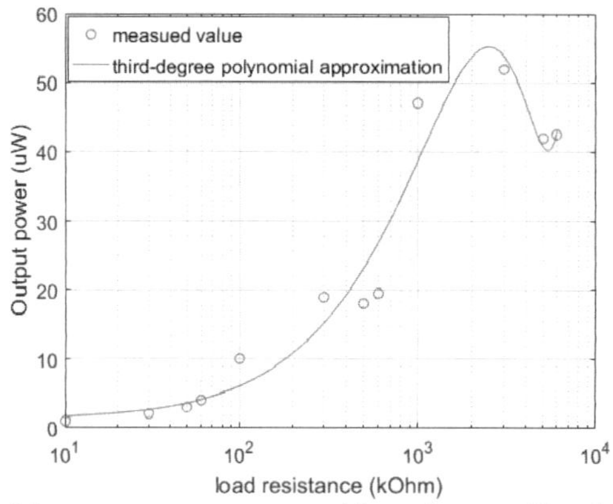

Figure 8. Measure output power with variance of load resistances.

4. Conclusion

In this paper, a flapping energy harvester capable of harvesting energy from water flow was studied. The structure was modelled in ANSYS and a prototype device was fabricated and tested to verify simulation results. It was found that the position and size of the bluff body and angle of attack have significant effect on the performance of such flapping energy harvesters. These key parameters were optimized in ANSYS to achieve the maximum lift force and oscillation displacement. By attaching a MFC patch to the cantilever beam, electrical energy can be generated while the beam is oscillating.

Although the study demonstrated the feasibility of using this structure for energy harvesting from water flow, the measured output power is relatively low. One of the reasons is that the resulting oscillation has a very low frequency (0.9 Hz) and the piezoelectric transducer is not particularly efficient in this frequency range. Further work will be done to implement methods to up-convert the operating frequency of this structure if piezoelectric transducer is still considered or to adopt electromagnetic transducers which works much better at lower frequencies.

5. References

[1] Wang D-A, Pham H-T, Chao C-W, Chen J M 2011 A Piezoelectric Energy Harvester Based on Pressure Fluctuations in Kármán Vortex Street World *Renewable Energy Congress 2011, 8-13 May 2011, linkoping, Sweden*

[2] Wang S H and Garcia R 2014 Development of a Digital and Battery-Free Smart Flowmeter *Energies 2014*, **7**, 3695-3709

[3] Boisseau S, Duret A-B, Perez M, Jallas E and Jallas E 2016 Water flow energy harvesters for autonomous flowmeters *Journal of Physics: Conference Series* **773** 012019

[4] D. Zhu, S. Beeby, M. J. Tudor, Mein M. White and N. R. Harris (2013) *Novel Miniature Airflow Energy Harvester for Wireless Sensing Applications in Building*. IEEE Sensors Journal, **13**, pp 691-700

[5] NASA (2017) Piezoelectric Fiber Composite Actuator Portfolio. https://technology.nasa.gov/patent/LAR-TOPS-17 (accessible on 27 September 2017)

Acknowledgments

Jorge Antonio Nieves Juárez would like to acknowledge the financial support from the Consejo Estatal de Ciencia y Tecnologia del Estado de Queretaro and the Technical University of Queretaro to complete his internship at the University of Exeter.

Metal Layer reinforced multilayer ferroelectret-based energy harvester

S.Yong[1], J.J.Shi[1], S. P. Beeby[1]

[1]Electronics and Computer Science, University of Southampton, Southampton, SO17 1BJ, UK

Email: sy1v16@soton.ac.uk

Abstract. This paper presents a flexible insole energy harvester made from multiple layers of ferroelectret polymer foams and nickel metal films. The ferroelectret stack has been fabricated by distributing a number of layers of nickel film through the ferroelectret. The addition of the thin metal films has improved the mechanical coupling to the ferroelectret foams within the stack and enhanced its voltage and energy output.

1. Introduction

A ferroelectret is a polymer foams with porous structure (void) filled with air and the elastic polymer-air composites has electric charge stored at its air-polymer interfaces around the void. When compressed, the ferroelectret polymer foam generates an electrical signal that can be used as a sensor output or energy harvester [1, 2]. When used as an energy harvester, the electrical output is typically rectified and stored in a capacitor. Polypropylene (PP) film is a flexible and porous polymer film with thickness between 70 to 80 μm, and young's modulus of about 1 MPa in the thickness direction, this soft and thin film is very suitable for an insole energy harvester providing a comfortable solution for users [3]. The energy harvesting ability of porous PP film has been investigated in our previous study, each single piece of PP ferroelectret can only generate 1.12 μJ under 800N of compressive force. When used as an energy harvester, the electrical output from a single piece of PP ferroelectret is not enough to power any commercial low-power chipsets at present. Therefore it is essential to stack up multiple number of parallel connected PP films.

Previous work showed that an 80-layer insole ferroelectret could switch on and maintain transmission a low-power ZigBee transmitter chipset [4]. However this work also showed that the energy output of the multilayer ferroelectret is not proportional to number of PP ferroelectret films parallel connected in the device. This is because when the stack is compressed, each individual pp layer experiences a different amount of stress. The stress experienced by the PP film located at the centre of the stack decreases as the number of layers increases.

This paper presents an investigation into the influence of additional thin metal films inside the multilayer energy harvester to improve the mechanical coupling to the ferroelectret foams within the stack. In this work all of the PP ferroelectret films were coated with silver layer on the top and bottom surfaces and connected in parallel. Three thin nickel films were placed in between the PP layers, distributed evenly through the stack and bonded together. The PP ferroelectret energy harvester stack was tested with quantified force and frequency to study their performance as energy harvester.

2. Concept

Figure 1 shows the schematic of the multilayer ferroelectret made with 10 PP films with three nickel films distribute through the stack symmetrically around the centreline. When force is applied to the stack, the voids deform and generate an electrical output due to the movement of individual electric dipoles. In comparison with the PP films located at the top and bottom of the ferroelectret stack, the PP films close to the centreline (PP layer 5 and 6) experience the least deformation and generate a lower output voltage. This lower voltage causes an uneven distribution of charge throughout the thickness of the stack reducing the net voltage and energy output.

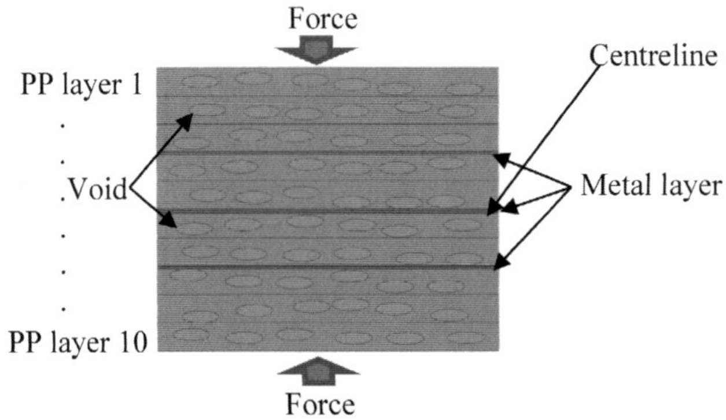

Figure 1 Schematic of the multilayer ferroelectret made with 10 PP films and three metal layers

In this design the Young's modulus of the metal layers are far greater than the Young's modulus of the PP films. When compressed, the metal films distribute the load more evenly across the stack resulting in a more consistent strain in each PP layer.

3. Experimental

The device presented here was fabricated using commercial available PP films (230 mm × 210 mm) purchased from Emfit Ltd. These films were cut into size of 45 mm × 70 mm and they were charged by the manufacturer. A silver paste (DuPont d5000) was printed onto each side of the PP film using a Dek 248 screen printer to form the top/bottom electrodes. Each silver electrode has an area size of 40 mm × 65 mm and the silver was cured in a box oven for 10 minutes at 50°C.

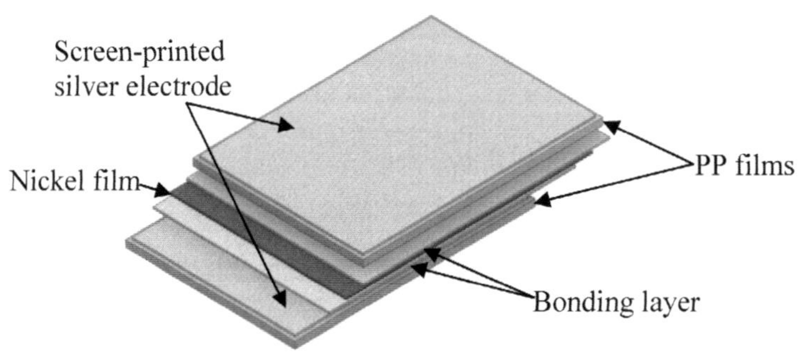

Figure 2. Device schematic showing a two PP layer and one nickel film stack with printed silver electrodes on each PP layer and with adhesive bonding layer.

The multilayer ferroelectret made by stacking 10 layers of silver coated PP films using adhesive films (Tesa® 64621). The adhesive film bonds layers together and provides insulation between layers that prevents self-discharge. All of the ferroelectret layers stacked in the device were connected in parallel. The three pieces of nickel films (40 μm thick each and Young's modulus of 1 GPa) were placed at the

3th, 6th and 8th bonding interface (figure 2). A cross section photograph of the multilayer ferroelectrets fabricated with and without nickel inserts are shown in figures 3(a) and 3(b). Both were fabricated simultaneously from the same PP sheets to enable comparison between them.

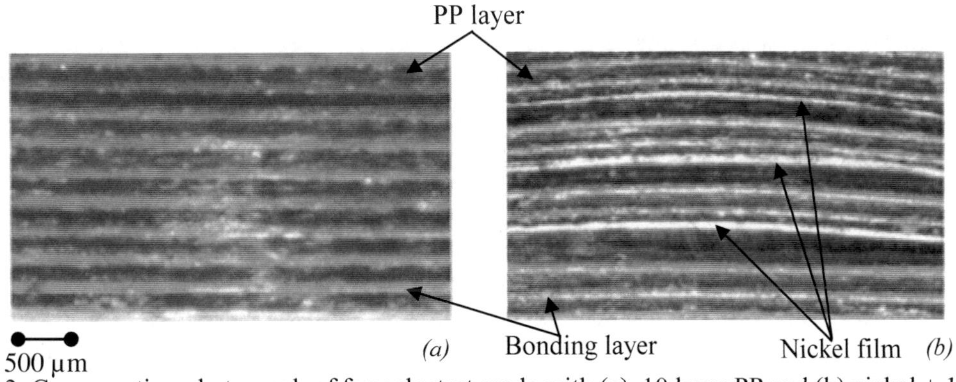

Figure 3. Cross-section photograph of ferroelectret made with (a), 10 layer PP and (b) nickel + 10 layer PP

The output voltage of the 10 layer PP and nickel + 10 layer PP ferroelectret were measured using an oscilloscope when the samples were compressed using an Instron electrodynamic instrument (ElectroPuls E1000, Instron Ltd). This applied a periodical square wave force of 800 N on the ferroelectret energy harvester samples at 1 Hz frequency. The electrical pulses generated from both ferroelectret stacks was rectified and stored on a 10 μF capacitor.

4. Result and discussion

The output voltage obtained from the ferroelectrets made with 10 layer PP and nickel + 10 layer PP are shown in figure 4(a). The maximum voltage output of the nickel + 10 layer PP ferroelectret (2.62 V) is 0.52 V higher than the 10 layer PP ferroelectret (2.1 V).

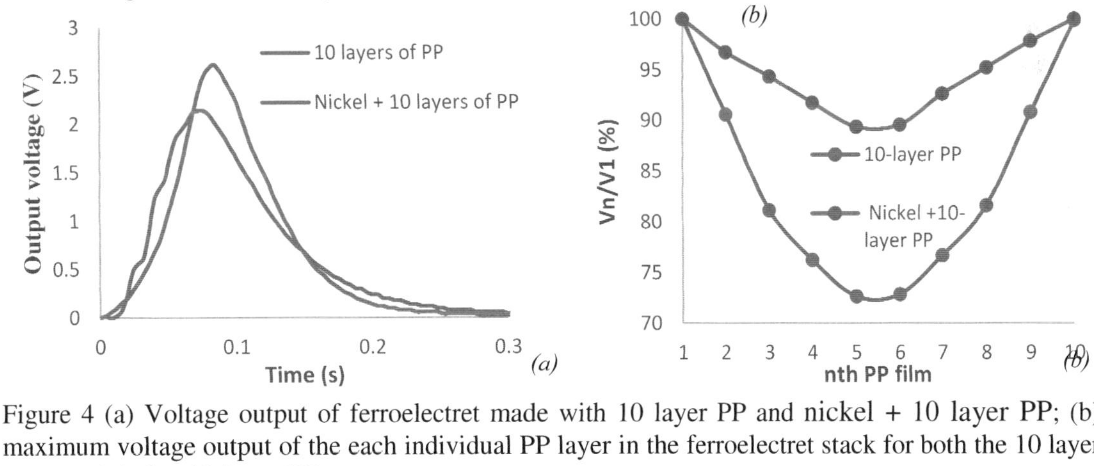

Figure 4 (a) Voltage output of ferroelectret made with 10 layer PP and nickel + 10 layer PP; (b) maximum voltage output of the each individual PP layer in the ferroelectret stack for both the 10 layer PP and nickel + 10 layer PP.

In figure 4(b) the maximum voltage output of the each polymer layer has been indexed to the voltage of the outer most layer (n = 1 and 10). The voltage output of layer n = 5 and 6 are only 72 % of V1. However, in nickel + 10 layer PP ferroelectret, the layers n = 5 and 6 achieve over 90 % of the maximum voltage output. These results indicate the additional nickel films improves the maximum voltage output deformation of the PP ferroelectret films close to the physical centre of the multilayer stack.

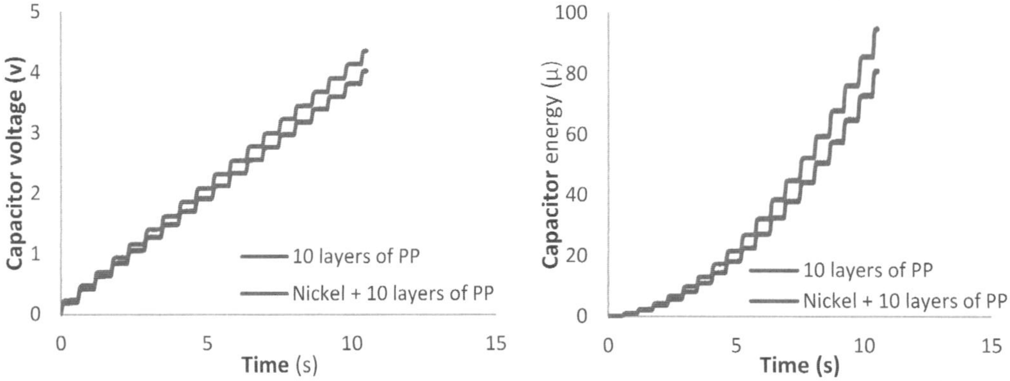

Figure 5 (a). Voltage charging curve of a 10 μF capacitor charged by ferroelectret made with 10 layer PP and 10 layer PP + Nickel, (b) Energy stored in a 10 μF capacitor charged by ferroelectret made with 10 layer PP and nickel + 10 layer PP

The charge output was stored on a capacitor to enable accurate calculation of the energy output. The capacitor voltage during the test was recorded and found to be 4.35 V for the nickel + 10 layer PP ferroelectret and 4.01V for the 10 layer PP ferroelectret after 10 mechanical cycles (figure 5(a)). Using the equation $E = 0.5CV2$, where C is the capacitance of the capacitor, V is the voltage of the capacitor, the energy E collected in the first 10 cycles by a nickel + 10 layer PP ferroelectret (94.6μJ) is 17.6% higher than a 10-layer PP ferroelectret (80.4 μJ) (figure 5(b)). It shows the additional nickel film increase not only the maximum voltage output of the multilayer PP ferroelectret but also improve its energy output.

5. Conclusion

This paper presents the design, assembly and test results of an energy harvester made of multilayer PP ferroelectrets and nickel metal films. Mechanical test indicates that the nickel layers couple the mechanical forces more effectively through the multilayer ferroelectret stack, in comparison with the multilayer PP ferroelectrets without the nickel metal films. The additional nickel layers increases the deformation to PP ferroelectrets at the centre of the energy harvester stack and this improves the voltage and energy output. It will enhance the performance of insole energy harvester, without affecting user comfort. Future work will focus on further improving the energy harvester voltage and energy output by using metal film with different thickness and Young's modulus.

Acknowledgements
This work was performed under the SPHERE IRC funded by the UK engineering and Physical Sciences Research Council (EPSRC), Grant EP/K031910/1.
DOI:10.5258/SOTON/D0267

References

[1] J Hillenbrand and G M Sessler, IEEE Trans. Dielectr. Electr. Insul 2000, 7, pp. 537-42
[2] G S Neugschwandtner, R Schwodiauer, S Bauer-Gogonea, S Bauer, M Paajanen, and J Lekkala, J. Appl. Phys 2001, 89 pp. 4503-11
[3] Z Luo, D Zhu, J Shi, S P Beeby, C Zhan, P Proynov and B Stark, IEEE Trans. Dielectr. Electr. Insul 2015. 22 pp. 1360-68
[4] Z Luo, D Zhu and S P Beeby, PowerMEMS 2015, pp. 660-65

OHA Ceramic Electret for Vibration Energy Harvesting

K Hakamata[1], T Miyoshi[2], C Itoga[1], Y Tanaka[1], and Y Suzuki[2]
[1] Department of Industrial Chemistry, Faculty of Engineering, Tokyo University of Science, 1-3 Kagurazaka, Shinjuku-ku, Tokyo 162-0826, Japan
[2] Department of Mechanical Engineering, The University of Tokyo, 7-3-1 Hongo, Bunkyo-ku, Tokyo 113-8656, Japan
E-mail: yutanaka@rs.tus.ac.jp

Abstract. With a motivation for developing oxy-hydroxyapatite (OHA) ceramic-electret applicable to vibration energy harvesting, the poling characteristics of OHA ceramics were investigated in terms of microstructure using thermally stimulated depolarization current (TSDC). OHA is a hydroxyapatite ($Ca_{10}(PO_4)_6(OH)_2$)-derivative material, in which the OH^- ions are partially substituted by O^{2-} ions and vacancies ($Ca_{10}(PO_4)_6(OH)_{2-x}O_{x/2}$). Patterned electret using 14-mm diameter OHA ceramic was prototyped with stripe-shaped electrodes, which serve as the electrostatic shield on the surfaces. The relationship between surface potential and the pattern width of the electrodes was experimentally examined, and the power generation characteristics were investigated.

1. Introduction

Energy harvesting is the technology to capture small amounts of ambient energy in order to realize maintenance-free standalone power source for low-power electronics. Its applications include automotive sensors such as tire pressure monitoring systems (TPMS), wireless sensors for structural health monitoring, and wearable devices [1]. At a low-frequency vibration range below 100 Hz, electret vibration energy harvester (EH) should have higher output power than electromagnetic or piezoelectric counterparts [2]. Recently, in addition to conventional electret materials such as polytetrafluoroethylene (PTFE) and SiO_2 [3], new polymer electret materials with higher surface charge density such as CYTOP-EGG [4, 5] and parylene AF4 [6] are proposed with a view of energy harvesting applications. On the other hand, inorganic electret has advantage for high thermal stability of charges. Tanaka et al. [7] proposed oxy-hydroxyapatite (OHA) ceramics as a promising electret material with extremely-high stored charges and high surface potential.

In the present study, we investigate the poling characteristics of OHA ceramics, and examine a patterning method of the surface potential for power generation of vibration EH prototype using OHA ceramic electret.

2. Poling characteristics of OHA

OHA is a hydroxyapatite ($Ca_{10}(PO_4)_6(OH)_2$)-derivative material, in which the OH^- ions are partially substituted by O^{2-} ions and vacancies ($Ca_{10}(PO_4)_6(OH)_2O_{x/2}$) [1]. Although OHA shows no piezoelectric property, our first-principle MD simulations demonstrated that the local flip of H^+ and the exchange of OH^-/O^{2-}, OH^-/vacancy and O^{2-}/vacancy occur in the OH^- sites linearly arranged along the c-axis in the apatite structure (Fig. 1), showing the possibility for inducing poling within the OHA ceramics.

OHA ceramic disks (ϕ 10 mm, t=1 mm) were prepared by sintering hydroxyapatite powder compacts at 1250-1400°C for 2 h under air. Sintering temperature affects the degree of grain growth; the grain size of OHA ceramics is monotonously increased from ca. 2 to 12 μm. As shown in Fig. 2a, the OHA ceramic "electrets" was prepared at 200°C for 1 h under the electrical field of 8 kV/cm. The field was kept during the cooling process to avoid the dipole relaxation [7]. The poling condition of the obtained OHA ceramics was investigated with a surface potential measurement and a thermally stimulated depolarization current (TSDC) technique (Fig. 2b).

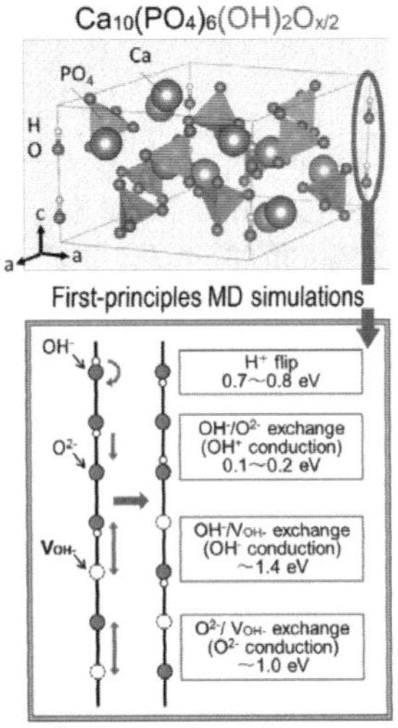

Figure 1. Crystal structure and carrier dynamics for OHA estimated by first-principle MD simulation.

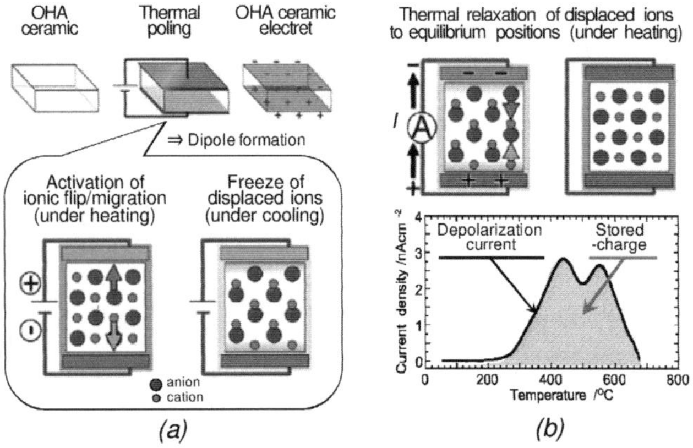

Figure 2. Schematic of (a) approach for preparing OHA ceramic electret and (b) TSDC technique.

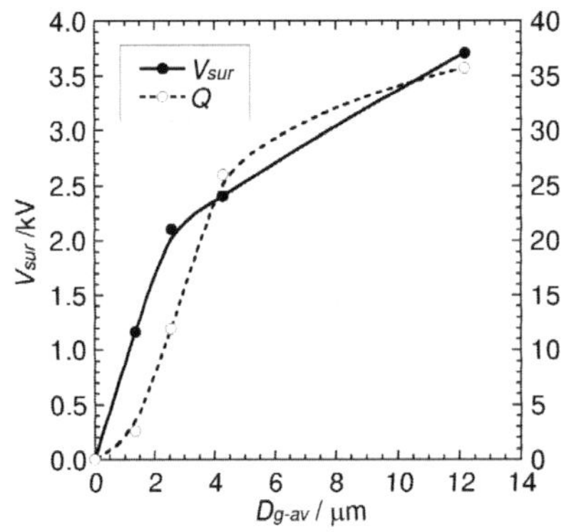

Figure 3. Effect of the grain size on surface potential (V_{sur}) and stored-charge (Q) of OHA electrets.

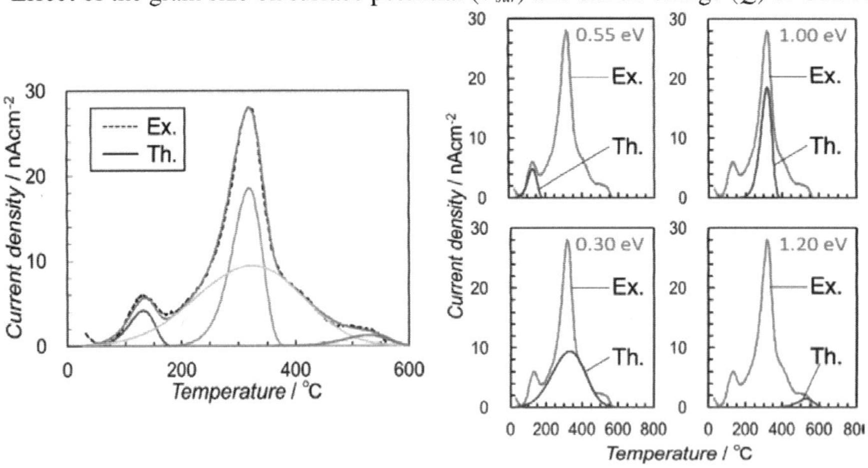

Figure 4. Experimentally obtained TSDC curve (Ex.) of OHA electrets with the theoretical curves (Th.) calculated using E_{dr} values of 0.3-1.2 eV.

3. TSDC Experiments

Figure 3 shows the surface potential V_{sur} and the stored charges Q versus the grain size. It can be seen that there is a positive correlation between V_{sur}, Q and the grain size, supporting the effect of the grain boundaries on the dipole formation in the OHA ceramics. For the detailed understanding of dipole formation/relaxation mechanism of the OHA ceramics, TSDC measurements were made and compared with the activation energy obtained in the MD simulation as shown in Fig. 4. The results show that four different kinds of poling modes with different activation energies E_{dr} exist in the OHA electret. The values of E_{dr} are in good agreement with the activation energy for flip/migration of ionic carries determined by the MD simulation.

4. Patterning OHA Electret

Figure 5 shows a schematic of an in-plane vibration energy harvester using the OHA electret. In order to increase the changing rate of overlapping area and thus the output power, strip patterns are usually employed for in-plane vibration EH. Because it is difficult to physically etch OHA, conductive strip patterns, which serve as the electrostatic shield on the surface, are formed on the surface of OHA. In the present study, poled OHA ceramic disks (ϕ 14 mm, 1 mm thickness) with the surface potential of about -2400 V are used, and 1-2 mm wide Au patterns were formed by sputtering through a hard mask. The surface electrodes are grounded to ensure zero potential on the electrodes. We employ an Electrostatic Force Microscopy (EFM) [8] to measure the surface potential profile in between the electrodes.

Figure 6 shows the surface potential distribution across the strip patterns on the OHA electrets. It is found that the surface potentials are much lower than that of the plain OHA ceramic disk (~-2400 V), and decreased with decreasing the width of the strip patterns. With the 2 mm-wide electrodes, the peak value of the surface potential near the center is -1400 V, but the potential is decreased toward the edge of the electrodes. Because thinner pattern is advantageous for vibration power generation, more detailed investigation is necessary to clarify this issue.

5. Power Generation Experiment

The OHA ceramic electrets with Au strip were bonded on a glass plate as shown in Fig. 7a. For power generation experiments, the OHA electrets is fixed onto a stage connected to an electromagnetic shaker, and in-plane sinusoidal motion is applied. A glass plate with current-collecting counter electrodes with the same electrode width with the electret is prepared. It is fixed onto a 5-axis stage to precisely adjust the alignment and the air gap between the surface of the OHA electret and the counter electrodes. The vibration frequency is 10 Hz. The peak-to-peak vibration amplitudes are set to 1 mm for the 1 mm-wide electret/electrode, and 2 mm for the 2 mm-wide electret/electrode. Figure 8 shows the output voltage waveforms across a 400 MΩ load, and the output power versus the load resistance with different air gap. The peak-to-peak voltage amplitude up to 50 V and the output power of 0.9 μW with the 200 MΩ load

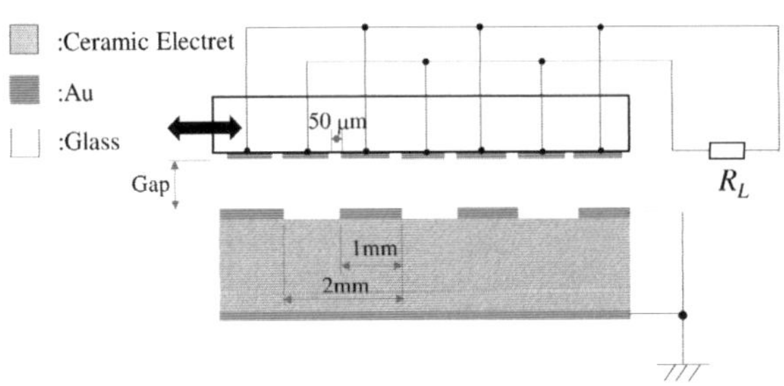

Figure 5. Schematic of the vibration electret EH using stripe-shaped Au electrodes as the electrostatic shield.

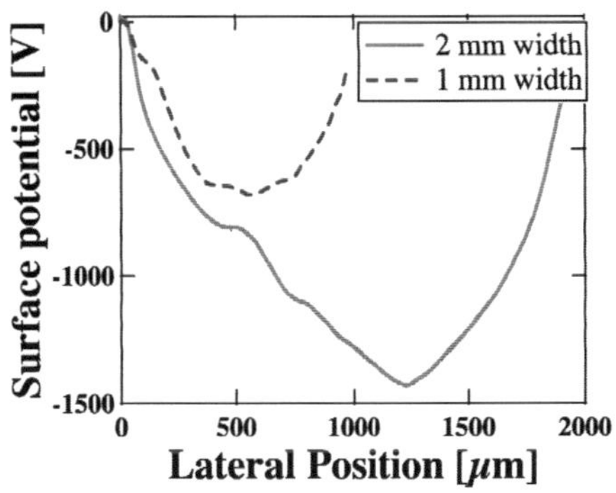

Figure 6. Surface potentials of OHA ceramic electrets with patterned Au electrodes on the surface.

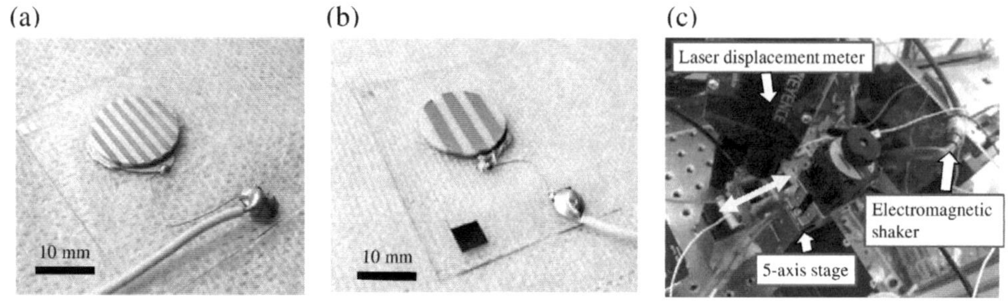

Figure 7. (a) OHA ceramic electret with 1 mm wide Au strip, (b) with 2 mm wide Au strip, and (c) Power generation experimental setup using an electromagnetic shaker.

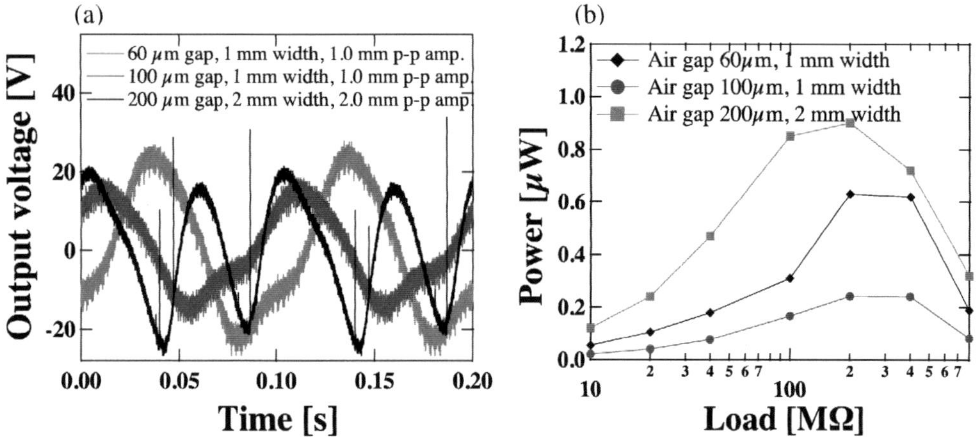

Figure 8. (a) Waveforms of output voltage, and (b) Output power versus external load.

is obtained with the 2-mm wide electret. Higher output power can be achieved by optimization of the strip width for electrostatic shielding and by using thinner air gap.

6. Conclusion

The poling characteristics of OHA ceramics were investigated in terms of microstructure using TSDC techniques. A patterning method of the surface potential of OHA electret is examined by using stripe-shaped electrodes on the surface. It is found that the surface potential becomes much lower than the plain electrode, and is decreased with the pattern width. Preliminary power generation experiment using in-plane vibration is made, and up to 0.9 μW output power is obtained with 10 Hz and 2 mm oscillation.

This work was supported by JST CREST Grant Number JPMJCR15Q3, Japan.

References

[1] S. Roundy et al., *Comp. Commu.*, **26**, pp. 1131-1144, 2003
[2] Y. Suzuki, *IEEJ Trans. Electr. Electr. Eng.*, **6**, pp. 101-111, 2011
[3] G.M. Sessler, *Electrets, 3rd ed.*, Laplacian Press, 1998
[4] Y. Sakane et al., *J. Micromech. Microeng.*, **21** 104011, 2008
[5] K. Kashiwagi et al., *J. Micromech. Microeng.*, **21** 125016, 2011
[6] H.-W. Lo, and Y.-C. Tai, *J. Micromech. Microeng.*,**18** 125016, 2008
[7] Y. Tanaka, et al., *J. Appl. Phys.*, **107** 014107, 2010
[8] D. Sato et al., *Imaging Conf. Jpn. 2014*, pp. 255-258, 2014

Electrode optimization of an electret-based vibration generator in slot-effect configuration

Cuong Phu Le and Einar Halvorsen

Department of Microsystems, University College of Southeast Norway, Campus Vestfold, Raveien 215, 3184 Borre, Norway

E-mail: `Einar.Halvorsen@usn.no`

Abstract. This paper offers insights into optimization of electret-based energy harvesters made of two patterned electrodes facing movable electret strips parallelized with a pitch p. The effective electrode dimension of such harvester structures is optimized to maximize power under given displacement amplitude and acceleration amplitude, while the length and the width of the design area are kept fixed. The geometrical optimization includes fringing field effects that may have significant impacts on the output power of the small-scale devices. The optimization problem is carried out through an established accurate lump-model represented by an equivalent circuit. The optimal electrode width that gives maximum power is proportional to the proof-mass displacement amplitude by an approximate formula $a_{\mathrm{opt}} \approx 1.45X_0$. Optimizing the pitch can further increase the maximum power as it simultaneously regulates both optimal ratios to the electrode width and the gap size.

1. Introduction

Small-scale vibration energy generators have promoted operation of wireless sensors nodes without the need of battery by converting kinetic energy from surrounding into electrical energy. Electret-based energy harvesting is the most favourable technique with respect to the feasibility of standard microfabrication process. The use of internal bias by an electret makes the harvesters more easily self-sufficient and integrated. Recently, there are a number of harvesters structured by an electret pattern on a movable substrate and a set of metal-strip electrodes on a stationary substrate. The two substrates are separated by a micro-gap. Such harvesters have experimentally demonstrated potential applications of monitoring car-tire and windmill [1, 2].

Because of significant fringing fields in the microscale devices, an accurate harvester-model is still a challenge. There are several works in modeling and design optimization to maximize harvester performance. Their models are based on either using an ideal field assumption or time-consuming FEM computation [3, 4]. So far, there may have only an effort that investigates electrode optimization for slot-effect configured generators [5]. However, the conformal mapping technique used for the model is limited by assumption on material and geometry.

This work presents an effort to overcome the aforementioned limitations. The transducer electrodes are dimensionally optimized for power enhancement. This is accomplished by a built analytic tool that includes all possible effects of geometry, materials and fringing fields. A standard equivalent-circuit is established to capture all electromechanical dynamics as well as to numerically solve the geometrical optimization problem, based on the time-domain circuit simulator SPICE.

2. Analysis

2.1. MEMS electret-based transducer

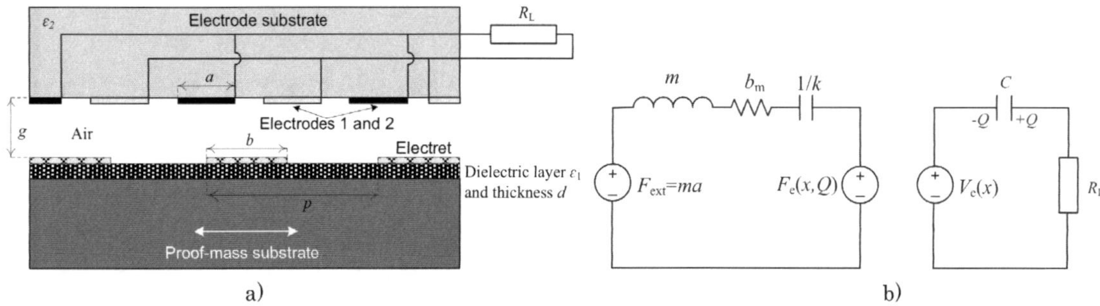

Figure 1. a) Simplified cross-section structure and b) equivalent-circuit model of an electret-based energy harvester.

A simplified cross-section of the electret-based energy transducer is shown in figure 1a. The top layer of the conductive proof-mass substrate consists of M parallel electret-strips with a width b deposited on a dielectric layer with thickness d and permittivity ε_1. The electret strips distributed to each other by a pitch p symmetrically faces a set of fixed metal-electrodes 1 and 2 with width a on the top substrate with permittivity ε_2 by an air gap g with permittivity ε_0. The proof mass m suspended by mechanical springs (not shown) with stiffness k has a relative displacement x to the two fixed-electrodes in the lateral direction. Thickness of each electrode/electret pattern is assumed negligible. The end-effects can be neglected as the period number M is considered to be sufficiently large, so that the charge distribution on each electret strip is assumed periodic and uniform. Under the proof-mass motion driven by a time-varying acceleration $a(t)$, the total fixed-charge Q_e over the entire electret induces charge variation between two fixed-electrodes, generating an electrical current through a resistive load R_L.

2.2. Equivalent-circuit model

The method used here to build an accurate model is established from our own previous work [6] that provides a fast and full exploration of design parameters. The calculation approach is essential to include all possible effects of fringing fields that may affect the output power. To this end, singular divergences of the surface charge density are taken into account at the edges of the metal strips in form of reciprocal square root. The method to calculate the all induced-charge coefficients γ_ij in the electrostatic system is considered as an adaption of previous boundary treatments applied for surface acoustic wave microstrip-devices [7, 8]. The total electrostatic energy stored in the system is

$$W_\mathrm{e} = \frac{1}{2\pi\varepsilon_0 ML}\left(\gamma_{12}Q_1Q_2 + \gamma_{13}Q_1Q_\mathrm{e} + \gamma_{23}Q_2Q_\mathrm{e}\right) + \frac{1}{4\pi\varepsilon_0 ML}\left(\gamma_{11}Q_1^2 + \gamma_{22}Q_2^2 + \gamma_{33}Q_\mathrm{e}^2\right) \quad (1)$$

where Q_1 and Q_2 are the total charges on the electrodes 1 and 2, and L is the electrode length. The induced-charge coefficients depend on the transducer geometry, the material properties and the relative displacement x. Further details of the coefficient calculation γ_ij can be found in [6].

The electromechanical transducer can be now represented by an equivalent-circuit based on the $e \rightarrow V$ convention [9, 10] in figure 1b, where b_m is the mechanical damping coefficient and $F_\mathrm{ext} = ma$ is the external force. In the circuit, three key elements $F_\mathrm{e}(x,Q)$, $V_\mathrm{e}(x)$ and C are

$$F_\mathrm{e}(x,Q) = \frac{\partial W_\mathrm{e}}{\partial x} = \frac{\gamma'_{33}(x) - \frac{4\pi d}{\bar{\varepsilon}_1 p(\gamma_{11}-\gamma_{12})}\left[\gamma'_{13}(x) + \gamma'_{23}(x)\right]}{4\pi\varepsilon_0 ML}Q_\mathrm{e}^2 + \frac{\gamma'_{13}(x) - \gamma'_{23}(x)}{2\pi\varepsilon_0 ML}Q_\mathrm{e}Q \quad (2)$$

and

$$V_e(x) = \frac{\gamma_{13}(x) - \gamma_{23}(x)}{2\pi\varepsilon_0 ML} Q_e, \quad C = \frac{\pi\varepsilon_0 ML}{\gamma_{11} - \gamma_{12}}. \tag{3}$$

where $Q = \frac{Q_1 - Q_2}{2}$. The force $F_e(x, Q)$ represents the electromechanical coupling between mechanical and electrical domains of the transducer, the displacement-dependent built-in source $V_e(x)$ is the open-circuit voltage when $Q = 0$ and C is the effective capacitance between the two electrodes 1 and 2.

3. Electrode optimization

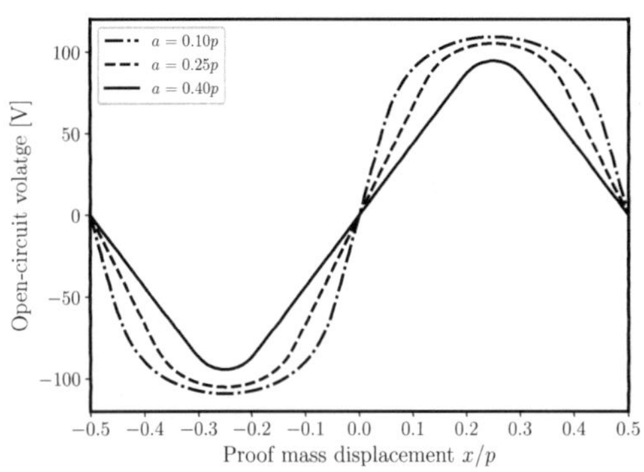

Figure 2. Open-circuit voltage vs. proof mass displacement for different metallization ratios a/p.

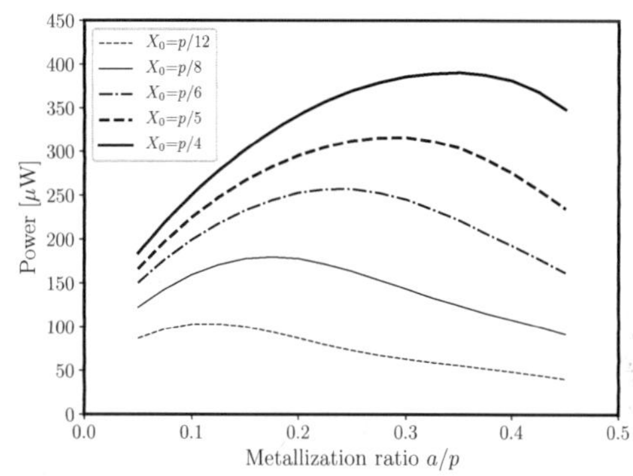

Figure 3. Output power as a function of metallization ratio for different amplitudes of proof mass displacement.

In order to optimize the electrode dimension of the transducer, the active design area is kept fixed during the optimization which seeks an effective electrode area that gives maximum power for given displacement amplitude X_0 or acceleration amplitude A. All design parameters are taken from the device example in [3], except a and p that are the dimension parameters to be optimized. The device had a set of Si_3N_4 electret strips on the SiO_2 dielectric layer deposited on the silicon proof mass substrate. The substrate for the metal-electrodes makes use of glass material. The design area is 10×10 mm^2.

Figure 2 shows open-circuit voltages for different metallization ratios $0 < a/p < 0.5$. The maximum voltage-amplitude at $|x/p| = 0.25$ is higher for the lower ratio a/p, but the linear range is extended for larger electrode. With short electrode, the voltage rate decreases for the displacement $|x/p| \to 0.25$. It can be explained that the induced charges on the two electrodes have less variation when one of the electrodes completely overlaps the electret pattern. The voltage amplitude increases for decreasing a/p, but the effective capacitance C is correspondingly reduced for example $C = 11.3$ pF for $a/p = 0.1$ and $C = 31.5$ pF for $a/p = 0.4$. The higher ratio a/p is required to maximize power for the larger amplitude X_0 under load optimization, see Figure 3. For $X_0 = p/12$, a maximum power is obtained at an optimal electrode $a_{\mathrm{opt}} = 0.12p$ while this value is of $a_{\mathrm{opt}} = 0.35p$ at $X_0 = p/4$. The relation of a_{opt} to X_0 can be approximated by a linear fit in figure 4. The optimal electrode width is approximately $a_{\mathrm{opt}} \approx 1.45X_0$, so that the optimization is achieved when the displacement reaches two extremes of the linear range. The device parameters used for the lumped-model here have $d/p = 2 \times 10^{-3}$ and $g/p = 2 \times 10^{-2}$, but this optimal condition can be generalized to various geometries with $d \ll g < p$ for the similar MEMS harvester structures.

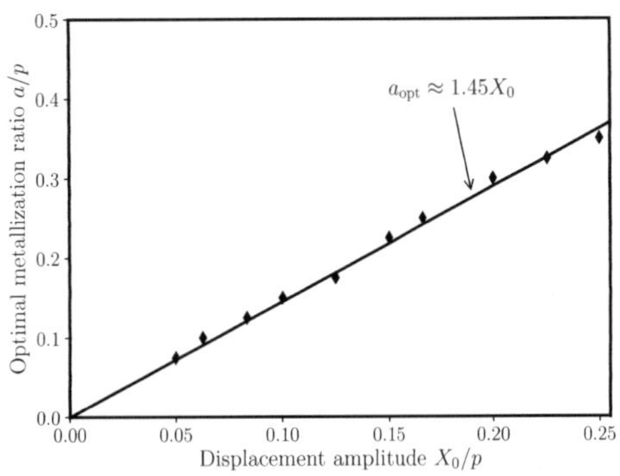

Figure 4. Relationship between optimal electrode width a_{opt} and displacement amplitude X_0.

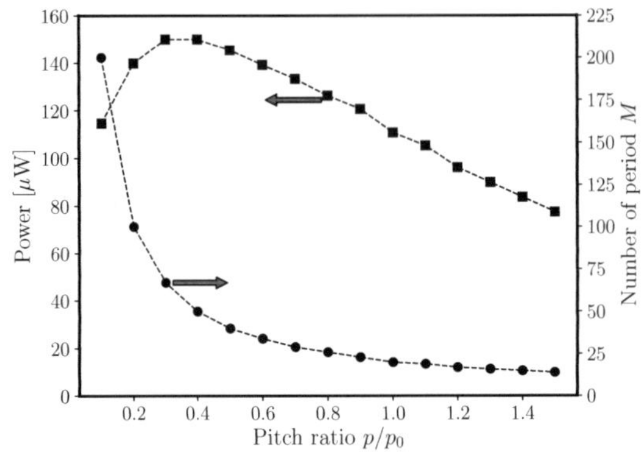

Figure 5. Output power and number of period as a function of pitch p for $p_0 = 500$ μm, $A = 2.0$ g and $a/p = 0.4$.

Figure 5 shows an alternative to optimize electrode area by varying the pitch p while keeping $a/p = 0.4$ fixed. Increase of M when scaling p down causes stronger electrostatic force, leading to smaller X_0 at the same acceleration amplitude. The maximum power achieved at $p/p_0 = 0.3$ is about 1.08 times higher than that obtained by only optimizing a/p ($a_{opt} = 0.15p$) at $p/p_0 = 1.0$. It is because the change of the pitch makes the ratio g/p further optimized while the condition $a = a_{opt}$ is still ensured by impedance matching which affects X_0. The same maximum power is obtained at optimal load when varying both a/p and p/p_0.

4. Conclusion

We investigated geometrical optimization for the electret-based vibration energy harvester configured in the slot-effect scheme. The optimization problem is to explore an effective electrode area for maximizing the harvested power with respect to the constraint of the total design area. To build a simulation tool, an accurate lumped-model that governs both the electromechanical dynamics and the fringing field effects was established in form of the standard equivalent circuit. The optimal electrode is found to be proportional to the proof-mass displacement amplitude by a linear approximation $a_{opt} \approx 1.45X_0$, while the pitch optimization is possible to enhance a higher maximum power.

Acknowledgments

This work was supported by the Research Council of Norway through the Grant no. 229716/E20.

References

[1] Renaud M, Altena G et al. 2013 *Proc. Transducer2013, (Barcelona, June)*, 693–696
[2] Perez M, Boisseau S et al. 2016 *Smart Mater. Struct.* **25** 045015
[3] Renaud M, Altena G et al. 2015 *Smart Mater. Struct.* **24** 085023
[4] Crovetto A, Wang F and Hansen O 2014 *J. Microelectromechanical Systems* **23** 1141-1155
[5] Okamoto H, Hamate Y and Kuwano H 2012 *Smart Mater. Struct.* **21** 065001
[6] Le C P and Halvorsen E 2017 *Smart Mater. Struct.* **26** 085042
[7] Gladwell G M L and Coen S 1975 *IEEE Trans. Microwave Theory Tech.* **23** 865–870
[8] Medina F and Horno M 1992 *IEEE Trans. Microwave Theory Tech.* **40** 1748–1756
[9] Tilmans H A C 1996 *J. Miromech. Miroeng.* **6** 157-176
[10] Jezierski E 2001 *Proc. Workshop Robot Motion and Control (Bukowy Dworek, Poland, October)* 181–188.

IOP Conf. Series: Journal of Physics: Conf. Series **1052** (2018) 012118 doi:10.1088/1742-6596/1052/1/012118

Comparative performance of voltage multipliers for MEMS vibration-based energy harvesters

Binh Duc Truong, Cuong Phu Le and Einar Halvorsen

Department of Microsystems, University College of Southeast Norway, Campus Vestfold, Raveien 215, 3184 Borre, Norway.

E-mail: `Einar.Halvorsen@usn.no`

Abstract. This paper investigates by numerical simulation the performance of an electrostatic vibration energy harvester when it is electrically configured in two different diode-capacitor multiplier topologies. The complete lumped-model of an overlap-varying generator along with power electronic interface circuit is constructed for analysis using a crcuit simulator. Parasitic capacitance of the transducers and nonideal diode traits such as leakage current and junction capacitance are incorporated. We find that both configurations are able to efficiently operate with a ratio of capacitance variation much lower than 2, which overcomes a challenging obstacle of MEMS-based devices. Other advantages and disadvantages of the two topologies are compared and discussed.

1. Introduction

There are circumstances in which sensing devices are positioned in inaccessible places such as the moving parts of machinery and vehicles, or where cabling is expensive or even impossible [1, 2]. In this scenario, energy harvesting can enable or lower the cost of an application by providing a means to implementt battery-less systems [3]. Many environmental energy sources can be exploited these purposes, for instance temperature gradient or mechanical vibrations [4]. This paper focuses on vibrational energy harvesters with electrostatic transduction.

The first report of a MEMS-based electrostatic generator that includes integrated electronics was presented in [5]. Since then, several studies on power electronic interface circuit for capacitive transducers in which the harvester is commonly coupled to AC-DC and/or DC-DC converters, have been published [6–8]. Complexity of control circuitry and power consumption are the main concerns of those architectures. Furthermore, in the majority of cases, those converters cannot start up below a certain input voltage [9] which is can be higher than a typical output voltage generated by the microharvesters [10]. Therefore, a processing circuit that scales up the voltage making it suitable for the following converters is sometimes required. Based on the ancient "doubler of electricity" invented by Bennet in the lare 18th century, de Queiroz *et al.* proposed a promising electronic version adapted to MEMS devices [11]. The circuit consists of two variable capacitors, three diodes/switches and one fixed capacitor where the harvested energy is stored. When the voltage across the storage capacitor reaches a certain maximum value, the accumulated charge is then dumped into a battery by using a simple buck converter [12]. An attempt to increase charging current for the doubler circuit was reported in [13]. The main disadvantage of such a configuration is its inability to operate efficiently with

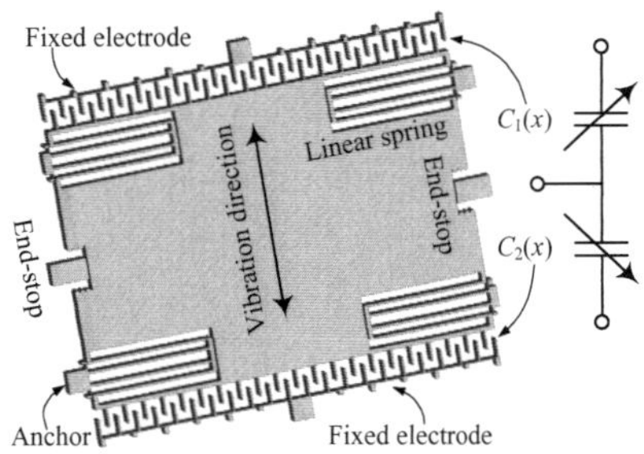

Figure 1: Overlap-varying electrostatic energy harvester (EH).

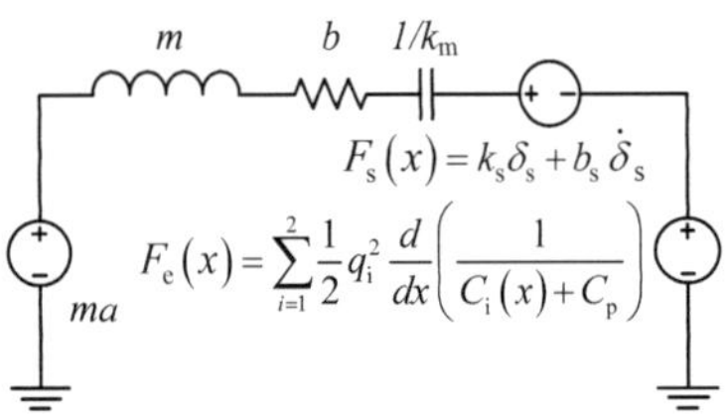

Figure 2: Equivalent circuit for mechanical domain.

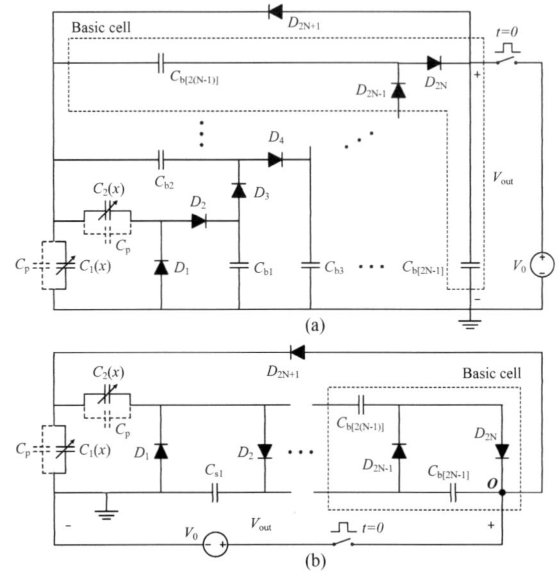

Figure 3: EH electrically configured as (a) Parallel-Series (PS) and (b) Cockcroft-Walton (CW) voltage multipliers.

transducers that have ratio of capacitance variation η lower than 2. This challenge is difficult to overcome in practice since MEMS harvesters tend to suffer from parasitic capacitances and the variable capacitance is small due to limitation of microfabrication processes and device size.

Recent work by Lefeuvre *et al.* [14] introduced a self-biased topology for electrostatic vibration energy harvesters. Based on theoretical analysis in the electrical domain using a rectangular Q-V cycle, the authors suggested that this circuit configuration can operate with $\eta < 2$. With an N-stage voltage multiplier, the ideal minimum required ratio of capacitance variation is $\eta_{\rm min}^{\rm ideal} = \frac{N+1}{N}$ [15]. In the same manner, this paper presents and compares two alternative electrical configurations: (i) A parallel-series multiplier adapted from the Greinacher voltage doubler [16] and (ii) a variation of Cockcroft-Walton multiplier [17]. Dynamic performances of both topologies are analyzed using circuit simulations.

2. Energy harvester model and voltage multiplier configurations

A sketch of the in-plane overlap-varying energy harvester is shown in Figure 1. The proof mass is suspended by linear, folded springs with total stiffness $k_{\rm m}$. The transducer capacitances as functions of the proof mass displacement x are $C_{1/2}(x) = C_0(1 \pm x/x_0)$, where C_0 and x_0 are the nominal capacitance and the nominal overlap respectively. The maximum displacement $X_{\rm max}$ is defined by rigid end-stops. The movable mass is subject to the fictitious force ma, the electrostatic force $F_{\rm e}$ and the impact force $F_{\rm s}$ from the end-stops. The equivalent circuit for the mechanical domain is presented in Figure 2, where b is the damping coefficient, $q_{1/2}$ are the charges on the two transducers, $C_{\rm p}$ is the parasitic capacitance and $\delta_{\rm s}$ is the relative displacement of the mass and the end-stops during contact at high input acceleration.

Two ordinary anti-phase transducers are electrically configured as N-stage voltage multiplier as shown in Figure 3. The Parallel-Series (PS) configuration was first proposed and thoroughly analyzed in [18], showing its superior performance over the circuit presented in [14]. The Cockroft-Walton (CW) topology is generalized from 2-stage CW generator introduced in [19]. The voltage source V_0 in series with a switch is used to pre-charge the capacitors to initiate

Table 1: Model parameters

Parameters	Value
Proof mass, m	0.79 mg
Spring stiffness, k_{m}	3.60 N/m
Thin-film air damping, b	3.48e-5 Ns/m
Nominal overlap, x_0	80 μm
Nominal capacitance, C_0	15 pF
Parasitic capacitance, C_{p}	7.5 pF
Contact stiffness, k_{s}	3.361 MN/m
Impact damping, b_{s}	0.435 Ns/m
Maximum displacement, X_{max}	≈ 31 μm
Relative contact displacement, δ_{s}	$\delta_{\mathrm{s}} = \lvert x\rvert - X_{\mathrm{max}}$
Multiplying capacitor, $C_{\mathrm{b[j]}}$	$C_{\mathrm{b}} = 1$ nF

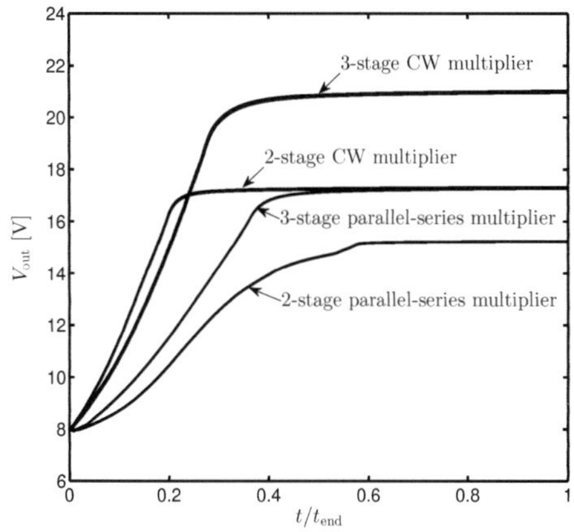

Figure 4: Time evolution of output voltage.

operation of the multiplier circuits in the simulations. How to secure such a pre-charge in a complete system is not adressed her. Output of the PS multiplier is the voltage across $C_{\mathrm{b[2N-1]}}$ while that of the CW circuit is the potential between the node O and ground. The dynamic model of the Schottky diode BAS716 is used for SPICE simulations incorporating junction capacitance and losses due to leakage current. All fixed capacitances $C_{\mathrm{b[j]}}$ are chosen equal for simplicity.

In order to investigate the potential merits of these two configurations, the transducers are designed so that the capacitance variation ratio is

$$\eta = \frac{C_{\mathrm{max}} + C_{\mathrm{p}}}{C_{\mathrm{min}} + C_{\mathrm{p}}} = 1.7 \tag{1}$$

where the minimum and maximum capacitances are $C_{\mathrm{min}} = 9.19$ pF and $C_{\mathrm{max}} = 20.81$ pF. The arameters are summarized in Table 1.

3. Simulation results

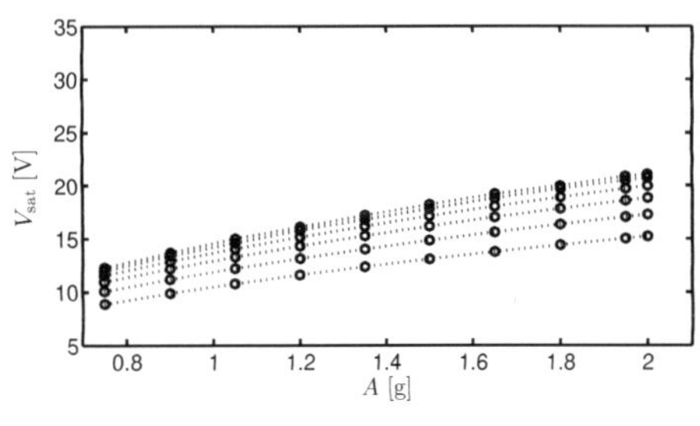

(a) Parallel-series multiplier.

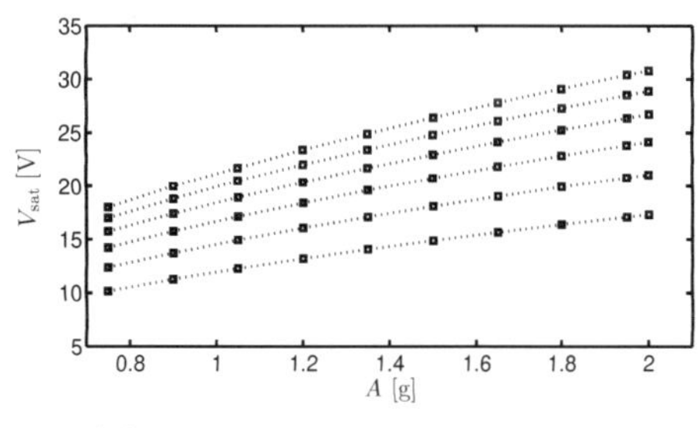

(b) Cockcroft-Walton multiplier.

Figure 5: Saturation voltage of the two configurations versus input acceleration amplitudes.

Figure 4 shows time evolution of the output voltage V_{out} for both PS and CW topologies with number of stages $N = 2$ and 3, initial bias $V_0 = 8$ V, input acceleration amplitude $A = 2$ g and drive frequency $f = f_0 = 1/2\pi\sqrt{k_{\mathrm{m}}/m}$. As long as V_0 is sufficient for operation of the circuits, V_{out} initially increases. After several transient cycles, a saturated output voltage is observed for all cases. It can be explained by an increase of electromechanical coupling.

Figure 5a and 5b present the saturation voltage V_{sat} of the two multiplier configurations over a range of acceleration amplitude $A \in [0.75\,\text{g}, 2\,\text{g}]$ for several numbers of stages $N \in \{2, \ldots, 7\}$. The simulation results show that both voltage multipliers give higher V_{sat} with increase of N and A. However, the CW multiplier performs better than the PS multiplier. In addition, it is worth mentioning that V_{sat} is independent on V_0. Therefore, the lowest possible value of pre-charging voltage is desirable. In particular, with $N = 2$, the minimum required initial bias for PS and CW multipliers are $(V_0)^{\text{PS}}_{\text{min}} = 5.5$ V and $(V_0)^{\text{CW}}_{\text{min}} = 4$ V respectively making the CW multiplier somewhat more attractive.

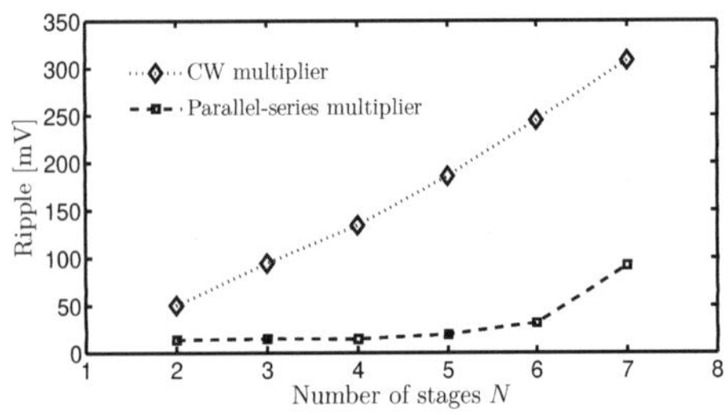

Figure 6: Comparison of ripple voltage between the two topologies.

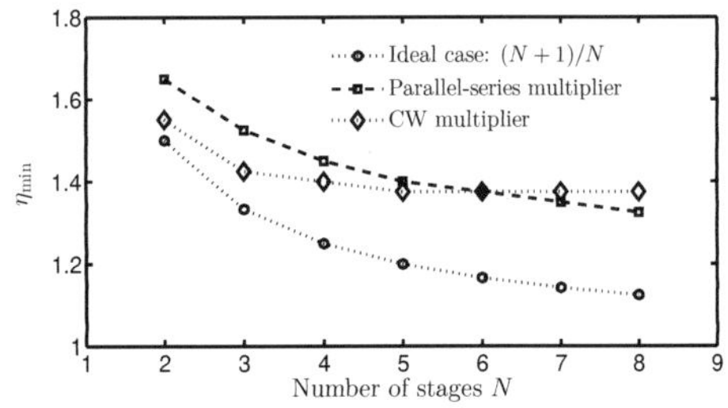

Figure 7: Minimum required capacitance variation ratio versus number of stages.

Due to the charge and discharge of capacitors, the output is distorted and shows some ripple superimposed on the DC saturation voltage. Figure 6 depicts peak-to-peak ripple of the two multipliers versus the number of stages N for $A = 2$ g and $f = f_0$. The comparison shows a potential advantage of the PS multiplier over the CW configuration if ripple is a concern.

To explore the benefits of N-stage multipliers, we choose to adjust X_{max} as a way to modify η while the other parameters in Table 1 are kept unchanged. Figure 7 reveals variations of the minimum required ratio of capacitance variation η_{min} when the number of stages N increases. When electrical losses are taken into account, η_{min} is a little higher than the threshold corresponding to the ideal case. A general trend for the PS topology is the continued decrease of η_{min} when adding more stages. However, for the CW configuration, this statement is only the case with $N \leq 5$. Higher N results in almost unchanged η_{min}. As a consequence, the η_{min}-curve of PS multiplier crosses that of CW circuit at $N = 6$. Hence, the PS multiplier can be advantageous when $N \geq 7$.

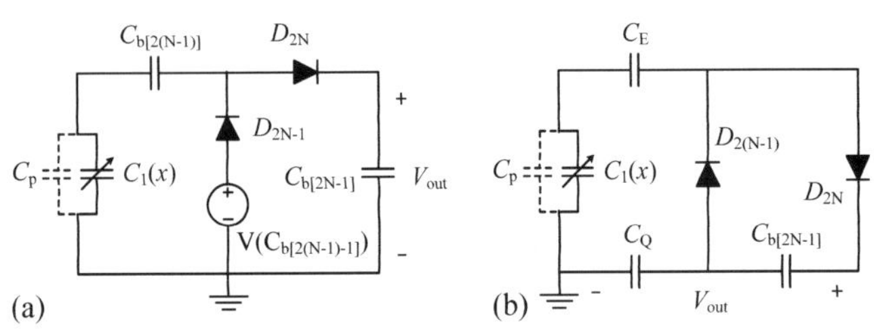

Figure 8: Equivalents for the final stages of (a) PS and (b) CW multipliers.

To explain the behavior for increasing N, we consider simplified operation of the two topologies at the final stage as shown in Figure 8, which directly affect the output performance of the harvesting system. For the PS multiplier, the voltage across capacitor $C_{\text{b[2(N-1)-1]}}$ can effectively be represented by a DC voltage source $V(C_{\text{b[2(N-1)-1]}})$. Meanwhile, two series connected capacitors of CW multiplier $\left(C_2(x), C_{\text{b2}}, C_{\text{b4}}, \ldots, C_{\text{b[2(N-1)]}}\right)$ and

$(C_{b1}, C_{b3}, C_{b5}, \ldots, C_{b[2N-3]})$ are presented by two equivalent capacitors C_E and C_Q respectively, where

$$C_E = \frac{C_b\big(C_2(x) + C_p\big)}{C_b + (N-1)\big(C_2(x) + C_p\big)}, \; C_Q = \frac{N-1}{C_b}. \tag{2}$$

The harvested energy of the PS topology is always stored in a capacitor whose capacitance is independent on N. For the CW topology, in contrary, capacitance of the equivalent storage capacitor can be treated as a function of N, i.e., $C_{out} = \big(C_Q^{-1} + C_{b[2N-1]}^{-1}\big)^{-1} = \frac{C_b}{N}$. Increase of N reduces C_Q, and therefore C_{out}. Hence, the voltage ripple of the latter significantly increases with N while that of the former is only changed negligibly.

Our simulations also show that increase of $V(C_{b[2(N-1)-1]})$ allows η_{min} of the PS multiplier to be reduced by increasing N. In this case, $C_{b[2(N-1)]}$ is not affected by N and is still satisfies the condition that it is relatively much larger than the transducer capacitance for operation of the circuit. While for the CW multiplier, the more stages, the smaller C_E and C_Q are obtained. Therefore η_{min} cannot decrease further at high N.

4. Conclusion

This paper presented a study of the performance of MEMS capacitive energy harvester when electrically configured as voltage multipliers. Two topologies were investigated and compared. Both are able to operate with the capacitance variation ratio η lower than 2 which is the threshold for operation of the Bennet's voltage doubler. The CW multiplier has the advantages of higher saturation voltage and lower required initial bias. However, the PS multiplier gives smaller ripple at the output. In addition, even though the minimum required capacitance ratio η_{min} of both configurations can be reduced by adding more stages, the PS topology benefits more from a high number of stages.

Acknowledgment

This work was supported by the Research Council of Norway through Grant no. 229716/E20.

References

[1] Wang L and Yan Y 2014 *Measurement Science and Technology* **25** 115101
[2] Ngigi R W, Pislaru C, Ball A and Gu F *Journal of Physics: Conference Series* **364** 012016
[3] Ferrari M, Ferrari V, Guizzetti M and Marioli D 2009 *Smart Materials and Structures* **18** 085023
[4] Vullers R, van Schaijk R, Doms I, Hoof C V and Mertens R 2009 *Solid-State Electronics* **53** 684 – 693
[5] Roundy S 2003 Ph.D. thesis University of California at Berkeley Berkeley CA
[6] Yen B C and Lang J H 2006 *IEEE Transactions on Circuits and Systems—Part I: Regular Papers* **53** 288–295
[7] Torres E O and Rincon-Mora G A 2010 *IEEE Journal of Solid-State Circuits* **45** 483–496
[8] Kempitiya A, Borca-Tasciuc D A and Hella M M 2013 *IEEE Transactions on Industrial Electronics* **60** 5639–5647
[9] Damaschke J M 1996 *Applied Power Electronics Conference and Exposition* pp 856–860 vol.2
[10] Hudak N S and Amatucci G G 2008 *Journal of Applied Physics* **103** 101301
[11] de Queiroz A C M and Domingues M 2011 *IEEE 54th International Midwest Symposium on Circuits and Systems (MWSCAS)* pp 1–4
[12] de Queiroz A C M and Domingues M 2011 *IEEE Transactions on Circuits and Systems II: Express Briefs* **58** 797–801
[13] Dragunov V and Dorzhiev V 2013 *Journal of Physics: Conference Series* **476** 012115
[14] Lefeuvre E, Risquez S, Wei J, Woytasik M and Parrain F *Journal of Physics: Conference Series* **557** 012052
[15] de Queiroz A C M 2016 *2016 IEEE International Symposium on Circuits and Systems (ISCAS)* pp 650–653
[16] Hauschild W and Lemke E 2014 *High-Voltage Test and Measuring Techniques* 1st ed (Springer-Verlag Berlin Heidelberg)
[17] Cockcroft J D and Walton E T S 1932 *Proceedings of the Royal Society of London A: Mathematical, Physical and Engineering Sciences* **136** 619–630
[18] Truong B D, Le C P and Halvorsen E *arXiv:1709.08759*
[19] Truong B D, Le C P and Halvorsen E *arXiv:1709.08754*

An electrostatic energy harvester with sandwiched structure of two electret layers

Yulong Zhang[1,2], Xinge Guo[1], Yushen Hu[1] and Fei Wang[1,2,3]

1 Department of Electrical and Electronic Engineering, Southern University of Science and Technology, Shenzhen 518055, China;
2 Shenzhen Key Laboratory of 3rd Generation Semiconductor Devices, Shenzhen 518055, China;
3 State Key Lab of Transducer Technology, Shanghai Institute of Microsystem and Information Technology, Chinese Academy of Sciences, Shanghai 200050, China.
E-mail: wangf@sustc.edu.cn

Abstract. In this paper, we proposed a fully packaged sandwich electrostatic energy harvester, which consists of two opposite charged electret layers to improve the output performance of the packaged device. The new device is based on the out-of-the-plane gap closing scheme, and the main components are similar to our previous research. This sandwich device mainly contains two variable capacitors, and the volume of the device is about $1.3 \times 1.8 \times 0.15$ cm^3. Because of the complementation between the two opposite charged variable capacitors, the output performance of the overall sandwich device is better than the single electret layer devices. We have achieved a maximum power output of 20.78 μW at the acceleration of 19.96 m/s^2, which is almost 1.4 times of the packaged single capacitance structure, without changing the overall volume of the packaged device. Also, this sandwich device shows good performance at random vibration.

1. Introduction

With the growth of the internet of things (IOT), the low power wireless sensors are used more and more frequently, and a perpetual power for the low power sensors is an urgent need. So, micro energy harvesters as the self-sustainable power sources for the low power systems have attracted much attention. The vibration energy harvester is one of the micro energy harvesters, and it can convert the kinetic energy from ambient environment into electric energy. According to the principle of energy conversion, different kinds of vibration energy harvesters have been designed and studied, like electromagnetic vibration energy harvester [1]-[4], piezoelectric vibration energy harvester [5]-[10], and electrostatic vibration energy harvester [11]-[22]. Among these vibration energy harvesters, the electrostatic energy harvester is easier to fabricate and to integrate with the micro systems through the simple Si-based MEMS technology. With the advanced MEMS technology, the electrostatic energy harvester takes the advantages of small size, high energy density, high conversion efficiency, and long-term output stability [23].

Up to now, there are a lot of reports about the electrostatic vibration energy harvesters which mainly focuses on three different configurations of variable capacitors by changing the gap distance [11]-[17], changing the overlap area [18], [19], or changing the counter electrodes [20]-[22]. In most of these studies, the main purpose is to improve the power output or bandwidth of the device to get a higher power density at a wider bandwidth. In our previous study, a four-wafer stacked electrostatic

IOP Conf. Series: Journal of Physics: Conf. Series **1052** (2018) 012119 doi:10.1088/1742-6596/1052/1/012119

energy harvester with out-of-the-plane gap closing scheme (changing the gap distance) has been developed [11]. In this device, there was only one electret layer, and the effective output was only from the top variable capacitor structure. The bottom cap is used for packaging and controlling the maximum distance of the proof mass. Now, in this paper, we have designed a fully packaged sandwich electrostatic energy harvester which contains three components and two opposite charged electret layers, as shown in Figure 1. The resonant structure is a proof mass which is suspended to a fixed frame through four spring structures. The top and bottom plates are the same, and two electret layers are deposited on the surface of the two cavities by spray coating [24]. Before packaging, the two electret layers are negatively and positively charged to about ±600 V, respectively. Then a sandwich device contained two variable capacitors is achieved. The volume of the final device is about 1.3 × 1.8 × 0.15 cm^3.

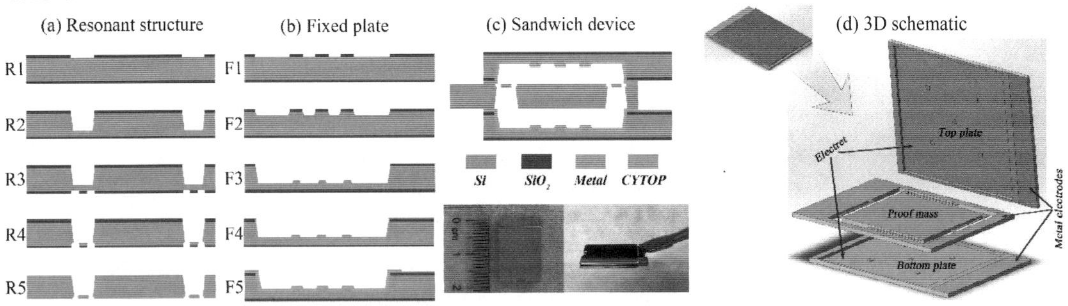

Figure 1. Fabrication process flow and 3D schematic of the sandwich electrostatic energy harvester.

2. Characterization

Figure 2 shows the circuit diagrams of two types of devices based on the sandwich structure. Based on the work principle of the pre-charged electret vibration energy harvester [16], [25], induced charges are obtained on both sides of the proof mass. When the proof mass is driven to move, there will be two kinds of current flow through the external load at the same time. Obviously, the output of the same charged sandwich device (Figure 2(a)) will be lower than that of the opposite charged device (Figure 2(b)), because of the opposite current flow through the external load. Therefore, we focus on the power output of the sandwich device with opposite charges here.

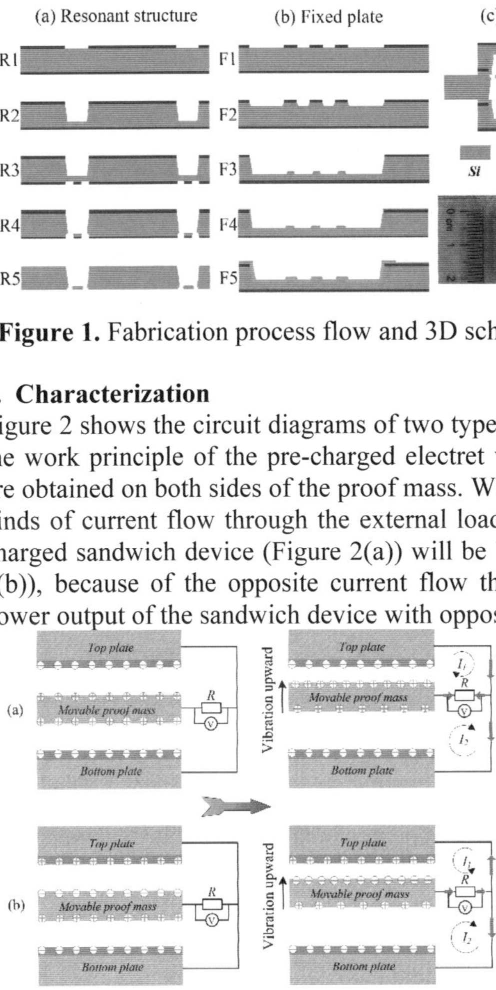

Figure 2. The power output principles of the packaged sandwich electrostatic energy harvesters with the same (a) and opposite (b) charged electret layers, respectively.

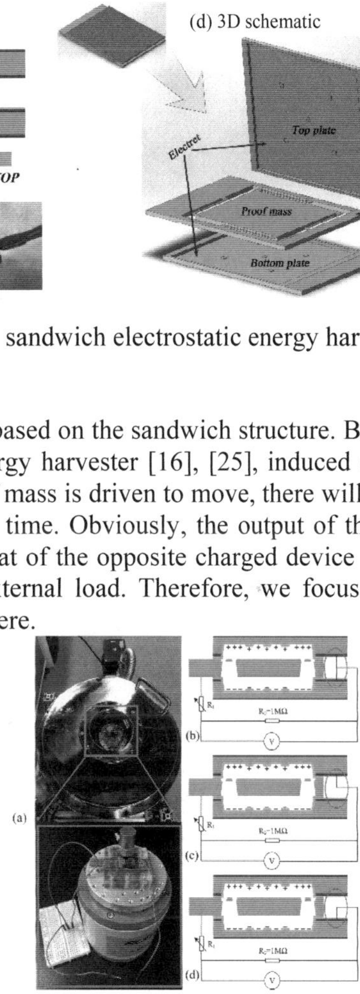

Figure 3. The test setup (a) and three test methods of the opposite charged sandwich electrostatic energy harvester. (b) is the overall sandwich device. (c) and (d) are the top and bottom single devices, respectively.

The output performance of the opposite charged electrostatic energy harvester was characterized using a shaker setup to mimic the external vibration source, as shown in Figure 3. The sandwich device was fixed on an acrylic stack mounted to the shaker as shown in Figure 3(a). And then we measured the power output of the sandwich device (Figure 3(b)), the single top device (Figure 3(c)), and the single bottom device (Figure 3(d)), respectively.

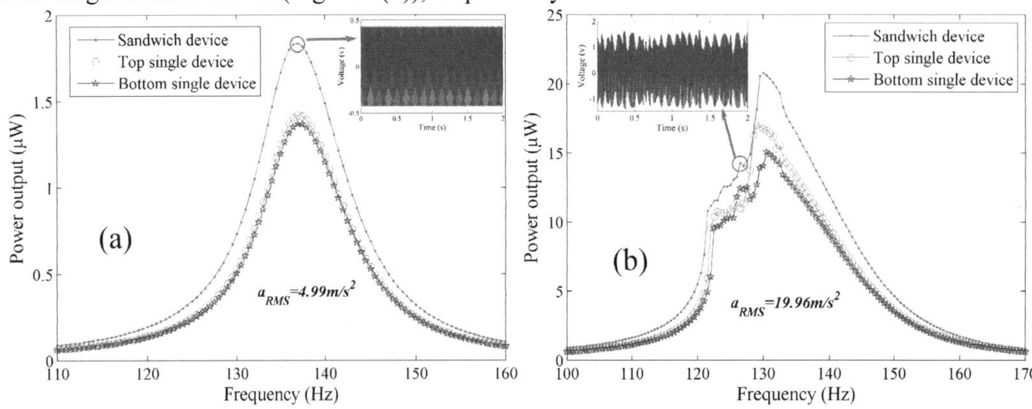

Figure 4. The experimental results of powers vs frequencies at different accelerations. (a) 4.99 m/s^2, (b) 19.96 m/s^2.

Figure 4 shows the swept sine tests for the overall sandwich device, the top device, and the bottom device at the accelerations of 4.99 m/s^2 and 19.96 m/s^2, respectively. Thanks to the complementation between the two opposite charged variable capacitors, the output performance of the overall sandwich device is higher than the single devices. The maximum power outputs of the overall sandwich device are 1.83 μW and 20.78 μW at the accelerations of 4.99 m/s^2 and 19.96 m/s^2, respectively, which are almost 1.4 times of the packaged single capacitor structure, without changing the overall volume of the packaged device. With the acceleration increased, the collision between the proof mass and the fixed plate occurs, which can be observed from the unstable voltage output in Figure 4(b). The collision indicates that the displacement of the proof mass has reached the limit, and it can broaden the bandwidth of the device.

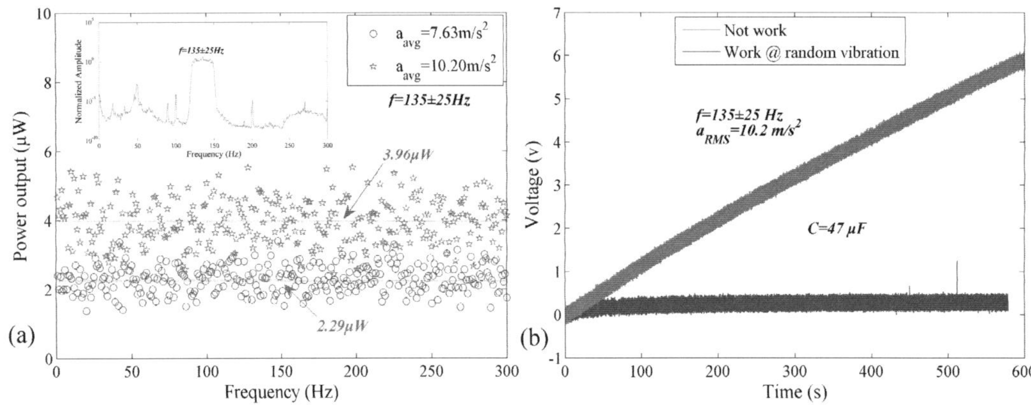

Figure 5. (a) The power output of the sandwich devices under random vibrations with bandpass filter (f=135±25Hz) at different average acceleration, (b) Voltage of a 47μF capacitor charged at 10.2 m/s^2 @ 135±25Hz.

The sandwich device with two electret layers was also tested when driven by a random vibration with the bandpass of 135±25 Hz at the average accelerations of 7.63 m/s^2 and 10.2 m/s^2, as shown in Figure 5(a). At 10.2 m/s^2, we got the output power of 3.96 μW, which is higher than the result in our previous single device [16] with the same device size. The new device can also work in a broader bandwidth (135±25 Hz) than before (160±12.5 Hz). It is well known that, it's easy to get an optimal power output at the resonant frequency of the device, but it's difficult in a wide bandwidth. Here, we have achieved a higher output in a broader bandwidth from random vibration. We have used the energy harvester to charge a capacitor of 47 μF as shown in Figure 5(b). In 10 minutes, the capacitor can be charged to about 5.9 V by the sandwich device which is driven by a vibration amplitude of 10.2 m/s^2 at random frequency range of 135±25 Hz at atmosphere.

3. Conclusion

We propose a new method to improve the output performance of the packaged electrostatic energy harvester, named sandwich device with two opposite charged electret layers. The volume of the new device is about $1.3 \times 1.8 \times 0.15$ cm^3. This new structure can effectively improve the power output at frequency sweep and random without changing the volume of the overall packaged device. This study provides a promising packaging method with enhanced performance for the electret based electrostatic energy harvester.

Acknowledgement

This work was supported by National Natural Science Foundation of China (Project No.: 51505209) and Shenzhen Science and Technology Innovation Committee (Projects Nos.: JCYJ20170412154426330 and KQTD2015071710313656). Fei Wang is also supported by Guangdong Natural Science Funds (Project Nos.: 2015A030313812 and 2016A030306042).

References

[1] J. Chen, D. Chen, T. Yua and X. Chen, *Appl. Phys. Lett.*, 100, 213509, 2012.
[2] H. Liu, K. H. Koh and C. Lee, *Appl. Phys. Lett.*, 104, 053901, 2014.
[3] Q. C. Tang and X. X. Li, *IEEE-ASME T. Mech.*, 20, 115-21, 2015.
[4] V. Nico, E. Boco, R. Frizzell and J. Punch, *Appl. Phys. Lett.*, 108, 013902, 2016.
[5] R. Elfrink, T. M. Kamel, M. Goedbloed, S. Matova, D. Hohlfeld, R. van Schaijk and R. Vullers, *J. Micromech. Microeng.*, 19, 094005, 2009.
[6] Q. C. Tang, Y. L. Yang and X. X. Li, Smart Mater. Struct., 20, 125011, 2011.
[7] S. Li, et.al., *Sensor. Actuat. A-Phys.*, 247, 547–554, 2016.
[8] S. Li, Z. Peng, A. Zhang, and F. Wang, *AIP Advances* 6, 015019 (2016).
[9] L. V. Blarigan, P. Danzl and J. Moehlis, *Appl. Phys. Lett.*, 100, 253904, 2012.
[10] Z. Yi, B. Yang, G. Li, J. Liu, X. Chen, X. Wang and C. Yang, *Appl. Phys. Lett.*, 111, 013902, 2017.
[11] F. Wang and O. Hansen, *Sensor. Actuat. A-Phys.*, 211, 131-7, 2014.
[12] C. Wang, L. Cai, Y. Feng, L. Chen, W. Yan, Q. Liu, T. Yao, F. Hu, Z. Pan, Z. Sun and S. Wei, *Appl. Phys. Lett.*, 104, 243112, 2014.
[13] T. Takahashi, M. Suzuki, T. Nishida, Y. Yoshikawa and S. Aoyagi, *MEMS 2015*, 1145-48.
[14] K. Tao, S. W. Lye, J. Miao, L. Tang and X. Hu, *J. Micromech. Microeng.*, 25, 104014, 2015.
[15] Y. Lu, F. Cottone, S. Boisseau, F. Marty, D. Galayko and P. Basset, *Appl. Phys. Lett.*, 107, 253902, 2015.
[16] Y. Zhang, et.al., *Review of Scientific Instruments*, 87, 125001 (2016).
[17] P. Pondrom, G. M. Sessler, J. Bos and T. Melz, *Appl. Phys. Lett.*, 109, 053906, 2016.
[18] S. Boisseau, G. Despesse and A. Sylvestre, *Smart Mater. Struct.*, 19, 075015, 2011.
[19] S. Risquez, M. Woytasik, J. Wei, F. Parrain and E. Lefeuvre, *DTIP 2015*, 1-5.
[20] Y. Naruse, N. Matsubara, K. Mabuchi, M. Izumi and S. Suzuki, *J. Micromech. Microeng.*, 19, 094002, 2009.
[21] A. Crovetto, F. Wang and O. Hansen, *J. Micromech. Microeng.*, 23, 114010, 2013.
[22] A. Crovetto, F. Wang and O. Hansen, *J. Microelectromech. S.*, 23, 1141-55, 2014.
[23] J. Cao, W. Wang, S. Zhou, D. Inman and J. Lin, *Appl. Phys. Lett.*, 107, 143904, 2015.
[24] Y. Xu, et.al., *Micro & Nano Letters*, 11, 640-644, 2016.
[25] F. Wang, and O. Hansen, *IEEE Electron Device Letters*, 34, 1047-1049, 2013.

IOP Conf. Series: Journal of Physics: Conf. Series **1052** (2018) 012120 doi:10.1088/1742-6596/1052/1/012120

A non-resonant rotational triboelectric energy harvester with high output performance

J Lin, H Liu*, T Chen, Z Yang and L Sun

School of Mechanical and Electric Engineering, Jiangsu Provincial Key Laboratory of Advanced Robotics, Collaborative Innovation Center of Suzhou Nano Science and Technology, Soochow University, Suzhou, 215123, China

Email: hcliu078@suda.edu.cn

Abstract. Energy harvesting technology has received significant attention around the globe over decades. However, it meets with difficulties of achieving high output performance under non-resonant frequency. In this paper, a non-resonant rotational triboelectric energy harvester with high output performance, which relied on the coupling of triboelectrification and electrostatic induction, was presented. The harvester introduced a cylindrical stator and disk-shaped rotor, which are made of NdFeB. Without complicated movement mechanism, the stator and rotor, of a higher pair have theoretical line contact by magnetic attractive force. Under a hand shake frequency of about 8Hz, the peak-peak value can reach to 140V. The maximum power of 0.31mW was obtained at a resistance of about $10^7\Omega$. In addition, the voltage can reach up to 154V when running at a speed of 8 km/h on the treadmill. The harvester can be potentially used as a power supply to intelligent wearable devices.

1. Introduction

Energy harvesting technology has received significant attention of the researchers around the globe over decades because of the rapid growth of mobile electronic devices [1-4]. Many energy harvesting devices based on different conversion mechanisms such as electromagnetic [5-6], piezoelectric [7-8], or triboelectric effects [9-10] have presented. Nanotechnology for harvesting mechanical energy using nanogenerator emerged recently [11]. For example, triboelectric nanogenerator(TENG) have been proved as an efficient, light, cost-effective and robust approach to harvest ambient mechanical energy [12]. In order to achieve high power efficiency, the natural frequency of an energy harvester should be designed to match with the environment vibration source [13-14]. It becomes a great challenge to design an energy harvester with high energy output and non-resonant frequency. The key solutions are to improve the energy output density and energy conversion efficiency of devices.

In this study, we demonstrated a non-resonant rotational triboelectric energy harvester with high output performance based on triboelectrification and electrostatic induction. Under an external excitation, the rotor magnet is able to rotate and impact/separate with two TENGs. The output performance of harvester can be improved significantly.

2. Device and working principle

The schematic diagram of the harvester is shown in Figure 1. It consists of a cover, two TENGs, a TENG fixer, a copper foil, a stator magnet, a rotor magnet, and a casing. TENG consists of one-piece membrane structure formed by Fluorinated Ethylene Propylene (FEP) and copper (Cu) electrode. The stator and rotor are both made of NdFeB magnet. The cylindrical stator magnet is fixed in the center of the casing, and the disk-shaped rotor magnet is attached on the edge of the stator by magnetic attractive force. Without assembling any shaft or bearing structure, the stator and rotor magnets formed a higher pair have theoretical line contact. It enables the rotor magnet easily rotate around the center stator magnet at low frequency vibration and avoids unnecessary surface friction. The copper foil was attached to the rotor magnet, so the foil can easily and directly impact against the FEP back and forth while rotating. Because of the triboelectrification, the AC voltage output can be generated from the TENG.

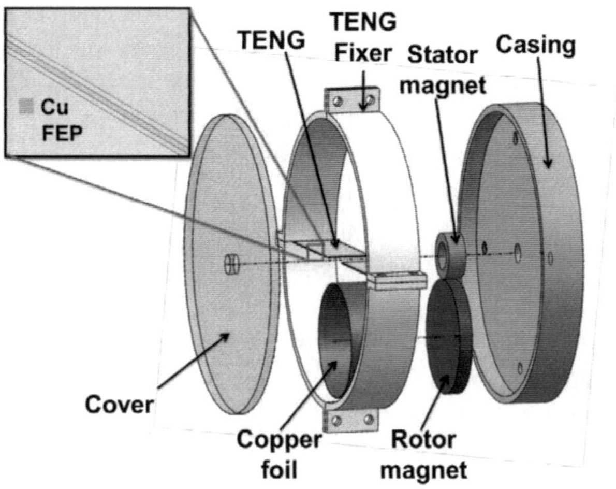

Figure 1. Schematic drawing of the triboelectric energy harvester.

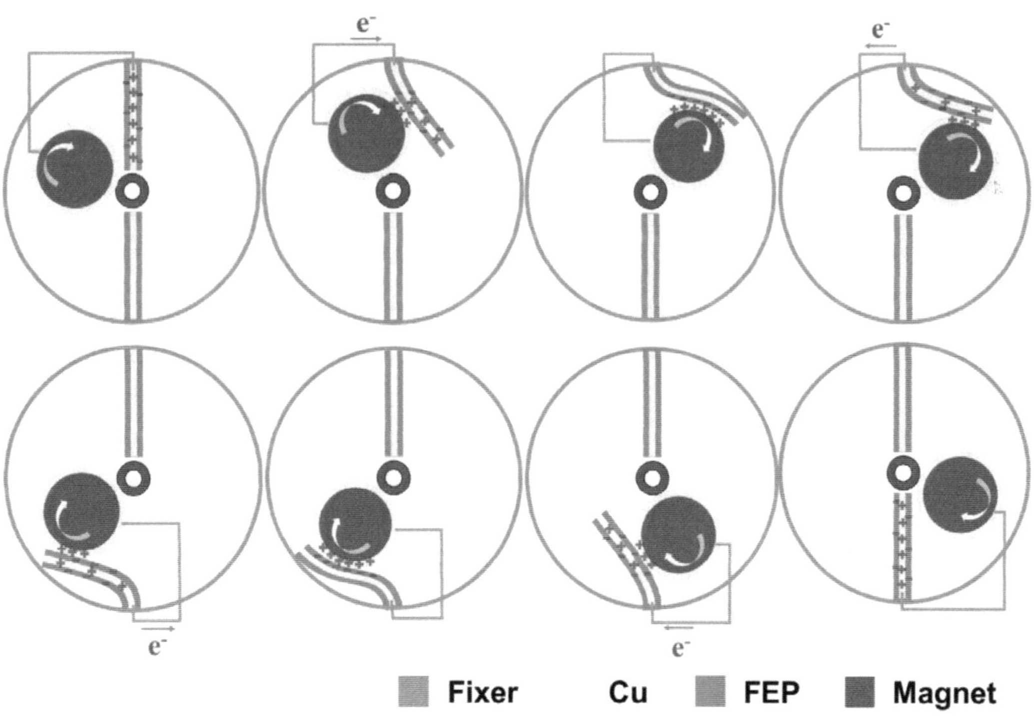

Figure 2. Schematic diagram of the working principle for the TENG

The working principle of TENG is based on the couple of electrification effect and electrostatic induction. The design of the TENG is based on the relative sliding motion of grated surfaces. Here, as schemed in Figure 2, the process of electricity generation is illustrated. At the first position, the sliding Cu foil and FEP film are completely isolated. Upon contact between the FEP film and the Cu foil, charge transfer takes place. Due to the difference on triboelectric polarities, electrons are injected from Cu into the surface of FEP. Since the two sliding surfaces are completely contacted, triboelectric charges with opposite polarities are fully balanced out, making no electron flow in the external circuit. Once a relative sliding occurs as a result of rotation, the negative charges on the FEP will drive free electrons on the back electrode to the sliding electrode through the external circuit due to electrostatic induction, neutralizing positive triboelectric charges on the sliding electrode and leaving positively induced charges behind on the back electrode. The flow of the induced electrons lasts until the two sliding surfaces completely separated. As the relative rotation continues, the Cu will come into contact with next FEP film. Therefore, in a cycle of electricity generation, AC electric output is generated.

3. Experiment and results

The experimental results in Figure 3(a) show the open-circuit voltage of the harvester when there is only one TENG is applied. Figure 3(b) shows the open-circuit voltage of two TENGs. The peak-peak value can reach to 140 V, which was a high output performance, at a hand shake frequency of about 8Hz. From Figure 3, it is known that the number of the TENG will influence the frequency of the voltage instead of the value of the voltage; also, the number of the TENG will influence the smoothness of the rotation. So two TENGs were applied in the following experiments.

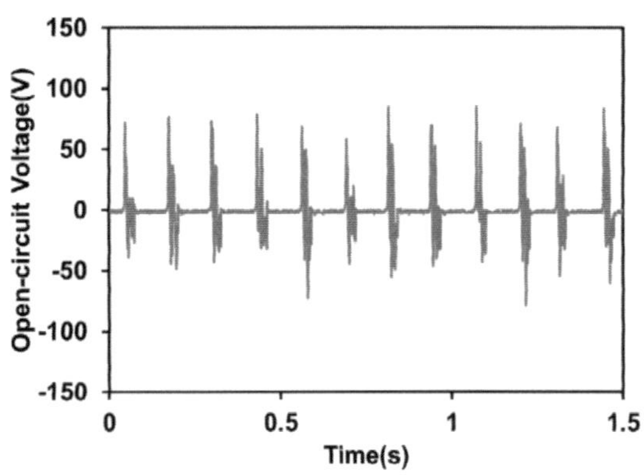

Figure 3(a) The open-circuit voltage of the harvester when one piece of TENG was applied.

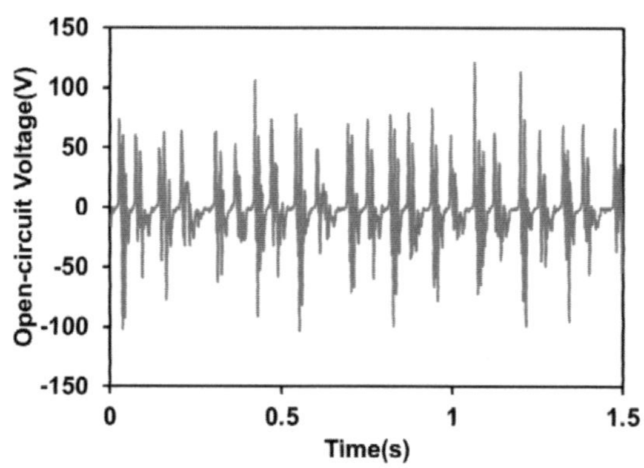

Figure 3(b) The open-circuit voltage of the harvester when two pieces of TENG were applied.

Figure 4 shows the peak voltage and power under different external load resistances with hand shaking vibration frequency of 8 Hz. The maximum power of 0.31 mW was obtained at a resistance of about 10^7 Ω. Figure 4 shows the open-circuit voltage of the harvester at different speeds. People swing hands when running. Figure 5 was measured on a treadmill, which the harvester was worn on the wrist. It can be seen that the harvester is non-resonant and low-frequency, the output was appreciable at any speeds, and the amplitude of the voltage is improved as the speed increasing, which means more energy is generated. The voltage can reach up to 154 V when running at a speed of 8 km/h.

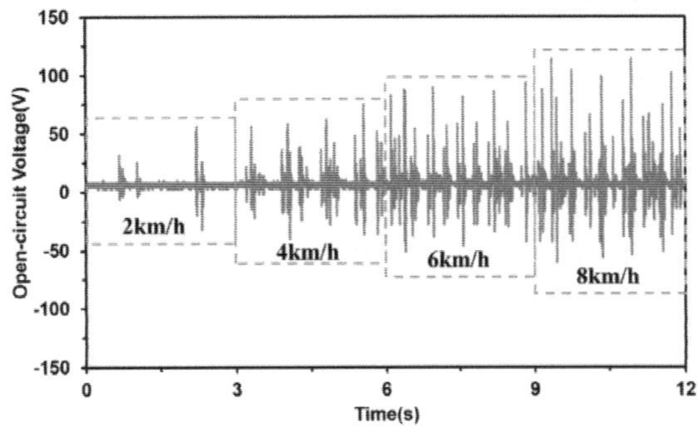

Figure 4. Open circuit voltage of the device at different speeds on the treadmill.

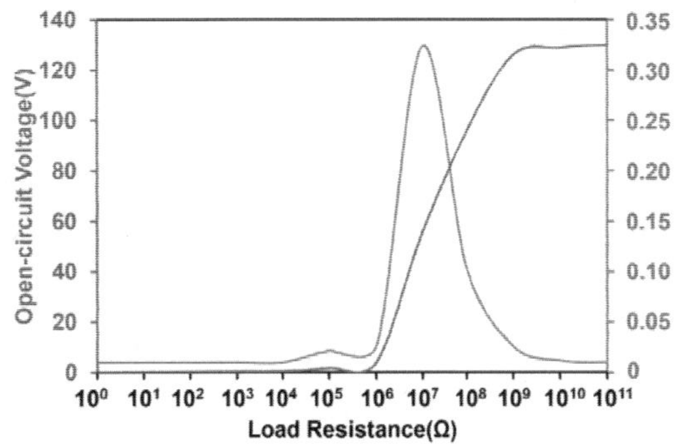

Figure 5. Peak voltage and power under different external load resistances with hand shaking vibration frequency of 8 Hz

4. Conclusion

In this paper, we presented a rotational triboelectric energy harvester, which can achieve high output performance under non-resonant frequency. Two TENGs are designed to impact with copper foil, resulting in a significant improvement of electrical output. Under a hand shake frequency of about 8 Hz, the peak-peak value can reach to 140 V. The maximum power of 0.31 mW was obtained at a resistance of about 10^7 Ω. When running on the treadmill, the voltage can reach up to 154 V at a speed of 8 km/h. The harvester is so sensitive to human motion that it can be potentially used as a power supply to intelligent wearable devices.

5.Acknowledgements

This work is partially funded by the National High Technology Research and Development Program of China (863 Program) (Grant No. 2015AA042601) and the National Natural Science Foundation of China (Grant No. 51405318).

References

[1] Ylli K, Hoffmann D, Folkmer B, et al. 2013 *Journal of Physics Conference Series* **476** 914-919.
[2] Zhong X, Yang Y, Wang X, et al. 2015 *Nano Energy* **13** 771-780.
[3] Beeby S P, Tudor M J and White N M 2006 *Meas. Sci. Technol.* **17** R175-R195.
[4] Zhang X S, Han M D, Meng B and Zhang H X 2015 *Nano Energy* **11** 304-322.
[5] Liu, H., Ji, Z., Chen, T., Sun, L., Menon, S. C., Lee, C. 2015 *IEEE Sensors Journal* **15** 4782-4790.
[6] Özge Zorlu, & Külah, H. 2013 *Sensors & Actuators A Physical* **202** 124-134.
[7] Jinsung, C., Young, L. K., Kang, C., Wha, K. M., Sang - Woo, K., & Min, B. J. 2014 *Advanced Functional Materials* **24** 2038-2043.
[8] Sultana, A., Alam, M. M., Garain, S., Sinha, T. K., Middya, T. R., & Mandal, D. 2015 *Acs Applied Materials & Interfaces* **7** 19091-7.
[9] Zhao, X., Wei, G., Li, X., Qin, Y., Xu, D., & Tang, W., et al. 2017 *Nano Energy* **34** 549-555.
[10] Lee, S., Lee, Y., Kim, D., Yang, Y., Lin, L., & Lin, Z. H., et al. 2013 *Nano Energy* **2** 1113-1120.
[11] Lee S, Lee Y, Kim D, et al, 2013 *Nano Energy* **2** 1113-1120.
[12] Ren X, Fan H, Wang C, et al, 2017 *Nano Energy* **35** 233-241.
[13] Williams C B and Yates R B 1996 *Actuat. A-Phys.***52** 8-11.
[14] Dhakar L, Tay F and Lee C 2015 *J. Microelectromech. S.* **24** 91-99.

Charge Doubler Vibration Energy Harvester Using Self-Synchronized mechanical switches

M. A. Ben Ouanes[1,2], **H. Samaali**[1,2], **P. Basset**[2] **and F. Najar**[1]

[1]Applied Mechanics and Systems Research Laboratory, Tunisia Polytechnic School, BP 743, La Marsa 2078, University of Crthage, Tunisia.
[2]Université Paris-Est, ESYCOM, ESIEE Paris, Noisy-le-Grand, France.

E-mail: benouanesaymen@gmail.com

Abstract. In this work, we study an electrostatic vibrational energy harvester e-VEH based on the charge doubler principle introduced by Bennet two centuries ago. We proof the advantages of using self-synchronized mechanical switches instead didoes, to avoid the problems of leakage current and threshold voltage. We made a new device as a proof of concept, which can work with mechanical switches or with diodes. To harvest the kinetic energy, an electrode moves between two fixed ones thanks to external vibration and creates an asymmetric variation of two variable capacitors. Due to his motion, this electrode also drives a sequence of contacts in the mechanical switch's configuration. A comparison between this design and the one using diodes in the same conditions shows that the output voltage increase with an exponentially form for both design. However, the one using mechanical switches is faster by a factor of 80%. This induce a time reduction of 99% of harvested power at 40nW.

1. Introduction

Nowadays, the use of vibrational energy is becoming a revolutionary source of energy. This latter, is directly extracted form the environment and contributes to the development of compact and self-contained devices like wireless sensors and portable devices.

There is 3 major kinds of transducers, which convert kinetic energy into useful electrical one. First, there is the electromagnetic transducer which is constituted by a wire coil and a magnet [1]. This one generates a high power in a large scale, but it is not adapted to the micro-scale. On the other hand, we find piezoelectric transducers [2]. It can leads to a high power generation at the small scale but the use of this technology induces special steps for fabrication with relatively a high price. Finally, there is the electrostatic transducers. Simple capacitive electrodes compose this kind of transducers [3]. It is easily implementable in small-scale but it generates high voltage and its conditioning electronics can be complex for minimazing the efficiency.

Among the conditioning circuits for electrostatic transducers, there is the charge doubler. It was developed by Bennet two centuries ago and consists of a sequence of contacts for grounding or connecting three conductive plates, which form two variable capacitors [4]. Recently, Queiroz used this technic of charge doubler to harvest vibration energy from the environment by replacing the contacts with diodes [5].

In this work, we have reintroduced the contacts for the purpose of kinetic energy harvesting. Self-synchronized mechanical switches replace the diodes to be ride of the issues induced by diode's threshold voltage and leakage current. We made a new device as a proof of concept based on the same principle that in [6] which can work with mechanical switches or diodes to compare the 2 approaches and prove the advantage of using mechanical switches.

2. The charge doubler for e-VEH with self-synchronized mechanical switches

The functional representation of our charge doubler device for e-VEH is shown in figure 1. It is composed by three electrodes (A, B and C). Thanks to external excitation, the central one B moves between the two others, which are fixed. This induces an asymmetric variation of the two variable capacitors C_{ab} which is formed between the electrodes A and B, and C_{bc} formed between the electrodes B and C. C_{ab} varies between 13pF and 68pF and C_{bc} varies between 25pF and 58pF. The mechanical switches (S_1, S_2 and S_3) play the role of the classically used diodes. S_1 and S_2 act as stoppers to limit the displacement of the central electrode. The third mechanical switch S_3 connects electrode A with electrode C, while S_1 and S_2 connect electrode B and C to the ground respectively. After a sequence of contacts by these mechanical switches and the asymmetric variation of the capacitance due to the oscillations of the central electrode, the charge increases at each cycle in the electrodes. Figure 2 represents the equivalent electrical circuit of our charge doubler device. We added another capacitor for storage C_{res} of $1nF$ and a high impedance follower at the output for the characterization, which add an additional $10pF$ in the circuit.

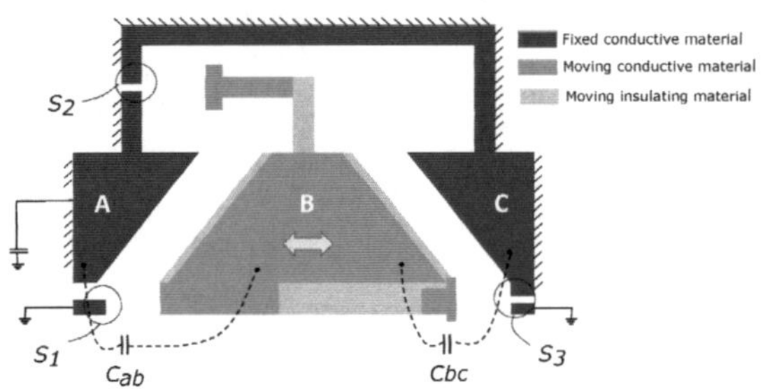

Figure 1. Functional representation of the e-VEH charge doubler device with mechanical switches.

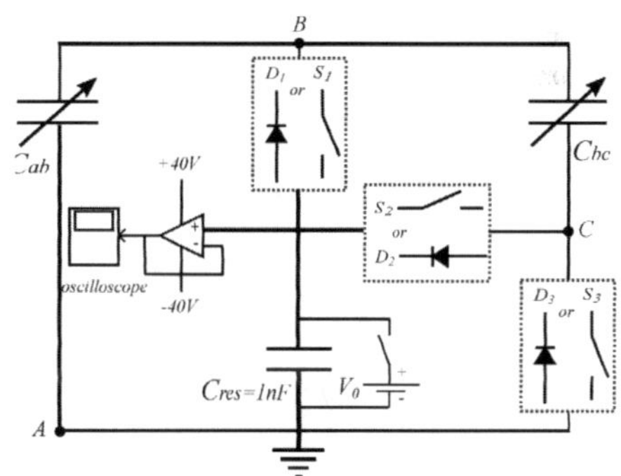

Figure 2. Equivalent electrical circuit of the charge doubler device with diodes or mechanical switches.

The sequence for harvesting the kinetic energy is the following. Initially, there is some charge on the electrodes due the ambient noise. The operation starts when the central electrode B is positioned near A while connected to the ground through S_1. This induces an increase of charge $-\Delta Q$ on B. In this configuration, all the capacitances are in parallel, while C_{ab} is at its maximum value and C_{bc} is at its minimum values. Then B moves towards C due to the external vibrations. When B reaches C, this one is grounded through S_2 and a charge ΔQ is induced on C. Now, all the capacitances are connected in series, C_{ab} turns to its minimum value and C_{bc} to its maximum values . Finally, the mobile electrode moves back. When B reaches A, B is grounded again through S_1 and A is connected to C through S_3 and then, the electrode A has a charge increase equal to $+2\Delta Q$. All the capacitances are again in parallel and a charge $-2\Delta Q$ is induced on B. So, the equivalent circuit oscillates between series and parallel configurations of capacitances [3], and at each vibration period the charge double on the electrodes.

3. Experimental validation

To proof the utility of the use of mechanical switches, we have tested our device with mechanical switches and with the diodes in the same conditions. The electrodes are made of steel. A shaker

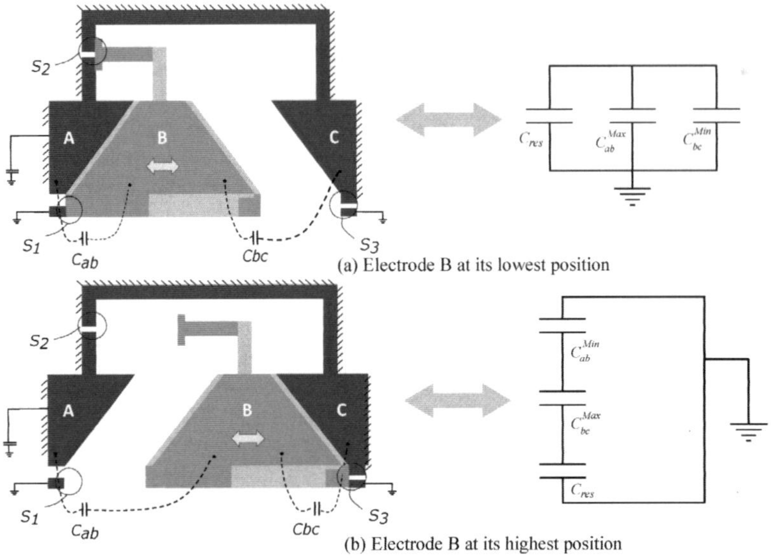

Figure 3. Operating cycles of the e-VEH charge doubler device with the corresponding equivalent circuit.

creates vibrations and allows the oscillation of the central electrode. The overall dimensions of the active part of the device is about $96cm^3$ against 315mg as the weight of the central electrode.

Figure 4. real charge doubler device mounted on the shaker.

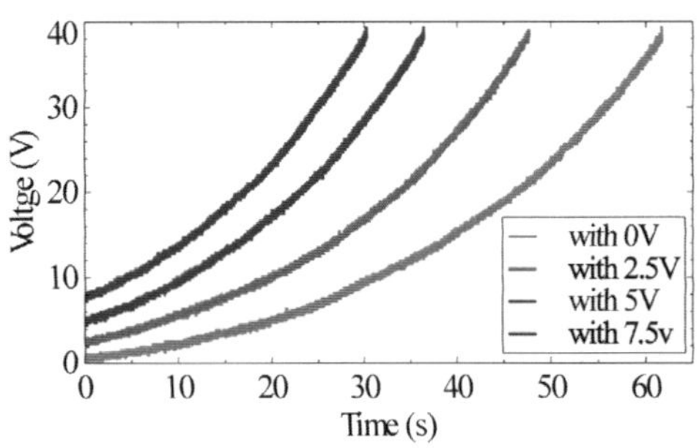

Figure 5. Voltage variation for different initial voltage, $8Hz$ and $0.5g_{rms}$.

The first set of experiments consists in varying the initial voltage across C_{res} with the device using the mechanical switches. The vibration conditions was set at 8Hz with $0.5g_{rms}$ and the experimental setup limits the measurement voltage across C_{res} at $40V$. We see in figure 5 that the slope of the voltage increase depends of the initial voltage across C_{res} : If the initial voltage decrease, the time to reach $40V$ increase. For example, with only ambient noise, 40V is reached after 62s against, 33s with 7,5V as initial voltage.

Then we compared the device using diodes and the one using mechanical switches with similare conditions (8Hz at $0.5g_{rms}$ and 1V as initial voltage). Results are shown in figure 6. Fig 7 shows the corresponding harvested power calculated by deriving the instantaneous energy in C_{res}, which is determinated by this equation:

$$E = \frac{1}{2} C_{res} V^2$$

It shows that the device with switches is much faster: $50nW$ of harvested power is reached in 45_s against 87_s with the diodes

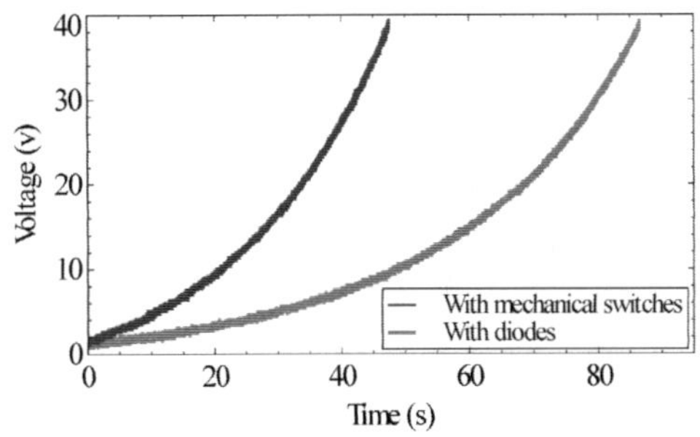

Figure 6. Voltage variation according to the time for the design with mechanical switches and with diodes with $8Hz$, $0.5g_{rms}$ and $1V$ as initial voltage.

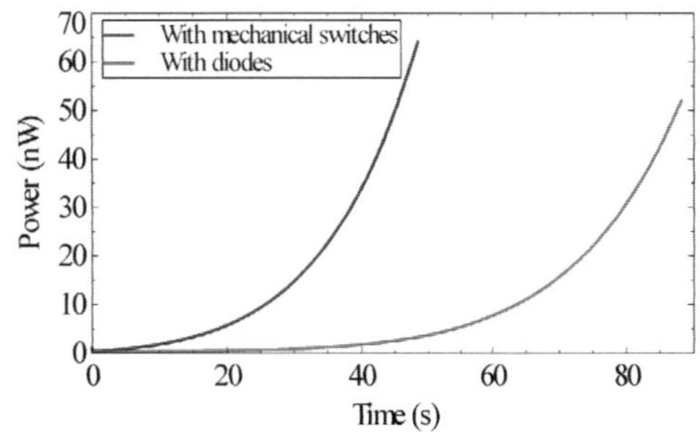

Figure 7. harvested power of charge doubler device with mechanical swithes and with diodes with $8Hz$ and $0.5g_{rms}$ and with $1V$ as initial voltage.

Table 1, we compares the time to reach 40nW for the two configurations for different initial pre-charges. We can see that the one with mechanical switches is always faster than one with diodes. In addition, we observe that the time gap to reach the 40nW for both design decreases with the growing of the initial voltage. This comes from the voltage drop across diodes which is more predonminent with low initial voltage.

Table 1. time to reach 40nW for different initial voltage using the design with mechanical switches and with diodes, and the time gap between them.

initial voltage	*with mechanical switches*	*with diodes*	*time gap*
1V	41.93s	83.73s	41.8s
2.5V	41.4s	75.24s	33.21s
5V	30.21s	54.53s	24.32s
7.5V	23.88s	37.66s	13.78s

4. Conclusion

Using the self-synchronized mechanical switches instead of diodes in the charge doubler device is more suitable for e-VEH because it supress the voltage drop and the reverse current across the diodes For instance, we showed that the design with mechanical switches is faster by a factor of 99% at 40nW. However, the mechanical switches can add reliability issues.

5. References

[1] Glynne-Jones, P., Tudor, M. J., Beeby, S. P., and White, N. M. 2004 electromagnetic, vibration-powered generator for intelligent sensor systems. Sensors and Actuators A: Physical,10 344-349..

[2] Marzencki M, Defosseux M and Basrour S 2009 *J. MEMS* **18** 1444-1453

[3] Basset, P., Galayko, D. and Blokhina, E., 2016 *Electrostatic kinetic energy harvesting.*

[4] Bennet A and Kaye R 1787 *Philosophical Trans. of the Royal Society of London* **77** 288-296

[5] de Queiroz A C M 2010 *53rd IEEE Int. Midwest Sym. on Circuits and Systems* Seattle WA 1-4 Aug.

[6] Ouanes M. B., Lu Y., Samaali H., Basset P., and Najar F, 2016 *Proc. PowerMEMS2016* Paris 6-9 DEC.

IOP Conf. Series: Journal of Physics: Conf. Series **1052** (2018) 012122 doi:10.1088/1742-6596/1052/1/012122

NON-LINEAR VIBRATION ELECTRET-HARVESTER WITH OPTIMIZED CURVED BEAM FOR LOW-FREQUENCY OPERATION

Koki Yamamoto[1], Adrien Badel[1], Fabien Formosa[1], Ludovic Charleux[1]
Takayuki Fujita[2], Kensuke Kanda[2], Kazusuke Maenaka[2]
[1]SYMME Laboratory, Université Savoie Mont Blanc, Annecy, France

[2]Graduate school of Engineering, University of Hyogo, Himeji, Japan

koki.yamamoto@univ-smb.fr

Abstract In this study, we report the design and fabrication results of an optimized curved beam structure for low frequency, under one-hundred Hz, and non-linear operation of MEMS (Micro-Electromechanical Systems) vibration energy harvester. The optimization procedure for shape and dimension of the curved beam were established combining a FEM and a numerical analysis. Simulated curved beam with bi-stable motion was evaluated for wide bandwidth operation of a vibration energy harvester and preliminary fabrication process was confirmed.

1. Introduction

Wireless sensor nodes (WSNs) become more important to monitor environmental parameters of human activities [1]. A typical WSN consists of many components e.g. sensors, integrated circuit, power management component, and battery. The battery is the largest problem for the spreading of WSNs. The energy harvesting technologies are expected to convert variety of environmental energy to electrical energy. We focused on vibration energy sources generated from a motor rotation and industrial equipment. It is difficult to apply typical resonant type energy harvester for environmental vibrations, because these vibrations have low frequency spectrum under 100 Hz and vibrate irregularly. In our previous device, a linear resonator structure with silicon-based MEMS (Micro Electromechanical Systems) processing was demonstrated [2]. The high mechanical quality factor of silicon MEMS structure is not only advantageous for sensors and actuators but also for energy harvesting devices for real vibration sources. In order to utilize the irregular vibration sources, we developed a bi-stable spring by MEMS fabrication [3]. The spring can collect energy from random vibration sources by using its nonlinearly. Most bi-stable harvesters are developed and studied at the mesoscale [4], but not at the MEMS scale. We reported the MEMS type bi-stable spring with curved parallel leaf spring and the combination of bi-stable and linear spring structure for low acceleration operation [5]. In this study, we established the optimization method by using the combination of two simulation software for our targeted applications.

2. Simulation and modelling of bi-stable spring

The mesoscale bi-stable spring is usually realized by applying an axial compression force to a clamped-clamped beam. However, it is difficult to apply such a force on MEMS devices. Thus, we investigate a bi-stable phenomenon in an initially curved beam design of a MEMS structure with various parameters. Figure 1 shows a half beam model of the proposed design in FEM analysis software (ANSYS workbench R17.0). The left-end is the clamped point and the right-end is a guided point for the half model. The snap-thorough force of the bi-stable spring is affected by the beam shape (*width*) and the initial deflection of the guided point (*amplitude*).

Figure 2 shows design optimization procedure by using ANSYS and a numerical simulation software (MATLAB2017a). By applying initial parameters e.g. beam length, width and deflection amplitude generated by MATLAB, ANSYS calculates the non-linear spring stiffness as a function of the deflection. Then the computed non-linear stiffness is automatically transferred to MATLAB SIMULINK where the frequency response of the harvester is computed when a 120 mg mass is attached between two non-linear springs. From the frequency response corresponding to a frequency sweep-up operation, the integrated area of the beam vibration amplitude and the frequency is output and can is used as an index of the bandwidth. SIMULINK also computes and outputs the number of equilibrium points as an index of bi-stability. An iterative and automatic simulation is performed to find the optimal design.

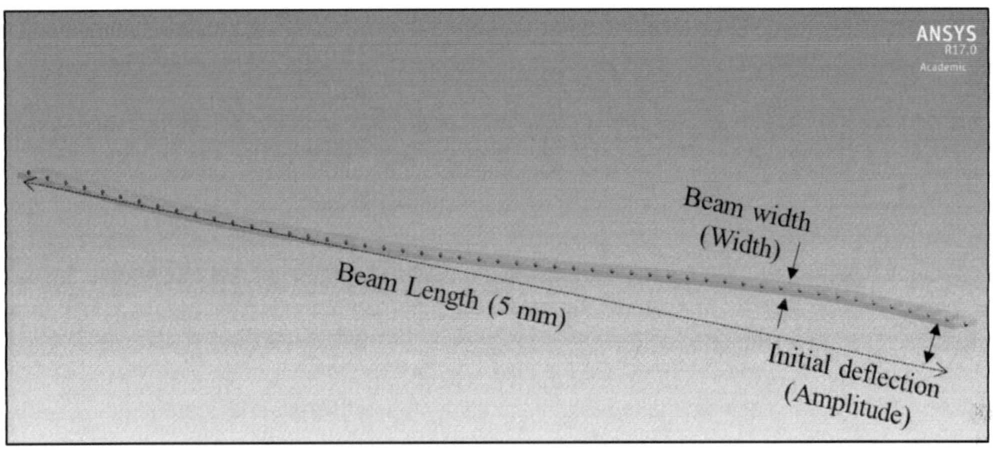

Figure 1. The simulation model of proposed curved beam in FEM software

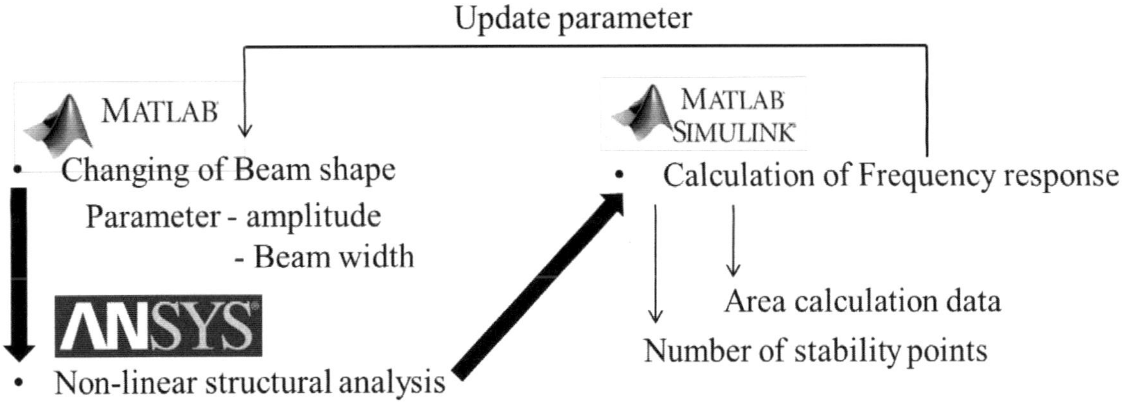

Figure 2. Computing for optimized design with ANSYS and MATLAB software

IOP Conf. Series: Journal of Physics: Conf. Series **1052** (2018) 012122 doi:10.1088/1742-6596/1052/1/012122

3. Simulation results

We set a target acceleration at 1 $G_{p\text{-}p}$ for bi-stable motion. The considered beam widths range from 20 to 50 μm with 5 μm step, the deflection amplitudes range from 50 to 500 μm with 50 μm step. The other parameters e.g. the total beam length of 10 mm, the beam thickness of 200 μm were fixed. The integrated area of the frequency response is shown in Fig. 3. The area size indicates easiness of bi-stable motion on specific acceleration of 1 G. The 150 μm amplitude and 30 μm width correspond to the peak value of the area. Figure 4 shows the number of equilibrium points in the nonlinear spring. The bi-stable motion has three equilibrium points (two stable, one unstable), and corresponds to the case where the spring can hold two stable different shapes after any external force is unloaded. The blue , where only one equilibrium exists corresponds to a mono-stable behaviour. By a combination of both simulation results, the optimized dimension is obtained as 150 μm in amplitude and 30 μm in width.

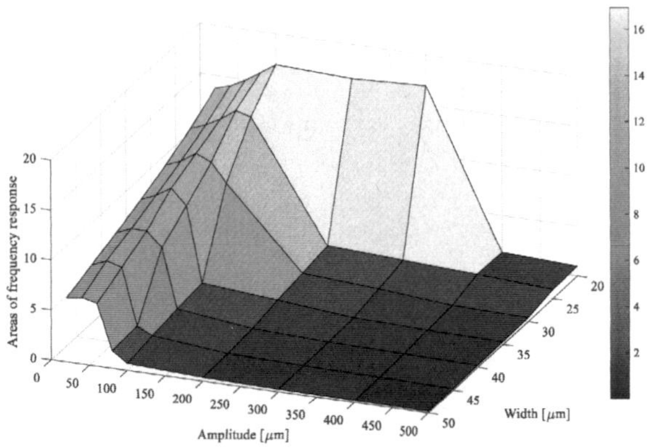

Figure 3. Integrated area of frequency responses for amplitude vs. width.

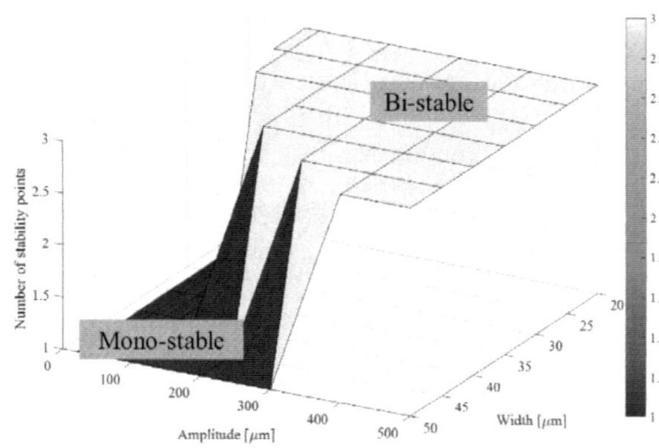

Figure 4. Number of equilibrium points for amplitude vs. width.

4. MEMS fabrication for the proposed beam

We fabricated the proposed curved beam in silicon MEMS process. The designed beam are included in a the electret-part of an electrostatic energy harvester. Figure 5 shows the fabrication process flow of the electret mass with curved spring. We used 200 μm thickness, 4 inch diameter single crystal silicon as a starting material. (1) The SiO_2 film is formed and patterned for mass and spring shape. (2-3) Aluminium film is deposited and patterned for electrodes. (4) The CYTOP for electret part is spin-coated and patterned. (5) Finally, the spring-mass structure is fabricated by Deep-RIE. The completed devices were picked up from substrate. Figure 6 shows fabricated device wafer (Fig. 6a), the close-up view of single chip (Fig. 6b) and the curved beam (Fig. 6c). The proof mass has a size of 12 × 10 mm^2. The curved beam has a width of 30 μm and an initial deflection of 150 μm.

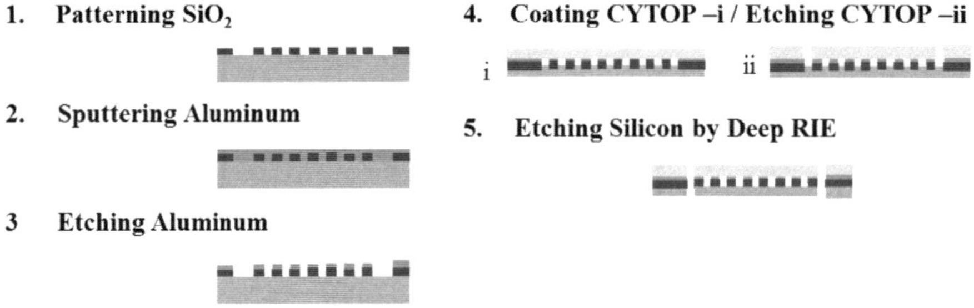

Figure 5. Processing flow of electret parts for the harvester

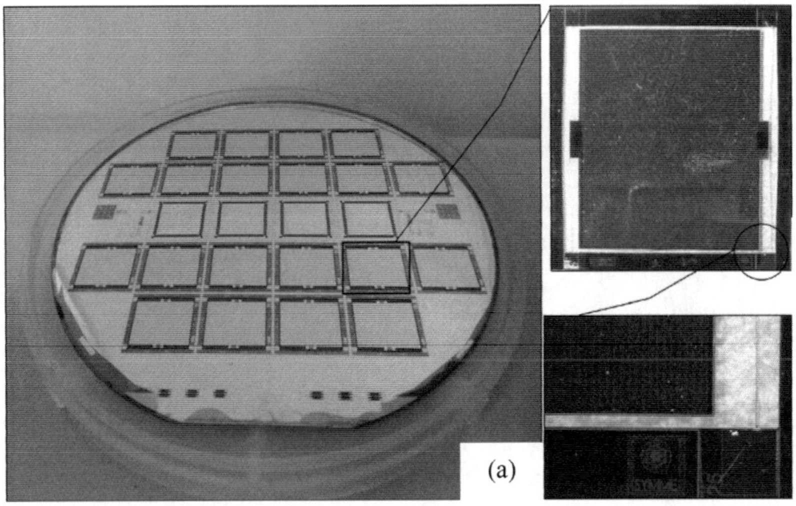

Figure 6. (a) Completed wafer after processing (b) The fabricated electret parts (c) Enlarged parts in the edge of the spring

5. Conclusion

We established the combination analysis method that can be effectively used to find the optimal geometry of a bi-stable spring. Samples for harvesting are being fabricated for experimentally verifying the proposed concepts.

Acknowledgement

This work was supported by JSPS (Japan Society for the Promotion of Science) KAKENHI Grant-in-Aid for Scientific Research(C) Number 16K06308.

References

[1] F. Shaikh et al., *Energy harvesting in wireless sensor networks*, Journal of Renewable and Sustainable Energy Reviews, 2016, Volume 55, pp. 1041-1054.

[2] N. Miwatani et al., *Design improvement for preventing discharge during fabrication of electrostatic energy harvester*. J. Physics, Conference Series, 2016, Volume 773, 012117.

[3] K. Yamamoto et al., *Vibration Energy Harvester with Bi-stable Curved Beam Spring Offset by Gravitational Acceleration*, Journal of Physics, Conference Series, Volume 660, 012127.

[4] W. Q Liu et al., *Novel piezoelectric bistable oscillator architecture for wideband vibration energy harvesting*, Journal of Smart Mater. Struct. 22, 2013, 035013

[5] B Vysotskyi et al, *Innovative Energy Harvester Design Using Bistable Mechanism With Compensational Springs In Gravity Field*, Journal of Physics, Conference Series, Volume 773, 012064

Probing thermal phonon mean free path using phononic crystal nanostructures

M Nomura[1,2], J Nakagawa[1], K Sawano[3], J Maire[1] and S Volz[1]

[1]Institute of Industrial Science, The University of Tokyo, Tokyo 153-8505, Japan
[2]PRESTO, Japan Science and Technology Agency, Saitama 332-0012, Japan
[3]Advanced Research Laboratories, Tokyo City University, Tokyo, 158-0082, Japan

E-mail: nomura@iis.u-tokyo.ac.jp

Abstract. We demonstrate an experimental method to obtain information of thermal phonon mean free path (MFP) spectra in thin films using hole based two dimensional phononic crystal nanostructures. The characteristic length of the system can be swept by changing the radius of circular holes of the phononic crystals. Then, thermal conductivities of single-crystalline Si, amorphous SiGe, and poly-SiGe thin films (150 nm thick) were measured for phononic crystals with different characteristic length between 20 nm and 200 nm. We observed different characteristic length dependence of the thermal conductivity among these three systems and the trend can be explained by the thermal phonon MFP spectra of the materials.

1. Introduction

Nanoscale heat transfer has been investigated because of its importance in both fundamental physics and other applications, such as thermoelectrics.[1] The thermal phonon MFP is an important physical parameter and determines the thermophysical property of a system. Some characteristic phonon transport phenomena have been reported in a variety of nanostructures.[2-9] Recent theoretical and experimental works have simulated or measured thermal conductivity accumulation distribution as a function of thermal phonon MFP.[10-14] It is known that the thermal phonon distributions span very broad frequencies and thus lead to broad thermal phonon MFP spectra. As thermal phonon MFP is directly connected to thermal conductivity, the impact of nanostructuring on thermal conductivity seems strongly dependent on the relation between the thermal phonon MFP spectrum and the characteristic length of the nanostructures. In this contribution, we report a method to obtain the information of thermal phonon MFP in solid by measuring in-plane thermal conductivity in 2D phononic crystal (PnC) nanostructures. In-plane thermal conductivity measurement was performed for a series of PnC nanostructures, which have different characteristic lengths, namely the shortest distance between two circular holes. The PnC nanostructures were fabricated in different materials; single-crystalline Si, amorphous and poly-SiGe, which have completely different MFP spectra. The observed different characteristic length dependence of the thermal conductivity among these three systems gives us some useful information on thermal phonon MFP spectrum in the thin film.

2. Sample and fabrication of phononic crystal nanostructures

Single-crystalline Si, amorphous SiGe, and poly-SiGe membranes were used for this study. For Si samples, we used a commercially available (100) nominally boron-doped silicon-on-insulator wafer

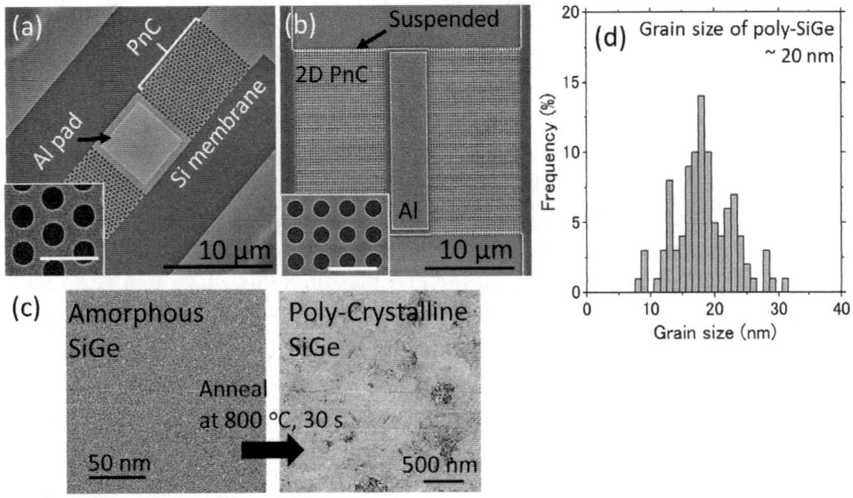

Figure 1. SEM images of the whole suspended structures: (a) hexagonal lattice Si PnC and (b) square lattice SiGe PnC structures. (c) TEM images of the surfaces of amorphous SiGe and poly-SiGe layers. (d) Histogram of the measured grain sizes of poly-SiGe on the surface.

with a 145 nm-thick upper Si layer and a 1 µm-thick buried SiO_2 layer. For SiGe samples, an amorphous $Si_{0.5}Ge_{0.5}$ layer with a thickness of 150 nm was deposited on the SiO_2/Si(001) substrate using solid-source molecular-beam-epitaxy at room temperature. A poly-SiGe sample was fabricated by annealing the amorphous SiGe sample at 800°C for 30 s in a N_2 atmosphere. Figures 1(a) and 1(b) shows scanning electron microscopy (SEM) images of the whole suspended Si and SiGe structures with two-dimensional PnC structures. An Al pad in the center enables microscale thermoreflectance (TDTR) measurements. The PnC structures were formed using circular holes aligned in periodic square or hexagonal lattice formations, as shown as the inset in Figs. 1(a) and 1(b). The PnC structures were fabricated with a fixed period, a = 300 nm, and a variety of radii. Figure 1(c) shows surface images, taken by transmission electron microscopy (TEM), of the amorphous and poly-SiGe membranes. The polycrystallization process formed crystal grains, which are expected to have changed the thermal phonon MFP spectrum. We analysed the TEM image to obtain a histogram of the grain size of the poly-SiGe [Fig. 1(d)]. It was found that the grain size is distributed between 10 and 30 nm in diameter. PnC nanostructures were fabricated with either single-crystalline Si, amorphous SiGe, or poly-SiGe thin films to change the characteristic length of the system. We can obtain some useful information about thermal phonon MFP spectra by measuring the sensitivity of thermal conductivity to the characteristic length. For example,

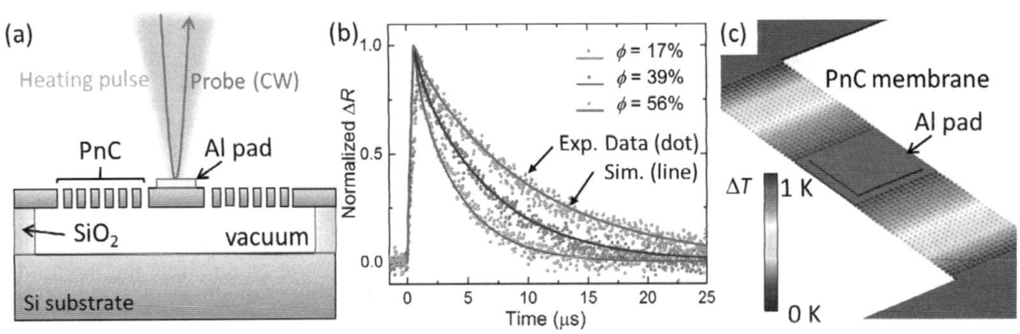

Figure 2. (a) Schematic of the µ-TDTR technique and cross section of the structure. (b) Measured (dot) and simulated (line) TDTR signal for a square-lattice PnC structure with various porosities. (c) Simulated temperature distribution 15 µs after heating for a SiGe PnC structure with r = 100 nm.

most of the thermal phonon in Si at room temperature have MFPs between 100 nm and 10 μm, resulting in Si showing a high sensitivity to PnC nanostructures with a characteristic size of around 100 nm.[15] The characteristic length was swept between 20 nm and 220 nm by changing the radius of the circular holes of PnC nanostructures. Here, the periodicity of the holes does not affect the thermal conductivity through the band-folding effect,[16] because phonon coherence is mostly lost at room temperature.

3. Simulation and experiment

To measure the in-plane thermal conductivities of the PnC nanostructures, a method using micro TDTR was developed. The Al pad was heated by a pulsed laser beam (wavelength λ = 642 nm) for 500 ns and the temporal evolution of the temperature of the Al pad was monitored using a continuous-wave laser beam (λ = 785 nm) as shown in Fig. 2(a). All the measurements were performed at room temperature. This well-defined system made it possible to obtain a highly reliable value for the thermal conductivity by analysing the temporal evolution of the TDTR signal. Figure 2(b) shows the measured TDTR signal (dots) and a simulated curve (solid line). The temperature evolution in the whole structure was simulated via the finite element method using exactly the same structural parameters as the experiment. Figure 2(c) shows the simulated temperature distribution, 15 μs after pulse heating, for a PnC structure with r = 100 nm. The value of thermal conductivity was obtained via the least squares method.

The thermal phonon MFP spectrum was investigated for the three thin films by comparing the thermal conductivities. The membranes have different thermal phonon MFP spectra owing to differences in their structures at atomic and nanometer scales. Therefore, the characteristic length dependence of thermal conductivity is also expected to be different owing to the different MFP spectra. The neck size is the characteristic length of the PnC nanostructure and is defined as the distance between two adjacent holes. Figure 3 shows the neck size dependences of the thermal conductivities of PnC nanostructures fabricated in the thin films. The thermal conductivities (dots) for the square-lattice PnC (κ_s) and hexagonal-lattice PnC (κ_h) nanostructures are plotted for a variety of radii. The orange and blue lines are the polynomial fits to the measured data to calculate the difference in thermal conductivity $D(\Phi)$, as defined by:

$$D(\Phi) = \frac{\kappa_s(\Phi) - \kappa_h(\Phi)}{[\kappa_s(\Phi) + \kappa_h(\Phi)]/2}. \qquad (1)$$

The PnC patterning reduces the thermal conductivity from the unpatterned membrane value (neck = 300 nm). We note that the thermal conductivity itself is reduced as a result of surface scattering at the sidewalls of the holes. For all the materials, the hexagonal-lattice PnC nanostructures showed lower thermal conductivity than the square-lattice PnC nanostructures. This indicates that the placement of the holes influences heat transfer with ballistic phonon transport properties. The reductions from the

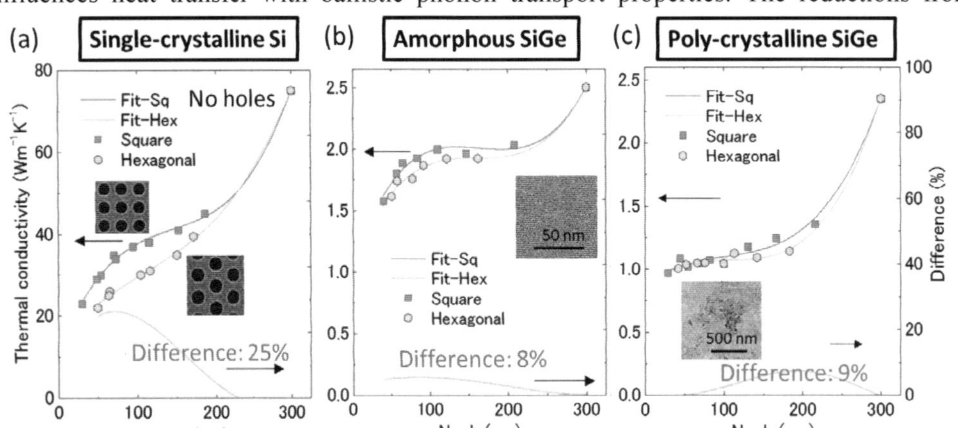

Figure 3. Measured in-plane thermal conductivity for PnC nanostructures with a variety of neck size at room temperature: (a) single-crystalline Si, (b) amorphous SiGe, and (c) poly-SiGe. The difference $D(\Phi)$ is shown by green lines (right vertical axis) at the bottom of figures.

membrane values are about 20% and 50% at the neck size of 200 nm for the amorphous and poly-SiGe, respectively. It is considered that the poly-SiGe has thermal phonons are distributed in very short (< 20 nm) and long (> 200 nm) MFP ranges. In amorphous SiGe, thermal conductivity sharply drops below the neck size of 70 nm. This can be explained by the fact that amorphous SiGe has a large thermal phonon MFP distribution in the range of tens of nanometers. In poly-SiGe, relatively weak dependence is observed between neck sizes of 20 nm and 200 nm, as shown in Fig. 3(c). It is considered that the transport of phonons with MFPs around tens of nanometers is impeded in poly-SiGe, owing to strong grain boundary scattering, which results in a poor contribution of these phonons to thermal conduction. This agrees with the grain size distribution shown in Fig. 1(d). This result suggests that nanostructures with dimensions comparable to the grain size do not reduce thermal conductivity effectively. Therefore, the characteristic length of the nanostructures should be chosen to avoid the overlap with other characteristic length. This series of experimental results indicates that the impact of nanostructuring on the thermal conductivity change is qualitatively explained by thermal phonon MFP spectra. Although the discussion is qualitative, the reported characteristic length dependences of the thermal conductivity with different MFP spectra may be useful to determine some parameters in the simulations for nanoscale heat transport.

4. Summary

The relation between thermal phonon MFP spectra and the impact of nanostructuring on thermal conductivity using Si, amorphous SiGe, and poly-SiGe membranes have been investigated. The experimentally obtained characteristic length dependences of thermal conductivity were qualitatively explained by thermal phonon MFP spectra. This thermal conductivity measurement, sweeping the characteristic length, can be used as thermal phonon MFP spectrum proving and gives useful information for designing nanostructures for highly efficient thermoelectric materials.

References

[1] Shi L 2012 *Eng. Nanoscale Microscale Thermophys.* **16**, 79
[2] Boukai A.I, Bunimovich Y, Tahir-Kheli J, Yu J.-K, Goddard A W, and Heath J R 2008 *Nature* **451** 168
[3] Maire J and Nomura M 2014 *Japan. J. Appl. Phys.* 53 06JE09
[4] Nakamura Y, Isogawa M, Ueda T, Yamasaka S, Matsui H, Kikkawa J, Ikeuchi S, Oyake T, Hori T, Shiomi J, and Sakai A 2015 *Nano Energy* **12**, 845
[5] Nomura M, Kage Y, Nakagawa J, Hori T, Maire J, Shiomi J, Anufriev R, Moser D, and Paul O 2015 *Phys. Rev.* B **91**, 205422
[6] Nomura M, Kage Y, Müller D, Moser D, and Paul O 2015 *Appl. Phys. Lett.* **106**, 223106
[7] Anufriev R, Maire J, Nomura M 2016 *Phys. Rev.* B 93, 045411
[8] Maire J, Anufriev R, Ramiere A, Yanagisawa R, Volz S, and Nomura M 2017 *Science Advances* 3, e1700027
[9] Anufriev R, Ramiere A, Maire J, and Nomura M 2017 *Nat. Commun.* 8, 15505
[10] Esfarjani K, Chen G, and Stokes H 2011 *Phys. Rev.* B **84**, 085204
[11] Yang F and Dames C 2013 *Phys. Rev.* B **87**, 035437
[12] Hori T, Shiomi J, and Dames C 2015 *Appl. Phys. Lett.* **106**, 171901
[13] Regner K.T, Sellan D.P, Su Z, Amon C.H, McGaughey A.J.H, and a Malen J 2013 *Nat. Commun.* **4**, 1640
[14] Minnich A, Johnson J, Schmidt A, Esfarjani K, Dresselhaus M, Nelson K, and Chen G 2011 *Phys. Rev. Lett.* **107**, 095901
[15] Nakagawa J, Kage Y, Hori T, Shiomi J, and Nomura M 2015 *Appl. Phys. Lett.* **107**, 023104
[16] Anufriev R and Nomura M 2015 *Phys. Rev.* B **91**, 245417

Acknowledgments
This work was supported by the Project for Developing Innovation Systems of the MEXT, Japan, Kakenhi (15H05869 and 15K13270), and the special fund of Institute of Industrial Science, The University of Tokyo.

Thermoelectric Properties of Size-Controlled Si and Metal Silicides Nanocomposites

S. Tanusilp[1], A. Yusufu[1, 2], and K. Kurosaki[1,3]

[1] Osaka University, 2-1 Yamadaoka, Suita, Osaka 565-0871, Japan
[2] University of Fukui, 1-2-4 Kanawa-cho, Tsuruga, Fukui 914-0055, Japan
[3] JST PRESTO, 4-1-8 Honcho, Kawaguchi, Saitama 332-0012, Japan

kurosaki@see.eng.osaka-u.ac.jp

Abstract. This paper describes our recent achievements of enhancement of the thermoelectric (TE) figure of merit *ZT* in nanocomposites of Si and silicides of various metals (V, Ni, Ti, and Co). We proposed a method of melt spinning followed by spark plasma sintering for synthesis of the nanocomposites. Effect of the size of the structure on the TE properties of the nanocomposites is discussed.

1. Introduction

The efficiency of thermoelectric (TE) devices is determined by the temperature gradient across the devices as well as the properties of TE materials called dimensionless figure of merit $ZT = S^2T\rho^{-1}\kappa^{-1}$ where S is the Seebeck coefficient, T is the absolute temperature, ρ is the electrical resistivity, and κ is the total thermal conductivity ($\kappa = \kappa_{lat} + \kappa_{el}$, where κ_{lat} and κ_{el} are the lattice and electronic contributions, respectively) [1]. Well-known TE materials such as Bi_2Te_3 [2] and PbTe [3] exhibit the *ZT* over unity, while the practical use of these materials is still limited as they consist of highly toxic and rare elements. Si-based TE materials provide desirable merits for the commercial utilization and mass production. Si exhibits good electrical properties (high $S^2\rho^{-1}$), but high κ_{lat}, resulting in a low *ZT* (~0.02 in maximum at room temperature) [4]. Nanostructuring has emerged as promising approach to decrease κ_{lat} and thus to enhance *ZT* of Si-based TE materials. The enhanced *ZT* of Si has been demonstrated in various nanostructures such as nanowires [5].

Recently, it has been revealed that nanocomposite structure reduces κ_{lat} without significant influences on the electronic properties. To create such structure, we have proposed a method combining melt spinning (MS) and spark plasma sintering (SPS) [6]. In the MS procedure, mixtures of Si and various metals (V [7], Ni [8], and so on) with an appropriate ratio are melted and rapidly cooled, then ribbon-state nanocomposites composed of Si and metal silicides are obtained. To create bulk samples from the ribbons, the ribbons are sintered by SPS with keeping the nanostructures. Here, we review the recent results on the microstructure characterization as well as the TE properties of such nanocomposites.

2. Experimental Procedures

The Si-VSi_2 and Si-$NiSi_2$ nanocomposites with the chemical compositions of $(Si_{100}P_3)_{97}V_3$ and $(Si_{100}P_3)_{98}Ni_2$ were synthesized from chunks of Si (11N), P (6N), and chunks of V (4N) or Ni (3N), where P acts as an electron dopant. The mixtures were placed into boron nitride nozzles and melted via induction heating in a flowing Ar atmosphere, and the molten alloys were rapidly solidified on a Cu roller. The rotation speed of the Cu roller with 30 cm diameter was set to be 2000, 4000, and 6000 rpm. The melt-spun ribbons obtained at each rotation speed were noted as MS (2000), MS (4000), and MS (6000). In order to sysnthesize the bulk samples from the MS ribbons, the SPS was carried out at 1473 K for 3 min for $(Si_{100}P_3)_{97}V_3$ and 1273 K for 4 min for $(Si_{100}P_3)_{98}Ni_2$. The sintered bulk samples prepared from the MS ribbons with each rotation speed are denoted as MS (2000)-SPS, MS (4000)-SPS, and MS (6000)-SPS.

The phase state and microstructure of the MS ribbons and the MS-SPS bulk samples were characterized by powder x-ray diffraction analysis (XRD; Ultima IV, Rigaku Co, Cu Kα) and scanning electron microscopy (FE-SEM, JSM-6500F, JEOL) at room temperature. The TE properties of the MS-SPS samples for $(Si_{100}P_3)_{97}V_3$ and $(Si_{100}P_3)_{98}Ni_2$ were measured from 300-1073 K. S and ρ were measured simultaneously using a commercial apparatus (ZEM-3, Ulvac). κ was evaluated from the relationshiop $\kappa = \alpha C_P d$, where α, C_P, and d are the thermal diffusivity, heat capacity, and density, respectively. α was measured using a laser flash apparatus (LFA-457, Netzsch). C_P was evaluated from those of pure Si and metal silicides based on the Kopp-Neumann rule. d was calculated from the weight and dimensions of the bulk samples. The Hall carrier concentration (n_H) and Hall mobility (μ_H) were measured using a Hall measurement system (Resitest8300, Toyo) at room temperature by the van der Pauw geometry in vacuum under applied magnetic field of 0.5 T.

3. Results and Disccussion

As shown in Fig. 1, the powder XRD patterns of all samples exhibit very sharp peaks of Si as the main phase. In the $(Si_{100}P_3)_{97}V_3$ system (Fig. 1 (a)), small peaks of VSi_2 [9] can be seen in both of the MS and MS-SPS samples. On the other hand, since $NiSi_2$ has very similar XRD patterns to Si [9] as they have the small lattice mismatch, only peaks for Si can be seen in the $(Si_{100}P_3)_{97}Ni_2$ system (Fig. 1(b)). The XRD patterns of the MS and MS-SPS samples are very similar to each other, meaning that the phase states do not change during SPS.

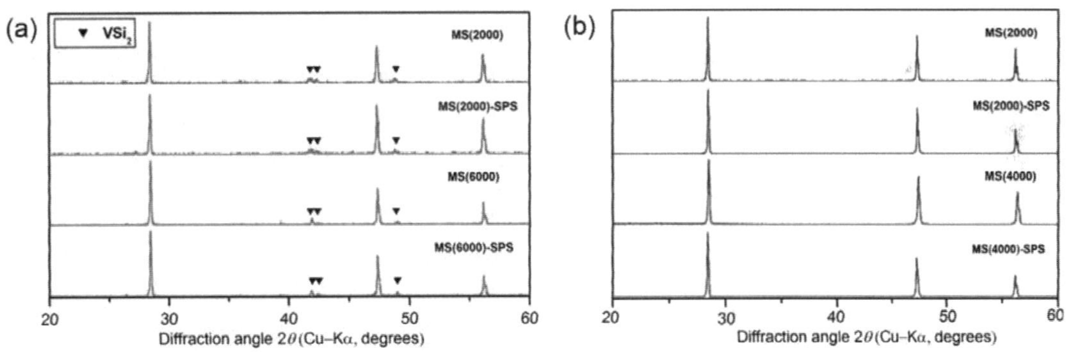

Figure 1. Powder XRD patterns of $(Si_{100}P_3)_{97}V_3$ and $(Si_{100}P_3)_{98}Ni_2$ synthesized by MS and MS followed by SPS.

In Fig. 2, the presence of two phases can be observed clearly, where white dots and dark area correspond to the metal silicide precipitates and Si matrix, respectively. In both the $(Si_{100}P_3)_{97}V_3$ and $(Si_{100}P_3)_{98}Ni_2$ systems, the nanocomposites containing a few dozen nm size precipitates were obtained in the MS ribbons, while the size increased to several μm size after SPS. The VSi_2 or $NiSi_2$ precipitates exist throughout the Si matrix and their size decreases with increasing the Cu roller speed. The estimated average sizes of the VSi_2 precipitates are 5 ± 2 μm, 2 ± 2 μm, and 0.5 ± 0.5 μm for MS (2000)-SPS, MS (4000)-SPS, and MS (6000)-SPS, respectively. On the other hand, the estimated average sizes of the $NiSi_2$ precipitates are 8 ± 2 μm and 1 ± 0.5 μm for the MS (2000)-SPS and MS (4000)-SPS, respectively.

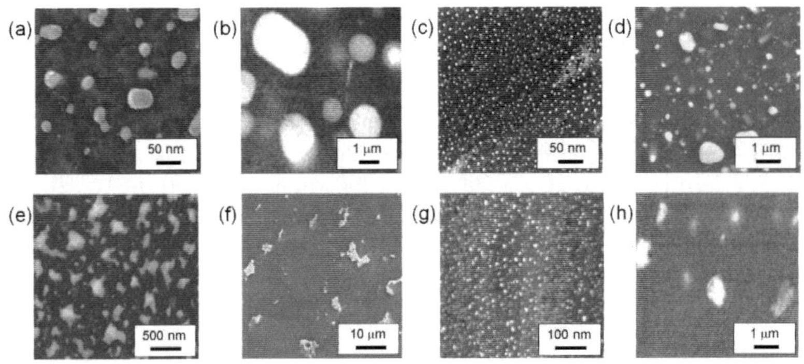

Figure 2. SEM images of $(Si_{100}P_3)_{97}V_3$. (a) MS (2000), (b) MS (2000)-SPS, (c) MS (6000), and (d) MS (6000)-SPS. SEM images of $(Si_{100}P_3)_{98}Ni_2$. (e) MS (2000), (f) MS (2000)-SPS, (g) MS (4000), and (h) MS (4000)-SPS.

As can be seen in Fig. 3, there are no clear trends between μ_H and size of the precipitates and matrix grains, while as the size of the precipitates and matrix grains decrease, the κ_{lat} decreases. As the results, the ratio of μ_H/κ_{lat} which is the most important parameter to determine the *ZT* increases with decreasing the size of both precipitates and Si matrix grains (Figs. 3(e) and 3(f)).

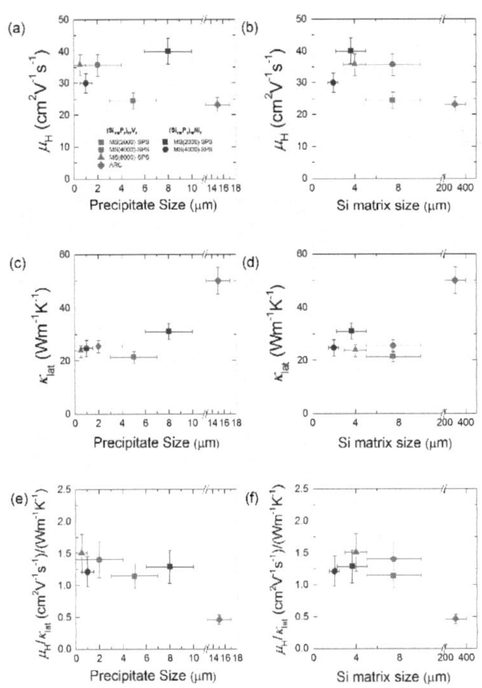

Figure 3. Effects of size of the silicide precipitates (VSi_2 and $NiSi_2$) and Si matrix grains on room-temperature values of the Hall mobility μ_H, lattice thermal conductivity κ_{lat}, and μ_H/κ_{lat} of MS-SPS bulk samples of $(Si_{100}P_3)_{97}V_3$ and $(Si_{100}P_3)_{98}Ni_2$: (a) μ_H vs. size of the metal silicide precipitates, (b) μ_H vs. size of the Si matrix grains, (c) κ_{lat} vs. size of the metal silicide precipitates, (d) κ_{lat} vs. size of the Si matrix grains, (e) μ_H/κ_{lat} vs. size of the metal silicide precipitates, and (f) μ_H/κ_{lat} vs. size of the Si matrix grains.

Temperature dependences of *S*, ρ, κ_{lat}, and *ZT* of the MS-SPS bulk samples of $(Si_{100}P_3)_{97}V_3$ [7] and $(Si_{100}P_3)_{98}Ni_2$ [8] are shown in Fig. 4. In these figures, the data for the samples with the same chemical compositions but synthesized by a traditional arc melting are shown for comparison. These samples are denoted as ARC. In both $(Si_{100}P_3)_{97}V_3$ and $(Si_{100}P_3)_{98}Ni_2$, the *S* values of the MS-SPS samples are almost the same with that of ARC. On the other hand, the ρ values of the MS-SPS $(Si_{100}P_3)_{97}V_3$ samples are slightly higher but those of the MS-SPS $(Si_{100}P_3)_{97}Ni_2$ samples are clearly lower than those of the ARC samples. In the $(Si_{100}P_3)_{97}V_3$ system, small size of the microstructure leads to increase in the carrier-grain boundary scattering, while in the $(Si_{100}P_3)_{97}Ni_2$ system, the $NiSi_2$ precipitates would have little influences on the carrier-grain boundary scattering due to the small lattice mismatch between Si and $NiSi_2$. As shown in Fig. 4 (c), the κ_{lat} of all MS-SPS samples are significantly lower than those of the

ARC samples, where the metal silicide precipitates scatter heat carrying phonons effectively. In both systems, the MS-SPS samples show higher ZT values than those of ARC samples. The best ZT value is 0.23 at 1073 K obtained for the $(Si_{100}P_3)_{98}Ni_2$ MS (2000)-SPS sample.

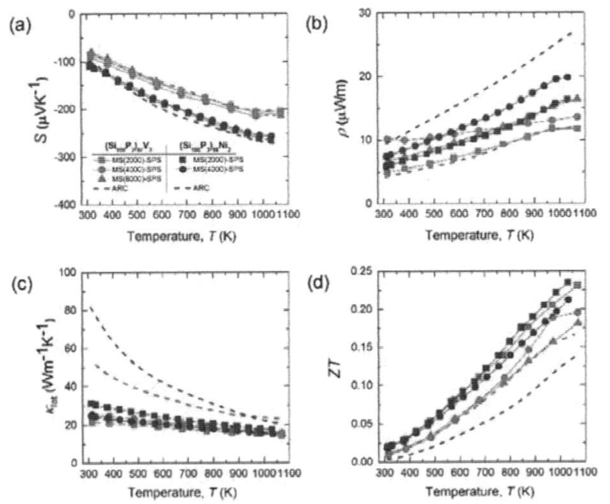

Figure 4. Temperature dependences of (a) Seebeck coefficient S, (b) electrical resistivity ρ, (c) lattice thermal conductivity κ_{lat}, and (d) dimensionless figure of merit ZT of MS-SPS bulk samples of $(Si_{100}P_3)_{97}V_3$ [7] and $(Si_{100}P_3)_{98}Ni_2$ [8]. Each TE property was repeatly measured a few times at each measurement temperature; the deviations for S, ρ, and α were estimated to be 2%, 3%, and 3%, respectively, resulting in approximately 10% deviation for ZT.

4. Conclusion

The Si-VSi_2 and Si-$NiSi_2$ nanocomposites with the chemical compositions of $(Si_{100}P_3)_{97}Ni_3$ and $(Si_{100}P_3)_{98}Ni_2$ were successfully synthesized by MS followed by SPS (MS-SPS method), where VSi_2 and $NiSi_2$ disperse homogeneously throughout Si matrix. The size of the metal silicide precipitates and Si matrix grains is much smaller than those of the ingot samples prepared by a conventional arc melting method and decreases with increasing the rotation speed of the Cu roller in the MS process. The MS-SPS samples exhibit significantly reduced κ_{lat} with keeping high μ_H. In both $(Si_{100}P_3)_{97}Ni_3$ and $(Si_{100}P_3)_{98}Ni_2$, the MS-SPS samples show higher ZT values than those of the ingot samples. The MS-SPS method has a great advantageous for suppressing κ_{lat} with keeping high μ_H, resulting in the enhancement of ZT.

References

[1] G. J. Snyder and E. S. Toberer, *Nat. Mater.* **7** (2008), 105-114.
[2] Y. Lan, A. J. Minnich, G. Chen, and Z. Ren, *Adv. Func. Mater.* **20** (2010), 357-376.
[3] J. P. Heremans, V. Jovovic, E. Toberer, A. Saramat, K. Kurosaki, A. Charoenphakdee, S. Yamanaka, and G. J. Snyder, *Science* **321** (2008), 554-557.
[4] G. H. Zhu, H. Lee, Y. Lan, X. Wang, G. Joshi, D. Wang, J. Yang, D. Vashaee, H. Guilbert, A. Pillitteri, M. Dresselhaus, G. Chen, and Z. Ren, *Phys. Rev. Lett.* **102** (2009), 196803.
[5] A. I. Hochbaum, R. Chen, R. D. Delgado, W. Liang, E. C. Garnett, M. Najarian, A. Majumdar, and P. Yang, *Nature* **451** (2008), 163-167.
[6] K. Kurosaki, A. Yusufu, Y. Miyazaki, Y. Ohishi, H. Muta, and S. Yamanaka, *Mater. Trans.* **57** (2016), 1018-1021.
[7] S. Tanusilp, K. Kurosaki, A. Yusufu, Y. Ohishi, H. Muta, and S. Yamanaka, *J. Electron. Mater.* **46** (2017), 3249-3255.
[8] P. Eaksuwanchai, K. Kurosaki, S. Tanusilp, Y. Ohishi, H. Muta, and S. Yamanaka, *J. Appl. Phys.* **121** (2017), 225110-1-225110-5.
[9] JCPDS Card Nos. 00-005-0565 (Si), 01-072-6183 (VSi_2), and 00-043-0989 ($NiSi_2$) (XRD).

Acknowledgement
This work was supported in part by JST, PRESTO Grant Number JPMJPR15R1.

IOP Conf. Series: Journal of Physics: Conf. Series **1052** (2018) 012125 doi:10.1088/1742-6596/1052/1/012125

Field-effect and chemical charge-type modulations of carbon nanotubes using functional polymers for thermoelectric energy harvesters

S Horike[1], T Fukushima[1], T Saito[2], Y Koshiba[1], M Morimoto[3] and K Ishida[1]

[1] Department of Chemical Science and Engineering, Kobe University, 1-1 Rokkocai-cho, Kobe 657-8501, Japan
[2] Nanomaterials Research Institute, National Intsitute of Afvanced Science and Technology, 1-1 Higashi, Tsukuba 305-8565, Japan
[3] Graduate School of Science and Engineering for Research, University of Toyama, 3190 Gofuku, Toyama 930-8555, Japan

horike@crystal.kobe-u.ac.jp

Abstract. As devices continue to shrink in size and grow in abundance, power supply methods using clean and abundant materials become increasingly important. Single-walled carbon nanotubes (SWCNTs) are promising materials as the thermoelectric (TE) energy-harvesting candidate due to their flexibility, nontoxicity, and relatively large TE power factors. This paper reviews our current studies on the charge-type conversion of SWCNTs for readily pairing several p- and n-type SWCNTs to improve the TE voltage output. First, we describe the polarity tuning of SWCNTs by electric dipoles of ferroelectric polymer. The sign of Seebeck coefficient of SWCNTs could be tuned by the direction of the electric dipoles adjacent to nanotubes in the ferroelectric-gate field-effect transistor-like device configuration. Second, we demonstrate the electron doping into SWCNTs by several versatile polymer-dopant charge transfer. The obtained negative Seebeck coefficients of SWCNTs were correlated to the highest occupied molecular orbital energy levels of the dopant polymers. Our findings enables the direct use of low-temperature waste heats around dwelling environments using eco-friendly materials and techniques.

1. Introduction

Seamlessly sensing every object around us and connecting their information through the Web systems offer us comfortable daily lives with accuracy, efficiency, and economic benefits. On the monitoring, vast of wearable sensing devices embedded in the objects will play a crucial role, connecting our physical world to the networks; the supply of electrical power to these sensors is a recent important problem. For automatic operations of these sensors over an extended period of time without any additional maintenance, energy harvesting technologies have emerged as the promising candidates. Of these, thermoelectric (TE) conversion devices, which generate electrical power from temperature differences, will provide a good solution because heat flows is quite common in dwelling environments.

Single-walled carbon nanotubes are considered the main candidate for this application due to their relatively large Seebeck coefficient, electrical conductivity, flexibility, nontoxicity, and printability onto plastics, papers, and textiles, efficiently harvesting the most ubiquitous energy source available to us: human body heats. In order for this technology to become truly pervasive, we should develop the way to generate sufficient power from the small temperature differences. Typically, any TE device should be modularized so that it contains many p- and n-type TE legs that are connected electrically in series and thermally in parallel, because this device architecture enables to directly add the TE voltages of each leg that is subject to the same temperature difference. Both the p- and n-type materials are therefore needed for developing the practical TE devices. However, SWCNTs generally show a p-type nature in ambient air conditions, as electrons are withdrawn by oxygen impurities; air-stable n-type conversion of SWCNTs is a current challenging issue for TE application. In this Article, we review our recent studies on the charge-type conversion of SWCNTs using field-effect-induced charge-carrier accumulation by electric dipoles of a ferroelectric polymer [1] and a charge-transfer interaction between nanotubes and versatile polymer dopants.[2]

2. Polarity tuning of SWCNTs by ferroelectric dipole

Conventional techniques of charge-carrier injection into the SWCNTs include (1) the lattice substituent of carbon networks by hetero atoms, (2) the charge-transfer interaction between nanotubes and dopants, and (3) the charge-carrier accumulation into the nanotube in the field-effect transistor (FET) configuration.[3–5] The lattice substituent enables the stable electron injection into nanotubes; however, the introduction of hetero atoms destroys the expanded sp^2 framework of nanotubes and causes the significant decrease in the charge-carrier mobility. Charge carriers could be conveniently injected via charge-transfer interaction between nanotubes and dopants physically adsorbed on the nanotube surfaces. However, the n-type SWCNTs are generally unstable in air due to autoxidation. The development of air-stable n-type SWCNTs via polymer-dopant charge transfer will be described in the next section.

In this section, we demonstrate the polarity (charge-type) tuning of the SWCNTs using (3) the charge-carrier accumulation in the FET configuration.[1] Several authors have reported an ambipolar charge transporting properties of SWCNTs-based FETs; SWCNTs could intrinsically accumulate and transport both holes and electrons.[5] Therefore, using field-effect-induced charge carrier modulation should be an effective way to tuning their polarities. However, this charge-type conversion, in principle, need a gate voltage in order to continuously accumulate the carriers. To this problem, we used a ferroelectric polymer (P(VDF/TrFE)) and applied the ferroelectric-gate FET (FeFET)-like device configuration as shown in figure 1 (a) and (b). This polymer holds a relatively large electric dipole field perpendicular to its carbon main chain. The dipole can be reversed and retained by the electric field; one can expect that the charge-carrier accumulation can be maintained once it is applied to a poling treatment.

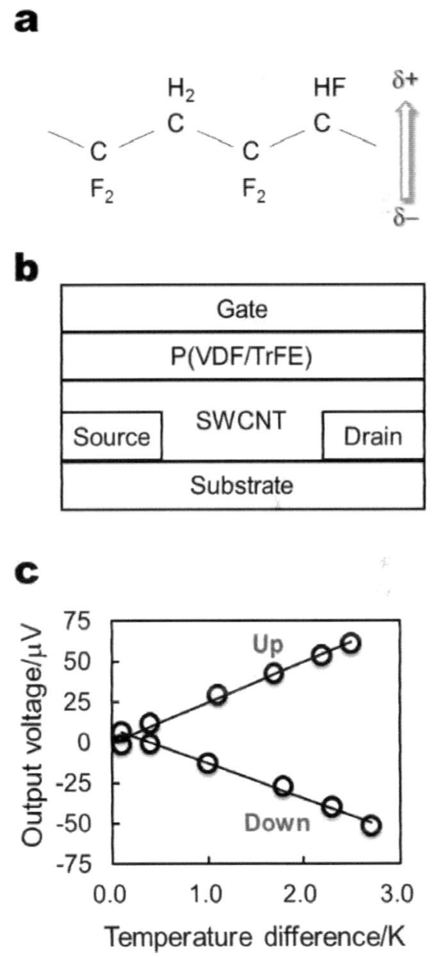

Figure 1. (a) Chemical structure of P(VDF/TrFE). (b) Device configuration. (c) Seebeck measurements under dipole.

The device fabrication procedure is described in our current paper. The charge-carrier type was evaluated by the TE charge-carrier determination after the poling treatments. Figure 1 (c) graphically shows the Seebeck coefficients of SWCNT thin films under ferroelectric dipoles. The sign of the coefficient could be tuned by the direction of the dipoles; positive and negative coefficients were observed in the "up" and "down" states, respectively. It was evident that the majority charge carrier changed according to the direction of the dipoles, as positive and negative coefficients directly reflect the hole and electron majority charge-carrier types, respectively. The Seebeck coefficients under the dipoles (±25 μV K^{-1}) decreased by half compared to that of a control SWCNT thin film (+50 μV K^{-1}). On the other hand, the electrical conductivities showed two-fold increases from 0.5 to 1.0 S cm^{-1} after the poling. These changes of Seebeck coefficient and electrical conductivity are straightforward according to the conventional theory in which these parameters are complementary regarding the carrier density. The carrier accumulation induced by the poling improved the carrier density of SWCNTs, thus increasing the conductivity but decreasing the Seebeck coefficients.

In general, TE power factor is defined as $P = S^2\sigma$ (S: Seebeck coefficient, σ: electrical conductivity). The power factors of SWCNT thin films under dipoles were about 0.05 μW K^{-2} m^{-1} in both the directions of dipoles. This value is about a half of that of control film (0.13 μW K^{-2} m^{-1}). However, it should be noted that the polarity-controlled SWCNTs can be connected as TE modules. To demonstrate this, we fabricated the TE module consisting the p- and n-type SWCNTs. The TE output voltage of 50 μV K^{-1} was obtained from one module; the TE voltages of the polarity controlled SWCNTs could be combined to improve the overall TE voltage.

3. N-type conversion of SWCNTs by doping of vinyl polymers

Next, we demonstrate the charge-transfer-induced electron injection into the SWCNTs via versatile polymer doping. [2] As mentioned above, charge transfer interaction can conveniently inject electrons into nanotubes. In early case, amine compounds such as poly(ethyleneimine) (PEI) were demonstrated as successful molecular dopants behaving as Lewis bases, where a lone pair (lp) of electrons on nitrogen was partially injected into conduction band of nanotubes via charge transfer interaction. However, it is known that the electron transport behaviour and negative Seebeck coefficient of these amine-doped SWCNTs gradually reverts into p-type upon exposure to air. It is when a new class of molecular dopants with versatility, stability, biocompatibility, and low-cost must be explored.

Herein, we describe our recent findings on the simple, cheap, non-toxic, chemically stable molecular dopants for nanotubes. It is reasonable to consider that the atom, which participates in the highest occupied molecular orbital (HOMO) and has high electron density, mainly contributes to the charge transfer interaction. In the PEI case, the nitrogen corresponds to such a kind of atom because (1) the lp on the nitrogen is not involved in the covalent bonding and thus should participate in the HOMO, and (2) the nitrogen has a relatively high electron density due to its large electronegativity. On the basis of these consideration, we selected poly(vinyl alcohol) (PVA) and poly(vinyl acetate) (PVAc) as the eco-friendly, cheap, and potential electron donors. The oxygen atoms of these polymers have an abundant lps and are negatively charged, thus providing a possibility to effectively donate electrons to nanotubes. To compare the electron donating ability, we also used poly(vinyl chloride) (PVC) and poly(vinyl pyrrolidone) (PVP) which have the identical carbon main chain and the different functional side groups.

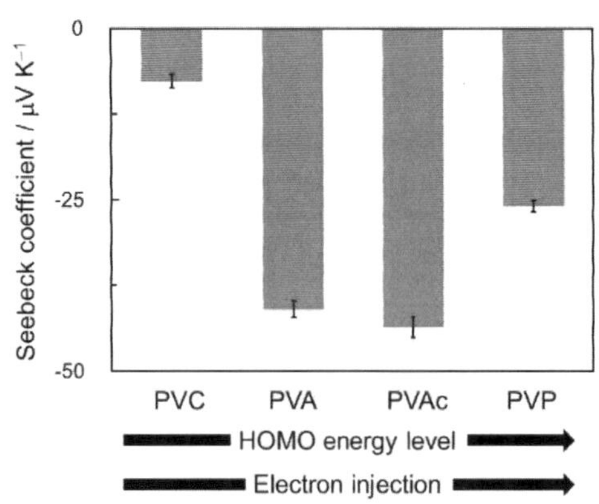

Figure 2. Seebeck coefficients of polymer-doped SWCNTs at 300 K.

The experimental details are described in our current paper. After depositing SWCNT thin films on the substrates, the polymers were doped by spin coating or drop casting directly onto the SWCNT thin films.

Figure 2 shows the Seebeck coefficients and power factors of the control and the doped SWCNTs. The positive coefficient of control SWCNTs indicate the p-type nature due to the hole injection from the oxygen impurities. The doping of the polymers changed this coefficient into negative. Of these, the largest negative values were obtained by doping of PVA and PVAc, matching with that of control SWCNTs. These differences in the values are suitably explained by the HOMO energy levels of the dopant polymers (PVC: −7.9, PVA: −7.1, PVAc: −7.0, PVP: −6.3 eV, calculated by density functional theory method at B3LYP/6-31G* levels on the monomer units). The higher HOMO energy level provides the more electron injection. The electron donation of PVC is relatively weak, and the electrons were mainly devoted to cancel out the original holes introduced from oxygen impurities, changing the Seebeck coefficient close to zero. The doping of PVA and PVAc injected adequate amount of electrons, thus recovering the coefficients to the matching levels to the control. The greater amount of electrons were injected from PVP so that the coefficient took smaller value than those of PVA- and PVAc-doped SWCNTs owing to the complicated behaviour of Seebeck coefficient and charge carrier density. The PVA-doped SWCNTs showed the largest TE power factor among the tested materials.

The stabilities of the negative coefficients of the SWCNTs in ambient air condition were evaluated. The PVC-doped SWCNTs showed poor stability to readily revert into p-type after 1 day. Although PVP has been considered a relatively stable electron donor; however, the doped SWCNTs indicated significant decrease in the negative coefficient. On the other hand, the PVA- and PVAc-doped SWCNTs showed excellent stabilities in air exceeding at least 3 weeks. We roughly attribute these differences in the stability to oxygen impurity blockage abilities of the dopant polymers; the low oxygen transmission coefficient of the polymer disturbed the adsorption of additional oxygen impurities and the hole injections by autoxidation. The higher polarity tends to provide the smaller oxygen transmission coefficient in general; this property is unique aspect to the polymers with oxygen atoms.

4. Conclusions and future perspectives

We reviewed our current studies on the charge-type conversion of SWCNTs. The ferroelectric dipoles of P(VDF/TrFE) effectively tuned the polarity of SWCNTs. Hybrid energy devices harvesting temperature differences, temperature changes, and vibrations may be developed in the future, as ferroelectric materials intrinsically hold pyroelectric and piezoelectric properties. The doping of versatile vinyl polymers successfully turned the charge type species of SWCNTs from hole into electron. The obtained negative Seebeck coefficients strongly related to the HOMO energy levels of the polymers. The largest coefficients and the considerable stabilities were obtained using oxygen abundant polymers such as PVA and PVAc. We expect that our findings will contribute to the more effective molecular dopants.

References

[1] Horike S, Misaki M, Koshiba Y, Morimoto M, Saito T and Ishida K 2016 *Appl. Phys. Express* **9** 081301
[2] Horike S, Fukushima T, Saito T, Kuchimura T, Koshiba Y, Morimoto M and Ishida K *Mol. Syst. Des. Eng.* (in press)
[3] Czerw R, Terrones M, Charlier J.-C, Blase X, Foley B, Kamalakaran R, Grobert N, Terrones H, Tekleab D, Ajayan P. M, Blau W, Rühle M and Carroll D. L. 2001 *Nano Lett* **1** 457
[4] Kong J and Dai H 2001 *J. Phys. Chem. B* **105** 2890
[5] Lin Y. M, Appenzeller J and Avouris Ph 2004 *Nano Lett* **4** 947

IOP Conf. Series: Journal of Physics: Conf. Series **1052** (2018) 012126 doi:10.1088/1742-6596/1052/1/012126

Thermoelectric performances in transparent ZnO films including nanowires as phonon scatterers

Takafumi Ishibe, Atsuki Tomeda, Kentaro Watanabe, and Yoshiaki Nakamura

Osaka University, Machikaneyamacho 1-3, Toyonaka, Osaka, Japan

E-mail: nakamura@ee.es.osaka-u.ac.jp

Abstract. We report the fabrication technique and electrical propertis of embedded-ZnO nanowires (NWs) structure for high-performance transparent thermoelectric materials. In fabrication technique, we revealed that the length of NW is less than 2 μm for embedding NWs. Our embedded-ZnO NWs structures showed a little smaller electrical conductivities than ZnO films (without ZnO NWs) at the carrier concentration of ~ 10^{18} cm^{-3}. However, at high carrier concentration, our embedded-ZnO NWs structures are expected to show high thermoelectric performances due to the drastic reduction of κ and smaller σ degradation, because there is the difference of MFP between electron and phonon. Our work showed the design guide to enhance the performance of transparent thermoelectric materials.

1. Introduction

Thermoelectric applications require materials with high value of $ZT = S^2\sigma T/\kappa$ (S: seebeck coefficient, σ: electrical conductivity, T: absolute temperature, and κ: thermal conductivity). However, these parameters are correlated with each other and it has been difficult to control them independently. Nanostructuring has recently been found to be an effective way to decrease κ [1] or to increase $S^2\sigma$ [2].

ZnO is a semiconductor composed of ubiquitous elements with wide energy band gap (3.3 eV). Therefore, ZnO-based materials are expected as transparent thermoelectric materials due to high $S^2\sigma$, which harvests the wasted heat from transparent products such as hot glass windows in the buildings [3]. However, their ZT is quite low at room temperature compared with the value for practical application ($ZT \sim 1$), because the κ is high (~ 45 $Wm^{-1}K^{-1}$ at room temperature) [4]. Here, we propose an embedded-ZnO nanowires (NWs) structure for high ZT thermoelectric ZnO material [5]. The design concept is that crystal interfaces of NWs scatter phonons to reduce κ, and power factor can also be increased due to one-dimensional carrier conduction [6] or energy filtering effect at the interfaces [2] (Figure 1). In this paper, we develop formation of embedded-ZnO NWs structure that enhances thermoelectric properties due to scattering of phonon and low energy electron.

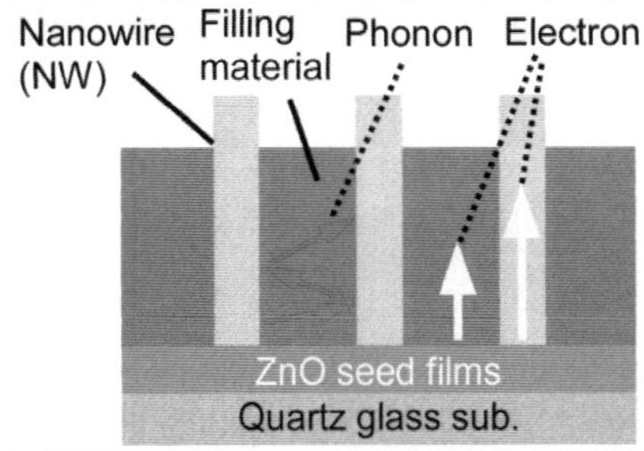

Figure 1. Design concept of embedded-ZnO nanowires (NWs) structure.

2. Sample preparation and measurements

ZnO films were formed on quartz glass substrates as a seed template at 400°C under an oxygen pressure of 1 Pa for 1 h by pulsed laser deposition (PLD) (Figure 2(a)). The targets were put 3 cm away from Si substrates and irradiated with 10 Hz pulsed laser (ArF excimer laser with 193 nm wavelength). Highly-oriented ZnO NWs were grown on the seed films under an Ar carrier gas of 20 sccm by physical vapor transport (Figure 2(b)). The temperature at the position of source material was increased to 780°C in 1 h and was maintained for 1 h. Finally, ZnO was deposited again by PLD over ZnO NWs to form embedded-ZnO NWs structure (Figure 2(c)).

The surface and cross-sectional morphologies of the samples were observed by scanning electron microscopy (SEM) with a 10 keV electron beam. Electrical properties of the samples were measured by van der Pauw method.

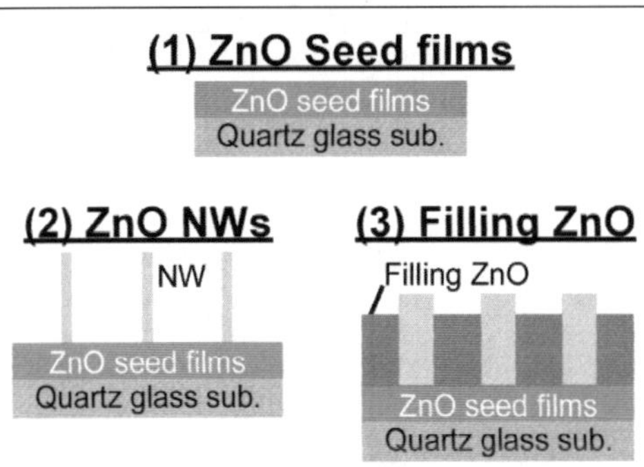

Figure 2. Fabrication process of embedded-ZnO NWs structure. (a) ZnO seed films, (b) ZnO NWs, (c) Filling ZnO.

3. Nanostructure formation and thermoelectric properties

Figures 3(a) and (b) show the cross-plane and plane-view scanning electron microscope images of highly-oriented ZnO NWs, respectively. We confirmed that NWs were formed with areal density of 4.4×10^9 cm^{-2} and the length of ~300 nm from these results. In the case of NWs with the length of > 2 μm, NWs were not embedded fully, because filling ZnO was trapped at the front edge of NWs. Therefore, it is important that NWs have the length less than 2 μm for forming embedded-ZnO NWs structure. After depositing ZnO for embedding, the structures of NWs were maintained as shown in Figure 3(c). Table 4 shows that σ of embedded-ZnO NWs structures were smaller than ZnO films (without ZnO NWs) at the carrier concentration of ~ 10^{18} cm^{-3}. The interfaces between NWs and films may scatter electrons, leading to reduction of σ. In general, thermoelectric materials show the best characteristics at high carrier concentration (~10^{20} cm^{-3}). At high carrier concentration, the NW interface scattering of electron is considered to be smaller due to shorter length of the mean free path (MFP) based on ionized impurity scattering process. Therefore, in the present case (low carrier concentration of ~10^{18} cm^{-3}), the NW interface scattering of electron may degrade σ, but its contribution become smaller at high carrier concentration, which indicates the difference of σ between embedded-ZnO NWs structures and ZnO films can become smaller.

On the other hands, the MFP of phonon is much longer than electron. Our previous work demonstrated ~200 times reduction of by introducing the nanoscale interfaces in Si films (Connected Si nanocrystals: CSN) [1]. Considering the present embedded-ZnO NWs structures, the NW structures can reduce κ, resulting in high thermoelectric performances.

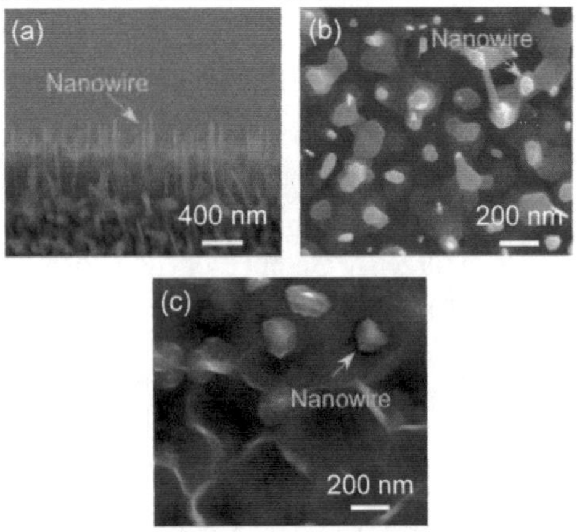

Figure 3 Scanning electron microscope images before ((a) cross-plane (b) plane-view) and after forming filling ZnO.((c) plane-view).

Table 1. Electrical conductivities of embedded-ZnO NWs structures (ZnO NWs) and ZnO films at the carrier concentration of ~10^{18} cm^{-3}

	ZnO NWs	ZnO films
σ ($\Omega^{-1}cm^{-1}$)	3.1	18.8

4. Conclusions

We developed embedded-ZnO NWs structures for high performance transparent thermoelectric materials. In fabrication technique, we revealed that the length of NW is less than 2 μm for embedding NWs. Our embedded-ZnO NWs structures showed a little smaller σ than ZnO films (without ZnO NWs) at the carrier concentration of ~ 10^{18} cm^{-3}. However, at high carrier concentration, our embedded-ZnO NWs structures can show high thermoelectric performances due to the drastic reduction of κ compared with the small reduction of σ, because there is the difference of MFP between electron and phonon.

References

[1] Y. Nakamura, M. Isogawa, T. Ueda, S. Yamasaka, H. Matsui, J. Kikkawa, S. Ikeuchi, T. Oyake, T. Hori, J. Shiomi, and A. Sakai, Nano Energy 12, (2015), pp. 845-851.
[2] S. V. Faleev and F. Leonard, Phys. Rev. B, 77 214304 (2008), pp. 1-9.
[3] J. Loureiro, N. Neves, R. Barros, T. Mateus, R. Santos, S. Filonovich, S. Reparaz, C. Torres, F. Wyczisk, L. Divay, R. Martinsa, and I. Ferreira, J. Mater. Chem. A, 2 (2014), pp. 6649-6655.
[4] K. P. Ong, D. J. Singh, and P. Wu, Phys. Rev. B 83 1155110 (2011), pp. 1-5

[5] T. Ishibe, A. Tomeda, K. Watanabe, J. Kikkawa, T. Fujita, and Y. Nakamura, J. Electron. Mater., 46 (2017), pp. 3020-3024.
[6] U. Ozgur, Y. I. Alivov, C. Liu, A. Teke, M.A. Reshchikov, S. Dogan, V. Avrutin, S. J. Cho, and H. Morkoc, J. Appl. Phys. 98 041301 (2005), pp. 1-103.

Acknowledgements
A part of this work was supported in part by a Grant-in-Aid for Scientific Research A (16H02078), for Exploratory Research (15K13276), and for JSPS Research fellow (17J00328). A part of PLD in this work was supported by Nanotechnology Platform Project of the Ministry of Education, Culture, Sports, Science, and Technology, Japan (S-17-OS-0025).

IOP Conf. Series: Journal of Physics: Conf. Series **1052** (2018) 012127 doi:10.1088/1742-6596/1052/1/012127

A fully integrated autonomous power management system with high power capacity and novel MPPT for thermoelectric energy harvesters in IoT/wearable applications

Hamed Osouli Tabrizi[1] , H M P C Jayaweera[1] and Ali Muhtaroğlu[1,2]

[1]Sustainable Environment and Energy Systems
[2]Dept. of Electrical-Electronics Engineering
Middle East Technical University Northern Cyprus Campus
Kalkanli, Guzelyurt, Mersin 10, Turkey
Email:{hamed.tabrizi, amuhtar}@metu.edu.tr

Abstract. This paper reports a fully integrated autonomous power management system for thermoelectric energy harvesting with application in batteryless IoT/Wearable devices. The novel maximum power point tracking (MPPT) algorithm does not require open circuit voltage measurement. The proposed system delivers 0.5 mA current with 1 V regulated output based on simulations, which is the highest output current for a fully integrated converter reported in the literature for ultra-low voltage applications, to the best knowledge of the authors. Regulated 1 V output can be achieved for load range >2 kΩ, and input voltage range >140 mV. The circuit has been implemented in UMC-180nm standard CMOS technology and simulated.

1. Introduction

Elimination of the battery is desirable in emerging applications such as internet of things (IoT) and wearable healthcare devices, due to considerable miniaturization and maintenance cost benefits [1]. Recent studies have shown energy storage devices can be replaced by ambient energy harvesters [2,3]. Among various ambient energy sources, thermal energy is promising for wearable computing and similar applications, due to the availability of hundreds of microwatts of harvestable power. Despite adequate power availability, maximum open circuit voltage generated by thermoelectric generators (TEGs) exposed to the typically attainable 2-5 K temperature gradient, cannot exceed 500 mV [3,4]. Requisites of a miniaturized low-cost batteryless power management system, therefore, include 1) Full integration, 2) low start-up voltage, 3) stepped-up and regulated output voltage, and 4) relatively high output power. Charge pump converters with LC-tank oscillators have been introduced [5-8] to overcome MOSFET threshold limitations associated with low input voltage. Our group has reported solutions with an input voltage as low as 200 mV, where the maximum output power of 400 μW is achieved with a 2.7 kΩ load, which is sufficient to meet the real-time demand of most ultra-low power IoT/Wearable circuits [7,8]. Real load dynamically varies in the above-mentioned applications, as does the TEG output due to alternating ambient temperature. Thus an autonomous system is required to ensure maximum power point tracking (MPPT) under various load and source conditions.

MPPT algorithm presented in [3,9] measures TEG open circuit voltage, V_{TEG}, for impedance matching, and thus causes system disruption when disconnected. The method does not work well when the converter efficiency varies with input impedance values, as is the case for the LC-tank coupled charge pump topology. The system proposed in this paper utilizes a novel MPPT that refrains from disconnecting the converter from source. The rest of the paper is organized as follows: The proposed system is discussed in section 2. In section 3, characteristics of the LC-tank charge pump circuit is presented. MPPT algorithm is discussed in Section 4. Section 5 presents the results, and the conclusion is provided in Section 6.

2. System Topology

2.1. TEG model and Power Efficiency

TEG first order model consists of a voltage source with open circuit voltage V_{TEG}, and a series resistance R_{TEG}, as shown connected to the power management system in figure 1. R_{TEG} restricts the maximum theoretical input power to $P_{in,max}$ as calculated from equation (1). Real input power is maximized when R_{TEG} is low and power management input impedance, Z is in the range of R_{TEG}.

$$P_{in,MAX} = \frac{(V_{TEG})^2}{4\,R_{TEG}} \tag{1}$$

Power efficiency of the system can then be calculated using equation (2) with respect to $P_{in,max}$.

$$\eta = \frac{P_{out}}{P_{in,max}} = \left(\frac{V_{out}}{V_{TEG}}\right)^2 \times \left(\frac{4\,R_{TEG}}{R_L}\right) \tag{2}$$

2.2. Power Management System

The proposed power management circuit (figure 2) is composed of an LC-tank oscillator connected to a charge pump with an adjustable number of stages, using a minimum number of PMOS switches. V_{st5} node supplies sufficient voltage to the digital MPPT even when V_{TEG} is low. Low-frequency square wave (clock) signal necessary for the MPPT block is generated using a thyristor-based ring oscillator [11] with ultra-low power consumption. The output voltage is regulated utilizing a low-dropout (LDO) regulator and a subthreshold voltage reference circuit [10]. When sufficient input power is available, op-amp and LDO start regulation.

3. LC-tank and Charge Pump Characteristics

The output voltage of the n-stage charge pump can be determined by equation (3) based on the lumped circuit model in figure 3 previously presented by our group [8]. $V_{C_{2n}}$, maximum voltage at charge pump capacitors of stage n, can be calculated in an iterative manner for n-1 stages. Input impedance of the DC-DC converter with LC-tank oscillator can be calculated using equation (4).

$$V_{out} = (V_{osc} + V_{C_{2n}})\,e^{-\frac{t}{(R_L+R_P)C}} \times \frac{R_L}{R_L+R_P}\ , t = \frac{T}{4},\quad where\quad V_{C_{2n}} = (n-1)V_{osc} + V_{in} - \frac{t i_n}{2C} \tag{3}$$

$$Y = \frac{1}{Z} = \frac{R_M C^2\omega^2}{1+R_M{}^2C^2\omega^2} + \frac{(R_M+R_P)C^2\omega^2}{4+(R_M+R_P)^2C^2\omega^2}(n-1) + \frac{R_LC_LC\omega^2(R_MR_LCC_L\omega^2-1)+C\omega^2(R_LC_L+R_MC+R_LC)}{(R_MR_LCC_L\omega^2-1)^2+(R_LC_L+R_MC+R_LC)^2\omega^2} \tag{4}$$

where R_P and R_M are average PMOS and NMOS average switch resistance in charge pump respectively, n is the number of stages, R_L represents load, i_n is the output current of stage n, V_{osc} is the LC-tank oscillation amplitude, V_C represents the peak voltage across the charge pump capacitors, T is the oscillation period, and V_{in} is the input voltage. Equation (3) establishes that V_{out} is directly related to n. From (4), converter input impedance has an inverse relation with n. V_{osc} also has an inverse relation with n due to the loading effect of charge pump on LC-tank.

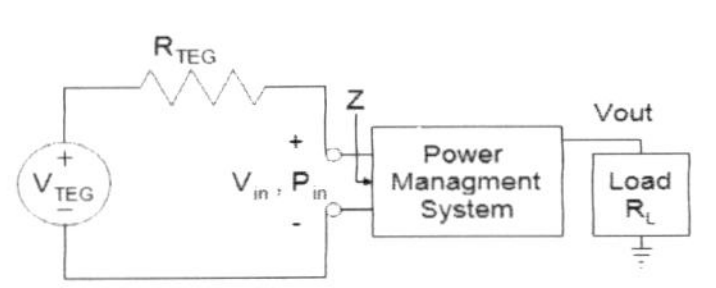

Figure 1. TEG first order model connected to the power management system.

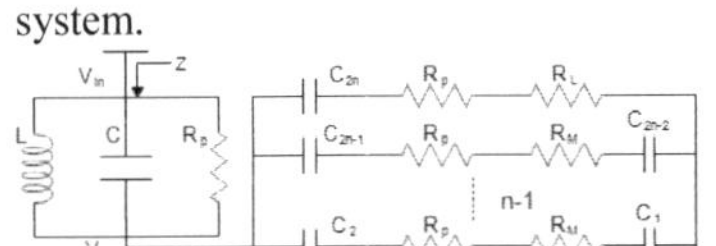

Figure 3. LC-tank and n stage charge pump circuit model proposed in [8].

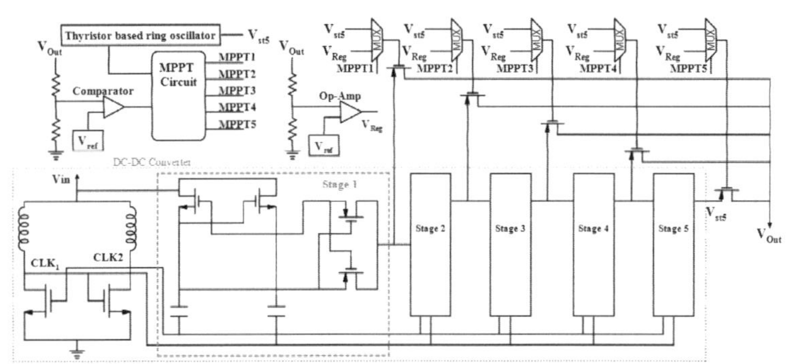

Figure 2. The proposed fully integrated power management system for thermoelectric energy harvesting.

4. MPPT Algorithm

MPPT algorithm maximizes η from equation (2) by maximizing V_{out} for a given R_{TEG}, V_{TEG}, and R_L. Therefore, regulation target is the maximum voltage that V_{out} can reach. Among the variables in equations (3), only n and C can be used to adjust V_{out}. Changing C requires an adjustable capacitor bank that significantly increases layout cost and charge redistribution losses (CRL). n is hence selected as the control parameter for MPPT. Stage 5 (last stage) is activated in the start-up state to maximize V_{out}. As long as V_{out} target is achieved, digital MPPT is inactive. When the output voltage drops below the target value, which can be a result of an increase in load or decrease in source voltage, MPPT block looks for a new optimum to reach the target voltage, by reducing the number of stages.

5. Results

R_{TEG} value varies for different products. 40 Ω is chosen in the simulations as a typical value for tiny TEGs [13]. DC-DC converter behaviour without voltage regulation is depicted in figure 4. Figure 4(a) shows that there is an optimal stage to maximize V_{out} for a given load. Loading on the LC-tank oscillator increases with the number of stages, resulting in decrease in the voltage swing as shown in figure 4(c). This also reduces input Z and V_{in} of the power management circuit, as depicted in figures 4(b) and 4(d) respectively. Therefore, output cannot be boosted up further by increasing the number of stages. Varying V_{TEG} has a similar effect as varying R_L, although this is not shown here for brevity: With fixed R_L, the optimal stage count varies with V_{TEG} value. Figure 5(a) depicts simulation results for the complete MPPT system with a given input voltage and load. Since 1 V output is not reached at start-up, MPPT block decreases the number of stages until a stage is found that delivers the target. Once target value is reached, MPPT stops. Figure 5(b) presents to a condition that leads to a lower efficiency despite increase in V_{TEG}. However, output power calculated from equation (1) is almost two times more compared to the one in figure 5(a). The power management design in this paper, which delivers maximum power to load, is practically more desirable than a design that would prioritize high efficiency over output power, as demonstrated here. Figure 6 shows the layout of the proposed system. Table 1 compares this work to the recent state-of-the art thermoelectric energy harvesting solutions in the literature. The solution in this paper provides lower cost in terms of process technology (0.18 μm), and is fully integrated. Circuit proposed in [9] is fully integrated, but use of ring oscillator has increased start-up time and limited output power. In [9], input impedance of the system is in the range of 300 Ω, which results in lower output power capacity, despite higher end to end efficiency.

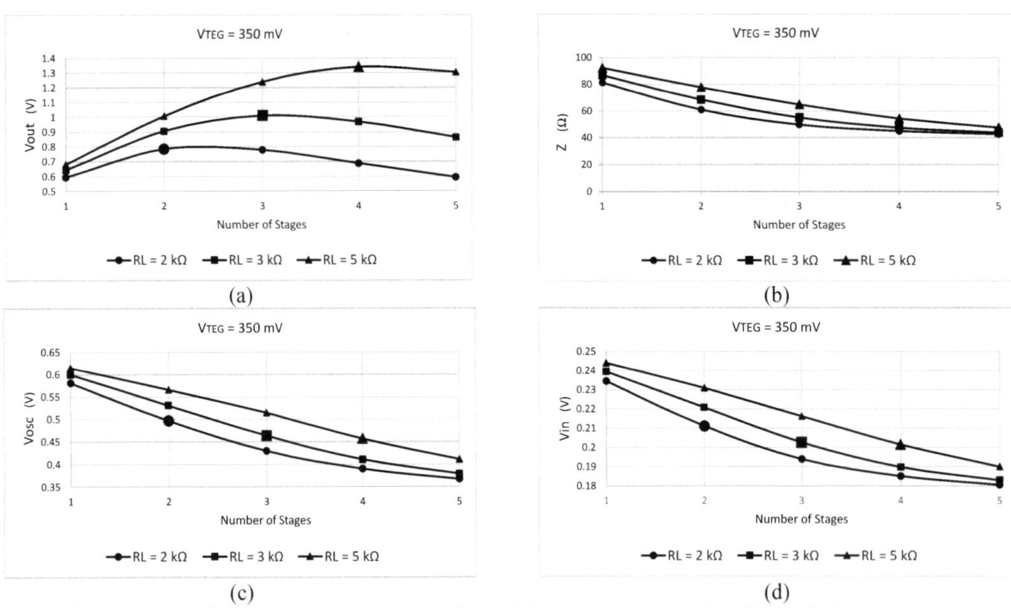

Figure 4: Optimization of charge pump stage count for each load, (a) to maximize V_{out} with fixed source voltage, and resulting change in (b) input impedance, Z, (c) LC-tank oscillation amplitude, (d) V_{in}.

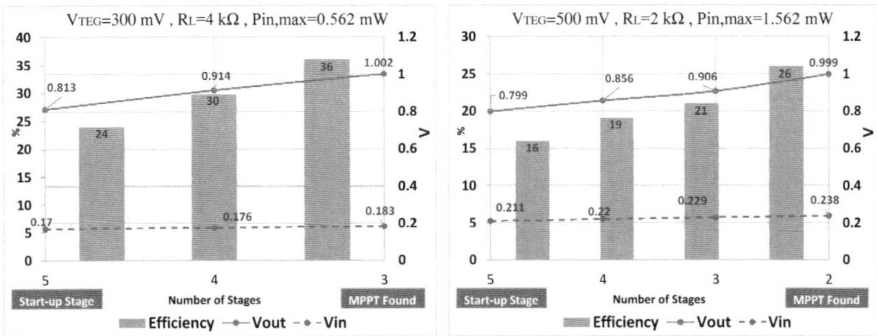

(a) (b)

Figure 5: Sample simulation results to demonstrate MPPT operation: (a) For V_{TEG}=300 mV and R_L=4 kΩ, MPPT finds optimal stage count as 3 for 1 V regulated output with 36% efficiency. (b) For V_{TEG}=500 mV and R_L=2 kΩ, MPPT find optimal stage count as 2 for 1 V regulated output with 26% efficiency.

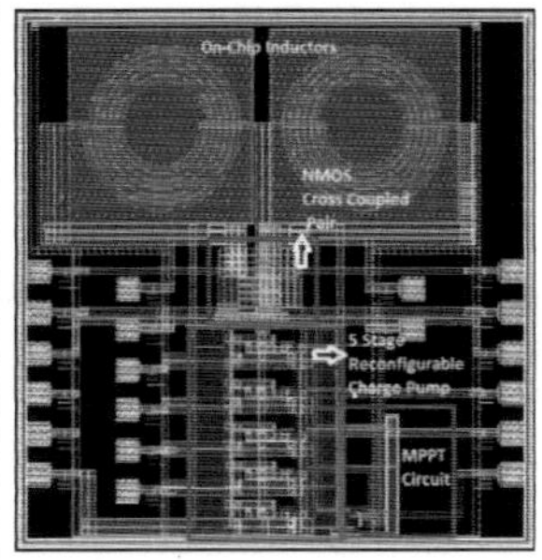

Figure 6: Chip layout: Inductors occupy 0.66 mm^2. Reconfigurable charge pump and MPPT take up 0.24 mm^2 and 0.05 mm^2 respectively.

6. Conclusion

A fully integrated autonomous power management system with high output power suitable for batteryless IoT/Wearable devices is introduced. The solution maximizes output power for a wide range of loads and input voltage variations through a novel MPPT, and delivers 0.5 mA current with 1 V regulated output at 220 mV input voltage, which is the highest output current for a fully integrated converter in the literature for ultra-low voltage applications to the best knowledge of the authors. Using this system, regulated 1 V output can be achieved for loads of > 2 kΩ and input voltage range of > 140 mV. Cold start-up from input voltages as low as 140 mV and high output power delivery are simultaneously attained in simulations due to the presence of the integrated LC-tank oscillator in the system.

Table 1. Comparison with state-of-the-art thermoelectric energy harvesters.

	[5]	[9]	[10]	[3]	[8]	This Work
Process	0.18 μm	65nm	0.18 μm	65 nm	0.18 μm	0.18 μm
Maximum Output Current	200 μA	730 μA	71 μA	358 μA	150 μA[a]	500 μA
Separate Start-Up Unit	No	Yes	No	Yes	No	No
Min V_{in} for start-up	150 mV	80 mV	350 mV	65 mV	360 mV	140 mV
Regulated V_{out}	1.8 V	0.7 V–1V	1.8V	1.8 V	1.2V	1 V
Peak End to End efficiency	N/A 73%	73%	N/A 80%	68%	43%[a]	36.7%
Off chip (L+C+R)	0+6+0	4+2+0	0+7+4	1+2+0	0+0+0	0+0+0
MPPT Tracking Time	NO MPPT	~20ms	< 180ms	< 20ms	NA	< 10ms

[a] Extracted for the V_{TEG}voltage range of < 500 mV

7. References

[1] Muhtaroğlu A 2017 *Energy Harvesting and Energy Efficiency* (Springer International Publishing) pp 63-85
[2] Balsamo D, Elboreini A, Al-Hashimi B and Merrett G 2017 Exploring ARM mbed support for transient computing in energy harvesting IoT systems. *Advances in Sensors and Interfaces (IWASI), 7th IEEE International Workshop on*
[3] Rozgić D and Marković D 2017 *IEEE Transactions on Biomedical Circuits and Systems* May 23
[4] Zoller T, Nagel C, Ehrenpfordt R and Zimmermann A 2017 *IEEE Transactions on Components, Packaging and Manufacturing Technology*. May 17
[5] Bassi G, Colalongo L, Richelli A and Kovacs-Vajna ZM 2013 *IET Power Electronics*. Jul 1 6(6) 1151-6
[6] Jayaweera HM, Pathirana WP and Muhtaroğlu A 2015 *J. of Phys: Conference Series* Vol. 660, No. 1 p 012017 IOP Publishing
[7] Jayaweera HM and Muhtaroğlu A 2016 *J. of Phys: Conference Series* Nov Vol 773, No. 1, p 012085 IOP Publishing
[8] Jayaweera HM and Muhtaroğlu A 2017 *Microelectronics Journal*. Jan 31;59:33-9
[9] Grezaud R and Willemin J 2013 *In New Circuits and Systems Conf (NEWCAS)*, Jun 16 pp 1-4
[10] Seok M, Kim G, Blaauw D and Sylvester D 2012 *IEEE Journal of Solid-State Circuits*. Oct;47(10):2534-45
[11] Mahato AK 2014 *Engineering and Computational Sciences (RAECS),* Mar 6 pp 1-5

IOP Conf. Series: Journal of Physics: Conf. Series **1052** (2018) 012128 doi:10.1088/1742-6596/1052/1/012128

Thermoelectric Properties of Chromium Selenides

Q Guo[1] and T Mori[1,2]

[1] WPI International Center for Materials Nanoarchitechtonics (MANA), National Institute for Materials Science (NIMS), Namiki 1-1, Tsukuba 305-0044, Ibaraki, Japan
[2] Graduate School of Pure and Applied Sciences, University of Tsukuba, Tennoudai 1-1-1, Tsukuba 305-8671, Japan

MORI.Takao@nims.go.jp

Abstract. Start your abstract here Magnetic semiconductors, such as $CuFeS_2$ and Mn-doped $CuGaTe_2$ with chalcopyrite structure, have received considerable attention due to their excellent thermoelectric properties [1, 2]. This paper reports the preparation and thermoelectric properties of a transition metal chalcogenides, chromium selenide Cr_2Se_3.

The chromium selenide Cr_2Se_3 adopts a rhombohedral structure with space group *R*3 based on NiAs-type structure. There is Cr vacancy ordering in every second metal layer. Rhombohedral Cr_2Se_3 is reported by Y.Adachi et al. to be antiferromagnetic with Néel temperature T_N = 43 K. [3] It was also found that Cr_2Se_3 is metallic when the temperature is over T_N, deducing from the temperature dependence of electrical resistivity. K.R.Pisharody prepared chromium sulfo-selenides $Cr2S_{3-x}Se_x$ ($0 \leq x \leq 1$) by reacting high purity in quartz tubes at 1273 K. All the samples are n-type materials showing negative Seebeck coefficient values and zT = 0.1 was achieved in Cr_2Se_3 at 500 K.

In this work, we have prepared $Cr_{2+x}Se_3$ ($-0.04 \leq x \leq 0.08$) by reacting the elements and investigated the effect of Cr content on thermoelectric properties. All the samples are indexed to be the Cr_2Se_3 phase, as observed from the powder X-ray diffraction (XRD) pattern. According to the onset temperature (T_{max}) of the intrinsic conduction and the Seebeck coefficient αmax at Tmax, The band gap values 0.22 eV for $Cr_{2.04}Se_3$ and 0.17 eV for $Cr_{2.08}Se_3$ were estimated using the equation $E_g = 2\alpha_{max}\ T_{max}$,[4] where T_{max} is the onset temperature of the intrinsic conduction and αmax the Seebeck coefficient peak αmax at T_{max}. The change in the conduction type from p-type to n-type was observed when the amount of Cr is over 2.04. For all the samples, the thermal conductivity increases with the increasing temperature. The highest zT = 0.22 was gained in *p*-type $Cr_{2.04}Se_3$ and zT = 0.12 for *n*-type $Cr_{2.08}Se_3$. Magnetic properties or Hall effect measurement remains to be performed.

References

[1] Ang R, Khan A U, Tsujii N, Takai K, Nakamura R and Mori T, 2015 *Angew. Chem. Int. Ed.* **54** 12909-13
[2] Ahmed F, Tsujii N and Mori T 2017 *J. Mater. Chem. A* **5** 7545-54
[3] Adachi Y, Yuzuri M, Kaneko T, Abe S and Yoshida H 1994 *J. Phys. Soc. Jpn.* **63** 369-70
[4] Goldsmid H J and Sharp J W 1999 *J. Electr. Mater.* **28** 869-72

Acknowledgments

This research was supported by JST CREST Grant Number JPMJCR15Q6, Japan.

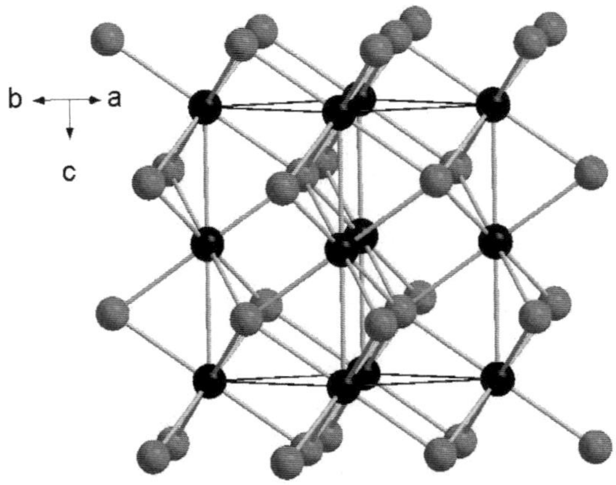

Figure 1. Crystal Structure of Cr_2Se_3. Black and red spheres represent Se and Cr, respectively. Cr vacancy ordering exists in every second metal layer.

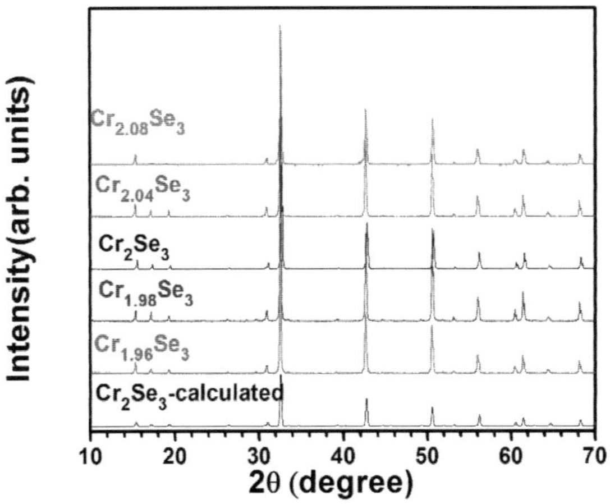

Figure 2. Powder X-ray diffraction (XRD) patterns for $Cr_{2+x}Se_3$ ($-0.04 \le x \le 0.08$).

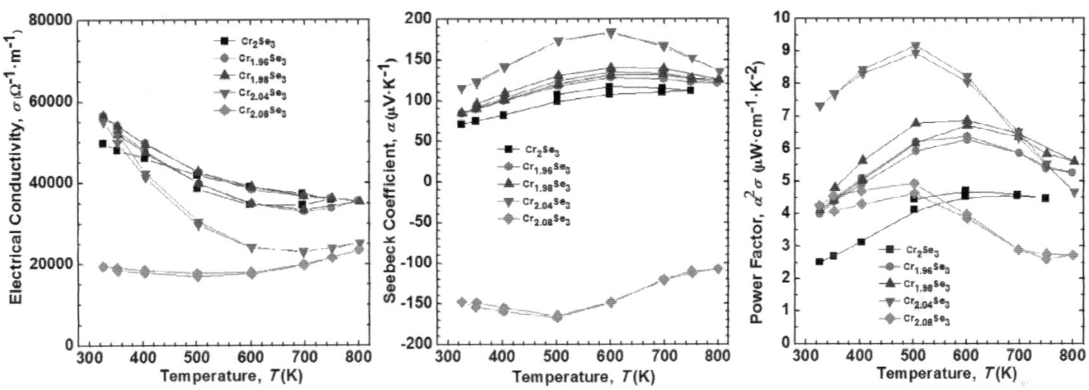

Figure 3. Temperature dependence of electrical properties for $Cr_{2+x}Se_3$ ($-0.04 \le x \le 0.08$).

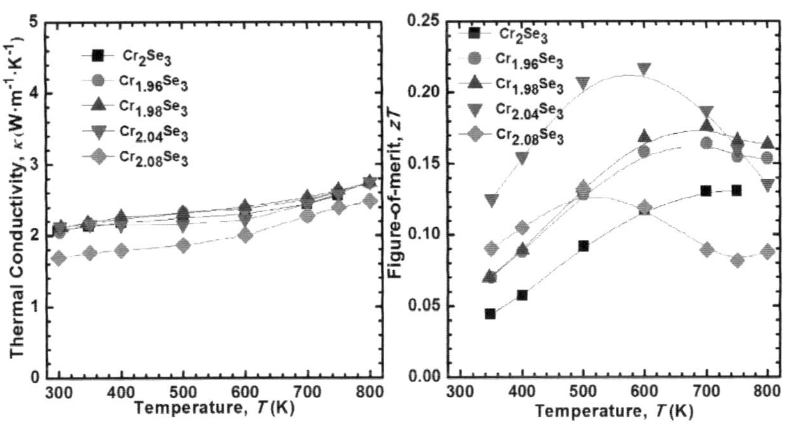

Figure 4. Temperature dependence of thermal conductivity and zT for $Cr_{2+x}Se_3$ ($-0.04 \le x \le 0.08$).

Fabrication and characterization of roll-type thin-film thermoelectric generators

J. Hamada [1]**, K. Yamamoto** [1]**, and M. Takashiri** [1]

[1] Department of Materials Science, Tokai University, 4-1-1 Kitakaname, Hiratsuka, Kanagawa 259-1292, Japan

E-mail: takashiri@tokai-u.jp

Abstract. We propose and demonstrate a facile fabrication method of roll-type thin-film thermoelectric generators to conveniently use a low electric power converted from ambient heat sources. The generators consisted of n-type Bi_2Te_3 and p-type Sb_2Te_3 thin films, which were deposited on a polyimide sheet by radio-frequency magnetron sputtering. Both types of thin films were thermally annealed to increase the thermoelectric properties. The annealed thin films were connected to each other using silver paste, with the p-n junctions connected electrically in series. The completed generator consisted of five p-n pairs was 625 mm long and 26 mm wide. To form the roll-type structure, the generator was bent so that the fold lines were located at the p-n junctions. The performance of the generators, such as the open circuit voltage (V_{oc}) and maximum output power (P_{max}), were measured by applying a temperature difference (ΔT) between the ends of the generator. As a result, the generator exhibited a V_{oc} of 20.1 mV and a P_{max} of 48.5 nW at a ΔT of 30 K. In addition, we calculated the performances of generators with different film thickness or number of p-n pairs based on the experimental results.

1. Introduction

The use of thermoelectric power generators is one of the key issues to make the internet of things (IoT) widespread. This is because IoT technology uses various sensors such as wireless sensor nodes [1,2], wearable sensors [3,4]. These sensors do not need high electric power, but do need batteries with long duration and high reliability. The thermoelectric generators satisfy these requirements. They directly generate a small amount of electrical energy from the ambient thermal energy in the surrounding space. In addition, the thermoelectric generators do not have moving parts or fluids because they use an all-solid-state technology, leading to an almost maintenance-free throughout the lifespan. Thin-film thermoelectric generators are quite favorable for the application of IoT. The thin-film thermoelectric generators are not able to produce higher electric power, but they are lightweight and can be attached on curved surfaces such as pipes or human bodies by using flexible substrates.

In general, thin-film thermoelectric generators have a lateral (in-plane) configuration. A rectangular meander made of many p- and n-type material pairs connected with metal electrodes is formed on a substrate [5-7]. This configuration can increase the output power as the number of pairs and the temperature difference (ΔT) are increased. However, the generation of ΔT in the in-plane direction is limited in application compared to that in the cross-plane direction. To extract the favorable features of the in-plane type of thin-film thermoelectric generators, those with roll-type structures have been proposed. Shiozaki et al. [8] fabricated flexible thermoelectric generators with a 3D woven shape, which were formed to stretch the cold and hot junction of the thermoelectric legs by mechanical force. This structure can make the ΔT appear to be in the cross-plane direction. To make the use of roll-type thin-film thermoelectric generators widespread, simple structure and facile fabrication methods, which are not used by MEMS techniques such as photolithography and precision machining, should be considered. In this study, we propose and demonstrate a simple structure of roll-type thin-film thermoelectric generators by a facile fabrication method.

2. Experimental setup

Figure 1 shows the illustration of the fabrication process of a roll-type thermoelectric thin-film generator. The thin films were deposited using RF magnetron sputtering (Tokuda CFS-8EP). The basic experimental setup was described in our previous publications [9,10]. Initially, we prepared ten pieces of polyimide sheets (26 mm × 76 mm × 25 μm). N-type Bi_2Te_3 thin films with a thickness of 0.8 μm were deposited on the five pieces of polyimide sheets, and p-type Sb_2Te_3 films with a thickness 0.8 μm were deposited on the remaining five pieces of polyimide sheets. The polyimide sheet was cut into 10 pieces because a longer polyimide sheet made it difficult to obtain a uniform film thickness due to the limitation of the deposition area in the sputtering equipment, and made it difficult to perform thermal annealing with a uniform temperature due to the limited soaking area in the electric furnace. After depositing the n- and p-type thin films, these films were thermally annealed at the optimized temperature. Then, the sample was assembled by electrically connecting the p-n junctions in series to form the generators. The p-type film was set to adhere to the n-type film, and connected by spraying silver paste. After baking the sample to dry the silver paste, the generator was bent like an accordion so that the fold lines were located at the p-n junctions. To measure the performance of the generators by applying ΔT between the ends of the generator, the lower half of the generator was dipped in hot water while the upper half was kept in the atmosphere.

3. Results and discussion

Figure 2 shows the performances of the generators, using both types of the thin films annealed at 300°C, as a function of ΔT; the ΔT at the ends was changed from 10 to 30 K. In Fig. 2(a), a V_{oc} of 11.2 mV was achieved at a ΔT of 10 K. With increasing the ΔT, V_{oc} linearly increased. A V_{oc} of 20.7 mV

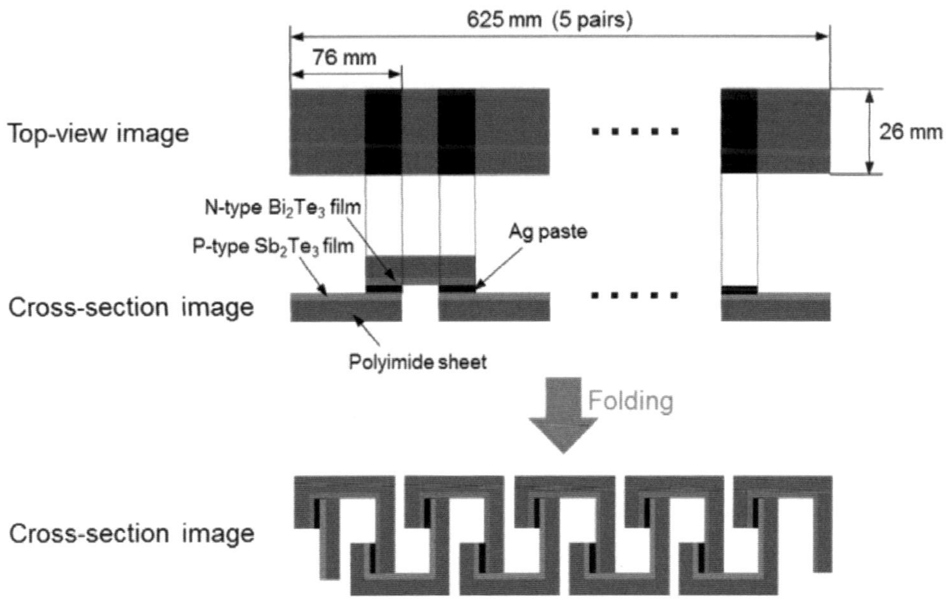

Figure 1. Illustration of the roll-type thermoelectric thin-film generator.

was achieved at a ΔT of 30 K. This result is in relatively good agreement with that of previous reports with similar device structures and the similar ΔT [6,11].

The maximum output power (P_{max}) is given by Eq. 1,

$$P_{max} = V_{oc}^2/4R_{total}, \tag{1}$$

where R_{total} is the measured total resistance of the generator. R_{total} of the generator was measured by a digital multimeter to be 2200 Ω. In Fig. 2(b), P_{max} was determined to be 14.4 nW at the ΔT of 10 K. P_{max} increased as the ΔT increased, and a P_{max} of 48.5 nW was achieved at the ΔT of 30 K. However, this P_{max} is not enough to activate sub-microwatt or microwatt wearable devices such as a temperature sensor [12] or CMOS image sensor [13].

Therefore, to apply the generators using a facile fabrication method into sub-microwatt wearable devices, we calculated the generator performance by changing the film thickness or the number of p-n pairs based on the experimental results. Figure 3(a) shows the calculated V_{oc} of the generator at ΔT = 30 K as a function of film thickness for various numbers of p-n pairs. An experimental result (5 p-n pairs with thickness 0.8 μm) is also plotted in this figure. The Seebeck coefficient shows no dependence on the film thickness, so that V_{oc} is only proportional to the number of p-n pairs. When the generator is composed of 20 p-n pairs, the V_{oc} is expected to be 82.6 mV.

To calculate P_{max}, R_{total} should be determined when the film thickness or the number of p-n pairs are changed. R_{total} is sum of the resistance of the n- and p-type thin films, R_{film}, and the contact resistance, R_c. R_{film} is given by Eq. 2,

$$R_{film} = n\left[\frac{1}{\sigma_{Bi-Te}}\left(\frac{l}{w \cdot d}\right) + \frac{1}{\sigma_{Sb-Te}}\left(\frac{l}{w \cdot d}\right)\right], \tag{2}$$

where n is the number of p-n junctions, σ_{Bi-Te} and σ_{Sb-Te} are the measured electrical conductivities of Bi_2Te_3 and Sb_2Te_3 thin films annealed at 300°C, respectively. The length (76 mm), width (26 mm), and thickness are represented by l, w, and d, respectively. On the other hand, R_c is given by Eq. 3,

$$R_c = (2n-1)R_{c0}, \tag{3}$$

where (2n-1) means the number of p-n junctions, and R_{c0} is the contact resistance per p-n junction. Based on the experimental result, R_c for 5 pairs of p-n junctions is calculated by substituting R_{film} (840 Ω) from R_{total} (2200 Ω), and was determined to be 1360 Ω. As a result, R_{c0} was determined to be 151

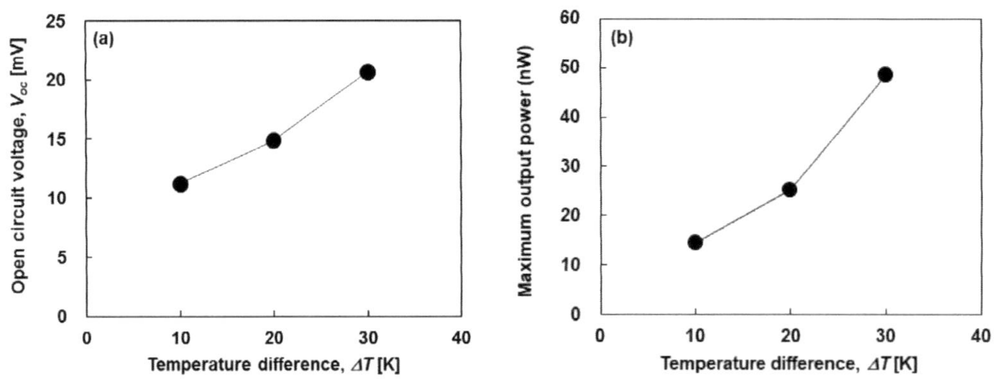

Figure 2. The performances of the generators, using both types of the thin films annealed at 300°C, as a function of ΔT; the ΔT at the ends was changed from 10 to 30 K.
(a) Open circuit voltage, (b) maximum output power.

Ω. We here assumed that the magnitude of R_{c0} is the same in each p-n junction. Figure 3(b) shows the calculated P_{max} of the generator at $\Delta T = 30$ K as a function of film thickness for various numbers of p-n pairs with the experimental result. The calculated P_{max} increased as the number of p-n pairs, regardless of the film thickness. The calculated P_{max} for all numbers of p-n pairs rapidly increased until the film thickness reached approximately 1.0 μm, because of the decrease in the resistance of the Bi_2Te_3 and Sb_2Te_3 thin films. On the other hand, the calculated P_{max} did not change so much when the film thickness was greater than 3 μm. Therefore, for example, to activate a CMOS image sensor at the consumption power of 226 nW [13] using the roll-type thin-film thermoelectric generator, the structure of the generator is expected to need 20 p-n pairs with 2.0 μm thick-film.

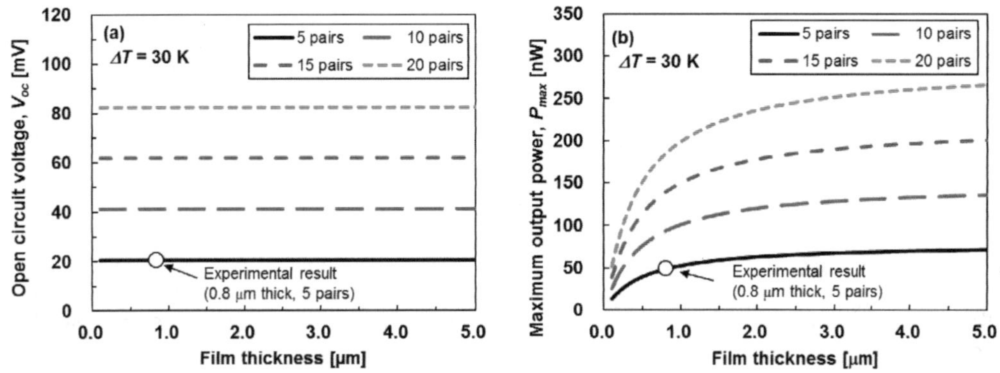

Figure 3. The calculated V_{oc} of the generator at ΔT = 30 K as a function of film thickness for various numbers of p-n pairs. (a) Open circuit voltage, (b) maximum output power.

4. Conclusion

The roll-type thin-film thermoelectric generators were prepared by RF sputtering method. By applying a ΔT between the ends of the generator, we confirmed that the generators produced an electric power.

Acknowledgments

This study was partly supported by JSPS KAKENHI Grant Number 16H04282.

References

[1] Roundy S, Wright K P, and Rabaey J 2003 *Comput. Commun.* **11** 1131
[2] Paradiso A J and Starner T 2005 *IEEE Pervasive Comput.* **4** 18
[3] Francioso L, Pascali D C, Farella I, Martucci C, Creti P, Siciliano P, and Perrone A 2011 *J. Power Sources.* **196** 3239
[4] Weber J, Potje K K, Haase F, Detemple P, Völklein F, and Doll T 2006 *Sens. Actuators.* A **132** 325
[5] Al Kadhim A H A, Abbas A K, and Hassan H A 2013 *Mater. Sci. Semicond. Process.* **16** 612
[6] Takayama K and Takashiri M 2017 *Vacuum.* **144** 261
[7] Siddique A R M, Mahmud S, and Van H B 2017 *Renewable Sustainable Energy Rev.* **73** 730
[8] Shiozaki M, Sugiyama S, Watanabe N, Ueno H, and Itoigawa K, 2006 *19th IEEE International Conference on Micro Electro Mechanical Systems* **10** 1109
[9] Hamada J and Takashiri M 2017 *J. Cryst. Growth.* **468** 188
[10] Yamauchi K and Takashiri M 2017 *J. Alloys Compd.* **698** 977
[11] Mizoshiri M, Mikami M and Ozaki K 2013 *J. Appl. Phys.* **52** 06GL07
[12] Chen P, Chen T K, Wang Y S, and Chen C C 2009 *IEEE Sens. J.* **9** 1639
[13] Hanson S, Seok M, Lin Y S, Foo Z Y, Kim D, Lee Y, Liu N, Ylvester S D, and Blaauw D A 2010 *IEEE J. Solid-State Circuits.* **45** 759

IOP Conf. Series: Journal of Physics: Conf. Series **1052** (2018) 012130 doi:10.1088/1742-6596/1052/1/012130

In-plane Thermoelectric Properties of Nano-TiS_2/CNT/PEDOT−PSS Hybrid Films

K Okamoto and H Anno

Department of Electrical Engineering, Tokyo University of Science, Yamaguchi, 1-1-1 Daigaku-Dori, Sanyo Onoda, Yamaguchi, Japan
E-mail: k.okamoto@rs.tusy.ac.jp

Abstract. We present the details of in-plane thermoelectric properties of nano-titanium disulfide (nano-TiS_2) / Carbon nanotube (CNT) / poly(3,4-ethylenedioxythiophene)−poly(styrenesulfonate) (PEDOT−PSS) hybrid films, in which PEDOT−PSS was used as a binder with high electrical conductivity. We attempted to reduce thermal conductivity, κ, of PEDOT−PSS/CNT hybrid film by enhancing the phonon scattering at the interface between CNT and nano-TiS_2 by the addition of nano-TiS_2. The Seebeck coefficient (S), electrical conductivity (σ), and power factor (PF) of a nano-TiS_2/CNT/PEDOT−PSS hybrid film at 50 °C for a weight ratio of 1:9:23 (CNT content 70 wt%) were S = 33.3 μV/K, σ = 406 S/cm, and PF = 45.0 μW/(mK^2), respectively. The in-plane thermal conductivity ($\kappa_{//}$) was 0.91 W/(mK) at room temperature for the hybrid film with a weight ratio of 1:9:23 (CNT content = 70 wt%). Moreover, the in-plane ZT of the hybrid film was 0.0163, which is about 4 times greater than that for PEDOT−PSS and CNT buckypaper (BP).

1. Introduction

Carbon nanotube (CNT) has potential application as electronic material, construction material, and energy material. It was reported that there is possibility of mass production of single-wall CNT by the super growth method [1]. In recent years, CNT has been noted as a thermoelectric material and reported to have a high Seebeck coefficient and high electrical conductivity [2]. However, the dimensionless thermoelectric figure of merit, ZT, is low because of high thermal conductivity [3]. Furthermore, poly(3,4-ethylenedioxythiophene)−poly(styrenesulfonate) (PEDOT−PSS) has been noted as a thermoelectric material with high electrical conductivity and low thermal conductivity [4]. In recent years, the improvement in Seebeck coefficient of PEDOT-PSS has improved the ZT [5]. We have investigated the improvement in thermoelectric properties and thermoelectric devices of PEDOT-PSS [6, 7]. However, thermoelectric materials require further improvement in thermoelectric properties. Furthermore, titanium disulfide nano-plates have demonstrated thermoelectric properties and are reported to have a high negative Seebeck coefficient and high electrical conductivity [8, 9].

In this study, we present the details of in-plane thermoelectric properties of nano-titanium disulfide (nano-TiS_2)/CNT/PEDOT−PSS hybrid films, in which PEDOT−PSS was used as a binder with high electrical conductivity. We attempted to reduce the κ of PEDOT−PSS/CNT hybrid film by enhancing the phonon scattering at the interface between CNT and nano-TiS_2 by the addition of nano-TiS_2.

2. Experimental Procedure

Nano-TiS_2 was synthesized by a process similar to that reported in reference [10]. A mixture of 5 vol% dimethyl sulfoxide (DMSO, Sigma-Aldrich) and PEDOT−PSS solution (PH-1000, Heraeus) was stirred for 1 h. Nano-TiS_2 and CNT (super-growth method, Zeon) were added to the PEDOT−PSS and DMSO solution and dispersed by sonication. The weight ratio of PEDOT−PSS:nano-TiS_2:CNT was 9:1:x, where x = 1−90. The prepared solution was cast on a polyimide substrate and dried at 60 °C in air for about 12 h to form a hybrid film. Nano-TiS_2 was evaluated by X-ray diffraction (XRD) and transmission electron microscopy (TEM). Seebeck coefficient and electrical conductivity were measured by a ZEM3 system (ADVANCE RIKO, Inc.), and the in-plane thermal diffusivity ($\alpha_{//}$) was measured by AC calorimetric method. The specific heat capacity (C_p) was measured by differential scanning calorimetry (DSC, NETZSCH), and the in-plane thermal conductivity was given by $\kappa_{//} = \alpha_{//} C_p \rho$, where ρ is the density.

3. Results and Discussion

Figure 1 shows the XRD pattern of the synthesized nano-TiS_2. The XRD data of the synthesized nano-TiS_2 agrees with that of PDF reference data (PDF No. 00-015-0853). The difference between the (001) peak (2θ = 15.558°) in the XRD pattern of the synthesized nano-TiS_2 and that of the PDF reference was thought to be large, because the synthesized nano-TiS_2 particles are disc-shaped, as observed by TEM. No impurity phases were detected in the profile. The lattice constants a and c were determined as 0.3405 nm and 0.5694 nm, respectively, for nano-TiS_2 and were consistent with those of bulk TiS_2.

Figure 2 shows the TEM image of the synthesized nano-TiS_2. TEM observation confirmed that the nano-TiS_2 particles are disc-shaped with an average diameter of 223 ± 57.9 nm and a thickness of about 14.6 ± 5.9 nm.

Figure 3 shows the dependence of thermoelectric properties on the CNT content for the nano-TiS_2/CNT/PEDOT−PSS hybrid films at 50 °C. The Seebeck coefficient increased monotonically with increasing CNT content. However, the electrical conductivity remained almost unchanged with the changing CNT content. The Seebeck coefficient and electrical conductivity of PEDOT-PSS were S = 16.2 μV/K and σ = 746 S/cm and for CNT buckypaper (BP) were S = 46.1 μV/K and σ = 291 S/cm, respectively. The Seebeck coefficients of the hybrid films were considered to be the characteristic of CNT. The electrical conductivity remained almost unchanged with the changing CNT content, suggesting that the dispersibility of CNT was not influenced by the addition of nano-TiS_2. The Seebeck coefficient, electrical conductivity, and power factor of a nano-TiS_2/CNT/PEDOT−PSS hybrid film at 50 °C for a weight ratio of 1:9:23 were S = 33.3 μV/K, σ = 406 S/cm, and PF = 45.0 μW/(mK^2), respectively.

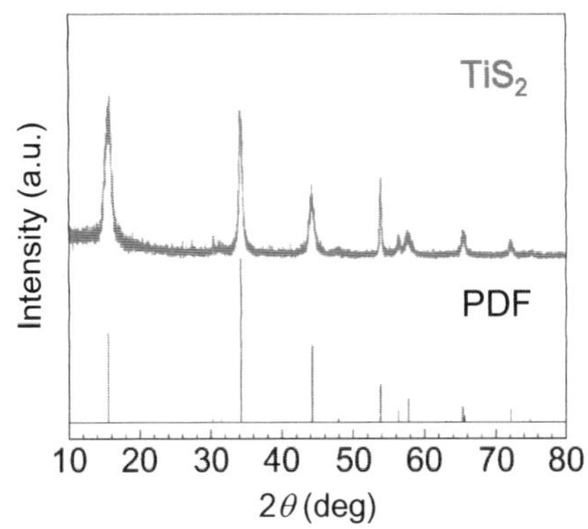

Figure 1. X-ray diffraction pattern of the synthesized nano-TiS_2.

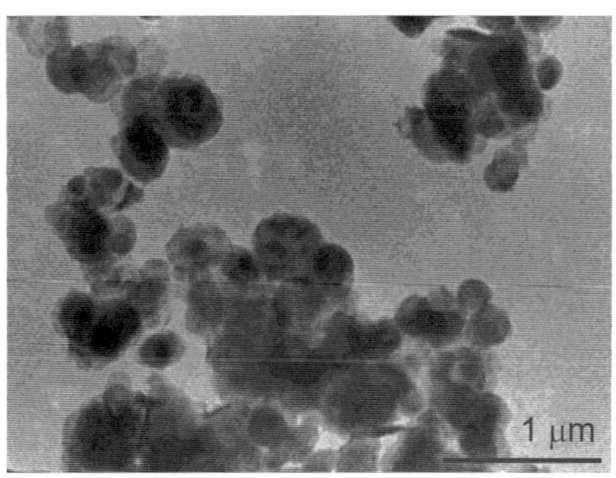

Figure 2. TEM image of the synthesized nano-TiS_2.

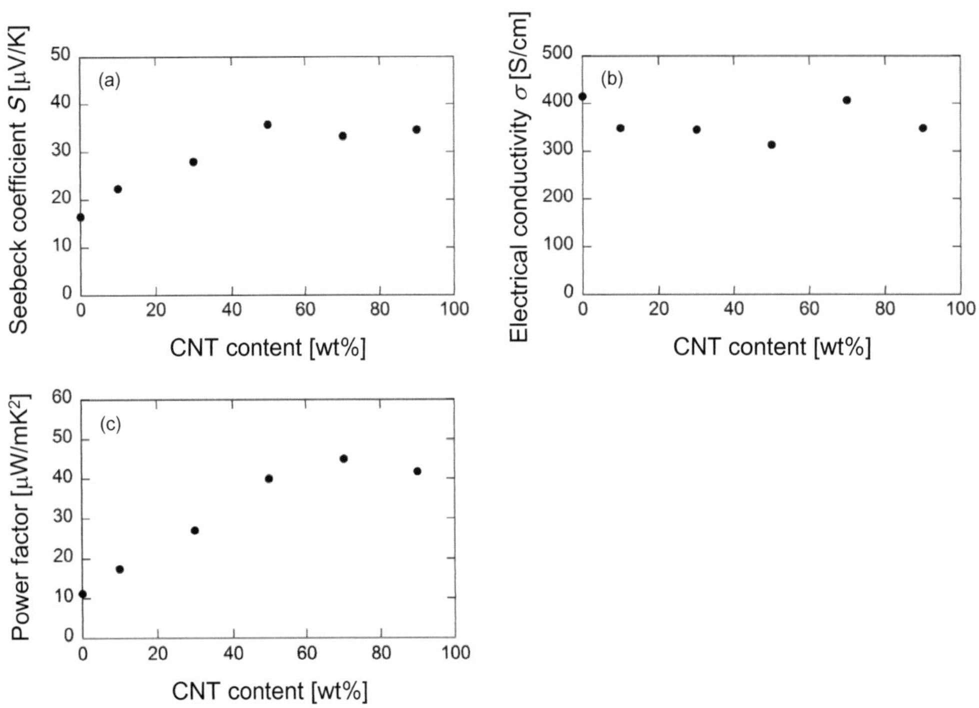

Figure 3. Dependence of thermoelectric properties on CNT content for nano-TiS_2 / PEDOT–PSS / CNT. (a) Seebeck coefficient; (b) electrical conductivity; (c) power factor.

Table 1. Thermoelectric properties of nano-TiS_2/CNT/PEDOT–PSS and reference samples.

	$\sigma_{//}$ [S/cm]	$S_{//}$ [μV/K]	$\kappa_{//}$ [W/mK]	$ZT_{//}$
Nano-TiS_2/PEDOT–PSS/CNT	406	33.3	0.91	0.0163
PEDOT–PSS	746	16.2	1.43	0.0045
CNT BP	291	46.1	5.95	0.0034

Table 1 shows the thermoelectric properties of nano-TiS_2/CNT/PEDOT–PSS hybrid films and reference samples. For thermal conductivity of the hybrid film, a CNT content of 70 wt% with the maximum power factor was measured. The in-plane thermal conductivity, $\kappa_{//}$, was 0.91 W/(mK) at room temperature for the hybrid film with a weight ratio of 1:9:23 (CNT content = 70 wt%). This value for the hybrid film is less than that of PEDOT–PSS and CNT BP. Figure 4 (a) and (b) show the schematic diagram of cross-sections of CNT/PEDOT–PSS and nano-TiS_2/CNT/PEDOT–PSS hybrid films. In the hybrid film, the nano-TiS_2 particles were considered to be dispersed in CNT covered with PEDOT–PSS. The reduction in κ can be explained by an enhancement in phonon scattering at the interface between CNT and nano-TiS_2 and the decrease in orientation of CNT by the addition of nano-TiS_2. The in-plane ZT for the hybrid film was 0.0163, which is about 4 times greater than that of PEDOT–PSS and CNT BP.

4. Conclusion

We prepared nano-TiS_2/CNT/PEDOT–PSS hybrid films and evaluated their in-plane thermoelectric properties. The nano-TiS_2/CNT/PEDOT–PSS hybrid films exhibited high Seebeck coefficient and electrical conductivity due to CNT and PEDOT-PSS, respectively. The thermal conductivity of the hybrid film was lower than that of PEDOT-PSS and CNT BP, which is possibly

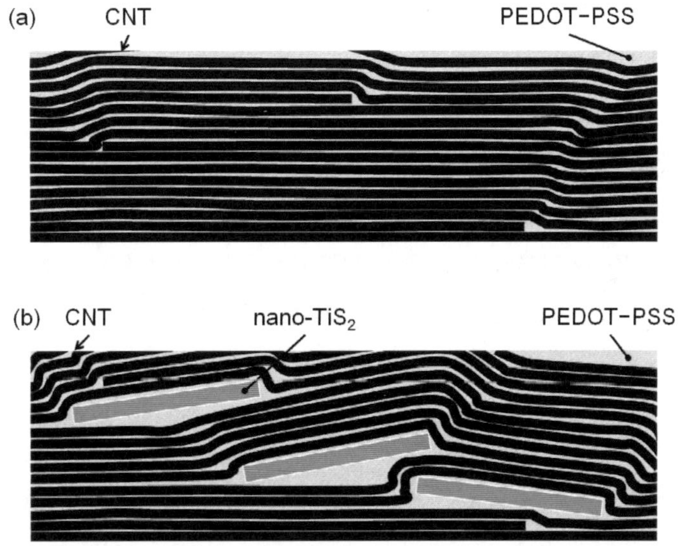

Figure 4. Schematic diagram of cross-section of (a) CNT/PEDOT−PSS and (b) nano-TiS_2/CNT/PEDOT−PSS hybrid film.

because of the effect of phonon scattering between CNT and nano-TiS_2, and the decrease in orientation of CNT due to the addition of TiS_2. The in-plane *ZT* for the hybrid film was 0.0163, which is about 4 times greater than that of PEDOT−PSS and CNT BP.

Acknowledgements

This work is based on the results obtained from a project commissioned by the New Energy and Industrial Technology Development Organization (NEDO). The authors would like to thank Zeon Corporation for joint research and support. The authors would also like to thank Ph. D. Yoshiaki Hirano, graduate student Keisuke Oshima, and Kenji Goto from Tokyo University of Science, Yamaguchi, for assisting with the experiments. The authors would like to thank Prof. Emer. Naoki Toshima and Prof. Yukihide Shiraishi from Tokyo University of Science, Yamaguchi, for useful discussions.

References

[1] Hata K, Futaba DN, Mizuno K, Namai T, Yumura M, and Iijima S 2004 *Science* **306** 1362−1364
[2] Toshima N, Oshima K, Anno H, Nishinaka T, Ichikawa S, Iwata A, and Shiraishi Y 2015 *Adv. Mater.* **27** 2246−2251
[3] Kim P, Shi L, Majumdar A, and McEuen P L 2001 *Phys. Rev. Lett.* **87** 215502
[4] Wei Q, Mukaida M, Kirihara K, and Ishida T 2014 *ACS Macro Lett.* **3** 948−952
[5] Kim G-H, Shao L, Zhang K, and Pipe K P 2013 *Nat. Mater.* **8** 719−723
[6] Hokazono M, Anno H, and Toshima N 2014 *J. Electron. Mater.* **43** 2196−2201
[7] Anno H, Nishinaka T, Hokazono M, Oshima N, and Toshima N 2015 *J. Electron. Mater.* **44** 2105−2112
[8] Wan C, Gu X, Dang F, Itoh T, Wang Y, Sasaki H, Kondo M, Koga K, Yabuki K, Snyder G J, Yang R, and Koumoto K 2015 *Nat. Mater.* **14** 622−627
[9] Wan C, Kidama Y, Kondo M, Sasai R, Zian X, Gu X, Koga K, Yabuki K, Yang R, and Koumoto K 2015 *Nano Lett.* **15** 6302−6308
[10] Sohee J, Dongwon Y, Jung-tak J, Minkyoung K, and Jinwoo C 2012 *J. Am. Chem. Soc.* **134** 18233−18236

IOP Conf. Series: Journal of Physics: Conf. Series **1052** (2018) 012131 doi:10.1088/1742-6596/1052/1/012131

Enhancement of thermoelectric power of a Si nanowire micro thermoelectric generator by improving the thermal conductivity of AlN thermally conductive film

T. Zhan[1], R. Yamato[1], S. Hashimoto[1], S. Oba[1], Y. Himeda[1], Y. Xu[2], T. Matsukawa[3], and T. Watanabe[1]

[1]Waseda University, 3-4-1 Ookubo, Shinjuku-ku, Tokyo 169-8555, Japan
[2]National Institute for Materials Science, 1-2-1 Sengen, Tsukuba 305-0047, Japan
[3]National Institute of Advanced Industrial Science and Technology, 1-1-1 Umezono, Tsukuba, Ibaraki 305-8560, Japan

zhantianzhuo@gmail.com

Abstract. We report the enhancement of thermoelectric power of Si nanowire micro thermoelectric generator (SiNW-μTEG) by improving the thermal conductivity of aluminum nitride (AlN) thermally conductive film. This work builds on the previous results that have been reported at SSDM 2017. [1] This work shows the measured thermal conductivities of the two different AlN films to understand the underlying mechanisms of the enhancement of the thermoelectric power.

1. Introduction
Energy harvesting devices that convert environmental heat energy into electrical energy using the Seebeck effect have attracted considerable attention for realizing a trillion-sensor network. [2] Recently, Si nanowires have been reported to be promising candidates for enhancing thermoelectric properties owing to the extremely low thermal conductivity. [3]

2. Sample preparation
Fig. 1 shows the schematic diagram of the fabricated SiNW-μTEG. SiNWs (8 μm in length; 100 nm in width) are fabricated on silicon-on-insulator (SOI) substrate by electron beam lithography and dry etching, and the surface is covered with a thermally grown SiO_2 film. P ions were implanted at a dose of 1.0×10^{15} ions/cm^{-2}. The Si pad and the both ends of the SiNWs are nickelized. AlN film (~550 nm thick) was deposited on the Si pad by sputtering as thermally conductive film. Two deposition methods were utilized to grow the AlN films: non-reactive RF sputtering method using AlN target in Ar gas and reactive RF sputtering method using Al target in Ar/N_2 mixed gas.

3. Results and discussion
Fig. 2 shows the X-ray diffraction (XRD) patterns of the two different AlN films. The full width half maximum (FWHM) results show that the reactively sputtered AlN film has a higher crystallinity than the non-reactively sputtered AlN film.

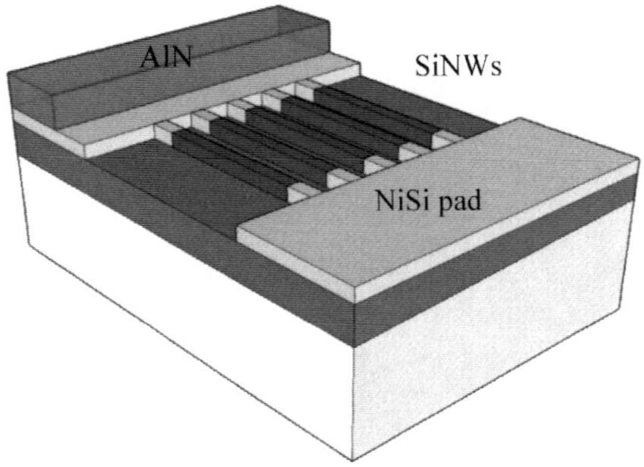

Fig. 1 Schematic diagram of the fabricated SiNW-μTEG.

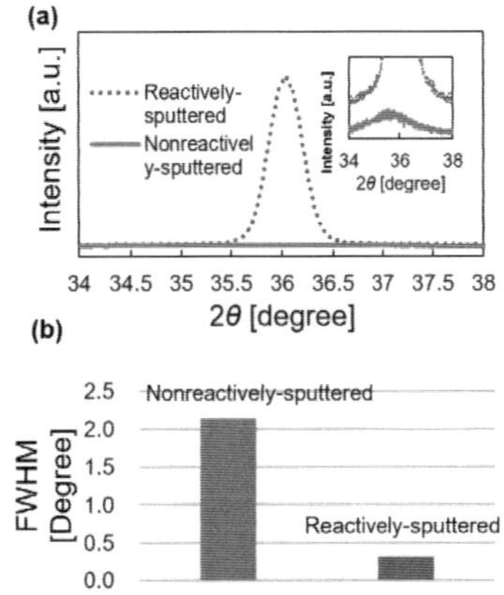

Fig. 2 (a) XRD patterns of the two different AlN films. (b) FWHM of the AlN (0002) peaks.[1]

The thermoelectric properties of the μTEGs were measured with a prober system with a micro IR thermography camera. The thermoelectric current (I_{TE}) and thermoelectric power density (P_{TE}) were measured by applying both the temperature difference and loading voltage (V_{load}) to the μTEGs. The maximum power densities of the SiNW-μTEGs with reactively- and non-reactively-sputtered AlN thermally conductive layers are 27.9 and 2.78 nW cm^{-2}, respectively. The thermoelectric power of the μTEG with the reactively sputtered AlN film is higher by an order of magnitude than that with the non-reactively sputtered AlN film. Fig. 3 shows the temperature distribution and temperature drop across the SiNWs of the μTEGs with two different AlN films. We measured the thermal conductivities of the AlN films by the ω method. [4] Fig. 4 shows that the reactively sputtered AlN film shows a higher thermal conductivity (3 W/mK) than the non-reactively sputtered AlN film (1.8 W/mK).

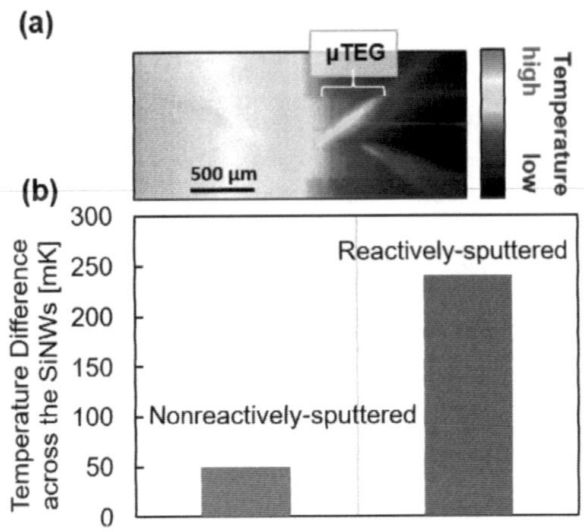

Fig. 3 (a) Thermography image of the μTEG. (b) Temperature difference across the SiNWs of the μTEGs with two different AlN films.[1]

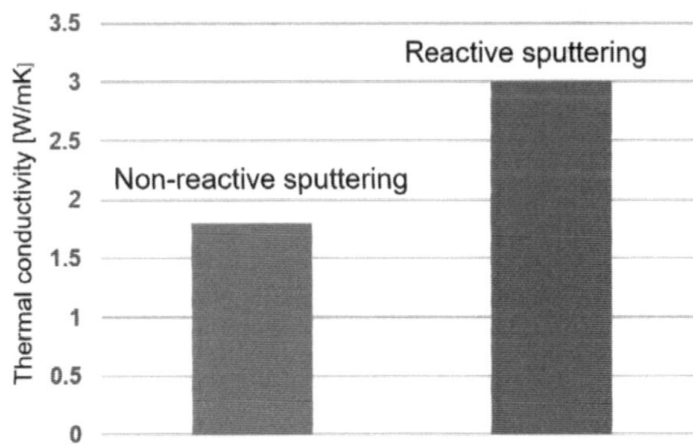

Fig. 4 Measured thermal conductivities of the two different AlN films.

4. Conclusions

We have found that the thermoelectric power of the μTEG with the reactively sputtered AlN film is higher by an order of magnitude than that with the non-reactively sputtered AlN film. The difference in the temperature drop across the SiNWs is attributed to the different parasitic thermal resistance in the two μTEGs. The results indicate that the improvement of the thermal conductivity of the thermally conductive film could enhance the thermoelectric power of μTEGs.

References

[1] R. Yamato, et al., 2017 International conference on solid state devices and materials.
[2] J. Gubbi, R. Buyya, S. Marusic, and M. Palaniswami, Future Gener. Comput. Syst. 29 (2013) pp. 1645-1660
[3] A. I. Boukai, et al., Nature, 451 (2008) pp. 168-171.
[4] R. Kato Y. Xu, and M. Goto, Jpn. J. Appl. Phys. 50 (2011) pp.106602A reference

Supramolecular Carbon Nanotube Films Adaptive to Thermoelectrics

Yoshiyuki Nonoguchi,[1,2] Tsuyoshi Kawai[1]

[1] Graduate School of Materials Science, Nara Institute of Science and Technology, Ikoma 630-0192, Japan,
[2] JST PRESTO, Kawaguchi 332-0012, Japan

nonoguchi@ms.naist.jp

Abstract. We present a rational route to stable carbon nanotubes, where supramolecular chemistry is important for stabilizing ionic bonding between charged nanotubes and counter ions. Developed SWNT materials, which are stable in air, at 100 °C, could readily be implemented to thermoelectric power generators.

1. Introduction

Some challenges in sustainable energy supply and powering remain to be solved in the near future. Thermoelectrics is a technique that immediately converts temperature difference into electric potential difference. Unlike conventional turbine generators with high pressure gases and large facilities, thermoelectric generators can produce power from small temperature differences and therefore can be widely applied in the recovery of waste heat from currently untapped sources such as commercial buildings and industrial plants. In addition, it can be expected that a lot of waste heat is potentially used as supplementary power for mobile vehicles (cars, buses, etc.) and information technology (IT) equipment including security and medical devices. In particular, the application of flexible and light-weight thermoelectric generators will be expanded in applications above. Importantly, for practical thermoelectric power conversion materials in a temperature range lower than 400 K, flexibility is crucial to cover and adhere closely to the heat sources that emit low thermal radiation.

Single-walled carbon nanotubes (SWNTs) are promising active thermoelectic materials owing to their narrow bandgap energy and high charge carrier mobility. Indeed, single-walled carbon nanotubes (SWNTs) show relatively high thermopower at room temperature. Thermoelectric modules mostly require both p-type and n-type materials. However, it has long been difficult to prepare air-stable n-type SWNTs because negatively-charged carbons, called carbanions, are highly reactive to electron acceptors such as oxygen. Additionally, the tuning of doping level is inevitable for optimizing thermoelectric properties. In order to solve these issues, here we introduce several approaches to the reliable doping of SWNTs.

2. Results and discussion

2.1. A general concept

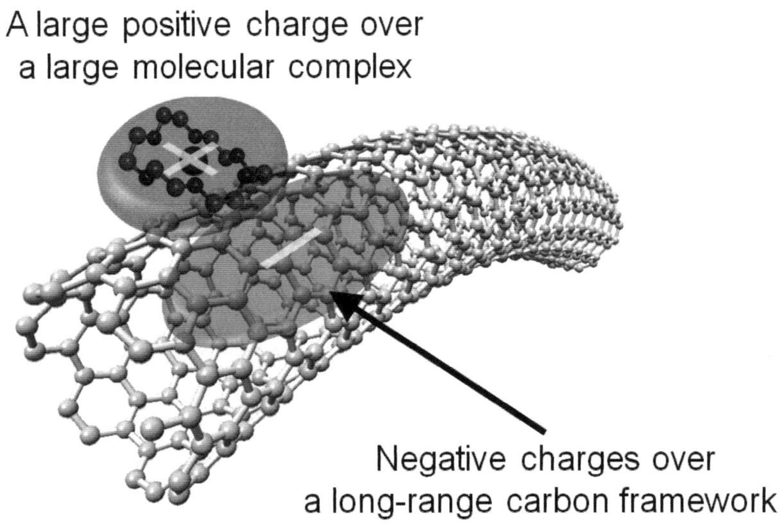

Figure 1. A schematic illustration of charge balanced n-type SWNT anion-metal cation complexes

We investigate the chemical doping of SWNTs in terms of non-covalent molecular interactions. Particularly, we attribute charged SWNTs to molecular ions, which enables the rational design of stable ion-pair complexes (Fig. 1) [2-6]. A charge could be delocalized over several carbon atoms through π-conjugation. Such charge is then compensated by a counter charge; for example, a negative charge doped into SWNTs can be stabilized by complexation with cations. In this situation, charge density balance between positive and negative charges is important for the stabilization; the "hard/soft acids/bases (HSAB)" rule is adopted [7].

2.2. Phosphine functionalization

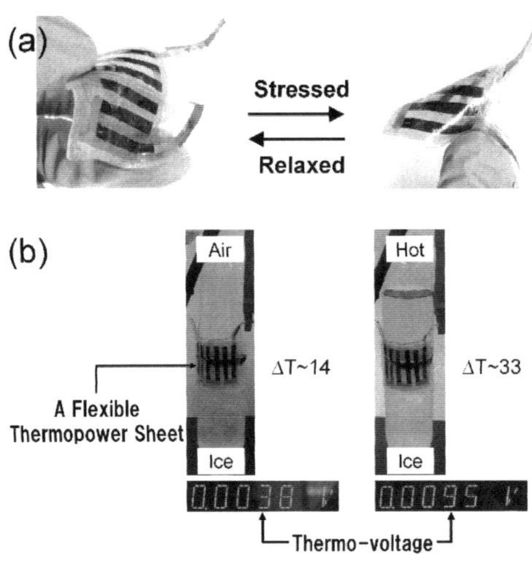

Figure 2. Prototype, flexible thermoelectric power generators equipped with p-type and n-type doped SWNT films. (a) flexibility. (b) Temperature-dependent thermovoltage generation.

Strategies for conventional molecular doping mostly rely on the use of amine-functional compounds. However, SWNTs functionalized with amines readily lose their n-type characteristics in air. We here used phosphines [1]; phosphonium generated after charge transfer could be a counter ion for n-type SWNTs. A series of phosphine derivatives successfully formed n-type SWNTs and their SWNT composites showed moderate air-stability. The encapsulation of molecular dopants inside the hollow inner space of SWNTs leads to the progression of n-type doping [3]. Furthermore, they can be integrated into flexible thermoelectrics that can recover any waste heat (Fig.2).

2.3. The functionalization of crown ether complexes

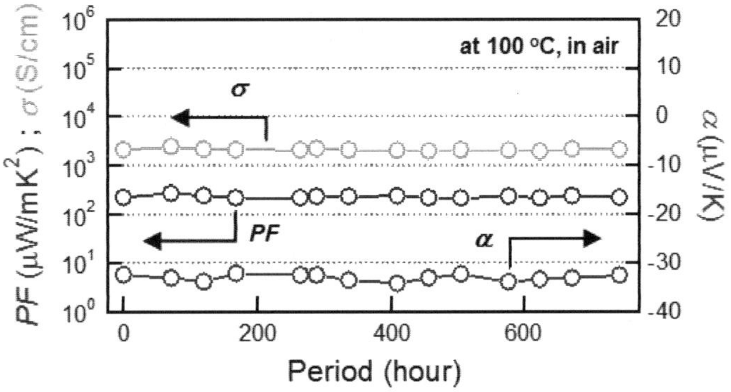

Figure 3. Time-course measurement of thermoelectric properties of n-type doped SWNTs with potassium (K^+)-benzo-18-crown complexes.

Molecularly-doped n-type SWNTs coordinated with large cationic complexes show unprecedented air- and thermal stability [2]. We used supramolecular complexes based on crown ethers as counter cations to n-type SWNTs; they spontaneously form complexes with alkali metal ions such as sodium (Na^+) and potassium (K^+) ions in polar solvent. Particularly, potassium complexes with benzo-18-crown ethers gave a SWNT composite showing air-stability above 100 °C (Fig. 3).

2.4. Organic dye supramolecular systems

Triarylmethane derivatives in their basic and reduced forms show excellent n-type doping ability against SWNTs, resulting in their stabilization in air [4,5].

2.5. Hole doping

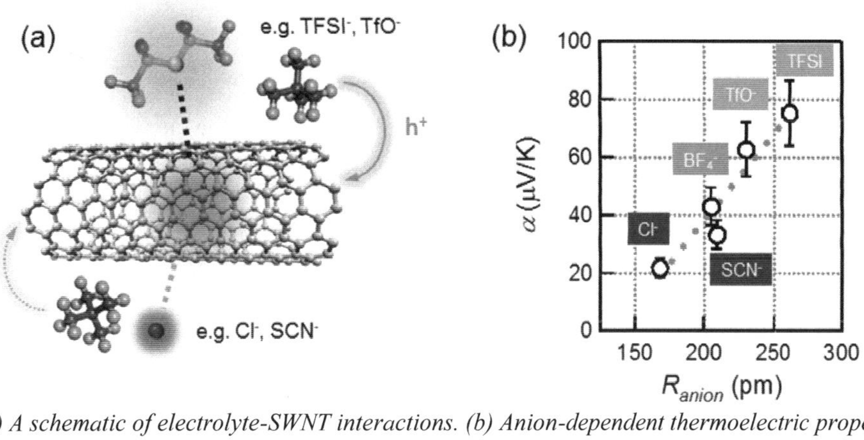

Figure 4. (a) A schematic of electrolyte-SWNT interactions. (b) Anion-dependent thermoelectric properties of polyelectrolyte-SWNT composites.

The p-type doping of SWNTs, and its stabilization are demonstrated by the complexation with charge-delocalized complex anions such as bis(trifluoromethanesulfonyl)imide (TFSI) (Fig. 4). The stable p-type doping of SWNTs is also developed using polyelectrolytes [6]; solid-state absorption and photoelectron spectroscopy reveal that the adsorption of electrolytes induces hole transfer to SWNTs.

3. Conclusion

We have demonstrated a rational strategy for successful amphoteric chemical doping to SWNTs. Additionally, these stable doped SWNTs were integrated into flexible and robust power modules enabling thermal-to-electricity conversion below 100 °C. The flexible thermoelectric power generator is, in near future, expected to harness electrical power from industrial plants and many other heat sources such as vehicles, and buildings, boosting energy-saving efforts in the fight against global warming.

4. acknowledgment

We thank financial supports from JSPS KAKENHI grant number JP 26790014, and JST-PRESTO grant number JPMJPR16R6.

References

[1] Y. Nonoguchi, K. Ohashi, R. Kanazawa, K. Ashiba, K. Hata, T. Nakagawa, C. Adachi, T. Tanase and T. Kawai, Sci. Rep., 3 (2013), article number 3344.
[2] Y. Nonoguchi, M. Nakano, T. Murayama, H. Hagino, S. Hama, K. Miyazaki, R. Matsubara, M. Nakamura and T. Kawai, Adv. Funct. Mater., 26 (2016), pp. 3021–3028.
[3] Y. Nonoguchi, Y. Iihara, K. Ohashi, T. Murayama, T. Kawai, Chem. Asian J., 11 (2016), pp. 2423–2427.
[4] Y. Nonoguchi, A. Tani, T. Ikeda, C. Goto, N. Tanifuji, R. M. Uda, T. Kawai, Small, 13 (2017), 1603420.
[5] Y. Nonoguchi, S. Sudo, A. Tani, T. Murayama, Y. Nishiyama, R. M. Uda, T. Kawai, Chem. Commun., 53 (2017), 10259–10262.
[6] M. Nakano, T. Nakashima, T. Kawai and Y. Nonoguchi, Small, 13 (2017), 1700804.
[7] R. G. Pearson, J. Am. Chem. Soc., 85 (1963), pp. 3533–3539.

Design and Performance of Transverse-Type Thin-Film Nano-Thermoelectric Generators

N Chiwaki[1]**, T Seino**[1] **and S Sugahara**[1]

Laboratory for Future Interdisciplinary Research of Science and Technology,
Tokyo Institute of Technology
4259-J3-14 Nagatsuta, Yokohama, 226-8503, Japan

E-mail: chiwaki.n.aa@m.titech.ac.jp

Abstract. Design methodology and performance of nano thermoelectric generators (nTEGs) using human body heat are investigated for wearable device applications. A transverse type of module suitable for a thin-film thermopile is employed for the nTEGs. It is revealed that the thermal and electrical insulation of the interspace of their modules is an important factor to determine the output power of the nTEGs. A new module structure using insulator/vacuum hybrid insulation with its design methodology that maximizes the output power is developed. Output power of a few milliwatts can be achieved by a wrist-band style mounting of the nTEG module with an optimized design.

1. Introduction

Internet of humans (IoH) is a new system for health and medical care using accumulated vital/medical data of humans, which enables us to give useful/important knowledges for improving/maintaining quality of life. Wearable devices play an essential role as a man-machine interface for IoH systems, and micro/nano thermoelectric generators (μ/nTEGs) [1, 2] can be applied to a power source of these wearable devices. The wearable devices for IoH postulate the following performances and features: (i) The devices need power of a few mW for short-distance wireless communication, since they are expected to be used with communication devices/facilities such as a smartphone or wireless LAN. (ii) A driving voltage of 0.5-1 V also needs to drive CMOS circuits in the wearable devices. (iii) A thin-film thermoelectric material is required for size/weight saving and cost reduction, which also gives rise to the usage of a micro- or nano-fabricated thermopile for μ/nTEGs. It is worthy to note that fabrication process technologies used for very-large-scale integrated circuits (VLSIs), such as lithography, can be diverted to μ/nTEGs fabrication, and these technologies highly enhances the degree of freedom of design for μ/nTEGs [2]. Also note that sophisticated minute-contact technologies used for VLSIs are also useful for forming thermal and electrical contacts of μ/nTEGs.

Recently, we proposed a transverse-type μTEG with highly integrated thin-film Seebeck elements and developed its design methodology [2]. Since the heat flow through the Seebeck elements are horizontal to the cold and hot plates of the module, high output power adaptable to wearable devices can be generated even by means of a thin-film thermopile. The thermal insulation of the interspace between the cold and hot plates of the module is one of the important factors to determine the output power, and thus the performance limits of the module can be obtained using the vacuum isolation [2].

In this paper, design methodology of transverse-type thin-film nTEGs with various module structures is developed and their performances are computationally analyzed. Although the thermal-electrical insulator of the interspace of the module severely degrades the performance, a newly proposed module structure using insulator/vacuum hybrid insulation that would be preferable to the device fabrication can achieve sufficient output power applicable to wearable devices.

Table 1. Physical constants used for module design.

Module size DxD	1 cmx1 cm	Temperature difference ΔT_S	10 K
Thermoelectric material (BiTe)		Interconnect/contact material (Cu)	
Thickness t_0	100 nm	Thermal conductivity λ_{Cu}	386 W/(m·K)
Seebeck coefficient $S = S_p - S_n$	434 μV/K	Electrical resistivity ρ_{Cu}	17 nΩm
Thermal conductivity $\lambda = (\lambda_p + \lambda_n)/2$	1.43 W/(m·K)	Interlayer insulator (Porous Silica)	
Electrical resistivity $\rho = (\rho_p + \rho_n)/2$	8.11 μΩm	Thermal conductivity λ_{PS}	35.7 mW/(m·K)

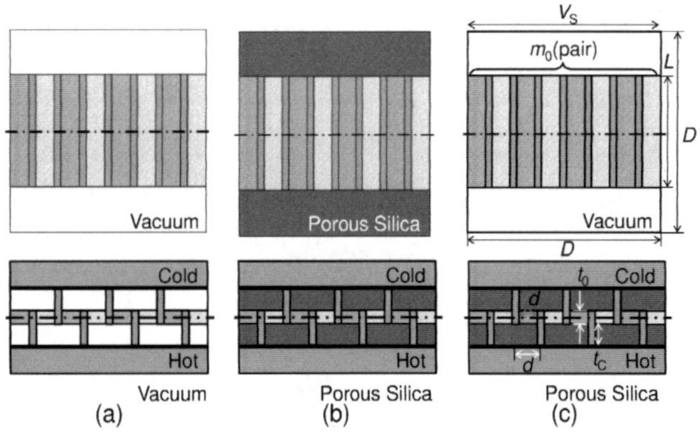

Fig.1. Schematics of transverse-type nTEGs using (a) V/V, (b) PS/PS and (c) PS/V modules.

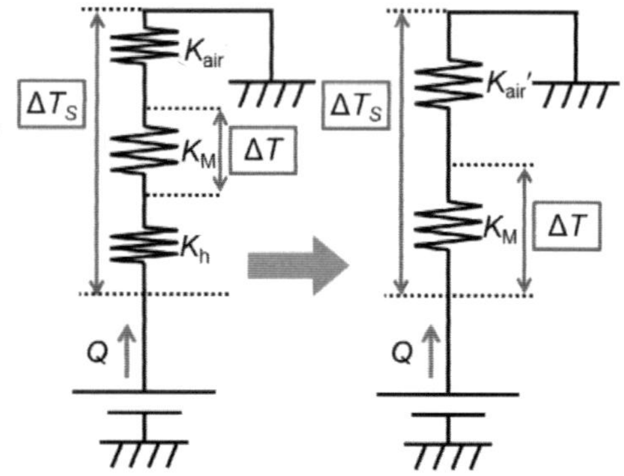

Fig.2. Homeothermic human model of a human body/wearable device/ambient system.

2. Module structures and design methodology

Figs. 1 (a)-(c) show schematics of various transverse-type nTEG modules. The Seebeck elements occupy the DxL region on the inside of the DxD module area for all the modules. The interspace between the hot and cold plates in the DxL region is thermally and electrically insulated by vacuum (V) isolation (Fig. 1 (a)) or porous silica (PS) filling (Figs. 1 (b) and (c)), and its outside region of the module is insulated by V isolation (Figs. 1 (a) and (c)) or PS filling (Fig. 1 (b)). Hereafter, these insulation structures shown in Figs. 1 (a), (b) and (c) are simply denoted by V/V, PS/PS and PS/V, respectively.

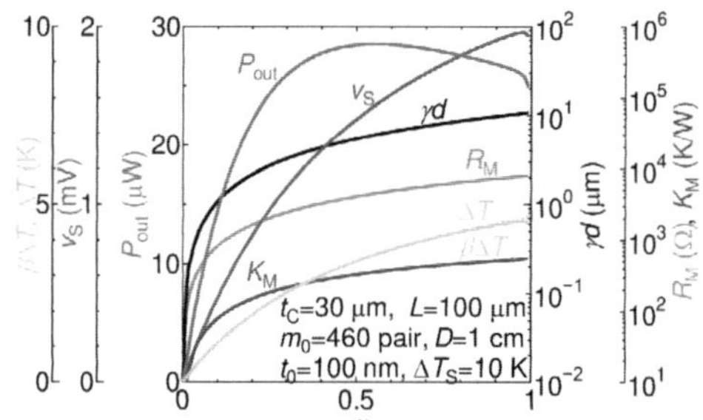

Fig. 3. P_{out} and design parameters/indices as a function of γ for the V/V insulation module.

Fig. 2 shows a homeothermic human model used for module design. The model includes heat flow limitation due to the thermogenic action of humans. In this system model, the maximum heat flow depends on the thermal resistance K_M of the module and K_{air}' that is a sum of the thermal resistance K_{air} from the module surface to ambient and the thermal resistance K_h of human body near the skin. Therefore, these (K_{air}' and K_M) are optimized to achieve maximized output power under the homeothermic human model. In this paper, a maximum heat flow of Q = 10 mW/cm^2 is assumed [3].

To maximize output power, the electrical resistance R_M of the module and K_M that are in the trade-off relation need to be optimized. For this purpose, a single trade-off parameter γ ($0<\gamma<1$) that represents a ratio of the width γd of a thermoelectric element to the unit area width d (see Fig.1) is introduced. The number and size of the thermoelectric elements and the resulting output voltage and power can be completely determined by γ instead of the thermal and electrical resistances [2]. The important design parameters are the width γd, the length L, the number m_0 of pair of the Seebeck element and the thickness t_C of the contact metal layer, and the useful design indices are the electromotive force v_S of each Seebeck element, the temperature difference ΔT of the module, the temperature difference $\beta\Delta T$ of each Seebeck element, K_M, K_{air}' and R_M. The physical constants used for the following designs are shown in Table 1. Thermal and electrical resistance network models are used to analyse the output characteristics of the nTEGs. Thermal resistance due to radiation heat and spreading thermal resistance are used for the V and PS insulation portions, respectively. The output power on electrical matched load, given by $P_{out} = V_s^2/4R_M = (m_0 \times v_s)^2/4R_M$, is used for the module designs, in which V_S (= $m_0 v_s$) represents the total electromotive force of the module.

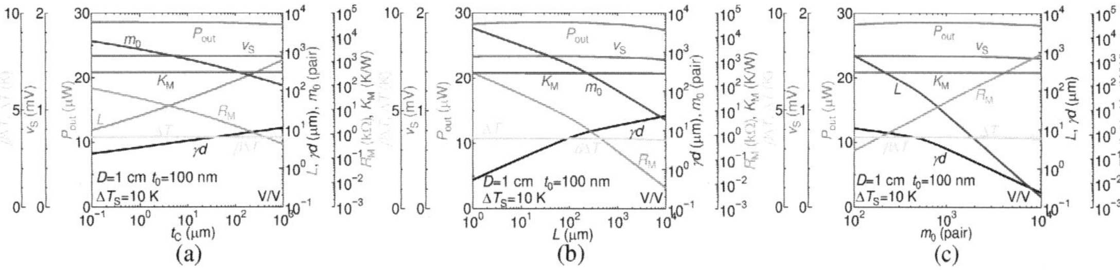

Fig. 4. P_{out} and design parameters/indices as a function of (a) t_C, (b) L and (c) m_0 for the V/V module, in which all the parameters/indices are optimized by γ. t_C is set to $t_C = (1+\gamma)d$ for (b) and (c).

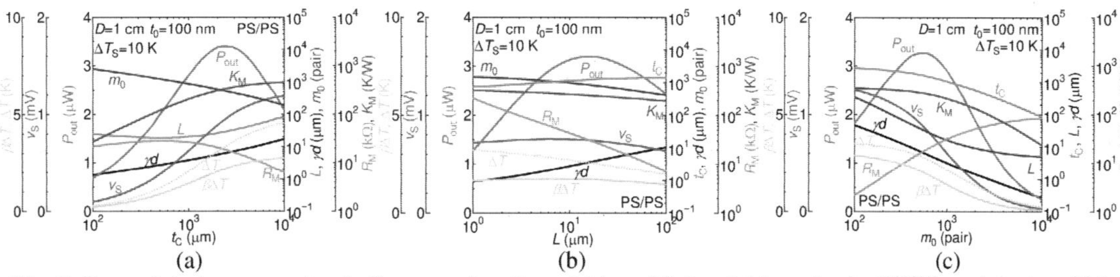

Fig. 5. P_{out} and design parameters/indices as a function of (a) t_C, (b) L and (c) m_0 for the PS/PS module, in which all the parameters/indices are optimized by γ.

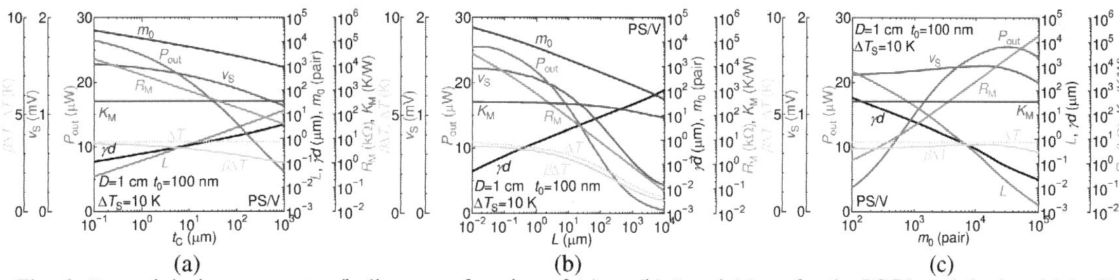

Fig. 6. P_{out} and design parameters/indices as a function of (a) t_C, (b) L and (c) m_0 for the PS/V module, in which all the parameters/indices are optimized by γ. t_C is set to $t_C = (1+\gamma)d$ for (b) and (c).

3. Design and performance

Firstly, the optimum design and resulting performance of the V/V module that gives the performance limit of the nTEGs are discussed. Fig. 3 shows P_{out}, v_S, ΔT, $\beta\Delta T$, γd, K_M, and R_M as a function of γ, in which t_C, L, m_0, and K_{air}' are fixed for simplicity (K_{air}' is adjusted to a particular value so that Q reaches 10 mW at the P_{out} peak). K_M and R_M increase with increasing γ owing to the increase in γd, and P_{out} shows the peak shape. Therefore, P_{out} is optimized by the single parameter γ. The γ value at the peak position represents the optimum condition of K_M and R_M.

Fig. 4 (a) shows P_{out} and the other design parameters/indices as a function of t_C for the V/V module. P_{out} is almost independent of t_C and remains constant at a maximized high value (~28 μW). Although v_S is maintained constant and m_0 decreases with increasing t_C, i.e., the total electromotive force V_S decreases with increasing t_C, R_M can be reduced (or designed) so that P_{out} is kept constant. The constant v_S can be achieved by the constant $\beta\Delta T$ (or K_M). Note that m_0 (which is a determination factor of V_S) can decreases with increasing L (which determines R_M) with K_M kept constant so that P_{out} remains constant at the highest value. Figs. 4 (b) and (c) show P_{out} and the other design parameters/indices as a function of L and m_0, respectively, for the V/V module. In both the cases, P_{out} also remains constant at 28 μW, since v_S is almost unchanged and the correlation between m_0 and L (or R_M) is substantively the same as the case in Fig. 4 (a). The maximized P_{out} value can be achieved by a number of combinations of L and m_0, and γd is determined by a selected m_0 (or L) value. The wide γd range between a few hundreds of nanometers to a few micrometers is applicable for the V/V module design.

The PS/PS module is preferable from the point of view of device fabrication. However, the output power is severely degraded. Fig. 5 (a) shows P_{out} and the other design parameters/indices as a function of t_C for the PS/PS module. P_{out} exhibits a single peak shape that results from the t_C-dependence on v_S, m_0, and R_M. Figs. 5 (b) and (c) show P_{out} and the other design parameters/indices as a function of L and m_0, respectively, for the PS/PS module. In both the cases, P_{out} achieves a peak. The maximum P_{out} value is severely lowered in comparison with the case of the V/V module. The PS/PS module requires the contact metal layer with thicker t_C in order to achieve a K_M value comparable to that of the V/V module. This is due to the high thermal conductivity of the PS insulation. In this situation, m_0 of the PS/PS module is smaller than that of the V/V module, and L (R_M) of the PS/PS module is reduced (enlarged) in comparison with the case of the V/V module. Therefore, P_{out} is degraded for the PS/PS module.

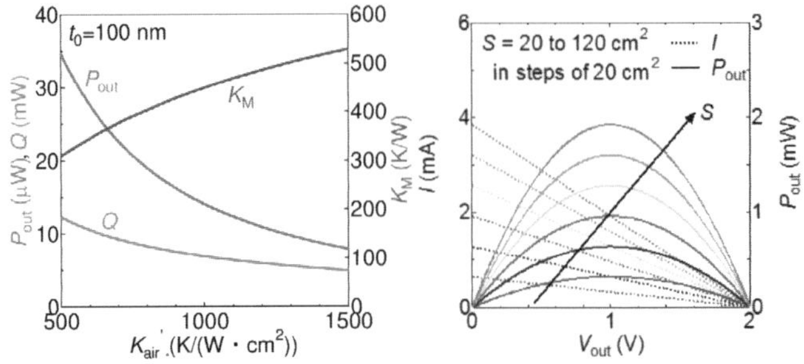

(a) (b)

Fig. 7. (a) P_{out}, Q and K_M as a function of K_{air}' for the PS/V module. (b) Output characteristics of the PS/V module, in which an optimum design at L=10 μm is used.

The PS/V module that would be relatively easy-to-fabricate insulation structure is effective at improving output characteristics of nTEGs. Fig. 6 (a) shows P_{out} and the other design parameters/indices as a function of t_C for the PS/V module. P_{out} is enhanced when the contact metal layer with thinner t_C is used. This is because v_S and m_0, i.e, V_S, can effectively increase with decreasing t_C, compared with the increase in R_M. Note that although m_0 decreases with increasing L so that K_M is kept constant, $\beta\Delta T$ (v_S) drops with increasing t_C owing to the effect of the parasitic thermal resistance of the metal contacts. Fig. 6 (b) shows P_{out} and the other design parameters/indices as a function of L for the PS/V module. P_{out} is strongly enhanced by reducing L. Owing to the correlation between m_0 and L, m_0 increases with decreasing L so that K_M is almost kept constant, although K_M drops for wider L values. Therefore, v_S and V_S are enhanced by reducing L, resulting in the increase in P_{out}. Fig. 6 (c) shows P_{out} and the other design parameters/indices as a function of m_0 for the PS/V module. L decreases with increasing m_0 with K_M held constant owing to the correlation between these parameters. In this case, v_S is successfully maintained constant. Therefore, V_S and thus P_{out} are enhanced with increasing m_0. Note that R_M increases with P_{out} and overhigh R_M suppresses the enhancement of P_{out}, as shown in Figs. 6 (a)-(c). It is worthy to note that the optimum feature size of the Seebeck elements is in 20-80 nm as shown in Fig. 6, i.e., the "nano" TEG structure is required for this module.

Fig. 7 (a) shows P_{out}, Q and K_M as a function of K_{air}' for the PS/V module, in which no restrictions are imposed on Q. P_{out} decreases with increasing K_{air}'. K_{air}' needs to be selected within the heat flow limit of humans. An alumite-treated fine-fin plate structure would allow K_{air}'=~700K/Wcm^2, and thus P_{out}> 20 μW/cm^2 can be achieved. This K_{air}' value also adapts the thermogenic ability of humans at room temperature. Fig. 7 (b) shows output characteristics of the PS/V insulation module, in which an optimum design at L=10 μm shown in Figs. 6 (b) and (c) is used and the module-mounting area is varied from 20 to 120 cm^2. When the module is mounted to a wrist-band style (100 cm^2), P_{out} of 1.6 mW can be achieved. This output value is sufficiently applicable to wearable devices with short-distance communication functionality.

Acknowledgment
This work was partly supported by Research and Development Project for Innovative Thermal Management Materials and Technologies from NEDO.

References

[1] Leonov V, Torfs T, Fiorini P and Van Hoof C 2007 *IEEE Sensors Journal* **7** pp 650-657
[2] Kondo T, Chiwaki N and Sugahara S 2017 *IEEE Electron Devices Technology and Manufacturing Conf.* (Toyama, Japan) pp 201-203
[3] Leonov V 2013 *IEEE Sensors Journal* **14** pp 2284-2291

IOP Conf. Series: Journal of Physics: Conf. Series **1052** (2018) 012134 doi:10.1088/1742-6596/1052/1/012134

Flexible thermoelectric system based on inorganic bulk materials

H.J Park, D.G Kim, Y.M Eom, W. Dimuthu, D.K Lee and W. Kim*

School of Mechanical Engineering, Yonsei University, Seoul 120-749, Republic of Korea

*woochul@yonsei.ac.kr

Abstract. This paper reports the flexible thermoelectric system based on inorganic bulk materials. Due to its structural design, while the bulk inorganic thermoelectric materials were used, it shows enough flexibility to be adopted on irregular surface such as human skin. To measure the thermoelectric performance, the refrigeration and power harvesting were conducted on human skin. It successfully refrigerate the human skin about 3.3 K power with the portable battery. And the 78-μW power output was achieved with 7.2-mV voltage on human skin.

1. Introduction

Thermoelectric phenomena is the energy conversion between thermal and electric energy and it has been used for waste heat recovery or refrigeration. But due to its structural limitation, its application has restricted to only flat surface so that the researches on flexible thermoelectric devices has been investigated. In case of mainly researched flexible thermoelectric devices, however, those are manufactured based on polymer materials[1, 2] or printed inorganic materials[3] for the flexibility. But due to its fabricated orientation, it is hard to make the temperature difference and shows the lower thermoelectric performance compared to bulk inorganic materials. In this investigation, flexible thermoelectric devices has been manufactured based on bulk inorganic materials which shows high thermoelectric performance with structural design and it shows high flexibility which can enhance its application fields.

2. Experiment

2.1. Material synthesis

Bi-Te based inorganic thermoelectric materials were synthesized at high temperature after sealed into the quartz tube with vacuumed. The melt grown materials were ball-milled for 1 h with 500 rpm. The ball-milled powder were sintered with using SPS (Spark Plasma Sintering) at 723 K for 10 min.

2.2. Device manufacturing

Sintered materials were polished in the shape of 5-mm cubic and electroplated with Ni /Au for the electrical contact. Each polished elements gets then stuck in holder made of Bakelite and holders are connected with commercial wires through holes on holders. And flexible wires are soldered as the electrodes for the flexibility as in Figure 1.

Figure 1. Schematic of flexible thermoelectric device

3. Result

3.1. Refrigeration on human skin

The flexible thermoelectric device was applied on human skin for the refrigeration integrated with cool-gel as the heat sink. The portable battery was used as the power source for the refrigeration in which 1.27 A current flew through the thermoelectric device. It successfully refrigerated the human skin about 3.3 K as shown in Figure 2. Since the commercial portable battery was used as the power source, it proposed the possibility for the mobile refrigerator for the human skin .

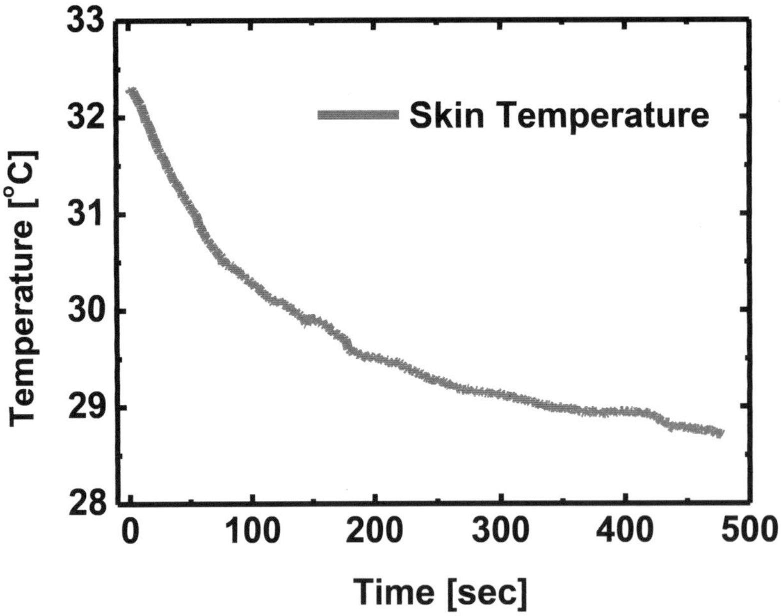

Figure 2. Refrigeration performance on human skin

3.2. Powergeneration on human skin
The flexible thermoelectric device was also applied on human skin for the power generation. Since the inorganic bulk thermoelectric materials ,which showed high thermoelectric performance near room temperature, were used, it produced 78-μW power output with 7.2-mV output voltage as shown in Figure 3. The output power density was about 4.8 μW/cm^2 which was high enough for the further application.

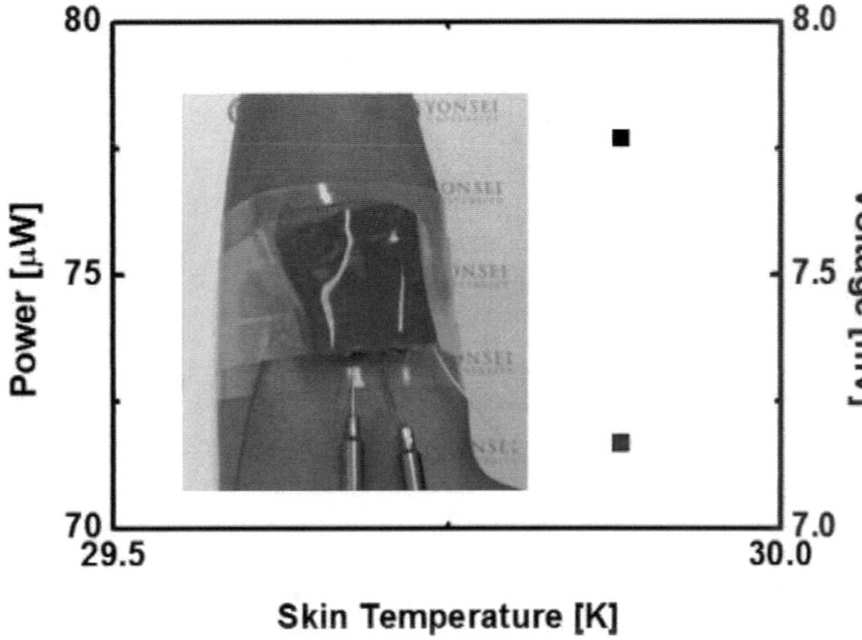

Figure 3. Output power with voltage from body heat harvesting

4. Conclusion
In this research, flexible thermoelectric system is manufactured by structural design based on inorganic bulk material which shows high thermoelectric performance and stability. From those characteristics, it is successfully applied on human skin for both energy harvesting and refrigeration for further applications. For the refrigeration, it successfully refrigerated the human skin with 3.3 K temperature drop powered by portable battery. Oriented from its structural design, the size of the thermoelectric device can be further enlarged for the application. For the body heat harvesting on human skin, it produced 78 μW power with 7.2 mV voltage output. Considering the size of the device, the output power density was about 4.8 μW/cm^2 which is the highest among the application of flexible thermoelectric device on human skin. The structural design proposed in this paper shows the possibility for the usage of the inorganic based thermoelectrice device to be enlarged.

Acknowledgement
This research was partially supported by the Graduate School of YONSEI University Research Scholarship Grants in 2017 and National Research Foundation of Korea (NRF) Grant funded by the Korean Government (MSIP) (NRF-2015R1A5A1036133)

References

[1] Bubnova O, Khan ZU, Malti A, Braun S, Fahlman M, Berggren M, et al. Optimization of the thermoelectric figure of merit in the conducting polymer poly(3,4-ethylenedioxythiophene). Nature Materials. 2011;10:429-33.
[2] Kim GH, Shao L, Zhang K, Pipe KP. Engineered doping of organic semiconductors for enhanced thermoelectric efficiency. Nature Materials. 2013;12:719-23.
[3] Lu ZY, Layani M, Zhao XX, Tan LP, Sun T, Fan SF, et al. Fabrication of Flexible Thermoelectric Thin Film Devices by Inkjet Printing. Small. 2014;10:3551-4.

Formation of various epitaxial nanodots in Si films for thermoelectric materials

S. Sakane[1], K. Watanabe[1,2], T. Fujita[2,3], N. Naruse[4], and Y. Nakamrua[1,2*]

[1]Osaka University, Japan, [2]CREST-JST, Japan, [3]Tohoku University, Japan, [4]Shiga university of Medical Science, Japan

*E-mail: nakamura@ee.es.osaka-u.ac.jp

Abstract. Nanostructures are expected to enhance thermoelectric properties of Si-based materials. For enhancement, the simultaneous achievement of low κ and high σ is required. In our previous study, we have demonstrated drastic reduction of κ in Si by introducing epitaxial connected Si NDs resulting in the smallest κ in Si. We have also reported that Si films containing Ge NDsexhibited the bulklike σ and the drastic κ reduction compared with that of bulk Si. However, S was not controlled in this nanostructure. In general, introduction of interfaces or high S materials can increase S of the total film. In addition to control of κ and σ, S can be controlled by tuning the material and size of NDs. In this work, we developed the formation method of the various material and size of introduced NDs.

1. Introduction

Thermoelectric conversion has been attracted as one of the energy harvestings. Thermoelectric conversion efficiency depends on dimensionless figure of merit ZT ($=S^2\sigma T/\kappa$). S is Seebeck coefficient, σ is electrical conductivity, κ is thermal conductivity and T is absolute temperature. For its enhancement, ZT has to be increased. The conventional approach to enhance ZT is to use heavy elements (Te, Pb, etc.) for κ reduction while maintaining high σ. However, these materials are often composed of rare metals, which are high cost and toxic. On the other hand, Si-based materials, which are eco-friendly and low cost materials , indicated high κ because they are composed of light elements , resulting in low ZT. Therefore, they are now very far from practical use.

Recently, introducing nanostructures has been expected to enhance thermoelectric performance of Si based materials. Nanostructuring was expected to cause phonon scattering, resulting in κ reduction. We have demonstrated drastic reduction of κ in Si by introducing epitaxial connected Si NDs resulting in the smallest κ in Si [1]. We have also reported that Si films containing Ge NDs exhibited the bulk-like σ and the drastic κ reduction compared with that of bulk Si. In this structure, epitaxial Ge NDs worked as phonon scattering centers and epitaxial Si layers worked as electric conduction layers. This result indicated the possibility of the independent control of the thermal and electric conductivities [2,3]. However, S was not controlled in this nanostructure. In general, introduction of interfaces or high S materials can increase S of the total film. In addition to control of κ and σ, S can be controlled by tuning the material and size of introduced NDs. In this work, we developed the formation method of the various material and size of NDs.

2. Experimental

We fabricated three types of Si films containing different NDs (Ge, Si, β-$FeSi_2$) on undoped Si substrates (>1000 Ωcm). Si substrates were introduced into molecular beam epitaxy chamber after chemical treatment. Clean Si surfaces were formed by depositing Si buffer layer at 550 °C. Ultrathin SiO_2 films were formed on clean Si surfaces by oxidizing at 600 °C at 2×10^{-4} Pa, and then nanowindows were formed in the ultrathin SiO_2 films by depositing 1 monolayer of Ge or Si ($Si + SiO_2 \rightarrow 2SiO\uparrow$ or $Ge + SiO_2 \rightarrow GeO\uparrow + SiO\uparrow$). Then, epitaxial Ge, Si or β-$FeSi_2$ NDs were formed by depositing Ge, depositing Si or codepositing Fe and Si (Fe : Si =1 : 2). [4-6]. Epitaxial Si layers were then formed on NDs by Si deposition. By repeating these formation processes: the ultrathin SiO_2 film,

NDs, and Si layers (= one formation cycle), we fabricated Si films containing various NDs. At these processes, we conducted *in-situ* observation of the surface structures by reflection electron diffraction (RHEED) and performed *ex-situ* observation of the samples by scanning transmission electron microscopy (STEM).

3. Results and Discussion

Figures 1 (a)-(c) show the RHEED patterns of Ge, Si and β-$FeSi_2$ NDs of first formation cycle, respectively. These patterns indicated epitaxial growth of $(001)_{Ge}$// $(001)_{Si}$, $(111)_{Si}$// $(111)_{Si}$ and $(110)_{\beta\text{-}FeSi2}$// $(111)_{Si}$ or $(101)_{\beta\text{-}FeSi2}$// $(111)_{Si}$. Ge, Si and β-$FeSi_2$ NDs samples were fabricated by repeating 8, 6 and 6 formation cycles. We confirmed epitaxial growth of these NDs continued.

We fabricated the larger size β-$FeSi_2$ NDs (36 ML) than our previous study (18 ML, 10 nm)[6]. The cross-sectional STEM image of β-$FeSi_2$ NDs (36 ML) sample was shown in Fig. 2. We found that multilayered structures of β-$FeSi_2$ NDs /Si layers were formed, and that the ND size was about 20 nm, which is larger than our previous study[6]. The epitaxial relationship of $(110)_{\beta\text{-}FeSi2}$// $(111)_{Si}$ or $(101)_{\beta\text{-}FeSi2}$// $(111)_{Si}$ was also confirmed by fast Fourier transformation analysis of STEM images.

In general, formation of stacking structure of NDs is difficult due to the silicidation. However, in the case of the present formation using ultrathin SiO_2 film, ultrathin SiO_2 films can work as barrier of intermixing, resulting in the success in the formation of stacking structure of various NDs.

4. Conclusion

We fabricated various epitaxial nanodots in Si films. The formation of Ge, Si and β-$FeSi_2$ NDs samples was confirmed by RHEED and STEM. Also, large size β-$FeSi_2$ NDs sample was controlled by large deposition amount.

These various structures can be expected to show various thermoelectric properties.

Acknowledgement

This work was supported by JST-CREST program, Grant-in-Aid for Scientific Research A (16H02078), Grant-in-Aid for Exploratory Research (15K13276).

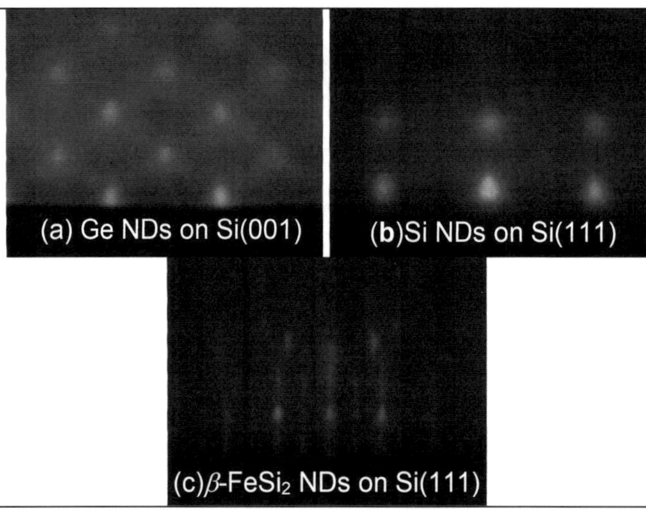

Figure 1. RHEED patterns of (a)Ge, (b)Si, and (c)β-$FeSi_2$ NDs at first cycle.

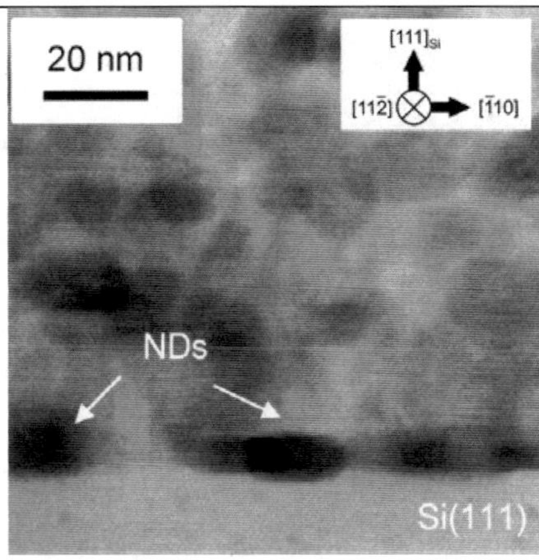

Figure 2. STEM image of β-$FeSi_2$ NDs (36 ML) samples

References

[1] Y. Nakamura, M. Isogawa, T. Ueda, S. Yamasaka, H. Matsui, J. Kikkawa, S. Ikeuchi, T. Oyake, T. Hori, J. Shiomi and A. Sakai 2015 *Nano Energy* **12**, 845.
[2] S. Yamasaka, Y. Nakamura, T. Ueda, S. Takeuchi, and A. Sakai 2015 *Sci. Rep.* **5**, 14490.
[3] S. Yamasaka, K. Watanabe, S. Sakane, S. Takeuchi, A. Sakai, K. Sawano, and Y. Nakamura 2016 *Sci. Rep.* **6**, 22838.
[4] Y. Nakamura, K. Watanabe, Y. Fukuzawa, and M. Ichikawa 2005 *Appl. Phys. Lett.* **87**, 133119.
[5] Y. Nakamura, Y. Nagadomi, S. P. Cho, N. Tanaka, and M. Ichikawa, 2006 *J. Appl. Phys.* **100**, 044313-1.
[6] S. Sakane, M. Isogawa, K. Watanabe, J. Kikkawa, S. Takeuchi, A. Sakai, and Y. Nakamura 2017 J. *Vac. Sci. Technol. A* **354**, 041402-1.

Development of stacking type thermoelectric power generation unit for potential waste heat recovery applications

A. Yamamoto[1] and H. Nishiate[1]

[1] National Institute of Advanced Industrial Science and Technology (AIST), Tsukuba, Ibaraki, 305-8568, Japan
e-mail : a.yamamoto@aist.go.jp

Abstract. A 10 W class thermoelectric power generation unit is designed and prototyped. The power unit is designed based on a concept of a compact heat exchanger integrated with thermoelectric module. The power unit specifically targeted hot water waste heat of which temperature is below 100 °C. The preliminary lab testing of the unit showed maximum power output of 10.6 W under conditions of T_h = 95 °C, T_c = 10 °C, the flow rate = 7 L/min. The size of unit is 5.4 x 11.2 x 14.4 cm and the calculated specific power of the unit is 12.1 W/L or 12 kW/m^3 which is much higher than a value ever reported. We conducted long term stability tests in three different sites; 1) hot spring, 2) waste heat from inductive heating furnace and 3) boiler blowdown water in addition to one preliminary test at other hot spring site. All collected data in the demonstrations described above proved good stability of the performance of the power unit. On-site, self-powered remote sensing is also demonstrated at the hot spring site and successful data acquisition of temperatures and water flow rates at every seconds were confirmed.

1. Introduction

The Fukushima nuclear accident in 2011 in Japan recalled us the importance of deployment of renewable energy and energy saving technology. Energy saving is the most important measure for social sustainability and utilizing wasted thermal energy is potential target to realize energy saving in large scale. Along this context, thermoelectric power generation from wasted heat has been re-focused and increasing number of demonstrative experiments have been reported [1][2].

In this study we report the stacking type thermoelectric power generation module which can be utilized in low grade waste heat such as hot water from factory, hot spring etc. [3][4][5][6][7]. The units were recently tested in various fields and showed reliable performance in those experiments.

2. Stacking Type Power Generation Unit

2.1. *Design Concept of the Unit*

A cost of initial installation is one of the most important part of deployment of new energy technology for any purpose. A binary turbine power generation can be a potential candidate technology for low grade waste heat recovery, but the initial cost still much higher than 0.5 million yen/kW. [8][9] Thermoelectric power generation has great potential in cost reduction since the devices used in the system is in principle semiconducting product. In this study we designed the power generation unit with conventional thermoelectric module integrated with plastic heat exchanger instead of expensive

IOP Conf. Series: Journal of Physics: Conf. Series **1052** (2018) 012136 doi:10.1088/1742-6596/1052/1/012136

metal based heat exchange. Ten 40 mm x 40mm thermoelectric modules were stacked with plastic water jacket and the piping was also made of plastics such as POM or PPS.

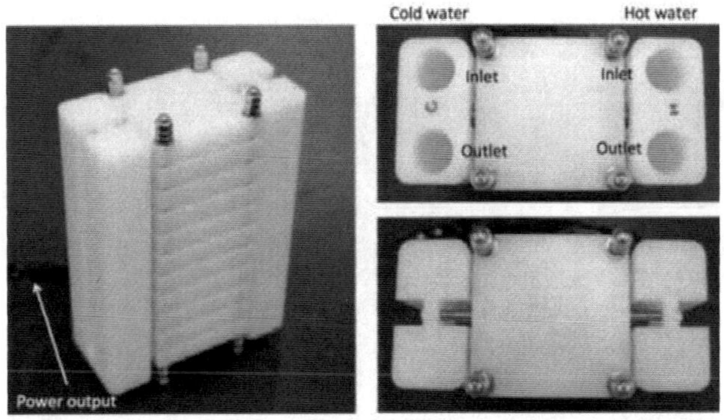

Figure 1. Pictures of prototype of the stacking type thermoelectric power generation unit.

2.2. *Basic performance of the Unit*

The power generation characteristics of the unit was examined in a laboratory test bench. Hot water and cold water were supplied by circulators and electrical power out put was measured as a function of the hot water temperature and the flow rate. Figure 2 shows the basic properties of the unit. The maximum power output was 10.4W at 95°C and 7 L/min.

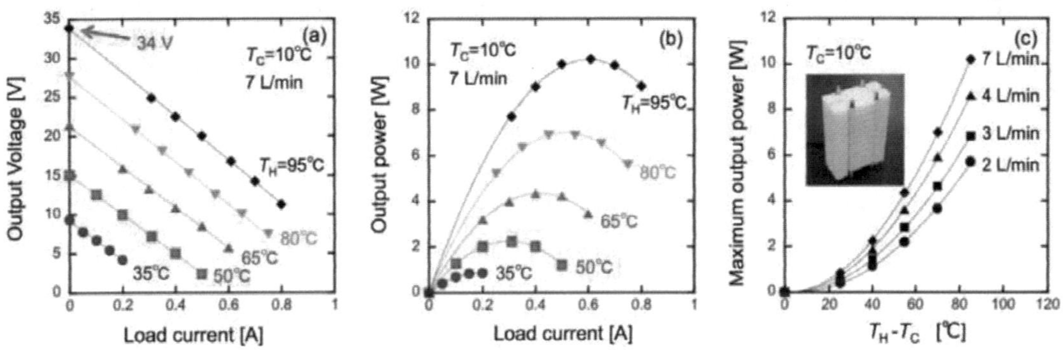

Figure 2. Basic power generation performance of the unit. (a) voltage - current relationship, (b) output power – current, (c) temperature difference – maximum output power.

3. Results of field test

The prototyped power generation units were then tested in three different sites to verify the durability in real use. The examined hot water are supplied from 1) cooling water of inductive heating furnace (SINTOKOGIO Ltd., Aichi) 2) secondary hot water in hot spring centre Yakushi-yu (Yumura hot spring, Hyogo) and 3) boiler blowdown water (The Japan Wool Textile Co., Ltd., Hyogo). The expected temperature of hot water are roughly 40 °C, 60 °C, and 80 °C, respectively

3.1. *Inductive heating furnace*

The cooling water of inductive furnace is designed to be cooled down by heat-exchanging with primary cooling water connected to a cooling tower. We modified the water pipings to bypass the fin and plate type heat exchanger so that the unit utilize hot water and cold water from existing system. Figure 3 shows result of contentious data acquisition for one month. The hot side temperature fluctuating since the furnace is often repeatedly operated during day and night. The flow rate of cooling water in figure 3 (b) showed gradual decrease and then completely clogged up at around three week. This was caused by sludge accumulated in the strainer located before the inlet of the power generation unit. Despite the lack of cooling water supply, the unit kept producing electricity as seen in figure 3 (c) for week3 and week4.

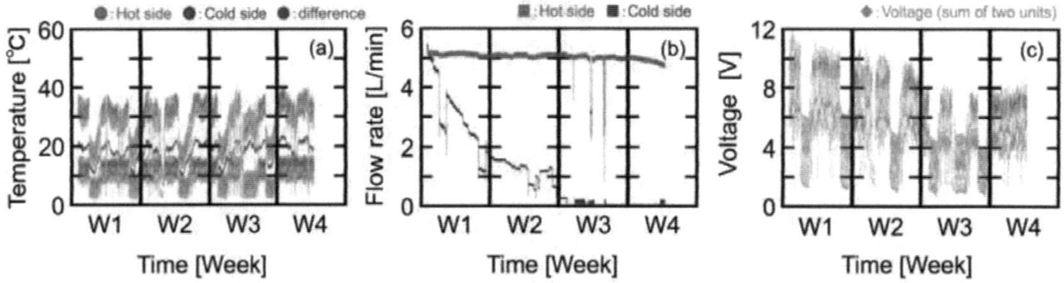

Figure 3. Time evolution of (a) hot and cold side temperature, (b) water flow output power – current, (c) temperature difference – maximum output power.

3.2. Hot spring

A demonstrative data acquisition and long distance transmission was performed using the power generation unit in hot spring center Yakushi-yu. The center provides Onsen bath service for public purpose and utilize the primal hot spring water (~ 98°C) for warming up the secondary clean water (~ 60°C) which is used for the bath service. We bypassed the piping to supply hot and cold water to the unit. The generated power was directly connected to a conventional battery charge controller that control 12V lead-acid battery. Data logger (Graphtec, GL-100-WL-4VT) monitored 5 channels (temperature and flow rate of hot water and cold water and output voltage) and transmitted the data via WiFi to rooter that is connected commercial 3G network. The data, updated every second was monitored by network PC which located in AIST in Tsukuba. All the electronics were operated by the generated power so that the system can be completely self-powered. Figure 4 shows one of example of remote sensing experiment. The flow rate of cold water is quite fluctuating, since we bypassed the cold water piping which is connected shower and toilet. The resulting power output also fluctuate depending on the cold water flow rate, however the hot water was relatively stable.

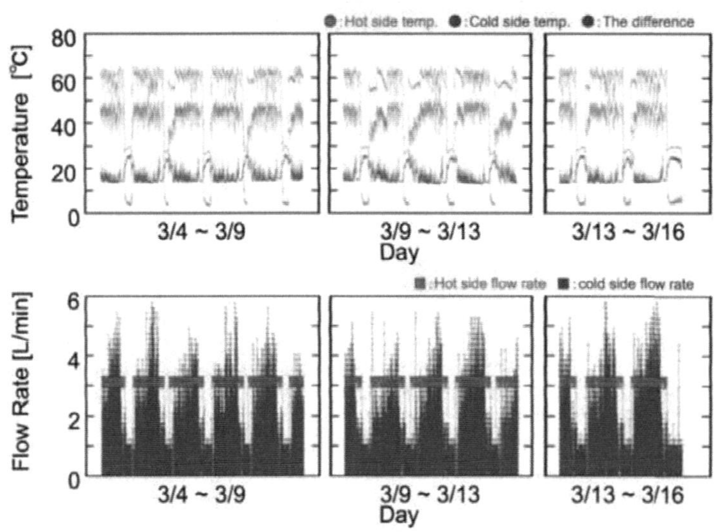

Figure 4. An example of the self-powered remote data acquisition results. Upper : water temperature and the difference, lower : water flow rates.

3.3. Boiler blowdown water

Innami factory of The Japan Wool Textile Co., Ltd. equips many through flow boiler to generate steam for factory operation and emits a lot of blowdown water. The water temperature is roughly 80°C and shows alkaline properties. We pumped up the blowdown water and directly used for the power generation. Two unit were connected in parallel and the data logger recorded the flow rate, the temperature and the output voltage of individual thermoelectric module every 10 sec. Total 50 million data was collected for the period from Oct. 2016 to June 2017. Figure 5 shows the time evolution plot of monitored data in the field test. The hot water clogged up repeatedly during the test as shown in figure 5 (b) and the reason is sludge accumulation at a strainer near the circulation pump for hot water. Frequent cleaning by an operator was needed to continue the experiment. We confirmed that the power generation unit itself was unchanged and quite stable after half year field test.

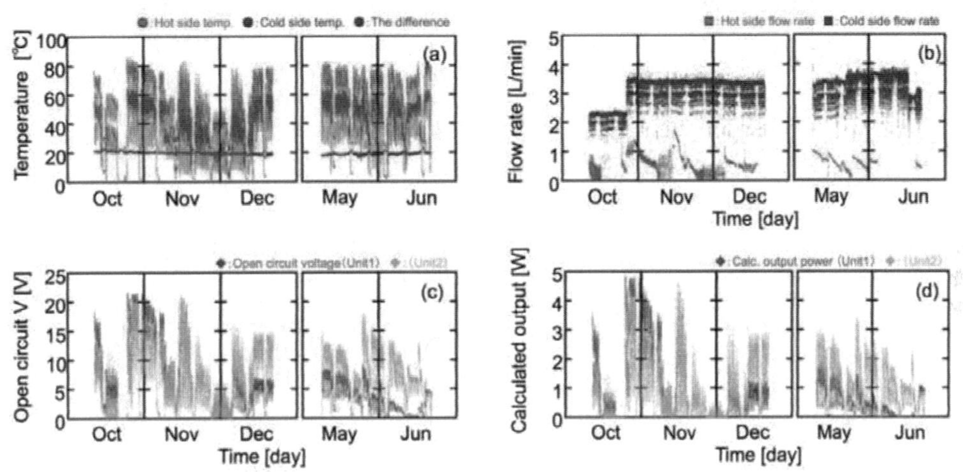

Figure 5. Time evolution data of (a) temperature of hot and cold water, (b) water flow rates, (c) output voltage, (d) generated power.

Summary

We designed and prototyped stacking type thermoelectric power generation unit and tested its performance in three field test. The unit produced electricity depending on the temperature and the flow rates of supplied hot water and cold water. The performance of the unit itself was unchanged and stable during the long term test. We need further verification of the life time of the unit through field demonstrations and need to find out some measures for the clogging at the strainer in the piping.

Acknowledgments

Authors would like to special thanks to S.Iijima of SINTOKOGIO Ltd., K.Taniguchi of Shin Onsen Town, and K.Togashi and Y.Konishi of The Japan Wool Textile Co., Ltd. for kind offers of field test sites and cooperation in the demonstrative experiments.

References

[1] K. Takahashi, *et al*, *Scientific Reports 3*, Article number: 1501 (2013)
[2] http://news.panasonic.com/jp/press/data/jn110620-2/jn110620-2.html, download on 2017/10/12
[3] H. Nishiate, et.al., *Proc. 11th Annual symposium of Thermoelectric Society of Japan* (TSJ 2014), p64
[4] H. Nishiate, et.al., *Proc. 12th Annual symposium of Thermoelectric Society of Japan* (TSJ 2015), p137
[5] H. Nishiate, et.al., *Proc. 12th Annual symposium of Thermoelectric Society of Japan* (TSJ 2015), p138
[6] H. Nishiate, et.al., *Proc. 13th Annual symposium of Thermoelectric Society of Japan* (TSJ 2016), p113
[7] Bertrand F.Tchanche, Gr.Lambrinos, A.Frangoudakis, G.Papadakis, *Renew. Sust. Energ. Rev.*, Vol.15, 8, October (2011) pp.3963-3979
[8] http://geothermal.jogmec.go.jp/report/file/023.pdf, download on 2017/10/12

IOP Conf. Series: Journal of Physics: Conf. Series **1052** (2018) 012137 doi:10.1088/1742-6596/1052/1/012137

A 3-way pushable electret-based energy harvester fabricated with 3d-printing and PDMS molding

Y. F. Chen[1], H. Honma[2], and H. Toshiyoshi[2]

1 Graduate Institute of Photonics and Optoelectronics, National Taiwan University, No. 1, Sec. 4, Roosevelt Rd., Taipei 10617, Taiwan
2 Institute of Industrial Science, The University of Tokyo, 4-6-1, Komaba, Meguro-ku, Tokyo, 153-8505, Japan

E-mail: d00941008@ntu.edu.tw

Abstract. An energy harvester device is developed with a PDMS chamber that contains upper and bottom electrodes (Al foil and electret) to form an air-dielectric variable capacitor, which can be pushed in three different manners to generate power through the up-and-down motion of the PDMS membrane. The 3-way mechanisms are experimentally compared..

1. Introduction

This paper reports the design, fabrication, and testing of a 3-way pushable energy harvester based on the three-dimensional (3D) printing and polydimethylsiloxane (PDMS) molding technology. The power is generated by the electrostatic inductance using the K^+-doping electret [1] due to the variance of the electrical capacitance between deformable electrodes. A similar device of PDMS-based flexible energy harvester with Parylene electret is reported in J. Micromech. Microeng. 2015 [2]. However, the method reported here is different from the prior work, and our device has better performance for power generation with the same effective active area of 1×1 cm^2.

2. Design and fabrication

Figure 1 shows a 3D model of the device structure, in which a 3D printed frame is used to create a PDMS mold. 3D printing has advantages such as low prototyping cost, fast, and clean, and PDMS device has characteristics of biocompatibility, non-reactivity, and mechanical flexibility. Containing all the properties of both technologies, a PDMS mold with a deformable thin membrane and a button-shaped chamber is created. Following the fabrication processes as shown in Fig. 2.

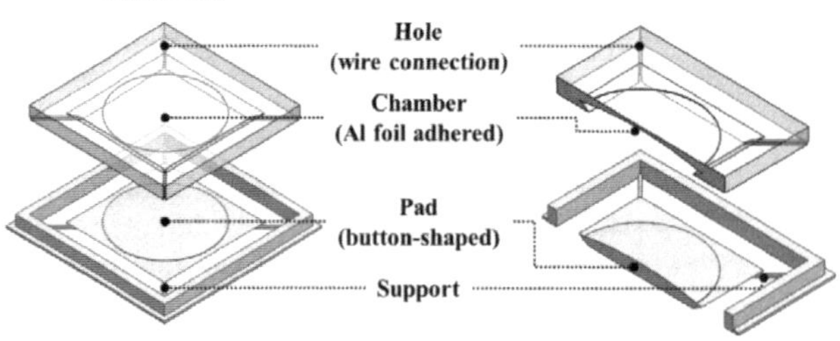

Figure 1. Illustration of 3D-printed frame and PDMS mold for fabricating device.

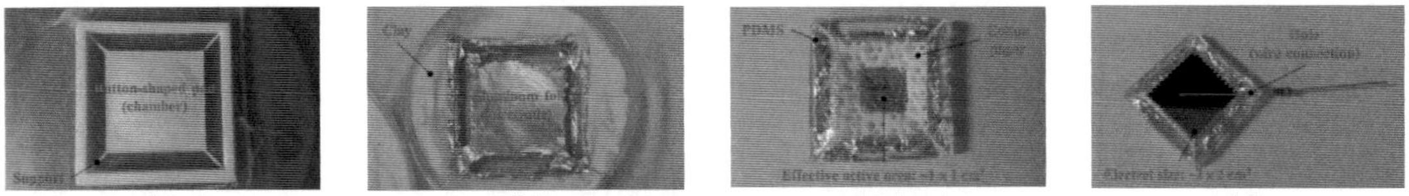

(a) (b) (c) (d)

Figure 2. The aluminum (Al) foil as an electrode layer was adhered onto the bottom side of the membrane during PDMS solidification, whereas a 2 × 2 cm^2 silicon electret is embedded into the chamber after PDMS molding. Also, a cotton paper is inserted as a spacer to form an air-dielectric variable capacitor; therefore, a 1 × 1 cm^2 effective active area of compact energy harvester device is created.

3. Power generation mechanisms and results of experiment

By using the identical energy harvester device, power can be generated due to capacitance changes through gap-closing, and three possible pushing mechanisms are used (Table 1). For Mechanism I (Finger tapping), the maximum output voltage of 15.6 V and the corresponding peak (rms) output power of 243.4 (54.0) μW/cm^2 with a 1 MΩ load of an internal resistance of the oscilloscope were obtained. This power may include triboelectric generation because the materials (PDMS, Al foil, and electret) may contact each other during pressing. For Mechanism II (Stationary stopper), the output voltages on the device at frequencies of 5, 10, 15, and 20 Hz were measured while the shaker was under controlled in the same displacement. The corresponding peak (rms) output power are 0.6 (0.1), 3.2 (0.4), 3.4 (0.6), and 4.8 (1.5) μW/cm^2. For Mechanism III (Vibrating mass), the output powers for the device loaded with a mass of 7.2 g at different frequencies were generated under the same acceleration of 1 m/s^2, and the peak (rms) output power of 12.5 (2.7) nW/cm^2 was measured at the frequency of 30 Hz. Compared with the above three mechanisms, the energy harvester device pushed by finger tapping can generate a much larger output power.

Table 1. Comparison of three possible ways to generate power.

Mechanism	Finger tapping	Stationary stopper	Vibrating mass
Photography of the device		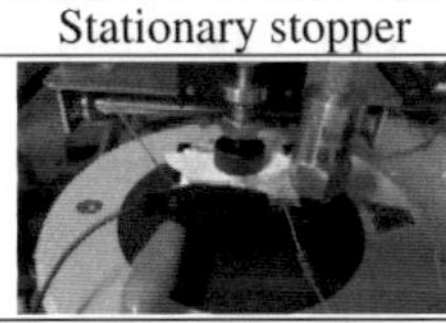	
Purpose	Buttons of keyboard or phone	Reciprocating motions	Machines or bridges
Figure of experimental result	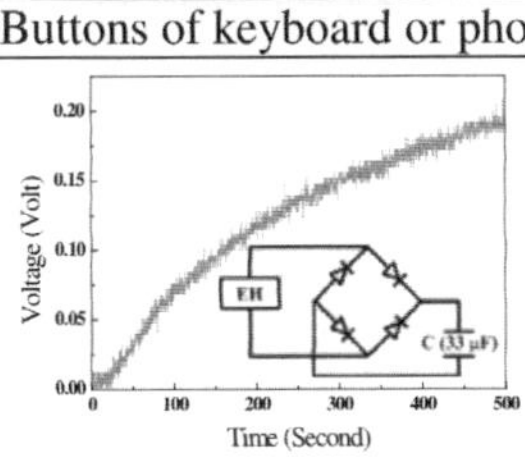 Stored voltage of 208 mV was collected in about 8 min by finger tapping.	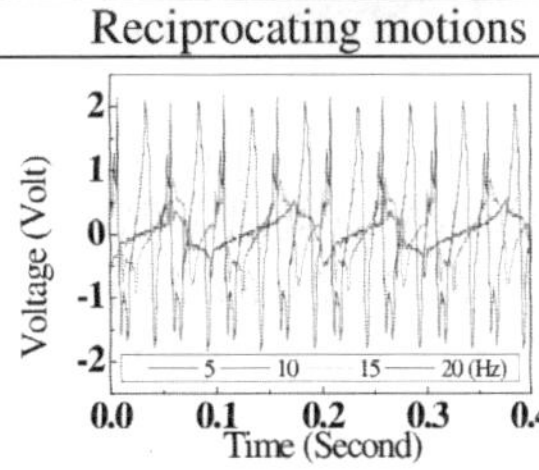 Output voltage on the energy harvester device at 5, 10, 15, and 20 Hz by a shaker.	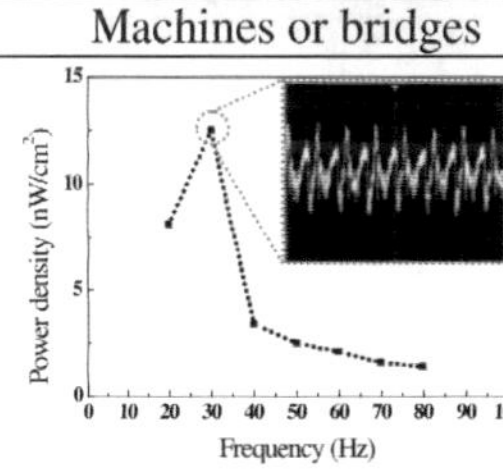 Output power vs. frequency on the energy harvester device with a mass loaded on top.
Maximum power generation	$P_{peak, output}$ = 243.4 μW/cm^2 $P_{rms, output}$ = 54.0 μW/cm^2	$P_{peak, output, 20 Hz}$ = 4.8 μW/cm^2 $P_{rms, output, 20 Hz}$ = 1.5 μW/cm^2	$P_{peak, output, 30 Hz}$ = 12.5 nW/cm^2 $P_{rms, output, 30 Hz}$ = 2.7 nW/cm^2

References

[1] Hashiguchi G *et al* 2016 *AIP Adv.* **6** 035004
[2] Chiu Y *et al* 2015 *J. Micromech. Microeng.* **25(10)** 104007

IOP Conf. Series: Journal of Physics: Conf. Series **1052** (2018) 012138 doi:10.1088/1742-6596/1052/1/012138

Thermoelectric nanogenerator networks: a viable source of power for autonomous wireless sensors

D. Tainoff, A. Proudhom, C. Tur, T. Crozes, S. Dufresnes, S. Dumont, D. Bourgault, O. Bourgeois

Institut Néel, CNRS, Université Grenoble Alpes, 25 Avenue de Martyrs, BP 166, 38042, Grenoble, France

olivier.bourgeois@neel.cnrs.fr

Abstract. Here we report the design, the elaboration and the measurement of an innovative planar thermoelectric (TE) devices made of large array of suspended nanogenerators [1]. The miniaturized TE generators are arranged in serial and/or in parallel depending on the expected final resistance adapted to the charge. Devices are elaborated using MEMS clean room processes. They can produce enough power (below 100 μWatt) to send a signal with common communication protocol and feed basic functions, for example a temperature measurement.

1. Introduction

With the emergence of low power communication protocols dedicated to the Internet of things (IoT), there is a real demand for autonomous micro-sources of energy that can supply batteries and deliver power on the order of 100 μW. That range of power can be supplied by standard TE module with a temperature gradient of few degrees under stationary operating conditions. However these conditions of use are not necessarily adapted for small wireless sensors especially when the energy source is intermittent or the volume available too limited.

2. Array of TE nanogenerators based on membrane

The expertise developed at the Institut Néel CNRS on the elaboration of suspended sensors dedicated to thermal measurements has been used to design suspended thermoelectric nano-generators [1,2]. The planar generators are made of suspended silicon nitride membranes functionalized with TE materials as described in the figure 1. Since membranes have a negligible mass, their temperature is free to fluctuate contrary to the bulk silicon frame that conducts well the heat. Thus a temperature gradient appears between the membranes and the bulk silicon frame when the whole setup is submitted to a heat source (air flow, radiation etc...). The proof of concept has been validated using Bi_2Te_3 TE thin films. Indeed this material has the best TE properties around 300 K with a conversion efficiency up to 10%.

For temperature gradient of 8 K the power generated by an individual membrane is around 0.3 nW. The duplication of this thermoelectric nano-generator thanks to standard MEMS clean room techniques enabled us to obtain few hundreds of membranes. In that case, a power of approximately 60μW is harvested for a temperature gradient of 10°C and a surface of 10 cm². A concentration of several thousands of nano-generators per cm² will be reached by using state of the art clean room technology. This may lead to power production of up to 750 μW for 10 cm², a power value sufficiently high to make sober sensors infinitely autonomous in energy. Prototypes including energy harvesting have been made and coupled to an ASIC in order to validate the possibility to feed a communicating temperature sensor.

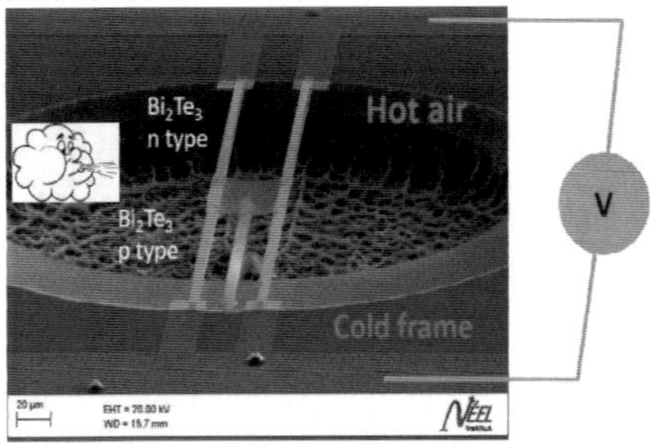

Figure 1. Scheme of a unique suspended membrane having two thermoelectric n and p junctions (Bi_2Te_3 materials), a nano-generator sensitive to heat flow.

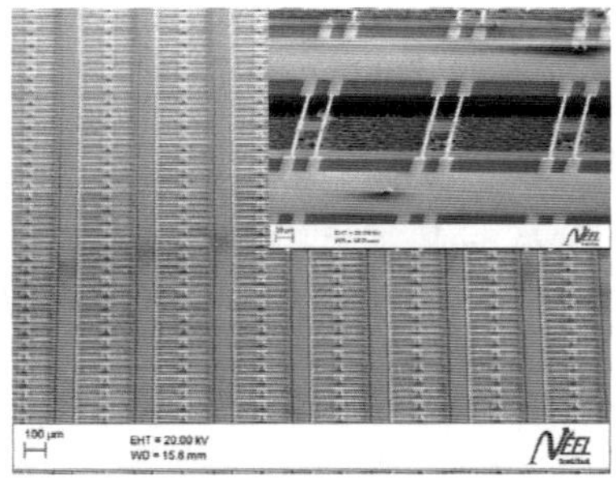

Figure 2. Array of thousands of membranes. The TE nano-generator are installed in series and parallel to adapt the internal resistance to the impedance of the charge (sensor to be feeded).

The patented design of the TE nano-generators array permits to arrange easily the micromodule in serial and/or in parallel depending of the expected final resistance [3]. The whole device is mechanically robust, versatile regarding the energy sources and not sensitive to aging having no moving part (see figures).

Figure 3. The ASIC made for the energy and power management of the array of TE micromodules.

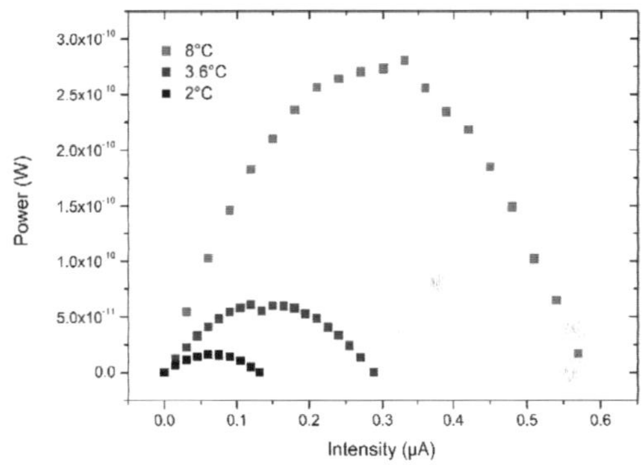

Figure 4. Power generated by a single membrane as the function of the bias current (in Peltier mode). The different temperature gradients used in the experiment are mentioned.

As a conclusion, planar suspended thermoelectric devices have been developed and patented by CNRS. This device permits to convert small intermittent gradient of heat into electrical power. The design of the device permits to adapt the resistance of the thermogenerator depending on the exact use case. The efficiency of the device is sufficient to expect its use in real IoT use cases.

References

[1] O. Bourgeois, D. Tainoff and D. Bourgault *Thermoelectric device*, patent 16/51336 owned by CNRS (2016).

[2] D. Tainoff, and O. Bourgeois *Gas sensor device*, patent owned by CNRS (2017).

[3] D. Tainoff, A. Proudhom, C. Tur, T. Crozes, S. Dufresnes, D. Bourgault, and O. Bourgeois. *Thermoelectric nanogenerator networks: a viable source of power for the autonomous connected sensors,* in prep. to be submitted to NanoEnergy (2017).

Development of electrical generator using ferromagnetic powders and non-magnetic fluid

Haruhiko SHIRAI[1], Hiromichi MITAMURA[1], Takuji NODA[2], Nobuaki ARAI[3]and Kazuyuki MORIYA[1]

[1]Graduate School of Infomatics,Kyoto University,
Japan Yoshida-Honmachi, Sakyo-ku, Kyoto, 606-8501,Japan
[2] The Institute of Statistical Mathematics, Tachikawa, Tokyo, 606-8502, Japan
[3] Field Science Education and Research Center, Kyoto University,
Japan Kitashirakawa Oiwake-cho, Sakyo-ku, Kyoto, 606-8502, Japan

E-mail: shirai-h@bre.soc.i.kyoto-u.ac.jp

Abstract. This paper reports a broadband vibration resonance model for vibration energy harvesting using ferromagnetic powders and suggested a non-resonance type device model by adding a non-magnetic fluid(water) to this device model to support frequencies lower than 10 Hz. The work builds on a vibration energy harvesting device reported at Mechanical Engineering Journal [1].In this study, because the electric generator can convert the reciprocating motion into electrical energy without rotary motion by ferromagnetic powders and non-magnetic fluid for the naturally occurring kinetic energy that is small such as the creature exercise, it may be said that this electricity generator is effective. We suggest that using ferromagnetic powders for broadband vibration and non-magnetic fluid for the low-frequency vibrations as devised in this study could be a breakthrough in solving the technical problems that arise in vibration energy harvesting.

1.Concept

It is difficult to generate energy from small, low-frequency (0.1–10 Hz) vibrations such as those produced by animals or natural phenomena by using the electromagnetic induction method. So far, authors have performed some basic study on the development of such a generation system [2, 3]. This generation system converts vibrational energy into electric energy using a permanent magnet and a pipe, which encloses ferromagnetic powders and non-magnetic fluid (water) (Fig. 1). The ferromagnetic powders form spike-shaped aggregates under the influence of the permanent magnet's magnetic field. When the pipe is vibrated by an external movement, the motion of these aggregates creates electromotive force because of the magnetic field crossing the coil by the right-hand rule.

This is a maintenance-free, durable, highly inexpensive generation system without moving parts. In this study, we have developed a generation system that is durable, inexpensive, and nearly maintenance-free due to removed mechanical parts such as sliding. In this system, a permanent magnet is placed outside a pipe filled with ferromagnetic powders and liquid, and electricity is produced by electromagnetic induction as a result of vibration of the system. The objective of this study is to examine the values of vibrational frequency and the magnetic substance to use and their relations with the efficiency of generation. An example of specifications is as follows. The vibration-driven generator was fixed to a shaker for the test tube to make it vibrate at 5.6 Hz (respective vibrational acceleration:2.728 G) with an amplitude of 40 mm. The electrical generator generated electricity in 185mVp-p, 2mA.The generation electricity per volume is 2.16[$\mathbf{\mu W}/cm^3$]. Furthermore, we are developing an efficient smaller electrical generator.

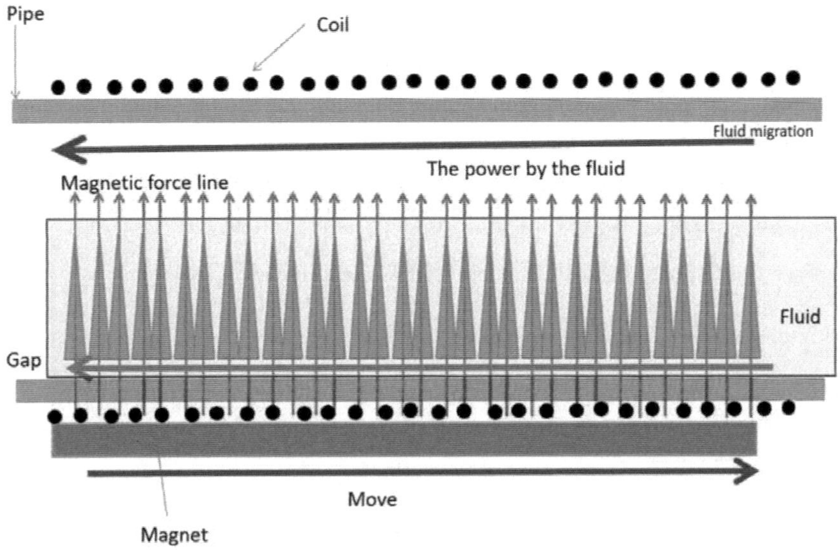

Fig. 1 Conceptual diagram of the developed system. Ferromagnetic powders and liquid are enclosed in a pipe. The ferromagnetic powders become a spikes-formed lump by the magnetic field of an outside permanent magnet.

2. Characteristic of the electrical generator

· We made many flexibility system vibration system by using ferromagnetic powders, and broadband of the vibration generation s, and broadband of the vibration generation enabled.

· The structure of the electrical generator is simple. Thus the electrical generator is adjustable and is adaptable itself for environment.

· We enabled low frequency vibration generation by using a nonmagnetic fluid.

Previously, we performed electromagnetic analysis using computer simulations because we conceived that the behavior of the fluid, direction of magnetic induction, and magnetic flux substantially contributed to our generator's electricity output. We were able to increase generation efficiency by analyzing the magnetic flux distribution of the generator's permanent magnet [1].Ferromagnetic powders used in this study are made of iron and iron compounds naturally available in our living environment. For example, ferromagnetic powders are included in magnetic tapes and groundwater and can harvest infinite, inexpensive energy out of industrial waste. It can be stated that this study "is the correct fusion of green energy and ecology."

References

[1] Shirai, H., Mitamura, H., Noda, T., Arai, N., Moriya, K., Study of the efficiency of electrical generator using ferromagnetic powders by electromagnetic analysis, Mechanical Engineering Journal, Vol. 3, No. 5, (2016).

[2] Shirai, H., Electrical generator with magnetic powders, International patent application PCT/JP2013/79357(2013).

[3] Shirai, H., Electrical generator with magnetic powders, Japanese patent disclosure 2014-93837(P2014-93837A)(2014). (in Japanese).

Vibration energy harvesting with piezoelectrets and electrets

X. Zhang[1] and G. M. Sessler[2]

[1]Tongji University, China, [2]Darmstadt University of Technology, Germany

x.zhang@tongji.edu.cn g.sessler@nt.tu-darmstadt.de

Abstract. We report on vibration energy harvesting with parallel tunnel *piezoelectret* films and negatively charged wavy shaped *electret* films made from fluoroethylene propylene (FEP).

1. Introduction

The large number of wireless communicating sensors used in the Internet of Things (IoT) makes the replacement of batteries impossible or very expensive as they are located in remote areas or harsh environments. A promising solution is to power the electronic devices by harvesting various ambient resources and converting their energy into useful electrical energy. Here, we report on vibration energy harvesting with parallel tunnel *piezoelectret* films and negatively charged wavy shaped *electret* films made from fluoroethylene propylene (FEP). Energy harvesters working in the thickness (33) and length extension (31) modes are discussed.

2. *Piezoelectret* energy harvesters

Piezoelectrets are cavity-polymer electrets with a strong piezoelectric effect. Due to large values of the figure of merit (FoM, $d_{33} \cdot g_{33}$ and $d_{31} \cdot g_{31}$) in parallel-tunnel FEP piezoelectret films, such films are very promising candidate as energy conversion material in the vibrational energy harvesters excited in 33 and 31 modes (Table I). In the vibrational energy harvesters using the longitude (g_{33}) piezoelectric effect in parallel tunnel FEP piezoelectret films, an output power of 6.1 μW was obtained at a vibrational acceleration of 1 *g* (9.81 m/s^2), a seismic mass of 54 g and a resonance frequency of 87 Hz (Figure 1). The output power has been significantly improved as the transverse (g_{31}) piezoelectric effect in parallel tunnel FEP piezoelectret films is utilized in miniaturized energy harvesters. Typically, with a total volume of 0.2 cm^3, a seismic mass of 0.09 g, a harvester resonance at 58 Hz, and a vibrational acceleration of 1 *g*, a peak output power at resonance of about 57 μW has been achieved (Figure 2).

3. *Electret* energy harvesters

Vibrational energy harvesters with electret layers are a kind of electrostatic energy harvester where the embedded electrical charges in electrets play the role of constant bias voltage to the variable capacitor transducer. In the 33 mode, an output power of 1.85 mW has been achieved in an energy harvester, consisting of a negatively charged wavy-shaped FEP electret film and a wavy-shaped counter electrode placed in a perpendicular way, at a vibrational acceleration of 1 *g*, a seismic mass of 80 g, and a resonance frequency of 90 Hz (Figure 3). In the 31 mode, the generated power across the optimal load resistor is about 360 μW at the resonance frequency of 22 Hz when using a seismic mass of 3 g and an input acceleration of 1 *g* (Figure 4).

IOP Conf. Series: Journal of Physics: Conf. Series **1052** (2018) 012140 doi:10.1088/1742-6596/1052/1/012140

A comparison of various energy harvesters based on piezoelectrets and electrets is given in Table II. The normalized power indicates that the harvesters operating in the 31 mode yield higher power output than those working in the 33 mode.

Table I: Piezoelectric coefficients of some materials. Literature values are in the audio and/or ultrasonic ranges. Values from this work are either static values or at frequencies up to 100 Hz.

Material	d_{33} [pC/N]	g_{33} [Vm/N]	FoM $d_{33} \cdot g_{33}$ [pC²m/N²F]	d_{31} [pC/N]	g_{31} [Vm/N]	FoM $d_{31} \cdot g_{31}$ [pC²m/N²F]
PZT-5H	640	0.021	13.44	283	0.0093	2.6
PVDF	33	0.33	10.9	23	0.22	5.0
PP piezoelectret	140	13*	1820	≈2	≈0.2	≈0.4
Parallel-tunnel FEP piezoelectret	1300	122*	159000	32**	3.0	96
					(Static values)	
					0.5~0.8	
		(10~35 Hz)			(10~100Hz)	

* Calculated g_{33} from $g_{33} = d_{33}/\varepsilon_0\varepsilon_r$, taking relative permittivity of 1.2

** Calculated d_{31} from $d_{31} = g_{31}\varepsilon_0\varepsilon_r$, taking relative permittivity of 1.2

Table II. Recent progress in piezoelectret energy harvesters. The normalized power is referred to an acceleration of 1 g = 9.81 m/s^2 and a seismic mass of 1 g. The adjustment is made by considering that the power output is proportional to a^2 and $m^{3/2}$.

Type of Harvester [Reference]	Normalized Power (μW)	Resonance Frequency (Hz)	Size of Active Area (cm²)
Piezoelectret g_{33} IXPP [a]	0.6	400	3
Piezoelectret g_{33} parallel tunnel FEP [this work]	0.02	87	3
Piezoelectret g_{31} parallel tunnel FEP [this work]	2000	58	0.4
Electret FEP (33 mode) [b]	20	35	4
Electret wavy FEP (33 mode) [this work]	2.6	90	4
Electret wavy FEP electret (31 mode) [this work]	100	30	3

[a] X. zhang, L. Wu, and G. M. Sessler, AIP Advances, 5, 077185(2015)
[b] P. Pondrom, G. M. Sessler, J. Bös, and T. Metz, Appl. Phys. Lett. 109, 053906(2016)

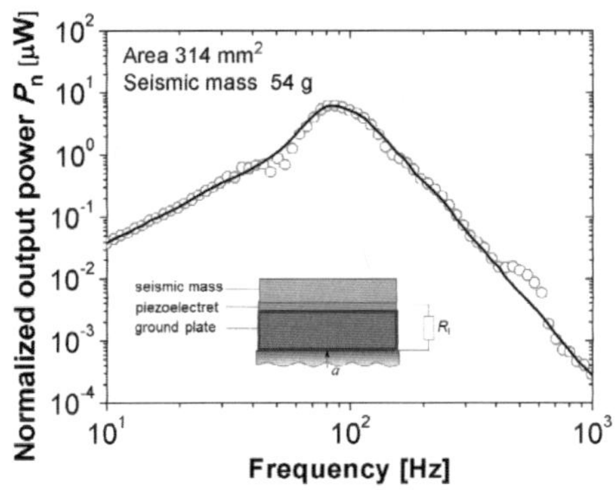

Figure 1. Output power as a function of frequency of Parallel tunnel FEP piezoelectret working in 33 mode.

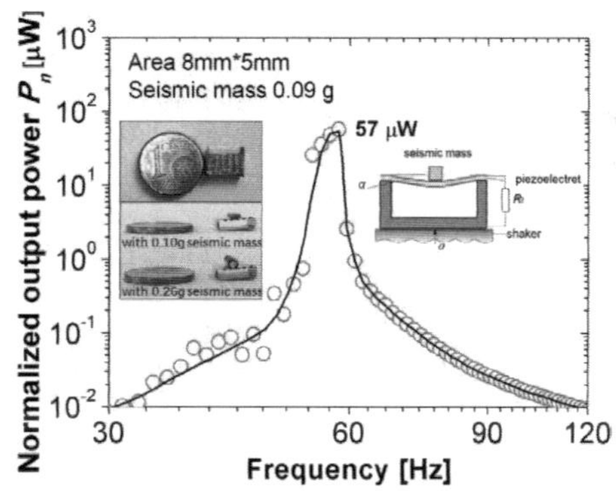

Figure 2. Output power as a function of frequency of Parallel tunnel FEP piezoelectret working in 31 mode.

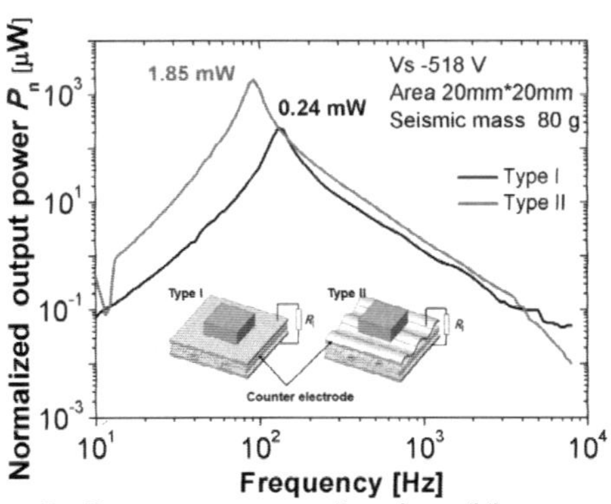

Figure 3. Output power as a function of frequency of wavy FEP electret working in 33 mode.

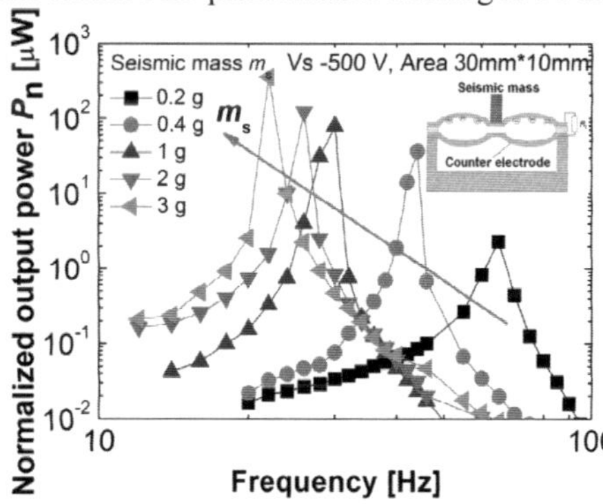

Figure 4. Output power as a function of frequency of wavy FEP electret working in 31 mode.

References

[1] S. Bauer, R. Gerhard-Multhaupt, G. M. Sessler, *Phys. Today* 57 (2004), 37.

[2] S. R. Anton, and K. M. Farinholt, In *Proc. Amer. Soc. Mechanical Eng,(ASME) Conf. Smart Materials, Adaptive Structures and Intelligent Systems (2012)*, pp. 929-937.

[3] X. Zhang, P. Pondrom, L. Wu, G. Sessler, *Appl. Phys. Lett.* 108 (2016), 193903.

[4] Y. Suzuki, *IEEJ Trans. Electr. Electr. Engin.*, 6 (2011), pp.101-111.

[5] P. Pondrom, J. Hillenbrand, G. M. Sessler, J. Bös, and T. Melz, *Appl. Phys. Lett.* 104 (2014), 172901.

IOP Conf. Series: Journal of Physics: Conf. Series **1052** (2018) 012141 doi:10.1088/1742-6596/1052/1/012141

THE DEVELOPPMENT OF THE CANTILEVER TYPED VIBRATION POWER GENERATION FLOOR

T. Yoshikawa

7-1 Kasugaoka Nabarishi-shi, Mie 518-0459, JAPAN

yoshikawa@ktc.ac.jp

Abstract. The purpose of out study is to establish the way for making vibration power generation floor. In the consideration of our study is composed of three processes. The first is to establish the design method obtaining the desirable maximum power from cantilever piezoelectric vibration unit. The 2nd is to construct the vibrating floor. The 3rd is to estimate the charging efficiency using the capacitor. After all we examine the average output power by the vibration power generation floor. The applications of the vibration power generation floor are assumed as power source of sensor network nodes and foot light at night and so on. Considering for the use of power supplier for sensor network node, the desirable generated power has been estimated more than 100uW averagely. That is our target. We have performed above three processes and succeeded in making the cantilever typed vibration power generating floor which can generate 120uW averagely.

1. Introduction

The power generating floor is very useful because the floor can be set everywhere and can continue to supply the power to low energy consumption information terminal (e.g. sensor network node). We tried to make the vibration power generating floor so as to apply to the power supplier of sensor network node. The target generation power is more than 100uW averagely for its application [1]. The conceptual view of the application of the vibration power generating floor is shown in Figure 1.

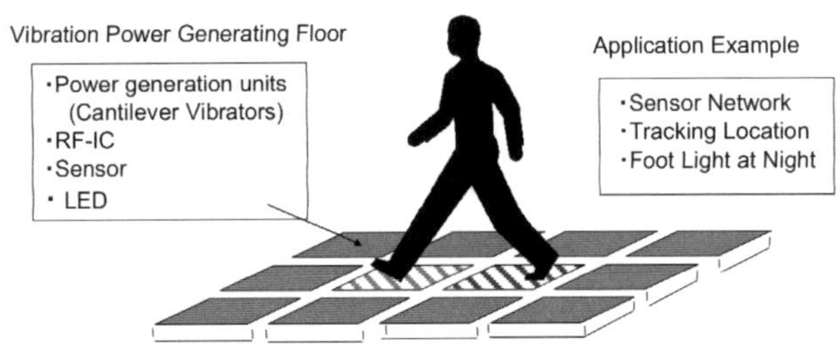

Figure 1. *Conceptual view of the application of the vibration power generation floor*

2. Cantilever Vibrating Unit

We have designed the output power of cantilever vibrator based on the equation of motion and piezoelectric foundations and material mechanics formulations [2]. Especially we focused on how much weight of additional mass is appropriate referring to our application. The basic excitation experiments (Figure 2) are performed and the results have been obtained (Figure 3). The maximum output power varies in inverse proportion to 5th power of resonant frequency when exciting acceleration is constant. Those results are in good agreement with the theoretical relationship.

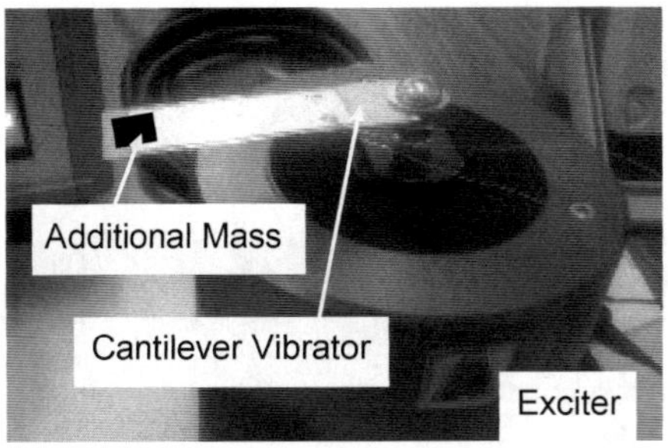

Figure 2. *The view of the basic excitation experiments of cantilever vibration unit*

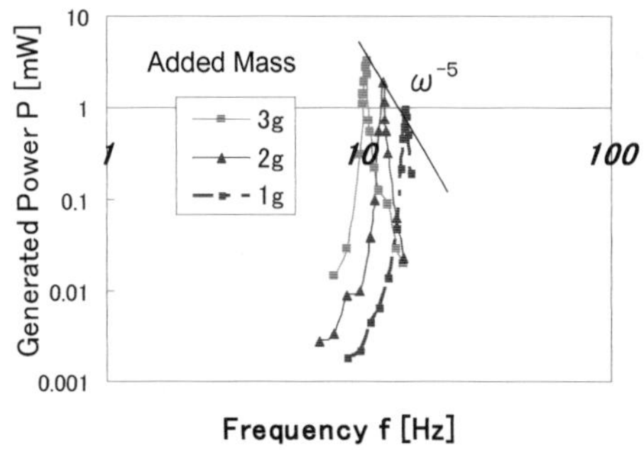

Figure 3. *The generated power characteristics under constant acceleration*

3. Cantilever Typed Vibration Power Generating Floor

Then we have made the vibration power generation floor with such evaluated vibrating unit (Figure 4). At first excitation force by human foot step is applied to upper plate, next the force is absorbed at the bottom buffer material and the burst vibration occurs at the upper rib which is the fulcrum of the cantilever vibrator unit. As the result, cantilever vibrator unit vibrates at that resonance and generate the power. We have measured the open Voltage and short current using the shunt resistance. Those results show the dumped vibration property (Figure 5). Then we have formulated the relationship between peak power and charged average power on a capacitor. This is expressed as following.

$$\text{Average power / Peak power} = X(1-\exp(-1/X)) \tag{1}$$

Here $X = \tau / T$, τ is damping time constant, T is step period. In addition to that, we have derived the theoretical value of the charging efficiency into capacitor the power of piezoelectric element. The value is expressed using the charging up start voltage and final voltage. Using above relationship, we have estimated the generated average power on the capacitor as 120uW. And we have measured the generating average power by the human walking steps using vibration power generating floor. The experimental value is the same as our estimated value. This result proofs our vibration power generating floor can be applied for the power supplier of sensor network node.

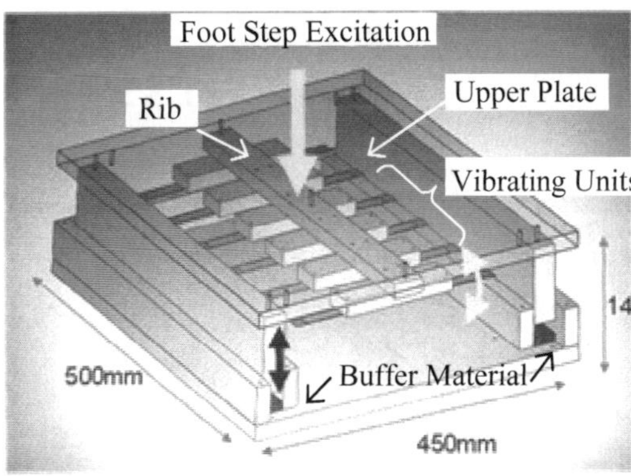

Figure 4. *3D view of power generating floor*

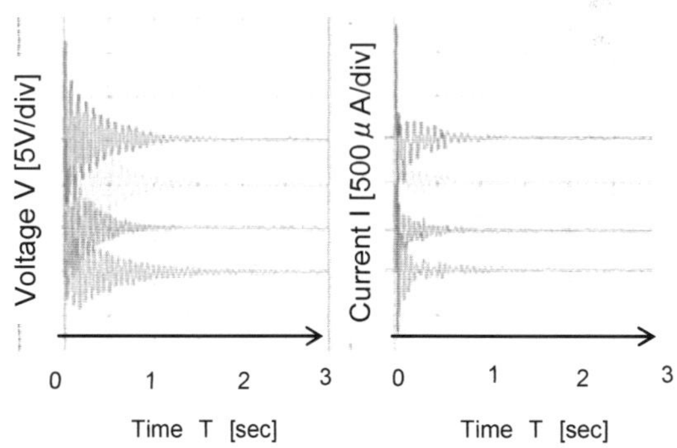

Figure 5. *The dumped vibration characteristics of Voltage and current*

References

[1] Takashi Yoshikawa, "Novel Concept for HEMS Apparatus", Elsevier Science Direct 2012 Energy Procedia 14, pp.1273-1279, 2012.

[2] Yoshio Kuriyama, et.al., "Study on the Cantilever Piezoelectric Plate Resonator ", Bulettin on Human Resources Development Vol.16 No.1, pp.57-63, 2004.

Sweat as energy source using an enzymatic microfluidic fuel cell

E. Ortiz-Ortega[1], R. A. Escalona-Villalpando[1], J. Galindo-de-la-Rosa[1], J. Ledesma-García[2], S. D. Minteer[3] and L.G. Arriaga[*1]

[1]Centro de Investigación y Desarrollo Tecnológico en Electroquímica, Pedro Escobedo, Qro., C.P. 76703, Mexico

[2]Facultad de Ingeniería, División de Investigación y Posgrado, Universidad Autónoma de Querétaro, Centro Universitario Cerro de las Campanas, Querétaro, Qro., C.P. 76010, México

[3]Department of Chemistry, University of Utah, 315 South 1400 East, Salt Lake City, UT, 84112, United States

*larriaga@cideteq.mx

Tel.: +52 442 211 6069

ABSTRACT.

An enzymatic microfluidic fuel cell was constructed to produce energy from lactate and oxygen preset in the sweat. In this case, laccase (Lc) connected to anthracene modified carbon nanotubes (An-MWCNT) to facilitate the direct electron transfer (DET) as the biocathode and lactate oxidase (LOx) as the bioanode, both enzymes were immobilized on Toray Carbon Paper (TCP). The oxygen reduction reaction (ORR) showed better results in a pH acid, which agrees with the slightly acid pH of sweat. On the other hand, the Lc showed to have good catalytic activity towards ORR in the presence of sweat, since only about 10% was inhibited after having added 660 µl of sweat. Finally, the enzymatic microfluidic fuel cell was capable to generate 64 µW cm-2 when sweat was fed into both electrodes, this confirms the high selectivity of the LOx and Lc towards lactate oxidation and oxygen reduction, respectively. These results are the first step to design an stamped device to produce energy in real time taking advantage of the sweat produced during workouts.

Keywords

Sweat, Lactate, Enzymes, Microfluidic Fuel Cell

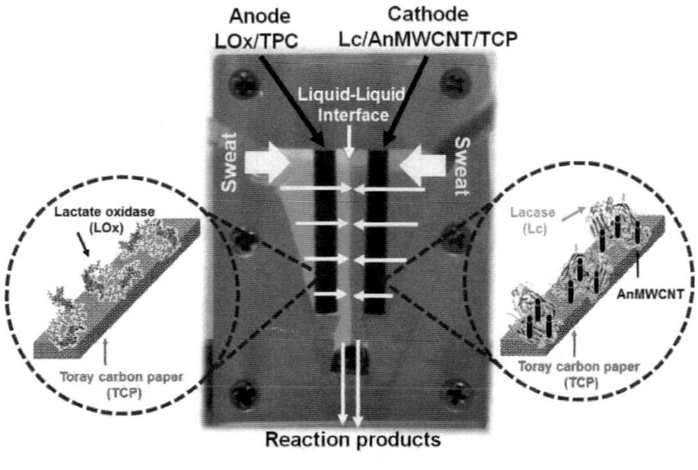

Figure 1. Enzymatic Microfluidic Fuel Cell.

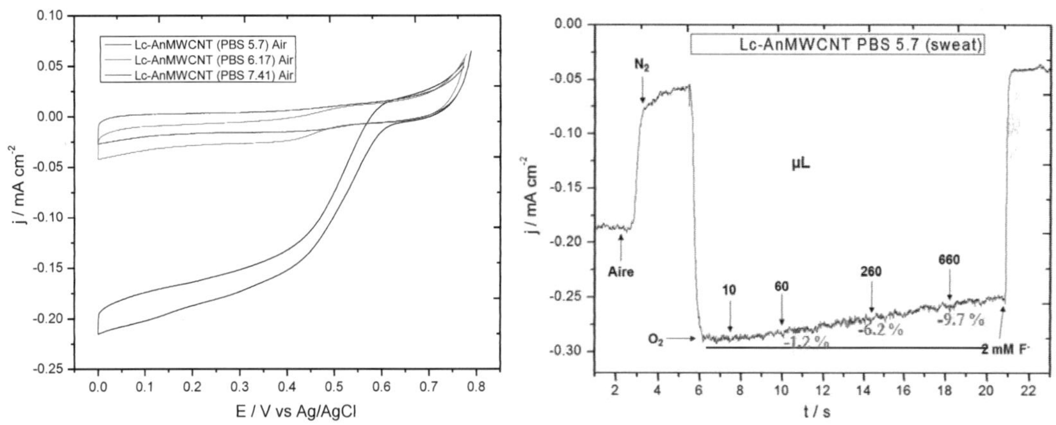

Figure 2. a) Cyclic Voltammetry of Lc/AnMWCNT in 0.05M PBS at different pH with scan rate 1 mV s^{-1} and b) Chronoamperometry curve of Lc/AnMWCNT in sweat at 0.2 V vs Ag/AgCl.

Table 1. Performance of an enzymatic microfluidic fuel cell using sweat.

Anode	Cathode	Anolyte	Catholyte	OCP / V	i_{JSC} / μA cm^{-2}
LOx	Lc-AnMWCNT	sweat	PBS pH 6.17	0.55	231.6
LOx	Lc-AnMWCNT	sweat	sweat	0.52	123.7
LOx	Pt/C	sweat	sweat	0.40	473.7

References

[1] R. A. Escalona-Villalpando, R. C. Reid, R. D. Milton, L.G. Arriaga, S. D. Minteer, J. Ledesma-García, Journal of Power Sources, 342 **(2017)** 546-552
[2] R.D. Milton, K. Lim, D.P. Hickey, S.D. Minteer, Bioelectrochemistry, 106 **(2015)** 56-63.
[3] M.T. Meredith, M. Minson, D. Hickey, K. Artyushkova, D.T. Glatzhofer, S.D. Minteer, ACS Cat., 1 **(2011)** 1683-1690.
[4] A. Mena-Bravo, M.D. Luque de Castro, Journal of Pharmaceutical and Biomedical Analysis, 90 **(2014)** 139–147
[5] C. J. Harvey, R. F. LeBouf, A. B. Stefaniak, Toxicology in Vitro, 24 **(2010)** 1790–1796.

IOP Conf. Series: Journal of Physics: Conf. Series **1052** (2018) 012143 doi:10.1088/1742-6596/1052/1/012143

Toward CMOS compatible wafer-scale fabrication of carbon-based microsupercapacitors for IoT

A D Smith[1], Q Li[1], A Anderson[2], A Vyas[1], V Kuzmenko[1], M Haque[1], L G H Staaf[1], P Lundgren[1] and P Enoksson[1]

[1] Chalmers University of Technology, Kemivägen 9, Göteborg 41258, Sweden
[2] University of Santa Barbara, Santa Barbara, CA 93106, USA

E-mail: smdavid@chalmers.se

Abstract. This work presents a wafer-scale method of microsupercapacitor (MSC) fabrication. Deposition of the electrode precursor, i.e. graphene oxide, is accomplished through spin-coating which allows for potential application in CMOS compatible processes for future integrated on-chip energy storage systems. Our MSCs have an areal capacitance of 0.4 mF/cm^2 at 10 μA, which is a very promising result. Further, the MSC has good rate capability as its capacitance decreases by only 0.03 mF/cm^2 when the current increases to 50 μA. The MSCs have a maximum energy density of 0.04 μWh/cm^2 and a maximum power density as high as 96 μW/cm^2. Additionally, the wafer-scale process demonstrates industrial viability.

1. Introduction

As microsystems shrink and become increasingly integrated (Internet of Things (IoT)), they have ever increasing power demands and can therefore greatly benefit from on-chip power supply and management systems. Microsupercapacitors (MSCs), combined with energy harvesting, offer the potential to alleviate these increasing power demands [1]–[6]. However, implementation of reliable wafer scalable and fully CMOS compatible MSC fabrication processes remains challenging. In this work, we present a novel, CMOS compatible fabrication method for development of wafer scalable MSC technology using rGO electrodes. This method has a potential to be incorporated back-end-of-the-line (BEOL) with other microsystems and many other carbon-based materials (such as, carbon nanotubes (CNTs), activated carbon (AC) and carbon nanofibers (CNFs)). These carbon materials show promising electrical characteristics as well as high surface to volume ratios – making them well-suited as MSC electrodes [7].

The MSC devices are comprised of metal current collectors. On top of these current collectors, GO is spin-coated and selectively etched to form positive and negative high surface area electrodes. GO is converted to rGO after annealing at high temperature. The porous nature of the rGO allows for the electrolyte to penetrate into the electrode and form a double layer capacitance at the electrode/electrolyte interface. Consequently, rGO offers tremendous capacitive properties due to its high surface area. A schematic of a typical MSC device fabricated in a wafer-scale process is shown in figure 1 with the silicon/SiO_2 substrate in black/grey, the current collectors in gold, the rGO in dark grey, and the electrolyte in green.

IOP Conf. Series: Journal of Physics: Conf. Series **1052** (2018) 012143 doi:10.1088/1742-6596/1052/1/012143

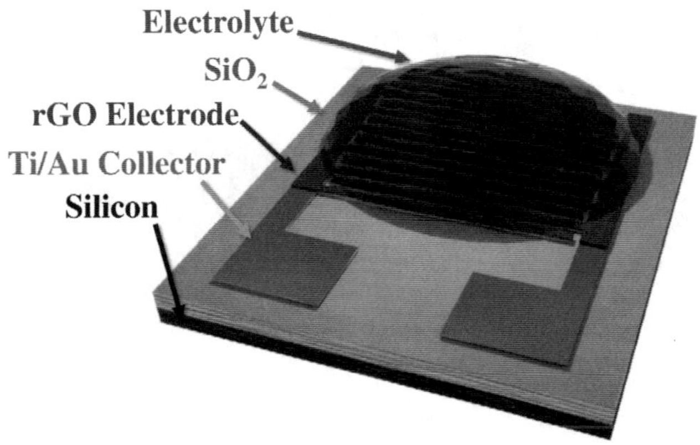

Figure 1. Graphical representation of a generic carbon-based MSC with interdigitated electrodes.

2. Device Fabrication

The wafer-scale fabrication procedure for MSCs is shown in Figure 2. A 2" silicon wafer with 400 nm of SiO_2 is employed as the substrate (figure 2a). Interdigitated Ti/Au (10/50 nm) current collectors are fabricated using metal evaporation and lift-off (figure 2b). Although gold is used for proof of concept, it can be replaced with CMOS compatible metals. Next, Graphene oxide (GO) is spin-coated (figure 2c). A 50 nm aluminum layer is then deposited and patterned to form a hard mask (figure 2d). This hard mask protects the GO electrode material during oxygen plasma etching at 100 W in a 100 mTorr (figure 2e). Aluminum is then etched away and GO is reduced (figure 2f) for 20 mins at 500 °C with hydrogen used as a reductant.

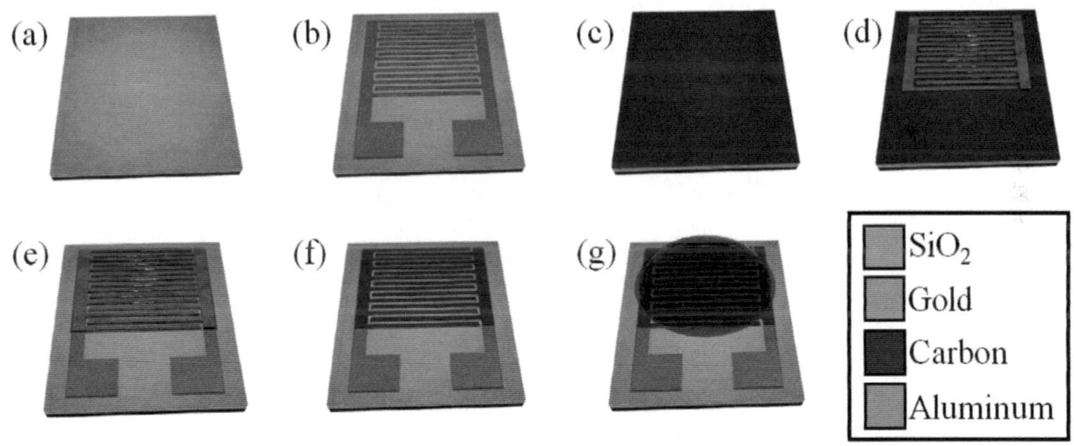

Figure 2. Schematic process plan for a ubiquitous carbon-based MSC.

Figure 3a displays rGO-based MSCs fabricated on a 2" silicon wafer (fabricated to step (e) in figure 2) where the inset presents an optical microscopy image of the interdigitated contact pattern – illustrating a very good etch profile from the Aluminum hard mask. Likewise, scanning electron microscopy (SEM) further confirms the presence of rGO (figure 3b).

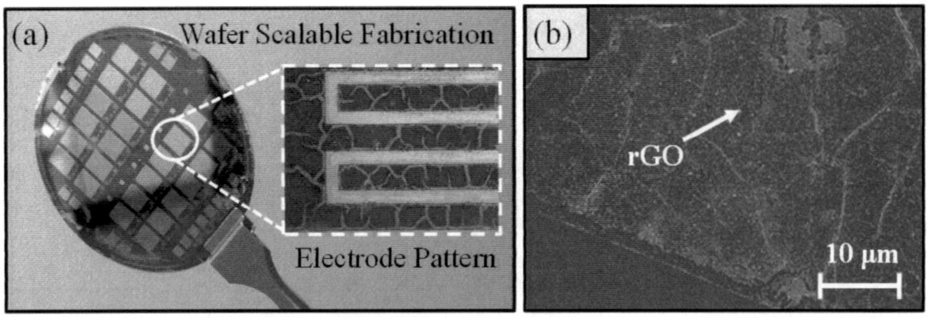

Figure 3. (a) Full size 2" wafer after graphene oxide etch, (inset) rGO interdigitated pattern post GO reduction. (b) SEM image rGO electrode after complete processing.

To evaluate the successful reduction of GO to rGO, Raman spectroscopy measurement was performed. The Raman spectrum of the material after the reduction process shows two major features, the G band due to E_{2g} symmetry of sp^2 carbon at 1587 cm^{-1} and the D band corresponding to breathing mode of A_{1g} symmetry at 1338 cm^{-1}. The intensity ratio of D to G band (I_D/I_G) is 1.39, similar to a previous report on rGO [8]. Finally, 6 M KOH electrolyte is applied onto the device (figure 2g). Although KOH is used for initial measurements, it can be replaced by a gel-based electrolyte (e.g. polyvinyl alcohol based electrolyte PVA/H_3PO_4) for CMOS compatibility.

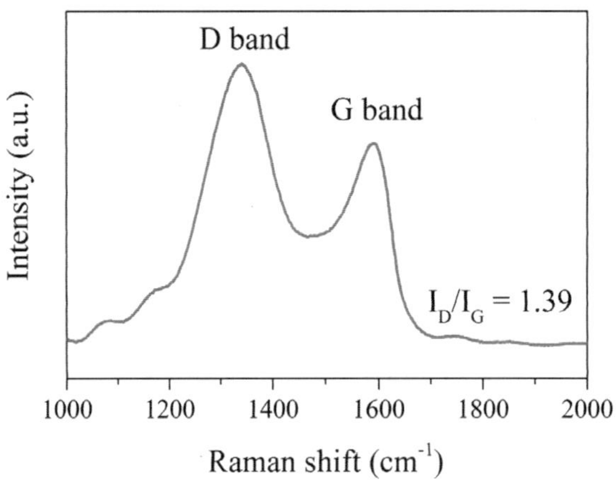

Figure 4. Raman spectroscopy of rGO with D and G bands. rGO is suggested by the intensity ratio of the two bands.

3. Results/Discussion

The electrochemical performance of the MSC is evaluated on Kari Suss PM 5 probe station coupled with Gamry Reference 3000AE potentiostat. Cyclic Voltammetry (CV) at scan rates of 1 – 10 V/s (figure 5a) show quasi-rectangular curves that indicate electric double layer (EDL) behavior dominates the storage mechanism, which is further evidenced by the linear Galvanostatic Charge/Discharge (GCD) curves at a current of 10, 20, 30 and 50 μA (figure 5b).

The following equations are applied for the calculation of capacitance: $C = I{\cdot}t \;/\; S{\cdot}\Delta V$, where C is the areal capacitance, I is the charge/discharge current, S is surface area of the device including the space between fingers and ΔV is the voltage range during discharge excluding IR drop. The energy density can be further calculated by $E = ½\, C{\cdot}\Delta V^2$, where E stands for energy density, and average power density by $P = E \;/\; t$. The calculation shows that the MSC has an areal capacitance of 0.4 mF/cm^2 at 10 μA. The value is comparable to 0.51 mF/cm^2 of a rGO based MSC fabricated by laser writing method [9]. The areal capacitance can be easily controlled by increase the thickness of active material layer by repeating the spin-coating step. The device also has excellent rate capability that capacitance drops by only approx. 0.3 mF/cm^2 when the current increases 5 times to 50 μA. The maximum energy of the device is around 0.04 μWh/cm^2 while the maximum power is as high as 96 μW/cm^2.

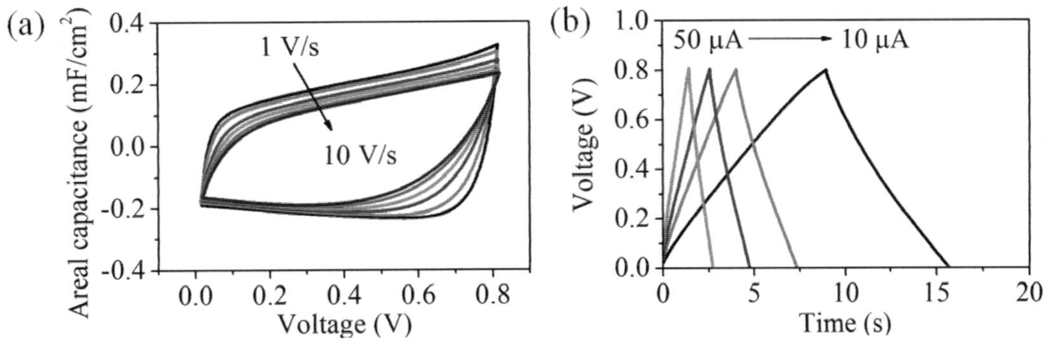

Figure 5. (a) Cyclic voltammetry, (b) Galvanostatic charge/discharge.

Although measurements for rGO are presented, the fabrication method is ubiquitously transferrable to other carbon materials. Due to the robust nature of the fabrication process, there is much room for performance enhancement with the resulting MSCs potentially being implemented on-chip with other microsystem technologies for future development of self-powering microsystems.

4. Conclusion

We present a MSC fabrication method which uses rGO as an electrode material. Deposition of rGO is accomplished through spin-coating and therefore offers potential CMOS compatibility. The process is scalable to wafer-scale sizes which allows for potential industrial viability. Our MSCs currently have an areal capacitance of 0.4 mF/cm^2 at 10 μA which is comparable to previous reports. Further, the MSC capacitance decreases by only approx. 0.03 mF/cm^2 when the current increases to 50 μA. The MSCs have a maximum energy density of approximately 0.04 μWh/cm^2 and a maximum power density as high as 96 μW/cm^2. The MSC fabrication method aims toward potential future application involving on-chip energy storage systems.

References

[1] Yun J, Kim D, Lee G, and Ha J S, 2014 *Carbon*, **79** pp 156–164
[2] Wang S, Hsia B, Carraro C, and Maboudian R 2014 *J. Mater. Chem. A* **2** 7997–8002
[3] Meng Q, Wu H, Meng Y, Xie K, Wei Z, and Guo Z 2014 *Adv. Mater* **26** 4100–6
[4] Yang Y, He L, Tang C, Hu P, Hong X, Yan M, Dong Y, Tian X, Wei Q, and Mai L 2016 *Nano Res* **9** 2510–19
[5] Wu Z-S, Feng X, and Cheng H-M 2014 *Natl. Sci. Rev.* **1** 277–92
[6] Zhai Y, Dou Y, Zhao D, Fulvio P F, Mayes R T, and Dai S 2011 *Adv. Mater.* **23** 4828–50
[7] Staaf L G H, Lundgren P, and Enoksson P 2014 *Nano Energy* **9** 128-41
[8] Das A K, Srivastav M, Layek R K, Uddin M E, Jung D, Kim N H, and Lee J H 2014 *J. Mater. Chem. A* **2** 1332-40
[9] Gao W, Singh N, Song L, Liu Z, A L M Reddy, Ci L, Vajtai R, Zhang Q, Wei B and Ajayan P M, *Nat. Nanotechnol.* **6** 496-500

IOP Conf. Series: Journal of Physics: Conf. Series **1052** (2018) 012144 doi:10.1088/1742-6596/1052/1/012144

Mask-programmable on-chip photovoltaic cell array

Y. Takeshiro[a], Y. Okamoto[a] and Y. Mita[a]

[a] Dept. EEIS, the Univ. of Tokyo, 7-3-1 Hongo, Bunkyo-ku, Tokyo (113-8656), Japan

E-mail: mems@if.t.u-tokyo.ac.jp

Abstract. We propose a gate-array style easy configuration method for voltage photovoltaic cell for rapid prototyping of on-chip IoT system. Using the same design chips fabricated by standard CMOS process and only changing the mask pattern for post processes, we successfully achieved two type chips: one generates 68 V open circuit voltage output photovoltaic cell and the other does 10 times higher short circuit current than it with 5.4 V open circuit voltage output.

1. Introduction

More and more applications of IoT devices are now proposed and each device requires different characteristic for power sources, and photovoltaic cells are widely used because of its high efficiency. A simple photovoltaic cell, however, can only generate low open circuit voltage (0.7 V ~ 1.1 V), which is not necessarily enough to drive the component of IoT devices, such as LSIs and sensors. Getting higher voltage without loss of area and energy, cells have to be connected serially corresponding to its purpose. To achieve this, serial connection of photovoltaic cells is employed and to connect these cells on a chip, some methods are reported (Ref.[1] and Ref.[2]). Among these methods, we proposed the photovoltaic diodes fabricated by post-CMOS process (Ref.[3]). This method can carry out the conventional bulk CMOS process without forming on SOI wafers, thus we could achieve high-quality photovoltaic cells with low cost. However, to modify the output characteristics, the method requires whole design modification with high cost and taking time. Therefore, simple modification method is needed, which is very attractive for fast prototyping of IoT devices.

In this paper, we propose gate-array style easy configuration method for getting required characteristic photovoltaic cells.

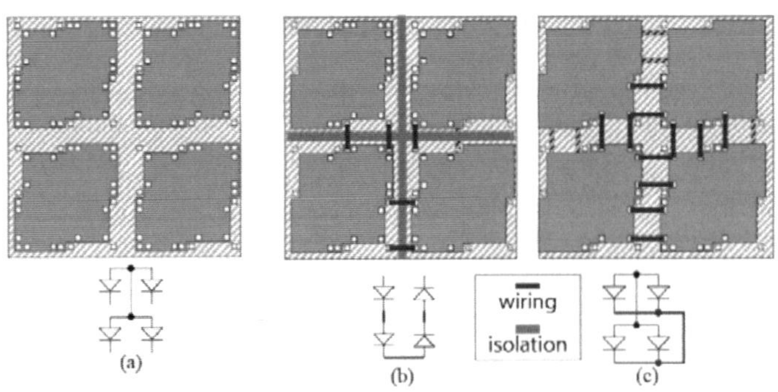

Figure 1. Connection and output characteristics configuration method.
(a)Before CMOS post-processed four diodes. Only anodes are connected by substrate.
(b) Serially connected four diodes for getting higher voltage. Each diode is mesa-isolated.
(c) Parallel connected four diodes for getting higher current.

Figure 2. SEM pictures of fabricated PV cells. (a) A part of serially connected photovoltaic cells. In this picture, four diodes are shown. (b)Enlarged view of a wire. Wires were released from the substrate to isolate diodes. diode design of mask-programmable PV cells.

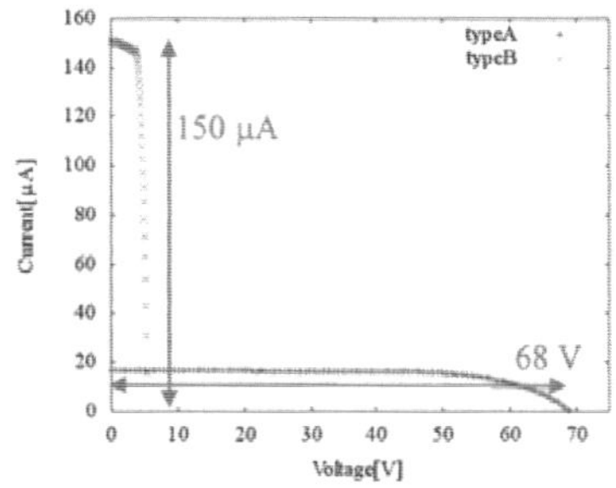

Figure 3. Measured current-voltage characteristics of fabricate two types PV cells. Type A achieved 68 V_{op}, while type B achieved 150 μA_{sc} .

2. Proposal Method

Figure 1 shows the configuration method of four diode connection. Each side of a diode has eight connectors for wiring with other diodes. After standard CMOS fabrication, diodes are connected as shown in **Fig. 1(a)**. All anodes are connected by substrate and cathodes are isolated. To connect these diodes serially for getting higher voltage (**Figure 1(b)**), these diodes are wired from anode side connector to cathode side one, and all diodes have to mesa-isolated. For getting higher current (**Figure 1(c)**), diodes are wired from cathodes connector to cathodes one, and these diodes are not mesa-isolated to keep their anodes connected.

3. Experiment

In this experiment, we designed 12 by 12 diodes on 7.5 mm by 7.5 mm chip for master slice. This design wafers are fabricated by foundry service on SOI wafers, whose thickness of device layer was 9 μm. For the demonstration, we fabricated two types design. One was that all diodes were connected in series for generating highest voltage which this design chip could generate (type A). The other was each row (12 diodes) were connected parallel, and columns were connected serially. This version is designed for driving LSIs, which require around 5 V (type B).

With those designs, two chips ware fabricated at the same time. Wires were patterned at first, and following process was almost the same as the previous research (Ref.[3]). **Figure 2(a)** shows the fabricated PV cell array (type A), and the close-up SEM image (**Figure 2(b)**) shows all the PV cells are completely isolated. These chips ware measured with light bulb and semiconductor device parameter analyzer (Keysight B1500A).

4. Results

Measurement results were illustrated as shown in **Figure 3**. Type A generated 68 V open circuit voltage with 15 μA short circuit current, and fill factor was 68%. Type B generated 5.4 V open circuit voltage with 150 μA short circuit current, and fill factor was 77%.

References

[1] H. Kwon, M. Nakada, Y. Hirabayashi A. Higo, M. Ataka, H. Fujita and H. Toshiyoshi 2007 *IEEE/LEOS Int. Conf. on Optical MEMS and Nanophotonics* pp. 49 – 50 (Hualien, Taiwan R.O.C.)

[2] C. L. Bellew, S. Hollar and K. S. J. Pister 2003 *Technical Digest of 12th Int. Conf. Solid State Sensors, Actuators and Microsystems* pp.1075–1079 (Boston, USA)

[3] I. Mori, M. Kubota, E. Lebrasseur and Y. Mita 2016 *IEEJ Trans. on Senors. and Micromachines*, **136** 2 pp. 24-30.

Power Electronics for Wireless Power Delivery in Synthetic Sensor Networks

P. D. Mitcheson, G. Kkelis, S. Aldhaher, J.M. Arteaga, D. C. Yates, D. Boyle, E. M. Yeatman

Department of Electrical and Electronic Engineering, Imperial College London, UK

E-mail: paul.mitcheson@imperial.ac.uk

Abstract. Energy harvesting is a very popular means of augmenting batteries or displacing their use for powering the wireless sensor networks; however, an alternative approach is to use wireless power delivery to energise the nodes. For a network occupying a well-known and confined physical geometry, it may be possible to achieve this via one or multiple sources in fixed locations; however, when the nodes cover a wide area, or are themselves mobile, fixed sources may not be practical. A possible solution is to use drones which can fly close to the sensor nodes, which then transfer power inductively from drone to node. In such systems the drone could also be used to gather data from the sensors nodes, reducing their power requirement for long distance data transmission. However, here we investigate the power electronics and energy storage solutions for this so-called *synthetic sensor network* (synthsense) setup. Here we describe the preliminary hardware subsystems which will allow the demonstration of synthsense.

1. Introduction

Inductive power transfer is becoming well known in short range, low power applications such as mobile phone charging, where the Qi standard has been adopted [1]. This technology offers an alternative solution to the problem addressed by energy harvesting: that of maintaining a remote power supply, typically of a sensor node, without the need for physical battery replacement or tethering through wires. One main advantage of harvesting is that it relies on ambient energy and so no additional physical infrastructure is required outside of the sensor node. Wireless power delivery on the other hand, requires an energy transmission source. In the case of wireless mobile phone charging, the use of a dedicated transmitter is straightforward: a charging pad on one's desk will suffice. However, for a large, distributed network of sensor nodes, perhaps fitted on a piece of civil infrastructure for health monitoring, or across a farm or vineyard to monitor crop or plant growth, installation of a fixed transmitter infrastructure may not be desirable or possible.

A possible way forward is to use a mobile vehicle as a power delivery source. Here we investigate the use of a drone, complete with a lightweight wireless power transmitter, capable of rapidly charging a storage element on the sensor node inductively, as the drone flies past.

2. Requirements

From an operational perspective, there are two main requirements for power delivery using this technique: the system must be capable of fast recharging thereby minimizing drone hovering time close to a sensor node; and the system must have the ability to efficiently transfer power over an air gap with variable geometry and over around 10 cm preventing the need for contact or docking between the drone

IOP Conf. Series: Journal of Physics: Conf. Series **1052** (2018) 012145 doi:10.1088/1742-6596/1052/1/012145

and the sensor node. From a technical perspective, the power source, mounted on the drone, must be light weight due to limited payload and must be efficient to transfer the maximum energy from the drone to the sensor node. We have previously demonstrated a drone charged above a charging pad and [2] and the solution for synthsense builds on this technology.

3. Solution and Results

The system uses high frequency inductive power delivery at 6.78 MHz. This contrasts with the Qi standard (which is based around 100 kHz). The reason to choose high frequency is two-fold: the efficiency of an inductive link is dependent on the product of the magnetic coupling between transmitter and receiver, k, and the Q-factors of the coils. High frequency allows high Q-factors and thus reduces dependence on k, increasing the effective range of the link. To achieve efficient power electronics in the MHz region, soft switching is mandatory, but it is hard to achieve the soft switching conditions when the magnetic link geometry is changing and the load is variable. A solution to this is to use the load independent class EF topology [3] which keeps soft switching as load conditions alter, using an aluminium PCB substrate for thermal management, Figure *2*.

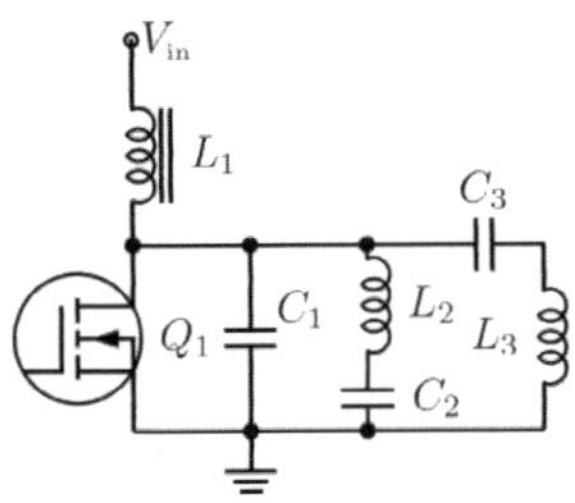

Figure 1 Load invariant class EF inverter

Figure 2 Inverter for mounting on drone using insulated metal substrate for thermal management (60 mmx70 mm)

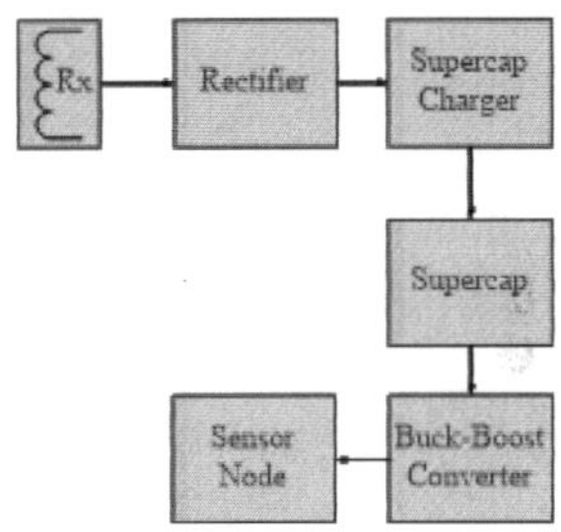

Figure 3 Receive side hardware for stepping up and down the capacitor voltage

On the receive side, the main challenge is to charge the sensor node as quickly as possible. This requires use of a supercapacitor, in this case an AVX 6 V, 5F device, which has shown a charge rate of 30 J/s, minimizing the hovering time of the drone when it is recharging a receiver.

4. Conclusions

We have discussed the requirements and presented preliminary hardware of the power electronics for the new paradigm for powering of sensor networks whereby a drone flies close to a sensor node to recharge it. The obstacles to overcome are achieving a light weight transmitter, achieved using MHz transmission and an IMS PCB, variable air gap geometry, achieved using a load independent inverter, rapid charging of the receiver, which is possible with a supercapacitor, and compact receiver dc-dc converter necessary to process the highly variable voltage output from the supercapacitor. We will show a working system in the poster presentation.

References

[1] Wireless Power consortium, "The Qi Wireless Power Transfer System Power Class 0 Spoecification," Wireless Power Consortium, 2016.

[2] S. Aldhaher, P. Mitcheson, J. Arteaga, G. Kkelis and D. Yates, "Light-Weight Wireless Power Transfer for Mid-Air Charging of Drones," in *European Conference on Antennas and Propogation*, paris, 2017.

[3] S. Aldhaher, P. D. Mitcheson and D. Yates, "Load-Indepdnent Class EF Inverters for Inductive Wireless Power Transfer," in *IEEE Wireless Power Transfer Conference*, Portugal, 2016.